D1744420

CURRENT DEVELOPMENTS IN
BIOTECHNOLOGY AND BIOENGINEERING

Series Editor

Ashok Pandey

Centre for Innovation and Translational Research, CSIR-Indian Institute of Toxicology Research, Lucknow, Uttar Pradesh, India
&
Sustainability Cluster, School of Engineering, University of Petroleum and Energy Studies, Dehradun, India

CURRENT DEVELOPMENTS IN BIOTECHNOLOGY AND BIOENGINEERING

Membrane Technology for Sustainable Water and Energy Management

Editors

XUAN-THANH BUI

Key Laboratory of Advanced Waste Treatment Technology & Faculty of Environment and Natural Resources, HCMUT, VNU-HCM, Ho Chi Minh City, Vietnam

WENSHAN GUO

Centre for Technology in Water and Wastewater, School of Civil and Environmental Engineering, University of Technology Sydney, Sydney, NSW, Australia

CHART CHIEMCHAISRI

Department of Environmental Engineering, Faculty of Engineering, Kasetsart University, Bangkok, Thailand

ASHOK PANDEY

Centre for Innovation and Translational Research, CSIR-Indian Institute of Toxicology Research, Lucknow, Uttar Pradesh, India & Sustainability Cluster, School of Engineering, University of Petroleum and Energy Studies, Dehradun, India

ELSEVIER

Elsevier
Radarweg 29, PO Box 211, 1000 AE Amsterdam, Netherlands
The Boulevard, Langford Lane, Kidlington, Oxford OX5 1GB, United Kingdom
50 Hampshire Street, 5th Floor, Cambridge, MA 02139, United States

Copyright © 2023 Elsevier Inc. All rights reserved.

No part of this publication may be reproduced or transmitted in any form or by any means, electronic or
mechanical, including photocopying, recording, or any information storage and retrieval system, without
permission in writing from the publisher. Details on how to seek permission, further information about the
Publisher's permissions policies and our arrangements with organizations such as the Copyright Clearance
Center and the Copyright Licensing Agency, can be found at our website: www.elsevier.com/permissions.

This book and the individual contributions contained in it are protected under copyright by the Publisher
(other than as may be noted herein).

Notices

Knowledge and best practice in this field are constantly changing. As new research and experience broaden
our understanding, changes in research methods, professional practices, or medical treatment may become
necessary.

Practitioners and researchers must always rely on their own experience and knowledge in evaluating and
using any information, methods, compounds, or experiments described herein. In using such information or
methods they should be mindful of their own safety and the safety of others, including parties for whom they have
a professional responsibility.

To the fullest extent of the law, neither the Publisher nor the authors, contributors, or editors, assume any
liability for any injury and/or damage to persons or property as a matter of products liability, negligence or
otherwise, or from any use or operation of any methods, products, instructions, or ideas contained in the
material herein.

ISBN: 978-0-443-19180-0

For information on all Elsevier publications
visit our website at https://www.elsevier.com/books-and-journals

Publisher: Susan Dennis
Editorial Project Manager: Catherine Costello
Production Project Manager: R.Vijay Bharath
Cover Designer: Vicky Pearson Esser

Typeset by STRAIVE, India

Working together
to grow libraries in
developing countries

www.elsevier.com • www.bookaid.org

Contents

6. Ultralow pressure membrane filtration for water and wastewater treatment

Chew Lee Leong, Muhammad Roil Bilad, Norazanita Shamsuddin, Hazwani Suhaimi, Nasrul Arahman, Adewale Giwa, and Ahmed Yusuf

7. Commercial scale membrane-based produced water treatment plant

Utjok W.R. Siagian, L. Lustiyani, K. Khoiruddin, I.N. Widiasa, Tjandra Setiadi, and I.G. Wenten

8. Membrane bioreactor for wastewater treatment: Fouling and abatement strategies

Shamas Tabraiz, Muhammad Zeeshan, Muhammad Bilal Asif, Uchenna Egwu, Sidra Iftekhar, and Paul Sallis

9. Membrane and filtration processes for microplastic removal

Linh-Thy Le, Xuan-Bui Bui, Cong-Sac Tran, Chart Chiemchaisri, and Ashok Pandey

10. Membrane bioreactor for municipal solid waste leachate treatment and organic micropollutant removals

Chart Chiemchaisri, Wilai Chiemchaisri, Varinthorn Boonyaroj, Anekpracha Kaewmanee, Chayanid Witthayaphirom, and Ngech Horng Heang

17. Membrane distillation-liquid desiccant air-conditioning for thermal comfort in buildings

Hung Cong Duong, Hai Thuong Cao, Long Duc Nghiem, Ashley Joy Ansari, Ngoc Lieu Le, Thi Oanh Doan, and Nguyen Cong Nguyen

18. Reverse osmosis (RO) membrane development and industrial applications

Nirenkumar Pathak, Umakant Badeti, Weonjung Sohn, Sherub Phuntsho, and Ho Kyong Shon

19. Potentialities of membrane distillation and membrane crystallization

E. Drioli, F. Alessandro, and F. Macedonio

20. Forward osmosis: Principle and applications in sustainable water and energy development

Duc-Viet Nguyen, Thanh-Tin Nguyen, Rusnang Syamsul Adha, Lei Zheng, Xuan-Thanh Bui, Xiaoli Ma, and Hoang Nhat Phong Vo

21. Application of forward osmosis membrane technology in nutrient recovery and water reuse

Hau Thi Nguyen, Nguyen Cong Nguyen, Shiao-Shing Chen, Huu Hao Ngo, Xuan-Thanh Bui, and Phuong-Thao Nguyen

Contributors

Zohaib Abbas Department of Environmental Sciences and Engineering, Government College University, Lahore, Pakistan

Amila Abeynayaka Institute for Global Environmental Strategies (IGES), Hayama, Kanagawa, Japan

Rusnang Syamsul Adha School of Earth Sciences and Environmental Engineering, Gwangju Institute of Science and Technology, Gwangju, Republic of Korea

F. Alessandro Institute on Membrane Technology, National Research Council of Italy (CNR-ITM), Rende, Italy

Ashley Joy Ansari Strategic Water Infrastructure Laboratory, School of Civil Mining and Environmental Engineering, University of Wollongong, Wollongong, NSW, Australia

Nasrul Arahman Department of Chemical Engineering, Engineering Faculty, Syiah Kuala University, Banda Aceh, Indonesia

Muhammad Bilal Asif Advanced Membranes and Porous Material Center, Physical Sciences and Engineering, King Abdullah University of Science and Technology (KAUST), Thuwal, Saudi Arabia; Institute of Environmental Engineering & Nano-Technology, Tsinghua-Shenzhen International Graduate School, Tsinghua University, Shenzhen, China

Umakant Badeti School of Civil and Environmental Engineering, University of Technology, Sydney, NSW, Australia

Muhammad Roil Bilad Faculty of Integrated Technologies, Universiti Brunei Darussalam, Gadong, Brunei

Varinthorn Boonyaroj Department of Environmental Engineering, Faculty of Engineering, Kasetsart University, Bangkok, Thailand

Xuan-Bui Bui Key Laboratory of Advanced Waste Treatment Technology, Ho Chi Minh City University of Technology (HCMUT); Vietnam National University Ho Chi Minh City (VNU-HCM), Ho Chi Minh City, Vietnam

Xuan-Thanh Bui Key Laboratory of Advanced Waste Treatment Technology & Faculty of Environment and Natural Resources, Ho Chi Minh City University of Technology (HCMUT); Vietnam National University Ho Chi Minh City (VNU-HCM), Ho Chi Minh City, Vietnam

Hai Thuong Cao Le Quy Don Technical University, Hanoi, Vietnam

Paula Carrera Centre for Environmental and Energy Research, Ghent University Global Campus, Incheon, Republic of Korea; Department of Green Chemistry and Technology, Faculty of Bioscience Engineering, Ghent University, Ghent, Belgium

Hicham Chatoui Laboratory of Biological Engineering, Faculty of Sciences and Techniques, Sultan Moulay Slimane University, Beni Mellal; Department of Health and Agro-Industry Engineering, School of Engineering and Innovation of Marrakech (E2IM), Private University of Marrakesh (UPM), Marrakech, Morocco

Guanghao Chen Department of Civil and Environmental Engineering, The Hong Kong University of Science & Technology, Hong Kong, China

Shiao-Shing Chen Institute of Environmental Engineering and Management, National Taipei University of Technology, Taipei, Taiwan, Republic of China

Chart Chiemchaisri Department of Environmental Engineering, Faculty of Engineering, Kasetsart University, Bangkok, Thailand

Wilai Chiemchaisri Department of Environmental Engineering, Faculty of Engineering, Kasetsart University, Bangkok, Thailand

Lijuan Deng Centre for Technology in Water and Wastewater, School of Civil and Environmental Engineering, University of Technology Sydney, Sydney, NSW, Australia

Thi Oanh Doan Hanoi University of Natural Resources and Environment, Hanoi, Vietnam

E. Drioli Institute on Membrane Technology, National Research Council of Italy (CNR-ITM), Rende, Italy

Hung Cong Duong Le Quy Don Technical University, Hanoi, Vietnam; Centre for Technology in Water and Wastewater, School of Civil and Environmental Engineering, University of Technology Sydney, Ultimo, NSW, Australia

Uchenna Egwu School of Engineering, Newcastle University, Newcastle, United Kingdom

Jauad El Kharraz Regional Center for Renewable Energy and Energy Efficiency (RCREEE), Cairo, Egypt

Alberto Figoli Institute on Membrane Technology, CNR, Rende, Italy

Francesco Galiano Institute on Membrane Technology, CNR, Rende, Italy

Adewale Giwa Department of Chemical Engineering, Khalifa University of Science and Technology, Abu Dhabi, United Arab Emirates

Wenshan Guo Centre for Technology in Water and Wastewater, School of Civil and Environmental Engineering, University of Technology Sydney, Sydney, NSW, Australia

Pri Januar Gusnawan Department of Chemical Engineering, Faculty of Industrial Technology, Institut Teknologi Bandung, Bandung, Indonesia

Tran Thi Thai Hang Master Program in Water Technology, Reuse & Management, Vietnamese German University, Binh Duong, Vietnam

Ngech Horng Heang Department of Environmental Engineering, Faculty of Engineering, Kasetsart University, Bangkok, Thailand

Seow Wah How Centre for Environmental and Energy Research, Ghent University Global Campus, Incheon, Republic of Korea; Department of Green Chemistry and Technology, Faculty of Bioscience Engineering, Ghent University, Ghent, Belgium

Sidra Iftekhar Department of Applied Physics, University of Eastern Finland, Kuopio, Finland

Jaewon Jang KEPCO Research Institute (KEPRI), Korea Electric Power Corporation (KEPCO), Naju-si, Jeollanam-do, Republic of Korea

Seongpil Jeong Korea Institute of Science and Technology (KIST) & KIST School, Korea University of Science and Technology (KUST), Seoul, Republic of Korea

Helen Julian Department of Chemical Engineering, Faculty of Industrial Technology, Institut Teknologi Bandung, Bandung; Department of Food Engineering, Faculty of Industrial Technology, Institut Teknologi Bandung, Sumedang, Indonesia

Anekpracha Kaewmanee Department of Environmental Engineering, Faculty of Engineering, Kasetsart University, Bangkok, Thailand

Chaeyeon Kang Centre for Environmental and Energy Research, Ghent University Global Campus, Incheon, Republic of Korea

Selim Karkour Beyond the Borders of Life Cycle Assessment, Tokyo, Japan

K. Khoiruddin Department of Chemical Engineering, Bandung Institute of Technology, Bandung, Indonesia

Hye-Won Kim Korea Institute of Science and Technology (KIST), Seoul, Republic of Korea

Linh-Thy Le Faculty of Public Health, University of Medicine and Pharmacy at Ho Chi Minh City (UMP); Key Laboratory of Advanced Waste Treatment Technology, Ho Chi Minh City University of Technology (HCMUT), Ho Chi Minh City, Vietnam

Ngoc Lieu Le International University; Vietnam National University, Ho Chi Minh City, Vietnam

Tran Le Luu Master Program in Water Technology, Reuse & Management, Vietnamese German University, Binh Duong, Vietnam

Chew Lee Leong Department of Chemical Engineering, Universiti Teknologi PETRONAS, Bandar Seri Iskandar, Malaysia

Jeng-Lung Lin Department of Environmental Engineering; Center for Environmental Risk Management, Chung Yuan Christian University, Taoyuan, Taiwan

L. Lustiyani Department of Chemical Engineering, Bandung Institute of Technology, Bandung, Indonesia

Xiaoli Ma Department of Materials Science and Engineering, University of Wisconsin-Milwaukee, Milwaukee, WI, United States

F. Macedonio Institute on Membrane Technology, National Research Council of Italy (CNR-ITM), Rende, Italy

Mohamed Merzouki Laboratory of Biological Engineering, Faculty of Sciences and Techniques, Sultan Moulay Slimane University, Beni Mellal, Morocco

Sla Min Centre for Environmental and Energy Research, Ghent University Global Campus, Incheon, Republic of Korea

Long Duc Nghiem Centre for Technology in Water and Wastewater, School of Civil and Environmental Engineering, University of Technology Sydney, Ultimo, NSW, Australia

Huu Hao Ngo Centre for Technology in Water and Wastewater, School of Civil and Environmental Engineering, University of Technology Sydney, Sydney, NSW, Australia

My Thi Tra Ngo Key Laboratory of Advanced Waste Treatment Technology & Faculty of Environment and Natural Resources, Ho Chi Minh City University of Technology (HCMUT); Vietnam National University Ho Chi Minh City (VNU-HCM), Ho Chi Minh City, Vietnam; Graduate School of Engineering, Nagasaki University, Nagasaki, Japan

Duc-Viet Nguyen Graduate School of Water Resources, Sungkyunkwan University, Suwon, Republic of Korea

Han Ngoc Mai Nguyen Key Laboratory of Advanced Waste Treatment Technology & Faculty of Environment and Natural Resources, Ho Chi Minh City University of Technology (HCMUT); Vietnam National University Ho Chi Minh City (VNU-HCM), Ho Chi Minh City, Vietnam

Hau Thi Nguyen Faculty of Chemistry and Environment, Dalat University, Dalat, Vietnam

Nguyen Cong Nguyen Faculty of Chemistry and Environment, Dalat University, Dalat, Vietnam

Phuong-Thao Nguyen Key Laboratory of Advanced Waste Treatment Technology & Faculty of Environment and Natural Resources, Ho Chi Minh City University of Technology (HCMUT); Vietnam National University Ho Chi Minh City (VNU-HCM), Ho Chi Minh City, Vietnam

Thanh-Tin Nguyen School of Earth Sciences and Environmental Engineering, Gwangju Institute of Science and Technology, Gwangju, Republic of Korea

Trung-Hieu Nguyen Department of Civil Engineering; Department of Environmental Engineering; Center for Environmental Risk Management, Chung Yuan Christian University, Taoyuan, Taiwan

Bing-Jie Ni Centre for Technology in Water and Wastewater, School of Civil and Environmental Engineering, University of Technology Sydney, Sydney, NSW, Australia

Ashok Pandey Centre for Innovation and Translational Research, CSIR-Indian Institute of Toxicology Research, Lucknow, Uttar Pradesh; Sustainability Cluster, School of Engineering, University of Petroleum and Energy Studies, Dehradun, India

Nirenkumar Pathak School of Civil and Environmental Engineering, University of Technology, Sydney, NSW, Australia

Minh-Thuan Pham Department of Civil Engineering; Department of Environmental Engineering; Center for Environmental Risk Management, Chung Yuan Christian University, Taoyuan, Taiwan

Sherub Phuntsho School of Civil and Environmental Engineering, University of Technology, Sydney, NSW, Australia

Dian Qoriati Department of Environmental Engineering; Center for Environmental Risk Management, Chung Yuan Christian University, Taoyuan, Taiwan

Chu Xuan Quang Center for Advanced Materials Technology, National Center for Technological Progress, Ha Noi, Vietnam

Safa Rachid Mining Environment and Circular Economy, Mohammed VI Polytechnic University (UM6P), Ben Guerir, Morocco

Paul Sallis School of Engineering, Newcastle University, Newcastle, United Kingdom

Tjandra Setiadi Center for Environmental Studies; Research Center for Bioscience and Biotechnology, Bandung Institute of Technology; Department of Chemical Engineering, Faculty of Industrial Technology, Institut Teknologi Bandung, Bandung, Indonesia

Norazanita Shamsuddin Faculty of Integrated Technologies, Universiti Brunei Darussalam, Gadong, Brunei

Ho Kyong Shon School of Civil and Environmental Engineering, University of Technology, Sydney, NSW, Australia

Utjok W.R. Siagian Department of Petroleum Engineering, Bandung Institute of Technology, Bandung, Indonesia

Muhammad Ahmar Siddiqui Department of Civil and Environmental Engineering, The Hong Kong University of Science & Technology, Hong Kong, China

Weonjung Sohn School of Civil and Environmental Engineering, University of Technology, Sydney, NSW, Australia

Hazwani Suhaimi Faculty of Integrated Technologies, Universiti Brunei Darussalam, Gadong, Brunei

Hismi Susane Department of Civil Engineering; Department of Environmental Engineering; Center for Environmental Risk Management, Chung Yuan Christian University, Taoyuan, Taiwan

Shamas Tabraiz School of Engineering, Newcastle University, Newcastle; Natural and Applied Sciences Section, School of Psychology and Life Sciences, Canterbury Christ Church University, Canterbury, United Kingdom

Nguyen Nhat Thoai Master Program in Water Technology, Reuse & Management, Vietnamese German University, Binh Duong, Vietnam

Tran Hung Thuan Center for Advanced Materials Technology, National Center for Technological Progress, Ha Noi, Vietnam

Cong-Sac Tran Key Laboratory of Advanced Waste Treatment Technology, Ho Chi Minh City University of Technology (HCMUT); Vietnam National University Ho Chi Minh City (VNU-HCM), Ho Chi Minh City, Vietnam

Duyen Phuc-Hanh Tran Department of Civil Engineering; Department of Environmental Engineering; Center for Environmental Risk Management, Chung Yuan Christian University, Taoyuan, Taiwan

La Vinh Trung Master Program in Water Technology, Reuse & Management, Vietnamese German University, Binh Duong, Vietnam

Nguyen Van Tuyen Center for Advanced Materials Technology, National Center for Technological Progress, Ha Noi, Vietnam

Hoang Nhat Phong Vo Queensland Alliance for Environmental Health Sciences (QAEHS), The University of Queensland, Woolloongabba, QLD, Australia

Qilin Wang Centre for Technology in Water and Wastewater, School of Civil and Environmental Engineering, University of Technology Sydney, Sydney, NSW, Australia

Ya-Fen Wang Department of Environmental Engineering; Center for Environmental Risk Management, Chung Yuan Christian University, Taoyuan, Taiwan

Wei Wei Centre for Technology in Water and Wastewater, School of Civil and Environmental Engineering, University of Technology Sydney, Sydney, NSW, Australia

I.G. Wenten Department of Chemical Engineering; Research Center for Bioscience and Biotechnology, Bandung Institute of Technology, Bandung, Indonesia

I.N. Widiasa Chemical Engineering Department, Diponegoro University, Semarang, Indonesia

Chayanid Witthayaphirom Department of Environmental Engineering, Faculty of Engineering, Kasetsart University, Bangkok, Thailand

Vita Wonoputri Department of Chemical Engineering, Faculty of Industrial Technology, Institut Teknologi Bandung, Bandung, Indonesia

Di Wu Centre for Environmental and Energy Research, Ghent University Global Campus, Incheon, Republic of Korea; Department of Green Chemistry and Technology, Faculty of Bioscience Engineering, Ghent University, Ghent, Belgium; Department of Civil and Environmental Engineering, The Hong Kong University of Science & Technology, Hong Kong, China

Sheng-Jie You Department of Environmental Engineering; Center for Environmental Risk Management, Chung Yuan Christian University, Taoyuan, Taiwan

Ahmed Yusuf Department of Chemical Engineering, Khalifa University of Science and Technology, Abu Dhabi, United Arab Emirates

Muhammad Zeeshan German Environment Agency, Section II 3.3; Department of Water Quality Control, Technical University of Berlin, Berlin, Germany

Lei Zheng Chongqing Institute of Green and Intelligent Technology, Chinese Academy of Sciences, Chongqing, China

Preface

The book titled *Membrane Technology for Sustainable Water and Energy Management* is a part of the Elsevier comprehensive book series Current Developments in Biotechnology and Bioengineering (Editor-in-Chief: Ashok Pandey). As water is an indispensable essential element in maintaining life and all human activities, meeting the demand for water in terms of both quality and quantity is a prerequisite for sustainable development. Since the beginning of the 20th century, global water consumption has increased sevenfold, mainly due to population growth and individual water needs. According to the current assessment regarding water resources, one-third of the countries in the world suffer from water shortage. By 2025, this number will increase to two-thirds, which means 35% of the world's population will face severe water scarcity. In addition, among the 17 Sustainable Development Goals (SDGs) of the United Nations, two goals are to ensure sustainable water resources and develop clean energy. Therefore, inexpensive, energy-saving, minimal or no chemical consumption water treatment technologies have been developed for adequate water supply and effective energy utilization as well as for sustainable water and energy management.

Due to an inextricable linkage between water and energy consumption, the augmentation of water supply must not come with a high cost or energy consumption. In recent years, membrane technology has attracted worldwide attention as an up-and-coming solution to improve the sustainability of water resources by exponential advantages, as follows:

- Achieves better contaminant removal efficiency compared to traditional activated sludge technology by blocking disease-causing bacteria and viruses, avoiding the risk of microbial outbreaks.
- Meets strict discharge regulations and treated wastewater is reusable for industrial, irrigation, and even potable purposes.
- Requires only half or less than the area of a conventional water treatment plant by modular design, allowing easy replication and scaling based on existing systems.
- Does not involve boiling, allowing a significant reduction in energy consumption, which is well suited for applications with renewable energy sources such as solar, wind, and tidal power.

This book consists of 21 chapters and are divided into three parts, focusing on the applications of membrane technology and energy management with inputs from leading experts in the field. Part A presents general problems regarding membrane properties, current membrane types available on the market (e.g., photocatalytic material-based membranes), and techniques used to control membrane fouling and desalination. In Part B, applications of membrane technologies (such as aerobic membrane bioreactors (MBR) and anaerobic MBR, ultralow-pressure membranes, and self-contained dynamic membrane bioreactors) to treat different types of waters and wastewaters are presented. Nutrient removal, microalgae harvesting, and the reduction of microplastic emissions to minimize their impacts on environment and human health are also covered in this part. Part C discusses the membrane processes for desalination, nutrient recovery, and water reuse, including forward osmosis membranes for sustainable water treatment and energy development, membrane distillation for concentrated brine treatment, transitional osmosis, and liquid desiccant air conditioners incorporating membrane distillation to provide energy-saving thermal comfort in buildings.

We express our sincere gratitude to the authors for contributing their high-quality work in a timely manner and revising it appropriately at short notice. We thank the reviewers, who reviewed and provided their valuable suggestions to the authors to improve the chapters for this book. We also thank and express our appreciation to Dr. Kostas Marinakis, former Senior Acquisitions Editor, Elsevier, the Netherlands;

Ms. Helena Beauchamp, and Ms. Catherine Costello, former and current editorial project managers, respectively, and the entire production team of Elsevier for supporting us constantly during the editorial process. Further, support from Ho Chi Minh City University of Technology (HCMUT) and Vietnam National University Ho Chi Minh (VNU-HCM) for the strong research group NCM2021-20-01 is highly appreciated.

We strongly believe that the information and technologies mentioned in this book can not only enhance the interdisciplinary scientific skills of the readers but also deepen their fundamental knowledge on membrane-based sustainable water and energy management.

Xuan-Thanh Bui

Wenshan Guo

Chart Chiemchaisri

Ashok Pandey

General on membrane, materials and application

1

Classification of membranes: With respect to pore size, material, and module type

Jaewon Jang

KEPCO Research Institute (KEPRI), Korea Electric Power Corporation (KEPCO), Naju-si, Jeollanam-do, Republic of Korea

1. Introduction

A membrane is a thin physical interface through which certain species are passed or filtered according to their physical/chemical properties. Membranes are divided into membranes that are found on the cell wall of an organism and the membranes that separate specific components in a solution or gas. This chapter intends to deal with the membranes that separate substances from nature, not the membranes related to life activities. In general, membranes are classified into two types: isotropic and anisotropic.

Isotropic membranes have a uniform composition and are classified into microporous and nonporous membranes (Baker, 2012). Microporous membranes have pore sizes in the range 0.01–10 µm, and there are microfiltration (MF) and ultrafiltration (UF) membranes according to the pore size. Microporous membranes are fabricated by phase inversion (controlling the polymer phase change reaction), track etching (irradiating a nonporous polymer membrane with heavy ions), and stretched polymer film (a technique in which a polymer is heated and stretched above its melting point) techniques (Lalia et al., 2013). Among them, phase inversion technique, a representative fabricating method, produces a polymer solution, casts it on a substrate, and causes a reaction by loading it with a nonsolvent on the polymer. Since the nonsolvent for most polymers is water, when the casted polymer film is immersed in water, the polymer precipitates and forms a membrane. Track-etched membranes are fabricated by attacking polymer chains with charged particles, irradiating them, and passing them through an etching solution. As the parts damaged by the particles in the membrane dissolve, pores

Copyright © 2023 Elsevier Inc. All rights reserved.

are formed. The stretched membrane is an oriented crystalline polymer membrane with pores created by extrusion and stretching processes. It is rapidly extruded by heating to a temperature close to the melting point of the polymer, then cooled and annealed to 300% of its initial state. In this process, slit-shaped pores (200–2500 Å) are formed. A nonporous (dense) membrane has pores with a diameter of several nanometers or less. In the process of using a nonporous membrane, substances can move according to diffusion by concentration difference, movement by pressure or electric field gradient. Therefore, the separation properties of the solutes in this situation are determined by differences in solubility for the membrane or differences in relative transport rates. Nonporous membranes are mainly used for nanofiltration (NF), forward osmosis (FO), reverse osmosis (RO), pervaporation (PV), and gas permeation.

Anisotropic membranes are nonuniform in composition or cross-sectional structure, and are composed of several layers. They are divided into two types: phase separation membranes (Loeb-Sourirajan membranes) and thin-film composite (TFC) membranes. The phase separation membranes have uniform chemical composition, but the pore sizes are different for each type of the membrane. On the other hand, TFC membranes are chemically and structurally nonuniform. In some cases, anisotropic membranes can also be fabricated by phase inversion method like the abovementioned isotropic membranes, but the formed pore sizes are nonuniform because different materials and process conditions are used. The anisotropic membrane is a structure in which a porous layer acts as a support, and a dense thin layer is formed on the surface. These layers are composed of different polymeric materials. The thin surface layer (active layer), which determines the performance of the TFC membrane, can be synthesized by various methods such as interfacial polymerization, solution coating, and plasma polymerization. The polymers such as cellulose acetate (CA), polyacrylonitrile (PAN), polyetherimide (PEI), polyethersulfone (PES), polyamide (PA), polycarbonate (PC), cross-linked polyether, polypropylene (PP), and polyvinylidene fluoride (PVDF), etc., are used (Ulbricht, 2006).

2. Types of membranes

Among the membranes that separate different substances in nature, a membrane for desalination or water treatment is a representative case. It basically serves to separate certain solutes and solvents from solutions. This kind of membrane can be classified into MF, UF, NF, and RO according to pore size (Fig. 1). The MF membrane can remove only large particles such as microorganisms, and since the UF membrane has slightly smaller pores than the MF membrane, microorganisms, bacteria, and proteins can be removed. The NF membrane is capable of removing tiny molecules and multivalent ions such as calcium, magnesium, iron, and manganese. The latest NF membrane technology has reached a level where it can reject up to 65% of NaCl from saline water (Fallahnejad et al., 2022; Park et al., 2012). The RO membrane can remove the solutes mentioned above, salt ions, and very small organic matter.

2.1 MF and UF membranes

The MF and UF processes are methods of filtering out specific impurities whose size is larger than the pores existing in the membrane. The MF membranes have pores larger

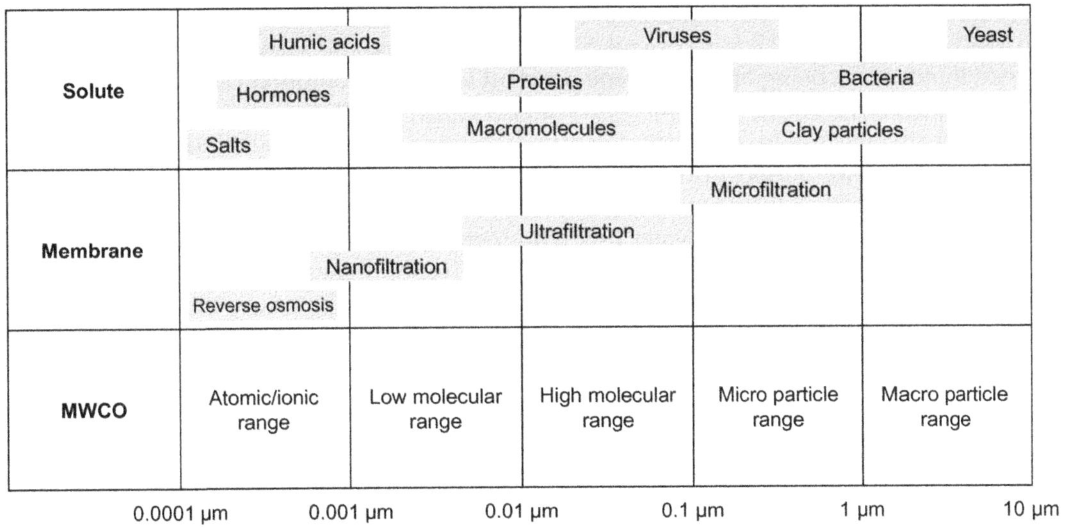

Solute	Humic acids		Viruses		Yeast
	Hormones		Proteins	Bacteria	
	Salts		Macromolecules	Clay particles	
Membrane				Microfiltration	
			Ultrafiltration		
		Nanofiltration			
	Reverse osmosis				
MWCO	Atomic/ionic range	Low molecular range	High molecular range	Micro particle range	Macro particle range

| 0.0001 μm | 0.001 μm | 0.01 μm | 0.1 μm | 1 μm | 10 μm |

FIG. 1   Classification of the membrane-based process according to molecular weight cut-off (MWCO).

(0.1–10 μm) than that of UF membranes and are usually able to reject large particles, asbestos, and various cellular materials such as red blood cells and bacteria. The UF membranes have pores sizes in the range 0.01–0.1 μm and can remove microorganisms, pyrogens, proteins, and viruses that are larger than the pores (Pellegrin et al., 2013). The two types of membranes are often fabricated from the same material, but some of the fabrication processes are different, resulting in different pore sizes. Materials such as PVDF, PAN, polysulfone (PSU), and poly (acrylonitrile)-poly(vinyl chloride) (PVC) copolymer have been used for the preparation of MF and UF membranes (Ulbricht, 2006). Cellulose acetate-cellulose nitrate blend, nylon, and poly(tetrafluoroethylene) are also used for the preparation of MF membrane. For preparing UF membranes, poly(ether sulfone) is also widely used (Baker, 2012).

The MF membrane can be used to separate particles larger than the colloid that do not dissolve in solution and is used in water and sewage treatment, desalination process pretreatment, food industry, biochemical application, and medical field. UF membranes are used in various areas such as water treatment, water remediation, surfactant recovery, food processing, protein separation, and gene engineering. Typically, UF membranes have anisotropic Loeb-Sourirajan structure. The first UF membranes were made from nitrocellulose in the 1900s. UF membranes developed rapidly in the 1960s with the introduction of CA-based products and began to be applied to industry in earnest in the 1970s (Saline Water Conversion—II, 1963). Recently, various materials such as TiO_2, Al_2O_3, ZrO_3, PAN, PES, and PVDF have been studied (Ulbricht, 2006).

2.2 NF membranes

NF membrane provides intermediate performance over UF and RO membranes. CA blends or polyamide composites are mainly used for preparing NF memebranes (Nunes and Peinemann, 2001). It has a nonporous structure and can filter out species with size greater than 0.001 μm. Accordingly, most organic molecules, viruses, and salts can be removed by NF process. In addition, using NF membrane divalent ions in the solution can be removed, so it can be used to convert hard water into soft water (Hilal et al., 2004). Because of these uses, NF membranes are essential in commercial desalination and water treatment industries and have been used for a long period in treating aqueous and organic solutions (Artuğ et al., 2007; Oatley-Radcliffe et al., 2017; Zhou et al., 2015). In general, NF operates at a much lower pressure than RO, and exhibits high water flux with high rejection for divalent ions. Commercial NF membranes include NF90, NF200, NF270 (Dow Filmtec, Dow Chemical Company, USA), K-SR2 (Koch Membrane Systems Inc., USA), ESNA1-LF2 (Hydranautics, USA), and NF99HF (Alfa Laval, USA) are representative products.

2.3 RO membranes

RO membrane was commercialized by Gulf General Atomics and Aerojet General by applying Loeb-Sourirajan CA membrane to produce spiral wound modules (Amjad et al., 1993). The CA membranes are made from acetylated cellulose, a polymer obtained naturally from cotton plant (Sourirajan, 1977). Cellulose acetylation proceeds through the following reaction (Fig. 2) (Stevens, 1999). The degree of acetylation can be defined by number of OH group pendants in cellulose substituted with acetyl groups. Acetylation has a significant influence on the properties of the CA membrane. The CA membranes with a low degree of acetylation tend to have high flux but low rejection. The acetylation levels can be classified from 0 to 3, and the maximally acetylated CA is called cellulose triacetate (CTA). The acetylation level of commercial CA membranes used in the RO process is usually 2.7 (Sagle and Freeman, 2004). This composition exerts the best balance between water flux and salt rejection. In addition, it is known that the greater the degree of acetylation, the greater the physical/chemical stability when the membrane is made (Sagle and Freeman, 2004; Yadav and Hakkarainen, 2021; Yang et al., 2018). When CA-based membrane is operated at 1500–2000 psig using NaCl feed solution with a concentration of 52,500 mg/L, a flux of about 5–11 gal per square foot per day (GFD) and a rejection of 99.5% are obtained (Saline Water Conversion—II, 1963). Using the CA membrane in an RO process has several advantages, such as ease of fabrication as a membrane module, excellent mechanical properties, and resistance to chlorine. In particular, it can withstand free chlorine up to 5 ppm without damaging the material (Sourirajan, 1977). A disadvantage of CA-based membranes is that their selectivity decreases as they become hydrolyzed over time (Sourirajan, 1977). In addition, it is very sensitive to changes in pH (stable only in the pH range 4–6), and there is a problem that the salt rejection decreases as the temperature increases (Baker, 2012).

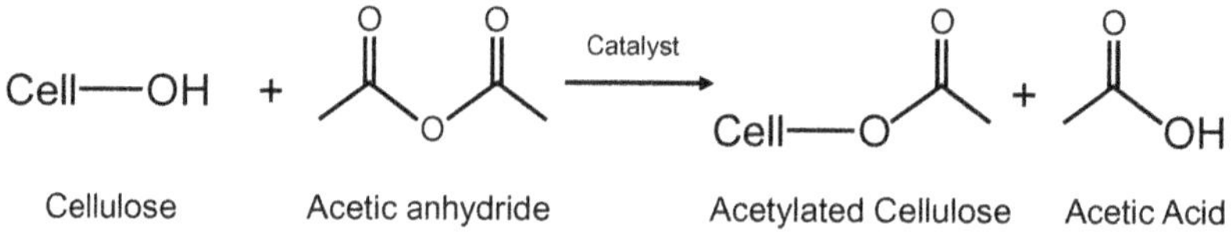

FIG. 2   Scheme of the cellulose acetylation reaction.

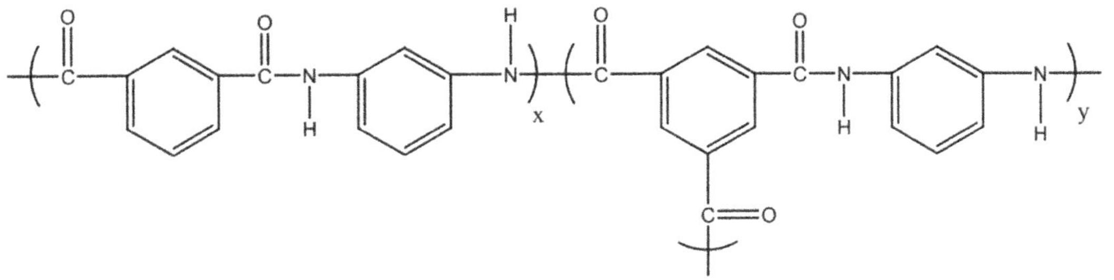

FIG. 3 Chemical structure of commercial polyamide selective layer in TFC membranes.

To overcome these problems, the TFC RO membrane was developed in 1972 and replaced the CA membrane. Usually, the TFC RO membrane is made of a polyamide selective layer (~200 nm) (Fig. 3), porous polysulfone layer (40–50 μm), and polyester support (120–150 μm). Unlike UF, MF, and NF membranes, the pores in the TFC RO membrane are very small (about 0.0001–0.001 μm), so it can be classified as a nonporous membrane. Accordingly, it exhibits a semipermeable property and can remove salt ions, minerals, metal ions, and organic molecules of low molecular weight. Due to these characteristics, the RO membrane is a very important element in the water treatment process, and the transport mechanism in the RO membrane is considered to be diffusion through the free volume area (Fritzmann et al., 2007; Greenlee et al., 2009; Lee et al., 2011; Peñate and García-Rodríguez, 2012; Van der Bruggen and Vandecasteele, 2002). In general, RO membranes are used to squeeze out water while leaving only salt in desalination process. In this case, a water pressure that can overcome the osmotic pressure of the brine should be applied to the concentrated side. Polyamide TFC membrane exhibits a flux of 27 GFD and rejection of 99.5% when an NaCl aqueous solution at a concentration of 2000 mg/L is supplied at temperature of 77°F and a pH of 7.5 at 225 psig. The degree of passage through the RO membrane depends on the solute during process operation because each solute has different solubility or mobility with respect to the RO membrane. The rejection properties of TFC membranes are determined by the surface layer of the cross-linked aromatic polyamide as the feed solution passes through the membrane (Baker, 2012). Unlike CA-based membranes, the TFC RO membranes show stable properties over a wide range of pH and temperature. However, one drawback is that it is very vulnerable to attacks by chlorine. When chlorine attacks polyamide, ring chlorination occurs, hydrogen bonding between chains are inhibited, and the polymer matrix is decomposed (Avlonitis et al., 1992). Owing to this phenomenon, the salt rejection of the TFC RO membrane is rapidly reduced when it comes into contact with chlorine. Therefore, when using the TFC RO membrane, a pretreatment process to remove chlorine must be performed before the feed solution is introduced.

3. Materials of membrane

When designing a membrane, materials, water flux, solute rejection, module configuration, mechanical/chemical/thermal/temporal stability, applicability to large-scale processes, and economic feasibility must be considered. Among these parameters, the basic performance of the membrane is mainly determined by the physical/chemical properties of the constituent materials and the structure of the pores. Thus, research to discover new materials and improve the membrane fabricating process is being actively conducted. Synthetic polymers such as PVDF, PSU, PAN, and PAN-PVC copolymers are the materials most commonly used for fabricating the UF, MF, NF, and RO membranes (Ulbricht, 2006). In addition, the membrane can be made of ceramic materials such as ceramics or zeolite, but these membranes have some demerits such as high cost for scale-up and high brittleness. In order to overcome the limitations of conventional materials, research to introduce novel materials or to improve the fabricating process has been actively conducted in recent years.

3.1 Organic membrane

A representative example of an organic-based membrane is a polymer membrane. Although thermal stability, chemical resistance, and stability to specific organic solvents are weaknesses that need to be overcome, many polymer membranes have been developed and actively used in various fields. The most intensively used polymer materials are PSf, PES, PVDF, PP, CA, polyvinyl alcohol (PVA), polytetrafluoroethylene (PTFE), and polyimide (PI) (Fig. 4) (Ahmad et al., 2013; Baker, 2012; Kim et al., 2002; Liu et al., 2011; Lohokare et al., 2008; Qin et al., 2003; Saljoughi and Mousavi, 2012; Shibutani et al., 2011; Soyekwo et al., 2014; Yuliwati and Ismail, 2011).

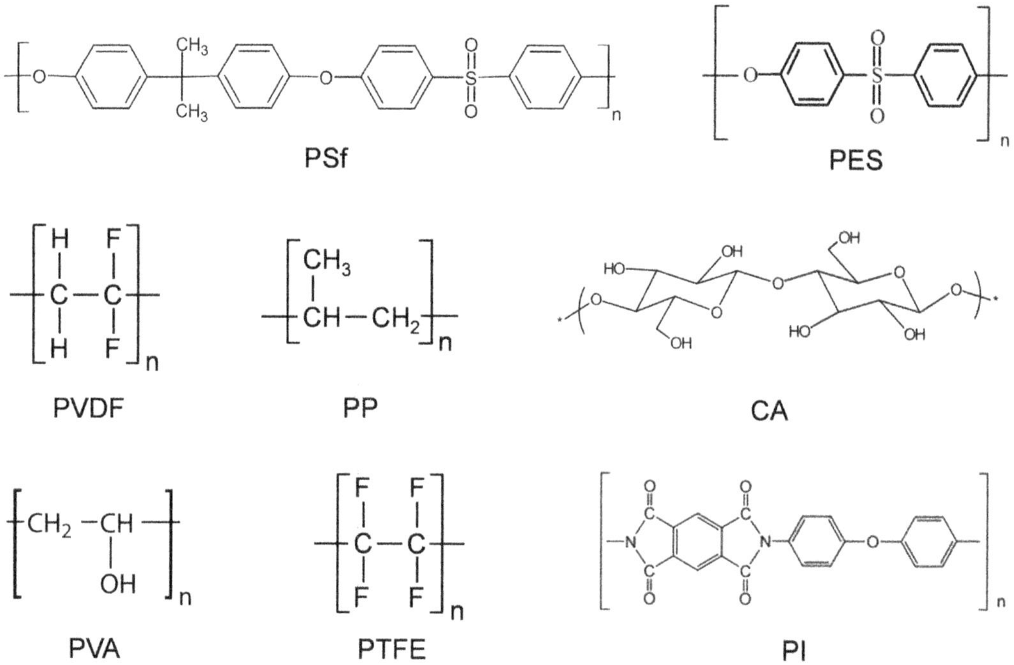

FIG. 4   Chemical structure of the polymers used in organic membrane.

Among them, PSf and PES are widely used as membranes for UF processes because they have excellent permeability, selectivity, mechanical stability, and chemical resistance, and are also used as membrane supports for NF and RO processes (Ahmad et al., 2013; Nady et al., 2011; Zhao et al., 2013). In the case of PP and PVDF, they have good chemical resistance and mechanical stability, and they were mainly used as membranes for MF processes because of their large pore size (Kang and Cao, 2014). The CA membranes are suitable for pressure-driven filtration processes because of their good thermal stability and water flux. They are easy to handle and can be sterilized by heating. Cellulose nitrate membranes are usually a mixture of cellulose nitrate and cellulose acetate. It can be used to recover biological samples from a solution due to its high protein adsorption property. PVA membrane is hydrophilic and has excellent mechanical properties. It can be used for various purposes, and it is also possible to use it as a selective layer by coating it on the surface of PVDF or PSf membrane. PTFE membrane has heat resistance, chemical resistance, and excellent mechanical properties and is used for liquid filters and air filters. In particular, it is used in high-tech applications such as a support layer for a proton exchange membrane in a fuel cell, a cable for the semiconductor device, a waterproof membrane for cell phones, and as a membrane for moisture venting in automobiles. The PI membrane has high thermal stability, chemical resistance, and gas selectivity. Although PI membranes can be applied in many fields, they are very useful as gas separation membranes.

Polymer-based membranes, despite their usefulness, have a disadvantage that they are susceptible to fouling due to their hydrophobic surfaces. This problem not only shortens the life of the membrane, but also deteriorates the process performance over time. Fouling is formed by the buildup of organic/inorganic substances, proteins, microorganisms, and microbial on membrane surfaces (Rana and Matsuura, 2010). Therefore, it is necessary to develop a polymer material that prevents the formation of fouling on the surface or a polymer membrane that is easy to clean with foulants.

3.2 Inorganic membranes

Inorganic materials are applied for preparing membranes because of their excellent chemical, thermal, and mechanical durability and the advantages of being easy to reuse after washing (Baker, 2012). Ceramic membranes, which are representative example of inorganic membranes, have better stain resistance and chemical durability than polymer-based membranes, so they are widely used in the water treatment field. For ceramic membranes, Al_2O_3 is most commonly used, and materials such as TiO_2, ZrO_2, ZnO, and SnO_2 are being studied (DeFriend et al., 2003). Among these metal oxide-based materials, TiO_2 has photocatalytic properties, thus TiO_2-incorporated membrane can decompose pollutants/organic species/microorganisms in addition to separating solute and solvent. This demonstrates the effect of delaying fouling on the membrane surface and the sterilization effect during water treatment. In addition to this, ceramic membranes that combine different materials such as TiO_2-SiO_2, TiO_2-ZrO_2, Al_2O_3-SiC, Ag-TiO_2, Zn-CeO_2, and zeolite are also being developed (Kumar et al., 2014; Mohmood et al., 2013). However, ceramic-based membranes are expensive and brittle compared to polymer-based membranes, which are the drawbacks that need to be overcome. Metal-based membranes are mainly made of stainless steel and have very fine pores. The main applications of these membranes are gas separation or used as a membrane support for high-temperature water treatment (Pellegrino and Sikdar, 2017).

3.3 Hybrid membranes

If a hybrid membrane is fabricated by mixing organic and inorganic materials, the performance that is difficult to obtain with a single material can be derived. When a membrane is made of a polymer-based composite material, mechanical strength, chemical/thermal stability, oxidation resistance, pH durability, and antifouling properties can be improved (Lee et al., 2016). Materials frequently combined with polymers include metal oxides such as Al_2O_3, ZnO, SiO_2, TiO_2, and Fe_2O_3, metals such as Cu and Ag, and carbon materials such as graphene and carbon nanotubes (Jeong et al., 2007; Majumder et al., 2005). When introducing these materials to the membrane, techniques such as phase inversion, interfacial polymerization, self-assembly, layer-by-layer, and surface grafting are adopted (Yin and Deng, 2015).

3.4 Novel materials for membrane

Recently, many research results using SiC as membrane support have been reported. For example, research on reducing the defect density of SiC by forming γ-Al_2O_3 nanocrystals on the SiC membrane via coating, drying, and sintering boehmite sol was reported (Facciotti et al., 2014). In addition, to solve issues such as low chemical stability and cost associated with SiC UF membranes, SiC membranes are produced in a one-step process by pyrolysis of allylhydrido polycarbosilane in the presence of α-SiC particles (König et al., 2014).

Research applying inorganic nanoparticles to water purification are also developing rapidly. Various nanoparticles such as gold, silver, copper, and core-shell nanocomposites have heavy metal adsorption properties, disinfection, and sterilization effects through photoelectric effect (Allred and Tost, 2014; Amin and Alazba, 2014; Ghasemzadeh et al., 2014; Li et al., 2015; Pendergast and Hoek, 2011; Ponder et al., 2000; Pradeep and Anshup, 2009; Sharma et al., 2008; Stoimenov et al., 2002; Thatai et al., 2014). Metal oxides such as Fe_3O_4, Al_2O_3, MgO, and TiO_2 are commonly used in this field. In addition to these, silver nanoparticles are also widely used because they have strong antibacterial properties. There are research cases where Ag NPs were applied directly to the membrane surface or combined with different materials to realize antibacterial properties (Cruz et al., 2015; Dumée et al., 2015; Mauter et al., 2011; Wehling et al., 2015).

In addition to this, many research results have been reported in which nanomaterials with two-dimensional shapes are applied to membranes (Huang et al., 2014; Jang et al., 2020, 2021, 2022; Liu et al., 2015). As a representative case, there are studies are conducted using graphene oxide (GO). GO nanosheets are hydrophilic due to the oxygen functional groups on the surface, and their controllable physicochemical properties, nanometer-scale pores, and excellent mechanical properties of individual nanosheets are beneficial for water treatment applications. GO can be deposited by dip coating or vacuum filtration on a ceramic or polymer support (Lou et al., 2014). Moreover, since composites can be synthesizes using GO with various materials, the properties can be engineered according to the purpose.

Therefore, various applications such as gas separation, oil/water separation, heavy metal removal, desalination, etc., are possible.

4. Membrane configurations and modules

Membrane modules are classified into four categories: plate-and-frame, tubular, hollow fiber, and spiral wound (Fig. 5). The plate-and-frame module is the simplest type of module and consists of two endplates, flat sheet membranes, supporting plates/spacers, and feed distribution plates. In this module, membranes and spacers are alternately stacked to form a sandwich structure (Yasukawa et al., 2018). The membranes constituting the module can be fabricated by interfacial polymerization method (Fig. 6). The plate is internally porous and has flow channels. This module has two modes: flow perpendicular to the membrane and cross-flow. In the case of flow perpendicular to the membrane, when the feed solution passes through the membrane plate, the cake layer builds up and the effective pore size decreases. However, in the case of cross-flow, membrane fouling is significantly reduced. Advantages of this module include good solute removal from water and easy membrane surface cleaning. The demerits are low packing density (100–$400\,\mathrm{m^2/m^3}$), high pressure drop, and frequent fouling formation (Balster, 2013; De, 2014). However, when the contamination of membrane becomes severe, the membrane can be easily replaced with a new one. The plate-and-frame module can be applied to rough-water purification prior to UF, NF, and RO process. In particular, it is very effective for PV applications. However, since this module has a small area per unit volume, a spiral wound or hollow fiber module is more effective for gas separation (Balster, 2013). Two limitations in utilizing plate-and-frame modules are (1) lack of optimized membrane

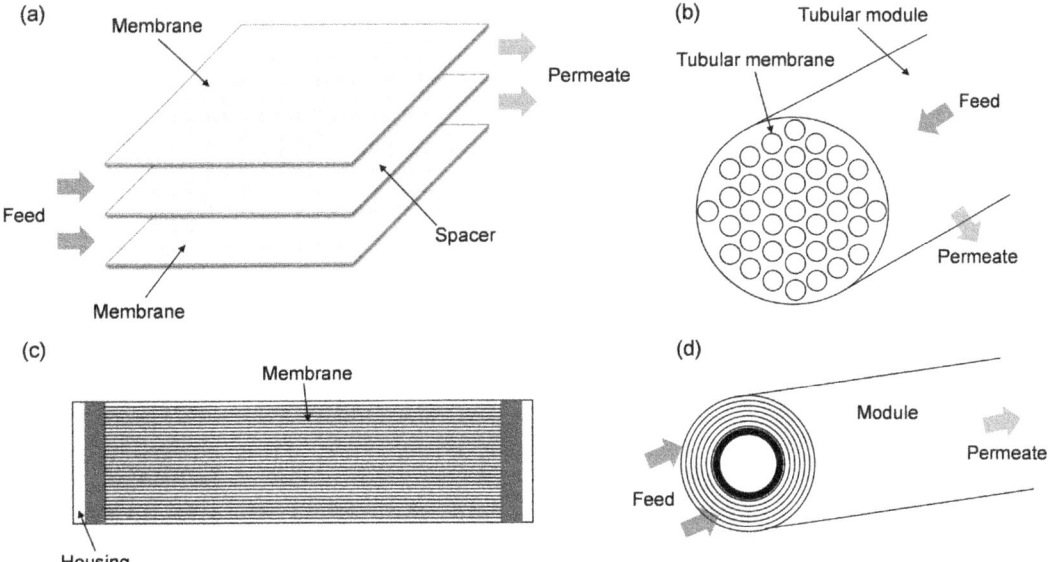

FIG. 5 Structure of (a) plate-and-frame, (b) tubular, (c) hollow fiber, and (d) spiral wound modules.

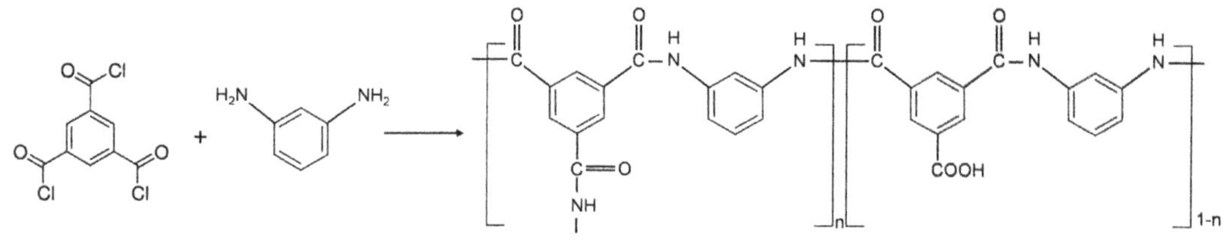

Trimesoyl chloride   n-phenylene diamine   Crosslinked structure   Linear structure

FIG. 6   Scheme of the interfacial polymerization reaction.

support for each system condition and (2) low packing density. Without proper membrane support, complicated process control is required, as it must operate at low water pressure and equalize the pressure across the membrane. In addition, lower packing density requires a larger system footprint and higher capital and operating costs.

In the case of the tubular module, membrane tubes with a diameter of 2–15 nm are packed in the shell. The tubular membrane is self-supporting, can withstand water pressure without deformation, and can be easily bundled directly inside the storage container. It has a packing density of 30–200 m^2/m^3 (Geankopolis, 2005; Pal, 2017; Seader et al., 2016). Ceramic-based tubular membranes are fabricated by slip casting or extrusion (Fan et al., 2016). Polymer-based tubular membranes can be fabricated by spinning technique, and an active layer is formed by interfacial polymerization (Emonds et al., 2021). The feed solution is pumped into the tubes and the permeate passes through the tubes and collects into the common header. However, this module also allows the liquid to flow freely on both sides of the membrane. The main difference between hollow fibers and tubular membranes is the flow area achievable on the bore side. Only laminar flow can be achieved in hollow fibers (internal diameter < 1 mm), thus limiting mixing at the membrane surface. With tubular membranes (inner diameter ≥ 2 mm), turbulence can be easily achieved, reducing concentration polarization, contamination, and scaling (Basile and Nunes, 2011). The tubular module is applicable to UF and RO fields. The relatively high cost is a disadvantage, but the advantage of resistance to membrane fouling due to good hydrodynamics makes up for the high cost.

Hollow fiber modules are fabricated by inserting a hollow fiber bundle into a pressure vessel. Packing density is in the range 500–9000 m^2/m^3 (Geankopolis, 2005; Seader et al., 2016). The hollow fiber membrane is a semipermeable membrane developed in the 1960s for RO applications. Since then, the scope of application has been expanded to include water treatment, desalination, cell culture, medical applications, and pharmaceuticals [Encyclopedia of Life Support Systems (Eolss), 2010]. There are four representative methods for fabricating hollow fiber membrane as follows:

(i)   Melt spinning: Molten thermoplastic polymer is extruded into the air through a spinneret.
(ii)   Dry spinning: Extrusion of the polymer dissolved in a suitable solvent into the air.
(iii)   Dry-jet wet spinning: After extruding the polymer dissolved in an appropriate solvent into the air, it is immersed in water.
(iv)   Wet spinning: Extrusion of the polymer dissolved in an appropriate solvent directly into the water.

The pore size and wall thickness of the fabricated hollow fiber membrane depend on variables such as the spinneret dimension, temperature, air gap, coagulant, dope and bore solution composition, and the speed of the motorized spool. A hollow fiber membrane is also classified as an isotropic or anisotropic membrane. An anisotropic hollow fiber membrane with a dense outer layer has been widely used industrially. This is commonly referred to as TFC hollow fiber membrane, and the performance is dependent on the thin and dense composite layer formed on the support. This surface layer can be controlled by various methods as follows.

(i) interfacial polymerization (Fig. 6),
(ii) dipping method,
(iii) plasma treatment,
(iv) chemical reaction; chemical vapor deposition, sputtering, and spray pyrolysis.

There is a method in which the feed solution passes through the fiber shell and exits through the end of the fiber, and another method in which the feed enters at the bore side and passes through the shell. The disadvantages of hollow fiber modules are that they have low fouling resistance and are difficult to clean. Thus, they need to be replaced frequently, but they are widely used in UF, RO, dialysis, and GP due to their low price and nice performance.

The spiral wound module consists of a spacer and membrane wound around a perforated central collection tube. The feed solution is passed axially under the module across the membrane envelope. Part of the feed penetrates into the membrane sheath, spirals toward the center, and exits through the collection tube (Pal, 2017). The module is inexpensive to construct and the hydrodynamic factors (flow rate, transmembrane pressure) can be adjusted by changing the spacer thickness to overcome concentration polarization and contamination. In a spiral-wound module-based process, recovery and flow velocity are determined by the characteristics of the membrane constituting the module and operating conditions. This module is most widely used in UF, MF, NF, RO, and dialysis membrane industry as well as in gas separation applications in the natural gas industry. Its packing density is in the range 200–800 m^2/m^3 (Geankopolis, 2005; Pal, 2017; Seader et al., 2016). The advantages of the spiral wound module are good fouling resistance, ease of cleaning, and low price. However, this system can be uneconomical for high-pressure applications as it requires a high-pressure vessel. In addition, nonuniform packaging of the spacers may result in feed bypass. The main characteristics of the four membrane modules are summarized in Table 1.

TABLE 1 Typical characteristics of membrane modules (Balster, 2013; Geankopolis, 2005; Seader et al., 2016).

	Plate and frame	Tubular	Hollow fiber	Spiral wound
Packing density (m^2/m^3)	100–400	30–200	500–9000	200–800
Fouling resistance	Good	Very good	Poor	Moderate
Ease of cleaning	Excellent	Good	Poor	Fair
Relative cost	High	High	Low	Low
Main application	MF, UF, NF, RO, PV, D	UF, RO	UF, RO, D, GP	MF, UF, NF, RO, D, GP

MF: microfiltration, UF: ultrafiltration, NF: nanofiltration, RO: reverse osmosis, D: dialysis, PV: pervaporation, GP: gas permeation.

5. Conclusions and perspectives

Membrane for water treatment or desalination has improved technological maturity through much research and contributed a lot to industrial development and improvement in humanity's quality of life. However, there are still areas to be improved, such as membrane fouling and chemical stability. If membrane fouling is reduced, it can become more cost effective because the membrane lifetime would be extended and the energy consumption reduced during operation. In addition, improvement in the chemical stability of the membrane also has great industrial significance. In particular, if the chlorine resistance of the polyamide TFC membrane, which is currently popularly used in the RO membrane, is improved, the overall process cost can be saved because there is no need to perform a pretreatment process to remove chlorine. In the future, research to develop more advanced membrane surface treatment technology or to find better materials for membranes is needed. Furthermore, technologically, if a method for more precisely and uniformly controlling the pore size in the membrane and a method for precisely controlling the passage channel for a specific component are developed, the performance of the membrane-based process will be dramatically improved.

References

Ahmad, A.L., Abdulkarim, A.A., Ooi, B.S., Ismail, S., 2013. Recent development in additives modifications of polyethersulfone membrane for flux enhancement. Chem. Eng. J. 223, 246–267. https://doi.org/10.1016/j.cej.2013.02.130.

Allred, B.J., Tost, B.C., 2014. Laboratory comparison of four iron-based filter materials for water treatment of trace element contaminants. Water Environ. Res. 86, 2221–2232. https://doi.org/10.2175/106143014X14062131178556.

Amin, M.T., Alazba, A.A., 2014. A review of nanomaterials based membranes for removal of contaminants from polluted waters. Membr. Water Treat 5, 123–146.

Amjad, Z., Workman, K.R., Castete, D.R., 1993. Reverse Osmosis: Membrane Technology, Water Chemistry, and Industrial Applications. Van Nostrand Reinhold, New York.

Artuğ, G., Roosmasari, I., Richau, K., Hapke, J., 2007. A comprehensive characterization of commercial nanofiltration membranes. Sep. Sci. Technol. 42, 2947–2986. https://doi.org/10.1080/01496390701560082.

Avlonitis, S., Hanbury, W.T., Hodgkiess, T., 1992. Chlorine degradation of aromatic polyamides. Desalination 85, 321–334. https://doi.org/10.1016/0011-9164(92)80014-Z.

Baker, R.W., 2012. Membrane Technology and Applications. John Wiley & Sons.

Balster, J., 2013. Plate and frame membrane module. In: Drioli, E., Giorno, L. (Eds.), Encyclopedia of Membranes. Springer, Berlin Heidelberg, Berlin, Heidelberg, pp. 1–3, https://doi.org/10.1007/978-3-642-40872-4_1584-1.

Basile, A., Nunes, S.P. (Eds.), 2011. Advanced Membrane Science and Technology for Sustainable Energy and Environmental Applications, Woodhead Publishing Series in Energy. Woodhead Publishing, Cambridge; Philadelphia.

Cruz, M.C., Ruano, G., Wolf, M., Hecker, D., Castro Vidaurre, E., Schmittgens, R., Rajal, V.B., 2015. Plasma deposition of silver nanoparticles on ultrafiltration membranes: antibacterial and anti-biofouling properties. Chem. Eng. Res. Des. 94, 524–537. https://doi.org/10.1016/j.cherd.2014.09.014.

De, S., 2014. Novel Separation Processes. National Programme on Technology Enhanced Learning (NPTEL).

DeFriend, K.A., Wiesner, M.R., Barron, A.R., 2003. Alumina and aluminate ultrafiltration membranes derived from alumina nanoparticles. J. Membr. Sci. 224, 11–28. https://doi.org/10.1016/S0376-7388(03)00344-2.

Dumée, L.F., He, L., King, P.C., Moing, M.L., Güller, I., Duke, M., Hodgson, P.D., Gray, S., Poole, A.J., Kong, L., 2015. Towards integrated anti-microbial capabilities: novel bio-fouling resistant membranes by high velocity embedment of silver particles. J. Membr. Sci. 475, 552–561. https://doi.org/10.1016/j.memsci.2014.10.051.

Emonds, S., Kamp, J., Borowec, J., Roth, H., Wessling, M., 2021. Polyelectrolyte complex tubular membranes via a salt dilution induced phase inversion process. Adv. Eng. Mater. 23, 2001401. https://doi.org/10.1002/adem.202001401.

Encyclopedia of Life Support Systems (Eolss), 2010. v.1: Desalination and Water Resources (Desware): Membrane Processes. EOLSS Publishers Co Ltd., Oxford.

Facciotti, M., Boffa, V., Magnacca, G., Jørgensen, L.B., Kristensen, P.K., Farsi, A., König, K., Christensen, M.L., Yue, Y., 2014. Deposition of thin ultrafiltration membranes on commercial SiC microfiltration tubes. Ceram. Int. 40, 3277–3285. https://doi.org/10.1016/j.ceramint.2013.09.107.

Fallahnejad, Z., Bakeri, G., Ismail, A.F., 2022. Performance of TFN nanofiltration membranes through embedding internally modified titanate nanotubes. Korean J. Chem. Eng. https://doi.org/10.1007/s11814-021-1036-5.

Fan, P.M., Zhen, K.F., Zan, Z.Y., Chao, Z., Jian, Z., Yun, J.Z., 2016. Preparation and development of porous ceramic membrane supports fabricated by extrusion technique. Chem. Eng. Trans. 55, 277–282. https://doi.org/10.3303/CET1655047.

Fritzmann, C., Löwenberg, J., Wintgens, T., Melin, T., 2007. State-of-the-art of reverse osmosis desalination. Desalination 216, 1–76. https://doi.org/10.1016/j.desal.2006.12.009.

Geankopolis, C.J., 2005. Transport Processes and Unit Operations. Prentice-Hall.

Ghasemzadeh, G., Momenpour, M., Omidi, F., Hosseini, M.R., Ahani, M., Barzegari, A., 2014. Applications of nanomaterials in water treatment and environmental remediation. Front. Environ. Sci. Eng. 8, 471–482. https://doi.org/10.1007/s11783-014-0654-0.

Greenlee, L.F., Lawler, D.F., Freeman, B.D., Marrot, B., Moulin, P., 2009. Reverse osmosis desalination: water sources, technology, and today's challenges. Water Res. 43, 2317–2348. https://doi.org/10.1016/j.watres.2009.03.010.

Hilal, N., Al-Zoubi, H., Darwish, N.A., Mohamma, A.W., Abu Arabi, M., 2004. A comprehensive review of nanofiltration membranes: treatment, pretreatment, modelling, and atomic force microscopy. Desalination 170, 281–308. https://doi.org/10.1016/j.desal.2004.01.007.

Huang, H., Ying, Y., Peng, X., 2014. Graphene oxide nanosheet: an emerging star material for novel separation membranes. J. Mater. Chem. A 2, 13772–13782. https://doi.org/10.1039/C4TA02359E.

Jang, J., Park, I., Chee, S.-S., Song, J.-H., Kang, Y., Lee, C., Lee, W., Ham, M.-H., Kim, I.S., 2020. Graphene oxide nanocomposite membrane cooperatively cross-linked by monomer and polymer overcoming the trade-off between flux and rejection in forward osmosis. J. Membr. Sci. 598, 117684. https://doi.org/10.1016/j.memsci.2019.117684.

Jang, J., Chee, S.-S., Kang, Y., Kim, S., 2021. MoS2-cysteine nanofiltration membrane for lead removal. ChemEngineering 5, 41. https://doi.org/10.3390/chemengineering5030041.

Jang, J., Kang, Y., Jang, K., Kim, S., Chee, S.-S., Kim, I.S., 2022. Ti3C2TX-ethylenediamine nanofiltration membrane for high rejection of heavy metals. Chem. Eng. J. 135297. https://doi.org/10.1016/j.cej.2022.135297.

Jeong, B.-H., Hoek, E.M.V., Yan, Y., Subramani, A., Huang, X., Hurwitz, G., Ghosh, A.K., Jawor, A., 2007. Interfacial polymerization of thin film nanocomposites: a new concept for reverse osmosis membranes. J. Membr. Sci. 294, 1–7. https://doi.org/10.1016/j.memsci.2007.02.025.

Kang, G., Cao, Y., 2014. Application and modification of poly(vinylidene fluoride) (PVDF) membranes—a review. J. Membr. Sci. 463, 145–165. https://doi.org/10.1016/j.memsci.2014.03.055.

Kim, I.-C., Yun, H.-G., Lee, K.-H., 2002. Preparation of asymmetric polyacrylonitrile membrane with small pore size by phase inversion and post-treatment process. J. Membr. Sci. 199, 75–84. https://doi.org/10.1016/S0376-7388(01)00680-9.

König, K., Boffa, V., Buchbjerg, B., Farsi, A., Christensen, M.L., Magnacca, G., Yue, Y., 2014. One-step deposition of ultrafiltration SiC membranes on macroporous SiC supports. J. Membr. Sci. 472, 232–240. https://doi.org/10.1016/j.memsci.2014.08.058.

Kumar, S., Ahlawat, W., Bhanjana, G., Heydarifard, S., Nazhad, M.M., Dilbaghi, N., 2014. Nanotechnology-based water treatment strategies. J. Nanosci. Nanotechnol. 14, 1838–1858. https://doi.org/10.1166/jnn.2014.9050.

Lalia, B.S., Kochkodan, V., Hashaikeh, R., Hilal, N., 2013. A review on membrane fabrication: structure, properties and performance relationship. Desalination 326, 77–95. https://doi.org/10.1016/j.desal.2013.06.016.

Lee, K.P., Arnot, T.C., Mattia, D., 2011. A review of reverse osmosis membrane materials for desalination—development to date and future potential. J. Membr. Sci. 370, 1–22. https://doi.org/10.1016/j.memsci.2010.12.036.

Lee, A., Elam, J.W., Darling, S.B., 2016. Membrane materials for water purification: design, development, and application. Environ. Sci. Water Res. Technol. 2, 17–42. https://doi.org/10.1039/C5EW00159E.

A. General on membrane, materials and application

Li, J., Zhao, T., Chen, T., Liu, Y., Ong, C.N., Xie, J., 2015. Engineering noble metal nanomaterials for environmental applications. Nanoscale 7, 7502–7519. https://doi.org/10.1039/C5NR00857C.

Liu, F., Hashim, N.A., Liu, Y., Abed, M.R.M., Li, K., 2011. Progress in the production and modification of PVDF membranes. J. Membr. Sci. 375, 1–27. https://doi.org/10.1016/j.memsci.2011.03.014.

Liu, H., Wang, H., Zhang, X., 2015. Facile fabrication of freestanding ultrathin reduced graphene oxide membranes for water purification. Adv. Mater. 27, 249–254. https://doi.org/10.1002/adma.201404054.

Lohokare, H.R., Muthu, M.R., Agarwal, G.P., Kharul, U.K., 2008. Effective arsenic removal using polyacrylonitrile-based ultrafiltration (UF) membrane. J. Membr. Sci. 320, 159–166. https://doi.org/10.1016/j.memsci.2008.03.068.

Lou, Y., Liu, G., Liu, S., Shen, J., Jin, W., 2014. A facile way to prepare ceramic-supported graphene oxide composite membrane via silane-graft modification. Appl. Surf. Sci. 307, 631–637. https://doi.org/10.1016/j.apsusc.2014.04.088.

Majumder, M., Chopra, N., Andrews, R., Hinds, B.J., 2005. Enhanced flow in carbon nanotubes. Nature 438, 44. https://doi.org/10.1038/438044a.

Mauter, M.S., Wang, Y., Okemgbo, K.C., Osuji, C.O., Giannelis, E.P., Elimelech, M., 2011. Antifouling ultrafiltration membranes via post-fabrication grafting of biocidal nanomaterials. ACS Appl. Mater. Interfaces 3, 2861–2868. https://doi.org/10.1021/am200522v.

Mohmood, I., Lopes, C.B., Lopes, I., Ahmad, I., Duarte, A.C., Pereira, E., 2013. Nanoscale materials and their use in water contaminants removal—a review. Environ. Sci. Pollut. Res. 20, 1239–1260. https://doi.org/10.1007/s11356-012-1415-x.

Nady, N., Franssen, M.C.R., Zuilhof, H., Eldin, M.S.M., Boom, R., Schroën, K., 2011. Modification methods for poly(arylsulfone) membranes: a mini-review focusing on surface modification. Desalination 275, 1–9. https://doi.org/10.1016/j.desal.2011.03.010.

Nunes, S.P., Peinemann, K. (Eds.), 2001. Membrane Technology: In the Chemical Industry, first ed. Wiley, https://doi.org/10.1002/3527600388.

Oatley-Radcliffe, D.L., Walters, M., Ainscough, T.J., Williams, P.M., Mohammad, A.W., Hilal, N., 2017. Nanofiltration membranes and processes: a review of research trends over the past decade. J. Water Process Eng. 19, 164–171. https://doi.org/10.1016/j.jwpe.2017.07.026.

Pal, P., 2017. Water treatment by membrane-separation technology. In: Industrial Water Treatment Process Technology. Elsevier, pp. 173–242, https://doi.org/10.1016/B978-0-12-810391-3.00005-9.

Park, S.-J., Cheedrala, R.K., Diallo, M.S., Kim, C., Kim, I.S., Goddard, W.A., 2012. Nanofiltration membranes based on polyvinylidene fluoride nanofibrous scaffolds and crosslinked polyethyleneimine networks. In: Diallo, M.S., Fromer, N.A., Jhon, M.S. (Eds.), Nanotechnology for Sustainable Development. Springer International Publishing, Cham, pp. 33–46, https://doi.org/10.1007/978-3-319-05041-6_3.

Pellegrin, M.-L., Aguinaldo, J., Arabi, S., Sadler, M.E., Min, K., Liu, M., Salamon, C., Greiner, A.D., Diamond, J., McCandless, R., Owerdieck, C., Wert, J., Padhye, L.P., 2013. Membrane processes. Water Environ. Res. 85, 1092–1175. https://doi.org/10.2175/106143013X13698672321940.

Pellegrino, J., Sikdar, S.K., 2017. Membrane technology fundamentals for bioremediation. In: Fundamentals and Applications. Routledge, pp. 457–509.

Peñate, B., García-Rodríguez, L., 2012. Current trends and future prospects in the design of seawater reverse osmosis desalination technology. Desalination 284, 1–8. https://doi.org/10.1016/j.desal.2011.09.010.

Pendergast, M.M., Hoek, E.M.V., 2011. A review of water treatment membrane nanotechnologies. Energ. Environ. Sci. 4, 1946. https://doi.org/10.1039/c0ee00541j.

Ponder, S.M., Darab, J.G., Mallouk, T.E., 2000. Remediation of Cr(VI) and Pb(II) aqueous solutions using supported, nanoscale zero-valent iron. Environ. Sci. Technol. 34, 2564–2569. https://doi.org/10.1021/es9911420.

Pradeep, T., Anshup, 2009. Noble metal nanoparticles for water purification: a critical review. Thin Solid Films 517, 6441–6478. https://doi.org/10.1016/j.tsf.2009.03.195.

Qin, J.-J., Li, Y., Lee, L.-S., Lee, H., 2003. Cellulose acetate hollow fiber ultrafiltration membranes made from CA/PVP 360 K/NMP/water. J. Membr. Sci. 218, 173–183. https://doi.org/10.1016/S0376-7388(03)00170-4.

Rana, D., Matsuura, T., 2010. Surface modifications for antifouling membranes. Chem. Rev. 110, 2448–2471. https://doi.org/10.1021/cr800208y.

Sagle, A., Freeman, B., 2004. Fundamentals of membranes for water treatment. Future Desalination Tex. 2, 137.

Saline Water Conversion—II, 1963. Advances in Chemistry. American Chemical Society, Washington, DC, https://doi.org/10.1021/ba-1963-0038.

Saljoughi, E., Mousavi, S.M., 2012. Preparation and characterization of novel polysulfone nanofiltration membranes for removal of cadmium from contaminated water. Sep. Purif. Technol. 90, 22–30. https://doi.org/10.1016/j.seppur.2012.02.008.

Seader, J.D., Henley, E.J., Roper, D.K., 2016. Separation Process Principles: With Applications Using Process Simulators. John Wiley & Sons.

Sharma, Y.C., Srivastava, V., Upadhyay, S.N., Weng, C.H., 2008. Alumina nanoparticles for the removal of Ni(II) from aqueous solutions. Ind. Eng. Chem. Res. 47, 8095–8100. https://doi.org/10.1021/ie800831v.

Shibutani, T., Kitaura, T., Ohmukai, Y., Maruyama, T., Nakatsuka, S., Watabe, T., Matsuyama, H., 2011. Membrane fouling properties of hollow fiber membranes prepared from cellulose acetate derivatives. J. Membr. Sci. 376, 102–109. https://doi.org/10.1016/j.memsci.2011.04.006.

Sourirajan, S., 1977. Reverse Osmosis and Synthetic Membrane. National Research Council of Canada.

Soyekwo, F., Zhang, Q.G., Deng, C., Gong, Y., Zhu, A.M., Liu, Q.L., 2014. Highly permeable cellulose acetate nanofibrous composite membranes by freeze-extraction. J. Membr. Sci. 454, 339–345. https://doi.org/10.1016/j.memsci.2013.12.014.

Stevens, M.P., 1999. Polymer Chemistry: An Introduction. N. Y. Oxf. Univ. Press Inc, pp. 3–34.

Stoimenov, P.K., Klinger, R.L., Marchin, G.L., Klabunde, K.J., 2002. Metal oxide nanoparticles as bactericidal agents. Langmuir 18, 6679–6686. https://doi.org/10.1021/la0202374.

Thatai, S., Khurana, P., Boken, J., Prasad, S., Kumar, D., 2014. Nanoparticles and core–shell nanocomposite based new generation water remediation materials and analytical techniques: a review. Microchem. J. 116, 62–76. https://doi.org/10.1016/j.microc.2014.04.001.

Ulbricht, M., 2006. Advanced functional polymer membranes. Polymer 47, 2217–2262. https://doi.org/10.1016/j.polymer.2006.01.084.

Van der Bruggen, B., Vandecasteele, C., 2002. Distillation vs. membrane filtration: overview of process evolutions in seawater desalination. Desalination 143, 207–218. https://doi.org/10.1016/S0011-9164(02)00259-X.

Wehling, J., Köser, J., Lindner, P., Lüder, C., Beutel, S., Kroll, S., Rezwan, K., 2015. Silver nanoparticle-doped zirconia capillaries for enhanced bacterial filtration. Mater. Sci. Eng. C 48, 179–187. https://doi.org/10.1016/j.msec.2014.12.001.

Yadav, N., Hakkarainen, M., 2021. Degradable or not? Cellulose acetate as a model for complicated interplay between structure, environment and degradation. Chemosphere 265, 128731. https://doi.org/10.1016/j.chemosphere.2020.128731.

Yang, S., Xie, Q., Liu, X., Wu, M., Wang, S., Song, X., 2018. Acetylation improves thermal stability and transmittance in FOLED substrates based on nanocellulose films. RSC Adv. 8, 3619–3625. https://doi.org/10.1039/C7RA11134G.

Yasukawa, M., Suzuki, T., Higa, M., 2018. Salinity gradient processes. In: Membrane-Based Salinity Gradient Processes for Water Treatment and Power Generation. Elsevier, pp. 3–56, https://doi.org/10.1016/B978-0-444-63961-5.00001-8.

Yin, J., Deng, B., 2015. Polymer-matrix nanocomposite membranes for water treatment. J. Membr. Sci. 479, 256–275. https://doi.org/10.1016/j.memsci.2014.11.019.

Yuliwati, E., Ismail, A.F., 2011. Effect of additives concentration on the surface properties and performance of PVDF ultrafiltration membranes for refinery produced wastewater treatment. Desalination 273, 226–234. https://doi.org/10.1016/j.desal.2010.11.023.

Zhao, C., Xue, J., Ran, F., Sun, S., 2013. Modification of polyethersulfone membranes—a review of methods. Prog. Mater. Sci. 58, 76–150. https://doi.org/10.1016/j.pmatsci.2012.07.002.

Zhou, D., Zhu, L., Fu, Y., Zhu, M., Xue, L., 2015. Development of lower cost seawater desalination processes using nanofiltration technologies—a review. Desalination 376, 109–116. https://doi.org/10.1016/j.desal.2015.08.020.

A. General on membrane, materials and application

Photocatalytic membrane reactors (PMRs) for hydrogen production

Duyen Phuc-Hanh Tran[a,b,c], Minh-Thuan Pham[a,b,c], Trung-Hieu Nguyen[a,b,c], Ya-Fen Wang[b], and Sheng-Jie You[b]

[a]Department of Civil Engineering, Chung Yuan Christian University, Taoyuan, Taiwan
[b]Department of Environmental Engineering, Chung Yuan Christian University, Taoyuan, Taiwan
[c]Center for Environmental Risk Management, Chung Yuan Christian University, Taoyuan, Taiwan

1. Introduction

The rapid development of industrialization and urbanization is accompanied by excessive consumption of fossil energy sources. This leads to the depletion of fossil resources along with serious environmental issues. Thus, it is really important to maximize energy-use efficiency and also develop renewable energy sources. The utilization of renewable energy sources such as hydro, wind, solar, biomass, geothermal, and hydrogen could be the key to overcome the challenges of meeting global energy demand as it releases less harmful pollutants into the atmosphere (Sharma et al., 2020). In recent years, most of the countries have started to implement laws committed to achieving net-zero carbon emission goals by 2030–2050 (Rogelj et al., 2018). This goal highlights the importance of utilizing sustainable resources to produce clean energy as an alternative to fossil fuels. Hydrogen (H_2) has drawn the attention of the scientific community as a sustainable alternative energy source due to its merits such as high calorific value ($122\,kJ\,g^{-1}$), freedom from pollution, and being a clean energy carrier (Carolin Christopher et al., 2021; Wang et al., 2021).

The photocatalysis process is known for its outstanding advantages such as low-energy consumption, easy operation at ambient temperature, low cost, and environmental friendliness (Liu et al., 2019a; Homocianu and Pascariu, 2022). The photocatalytic water-splitting process is one of the most viable methods to produce hydrogen and oxygen using clean and renewable sources in the presence of a semiconductor photocatalyst. It offers an eco-green

Copyright © 2023 Elsevier Inc. All rights reserved.

hydrogen production method, which is really important to reach fully sustainable energy generation (Singla et al., 2021). However, photocatalyst suspension separation is an obstacle hindering photocatalytic hydrogen production. In connection with this, a combination of membrane and photocatalytic technology is considered to be effective to achieve better hydrogen production efficiency and easy separation of photocatalysts. This technology allows photocatalyst recovery, reaction, and product separation to occur simultaneously through an integrated membrane and photocatalytic system (Molinari et al., 2019; Molinari et al., 2020). In addition, the photocatalytic membrane reactor (PMR) offers higher energy efficiency, modularity, and scaling-up capability compared to the conventional photocatalytic reactor (Chen et al., 2022).

Many efforts have been made in recent years to explore the improvement possibilities of PMR. A variety of photocatalysts ranging from conventional (e.g., TiO_2, ZnO, and ZnS) to metal-free catalysts, carbon-based nanomaterials, and other types have been investigated for the photocatalysis experiment (Monga et al., 2021; Singla et al., 2021). Present research results showed that modification methods such as doping or impregnation could improve the performance of photocatalyst in regard to reducing the bandgap value. These results highlight the importance of the material modification process as an effort in building up a full-scale PMR facility. Apart from photocatalyst properties, there are still numerous kinds of factors that have been proven to possess effects toward PMR efficiency. Operating conditions, membrane properties, and the feed water matrix directly influence the efficiency of PMR in producing hydrogen fuel (Argurio et al., 2018).

The main objective of this chapter is to focus on discussions related to the general overview of the PMR principle, updates on photocatalyst modification methods, and the factors affecting PMR efficiency. Moreover, this chapter elucidates how different types of light source and different photocatalyst properties affect the photocatalytic process as a part of PMR. The effects of photocatalyst material selection, operation mode, and light source on the hydrogen production performance of PMRs are also the main highlight of this chapter's content.

2. Photocatalytic material-based membranes

2.1 Definition

The membrane concept was originally named for thin films used in the filtration and separation of water and wastewater. There are different types of membranes based on their pore sizes such as microfiltration, nanofiltration, ultrafiltration, or reverse osmosis membrane. Nowadays, membranes can be applied in many other fields such as gas separation, ion exchange, etc. The membrane is a relatively thin film of inorganic or organic polymer that is fabricated by different methods and possesses certain porosity, pore size, and thickness. These are important parameters that determine the filtration and separation performances of membranes. The concept of "photocatalytic material-based membranes" can also be understood as "photocatalytic membranes." These membranes are the result of the incorporation of photocatalytic materials into or onto the surface of conventional membranes. This combination not only preserves the filtration and separation capabilities of conventional membranes but also adds photocatalytic activities to these membranes. The fabrication methods, as well as the advantages and disadvantages of different types of photocatalytic membranes, are presented in the following sections.

2.2 Photocatalytic materials

In choosing optimum photocatalytic materials for PMR, the electrochemical and optical properties of the materials need to be considered by the characteristic and experimental evidence. Regarding electrochemical properties, redox potentials are often used to describe the electric band position of the materials. In addition to the optical properties, the diffuse reflectance spectroscopy (DRS) spectra are often used to determine the bandgap of the materials (Mishra et al., 2018; George and Chowdhury, 2019). The DRS results include the light reflectance and light absorbance of the materials. To determine the bandgap materials, first, the wavelength transform to energy is first calculated by Eq. (1). Then, based on the results of DRS spectra (Fig. 1A), both the indirect and direct bandgap of the materials can be determined by the Tauc equation (Eq. 2) and Kubelka Munk (Eq. 3) (Makuła et al., 2018; Johannes et al., 2020). As shown in Fig. 1B–D, the indirect bandgaps are always higher than the indirect bandgaps. The bandgap is determined to understand the light absorption of the materials, which supports choosing the suitable light source and materials for the photocatalytic reaction (Bui et al., 2021). Based on the electrochemical and optical properties of the materials, the

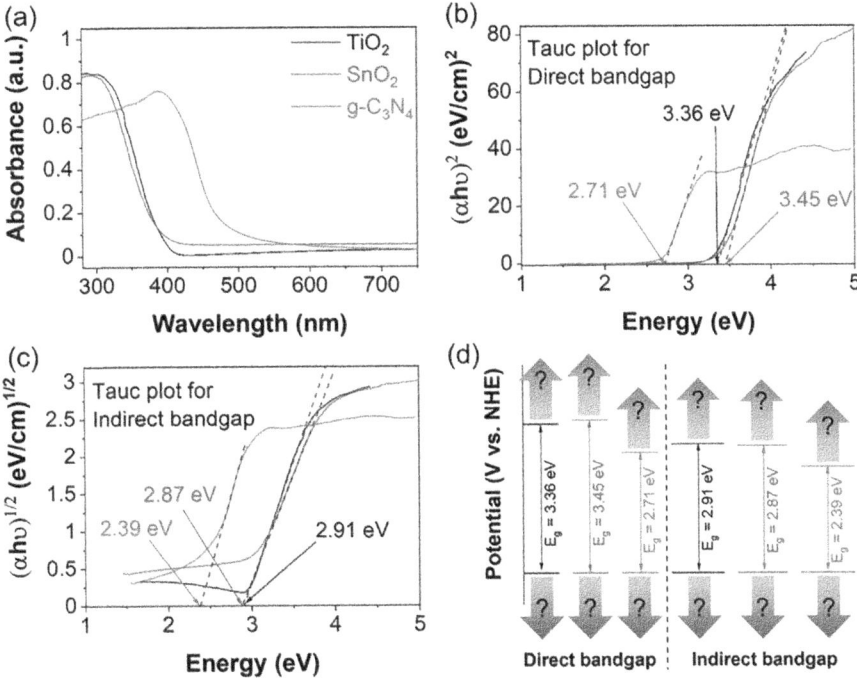

FIG. 1   DRS spectra (A). Tauc plots of $(\alpha h\upsilon)^2$ vs energy (B) and $(\alpha h\upsilon)^{1/2}$ vs energy (C) of TiO_2, SnO_2, and g-C_3N_4. Direct and indirect bandgaps of materials (D). DRS spectra cannot determine the positions of the CB and VB of materials. CC BY 4.0 (Bui et al., 2021).

most commonly used photocatalyst in PMR can be divided into two, including pure semiconductors and composite materials. The detailed characteristics of these materials are discussed in the following sections.

$$E = h\nu = \frac{hc}{\lambda} \qquad (1)$$

$$(\alpha h\nu)^r = B(h\nu - E_g) \qquad (2)$$

$$(F(R)h\nu)^r = B(h\nu - E_g) \qquad (3)$$

where E is the Photon energy (eV), h is the Planck's constant ($4.132 \times 10^{-15}\,eV\,s$), ν is the frequency (s^{-1}), c is the photon velocity ($3 \times 10^8\,m/s$), λ is the wavelength (nm), α is the absorption coefficient, B is the constant, E_g is the bandgap (eV), and r is equal to ½ for indirect bandgaps and 2 for direct bandgaps.

On the other hand, the Mott-Schottky plot techniques were invested in determining the redox ability, which included the conduction band (CB) positions and the valence band (VB) positions of the materials (Gelderman et al., 2007). The Mott-Schottky plots of three kinds of materials are shown in Fig. 2A and B. The CB maximum of materials is various with different reference electrodes and conductive media. Based on the mentioned information about the electrochemical and optical properties above, the band alignment and photocatalytic

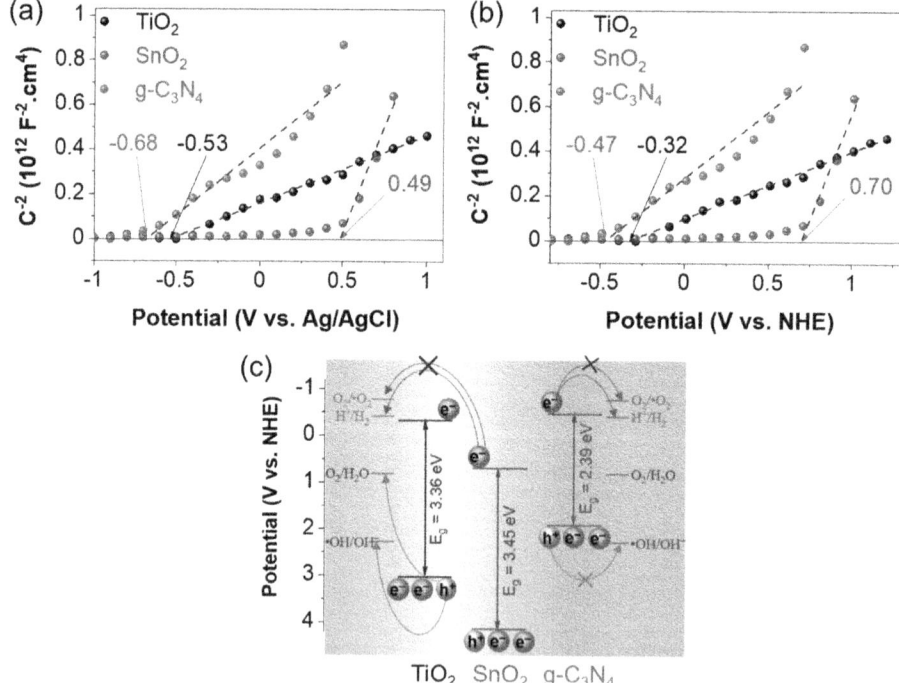

FIG. 2   Mott-Schottky plots vs Ag/AgCl reference electrode (A) and vs normal hydrogen electrode (B) of TiO₂, SnO₂, and g-C₃N₄. Band positions of materials (C). CC BY 4.0 (Bui et al., 2021).

mechanism of the materials were illustrated and shown in Fig. 2C. First, the excited electron (e^-) moves from the VB to the CB of the materials to reduce oxygen (O_2), hydrogen (H_2), O_2 radical ($^{\cdot}O_2$), and H^+. On the other hand, the hole (h^+) from VB oxidates the absorbed water to generate OH radicals. Because the CBM of SnO_2 is lower than the redox potential of H^+/H_2 and $O_2/^{\cdot}O_2$ (Fig. 2C), the SnO_2 cannot produce $^{\cdot}O_2$ and H_2. In contrast, the hydroxyl radical ($^{\cdot}OH$) cannot be produced by $g\text{-}C_3N_4$ because the VBM of $g\text{-}C_3N_4$ is higher than the redox potential of $^{\cdot}OH/OH^-$ (Li et al., 2016; Fu et al., 2020; Reischauer and Pieber, 2021).

2.2.1 Pure semiconductors

Pure semiconductors are often used due to their electrochemical properties and abundant availability. Some of the mostly used pure semiconductors include TiO_2, SnO_2, ZnO, Fe_2O_3, ZnS, etc. (Mills and Hunte, 2000; Liu et al., 2003; Lee and Wu, 2017; Al-hakkani et al., 2021; Rai and Kondal, 2021; Sun et al., 2022). In a typical semiconductor photocatalytic process, a semiconductor has an energy band structure consisting of a low-energy valence band (VB) and a high-energy conduction band (CB). Under light irradiations, electron (e^-) from the VB moves to the CB of the semiconductor. Then the free e^- reduces H_2O to generate H_2 (Fig. 3). For the PMRs process, the conduction band of the semiconductor is also an important factor that affects the hydrogen yield rate with the redox potential of H^+/H_2 ($-0.41\,V$ vs NHE). Unfortunately, the current problem that pure semiconductor materials have is regarding the relatively large bandgap, making them only capable of absorbing light in the ultraviolet light region (Fu et al., 2018; Lee et al., 2019). Besides, the CB and VB edges of some pure semiconductors are higher than the reference redox potential of H^+/H_2. This limitation is considered as an obstacle in the practical application of photocatalytic technology such as low efficiency, not durability, high-energy consumption. The materials used in photocatalytic hydrogen production must meet the small bandgap and the CB edge lower than the redox potential of H^+/H_2 ($-0.41\,V$ vs NHE).

In recent years, new semiconductors possessing smaller bandgaps have been fabricated to overcome the optical limitations of classic semiconductors. They include Ag_3PO_4, CdS, $g\text{-}C_3N_4$, Bi_2S_3, $SrTO_3$, etc. (Wu et al., 2014; Al et al., 2019; Gastiasoro et al., 2020; Dabhane

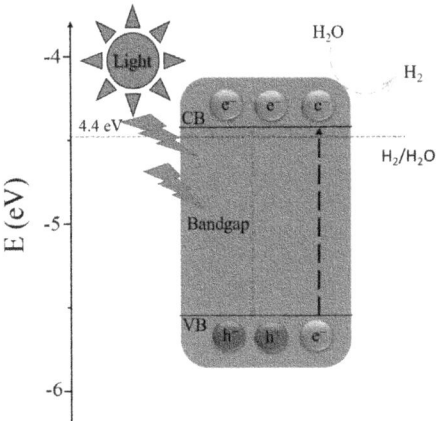

FIG. 3   Photocatalytic hydrogen generation of pure semiconductors.

et al., 2021; Pham et al., 2021). This small bandgap advantage could help semiconductors to absorb both UV light and visible light. However, the small bandgap makes the recombination ability between the electron–hole pairs (e⁻–h⁺) faster, thereby bandgap reducing the photocatalytic efficiency. A detailed bandgap value of typical pure semiconductor materials is shown in Table 1.

2.2.2 Composite

Nowadays, composite photocatalytic materials have been intensively studied to overcome the limitations of pure semiconductors, which have been mentioned in the previous sections. These newly composite materials commonly have a smaller bandgap and better properties

TABLE 1 The bandgap value of semiconductors photocatalyst (Babu et al., 2015; Opoku et al., 2017).

Photocatalytic material	Band levels with respect to NHE (eV)	
	Bandgap energy (eV)	CB
Si	1.1	−0.3
WO_3	2.8	0.71
Fe_2O_3	2.2	0.73
SiC	3.0	−0.46
$BaTiO_3$	3.3	0.55
CdO	2.1	0.74
CdS	2.4	−0.52
CdSe	1.7	−0.54
Fe_2O_3	3.1	0.73
TiO_2 Rutile	3.0	−0.05
TiO_2 Anatase	3.2	−0.25
$SrTiO_3$	3.4	−0.81
SnO_2	3.5	0.19
GaAs	1.4	−0.4
GaP	1.23	−0.97
ZnS	3.7	−0.91
ZnO	3.2	0.15
g-C_3N_4	2.7	−1.3
$CuTiO_3$	3.18	0.19
$FeTiO_3$	2.9	0.1
$KTaO_3$	3.02	−0.48

compared to pure catalysts. Several modification processes such as noble metal doping, ion doping, dye photosensitization, and photocatalyst mixing have been done to produce optimum photocatalytic materials. Noble metal-doped photocatalysts are potentially able to improve the electron–hole separation process by the ability of noble metals to trap electrons.

Moreover, noble metals such as Pt, Pd, Rh, Ni, Cu, Ai, and Ag could extend the light absorption level until visible range and alter the surface properties of photocatalytic materials. From the view of commercial application, gold has been intensively used as cocatalyst materials for photocatalytic reaction. As one of the examples, Primo and coworkers successfully synthesized an Au-doped Ceria photocatalyst, which can outperform the photocatalytic activity of tungsten trioxide (WO_3) under UV light irradiation (Primo et al., 2011). When reacting with visible light, the photocatalytic materials could produce O_2 from water ($10.5\,\mu mol\,h^{-1}$), which is larger than common WO_3 ($1.7\,\mu mol\,h^{-1}$) even under UV irradiation ($9.5\,\mu mol\,h^{-1}$). The mechanism of metal doping on a typical semiconductor photocatalyst for hydrogen production is shown in Fig. 4.

On the other hand, the ion doping method has also been investigated in pursuit of decreasing the wide bandgap value of photocatalysts. Therefore, the light absorption region of the material extends to the visible-light region (wavelength (λ) > 400 nm). Enhancing light absorption can increase photocatalytic efficiency and improve the applicability of materials. Some of the examples are doped TiO_2 (Chen et al., 2018), doped ZnO (Khan et al., 2021), doped $SrTiO_3$ (Zwara et al., 2019), etc. The main principle of ion doping is to provide dopant ions, which can function as hole and electron traps or mediate interfacial charge transfer. The dopant ions could also possibly be integrated with semiconductor catalysts through adsorption, oxide phases formation, and interior incorporation. The decoration of the metal onto the semiconductor is one of the best ways to reduce the bandgap of the semiconductor through the surface plasmon resonance (SPR) effect (Reddy et al., 2019).

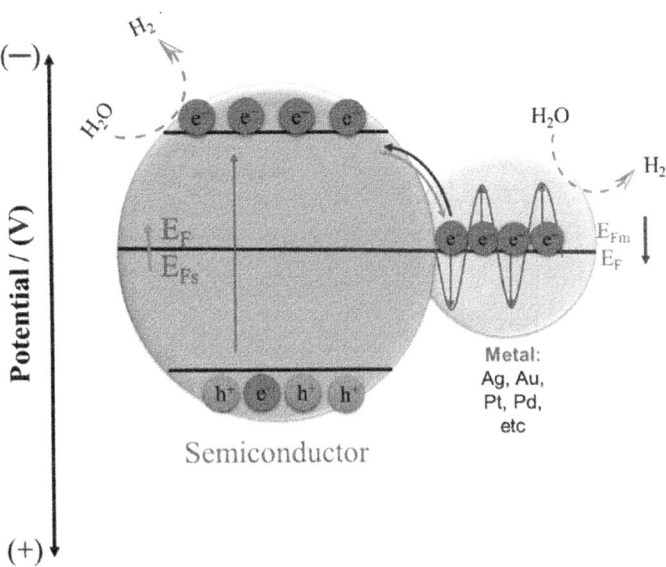

FIG. 4 Diagrams of SPR on the photocatalytic H_2 production.

In addition, a unique composite photocatalyst could be produced through the combination of different semiconductor photocatalyst types. Herein, the construction of heterojunction structures by combining two or more different types of semiconductors with other electronic band structures is considered one of the most popular and effective methods to increase the lifetime of the electron–hole pairs or change the bandgap of the mixture since the photocatalytic efficiency will be significantly increased. The heterojunction structure includes type-I, II, III, Z-scheme, S-scheme, and also p-n heterojunction (Moniz et al., 2015; Xu et al., 2020) (Fig. 5). For the three types of heterojunction and Z-scheme structure, the combination of two or three semiconductors together could reduce the bandgap and increase the lifetime of the e^-–h^+ pairs.

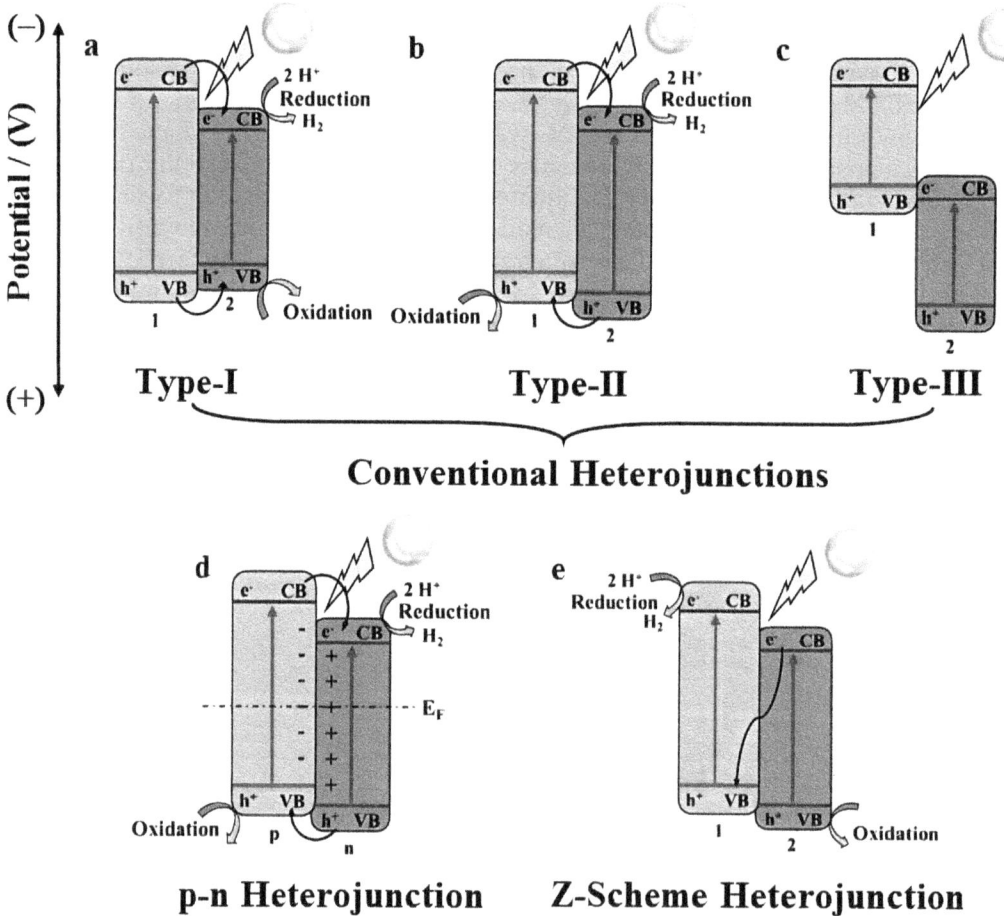

FIG. 5   Diagrams of type-I (A); type-II (B); type-III (C), p–n heterojunction (D), Z-scheme (E) heterojunction. 1, 2, p, and n refer to semiconductor 1, semiconductor 2, p-semiconductor, and n-semiconductor, respectively. *With permission from Lu, L., Wu, B., Shi, W., 2019. Metal–organic framework-derived heterojunctions as nanocatalysts for photocatalytic hydrogen production. Inorg. Chem. Front. 6, 3456–3467. https://doi.org/10.1039/c9qi00964g.*

Taking CdS/TiO_2 as an example for a catalyst made from the Z-scheme, the process of generating hydrogen of CdS/TiO_2 was $51.4\,\mu mol\,h^{-1}$ higher than that of the pure TiO_2 ($1.8\,\mu mol\,h^{-1}$) and the CdS ($0\,\mu mol\,h^{-1}$) (Fig. 6) (Meng et al., 2017). In addition, the DRS spectra of the materials are shown in Fig. 7A, TiO_2 shows a strong absorption in the UV light range. In contrast, the CdS/TiO_2 composite shows stronger absorption in the visible range. The bandgap of the TiO_2 decreased significantly from 3.2 to 2.3 eV by combining TiO_2 and CdS. On the other hand, the photocurrent-potential curves of the materials are shown in Fig. 7B. As shown in Fig. 7B, the current density of the CdS/TiO_2 composites is higher than that of TiO_2 and CdS. These results show the enhancement in photocatalytic efficiency of the CdS/TiO_2 composites. Furthermore, the photocatalytic mechanism of the CdS/TiO_2 composites is shown in Fig. 8. The e^- moved from the VB to the CB of TiO_2, then moved to the CB and VB of CdS, which means that the light time of the $e^- - h^+$ pair is longer. This is attributed to the Z-scheme photocatalytic reaction mechanism that effectively improves the separation and migration of electrons, which extend the photocatalytic reaction to photocatalysts in the visible region. The construction of the heterostructure not only improves light absorption,

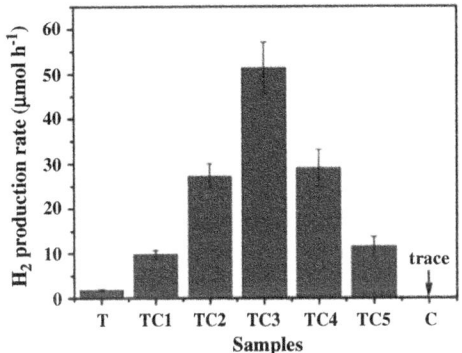

FIG. 6 Comparison of the photocatalytic hydrogen production rates of a TiO_2/CdS photocatalyst loaded with different amounts of CdS in methanol aqueous solution under Xenon irradiation for 1h. *With permission from Meng, A., Zhu, B., Zhong, B., Zhang, L., Cheng, B., 2017. Direct Z-scheme TiO2/CdS hierarchical photocatalyst for enhanced photocatalytic H2-production activity. Appl. Surf. Sci. 422, 518–527. https://doi.org/10.1016/j.apsusc. 2017.06.028.*

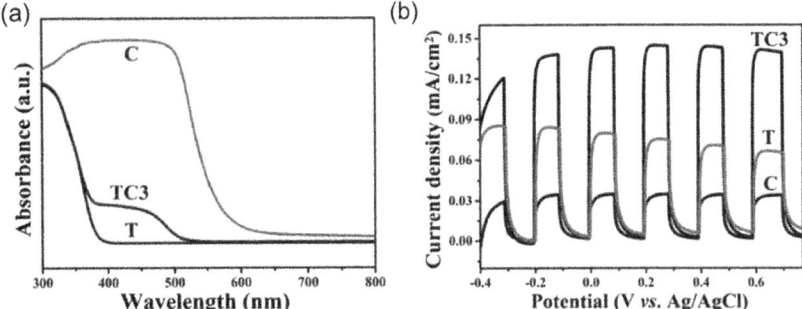

FIG. 7 (A) UV–vis diffuse reflectance spectra of T (black line), C (blue line, light gray in print version), and TC3 (red line, gray in print version) and (B) photocurrent-potential curves of T, C, and TC3 samples. *With permission from Meng, A., Zhu, B., Zhong, B., Zhang, L., Cheng, B., 2017. Direct Z-scheme TiO2/CdS hierarchical photocatalyst for enhanced photocatalytic H2-production activity. Appl. Surf. Sci. 422, 518–527. https://doi.org/10.1016/j.apsusc.2017.06.028.*

FIG. 8   The potential positions of TiO$_2$ and CdS band edges and schematic illustration of the direct Z-scheme photocatalytic mechanism for a TiO$_2$/CdS photocatalyst. *With permission from Meng, A., Zhu, B., Zhong, B., Zhang, L., Cheng, B., 2017. Direct Z-scheme TiO2/CdS hierarchical photocatalyst for enhanced photocatalytic H2-production activity. Appl. Surf. Sci. 422, 518–527. https://doi.org/10.1016/j.apsusc.2017.06.028.*

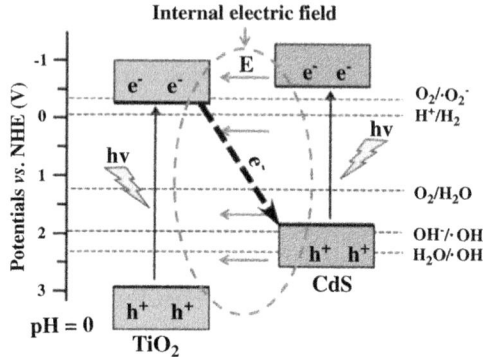

but also limits the recombination of charges, enhancing the efficiency of hydrogen separation from water. This is considered a promising strategy toward practical applications.

3. Membrane-based photocatalytic system

3.1 Hydrogen production with membrane filtration

In general, the Z-scheme approach is used in photocatalysis for the production of hydrogen and oxygen through oxidation and reduction reactions. Two photocatalysts must absorb light and drive oxidation and reduction processes on their surfaces for this procedure. Hydrogen generation through water splitting requires photocatalysts that possess a conduction band level lower than the water reduction potential and a valence band level higher than the water oxidation potential. The typical procedure is described through the following reactions (Eqs. 4–7):

$$2\lambda \rightarrow 2e_{CB}^- + 2h_{VB}^+ \quad \text{(Photoinduced } e^-/h^+ \text{ generation)} \tag{4}$$

$$H_2O + 2h^+ \rightarrow 2H^+ + \frac{1}{2}O_2 \quad \text{(Water oxidation half reaction)} \tag{5}$$

$$2H^+ + 2e^- \rightarrow H_2 \quad \text{(Proton reduction half reaction)} \tag{6}$$

$$H_2O + 2\lambda \rightarrow H_2 + \frac{1}{2}O_2 \quad \text{(Overall water splitting)} \tag{7}$$

However, producing pure hydrogen and oxygen and separating a suspended photocatalyst from this process are considered big challenges. Hence, the application of membrane technology aids in the resolution of this issue. In this way, the photocatalyst is retained on the membrane surface through filtration, while the membrane acts as a selective barrier to degraded compounds (Tsydenov and Vorontsov, 2015). Membrane photocatalysis involves using the same photocatalyst to execute oxidation and reduction on both sides of the membrane (Fig. 9). Electrons must travel from one side of the membrane to the other side, thus requiring four quanta to produce 2 H$_2$ and O$_2$. Ion-exchange membranes are commonly used

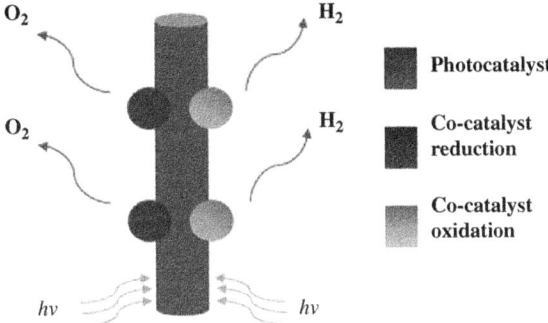

O_2

O_2

H_2

H_2

■ Photocatalyst

■ Co-catalyst reduction

■ Co-catalyst oxidation

hv

hv

FIG. 9 Schematic diagram of a photocatalytic membrane for hydrogen production.

for the separation of hydrogen from water. This type of membrane only enables cations to pass through, allowing the charge balance between the two compartments to be maintained and the reactants to be transported to the reaction site. Usually, these cations can be cations of reversible redox mediator pairs such as Ce^{4+}/Ce^{3+} (Kozlova et al., 2009), Fe^{3+}/Fe^{2+} (Primo et al., 2011), Co^{3+}/Co^{2+} (Sasaki et al., 2013), etc., or protons. These pairs of reversible redox mediators are responsible for transporting electrons from the oxidation catalysts (which produce oxygen) to the reduction catalysts (which produce hydrogen) through cyclic redox processes. In the case of protons, the protons produced by the oxidation of water to oxygen will move through the ion-exchange membrane to the compartment containing the reduction catalyst, accepting electrons and forming hydrogen. However, for photocatalytic membrane systems designed with photoanode and photocathode electrodes and an external circuit conductor, the conductor serves as the electron transfer between the two types of photocatalysts and the ion-exchange membrane acts as the proton transporter between the two compartments. Regardless of configuration, hydrogen-generating photocatalytic systems using selective membrane filtrations proved to have significant advantages over photocatalytic systems without ion-exchange membranes.

3.2 Photocatalytic membranes

Many factors need to be considered and improved for photocatalytic reactors to reach commercial application on an industrial scale. An important factor is related to the status of the photocatalyst in the reactor, i.e., suspended or immobilized photocatalysts. When the photocatalyst is distributed throughout in the reaction solution, the contact area between it and the reactant is very large and the reaction rate is high. However, this has the disadvantage of requiring further steps for photocatalyst recovery and reuse. In addition, suspended photocatalysts at high concentrations cause light scattering, which reduces the light absorption efficiency of the photocatalyst in the reaction solution. In the case where the photocatalyst is immobilized on solid substrates such as glass, cellulose fibers, or membranes, although its contact area is not high, it overcomes the disadvantages of suspended photocatalysts. Depending on the position of the photocatalyst on the membrane, photocatalytic membranes are classified into three types, including photocatalyst-coated membranes, photocatalyst-blended membranes,

TABLE 2   Comparison of different photocatalytic membranes.

Photocatalytic membranes	Advantages	Disadvantages
Photocatalyst-coated membranes	Time-saving, cheap, easy to use, flexible	Weak bonding force between photocatalyst and membrane
Photocatalyst-blended membranes	Minimum catalyst leaching Strong bonding force between photocatalyst and membrane	Time consuming, ease of agglomeration of the catalyst
Free-standing photocatalytic membranes	Minimum catalyst leaching	Light exposure in the membrane interior hindered, nonflexible, expensive

and free-standing photocatalytic membranes. The advantages and disadvantages of these types of photocatalytic membranes are described in Table 2.

3.2.1 Photocatalyst-coated membranes

Dip-coating, electrospraying, magnetron sputtering, or gas-phase deposition can all be used to deposit one or more photocatalyst coating layers on the membrane surface. The stability, filtration, and photocatalytic properties of the resulting photocatalytic membrane need to be considered.

The dip-coating method is considered the simplest method as it does not require much complicated equipment. Usually, the photocatalyst is dispersed in an aqueous medium or as a sol–gel, followed by dipping of the membrane into the mixture at a controlled rate and time. However, this method is difficult to control the uniform distribution and exact content of the photocatalyst on the membrane surface. The adhesion of the photocatalyst to the membrane surface is also noteworthy because it is usually due to weak electrostatic or van der Waals forces. However, this adhesion will be significantly improved in the case of chemical bond formation. Ceramic membranes have better adhesion properties to inorganic photocatalysts than organic membranes (Anwar et al., 2020). For organic membranes, an intermediate binder layer is precoated onto the membrane surface before the membrane is immersed in the dispersed mixture of the photocatalyst. This intermediate layer can be polymers such as polydopamine (PDA), polyvinyl alcohol (PVA), polyacrylic acid (PAA), etc. In addition, to form a chemical bond between the membrane surface and the intermediate layer, membrane surface pretreatment methods can be used such as cold plasma, UV light, electron beam, etc.

In the electrospraying method, high voltage is applied between the injector and the counter electrode with the desired substrate attached. This high voltage facilitates the formation of nanoparticles from the precursor mixture and helps in the adhesion of the nanoparticles to the substrate. The electrospinning method is also applied to a similar high voltage and is developed from the electrospraying technique. The electrospinning method used to generate nanofiber membranes can be used to alternately combine with electrospraying of photocatalyst nanoparticles. In addition, electrospinning a dope solution of polymer and photocatalyst produces nanofiber photocatalytic membranes (Daels et al., 2014; Nor et al., 2016).

In sputtering, microscopic particles of matter are ejected from the surface of the material by the bombardment of high-energy particles. This technique can create an extremely thin layer of one material onto the surface of another substrate. The commonly used photocatalyst for this technique is TiO_2 followed by anodization of the TiO_2-sputtered membrane to form TiO_2 nanotubes on the membrane surface. TiO_2 nanotubes have photocatalytic activity that is several times higher than that of sputtered TiO_2 (Fischer et al., 2014).

In a gas-phase deposition, a vacuum is utilized to help in the production of high-quality, high-performance solid materials. A substrate such as a membrane is placed in a vacuum containing volatile precursors; these precursors decompose or react on the membrane surface to form a thin layer of solid. The photocatalytic membrane containing a thin film of Pt/TiO_2 produced by this technique shows a higher rate of hydrogen generation than the membranes fabricated from traditional techniques (Fischer et al., 2014).

3.2.2 Photocatalyst-blended membranes

In terms of photocatalyst-blended membranes, the photocatalyst is often combined with a polymer solution to form the desired membrane structure. The composition ratio of the photocatalyst, polymer, and solvent utilized is considered an important factor affecting the produced membrane (Nguyen et al., 2022a). The procedure begins with a polymer solution that is thinly spread on a glass-style stand or any other supporting material such as nonwoven fiber before being submerged in a coagulation bath to harden the film that has been shaped. In the blending method, if there is an incompatibility between the membrane-forming material and the photocatalyst, it is necessary to modify the polymer or the photocatalyst to ensure uniform dispersion of the photocatalyst in the matrix. However, the ease of agglomeration of the catalyst on the membrane surface limits its potential application on an industrial scale (Homayoonfal et al., 2014).

3.2.3 Free-standing photocatalytic membranes

For free-standing photocatalytic membranes, the photocatalyst alone is fabricated as a membrane without the presence of any other substrate material. In terms of photocatalyst efficiency per unit mass of photocatalyst used, this type of membrane will not give high efficiency because light exposure in the membrane interior is hindered. In addition, this type of membrane does not have the flexibility of the polymer component, which limits its stability. In terms of cost, free-standing photocatalytic membranes will generally be much more expensive than the two types of photocatalytic membranes mentioned above. However, its advantage is to limit the phenomenon of leaching out of the photocatalyst during operation in an aqueous solution (Argurio et al., 2018; Nguyen et al., 2022c). Because there is a weak interaction between the inorganic photocatalyst and the organic membrane in the coated photocatalytic membrane; meanwhile, there is a strong interaction of the same nature of photocatalyst particles (referred to as ceramic membranes) in free-standing photocatalytic membranes. These free-standing photocatalytic membranes are usually inorganic catalysts such as pure TiO_2 and doping or composite forms of TiO_2. The introduction of cocatalysts such as Pt, carbon nanotubes, CdS, etc. will greatly improve the photocatalytic performance of the resulting membrane.

3.3 Configuration and operation of photocatalytic membrane reactors

The conventional photocatalytic reactor has limitations such as the difficult reuse and recovery of the photocatalyst and the product and reactant coexisting in the solution. Therefore, the combination of photocatalytic reactor and membrane filtration provides synergistic effects to overcome the above limitations. Depending on the location of the photocatalyst, the photocatalytic membrane reactor (PMR) consists of two main configurations: slurry (suspended) PMRs and immobilized PMRs. In slurry PMRs, the photocatalytic reaction occurs at the surface of the dispersed catalyst in solution and filtration occurs at the membrane. The membrane and lights can be placed inside or outside of the reaction solution. In immobilized PMRs, photocatalytic membranes are designed to perform the dual functions of filtration and photocatalysis (Nguyen et al., 2022b).

In most studies, the combination of photocatalysts and Nafion membranes is widely used due to the unique chemical structure of the membrane (Tsydenov and Vorontsov, 2015). Nafion is a thermoplastic resin made from tetrafluoroethylene and perfluoro (4-methyl-3, 6-dioxa-7-octene-1-sulfonyl-fluoride). Therefore, the fabrication process of the Nafion membrane is simple and may be done in a variety of shapes. The fluorocarbon composition in this copolymer makes it possess a higher chemical and thermal stability than other polymers (Dong et al., 2022; Tang et al., 2022). When Nafion is immersed in an acidic solution, the sulfonyl fluoride functional group converts to a sulfonic acid functional group, and the copolymer becomes saturated with protons. This sulfonic acid functional group enables the Nafion copolymer to easily exchange cations with the surrounding solution. Nafion membranes have a very wide range of applications based on this unique cation exchange property. In many researches, photocatalyst nanoparticles or Pt placed into or onto Nafion membranes have been investigated.

Combining the membrane inside the two-compartment reactor possesses many outstanding advantages such as limiting the backward of hydrogen and oxygen, producing high-purity hydrogen, and lowering the risk of fire and explosion. For example, Fujihara et al. (1998) investigated the performance of the Nafion membrane in hydrogen separation by using TiO_2 and Pt/TiO_2 as photocatalysts dispersed in two compartments with Pt electrodes. The TiO_2 photocatalyst catalyzes the reaction that produces oxygen and protons in the presence of Fe^{3+}/Fe^{2+} mediator pairs to transfer the generated electrons to the Pt electrode. The proton generated in this compartment will move through the Nafion membrane to the other compartment. In the other compartment, the Br_2/Br^- mediator pair will transfer electrons from the Pt electrode to the Pt/TiO_2 catalyst to cause a deprotonation reaction to form hydrogen (Fig. 10).

The application of the Z-scheme in a dual reactor system to generate hydrogen under visible light was studied by Yu and coworkers (Yu et al., 2011). In this study, $Pt/SrTiO_3$:Rh and $BiVO_4$ were utilized as the H_2-photocatalyst and the O_2-photocatalyst, respectively. The electron transfer mediators used in reactors consisted of only one Fe^{3+}/Fe^{2+} pair that could move across the Nafion membrane. After 6h irradiation, the average H_2 and O_2 generation rates were 0.65 and $0.32 \, mol \, g^{-1} \, h^{-1}$, respectively. This result revealed that the membrane resistance in the twin reactor would not interfere with the photocatalytic water-splitting process.

The second configuration of the photocatalytic reactor is immobilized PMRs, in which the photocatalyst is immobilized on the membrane or other substrate in the form of thin films. There have been studies combining polymer membranes, electrodes, and photocatalysts into

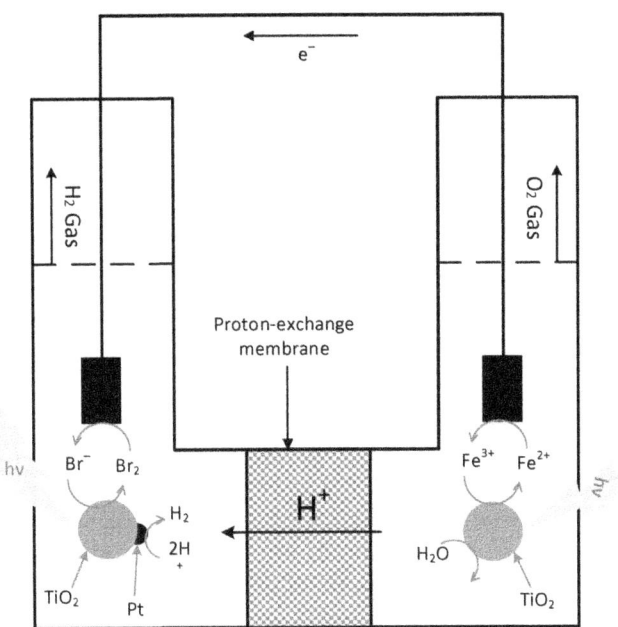

FIG. 10 An illustration of two-compartment photocatalytic reactors. *Reproduced with permission from Fujihara, K., Ohno, T., Matsumura, M., 1998. Splitting of water by electrochemical combination of two photocatalytic reactions on TiO2 particles. J. Chem. Soc. Faraday Trans. 94, 3705–3709. https:// doi.org/10.1039/A806398B.*

a single unit known as the polymer membrane electrode assembly (MEA). Seger and Kamat (2009), TiO_2/Nafion/Pt photocatalytic membrane was fabricated to catalyze the water-splitting reaction to produce hydrogen gas. In addition, on this membrane, there are two electrode layers made of a graphite flow plate between TiO_2 particles and the Nafion membrane and between the Nafion membrane and Pt nanoparticles used to conduct electrons between two photoanode and photocathode electrodes (Fig. 11). The reduction of H^+ at a Pt cathode under no applied bias was confirmed from the evolution of hydrogen with a rate of $69 \, \mu L \, h^{-1} \, cm^{-2}$.

However, the fabrication of multiple layers on the membrane surface can affect the selective filtration and separation performance of ion-exchange membranes. In addition, the electron

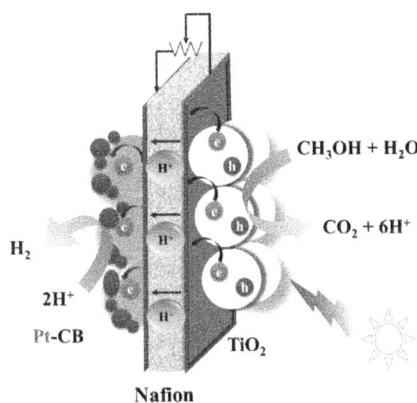

FIG. 11 Polymer membrane electrode assembly used in hydrogen generation. *Reproduced with permission from Seger, B., Kamat, P. V., 2009. Fuel cell geared in reverse: photocatalytic hydrogen production using a TiO2/ Nafion/Pt membrane assembly with no applied bias. J. Phys. Chem. C 113, 18946–18952. https://doi.org/10.1021/jp907367k.*

conduction on the two graphite electrodes could be significantly improved if the two catalysts are coated on two surfaces of the same metal plate. In this case, the photoelectrodes and the ion-exchange membrane are separated as two separate components in the two-compartment photocatalytic reactor. Kitano and coworkers developed a Pt-loaded vis TiO_2 thin film for the decomposition of water into H_2 and O_2 (Fig. 12) (Kitano et al., 2006). Under light irradiation of wavelengths longer than 390 nm, these Pt-loaded vis-TiO_2 thin films were reported to break down pure water stoichiometrically into H_2 and O_2. As a result, after 7 h of irradiation, the total amounts of H_2 and O_2 evolved were 22.1 and 9.48 mmol, respectively. It also revealed that separate generation of H_2 and O_2 from water in an H-type glass and the proton-exchange membrane was achievable under visible/solar light irradiation.

4. Factors affecting the hydrogen production performance of PMRs

4.1 Photocatalyst material characteristics

Photocatalyst characteristics play an important role in improving the hydrogen production rates in PMRs. An effective photocatalyst depends on many factors such as bandgap energy, the specific surface area and particle size distribution, the crystallographic structure, the components, etc. (Retamoso et al., 2019; Estrada-Flores et al., 2020). The bandgap energy plays an important role in the water-splitting reaction; specifically, semiconductors possessing E_g greater than 1.23 eV exhibit water-splitting activity (i.e., TiO_2, $SrTiO_3$, g-C_3N_4, etc.) (Liu et al., 2019b). However, TiO_2 has limited visible-light absorption due to its wide bandgap (3.2 eV) (Abdullah et al., 2017). Therefore, much efforts should be made (i.e., doping with metals, nonmetals, cocatalysts, etc.) to improve the photocatalytic efficiency of TiO_2. In contrast, g-C_3N_4 with a smaller bandgap (2.7 eV) possesses high absorption of visible light and it is potential for increasing the rate of hydrogen production (Fajrina and Tahir, 2019). The study

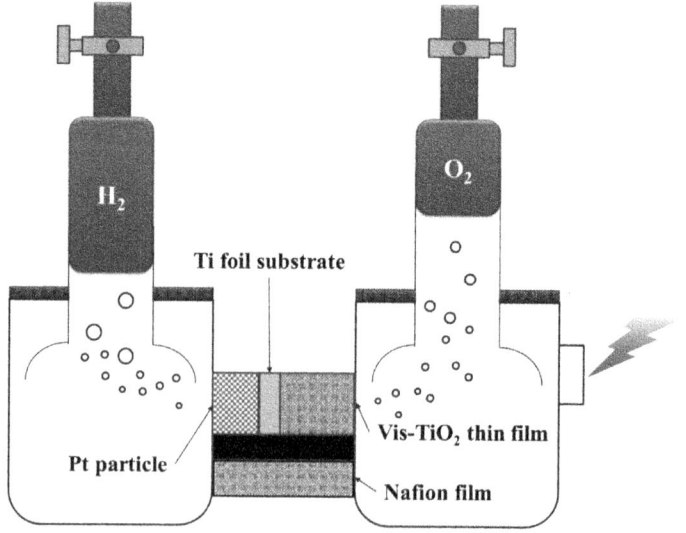

FIG. 12 H-type photocatalytic reactor for water splitting. *Reproduced with permission from Kitano, M., Tsujimaru, K., Anpo, M., 2006. Decomposition of water in the separate evolution of hydrogen and oxygen using visible light-responsive TiO2 thin film photocatalysts: effect of the work function of the substrates on the yield of the reaction. Appl. Catal. A. Gen. 314(2), 179–183. https://doi.org/10.1016/j.apcata.2006.08.017.*

of Mo and coworkers demonstrated the hydrogen production potential of two-dimensional (2D) MnO_2/monolayer g-C_3N_4 up to $28.0\,mmol\,g^{-1}h^{-1}$ (Mo et al., 2019). In addition, zinc-doped g-C_3N_4 is also found to be effective for hydrogen production (Qin et al., 2018). As a result, it can be seen that visible-light photocatalysts have a high potential for large-scale hydrogen production in PMRs.

In addition, membrane fouling in hydrogen-producing PMRs systems is closely related to the photocatalyst material and its concentration even its low contaminations. PMRs with high catalyst loading increase hydrogen absorption and conversion. However, overload of the photocatalyst also has an adverse effect on the overall efficiency of the process. In particular, high catalytic loads can cause an increase in the permeability flux leading to high membrane fouling rates (Sabouni and Gomaa, 2019). At the same time, the scattering/reflecting ability of the UV lamp is hindered, which reduces the reaction rate (Zheng et al., 2017). Therefore, optimal photocatalytic loading is necessary when taking into account the hydrogen evolution efficiency.

In addition, variables such as environmental impact and economic visibility have a considerable impact on photocatalyst material selection. In terms of environmental concerns, the material's toxicity is taken into account to provide a green path. Presently, some researches have also fully shown the nontoxicity of some photocatalysts such as TiO_2 (Gao et al., 2020), ZnO (Karamian and Sharifnia, 2018; Chen et al., 2021), CeO_2 (Hezam et al., 2020), g-C_3N_4 (Yang et al., 2020), etc. Despite that, there are still many studies which use allegedly high-toxicity nanomaterials due to its unreplaceable performance. Furthermore, the selected materials must provide market abundance and availability, which is required for close price correlation. Furthermore, the cost of the material is a significant consideration in this process. There are no standards by which to compare the variances in the permitted material selection. As a result, numerous attempts are made to test the efficiency of various photocatalysts.

4.2 Operation mode

Identical to other common filtration technologies, operation modes are also considered as the key factors which affect PMRs performance. Typically, there are two modes, including dead-end and cross-flow, that can be applied in PMRs operation (Wang et al., 2008; Song et al., 2012). In the dead-end mode, feed water is directly contacted to the membrane filter. The difference between the dead-end mode and the cross-flow mode is the relative direction of the feed flow and the membrane. In the dead-end mode, the feed flow is perpendicular to the membrane, while in the cross-flow mode they are parallel. The substrates are trapped and aggregated on the surface of the membrane, resulting in the formation of a cake layer. This layer decreases the membrane permeability and indirectly reduces the photocatalytic performance of PMRs (Ahmad et al., 2017). This cause can be attributed to the existence of some impurities in water that cannot penetrate the membrane that accumulates on the membrane surface and causes cake formation. During the cross-flow mode, the feed water is introduced through tangential flow, which leads to periodic removal of deposited particles on the membrane surface. Typically, the permeate perpendicularly flows across the membrane, while the retentate is recirculated in the feed basin. The high recirculation rate in the cross-flow mode effectively promotes mass transfer from the feed solution into the surface membrane, which technically increases the photocatalytic reaction rate (Argurio et al., 2018). Reflecting on those reasons, it can be suggested that the cross-flow mode serves as a more viable option to be applied in a bigger-scale PMRs process (Mozia et al., 2014).

Apart from the filtration modes, the flow rate value also becomes a core operation parameter for the PMRs process. That is due to its having the ability to dictate the contact periods between the photocatalyst and the reagents and products (Molinari et al., 2019). In an ideal condition, the mass transfer process between the reagent and the photocatalyst needs to happen fast enough to prevent reaction limitation. In addition, the contact time for the catalyst/reagent should also be set in an appropriate range to maintain the photocatalytic reaction selectivity. These two reasons are enough to highlight the importance of flow rate in operating PMRs instruments.

4.3 Light source

The photocatalytic hydrogen conversion efficiency of the PMR mostly depends on the coated materials. Based on the bandgap structure of the material, a suitable excitation light source could be determined in a photocatalytic process. The optimum excitation light source may affect the performance of photocatalytic activity (Yan et al., 2013; Bui et al., 2021). Thus, it is important to calculate the minimum excitation light source for semiconductor materials. Eq. (7) can be used to calculate the minimum excitation light source. Therefore, the choice of an appropriate light source also affects the operating cost of the photocatalytic hydrogen evolution system. Currently, there are many different light sources and materials that make up the light source, which dramatically affects the photocatalytic efficiency and directly affects the operating cost. In this case, the apparent quantum efficiency (AQE) is invested to calculate the effect of the light source on the photocatalytic materials and system. Herein, the AQE result can evaluate the optimization of utilizing photons for photocatalytic reactions (Eq. 8) (Ohtani, 2011).

$$\varphi = \frac{N_A \int_0^t (C_0 - C_t) V_t}{\text{Photon Flux} \times \text{Irradiation Area} \times 1000 M_{\text{reagents}}} \tag{8}$$

where N_A is the Avogadro constant ($6.022 \times 10^{23} \, \text{mol}^{-1}$), C_0 is the initial concentration of the reagents ($\text{mg} \, \text{L}^{-1}$), C_t is the concentration of reagents at time t ($\text{mg} \, \text{L}^{-1}$), V is the flow rate ($\text{L} \, \text{min}^{-1}$), and M_x is the molar reagents ($\text{g} \, \text{mol}^{-1}$). The photon flux of the lamp depends on the wattage (W) of the lamp, and the irradiation area depends on the photocatalytic reactor or the area of membranes.

In addition, the energy consumption and light intensity are calculated by Eqs. (9) and (10), respectively (Swan and Ugursal, 2009; Rohit et al., 2020).

$$\text{Energy consumption} = \text{Energy cost} \left(\frac{\text{money}}{W} \right) \times P \, (W) \times t \, (h) \tag{9}$$

$$I = \frac{P(W}{A \, (m^2)} \tag{10}$$

where P is the light source power (W), t is the operating time (h), and A is the reaction area (m^2).

5. Conclusions and perspectives

In the span of recent years, various new methods and improvements have been discovered in the realm of the PMRs process. Different types of configurations, namely coated-, blended-, and free-standing photocatalytic membranes have been identified to significantly improve the performance under different circumstances with the expense of several disadvantages that have been discussed in this chapter. Apart from that, the recent modifications on semiconductor catalysts such as TiO_2, SnO_2 have successfully improved the materials quality and overcome common disadvantages such as poor applicability and low production capability which often lead to an inefficient PMRs process.

However, the practical application of photocatalyst technology still has many obstacles. Herein, material type, material structure, and excitation light source are the biggest impeding factors of photocatalysis technology. Pure semiconductors often have a high bandgap, which makes them only able to absorb UV light; in addition, the CB and VB edges of some pure semiconductors are not suitable for H_2 production. The heterojunction structure includes combining the semiconductors and SPR effect, enhancing the photocatalytic performance of materials by controlling the electrochemical and optical properties of the materials. The DRS and Mott-Schottky is the typical technique to determine the bandgap and the CB and VB positions of the materials. Leading to these results, the challenge of light absorption and the CB and VB edges of the materials can be controlled. From there, a suitable light source will be identified and used in the H_2 production process; choosing an appropriate light source helps save energy and optimize H_2 production efficiency. In addition, operation mode and different membrane types also have a great influence on the photocatalyst efficiency; choosing a suitable membrane under a suitable reaction condition not only improves the photocatalyst efficiency, but also increases the durability and applicability of PMRs. Moreover, the selectivity of the reaction also needs to be considered through flow rate optimization. The creation of a highly efficient sunlight-responsive photocatalyst should be the focus of additional study in order to create a bigger yield of hydrogen via the photocatalytic process. Overall, the energy crisis and global warming could be tackled through combining green technology with another green technology, which in this case are photocatalytic and membrane reaction. Photocatalyst technology in general and PMRs in particular is still being hoped to be one of the green technologies of the future.

Acknowledgments

This research was financially supported by the Chung Yuan Christian University, Taiwan (Project No: 109609432). This work was supported by the Department of Civil Engineering Chung Yuan Christian University, Taiwan; Department of Environmental Engineering, Chung Yuan Christian University, Taiwan; Center for Environmental Risk Management, Chung Yuan Christian University, Taoyuan 32023, Taiwan.

References

Abdullah, H., Ismail, N.A., Yaakob, K., Khan, M.R., Rahim, S.A., 2017. CeO2-TiO2 for photoreduction of CO2 to methanol under visible light: effect of ceria loading. Malaysian J. Anal. Sci. 21 (1), 166–172. https://doi.org/10.17576/mjas-2017-2101-19.

A. General on membrane, materials and application

Ahmad, R., Kim, J.K., Kim, J.H., Kim, J., 2017. Nanostructured ceramic photocatalytic membrane modified with a polymer template for textile wastewater treatment. Appl. Sci. 7 (12). https://doi.org/10.3390/app7121284.

Al, M., Sen, S., Chakrabortty, D., 2019. Ag3PO4-based nanocomposites and their applications in photodegradation of toxic organic dye contaminated wastewater: review on material design to performance enhancement. J. Saudi Chem. Soc. 24 (1), 20–41. https://doi.org/10.1016/j.jscs.2019.09.001.

Al-hakkani, M.F., Gouda, G.A., Hassan, S.H.A., 2021. Heliyon review article a review of green methods for phyto-fabrication of hematite (α-Fe2O3) nanoparticles and their characterization, properties, and applications. Heliyon 7 (1), e05806. https://doi.org/10.1016/j.heliyon.2020.e05806.

Anwar, K., Said, M., Fauzi, A., Abdul, Z., Sohaimi, M., Usman, J., Olabode, Y., 2020. Innovation in membrane fabrication : magnetic induced photocatalytic membrane. J. Taiwan Inst. Chem. Eng. 113, 372–395. https://doi.org/10.1016/j.jtice.2020.08.014.

Argurio, P., Fontananova, E., Molinari, R., Drioli, E., 2018. Photocatalytic membranes in photocatalytic membrane reactors. Processes 6 (9), 162. https://doi.org/10.3390/pr6090162.

Babu, V.J., Vempati, S., Uyar, T., Ramakrishna, S., 2015. Review of one-dimensional and two-dimensional nanostructured materials for hydrogen generation. Phys. Chem. Chem. Phys. 17 (5), 2960–2986. https://doi.org/10.1039/C4CP04245J.

Bui, D., Pham, M., Tran, H., Nguyen, T., Cao, T.M., Pham, V.V., 2021. Revisiting the key optical and electrical characteristics in reporting the photocatalysis of semiconductors. ACS Omega 6, 27379–27386. https://doi.org/10.1021/acsomega.1c04215.

Carolin Christopher, F., Kumar, P.S., Vo, D.-V.N., Joshiba, G.J., 2021. A review on critical assessment of advanced bioreactor options for sustainable hydrogen production. Int. J. Hydrogen Energy 46 (10), 7113–7136. https://doi.org/10.1016/j.ijhydene.2020.11.244.

Chen, W., Chan, A., Sun-waterhouse, D., Llorca, J., Idriss, H., Waterhouse, G.I.N., 2018. Performance comparison of Ni/TiO2 and Au/TiO2 photocatalysts for H2 production in different alcohol-water mixtures. J. Catal. 367, 27–42. https://doi.org/10.1016/j.jcat.2018.08.015.

Chen, C., Jin, J., Chen, S., Wang, T., Xiao, J., Peng, T., 2021. In-situ growth of ultrafine ZnO on g-C3N4 layer for highly active and selective CO2 photoreduction to CH4 under visible light. Mater. Res. Bull. 137, 111177. https://doi.org/10.1016/j.materresbull.2020.111177.

Chen, L., Xu, P., Wang, H., 2022. Photocatalytic membrane reactors for produced water treatment and reuse : Fundamentals, affecting factors, rational design, and evaluation metrics. J. Hazard. Mater. 424 (PB), 127493. https://doi.org/10.1016/j.jhazmat.2021.127493.

Dabhane, H., Ghotekar, S., Tambade, P., Pansambal, S., Murthy, H.C.A., Oza, R., Medhane, V., 2021. A review on environmentally benevolent synthesis of CdS nanoparticle and their applications. Environ. Chem. Ecotoxicol. 3, 209–219. https://doi.org/10.1016/j.enceco.2021.06.002.

Daels, N., Radoicic, M., Radetic, M., Van Hulle, S.W.H., De Clerck, K., 2014. Functionalisation of electrospun polymer nanofibre membranes with TiO2 nanoparticles in view of dissolved organic matter photodegradation. Sep. Purif. Technol. 133, 282–290. https://doi.org/10.1016/j.seppur.2014.06.040.

Dong, R., Du, T., Dong, S., Zhao, X., Ma, R., Du, A., Fan, Y., Cao, X., 2022. A weather-resistant daytime radiative cooler based on fluorocarbon resin. Sol. Energy Mater. Sol. Cells 235, 111486. https://doi.org/10.1016/j.solmat.2021.111486.

Estrada-Flores, S., Martínez-Luévanos, A., Perez-Berumen, C.M., García-Cerda, L.A., Flores-Guia, T.E., 2020. Relationship between morphology, porosity, and the photocatalytic activity of TiO2 obtained by sol–gel method assisted with ionic and nonionic surfactants. Bol. Socied. Esp. Cerám. Vidrio 59 (5), 209–218. https://doi.org/10.1016/j.bsecv.2019.10.003.

Fajrina, N., Tahir, M., 2019. A critical review in strategies to improve photocatalytic water splitting towards hydrogen production. Int. J. Hydrogen Energy 44 (2), 540–577. https://doi.org/10.1016/j.ijhydene.2018.10.200.

Fischer, K., Gläser, R., Schulze, A., 2014. Nanoneedle and nanotubular titanium dioxide – PES mixed matrix membrane for photocatalysis. Appl. Catal. Environ. 160–161, 456–464. https://doi.org/10.1016/j.apcatb.2014.05.054.

Fu, C.-F., Wu, X., Yang, J., 2018. Material design for photocatalytic water splitting from a theoretical perspective. Adv. Mater. 30 (48), 1802106. https://doi.org/10.1002/adma.201802106.

Fu, J., Jiang, K., Qiu, X., Yu, J., Liu, M., 2020. Product selectivity of photocatalytic CO2 reduction reactions. Mater. Today 32, 222–243. https://doi.org/10.1016/j.mattod.2019.06.009.

A. General on membrane, materials and application

Fujihara, K., Ohno, T., Matsumura, M., 1998. Splitting of water by electrochemical combination of two photocatalytic reactions on TiO2 particles. J. Chem. Soc. Faraday Trans. 94, 3705–3709. https://doi.org/10.1039/A806398B.

Gao, Y., Qian, K., Xu, B., Li, Z., Zheng, J., Zhao, S., Ding, F., Sun, Y., Xu, Z., 2020. Recent advances in visible-light-driven conversion of CO2 by photocatalysts into fuels or value-added chemicals. Carbon Resour. Convers. 3, 46–59. https://doi.org/10.1016/j.crcon.2020.02.003.

Gastiasoro, M.N., Ruhman, J., Fernandes, R.M., 2020. Superconductivity in dilute SrTiO3: a review. Ann. Phys. 417, 168107. https://doi.org/10.1016/j.aop.2020.168107.

Gelderman, K., Lee, L., Donne, S.W., 2007. Flat-band potential of a semiconductor: using the Mott–Schottky equation. J. Chem. Educ. 84 (4), 685. https://doi.org/10.1021/ed084p685.

George, P., Chowdhury, P., 2019. Complex dielectric transformation of UV-vis diffuse reflectance spectra for estimating optical band-gap energies and materials classification. Analyst 144 (9), 3005–3012. https://doi.org/10.1039/C8AN02257G.

Hezam, A., Namratha, K., Drmosh, Q.A., Ponnamma, D., Wang, J., Prasad, S., Ahamed, M., Cheng, C., Byrappa, K., 2020. CeO2 nanostructures enriched with oxygen vacancies for photocatalytic CO2 reduction. ACS Appl. Nano Mater. 3, 138–148. https://doi.org/10.1021/acsanm.9b01833.

Homayoonfal, M., Reza, M., Shariaty-niassar, M., Akbari, A., 2014. A comparison between blending and surface deposition methods for the preparation of iron oxide/polysulfone nanocomposite membranes. Desalination 354, 125–142. https://doi.org/10.1016/j.desal.2014.09.031.

Homocianu, M., Pascariu, P., 2022. High-performance photocatalytic membranes for water purification in relation to environmental and operational parameters. J. Environ. Manage. 311, 114817. https://doi.org/10.1016/j.jenvman.2022.114817.

Johannes, A.Z., Pingak, R.K., Bukit, M., 2020. Tauc plot software: calculating energy gap values of organic materials based on ultraviolet-visible absorbance spectrum. IOP Conf. Ser.: Mater. Sci. Eng. 823 (1), 12030. https://doi.org/10.1088/1757-899x/823/1/012030.

Karamian, E., Sharifnia, S., 2018. Enhanced visible light photocatalytic activity of BiFeO3-ZnO p-n heterojunction for CO2 reduction. Mater. Sci. Eng. B: Solid-State Mater. Adv. Technol. 238–239, 142–148. https://doi.org/10.1016/j.mseb.2018.12.023.

Khan, S., Je, M., Nu, N., Ton, T., Lei, W., Taniike, T., 2021. C-doped ZnS-ZnO/Rh nanosheets as multijunctioned photocatalysts for effective H2 generation from pure water under solar simulating light. Appl. Catal. Environ. 297, 120473. https://doi.org/10.1016/j.apcatb.2021.120473.

Kitano, M., Tsujimaru, K., Anpo, M., 2006. Decomposition of water in the separate evolution of hydrogen and oxygen using visible light-responsive TiO2 thin film photocatalysts: effect of the work function of the substrates on the yield of the reaction. Appl. Catal. A. Gen. 314 (2), 179–183. https://doi.org/10.1016/j.apcata.2006.08.017.

Kozlova, E.A., Korobkina, T.P., Vorontsov, A.V., 2009. Overall water splitting over Pt/TiO2 catalyst with Ce 3D/Ce 4D shuttle charge transfer system. Int. J. Hydrogen Energy 34 (1), 138–146. https://doi.org/10.1016/j.ijhydene.2008.09.101.

Lee, G., Wu, J.J., 2017. Recent developments in ZnS photocatalysts from synthesis to photocatalytic applications—a review. Powder Technol. 318, 8–22. https://doi.org/10.1016/j.powtec.2017.05.022.

Lee, K., Yoon, H., Ahn, C., Park, J., Jeon, S., 2019. Strategies to improve the photocatalytic activity of TiO2: 3D nanostructuring and heterostructuring with graphitic carbon nanomaterials. Nanoscale 11 (15), 7025–7040. https://doi.org/10.1039/C9NR01260E.

Li, X., Yu, J., Jaroniec, M., 2016. Hierarchical photocatalysts. Chem. Soc. Rev. 45 (9), 2603–2636. https://doi.org/10.1039/C5CS00838G.

Liu, H., Ma, H.T., Li, X.Z., 2003. The enhancement of TiO2 photocatalytic activity by hydrogen thermal treatment. Chemosphere 50 (1), 39–46. https://doi.org/10.1016/s0045-6535(02)00486-1.

Liu, W., Li, Y., Liu, F., Jiang, W., Zhang, D., Liang, J., 2019a. Visible-light-driven photocatalytic degradation of diclofenac by carbon quantum dots modified porous g-C3N4: mechanisms, degradation pathway and DFT calculation. Water Res. 151, 8–19. https://doi.org/10.1016/j.watres.2018.11.084.

Liu, G., Sheng, Y., Ager, J.W., Kraft, M., Xu, R., 2019b. Research advances towards large-scale solar hydrogen production from water. EnergyChem 1 (2), 100014. https://doi.org/10.1016/j.enchem.2019.100014.

Makuła, P., Pacia, M., Macyk, W., 2018. How to correctly determine the band gap energy of modified semiconductor photocatalysts based on UV–vis spectra. J. Phys. Chem. Lett. 9 (23), 6814–6817. https://doi.org/10.1021/acs.jpclett.8b02892.

A. General on membrane, materials and application

Meng, A., Zhu, B., Zhong, B., Zhang, L., Cheng, B., 2017. Direct Z-scheme TiO2/CdS hierarchical photocatalyst for enhanced photocatalytic H2-production activity. Appl. Surf. Sci. 422, 518–527. https://doi.org/10.1016/j.apsusc.2017.06.028.

Mills, A., Hunte, S.L., 2000. An overview of semiconductor photocatalysis. J. Photochem. Photobiol. A Chem. 108 (1), 1–35. https://doi.org/10.1016/S1010-6030(97)00118-4.

Mishra, V., Warshi, M.K., Sati, A., Kumar, A., Mishra, V., Sagdeo, A., Kumar, R., Sagdeo, P.R., 2018. Diffuse reflectance spectroscopy: an effective tool to probe the defect states in wide band gap semiconducting materials. Mater. Sci. Semicond. Process. 86, 151–156. https://doi.org/10.1016/j.mssp.2018.06.025.

Mo, Z., Xu, H., Chen, Z., She, X., Song, Y., Lian, J., Zhu, X., Yan, P., Lei, Y., Yuan, S., Li, H., 2019. Construction of MnO2/monolayer g-C3N4 with Mn vacancies for Z-scheme overall water splitting. Appl. Catal. Environ. 241, 452–460. https://doi.org/10.1016/j.apcatb.2018.08.073.

Molinari, R., Lavorato, C., Argurio, P., Szymański, K., 2019. Overview of photocatalytic membrane reactors in organic synthesis, energy storage and environmental applications. Catalysts 9 (239), 1–39. https://doi.org/10.3390/catal9030239.

Molinari, R., Lavorato, C., Argurio, P., 2020. Visible-light photocatalysts and their perspectives for building photocatalytic membrane reactors for various liquid phase chemical conversions. Catalysts 10 (11). https://doi.org/10.3390/catal10111334.

Monga, D., Sharma, S., Shetti, N.P., Basu, S., Raghava, K., Aminabhavi, T.M., 2021. Advances in transition metal dichalcogenide-based two-dimensional nanomaterials. Mater. Today Chem. 19, 100399. https://doi.org/10.1016/j.mtchem.2020.100399.

Moniz, S.J.A., Shevlin, S.A., Martin, D.J., Guo, Z.-X., Tang, J., 2015. Visible-light driven heterojunction photocatalysts for water splitting—a critical review. Energ. Environ. Sci. 8 (3), 731–759. https://doi.org/10.1039/C4EE03271C.

Mozia, S., Darowna, D., Orecki, A., Wróbel, R., Wilpiszewska, K., Morawski, A.W., 2014. Microscopic studies on TiO2 fouling of MF/UF polyethersulfone membranes in a photocatalytic membrane reactor. J. Membr. Sci. 470, 356–368. https://doi.org/10.1016/j.memsci.2014.07.049.

Nguyen, H.T., Bui, H.M., Wang, Y.-F., You, S.-J., 2022a. Antifouling catalytic mixed-matrix membranes based on polyethersulfone and composition-optimized Zn-cu-Fe-O CWAO catalyst under dark ambient conditions. Environ. Technol., 1–17. https://doi.org/10.1080/09593330.2022.2041106.

Nguyen, H.T., Pham, M.-T., Nguyen, T.-M.T., Bui, H.M., Wang, Y.-F., You, S.-J., 2022b. Modifications of conventional organic membranes with photocatalysts for antifouling and self-cleaning properties applied in wastewater filtration and separation processes: a review. Sep. Sci. Technol. 57 (9), 1471–1500. https://doi.org/10.1080/01496395.2021.1982981.

Nguyen, H.T., Guo, S.-Y., You, S.-J., Wang, Y.-F., 2022c. Visible light driven photocatalytic coating of PAA plasma-grafted PVDF membrane by TiO2 doped with lanthanum recovered from waste fluorescent powder. Environ. Eng. Res. 27 (3), 210140–210144. https://doi.org/10.4491/eer.2021.144.

Nor, N.A.M., Jaafar, J., Ismail, A.F., Mohamed, M.A., Rahman, M.A., Othman, M.H.D., Lau, W.J., Yusof, N., 2016. Preparation and performance of PVDF-based nanocomposite membrane consisting of TiO2 nanofibers for organic pollutant decomposition in wastewater under UV irradiation. Desalination 391, 89–97. https://doi.org/10.1016/j.desal.2016.01.015.

Ohtani, B., 2011. Chapter 10—Photocatalysis by inorganic solid materials: Revisiting its definition, concepts, and experimental procedures. In: van Eldik, R., Stochel, G. (Eds.), Inorganic Photochemistry. vol. 63. Academic Press, pp. 395–430, https://doi.org/10.1016/B978-0-12-385904-4.00001-9.

Opoku, F., Govender, K.K., van Sittert, C.G.C.E., Govender, P.P., 2017. Recent Progress in the development of semiconductor-based photocatalyst materials for applications in photocatalytic water splitting and degradation of pollutants. Adv. Sustain. Syst. 1 (7), 1700006. https://doi.org/10.1002/adsu.201700006.

Pham, M., Hussain, A., Bui, D., 2021. Environmental technology and innovation surface plasmon resonance enhanced photocatalysis of Ag nanoparticles-decorated Bi 2 S 3 nanorods for NO degradation. Environ. Technol. Innov. 23, 101755. https://doi.org/10.1016/j.eti.2021.101755.

Primo, A., Marino, T., Corma, A., Molinari, R., García, H., 2011. Efficient visible-light photocatalytic water splitting by minute amounts of gold supported on nanoparticulate CeO2 obtained by a biopolymer templating method. J. Am. Chem. Soc. 133 (18), 6930–6933. https://doi.org/10.1021/ja2011498.

Qin, Z., Fang, W., Liu, J., Wei, Z., Jiang, Z., Shangguan, W., 2018. Zinc-doped g-C3N4/BiVO4 as a Z-scheme photocatalyst system for water splitting under visible light. Chin. J. Catal. 39 (3), 472–478. https://doi.org/10.1016/S1872-2067(17)62961-9.

Rai, H., Kondal, N., 2021. A review on defect related emissions in undoped ZnO nanostructures. Mater. Today: Proc. 4–8. https://doi.org/10.1016/j.matpr.2021.08.343.

Reddy, N.L., Rao, V.N., Vijayakumar, M., Aminabhavi, T.M., 2019. A review on frontiers in plasmonic nano-photocatalysts for hydrogen production. Int. J. Hydrogen Energy 44 (21), 10453–10472. https://doi.org/10.1016/j.ijhydene.2019.02.120.

Reischauer, S., Pieber, B., 2021. Emerging concepts in photocatalytic organic synthesis. IScience 24 (3). https://doi.org/10.1016/j.isci.2021.102209.

Retamoso, C., Escalona, N., González, M., Barrientos, L., Allende-González, P., Stancovich, S., Serpell, R., Fierro, J.L.G., Lopez, M., 2019. Effect of particle size on the photocatalytic activity of modified rutile sand (TiO2) for the discoloration of methylene blue in water. J. Photochem. Photobiol. A Chem. 378, 136–141. https://doi.org/10.1016/j.jphotochem.2019.04.021.

Rogelj, J., Shindell, D., Jiang, K., Fifita, S., Forster, P., Ginzburg, V., Handa, C., Kheshgi, H., Kobayashi, S., Kriegler, E., Mundaca, L., Séférian, R., Vilariño, M.V., 2018. Mitigation pathways compatible with 1.5°C in the context of sustainable development. In: Global Warming of 1.5°C. An IPCC Special Report on the impacts of global warming of 1.5°C above pre-industrial levels and related global greenhouse gas emission pathways. IPCC, Geneva, Switzerland.

Rohit, S., Abhijit, S.R., Eberhard, S., 2020. A simple calculator to decide UVA "on" time for cross-linking of thin Keratoconic corneas (<400 μm). J. Refract. Surg. 36 (10), 707. https://doi.org/10.3928/1081597X-20200729-01.

Sabouni, R., Gomaa, H., 2019. Photocatalytic degradation of pharmaceutical micro-pollutants using ZnO. Environ. Sci. Pollut. Res. Int. 26 (6), 5372–5380. https://doi.org/10.1007/s11356-018-4051-2.

Sasaki, Y., Kato, H., Kudo, A., 2013. Water splitting under sunlight irradiation using z-scheme photocatalyst system. J. Am. Chem. Soc. 135 (14), 5441–5449. https://doi.org/10.1021/ja400238r.

Seger, B., Kamat, P.V., 2009. Fuel cell geared in reverse: photocatalytic hydrogen production using a TiO2/Nafion/Pt membrane assembly with no applied bias. J. Phys. Chem. C 113, 18946–18952. https://doi.org/10.1021/jp907367k.

Sharma, S., Basu, S., Shetti, N.P., Kamali, M., Walvekar, P., Aminabhavi, T.M., 2020. Waste-to-energy nexus: a sustainable development. Environ. Pollut. 267, 115501. https://doi.org/10.1016/j.envpol.2020.115501.

Singla, S., Sharma, S., Basu, S., Shetti, N.P., Aminabhavi, T.M., 2021. Photocatalytic water splitting hydrogen production via environmental benign carbon based nanomaterials. Int. J. Hydrogen Energy 46 (68), 33696–33717. https://doi.org/10.1016/j.ijhydene.2021.07.187.

Song, H., Shao, J., He, Y., Liu, B., Zhong, X., 2012. Natural organic matter removal and flux decline with PEG-TiO2-doped PVDF membranes by integration of ultrafiltration with photocatalysis. J. Membr. Sci. 405–406, 48–56. https://doi.org/10.1016/j.memsci.2012.02.063.

Sun, C., Yang, J., Xu, M., Cui, Y., Ren, W., Zhang, J., Zhao, H., Liang, B., 2022. Recent intensification strategies of SnO2-based photocatalysts: a review. Chem. Eng. J. 427, 131564. https://doi.org/10.1016/j.cej.2021.131564.

Swan, L.G., Ugursal, V.I., 2009. Modeling of end-use energy consumption in the residential sector: a review of modeling techniques. Renew. Sustain. Energy Rev. 13 (8), 1819–1835. https://doi.org/10.1016/j.rser.2008.09.033.

Tang, L., Shao, S., Wang, A., Tian, C., Luo, F., Li, J., Li, Z., Tan, H., Zhang, H., 2022. Influence of fluorocarbon side chain on microphase separation and chemical stability of silicon-containing polycarbonate urethane. Polymer 242, 124538. https://doi.org/10.1016/j.polymer.2022.124538.

Tsydenov, D.E., Vorontsov, A.V., 2015. Influence of Nafion loading on hydrogen production in a membrane photocatalytic system. J. Photochem. Photobiol. A Chem. 297, 8–13. https://doi.org/10.1016/j.jphotochem.2014.09.014.

Wang, W.Y., Irawan, A., Ku, Y., 2008. Photocatalytic degradation of Acid Red 4 using a titanium dioxide membrane supported on a porous ceramic tube. Water Res. 42 (19), 4725–4732. https://doi.org/10.1016/j.watres.2008.08.021.

Wang, Q., Guo, R., Wang, Z., Shen, D., Yu, R., Luo, K., Wu, C., Gu, S., 2021. Progress in carbon-based electrocatalyst derived from biomass for the hydrogen evolution reaction. Fuel 293, 120440. https://doi.org/10.1016/j.fuel.2021.120440.

Wu, H., Liu, L., Zhao, S., 2014. The effect of water on the structural, electronic and photocatalytic properties of graphitic carbon nitride. Phys. Chem. Chem. Phys. 16, 3299–3304. https://doi.org/10.1039/c3cp54333a.

Xu, Q., Zhang, L., Cheng, B., Fan, J., Yu, J., 2020. S-Scheme heterojunction photocatalyst. Chem 6 (7), 1543–1559. https://doi.org/10.1016/j.chempr.2020.06.010.

Yan, H., Wang, X., Yao, M., Yao, X., 2013. Band structure design of semiconductors for enhanced photocatalytic activity: the case of TiO2. Progr. Nat. Sci.: Mater. Int. 23 (4), 402–407. https://doi.org/10.1016/j.pnsc.2013.06.002.

A. General on membrane, materials and application

Yang, X., Tian, Z., Chen, Y., Huang, H., Hu, J., 2020. One-pot calcination preparation of graphene/g–C3N4–co photocatalysts with enhanced visible light photocatalytic activity. Int. J. Hydrogen Energy 45 (23), 12889–12902. https://doi.org/10.1016/j.ijhydene.2020.03.028.

Yu, S., Huang, C., Liao, C., Wu, J.C.S., Chang, S., Chen, K., 2011. A novel membrane reactor for separating hydrogen and oxygen in photocatalytic water splitting. J. Membr. Sci. 382 (1–2), 291–299. https://doi.org/10.1016/j.memsci.2011.08.022.

Zheng, X., Shen, Z.-P., Shi, L., Cheng, R., Yuan, D.-H., 2017. Photocatalytic membrane reactors (PMRs) in water treatment: configurations and influencing factors. Catalysts 7 (8). https://doi.org/10.3390/catal7080224.

Zwara, J., Paszkiewicz-gawron, M., Łuczak, J., Pancielejko, A., Lisowski, W., Trykowski, G., Zaleska-Medynska, A., Grabowska, E., 2019. The effect of imidazolium ionic liquid on the morphology of Pt nanoparticles deposited on the surface of SrTiO3 and photoactivity of Pt-SrTiO3 composite in the H2 generation reaction. Int. J. Hydrogen Energy 44 (48), 26308–26321. https://doi.org/10.1016/j.ijhydene.2019.08.094.

3

In situ real-time monitoring technologies for fouling detection in membrane processes

Seongpil Jeong[a] and Hye-Won Kim[b]

[a]Korea Institute of Science and Technology (KIST) & KIST School, Korea University of Science and Technology (KUST), Seoul, Republic of Korea [b]Korea Institute of Science and Technology (KIST), Seoul, Republic of Korea

1. Introduction

Fouling is the major issue inhibiting the stable operation of the membrane process. In order to deal with the fouling problem, several approaches have been proposed such as: (1) understanding the fouling mechanism by using experimental tests or theoretical or empirical modeling approaches and (2) development of removal or mitigation processes to mitigate fouling on membranes. The foulants are commonly classified into four categories including particulates, organics, inorganics, and microbes. The fouling phenomena are different according to the (1) size of the membrane pores such as microfiltration (MF), ultrafiltration (UF), nanofiltration (NF), and reverse osmosis (RO), (2) type of the membrane such as flat-sheet, tubular, hollow fiber, and spiral wound, (3) materials of membranes such as polymer [polyvinylidene fluoride (PVDF), polypropylene (PP), polytetrafluoroethylene (PTFE), cellulose acetate (CA), polyamide (PA), etc.], ceramic (Al or Zr oxides) or carbon [carbon nanotubes (CNT) and graphene], and (4) operation types such as pressurized (MF/UF/NF/RO) or nonpressurized ones [forward osmosis (FO), membrane distillation (MD), electrodialysis (ED), and capacitive deionization (CDI)]. Various studies on fouling have been conducted on a laboratory scale and on pilot scale.

Fouling studies have focused on the observation of fouling parameters such as decrease in flux and increase in transmembrane pressure (TMP). Foulants in the feed solution for the membrane processes were also measured to guess how much fouling could be formed on

Copyright © 2023 Elsevier Inc. All rights reserved.

the membrane surfaces. By using the concentrations of foulants, the number of microbes, and saturation index (SI), various fouling models have been developed to understand various fouling patterns such as standard blocking, complete blocking, intermediate blocking, and cake formation. Fouling indexes also have been suggested from the silt density index (SDI), membrane fouling index (MFI), etc. However, the fouling mechanisms of the membrane process have not been clearly proven yet.

Recently, novel approaches for in situ fouling measurement have been suggested. Some of the in situ monitoring technologies can obtain data in real time. Prompt response is essential for membrane operation because once severe membrane fouling occurs, it is hard to recover the initial flux of the membrane due to formation of the irreversible fouling. Therefore, warning systems to prevent severe fouling can be suggested if the in situ real-time monitoring technology is developed.

In this chapter, conventional and novel fouling observation technologies will be introduced.

2. Foulants and their conventional quantifications in water samples

Before discussing foulants, it is required to understand the aspects of the foulants. The foulants can be classified into four major components such as particles, organics, inorganics, and microbes (Warsinger et al., 2015). In most environmental conditions, it is hard to separately understand mixed foulants' characteristics because four major foulants coexist on the membrane surface (Wang et al., 2020a). Therefore, many studies have discussed separation of foulants and analysis of their effects on fouling one by one (Kim et al., 2022). The foulants were detached from the membrane surface by the deionized water, acidic, or base solutions and analyzed from the collected water samples as the following (Lee et al., 2020).

2.1 Particles

Particles are materials having their size more than 0.45 µm; therefore, they settle down in water solutions with time. MF/UF membranes are usually used to remove particles, therefore, particulate foulants are usually formed on the MF/UF membrane surfaces. Particles in water solutions have been measured by using filtration and weighing to measure the suspended solids (Charfi et al., 2015), turbidity (Jeong et al., 2011), and particle counters (Lay et al., 2022). The turbidity meter and particle counter have been applied before and after membrane filtration to monitor the particle concentrations in the feed and permeate solutions in the desalination pilot (Jeong et al., 2011) and water treatment plants (Safaee et al., 2022), respectively.

2.2 Organics

Organic matters can (1) attach to membranes such as transparent exopolymer particles [TEPs, precursor organic foulant (Alayande et al., 2022)] and form a cake layer (Gong et al., 2017), (2) block the pores with the particulate fouling (Silalahi and Leiknes, 2009), or

(3) interact with divalent cations to form a thick and dense fouling layer (Kim et al., 2019; Nguyen et al., 2017). The organic matters could be used as substrates for biofouling (microbes) (Abushaban et al., 2022; Botton et al., 2012) or inhibitors if they are toxic to the microbes such as micropollutants (Besha et al., 2017). The effects of organics on the fouling layer vary due to their diversities; therefore, different analytical methods have been used for organic matter analysis in water solutions. Volatile suspended solids (VSS) can be used as parameters for organic matters (Deb et al., 2022). The total organic carbon (TOC) analyzer also can report the concentration of organic matters in the water solutions (Tibi et al., 2021). Liquid chromatography (LC) and gas chromatography (GC) coupled with mass spectrometry or organic carbon detectors such as LC-MS (Enfrin et al., 2021), LC-MSMS (Jamil et al., 2021), LC-OCD (Jeong et al., 2021), and GC-MS (Park et al., 2019) which can identify trace levels of organic matters are also used to analyze organic foulants. Fluorescence excitation emission matrix (FEEM) spectroscopy also has been applied to understand the detailed characteristics of the organic matters (Jeong et al., 2021).

2.3 Inorganics

Major inorganic foulants in water treatment are scalants such as $CaCO_3$ and $CaSO_4$ (Kim et al., 2020). The scaling formation on the membrane surface is also affected by the composition of ions (Kim et al., 2020). Moreover, high ionic strength ($>150\,mM$) in protein solution decreased biofouling formation on the membrane surface (Kilmer et al., 2021). Some inorganic ions such as metals (Cu, Zn, and Ag) can affect the inhibition of biofouling formation (Kim et al., 2021). The foulant could be washed by deionized water or selected solutions and collected as water samples. The ion concentrations in the feed and permeate solutions can be also analyzed by using the instrumental analysis devices such as ion chromatography (IC), atomic absorption spectrometry (AAS), and inductively coupled plasma spectrometry (ICP) (Kim et al., 2020, 2019).

2.4 Microbes

Biofouling is the most common fouling for the RO membrane because particulate fouling is mitigated by the pretreatment system (Bae et al., 2011). Biofouling is unique because it can grow on the membrane surface. It has been reported that various bacterial groups could be shifted or stabilized on the membrane surface during the operation of the membrane process (Bae et al., 2011). In order to check the concentration of the microbes, several methods can be used. Optical density (OD) is a simple measurement performed by using a spectrophotometer to measure microbial concentrations in the water sample (Lee et al., 2021a). If it is required to identify microbes from the biofouling, the collected microbial sample could be incubated on the specific culture media (Lee et al., 2021a). Microbial concentration can be also analyzed by cell counting after incubation (Lee et al., 2021a). The cell concentration in the water sample can be directly measured by using flow cytometry (Nguyen et al., 2020, 2021). Indirectly, adenosine triphosphate (ATP) and extracellular polymeric substances (EPS) can be detected on the membrane samples to measure the microbial activities (Lee et al., 2021a).

3. Conventional technologies for detecting fouling

3.1 Membrane autopsy

The membrane autopsy is the most commonly used method for fouling analysis in order to check the fouled membrane surface starting from a bare-eye observation (Lee et al., 2021b). The color or shape of the fouling area could be analyzed based on the operating conditions of the membrane process such as slimy biofouling or organic fouling, which could be easily distinguished from the scaling, which has crystal formation (Fortunato et al., 2020). Moreover, the areas that have high or low fouling potentials could be shown visually and the washing or cleaning tests can be conducted for the used membranes (Jung et al., 2018). The foulants on the membrane surface could be collected by scratching and be further analyzed using various instruments (Cho et al., 2016).

3.2 Confocal laser scanning microscopy (CLSM)

CLSM could provide three-dimensional (3D) images of the live or dead cells on membrane surfaces (Tian et al., 2021). The live and dead cells are shown as green and red color due to the attached dye on the cells. The 3D structural image also can show the location of the live and dead cells on the membrane surface under given experimental conditions. The thickness of the biofilm is commonly analyzed to understand biofouling behaviors. The CLSM images have been used to compare the fouled membrane surface (biovolume of foulants) with each other according to the operational conditions (Ng and Ng, 2010) or cleaning conditions (Guerrero-Navarro et al., 2020).

3.3 Atomic force microscopy (AFM)

AFM measures the force between the cantilever and the membrane surface (Chen et al., 2022). The measure attraction force between the cantilever and the membrane surface could be converted into the surface roughness of the membrane. The surface roughness of the fouled RO membrane was higher compared to that of the virgin RO membrane (Powell et al., 2017). When the modified membrane has low surface roughness, the flux was maintained showing a good antifouling property (Wang et al., 2021). Moreover, the chemically modified tips of the AFM were also used to measure the interactions between the organics and the RO membrane (Lei et al., 2018).

3.4 Scanning electron microscope/energy-dispersive X-ray spectroscopy (SEM-EDX)

SEM images are commonly used to check the morphology of the membrane surface (Kim et al., 2019). Inorganic foulants such as $CaSO_4$ and $CaCO_3$ could be distinguished according to their characteristics due to their shape and fractions of ion compositions by using SEM-EDX analysis (Kim et al., 2020, 2019). Comparisons of the ion compositions between the pristine membrane and fouled membrane can provide foulant characteristics additionally.

Cross-sectional images of the fouled membrane can show the thickness of membrane and the fouled layer (Kim et al., 2019). The fouled area could be further separated by using the mapping tool of the SEM-EDX device according to their ion compositions (Kim et al., 2019). The effective scanning depth of the SEM-EDX analysis is around 1–2 μm; therefore, if pristine membrane is covered by the thick cake layer, the atomic fraction of the foulant will be emphasized compared to that of the membrane.

3.5 Fourier transform infrared spectroscopy (FT-IR)

FT-IR analysis can be used to identify the new molecular bonding formation or removal during membrane fouling (Guo et al., 2019). Some studies suggested correlations between the FT-IR result and operational conditions such as pH, ionic strength, and temperature (Li et al., 2011) of the membrane processes. Moreover, recently two-dimensional (2D)-based FT-IR mapping technology has been applied to membrane fouling studies (Benavente et al., 2016).

3.6 Contact angle

The contact angle of the membrane surface shows the hydrophilic (contact angle <90 degrees) or hydrophobic (contact angle >90 degrees) characteristics. Because foulants (organic matters and biofoulant) are hydrophobic, hydrophobicity is the important factor for membrane fouling (Kim et al., 2020, 2019). Contact angle analysis is conducted for the modified membrane or contact angle variation is also monitored during membrane fouling (Nguyen et al., 2017). Moreover, contact angle hysteresis (which can be measured by a dynamic contact angle analyzer) can also provide information about surface homogeneity (Jeong et al., 2016).

4. Novel technologies for detecting fouling

The membrane process could be classified into pressurized (MF/UF/NF/RO) and nonpressurized systems (FO/MD/ED/CDI). In this chapter, various in situ monitoring systems, which have been applied to pressurized systems, will be introduced. There is a distinct difference between MF/UF and NF/RO processes due to their separation mechanisms such as size exclusion and charge repulsion. MF/UF membrane processes are generally used as pretreatment processes of the NF/RO processes. Therefore, in the membrane process operations under environmental conditions, the particulate, organic, and biofoulings are commonly observed on MF/UF membrane surfaces, while organic, inorganic, and biofoulings occur on NF/RO membrane surfaces.

4.1 Fouling observation on pressurized microfiltration/ultrafiltration (MF/UF) membrane systems

The foulings that occur mainly on the MF/UF membrane surface are mixtures of particles, organics, inorganics, and microbes. MF/UF systems usually used for the removal of particles and microbes are pretreatment of RO systems or the main process of the water production plant. Therefore, the main retentates on the membrane surfaces are particles and microbes. And in the permeate, organic matters and inorganic ions exist. The in situ monitoring technology analyzes the retentates on the membrane surface and transferred matters across the membrane.

4.1.1 Optical coherence tomography (OCT) technology

OCT technology could monitor the 2D cross-sectional image of the fouling layer forming on the membrane surface. The thickness and morphology of the fouling layer could be analyzed in real time. As one of the examples, the surface of MF membrane of the MBR system using real wastewater from the real-scale plant was observed up to 7 days under three different draw solute conditions (Fig. 1). OCT technology could show the morphology variations with time and the morphology variation could be used to understand as an additional tool complex cake layer formation on the membrane surface (Pathak et al., 2018). And the OCT technology is a nondestructive method, which does not affect the fouling layer when the image is captured. Therefore, continuous observation of the cross-sectional image for the cake layer is available by using OCT technology (Fortunato et al., 2016). The obtained images could be further analyzed by using image analysis tools such as MATLAB code (Fortunato and Leiknes, 2017). The combined analysis by using both experimental data and OCT images could provide comprehensive understanding of the membrane fouling by using modeling tools (Fortunato and Leiknes, 2017; Li et al., 2017; Pathak et al., 2018). And if the fouling mechanism is proven well, mitigation or removal of the fouling layer is effectively conducted.

4.1.2 In situ EEM (solid-phase fluorescence EEM, SPF-EEM)

The FEEM device has been used to identify different organic fractions according to the organic characteristics such as tryptophan-like or aromatic protein-like organics, etc., and no further pretreatment process was required (Henderson et al., 2009). The EEM technology has been used for the analysis of organic matters in the water sample; however, Yamamura et al. (2019) suggested in situ solid-phase fluorescence EEMs for membrane fouling

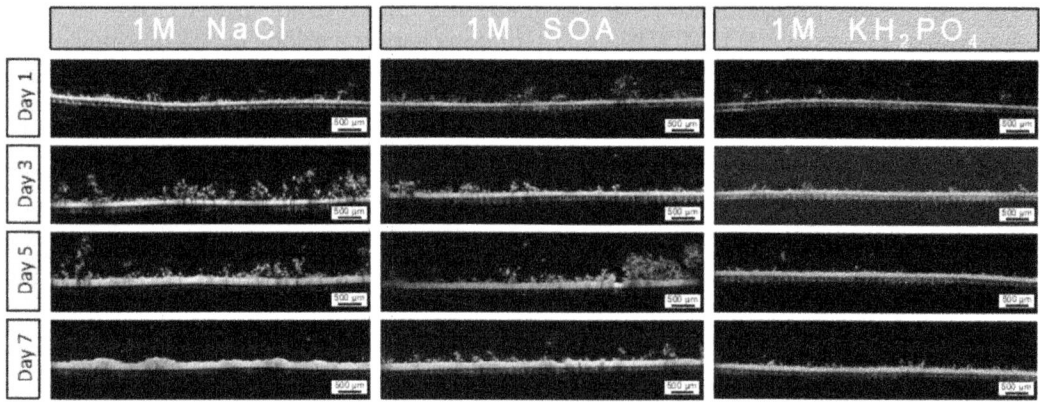

FIG. 1   The collected OCT scan image for the fouling layer on the membrane surface with time (Pathak et al., 2018).

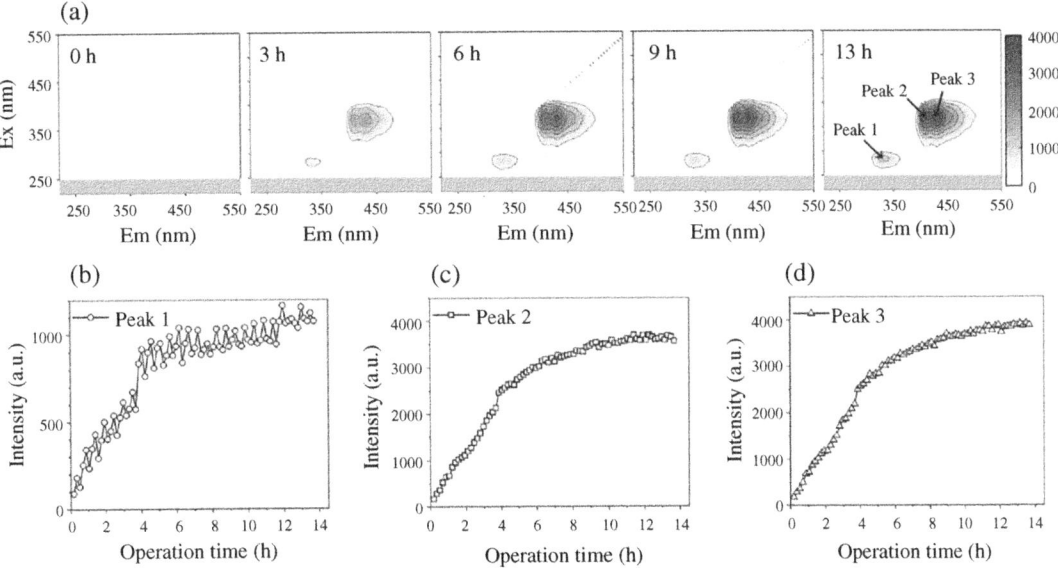

FIG. 2   In situ SPF-EEM monitoring results (Yamamura et al., 2019).

monitoring for the hollow fiber PVDF membrane system treating secondary effluent of the wastewater treatment plant (Yamamura et al., 2019). The SPF-EEM device has optical fibers connected to the EEM device. From the optical fiber, EEM spectrum released to the membrane surface and the fluorescence from the membrane fiber having solid-phase gels and liquid-phase proteins was obtained by the detector. During the filtration of the hollow fiber membrane system, the in situ EEM could acquire EEM results every 13 min. As shown in Fig. 2, the accumulation of organic matter on the membrane surface could be clearly detected with time. Moreover, the possible organic candidates such as proteinaceous substances or hydrophilic and transphilic organics could be identified by using the EEM peaks. By using the increase in peak areas, the proper time for membrane cleaning could be suggested which can mitigate irreversible fouling formation.

4.1.3 In situ real-time investigations by using the quartz crystal microbalance with dissipation monitoring (QCM-D)

The quartz crystal microbalance with dissipation monitoring (QCM-D) detects variations in acoustic waves in frequency and dissipation by determining deposition and viscoelastic properties using quartz crystal sensors. This technology has been applied to sense organic matters (proteins and polysaccharides) in the food industry. Rudolph et al. (2021) recently applied this in situ real-time monitoring methods to identify the fouling for the ceramic tubular MF membrane to treat thermomechanical pulping process water, which include ash, total sugars, acid-insoluble solids, acid-soluble lignin, total lignin, and hydrophobic extractives (Rudolph et al., 2021).

The QCM-D device measures the shift in frequency (Δf) and converts it to the changes in mass (Δm) by using the Sauerbrey relationship including measured overtone (n) and calculated constant (C) by using the thickness (t), density (ρ), and resonant frequency (f_0) of the crystal sensor (Sauerbrey, 1959).

$$\Delta m = -\frac{C}{n}\Delta f$$

$$C = \frac{t\rho}{f_0}$$

In addition, dissipation that can be measured by the relationship between lost energy during the one oscillation cycle (E_{lost}) and total stored energy in the oscillator (E_{stored}) was monitored to complement monitoring integrity in a liquid environment (Sauerbrey, 1959).

$$D = \frac{E_{lost}}{2\pi E_{stored}}$$

As shown in Fig. 3, the measured frequency and dissipation shifts were monitored during the fouling and rinsing processes for process water, retentate, and permeate from the MF membrane system. The frequency shifts decreased with increasing fouling and the decreased frequency shifts were restored during the rinsing process. The dissipation shifts showed opposite trends with frequency shifts. Therefore, by monitoring the frequency and dissipation shifts, it was confirmed that the foulant mass could be identified in real time.

4.1.4 Electrochemical impedance spectroscopy (EIS)

Electrochemical impedance spectroscopy (EIS) can show various layered structures including the fouling and membrane structures. Jing et al. (2016) applied EIS technology to check the fouling layers on the ceramic UF reactive electrochemical membrane (Jing et al., 2016). And chemical-free electrochemical regeneration (CFER) technology was used to

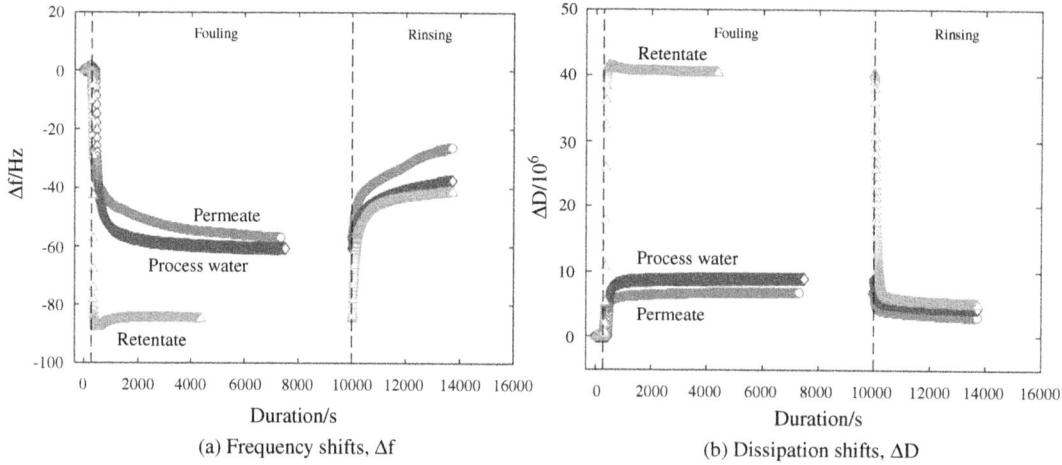

(a) Frequency shifts, Δf (b) Dissipation shifts, ΔD

FIG. 3 The frequency and dissipation shift of the QCM-D monitoring method for the fouling and rinsing of the MF membrane treating the thermomechanical pulping process water (Rudolph et al., 2021).

mitigate membrane fouling. The model organic foulants were humic acids (HA) and polystryrene (PS) microspheres (1.58 μm) to simulate natural organic matter (NOM).

The membrane flux and EIS spectra were monitored according to the different operational conditions of the UF membrane fouling test by using HA as shown in Fig. 4. The EIS spectra reflected the decreased flux by showing relatively higher imaginary impedance compared to the real impedance in the frequency of 10 mHz.

4.2 Fouling observation on the pressurized nanofiltration/reverse osmosis (NF/RO) membrane systems

The fouling on the NF/RO membrane is different from that on the MF/UF membrane due to several reasons such as higher operating pressure, cross-flow filtration mode for low fouling, and existence of the pretreatment process. Biofouling is the most commonly observed foulant in the NF/RO process; however, inorganic fouling (scaling) also occurs on the membrane surface in the desalination process.

4.2.1 Adenosine triphosphate (ATP) measurement

Biofouling on the RO membrane can be monitored indirectly by measuring ATPs in the feed and permeate of the RO plant. Nakaya et al. (2021) conducted ATP monitoring for 5 months at the RO pilot plant by collecting seawater samples every week and analyzing ATP concentrations by using the luciferin-luciferase bioluminescent assay (Nakaya et al., 2021). Furthermore, ATP was classified as the intercellular (ATP_{int}) and extracellular (ATP_{ext}) ATPs.

There were temporal variations of the intercellular and extracellular ATPs with the operational time of the RO pilot plant. The increased differential pressure ratio ($R_{\Delta P}$) working as the fouling index was observed after the increase in ΔATP_{int} and ΔATP_{ext}, which showed that the ΔATP_{int} and ΔATP_{ext} represented biofouling accumulation during the RO pilot plant operation.

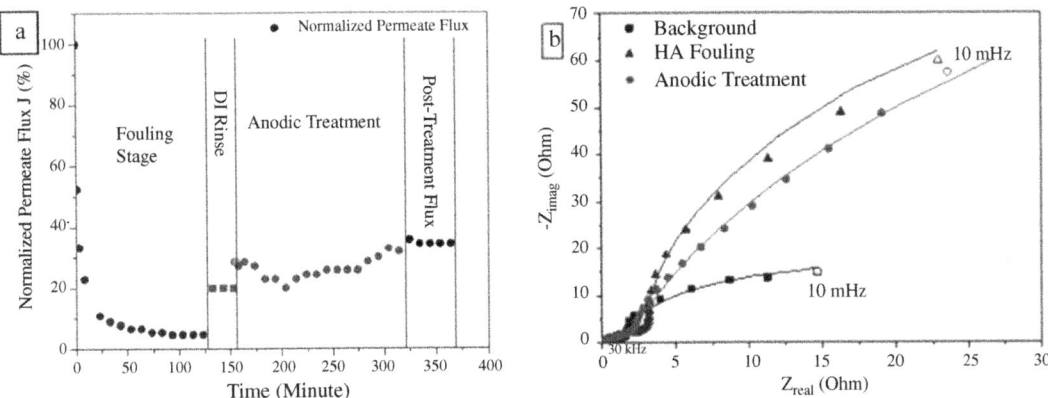

FIG. 4 The flux and EIS spectra of the UF membrane for the fouling test by using humic acids according to the operational conditions (anodic treatment) (Jing et al., 2016).

ΔATP_{int} and ΔATP_{ext} were calculated by using the following equation:

$$\Delta ATP_{int\ or\ ext} = \frac{\left(ATP_E - ATP_D \times \frac{EC_E}{EC_D}\right)}{ATP_D} \times 100\ (\%)$$

where ATP_D and ATP_E are the ATP concentration before and after the RO process. EC_D and EC_E are the electronic conductivities (EC) before and after the RO process.

Consequently, by using the ΔATP_{int} and ΔATP_{ext} values, the biofouling stages could be estimated such as (1) normal operating condition, (2) bacterial cell adhesion and biofilm growth ($\Delta ATP_{int} < 0$ and $\Delta ATP_{ext} < 0$), and (3) bacteria cell detachment from the biofilm ($\Delta ATP_{int} > 0$ and $\Delta ATP_{ext} > 0$).

4.2.2 Electrochemical impedance spectroscopy (EIS)

EIS technology has been applied to the RO fouling studies by research groups. The in situ EIS monitoring method for the RO membrane was suggested by Ho et al. (2016). The RO fouling was tested coupled with the RO-EIS module by using the incubated model microbe [*Pseudomonas aeruginosa* PA01 (ATCC, BAA-47)] in the 2000 mg/L of NaCl salt solution (brackish water level) according to the operating flux in the range from 8 to 40 (L/m²h). In addition, confocal laser scanning microscopy (CLSM) analysis was also conducted to show the live and dead cells on the RO membrane with time. It is worth mentioning that the biofouling stage on the RO membrane can be monitored and evaluated by using the EIS coupled with the RO system. Comprehensive understanding on the biofouling of the RO membrane was available by using operational data such as transmembrane pressure (TMP) and instrumental analysis data such as a CLSM image.

4.2.3 Real-time computational imaging by using a digital camera

The digital camera-based scaling observation method was conducted to confirm scaling formation of the RO membrane by Sarker and Bilton (2021). The plate-and-frame-type RO module was connected to the digital camera. The feed solution was synthesized brackish water, which had an electronic conductivity of 3250 μS/cm. The dominant expected scalant was $CaSO_4$. An RO test was conducted in the lab-scale RO device by simulating the intermittently operated PV-powered RO system established in developing countries. The digital camera collected the images of the RO membrane and hotspots were identified after the postimage processing. The morphology of the $CaSO_4$ scalant on the RO membrane was further compared by using SEM-EDS analysis. By using the image analysis tool, the scalant (red color) was clearly expressed in Fig. 5.

4.2.4 Excitation emission matrix—Parallel factor analysis (EEM-PARAFAC)

In order to treat landfill leachate, the NF membrane is the commonly applied method to remove organics and salts. The real landfill leachate was used as the feed solution for the NF membrane process coupled with the combinations of the various pretreatment methods including coagulation/flocculation (C/F) using alum, magnetic ion exchange resin (MIEX), and granular-activated carbon (GAC) adsorption. The NF membrane fouling test was conducted by using the dead-end filtration cell. The organic fractions were measured from

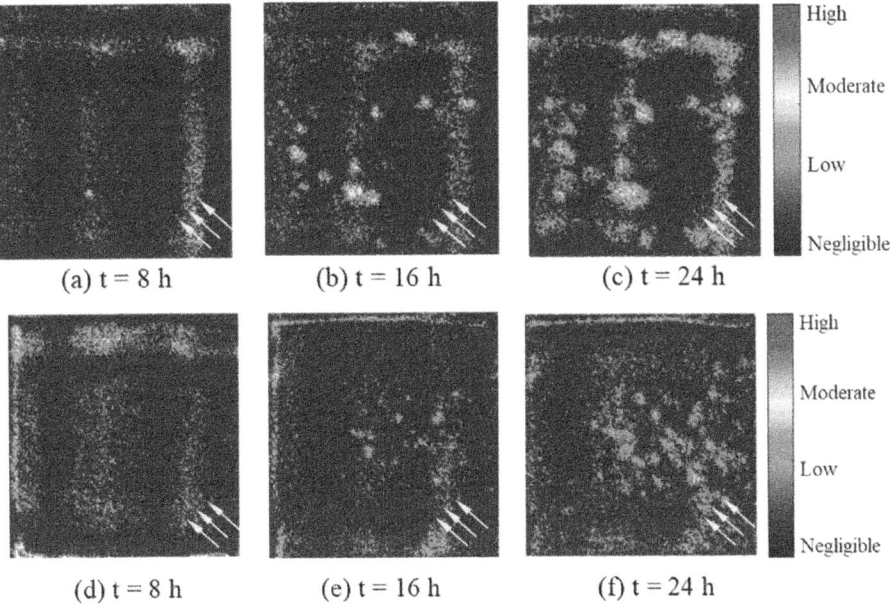

(a) t = 8 h   (b) t = 16 h   (c) t = 24 h

High
Moderate
Low
Negligible

(d) t = 8 h   (e) t = 16 h   (f) t = 24 h

High
Moderate
Low
Negligible

FIG. 5 Collected and postimage processed images by a digital camera (Sarker and Bilton, 2021).

the water solutions including dissolved organic carbons after the various pretreatments by using EEM-PARAFAC (Aftab et al., 2020).

The three different organic fractions were classified by using the EEM-PARAFAC analysis such as C1: fulvic-like component, C2: tryptophan-like component, and C3: humic-like component. The organic fractions (C1, C2, and C3) of the untreated and pretreated water solutions according to the pretreatment options were presented. The identified relationships between the pretreatments and organic matters by using the EEM-PARAFAC method can be used further to optimize proper pretreatment according to the different feed water conditions. Moreover, proper cleaning methods for the NF membrane could be suggested.

4.3 Fouling detection on the nonpressurized forward osmosis (FO) system

FO is a process using osmotic pressure in which water moves due to the difference in chemical potential between the feed and draw solution (Cath et al., 2006). In the FO process, silica, calcium carbonate, and calcium sulfate are mainly reported as major inorganic foulants (Mi and Elimelech, 2013). The effects of major foulants such as humic substances, polysaccaharides, and proteins on organic fouling have been reported in FO. The hydrophobic humic acid did not penetrate into the pores of the membrane showing relatively little effect (Parida and Ng, 2013), and hydrophilic polysaccharide was reported as the most significant foulant in the FO process (Shon et al., 2006). Also, reversible biofouling and irreversible biofouling are formed in stages (Goulter et al., 2009). Reversible fouling can be recovered by backwashing or fluid flow, but irreversible fouling is not recovered due to the cohesion of the negatively charged membrane and cells. In addition, biofouling may be aggravated by bonding with other types of fouling, and it is difficult to remove due to its strong adhesion to the membrane (Abid et al., 2017).

Various real-time in situ monitoring techniques such as direct observation, optical coherence tomography (OCT), UTDR, NMR, and EIS, which have been applied in advanced technologies such as NF, MF, and RO processes, were applied to the FO process to understand the mechanism of the fouling phenomenon (Ibrar et al., 2019).

4.3.1 Direct techniques

The direct observation technique analyzes the effect of fouling on process performance by directly observing the formation of fouling using a camera or a microscope. According to a previous study (Wang et al., 2010), Wang applied an optical microscope to the FO process for the first time and verified the effects of membrane orientation and spacers. In the study, the AL-FS mode may produce less flux than the AL-DS mode, but the fouling tendency is low. In addition, although the use of spacers was attributed to maintaining the stable flux, particles were found to accumulate in the spacer filament.

In addition, Tow (Tow et al., 2016) studied the cleaning mechanism of organic fouling by applying a camera to the FO process. In this study, the cleaning mechanism by backwashing was identified by visualizing the process in which the cake layer formed on the membrane surface is removed through swelling and wrinkling (Fig. 6).

4.3.2 Optical coherence tomography (OCT)

Optical coherence tomography (OCT) is an in situ real-time monitoring technique where it is possible to observe the process of accumulation of fouling on membranes in the cross section. Therefore, real-time information can be used to determine the strategies for fouling control and cleaning. Recently, the OCT technique has been applied to the FO process for fouling monitoring. In a previous study, Im et al. (2021)) used the OCT technique to visualize the fouling morphology and to develop a deep learning model for fouling prediction. In addition, Fortunato et al. (2017) proposed a novel 3D OCT for scanning biomass structure.

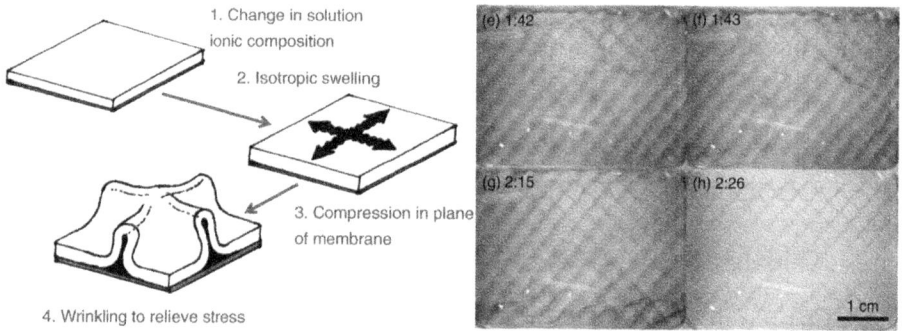

FIG. 6   Schematic of (1) cleaning mechanism and (2) video stills of wrinkles in the FO process (Tow et al., 2016).

4.3.3 Confocal laser scanning microscopy (CLSM)

Nondestructive and real-time monitoring for biofouling in forward osmosis was conducted using confocal laser scanning microscopy (CLSM). Mukherjee et al. (2016) developed biofilm flow cells that can observe the biofouling dynamic during the FO operation in real time, nondestructively. Information on area, thickness, and volume of biofilm was provided using this flow cell coupled with CLSM.

4.4 Fouling detection on the nonpressurized membrane distillation (MD) system

Membrane distillation is a thermal desalination process driven by temperature differences between feed and permeate sides (Lawson and Lloyd, 1997). The big difference of the MD process with respect to other membrane-based desalination processes is using the hydrophobic and porous membrane. Fouling and wetting are regarded as obstacles hindering upscale of the MD process (Naidu et al., 2016; Tijing et al., 2015; Warsinger et al., 2015). Wetting is a phenomenon in which the feed solution penetrates the pores of hydrophobic membrane (Franken et al., 1987). The degree of wetting is classified into nonwetting, surface wetting, partial wetting, and full wetting (Chamani et al., 2021). If wetting occurs, treated water quality is contaminated by the direct passing of the feed solution, and recovery of membrane and permeate system would be necessary.

In the MD process, inorganic fouling is serious because it mainly treats highly concentrated brine such as seawater or RO concentrates. Inorganic fouling was deposited not only on the membrane surface but also inside of the pores, causing wetting by deforming the membrane structure. In addition, NOM forms a gel-like layer with other inorganic materials and particles on the membrane surface, thereby reducing the permeability of the membrane (Yuan and Zydney, 1999). Biofouling in the MD process has little effect at an operating temperature of $40°$–$60°$ (Gryta, 2002). However, the effects of module types and temperature on the formation of biofouling are still necessary to study in the MD process (Costa et al., 2021). In this section, studies on application of in situ monitoring systems to MD processes are introduced.

4.4.1 Direct observation

Direct observation using a camera and microscopy is a simple but effective technique where it is possible to understand fouling and wetting mechanisms. In general, direct observation techniques are effective with transparent membranes. Although the MD membrane is opaque, a visualization system for detecting membrane wetting has been proposed using the phenomenon that the membrane becomes transparent when wetting occurs (Jacob et al., 2020; Kiefer et al., 2019; Kim et al., 2019). Kim first proposed a visualization system that detects the occurrence of fouling and wetting by observing the change in light intensity on the membrane surface using a CCD camera and LED lighting (Fig. 7). The system provided evidence that wetting occurred at the point-of-scale formation.

4.4.2 Optical coherence tomography (OCT)

In addition, the OCT scan is used to observe fouling deposition in MD studies. Lee et al. (2018) first applied the OCT technique to an MD fouling study in which different fouling

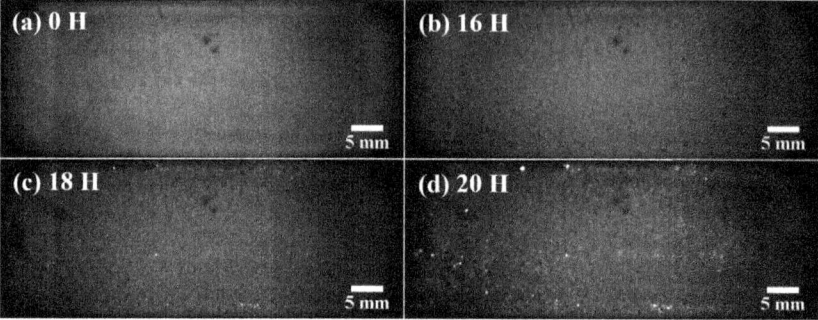

FIG. 7   Images obtained from a visualization system depending on operation time (Kim et al., 2019).

tendencies were identified depending on the scalant types. Also, Fortunato et al. (2018) suggested that OCT images could provide prior information for membrane autopsy. Recently, Wong et al. (2021) developed OCT application from the 2D to the 3D technique in the MD process, which makes it possible to quantify the fouling formation by separating the fouling layer from the membrane (Fig. 8).

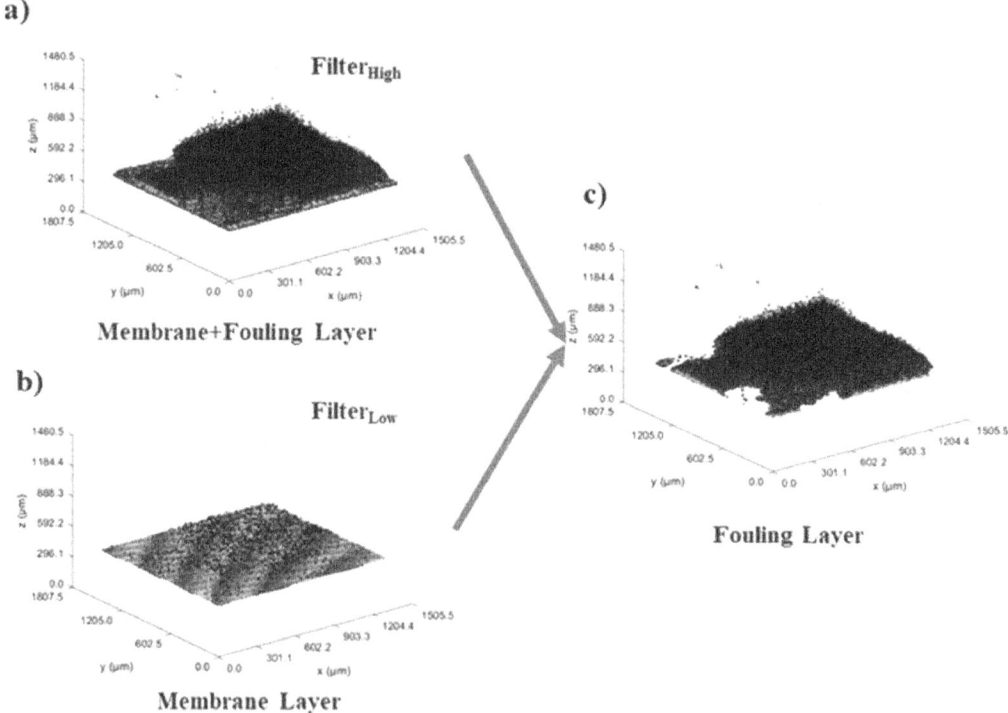

FIG. 8   3D OCT scans showing (A) membrane and fouling layer, (B) membrane layer, and (C) fouling layer (Wong et al., 2021).

4.4.3 *Electrical impedance spectroscopy (EIS)*

Electrical impedance spectroscopy (EIS) is an emerging in situ monitoring technique. It sensitively detects the electrical variation on the membrane surface in real time by measuring impedance, which makes it possible to observe the occurrence of fouling and wetting in the MD process. Ahmed et al. (2018) first applied the EIS system to the MD process for fouling monitoring, and verified that early detection of fouling is possible before the flux reduction and electrical conductivity increase. In addition, Chen et al. (2017) first proposed an impedance-based wetting monitoring system, and the ability to detect partial wetting in the early stage was defined. It means that stable performance of MD membranes can be maintained from occurrence of wetting.

4.5 Fouling observation on the nonpressurized electrodialysis (ED) system

Electrodialysis (ED) is a separation process of ionic species using an ionic exchange membrane (IEM) (Mohammadi et al., 2005). The ED system consists of two electrodes, a positively charged electrode and a negatively charged electrode, and a series of anion exchange membranes (AEM) and cation exchange membranes (CEM), noted to ED stack, between the electrodes on both sides (Li et al., 2021). IEM provides special ionic selectivity that allows for demineralization and concentration, and for the extraction of the ionic, organic species from the solution containing complex components.

An ED system has been developed by improving the properties of IEMs in both electrochemical and physicochemical terms. However, fouling on the surface of the ion exchange membrane (IEM) is a major challenge in the electrodialysis (ED) process, which hinders scale up (Al-Amshawee et al., 2020). An anion exchange membrane (AEM) has a positive charge on the surface, so colloidal substances or organic substances with a negative charge are adsorbed (Sosa-Fernandez et al., 2021; Zhao et al., 2017). A cation exchange membrane (CEM) has a negative charge on the surface, so inorganic substances having a positive charge such as calcium, magnesium, phosphate, and carbonate are adsorbed onto the surface of the IEM (Andreeva et al., 2017; Cifuentes-Araya et al., 2012).

4.5.1 *Electrical impedance spectroscopy (EIS)*

EIS detects microscale changes in concentration by distinguishing bulk and boundary layers, thereby providing information on the formation of a fouling layer on the membrane surface. In a study (Zhang et al., 2020), an in situ monitoring method was devised by improving the ex situ monitoring method, where it was difficult to monitor IEM fouling and concentration polarization (CP). Ion mass transfer was monitored using in situ EIS, which provided information on fouling and CP profile during the operation. It was reported that fouling improved the ion transport resistance at the surface, and the concentration polarization was intensified. Also, the performance of chemical cleaning was evaluated by monitoring impedance changes. As a result, BSA was found to cause irreversible fouling by blocking the membrane pores, which was confirmed by ion flux and impedance changes in EIS.

Xiang et al. (2021) applied EIS to the ED process to investigate the fouling mechanism of AEM caused by APAM. In this study, the resistance of membrane and solution at high frequency and electrochemical information of the boundary layer at low frequency were obtained through EIS, which was used to compare the tendency of the fouled membrane compared to the pristine membrane. EIS successfully distinguished the electrochemical properties of bulk and boundary layers. In addition, EIS has been shown to detect membrane fouling more sensitively compared to changes in conductivity and pH.

4.5.2 *Transmembrane electric potential (TMEP)*

TMEP, which represents voltage drop across the membrane, can represent a change in membrane resistance that is related to the degree of membrane fouling. Xiang (Xiang et al., 2021) proposed the mechanism of IEM fouling caused by APAM by applying TMEP and EIS, introduced in Section 4.5.1, to the ED process (Fig. 9).

4.6 Fouling observation on the nonpressurized capacitive deionization (CDI) system

CDI is an electrosorption process separating ion substances from the solution using porous carbon electrodes (Tanaka, 2003). Electrical potential is applied to the negatively charged electrode and the positively charged electrode (AlMarzooqi et al., 2014). Membrane capacitive deionization (MCDI) has improved the CDI process by increasing the selective separation efficiency of ions and enabling energy recovery during the regeneration process.

Fouling in the CDI process can be divided into fouling directly attached to the carbon electrode and fouling on the ion exchange membrane of MDCI. According to previous studies (Wang et al., 2020b), it is reported that organic matters block the pores of the carbon electrode, thereby reducing ion removal efficiency and increasing energy consumption. In addition, in the case of inorganic fouling, the influence of Ca and Mg was insignificant, and the influence of the Fe ion was relatively large (Mossad and Zou, 2013). In addition, in the MCDI process,

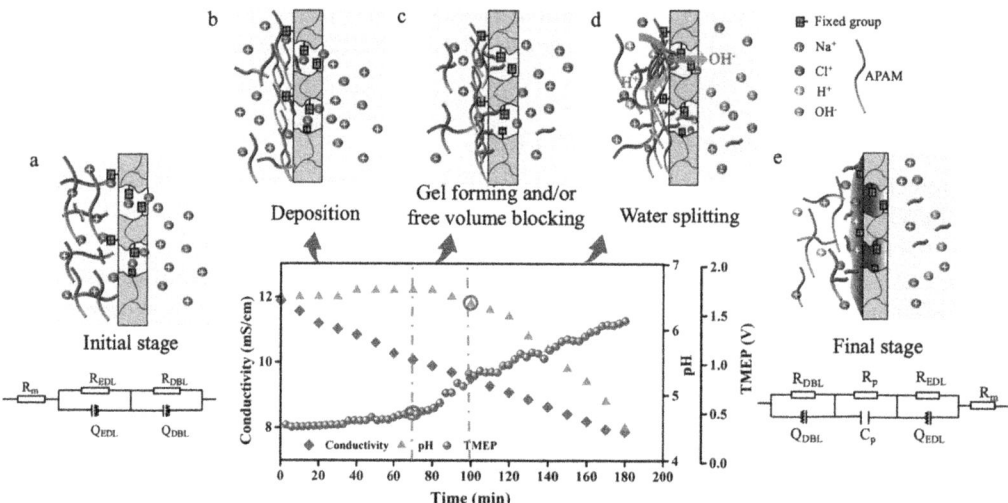

FIG. 9 Schematic of fouling deposition on AEM and comparison of conductivity, pH, and TMEP according to operation time (Xiang et al., 2021).

the selectively adsorbed ions pass through the pores and move to the electrode; so, fouling is formed inside the membrane pores as well as on the surface (Ali et al., 2021). However, research on fouling of the ion exchange membrane in CDI is lacking, due to the fact that the fouling of the ion exchange membrane is shown to be insignificant compared to the fouling of the electrode. Therefore, additional research on fouling in MCDI is required.

4.6.1 Electrical impedance spectroscopy (EIS)

It is reported that the ferric iron is accumulated in the carbon electrode in the CDI process, which greatly reduces the performance. Therefore, Wang et al. (2020c) monitored electrode resistance using EIS to evaluate the effect of iron on CDI system performance in brackish water (Fig. 10). As a result of EIS analysis, Fe scales interrupted the movement of ions in the electrode, and when both NOM and Fe scales were accumulated, the internal resistance of CDI increased. As such, EIS contributed to explain the negative effects of Fe scales and NOM on the CDI performance of the electrode.

5. Conclusions and perspectives

In this chapter, various methods to measure the foulants (particles, organics, inorganics, and microbes) on the membrane process were introduced. Moreover, in situ real-time monitoring methods to understand the formation of foulant on the membrane surface suggested by many research groups were also introduced. It is important to understand fouling properly and to observe fouling promptly to decrease the operational cost of membrane processes by mitigating or removing the fouling. Therefore, a small or simple in situ device that can monitor the key factor on membrane fouling could be used on the water-related market with respect to water treatment, wastewater treatment, desalination, and water reuse. Especially,

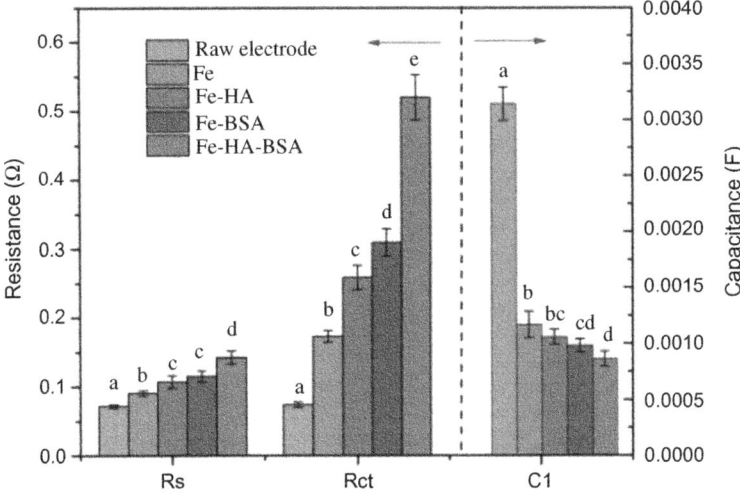

FIG. 10 Electrochemical impedance spectroscopy of CDI (Wang et al., 2020c).

membrane processes in an isolated area or which only operated in emergency will require a self-checking system for sustainable operation. Therefore, in situ real-time monitoring methods can be used as sensors for automated membrane processes in the future. Moreover, direct observation of the membrane surface could also help in understanding the fouling formation mechanism.

References

Abid, H.S., Johnson, D.J., Hashaikeh, R., Hilal, N., 2017. A review of efforts to reduce membrane fouling by control of feed spacer characteristics. Desalination 420, 384–402.

Abushaban, A., Salinas-Rodriguez, S.G., Philibert, M., Le Bouille, L., Necibi, M.C., Chehbouni, A., 2022. Biofouling potential indicators to assess pretreatment and mitigate biofouling in SWRO membranes: a short review. Desalination 527, 115543.

Aftab, B., Cho, J., Shin, H.S., Hur, J., 2020. Using EEM-PARAFAC to probe NF membrane fouling potential of stabilized landfill leachate pretreated by various options. Waste Manag. 102, 260–269.

Ahmed, F.E., Hilal, N., Hashaikeh, R., 2018. Electrically conductive membranes for in situ fouling detection in membrane distillation using impedance spectroscopy. J. Membr. Sci. 556, 66–72.

Al-Amshawee, S., Yunus, M.Y.B.M., Azoddein, A.A.M., Hassell, D.G., Dakhil, I.H., Hasan, H.A., 2020. Electrodialysis desalination for water and wastewater: a review. Chem. Eng. J. 380, 122231.

Alayande, A.B., Yun, E.-T., Hong, S., 2022. Mechanistic insights into the potential applicability of a sulfate-based advanced oxidation process for the control of transparent exopolymer particles in membrane-based desalination. Desalination 522, 115437.

Ali, A., Quist-Jensen, C.A., Jørgensen, M.K., Siekierka, A., Christensen, M.L., Bryjak, M., Hélix-Nielsen, C., Drioli, E., 2021. A review of membrane crystallization, forward osmosis and membrane capacitive deionization for liquid mining. Resour. Conserv. Recycl. 168, 105273.

AlMarzooqi, F.A., Al Ghaferi, A.A., Saadat, I., Hilal, N., 2014. Application of capacitive deionisation in water desalination: a review. Desalination 342, 3–15.

Andreeva, M., Gil, V., Pismenskaya, N., Nikonenko, V., Dammak, L., Larchet, C., Grande, D., Kononenko, N., 2017. Effect of homogenization and hydrophobization of a cation-exchange membrane surface on its scaling in the presence of calcium and magnesium chlorides during electrodialysis. J. Membr. Sci. 540, 183–191.

Bae, H., Kim, H., Jeong, S., Lee, S., 2011. Changes in the relative abundance of biofilm-forming bacteria by conventional sand-filtration and microfiltration as pretreatments for seawater reverse osmosis desalination. Desalination 273 (2–3), 258–266.

Benavente, L., Coetsier, C., Venault, A., Chang, Y., Causserand, C., Bacchin, P., Aimar, P., 2016. FTIR mapping as a simple and powerful approach to study membrane coating and fouling. J. Membr. Sci. 520, 477–489.

Besha, A.T., Gebreyohannes, A.Y., Tufa, R.A., Bekele, D.N., Curcio, E., Giorno, L., 2017. Removal of emerging micropollutants by activated sludge process and membrane bioreactors and the effects of micropollutants on membrane fouling: a review. J. Environ. Chem. Eng. 5 (3), 2395–2414.

Botton, S., Verliefde, A.R., Quach, N.T., Cornelissen, E.R., 2012. Influence of biofouling on pharmaceuticals rejection in NF membrane filtration. Water Res. 46 (18), 5848–5860.

Cath, T.Y., Childress, A.E., Elimelech, M., 2006. Forward osmosis: principles, applications, and recent developments. J. Membr. Sci. 281 (1–2), 70–87.

Chamani, H., Woloszyn, J., Matsuura, T., Rana, D., Lan, C.Q., 2021. Pore wetting in membrane distillation: a comprehensive review. Prog. Mater. Sci., 100843.

Charfi, A., Yang, Y., Harmand, J., Amar, N.B., Héran, M., Grasmick, A., 2015. Soluble microbial products and suspended solids influence in membrane fouling dynamics and interest of punctual relaxation and/or backwashing. J. Membr. Sci. 475, 156–166.

Chen, Y., Wang, Z., Jennings, G.K., Lin, S., 2017. Probing pore wetting in membrane distillation using impedance: early detection and mechanism of surfactant-induced wetting. Environ. Sci. Technol. Lett. 4 (11), 505–510.

Chen, Y., Zhang, J., Cohen, Y., 2022. Fouling resistant and performance tunable ultrafiltration membranes via surface graft polymerization induced by atmospheric pressure air plasma. Sep. Purif. Technol., 120490.

Cho, K., Jeong, S., Kim, H., Choi, K., Lee, S., Bae, H., 2016. Simultaneous dechlorination and disinfection using vacuum UV irradiation for SWRO process. Desalination 398, 22–29.

Cifuentes-Araya, N., Pourcelly, G., Bazinet, L., 2012. Multistep mineral fouling growth on a cation-exchange membrane ruled by gradual sieving effects of magnesium and carbonate ions and its delay by pulsed modes of electrodialysis. J. Colloid Interface Sci. 372 (1), 217–230.

Costa, F.C., Ricci, B.C., Teodoro, B., Koch, K., Drewes, J.E., Amaral, M.C., 2021. Biofouling in membrane distillation applications-a review. Desalination 516, 115241.

Deb, A., Gurung, K., Rumky, J., Sillanpää, M., Mänttäri, M., Kallioinen, M., 2022. Dynamics of microbial community and their effects on membrane fouling in an anoxic-oxic gravity-driven membrane bioreactor under varying solid retention time: a pilot-scale study. Sci. Total Environ. 807, 150878.

Enfrin, M., Hachemi, C., Callahan, D.L., Lee, J., Dumée, L.F., 2021. Membrane fouling by nanofibres and organic contaminants–Mechanisms and mitigation via periodic cleaning strategies. Sep. Purif. Technol. 278, 119592.

Fortunato, L., Leiknes, T., 2017. In-situ biofouling assessment in spacer filled channels using optical coherence tomography (OCT): 3D biofilm thickness mapping. Bioresour. Technol. 229, 231–235.

Fortunato, L., Jeong, S., Wang, Y., Behzad, A.R., Leiknes, T., 2016. Integrated approach to characterize fouling on a flat sheet membrane gravity driven submerged membrane bioreactor. Bioresour. Technol. 222, 335–343.

Fortunato, L., Bucs, S., Linares, R.V., Cali, C., Vrouwenvelder, J.S., Leiknes, T., 2017. Spatially-resolved in-situ quantification of biofouling using optical coherence tomography (OCT) and 3D image analysis in a spacer filled channel. J. Membr. Sci. 524, 673–681.

Fortunato, L., Jang, Y., Lee, J.-G., Jeong, S., Lee, S., Leiknes, T., Ghaffour, N., 2018. Fouling development in direct contact membrane distillation: non-invasive monitoring and destructive analysis. Water Res. 132, 34–41.

Fortunato, L., Alshahri, A.H., Farinha, A.S., Zakzouk, I., Jeong, S., Leiknes, T., 2020. Fouling investigation of a full-scale seawater reverse osmosis desalination (SWRO) plant on the Red Sea: membrane autopsy and pretreatment efficiency. Desalination 496, 114536.

Franken, A., Nolten, J., Mulder, M., Bargeman, D., Smolders, C., 1987. Wetting criteria for the applicability of membrane distillation. J. Membr. Sci. 33 (3), 315–328.

Gong, H., Jin, Z., Wang, Q., Zuo, J., Wu, J., Wang, K., 2017. Effects of adsorbent cake layer on membrane fouling during hybrid coagulation/adsorption microfiltration for sewage organic recovery. Chem. Eng. J. 317, 751–757.

Goulter, R., Gentle, I., Dykes, G., 2009. Issues in determining factors influencing bacterial attachment: a review using the attachment of Escherichia coli to abiotic surfaces as an example. Lett. Appl. Microbiol. 49 (1), 1–7.

Gryta, M., 2002. The assessment of microorganism growth in the membrane distillation system. Desalination 142 (1), 79–88.

Guerrero-Navarro, A.E., Ríos-Castillo, A.G., Ripolles-Avila, C., Felipe, X., Rodríguez-Jerez, J.J., 2020. Microscopic analysis and microstructural characterization of the organic and inorganic components of dairy fouling during the cleaning process. J. Dairy Sci. 103 (3), 2117–2127.

Guo, H., Tang, X., Ganschow, G., Korshin, G.V., 2019. Differential ATR FTIR spectroscopy of membrane fouling: contributions of the substrate/fouling films and correlations with transmembrane pressure. Water Res. 161, 27–34.

Henderson, R.K., Baker, A., Murphy, K., Hambly, A., Stuetz, R., Khan, S., 2009. Fluorescence as a potential monitoring tool for recycled water systems: a review. Water Res. 43 (4), 863–881.

Ho, J.S., Low, J.H., Sim, L.N., Webster, R.D., Rice, S.A., Fane, A.G., Coster, H.G., 2016. In-situ monitoring of biofouling on reverse osmosis membranes: detection and mechanistic study using electrical impedance spectroscopy. J. Membr. Sci. 518, 229–242.

Ibrar, I., Naji, O., Sharif, A., Malekizadeh, A., Alhawari, A., Alanezi, A.A., Altaee, A., 2019. A review of fouling mechanisms, control strategies and real-time fouling monitoring techniques in forward osmosis. Water 11 (4), 695.

Im, S.J., Fortunato, L., Jang, A., 2021. Real-time fouling monitoring and membrane autopsy analysis in forward osmosis for wastewater reuse. Water Res. 197, 117098.

Jacob, P., Dejean, B., Laborie, S., Cabassud, C., 2020. An optical in-situ tool for visualizing and understanding wetting dynamics in membrane distillation. J. Membr. Sci. 595, 117587.

Jamil, S., Loganathan, P., Kandasamy, J., Ratnaweera, H., Vigneswaran, S., 2021. Comparing nanofiltration membranes effectiveness for inorganic and organic compounds removal from a wastewater-reclamation plant's micro-filtered water. Mater. Today: Proc. 47, 1389–1393.

Jeong, S., Park, Y., Lee, S., Kim, J., Lee, K., Lee, J., Chon, H.-T., 2011. Pre-treatment of SWRO pilot plant for desalination using submerged MF membrane process: trouble shooting and optimization. Desalination 279 (1–3), 86–95.

A. General on membrane, materials and application

Jeong, S., Shin, B., Jo, W., Kim, H.-Y., Moon, M.-W., Lee, S., 2016. Nanostructured PVDF membrane for MD application by an O2 and CF4 plasma treatment. Desalination 399, 178–184.

Jeong, S., Song, K.G., Kim, J., Shin, J., Maeng, S.K., Park, J., 2021. Feasibility of membrane distillation process for potable water reuse: a barrier for dissolved organic matters and pharmaceuticals. J. Hazard. Mater. 409, 124499.

Jing, Y., Guo, L., Chaplin, B.P., 2016. Electrochemical impedance spectroscopy study of membrane fouling and electrochemical regeneration at a sub-stoichiometric TiO2 reactive electrochemical membrane. J. Membr. Sci. 510, 510–523.

Jung, J., Ryu, J., Choi, S.Y., Park, K.Y., Song, W.J., Yu, Y., Jang, Y.-S., Park, J., Kweon, J., 2018. Autopsy study of irreversible foulants on polyvinylidene fluoride hollow-fiber membranes in an immersed microfiltration system operated for five years. Sep. Purif. Technol. 199, 1–8.

Kiefer, F., Präbst, A., Rodewald, K.S., Sattelmayer, T., 2019. Membrane scaling in Vacuum Membrane Distillation-Part 1: in-situ observation of crystal growth and membrane wetting. J. Membr. Sci. 590, 117294.

Kilmer, N.T., Huss, R.L., George, C.C., Stennett, E.M., 2021. The influence of ion identity and ionic strength on membrane biofouling of a binary protein solution. Sep. Purif. Technol. 255, 117769.

Kim, H.-W., Yun, T., Kang, P.K., Hong, S., Jeong, S., Lee, S., 2019. Evaluation of a real-time visualization system for scaling detection during DCMD, and its correlation with wetting. Desalination 454, 59–70.

Kim, H.-W., Yun, T., Hong, S., Lee, S., Jeong, S., 2020. Retardation of wetting for membrane distillation by adjusting major components of seawater. Water Res. 175, 115677.

Kim, A., Kim, J.H., Patel, R., 2021. Modification strategies of membranes with enhanced anti-biofouling properties for wastewater treatment: a review. Bioresour. Technol. 126501.

Kim, J., Kim, H.-W., Tijing, L.D., Shon, H.K., Hong, S., 2022. Elucidation of physicochemical scaling mechanisms in membrane distillation (MD): implication to the control of inorganic fouling. Desalination 527, 115573.

Lawson, K.W., Lloyd, D.R., 1997. Membrane distillation. J. Membr. Sci. 124 (1), 1–25.

Lay, H.T., Wang, R., Chew, J.W., 2022. Influence of foulant particle shape on membrane fouling in dead-end microfiltration. J. Membr. Sci. 120265.

Lee, J.-G., Jang, Y., Fortunato, L., Jeong, S., Lee, S., Leiknes, T., Ghaffour, N., 2018. An advanced online monitoring approach to study the scaling behavior in direct contact membrane distillation. J. Membr. Sci. 546, 50–60.

Lee, W.J., Ng, Z.C., Hubadillah, S.K., Goh, P.S., Lau, W.J., Othman, M., Ismail, A.F., Hilal, N., 2020. Fouling mitigation in forward osmosis and membrane distillation for desalination. Desalination 480, 114338.

Lee, S., Xu, H., Rice, S.A., Chong, T.H., Oh, H.-S., 2021a. Development of a quorum quenching-column to control biofouling in reverse osmosis water treatment processes. J. Ind. Eng. Chem. 94, 188–194.

Lee, Y.-G., Kim, S., Shin, J., Rho, H., Kim, Y.M., Cho, K.H., Eom, H., Oh, S.-E., Cho, J., Chon, K., 2021b. Sequential effects of cleaning protocols on desorption of reverse osmosis membrane foulants: autopsy results from a full-scale desalination plant. Desalination 500, 114830.

Lei, H., Cheng, N., Zhao, J., 2018. Interaction between membrane and organic compounds studied by atomic force microscopy with a tip modification. J. Membr. Sci. 556, 178–184.

Li, H., Lin, Y., Yu, P., Luo, Y., Hou, L., 2011. FTIR study of fatty acid fouling of reverse osmosis membranes: effects of pH, ionic strength, calcium, magnesium and temperature. Sep. Purif. Technol. 77 (1), 171–178.

Li, M., Idoughi, R., Choudhury, B., Heidrich, W., 2017. Statistical model for OCT image denoising. Biomed. Opt. Express 8 (9), 3903–3917.

Li, C., Ramasamy, D.L., Sillanpää, M., Repo, E., 2021. Separation and concentration of rare earth elements from wastewater using electrodialysis technology. Sep. Purif. Technol. 254, 117442.

Mi, B., Elimelech, M., 2013. Silica scaling and scaling reversibility in forward osmosis. Desalination 312, 75–81.

Mohammadi, T., Moheb, A., Sadrzadeh, M., Razmi, A., 2005. Modeling of metal ion removal from wastewater by electrodialysis. Sep. Purif. Technol. 41 (1), 73–82.

Mossad, M., Zou, L., 2013. Study of fouling and scaling in capacitive deionisation by using dissolved organic and inorganic salts. J. Hazard. Mater. 244, 387–393.

Mukherjee, M., Menon, N.V., Liu, X., Kang, Y., Cao, B., 2016. Confocal laser scanning microscopy-compatible microfluidic membrane flow cell as a nondestructive tool for studying biofouling dynamics on forward osmosis membranes. Environ. Sci. Technol. Lett. 3 (8), 303–309.

Naidu, G., Jeong, S., Vigneswaran, S., Hwang, T.-M., Choi, Y.-J., Kim, S.-H., 2016. A review on fouling of membrane distillation. Desalin. Water Treat. 57 (22), 10052–10076.

Nakaya, S., Yamamoto, A., Kawanishi, T., Toya, N., Miyakawa, H., Takeuchi, K., Endo, M., 2021. Detection of dynamic biofouling from adenosine triphosphate measurements in water concentrated from reverse osmosis desalination of seawater. Desalination 518, 115286.

A. General on membrane, materials and application

Ng, T.C.A., Ng, H.Y., 2010. Characterisation of initial fouling in aerobic submerged membrane bioreactors in relation to physico-chemical characteristics under different flux conditions. Water Res. 44 (7), 2336–2348.

Nguyen, Q.-M., Jeong, S., Lee, S., 2017. Characteristics of membrane foulants at different degrees of SWRO brine concentration by membrane distillation. Desalination 409, 7–20.

Nguyen, H.T., Kim, Y., Choi, J.-W., Cho, K., Jeong, S., 2020. Assimilable organic carbon removal strategy for aquifer storage and recovery applications. Environ. Res. 191, 110033.

Nguyen, H.T., Kim, Y., Choi, J.-W., Jeong, S., Cho, K., 2021. Soil microbial communities-mediated bioattenuation in simulated aquifer storage and recovery (ASR) condition: long-term study. Environ. Res. 197, 111069.

Parida, V., Ng, H.Y., 2013. Forward osmosis organic fouling: effects of organic loading, calcium and membrane orientation. Desalination 312, 88–98.

Park, S., Nam, T., You, J., Kim, E.-S., Choi, I., Park, J., Cho, K.H., 2019. Evaluating membrane fouling potentials of dissolved organic matter in brackish water. Water Res. 149, 65–73.

Pathak, N., Fortunato, L., Li, S., Chekli, L., Phuntsho, S., Ghaffour, N., Leiknes, T., Shon, H.K., 2018. Evaluating the effect of different draw solutes in a baffled osmotic membrane bioreactor-microfiltration using optical coherence tomography with real wastewater. Bioresour. Technol. 263, 306–316.

Powell, L., Hilal, N., Wright, C., 2017. Atomic force microscopy study of the biofouling and mechanical properties of virgin and industrially fouled reverse osmosis membranes. Desalination 404, 313–321.

Rudolph, G., Hermansson, A., Jönsson, A.-S., Lipnizki, F., 2021. In situ real-time investigations on adsorptive membrane fouling by thermomechanical pulping process water with quartz crystal microbalance with dissipation monitoring (QCM-D). Sep. Purif. Technol. 254, 117578.

Safaee, H., Bracewell, A., Safarik, J., Plumlee, M.H., Rajagopalan, G., 2022. Online colloidal particle monitoring for controlled coagulation pretreatment to lower microfiltration membrane fouling at a potable water reuse facility. Water Res. 118300.

Sarker, N.R., Bilton, A.M., 2021. Real-time computational imaging of reverse osmosis membrane scaling under intermittent operation. J. Membr. Sci. 636, 119556.

Sauerbrey, G., 1959. Verwendung von Schwingquarzen zur Wägung dünner Schichten und zur Mikrowägung. Z. Phys. 155 (2), 206–222.

Shon, H., Vigneswaran, S., Kim, I.S., Cho, J., Ngo, H., 2006. Fouling of ultrafiltration membrane by effluent organic matter: a detailed characterization using different organic fractions in wastewater. J. Membr. Sci. 278 (1–2), 232–238.

Silalahi, S.H., Leiknes, T., 2009. Cleaning strategies in ceramic microfiltration membranes fouled by oil and particulate matter in produced water. Desalination 236 (1–3), 160–169.

Sosa-Fernandez, P., Miedema, S., Bruning, H., Leermakers, F., Post, J., Rijnaarts, H., 2021. Effects of feed composition on the fouling on cation-exchange membranes desalinating polymer-flooding produced water. J. Colloid Interface Sci. 584, 634–646.

Tanaka, Y., 2003. Mass transport and energy consumption in ion-exchange membrane electrodialysis of seawater. J. Membr. Sci. 215 (1–2), 265–279.

Tian, M., Xu, H., Yao, L., Wang, R., 2021. A biomimetic antimicrobial surface for membrane fouling control in reverse osmosis for seawater desalination. Desalination 503, 114954.

Tibi, F., Charfi, A., Cho, J., Kim, J., 2021. Effect of interactions between ammonium and organic fouling simulated by sodium alginate on performance of direct contact membrane distillation. Sep. Purif. Technol. 278, 119551.

Tijing, L.D., Woo, Y.C., Choi, J.-S., Lee, S., Kim, S.-H., Shon, H.K., 2015. Fouling and its control in membrane distillation—a review. J. Membr. Sci. 475, 215–244.

Wang, Y., Wicaksana, F., Tang, C.Y., Fane, A.G., 2010. Direct microscopic observation of forward osmosis membrane fouling. Environ. Sci. Technol. 44 (18), 7102–7109.

Tow, E.W., Rencken, M.M., Lienhard, J.H., 2016. In situ visualization of organic fouling and cleaning mechanisms in reverse osmosis and forward osmosis. Desalination 399, 138–147.

Wang, S., Huang, X., Elimelech, M., 2020a. Complexation between dissolved silica and alginate molecules: implications for reverse osmosis membrane fouling. J. Membr. Sci. 605, 118109.

Wang, T., Liang, H., Bai, L., Zhu, X., Gan, Z., Xing, J., Li, G., Aminabhavi, T.M., 2020b. Adsorption behavior of powdered activated carbon to control capacitive deionization fouling of organic matter. Chem. Eng. J. 384, 123277.

Wang, T., Zhang, C., Bai, L., Xie, B., Gan, Z., Xing, J., Li, G., Liang, H., 2020c. Scaling behavior of iron in capacitive deionization (CDI) system. Water Res. 171, 115370.

A. General on membrane, materials and application

Wang, C., Park, M.J., Seo, D.H., Shon, H.K., 2021. Inkjet printing of graphene oxide and dopamine on nanofiltration membranes for improved anti-fouling properties and chlorine resistance. Sep. Purif. Technol. 254, 117604.

Warsinger, D.M., Swaminathan, J., Guillen-Burrieza, E., Arafat, H.A., 2015. Scaling and fouling in membrane distillation for desalination applications: a review. Desalination 356, 294–313.

Wong, P.W., Guo, J., Khanzada, N.K., Yim, V.M.W., Kyoungjin, A., 2021. In-situ 3D fouling visualization of membrane distillation treating industrial textile wastewater by optical coherence tomography imaging. Water Res. 205, 117668.

Xiang, W., Han, M., Dong, T., Yao, J., Han, L., 2021. Fouling dynamics of anion polyacrylamide on anion exchange membrane in electrodialysis. Desalination 507, 115036.

Yamamura, H., Ding, Q., Watanabe, Y., 2019. Solid-phase fluorescence excitation emission matrix for in-situ monitoring of membrane fouling during microfiltration using a polyvinylidene fluoride hollow fiber membrane. Water Res. 164, 114928.

Yuan, W., Zydney, A.L., 1999. Humic acid fouling during microfiltration. J. Membr. Sci. 157 (1), 1–12.

Zhang, L., Jia, H., Wang, J., Wen, H., Li, J., 2020. Characterization of fouling and concentration polarization in ion exchange membrane by in-situ electrochemical impedance spectroscopy. J. Membr. Sci. 594, 117443.

Zhao, Z., Shi, S., Cao, H., Li, Y., 2017. Electrochemical impedance spectroscopy and surface properties characterization of anion exchange membrane fouled by sodium dodecyl sulfate. J. Membr. Sci. 530, 220–231.

Life-cycle assessment of membrane-based desalination technologies and alternatives

Safa Rachid[a], Selim Karkour[b], Amila Abeynayaka[c], Hicham Chatoui[f,g], Mohamed Merzouki[d], and Jauad El Kharraz[e]

[a]Mining Environment and Circular Economy, Mohammed VI Polytechnic University (UM6P), Ben Guerir, Morocco [b]Beyond the Borders of Life Cycle Assessment, Tokyo, Japan [c]Institute for Global Environmental Strategies (IGES), Hayama, Kanagawa, Japan [d]Laboratory of Biological Engineering, Faculty of Sciences and Techniques, Sultan Moulay Slimane University, Beni Mellal, Morocco [e]Regional Center for Renewable Energy and Energy Efficiency (RCREEE), Cairo, Egypt [f]Laboratory of Biological Engineering, Faculty of Sciences and Techniques, Sultan Moulay Slimane University, Beni Mellal, Morocco [g]Department of Health and Agro-Industry Engineering, School of Engineering and Innovation of Marrakech (E2IM), Private University of Marrakesh (UPM), Marrakech, Morocco

1. Introduction

Over the last decade, desalination has become a reliable option to provide fresh water, especially with respect to climate change and its direct impact on water resources. The desalination capacity increased at a rate of 7% per year between 2010 and 2019 (Eke et al., 2020). The desalination technology transforms water from some of the nontraditional water sources such as brackish water (1000–10,000 mg/L of salinity) or seawater (30,000–44,000 mg/L of salinity) into potable water in a sustainable way. Albeit brackish water is considered a low-cost desalinated water, this limited source represents less than 1% of global resources (Proskynitopoulou and Katosoyiannis, 2018).

Copyright © 2023 Elsevier Inc. All rights reserved.

As the volume of oceans seems limitless, there is a shift in the water industry that is looking for fresh water in this huge amount of saline water. According to the International Desalination Association (IDA), the global installed capacity was distributed on 18,500 plants in 150 countries while the global capacity attained 99.8 million m^3/day in 2017 (Diaz, 2021). A substantial increase in desalination installed capacity has been noticed in the regions of Europe and Africa, which was not the case a few years ago (Eke et al., 2020).

Worldwide, 99% of water consumption is for agricultural applications. With a global population growth projected to reach 9.8 billion by 2050, agricultural production needs to be increased by almost 50% by 2050 as compared to 2012 to meet the rising demand for food, fiber, and biofuels. Thus, the permanently rising demand for water will affect its availability and put this limited resource in severe scarcity (Mekonnen and Gerbens-Leenes, 2020).

The majority of the increase in agricultural production is anticipated to take place in sub-Saharan Africa and South Asia, as it would need to more than double by 2050 to cope with the upsurge in demand (Cervantes-Godoy et al., 2014). Besides this, because of climate change and land-use change, the water footprint is expected to expand by 22% by 2090 (Mekonnen and Gerbens-Leenes, 2020). This calls for obligatory sustainable management of water resources. In regions where water scarcity has already reached an acute status, decision-makers decided to invest in water desalination as an approach that promises fewer objections to fulfill the present or future gaps between water demand and supply (Archer et al., 2010).

Life-cycle assessment (LCA) seems to be a suitable method to study and assess the environmental footprint of this technology that has been increasingly adopted year after year. LCA is known to be an efficient tool to assess the environmental burden of any service or product and provide comprehensive results for both policy-makers and decision-makers. It is a methodology that consists of four stages: goal and scope, life-cycle inventory, life-cycle impact assessment, and interpretation. The standardized approach is to give a holistic science-based environmental impact assessment following the ISO 14044. This method started to receive attention from scientists globally, which made it the most used method to level up the quantitative sustainability measurements worldwide. LCA also gives a very valuable possibility of choosing the most eco-efficient way of achieving a specific functionality or service.

The peer-reviewed articles analyzed and studied to generate metadata in this chapter showed the different technologies assessed in each article, e.g., reverse osmosis (RO), multistages flash (MSF), multieffect distillation (MED), membrane distillation (Memstill), and microbial desalination cell/microbial fuel cell (MFC/MDC) as well as the system boundary considered in each case. Concerning the system boundaries, 73% of the studies considered the cradle-to-gate approach while only 13% considered the whole life cycle from cradle to grave. Most of the studies highlighted that the dismantling phase has minimal impacts as compared to the manufacturing and the operational phase. The share of the technologies assessed is 55% for the RO, 19% for MSF, 16% used MED, and 3% for capacitive deionization (CDI), Memstill, and MFC/MDC.

It should be noted that in 31% of the articles analyzed the scale of the desalination technology was not indicated (desalination capacity). The large-scale plants, i.e., >100,000m^3/d, were assessed the most in 25% of articles, the medium-scale plants (10,000–100,000m^3/d) in 21%, the pilot-scale plants (10,000–100m^3/d) in 9%, and the small-scale (10–100m^3/d) and laboratory-scale plants (<10m^3/d) together in 5% of articles. This shows the trend for researchers to perform the classic LCA by applying the LCA tool to the existing large-scale plants. Hence, innovative approaches are still needed to test and provide new conclusions and results for the discovery of new green technologies to optimize the desalination impacts on the environment.

2. Desalination technologies

Between 1990 and 2020, the cumulative installed capacity of desalination plants has increased by about 813% (Eke et al., 2020). The major desalination technologies are either membrane-based [i.e., reverse osmosis (RO), electrodialysis (ED), capacitive deionization (CDI), etc.] or thermal-based [i.e., multistages flash (MSF), multieffect distillation (MED), membrane distillation (MD), etc.], the thermal-based technologies are considered as intensive energy-demanding technologies. Taking into consideration that the majority of African countries are generating electricity from coal and oil [international energy agency (IEA)], in addition to the fact that brine (a by-product of desalination) has a negative impact on marine ecosystem, policy-makers are urged to move forward with sustainable management of desalination. In this section, the technical process of different membrane-based desalination technologies and some alternatives are described.

2.1 Reverse osmosis (RO) desalination technology

Reverse osmosis is a membrane desalination technology that has set the trend worldwide. This technology consists of using a semipermeable membrane that filters water when a force applied by a high-pressure pump and the salt with impurities is collected to restrain their penetration. Thus, purified water is obtained (see Figs. 1 and 2). Generally, the source of energy can be either thermal or electric and the consumption will vary on the scale or the volume of the water treated. The use of renewable energy causes 65–70 times lower airborne

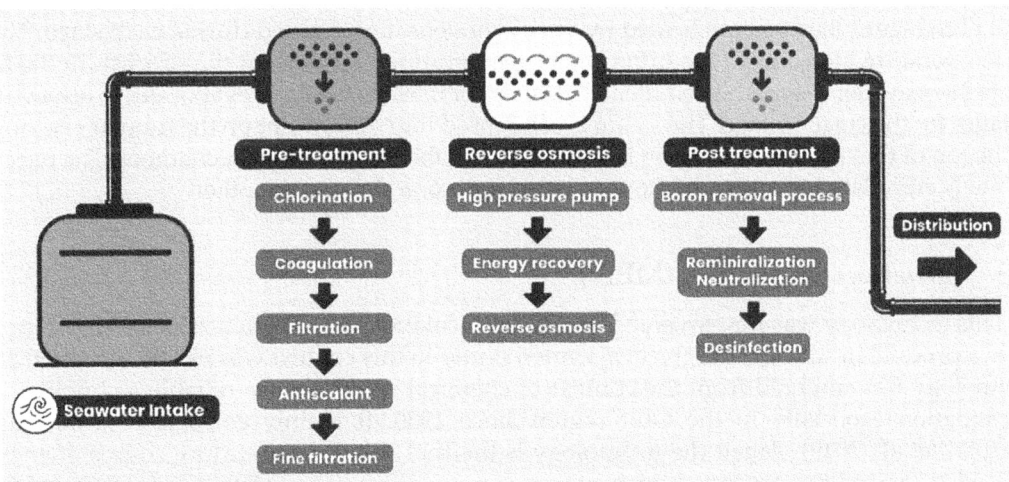

FIG. 1   Reverse osmosis detailed process.

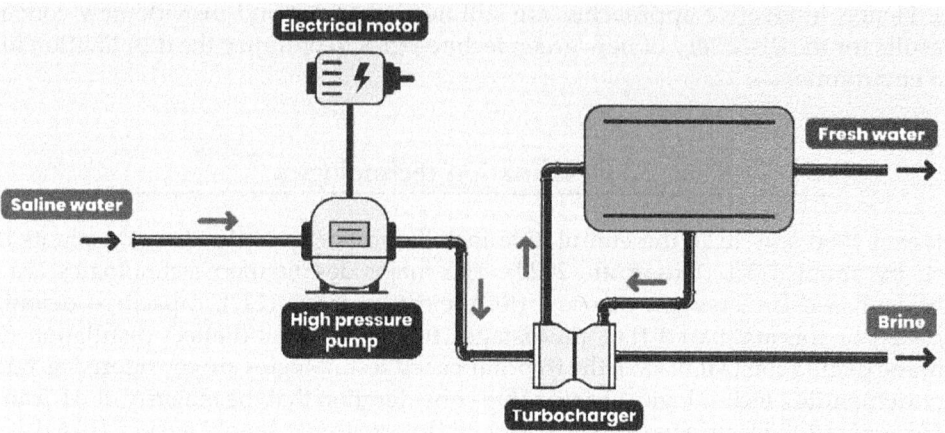

FIG. 2   Reverse osmosis process.

emissions than the electricity grid mix based on fossil fuel (Raluy et al., 2005b). Some factors contribute to the scaling of the membrane, for example, contaminants such as insoluble particles and ionic contamination. Accordingly, this imposes some maintenance operations frequently. However, the key to minimizing the cost or time, organic polymers and biocides can play the role of antiscaling and antifouling inhibitors. The reason behind its increasing adoption as a main technology worldwide lies in the fact that it delivers long-term high performance, high solute rejection, and is low cost (Eke et al., 2020).

2.2 Multistages flash (MSF)

MSF is a thermal technology where seawater flows through an evaporator in many stages or chambers maintained at lower pressure from hot stage to cold stage. This water passes through heat exchanger tubes that are heated by vapor condensation emitted during each stage. Thus, its temperature increases to the brine heater temperature required for the process. In the first stage the seawater is superheated then it "flashes" or releases heat and evaporates to achieve the balance for the current stage. The steam is condensed into fresh water on the tubular exchanger at the top of the cell. This operation is repeated until the last and coldest chamber. The rejected solute is cumulated through the flow of brine from one stage to the other.

2.3 Multieffect distillation (MED)

This technology was first invented in the chemical industry. There was needed to concentrate a product in solution but the evaporated liquid in this context was not the product. This technology has emerged from the context of chemical industry as a reliable technology for desalination especially in the Gulf region since 1950. It mainly consists of 2–16 stages (Tarpani et al., 2019). Albeit the technology is thermal, a low-temperature source of energy is used to power the process, which makes it more efficient in terms of energy compared to other thermal technologies. The heated seawater is evaporated by the inlet vapor. Tertiary steam is generated by the secondary steam at a lower pressure. This process is repetitive and the product water is the condensate that accumulates from stage to stage.

2.4 Capacitive deionization (CDI)

Capacitive deionization (CDI) is a process system that removes charged particles from water, using an electrical potential difference (electrical driving force on the ions) between two electrodes often made of porous carbon. One electrode (positively charged) adsorbs anions (negatively charged ions) and the other electrode (negatively charged) adsorbs cations (positively charged ions).

The fouling is controllable as there is no hydraulic pressure and a relatively low voltage (<1.8 V) is required. Furthermore, feed particulate characteristics, operating conditions, and electrode properties are among the main determinants of removal selectivity (Yu et al., 2016).

2.5 Membrane distillation (MD/Memstill)

It is a thermally driven separation technology in which a hydrophobic (permeable for water vapor and not for liquid water) microporous membrane, which is liquid water-resistant, allows the vapors to pass through (Fig. 3). The impetus of the process lies in the steam force impulsed by the temperature difference between the cold and the hot flow (Fig. 4).

The main application of membrane distillation is to purify and demineralize the different qualities of input feed water (Garcia-Suarez et al., 2019).

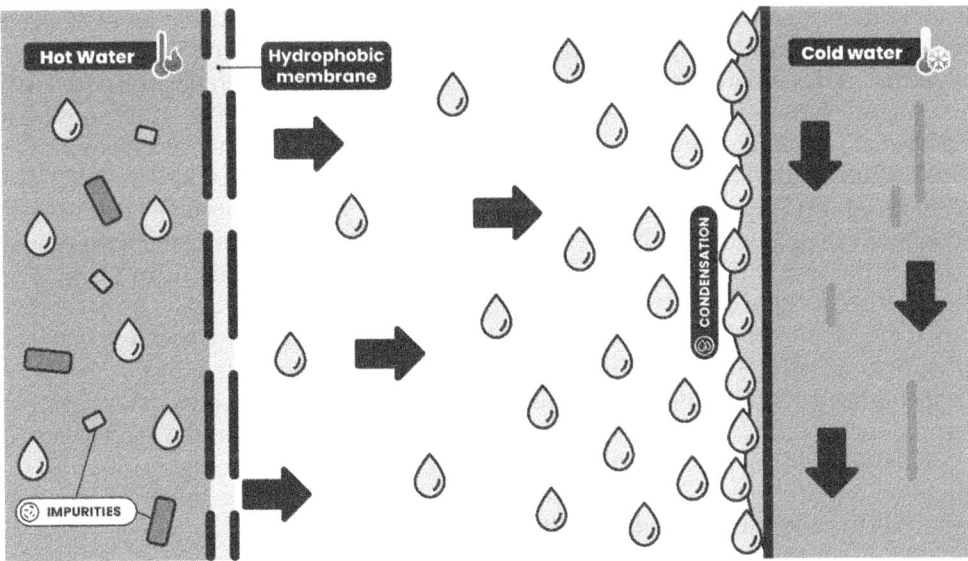

FIG. 3   Membrane distillation process layout.

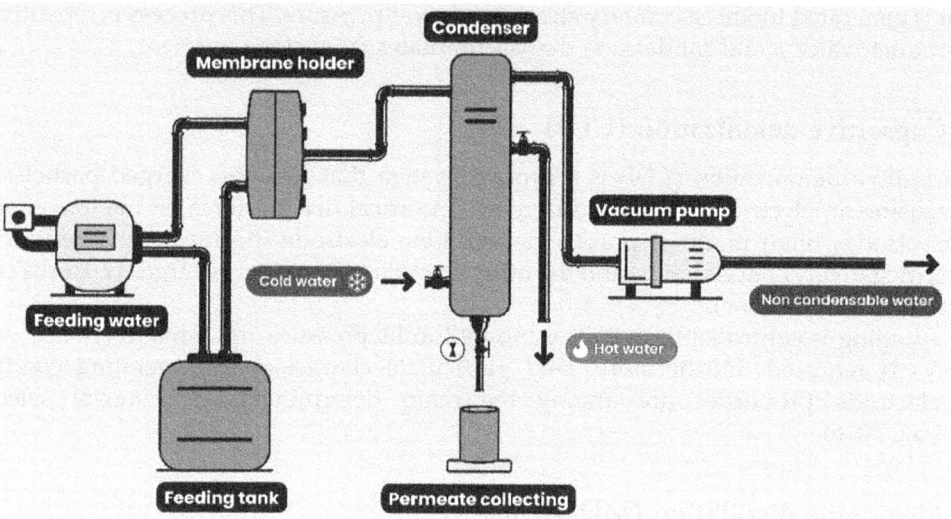

FIG. 4 Membrane distillation detailed process.

3. Life-cycle environmental impacts of desalination

3.1 Life-cycle inventory (LCI) review

A summary of the data inputs in di consumed different studies is provided in this section (Table 1). It should be noted that not all the research papers presented in this chapter clarified this step. Therefore several previous assessments lack transparency. Three key performance indicators can be identified from these studies: the water input (m^3), the amount of electricity consumed (kWh), and the amount of heat consumed (MJ). Concerning the amount of water input per m^3 of desalinated water, the average amount was found to be around 2.1 m^3 with values ranging between nearly 1 for RO/MED (Li et al., 2016) and 10 m^3 for Memstill (Tarnacki et al., 2012). The amount of electricity consumed was around 4.1 kWh per m^3 of desalinated water with values ranging between nearly 0.1 for RO (Shahabi et al., 2015b) and 28.6 kWh for MED (Tarpani et al., 2019). The lowest value (500 m) is probably due to the small distance between the plant and the ocean (Shahabi et al., 2017). The highest value is explained probably by the size of the desalination plant, relatively small in Chile (10 m^3/day) whereas for the other studies the capacity was in the range of 7570–320,000 m^3/day. It was difficult to confirm if the salinity of the water had an impact on electricity consumption due to the lack of samples for brackish water, indeed several previous studies highlighted the fact that seawater due to its higher mineral loads is more energy intensive to treat than brackish water. Finally, concerning the amount of heat consumed per m^3 of desalinated water, the values were in the range of 140–269 MJ/m^3. As expected, MED technology showed the highest inputs.

The other inputs provided by the studies mainly included the infrastructure usually expressed as the quantity of steel or concrete needed per m^3 of desalinated water (range between 1.85E – 06 and 1.80E – 02 kg of steel) and the consumption of raw materials linked to the membrane. Usually, the main material found in the RO membrane is polyamide (with values

TABLE 1 Summary of inputs provided in the different studies (per m³ of desalinated water) (LCI).

Reference	Technology	Type of water	Structure — Steel (kg)	Structure — Concrete (kg)	Tanks feed — Glass (kg)	KPI1 Water input (m³)	KPI2 Electricity (kWh)	KPI3 Heat (MJ)	Disinfection — Chlorine (kg)	Membrane — Polyamide (kg)	Antiscalant — Antiscaling (kg)	pH adjustment — Sulfuric acid (kg)	Membrane cleaning — Sodium hypochlorite (kg)	Coagulant/flocculation — Ferric chloride (kg)	Cleaning — Hydrochloric acid (kg)	Cleaning/neutralization — Caustic soda (kg)	Country
Munoz and Fernández-Alba (2008)	RO	Brackish	2.9E−03	3.7E−02	–	1.6	2.0	–	3.0E−06	4.4E−04	4.5E−03	–	–	–	–	–	Spain
Lyons et al. (2009)	RO	–	–	–	–	–	6.7	–	–	–	–	–	–	–	–	–	USA
Stokes and Horvath (2006)	RO	Brackish	–	–	–	–	0.2	–	–	–	–	–	–	–	–	–	USA
Stokes and Horvath (2006)	RO	Seawater	–	–	–	–	0.7	–	–	–	–	–	–	–	–	–	USA
Stokes and Horvath (2006)	RO	Seawater	–	–	–	–	0.7	–	–	–	–	–	–	–	–	–	USA
Jijakli et al. (2012)	RO	–	–	–	–	–	–	–	–	–	–	–	–	–	–	–	UAE
Tarnacki et al. (2012)	RO	Seawater	8.91E−04	–	–	2.4	3	–	7.1E−03	7.3E−05	2.5E−03	5.9E−03	5.9E−02	7.1E−03	–	–	
Tarnacki et al. (2012)	Memstill	Seawater	8.91E−04	–	–	10.0	–	140	–	–	–	–	–	–	–	–	
Hancock et al. (2012)	RO	Seawater	–	–	–	–	2.5	–	–	1.4E−04	–	8.0E−03	–	–	–	–	
Godskesen et al. (2013)	RO	Brackish	1.75E−02	4.58E−02	–	–	7.5	–	–	–	–	–	–	–	–	–	Denmark
Al-Sarkal and Arafat (2013)	RO	–	–	–	–	–	–	–	–	–	–	9.0E−07	7.4E−07	–	–	–	UAE
Amores et al. (2013)	RO	–	–	–	–	–	–	–	–	–	–	–	–	–	–	–	Spain
Del Borghi et al. (2013)	TVC-MED	–	–	–	–	–	3.9	217	–	–	–	–	2.6E−02	–	–	–	Italy
Del Borghi et al. (2013)	MSF/RO	–	–	–	–	–	4.3	263	–	–	–	2.5E−01	–	1.4E−02	–	–	Italy
Shahabi et al. (2015b)	RO	Seawater	–	–	–	2.2	3.1	–	–	1.4E−04	–	6.9E−04	1.9E−03	–	–	–	Australia

Continued

TABLE 1 Summary of inputs provided in the different studies (per m³ of desalinated water) (LCI)—cont'd

Reference	Technology	Type of water	Structure Steel (kg)	Structure Concrete (kg)	Tanks feed Glass (kg)	KPI1 Water input (m³)	KPI2 Electricity (kWh)	KPI3 Heat (MJ)	Disinfection Chlorine (kg)	Membrane Polyamide (kg)	Antiscalant Antiscaling (kg)	pH adjustment Sulfuric acid (kg)	Membrane cleaning Sodium hypochlorite (kg)	Coagulant/flocculation Ferric chloride (kg)	Cleaning Hydrochloric acid (kg)	Cleaning/neutralization Caustic soda (kg)	Country
Shahabi et al. (2015a)	RO open-intake	Seawater	—	—	—	2.3	0.1	—	—	1.4E−04	—	6.9E−04	—	1.6E−03	—	1.6E−03	Australia
Shahabi et al. (2015a)	RO beach well	Seawater	—	—	—	2.0	0.2	—	—	—	—	—	—	—	—	—	Australia
Linares et al. (2016)	RO	Seawater	—	—	—	—	3.5	—	—	—	—	—	—	—	—	—	—
Li et al. (2016)	LT-MED	Seawater	3.96E−06	—	—	0.0	1.5	251	—	—	—	—	—	—	—	—	China
Li et al. (2016)	LT-MED	Seawater	1.85E−06	—	—	0.0	2.1	269	—	—	—	—	—	—	—	—	China
Li et al. (2016)	RO	Seawater	2.46E−06	—	—	0.0	4.4	—	—	—	—	—	—	—	—	—	China
Li et al. (2016)	RO	Seawater	2.47E−06	—	—	0.0	5.0	—	—	—	—	—	—	—	—	—	China
Yu et al. (2016)	CDI	Brackish	1.80E−02	—	6.65E−02	1.7	0.4	—	—	—	—	—	—	—	—	—	Taiwan
Kim et al. (2017)	RO	—	—	—	—	—	—	—	—	1.2E−03	3.3E−05	—	1.2E−04	—	1.6E−03	5.3E−04	—
Aleisa and Al-Shayji (2018)	MSF	—	—	—	—	—	—	—	—	—	2.5E−07	7.3E−03	—	—	3.7E−06	3.0E−03	Kuwait
Aleisa and Al-Shayji (2018)	RO	—	—	—	—	—	—	—	—	—	5.0E−03	6.9E−03	1.6E−03	9.0E−05	5.9E−05	6.4E−05	Kuwait
Tarpani et al. (2019)	MED	Brackish	9.61E−03	—	—	—	14.29	—	—	—	—	1.4E−06	—	—	—	—	Chile
Tarpani et al. (2019)	MED	Brackish	9.61E−03	—	—	—	28.57	—	—	—	—	1.4E−06	—	—	—	—	Chile
Tarpani et al. (2019)	MED	Brackish	9.61E−03	—	—	—	—	—	—	—	—	1.4E−06	—	—	—	—	Chile
Tarpani et al. (2019)	MED	Brackish	9.61E−03	—	—	—	—	—	—	—	—	1.4E−06	—	—	—	—	Chile
Hsien et al. (2019)	RO	—	—	—	—	2.7	3.5	—	1.3E−03	—	—	7.3E−04	1.7E−03	—	—	2.9E−05	Singapore
Hsien et al. (2019)	RO	—	—	—	—	2.7	3.5	—	1.3E−03	—	—	—	—	8.0E−04	—	2.9E−05	Singapore
Hsien et al. (2019)	RO	—	—	—	—	2.7	3.5	—	1.3E−06	—	—	7.3E−04	3.5E−03	7.5E−04	—	2.9E−05	Singapore

close to 0.0001 kg material/m^3 of desalinated water). This membrane needs to be disinfected therefore the inputs also include some chemicals such as chlorine, sodium hypochlorite, or caustic soda. The amount of sulfuric acid for pH adjustment should also be noted.

A final comment can be made for these inputs, it is not always clear whether the common functional unit adopted (1 m^3 of desalinated water) includes or not the distribution until the consumer (i.e., do the boundaries are limited to the plant gate or the consumer gate?).

3.2 Global warming potential

The operation phase has been proved to be the main phase to impact the GWP results for brackish water reverse osmosis desalination as it was estimated to be about 60%–91% of the whole life cycle GWP (Stokes and Horvath, 2006). It was proved that RO technologies have lower CO$_2$ emissions than thermal desalination and the carbon footprint of seawater desalination is higher than that of brackish water (Cornejo et al., 2014). T wide range of outlined values have been linked to different parameters (e.g., location, technologies, life cycle stages, and estimation tools).

3.2.1 RO

For reverse osmosis (RO), the global warming potential was revealed to be in the range of 1.8–5.4 kg CO$_2$ eq/m^3 (Uche et al., 2013; Tarnacki et al., 2012; Raluy et al., 2005a; Meron et al., 2020; Meneses et al., 2010; Marín et al., 2012), which highlights the crucial impact of climate change due to desalination. It was found that raw materials and manufacturing of the main consumables and electricity for operation are the main contributors to global warming potential (Garcia-Suarez et al., 2019). It was concluded that electricity use represents almost 89%–99% of the GWP resulting from this desalination technology (Aberilla et al., 2020), which was also affirmed (Goga et al., 2019) in the case of South Africa. The main finding of the impact of climate change due to desalination was explained by the consumption of electricity and its emissions in a range of 3–10 kWh when reverse osmosis is applied (Meneses et al., 2010).

Mannan et al. (2019) consider thermal energy as the main contributor to this environmental impact.

3.2.2 MED

Concerning MED (multieffect distillation), the impact resulting from GWP varied from 1.8 to 47 kg CO$_2$ eq. The principal contribution to intensive energy use is steam consumption (Li et al., 2016). Almost all LCA analyses for desalination stressed the substantial impact of energy use on climate, which makes the electricity grid mix one of the most influential parameters, and it can play a major role to reduce the GWP to 99% (Tarnacki et al., 2012).

3.2.3 MSF

The range of impact on climate change is between 3.912 and 18.69 kg CO$_2$ eq. An LCA analysis concerning fossil fuel impacts in water desalination plants in Kuwait was performed, which showed that crude oil used for electricity generation is responsible for approximately 4 times more GWP when compared with other fossil fuels like heavy fuel oil, diesel, and natural gas. Hence, even if only 12.2% of Kuwait's electrical energy is generated from crude oil (CR), it contributes to 63% of Kuwait's GWP related to electricity (Al-Shayji and Aleisa, 2018).

3.2.4 CDI

The global warming potential was estimated at 1.43 kg CO2 eq. Meanwhile, for technology like CDI, electricity consumption was estimated to be relatively lower due to the low salinity of treated brackish water (Yu et al., 2016), then the highest impact on GWP was related mainly to activated carbon which is the main component of electrodes.

It is worth mentioning that the quality of feed water plays a remarkable role in the GWP of membrane technologies. As proved by Tarnacki et al. (2012) the electricity needed for water desalination using RO depends on the salinity of feed water because the power needed to create the suitable pressure for reverse osmosis technology increases with increase in salinity of water. Contrary to other impact categories, GWP has received the attention of most authors in those selected papers. About 56% of them have assessed the global warming potential of desalination, mainly because this determining factor (electricity/heat) has a direct impact on climate change category (Cetinkaya and Bilgili, 2019).

3.3 Average of environmental impacts of different desalination technologies

A global view of the GWPs resulting from the different desalination technologies was compiled using the research publications listed in the supplementary information table (Table 2). As different methods have been used for calculating LCIA in the literature, specific units (most commonly used: kg CO_2 eq) were chosen for the GWP and to perform calculation. The results are displayed in box plots to represent the minimum and maximum values, the first quartile (25th percentile) median, and the third quartile (75th percentile).

The data used in this comparison were selected according to a unified unit which limited the data processed. As shown in Fig. 5, RO has a very centered range, which reveals a minimum spread of data compared to other technologies. Also, this technology shows minimum value compared to all the other technologies. While it is remarkable that for MED (thermal technology) the range is wider as well as the spread of data even with the very limited values and papers for this technology compared to RO. This reveals the difference in the MED effects caused by different scales or geographic regions (countries). Also, the specificity for MED in our data collection is the difference in subtechnologies.

The lowest values concerning multistages distillation were associated with a global scale study (Aleisa and Al-Shayji, 2018) based on the integration of solar energy and the highest values were obtained for the Chilean study in which the diesel was used to generate the electricity for a small plant (Tarpani et al., 2019). To give an overview of the different desalination plant scales used in previous LCA studies, Table 2 shows the characterizations of each selected study. Table 2 reveals that the study with the smallest plant capacity (MED) has the highest impact compared to other studies with more capacity.

TABLE 2 Reported LCAs of desalination systems described in articles considered in this chapter.

Reference	Country	FU[a]	System boundary					Technology						Scale[b]					
			C.	C.G.G	C. Gr	C·D	C·Gt	RO	MSF	MED	Memstill	MFC/MDC	CDI	Lab	S	P	M	L	N/A
Lundie et al. (2004)	Sweden	NA	×																×
Raluy et al. (2004)	Spain	45,500 m³ of potable water/1m³ produced water		×		×		×	×	×									×
Raluy et al. (2005a)	Spain	1 m³ of produced water				×		×								×			
Raluy et al. (2005b)	Spain	45,500 m³ of potable water/1m³ desalted water				×		×	×	×						×			
Raluy et al. (2005c)	Spain	25,000 h m³/1m³				×		×											×
Stokes and Horvath (2006)	USA	123 million liters of water				×		×											×
Raluy et al. (2006)	Spain	45,500m3 of potable water		×		×		×	×	×						×			
Vince et al. (2008)	France	1 m³ of water				×		×	×										×
Munoz and Fernández-Alba (2008)	Spain	1 m³ of desalinated water				×		×	×							×			
Lyons et al. (2009)	USA	446 m³ of water				×		×											×
Stokes and Horvath (2009)	Australia	1 GL of desalinated water				×		×											×
Munoz et al. (2009)	USA	1 m³ of water	×													×			
Pasqualino et al. (2011)	Spain	1 m³ of wastewater	×												×				
Meneses et al. (2010a, b)	Spain	1 m³ of desalinated water				×	×	×								×			
Beery et al. (2011)	Germany	1 m³ of pretreated seawater				×		×											×
Zhou et al. (2011)	USA, Singapore, Spain	1 m³ of desalinated water				×		×											×

Continued

TABLE 2 Reported LCAs of desalination systems described in articles considered in this chapter—cont'd

Reference	Country	FU[a]	System boundary					Technology						Scale[b]					
			C. C	G.G	C. Gr	C-D	C.Gt	RO	MSF	MED	Memstill	MFC/MDC	CDI	Lab	S	P	M	L	N/A
Tarnacki et al. (2011)	Spain	1 m³ of desalinated water				×													×
Salcedo et al. (2012)	Spain	1 m³ of freshwater produced				×													×
Jijakli et al. (2012)	UAE	The supply of 1250L/d of clean water				×	×								×				
Tarnacki et al. (2012)	Spain	1 m³ of desalted water				×	×				×				×				
Hancock et al. (2012)	USA	1 m³ of water produced				×									×				
Norwood and Kammen (2012)	USA	1 kWh of energy				×													×
Godskesen et al. (2013)	Denmark	1 m³ of water				×	×												×
Antipova et al. (2013)	Spain	NR/ No characterization				×													×
Zhou et al. (2013)	Singapore	1 m³ of desalination brine				×													×
Al-Sarkal Arafat (2013)	UAE	1 m³ of potable water				×												×	
Amores et al. (2013)	Spain	1 m³ of potable water				×													×
Del Borghi et al. (2013)	Italy	1 m³ of treated water			×												×		
Shahabi et al. (2015a)	Australia	1 m³ of water supplied				×											×		
Shahabi et al. (2015b)	Australia	1 m³ of water, treated and distributed to a population center				×												×	

Reference	Country	Functional unit							
Lane et al. (2015)	Australia	1 year			×				×
Coday et al. (2015)	USA	1 barrel of O&G pit water		×					×
Shahabi et al. (2015b)	Australia	1 m³ of desalinated water			×	×			×
Shahabi et al. (2015c)	Australia	1 m³ of desalinated water			×				×
Karami et al. (2017)	Iran	1 ha of tomatofarm			×				×
Linares et al. (2016)	Global	1 m³ of water produced			×				×
Li et al. (2016)	China	1 m³ of water produced			×		×		×
Yu et al. (2016)	Taiwan	1 m³ of desalinated water			×			×	×
Cherchi et al. (2017)	USA	Per unit of energy consumed			×				×
Kim et al. (2017)	Australia	100,000 m³ of reusable water produced			×				×
Shahabi et al. (2017)	Australia	1m3 of desalinated water			×		×		×
Zhang et al. (2018)	USA	1 L of water being treated		×				×	
Al-Shayji and Aleisa (2018)	Kuwait	1 ton of desalinated water		×	×		×		×
Aleisa and Al-Shayji (2018)	Kuwait	1 m³ of desalinated water			×				×
Opher et al. (2019)	Israel	1 year of supply, reclamation and reuse of water	×						×
Ronquim et al. (2020)	Brazil	1 m³ of water							×
Al-Kaabi and Mackey (2019)	Arabian gulf	1 m³ of desalinated water			×		×		×

Continued

TABLE 2 Reported LCAs of desalination systems described in articles considered in this chapter—cont'd

Reference	Country	FU[a]	System boundary					Technology							Scale[b]				
			C.C	G.G	C.Gr	C.D	C.Gt	RO	MSF	MED	Memstill	MDC	MFC/CDI	Lab	S	P	M	L	N/A
Tarpani et al. (2019)	Chile	1 m³ of distillate	X																X
Mannan et al. (2019)	Qatar	1 m³ of potable water				X			X									X	
Alhaj and Al-Ghamdi (2019)	Kuwait. Algeria, Abu…	1 m³ of desalted water					X			X						X			
Zhang et al. (2019)	Qatar	1 L of incoming water	X											X					
Goga et al. (2019)	South Africa	1 m³ of potable water	X					X									X		
Hsien et al. (2019)	Singapore	1 m³ of water	X																

[a] FU = functional unit; C.C: cradle to cradle; G.G: gate to gate; C.Gr: cradle to grave; C.D: cradle to distribution; C.Gt: cradle to gate.

[b] Scales of the systems considered in particular studies:: Lab: <10 m³/d; S: (10–100 m³/d); P: (10,000–100 m³/d); M: (10,000–100,000 m³/d); L: (>100,000 m³/d).

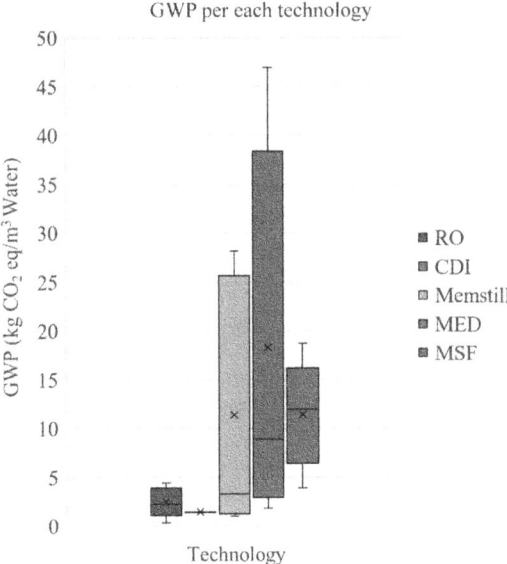

FIG. 5 Global warming potential of different desalination technologies.

4. Carbon footprint and correlation to the geography of first ranked countries in desalination capacities

Accurate estimations of GWP due to desalination depends on the availability of LCI and the LCAs conducted. However, is there sufficient specified information from those regions? The following section discusses the relationship between the published LCAs, GWP, and the desalination capacity.

4.1 The number of peer-reviewed scientific publications correlated to the capacity of desalination plants

A correlation between the countries having the highest desalination capacity in the world and the number of LCA articles focusing on desalination in those countries is presented. The results reveal a gap as in some cases no LCA study related to desalination was found in some countries where the desalination potential is high (e.g., Saudi Arabia has a desalination capacity of around 11,947,385 m^3/day). This highlights a gap between the academic interest and the actual need for desalination, especially in developing countries (Fig. 6).

In addition, it is observed that a good number of papers did not mention the scale of the desalination plant. Meanwhile, it is a substantial parameter for the transparency of their study to allow a clear understanding of the results. Accordingly, it can be also pointed out that only a few authors used primary data to perform the calculations and in some cases the life-cycle inventory (LCI) was not provided.

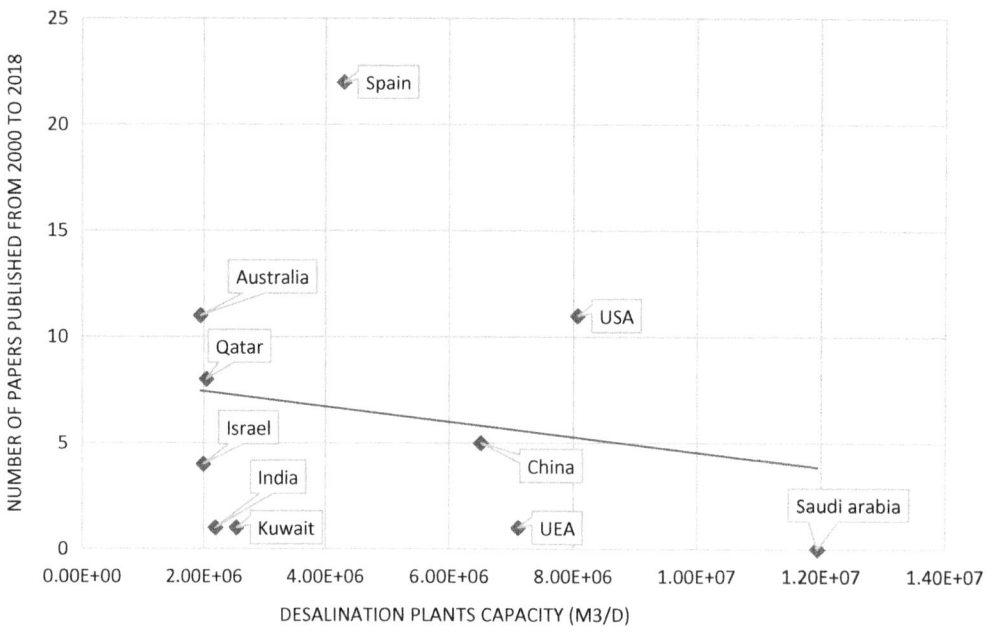

FIG. 6   Countries' existing desalination capacity and the number of research publications.

4.2 The impact of energy grid mix on the carbon footprint in desalination

The average value of electricity consumption has been calculated for each technology and the results are presented in Table 3. Besides electricity, heat is a major contributor to the global warming potential of thermal technologies (e.g., Memstill, MED, MSF). The latter is still mostly used in countries where the principal fuel is fossil fuel for either heat or electricity generation (Cetinkaya and Bilgili, 2019). This exacerbates the environmental burden and by using renewable energy this impact could be reduced by up to 99% (Mannan et al., 2019).

Airborne emission (NO_x, NMVOC, SO_x) data were compiled for three technologies: RO, MSF, and MED. The results show that MSF has the highest impact compared to the other two technologies. It generates the highest values for NO_x and NMVOC emissions while MED has the highest value for SO_x emissions. This is mainly related to emissions from heat and electricity consumption in those technologies.

Fig. 7 provides an estimate of the annual potential GHG emissions from water desalination. The top 10 countries that have the highest desalination capacity were picked for this estimation. The average carbon footprint for $1\,m^3$ of desalted water was obtained from this review. The figure shows that desalination could have a serious impact in China as the country is

TABLE 3   Average electricity consumed per $1\,m^3$ in the different desalination technologies.

Technology	RO	MSF	Memstill	CDI
Average electricity consumed per $1\,m^3$ desalinated water (kWh)	3.52	2–4	2–4	0.4

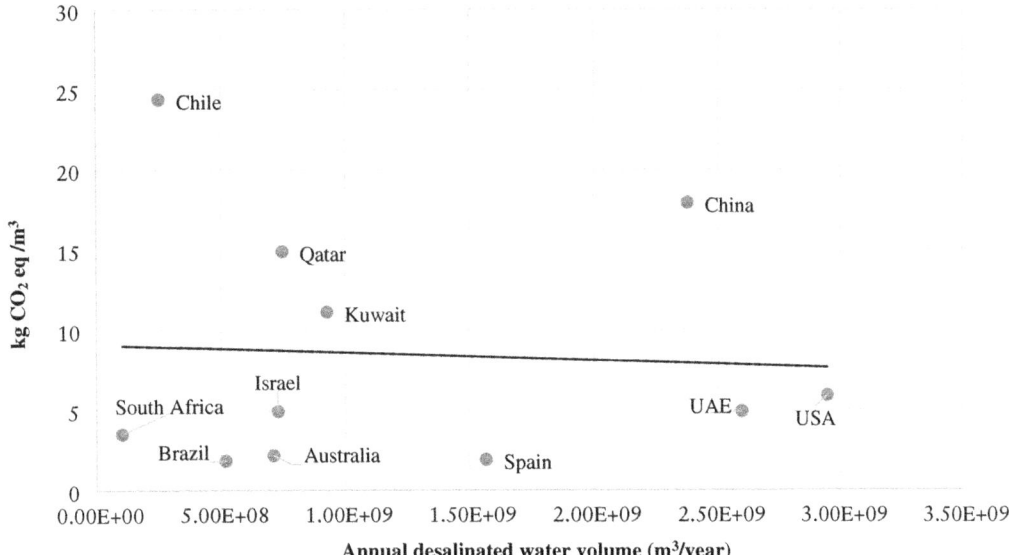

FIG. 7   Potential GHG emissions from water desalination.

relying on thermal technologies for desalination. On the other hand, in Spain, it is noteworthy to mention that even though its desalination capacity is relatively high, it keeps a low carbon footprint (44.33%) and this could be due to the generation of energy from green sources in Spain. This is mostly due to the fact that energy is the main factor that increases GWP (Tarnacki et al., 2012). From the graph, it can also be concluded that there is a poor correlation between the carbon footprint of each country and the water desalination capacity.

5. Techno-economic assessment (TEA) of different desalination technologies

There exists a recent trend toward techno-economic assessment as a complementary tool to LCA. This method insists on including the technical and economical aspects of an industrial product, service or process compared. In most of the papers discussing the TEA of the various desalination technologies, there was a noticeable focus on the RO, MED, as well as emerging technologies, e.g., electrodialysis (ED), forward osmosis (FO), and CDI.

Forward osmosis is a membrane technology driven by natural osmotic pressure to separate water from the dissolved solute via a semipermeable membrane. On the other hand, electro-dialysis is a technology where an electric potential difference is applied, the salt ions are transported from one solution through an ion exchange membrane so the dissolved species are separated from the water. FO-low-pressure reverse osmosis was proved to have a 16% cost reduction compared to the benchmark for desalination mainly seawater reverse osmosis desalination.

In a paper where ED and RO were compared, RO had two scenarios [with and without the pressure exchanger (PX)] (e.g., a device that can recover the brine's pressure energy), it was concluded that for a high salinity feedwater RO with PX was the optimal choice in terms of cost. ED was the optimal option concerning the low salinity feed water (<7 mg/L). RO without PX was highly dependent on the salinity of feed water.

RO and Ed technologies have specific parameters that fluctuate their financial cost. It was demonstrated that membrane quantity per unit, linear flow velocity, current efficiency, safety factor, volume fraction, and salinity control the increase or decrease of energy consumption and water production cost. To improve the ED energy consumption and process cost, coupling high counter-ion salt transport with low-cost ion-exchange membranes is an efficient technique (Generous et al., 2021). This could be achievable by different technologies: resin filling in the spaces to facilitate the transport. Also, the intake and depth of source water are critical as it adds additional energy consumption.

Consider MED as an example of thermal technologies. From a technico-economic perspective, it was proved that using polymer composites in MED plants instead of titanium tubes reduces the carbon footprint by 35% and the economic cost by 40% (Tahir et al., 2021). Moreover, the LCA analysis of the polymer scenario transpired favorable results compared to the titanium scenario. In this concern, Sohani et al. (2021) studied the technico-economic assessment of innovative design for solar still desalination technologies. This study concluded that using a combination of sun tracking and side mirrors as a design strategy increases freshwater production by 34.3% during active operation. Also, it reduces the cost to 0.0225 $/L.

From Fig. 8, it can be concluded that the ED is a cost-effective treatment when it comes to low salinity feed water (5 mg/L) and RO with a pressure exchanger for high salinity feed water (35 mg/L). Nevertheless, many other parameters are included in the process which contributes to the decision of implementing one technology over another. It is difficult to judge one technology as the most efficient technology only based on the cost (Tahir et al., 2021; Sohani et al., 2021; Generous et al., 2021).

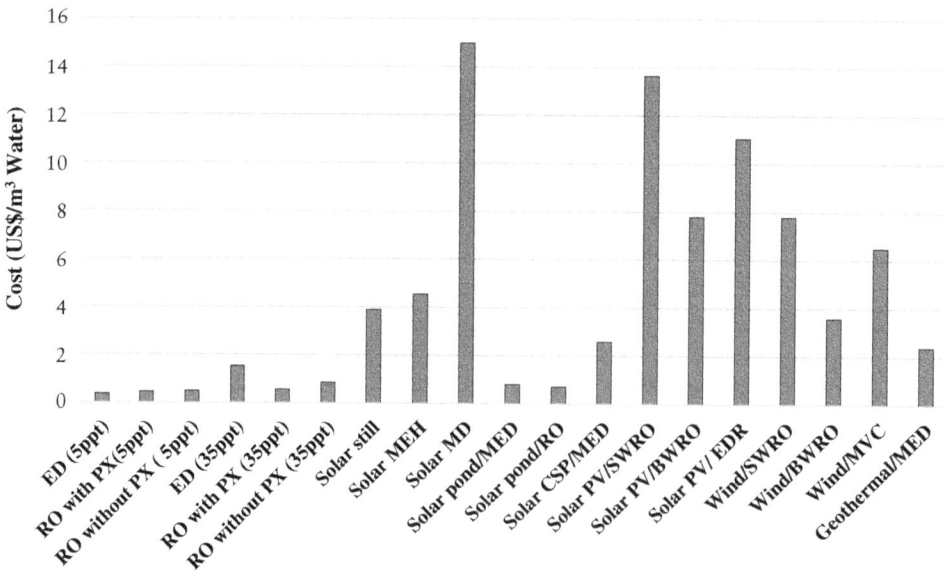

FIG. 8   Cost per unit volume of water (Generous et al., 2021; Sohani et al., 2021).

TEA gives a global overview from the technico-economic perspective in parallel with the environmental one. This kind of research is still rare which demands to shed light on assessing more technologies and scenarios either in design and conception, end-life scenarios, emerging technologies, etc. As a consequence, this helps in the engineering of new and improved technologies that reduce the environmental burden as well as being economically efficient.

6. Conclusions and perspectives

The water desalination community has shown a growing interest in the LCA method over the last few decades. Still, the trend is more for the classic LCA application with a high relaying on the background data from secondary sources such as literature. It is recommended to start adopting innovative approaches and consider the emerging green technologies for desalination. Based on the analysis of the results exposed in this chapter, it is worth noting that energy is considered as the most influential parameter. As for RO-based systems, the electricity consumption is between 3 and $10\,kWh$ while for thermal technologies it can reach even $20\,kWh$ per $1\,m^3$ including the thermal energy. This is the main reason behind the new tendency to study the green alternatives of energy sources coupling renewable energy with desalination plants.

When comparing the different technology's impact categories, RO has the minimal value in all the impact categories except abiotic depletion. RO had been estimated at $2.27\,E - 03\,kg$ SO_2 eq, $1E - 03\,kg^-$ eq, and $0.335\,kg\,CO_2$ eq in acidification, eutrophication, and GWP, respectively. While thermal technology has shown the maximal values in all impacts categories as MED has scored the highest values $3.84E - 1\,kg\,SO_2$ eq and $4.7\,kg\,CO_2$ eq in acidification and GWP, respectively. MSF has the highest impact on eutrophication with a maximum value of $3.3E - 02\,kg$ PO4 eq. It was observed that in the estimation of carbon footprint for each of the top 11 water desalination producing nations worldwide, the highest value was been estimated to be $1.74E + 10\,kg\,CO_2$ eq annually in Saudi Arabia principally due to the highest amount of desalinated water produced in addition to the adoption of thermal technologies while the lowest carbon footprint was $3.65E + 08$ in South Africa.

A gap was spotted between technology implementation and academia to develop the scientific evidence for decision-making. Countries with high desalination capacity (macro field of research) have not invested much in assessing the environmental impact of this operation. In addition, a review of different technico-economic assessment (TEA) papers revealed a poor interest in applying this method to desalination technologies even though it could lead to a better understanding of the technical and economic aspects. This provides a more holistic perspective to help decision-makers adopt the most efficient technologies that reduce the cost and environmental burdens.

References

Aberilla, J.M., Gallego-Schmid, A., Stamford, L., Azapagic, A., 2020. Environmental assessment of domestic water supply options for remote communities. Water Res. 175, 115687.
Aleisa, E., Al-Shayji, K., 2018. Ecological–economic modeling to optimize a desalination policy: case study of an arid rentier state. Desalination 430, 64–73.

A. General on membrane, materials and application

Alhaj, M., Al-Ghamdi, S.G., 2019. Integrating concentrated solar power with seawater desalination technologies: a multi-regional environmental assessment. Environ. Res. Lett. 14 (7), 074014.

Al-Kaabi, A.H., Mackey, H.R., 2019. Life-cycle environmental impact assessment of the alternate subsurface intake designs for seawater reverse osmosis desalination. In: Computer Aided Chemical Engineering. 46. Elsevier, pp. 1561–1566.

Al-Sarkal, T., Arafat, H.A., 2013. Ultrafiltration versus sedimentation-based pretreatment in Fujairah-1 RO plant: environmental impact study. Desalination 317, 55–66.

Al-Shayji, K., Aleisa, E., 2018. Characterizing the fossil fuel impacts in water desalination plants in Kuwait: a life cycle assessment approach. Energy 158, 681–692.

Amores, M.J., Meneses, M., Pasqualino, J., Antón, A., Castells, F., 2013. Environmental assessment of urban water cycle on Mediterranean conditions by LCA approach. J. Clean. Prod. 43, 84–92.

Antipova, E., Boer, D., Cabeza, L.F., Guillén-Gosálbez, G., Jiménez, L., 2013. Uncovering relationships between environmental metrics in the multi-objective optimization of energy systems: a case study of a thermal solar Rankine reverse osmosis desalination plant. Energy 51, 50–60.

Archer, D.R., Forsythe, N., Fowler, H.J., Shah, S.M., 2010. Sustainability of water resources management in the Indus Basin under changing climatic and socio economic conditions. Hydrol. Earth Syst. Sci. 14, 1669–1680.

Beery, M., Hortop, A., Wozny, G., Knops, F., Repke, J.U., 2011. Carbon footprint of seawater reverse osmosis desalination pre-treatment: initial results from a new computational tool. Desalin. Water Treat. 31 (1–3), 164–171.

Cervantes-Godoy, D., Dewbre, J., PIN, Amegnaglo, C.J., Soglo, Y.Y., Akpa, A.F., Bickel, M., Sanyang, S., Ly, S., Kuiseu, J., Ama, S., Gautier, B.P., Officer, E.S., Officer, E.S., Eberlin, R., Officer, P., Branch, P.A., Oduro-ofori, E., Anokye, P.A., Acquaye, N.E.A., Dandelar, V.M., Mineo, J., Fao, F., U. Agriculture Organization of the, Ifad, Wfp, A. K. B. Fadipe A.E.A, Afun, O.O., Ganiev, I., Maguire, C.J., Hoover, T.S., Scanlon, D.C., Fao, F., U. Agriculture Organization of the, Singh, K.M., Meena, M.S., Swanson, B.E., 2014. The future of food and agriculture: trends and challenges.

Cetinkaya, A.Y., Bilgili, L., 2019. Life cycle comparison of membrane capacitive deionization and reverse osmosis membrane for textile Wastewater treatment. Water Air Soil Pollut. 230 (7), 1–10.

Cherchi, C., Badruzzaman, M., Becker, L., Jacangelo, J.G., 2017. Natural gas and grid electricity for seawater desalination: an economic and environmental life-cycle comparison. Desalination 414, 89–97.

Coday, B.D., Miller-Robbie, L., Beaudry, E.G., Munakata-Marr, J., Cath, T.Y., 2015. Life cycle and economic assessments of engineered osmosis and osmotic dilution for desalination of Haynesville shale pit water. Desalination 369, 188–200.

Cornejo, P.K., Santana, M.V.E., Hokanson, D.R., Mihelcic, J.R., Zhang, Q., 2014. Carbon footprint of water reuse and desalination: a review of greenhouse gas emissions and estimation tools. Journal of Water Reuse and Desalination 4, 238–252.

Del Borghi, A., Strazza, C., Gallo, M., Messineo, S., Naso, M., 2013. Water supply and sustainability: life cycle assessment of water collection, treatment and distribution service. Int. J. Life Cycle Assess. 18, 1158–1168.

Diaz, J., 2021. Reverse Osmosis. Scholarly Commun. (encyclopedia).

Eke, J., Yusuf, A., Giwa, A., Sodiq, A., 2020. The global status of desalination: an assessment of current desalination technologies, plants and capacity. Desalination 495, 114633.

Garcia-Suarez, T., Kulak, M., King, H., Chatterton, J., Gupta, A., Saksena, S., 2019. Life cycle assessment of three safe drinking-water options in India: Boiled water, bottled water, and water purified with a domestic reverse-osmosis device. Sustainability 11 (22), 6233.

Generous, M.M., Qasem, N.A.A., Akbar, U.A., Zubair, S.M., 2021. Techno-economic assessment of electrodialysis and reverse osmosis desalination plants. Sep. Purif. Technol. 272, 118875.

Godskesen, B., Hauschild, M., Rygaard, M., Zambrano, K., Albrechtsen, H.J., 2013. Life-cycle and freshwater withdrawal impact assessment of water supply technologies. Water Res. 47, 2363–2374.

Goga, T., Friedrich, E., Buckley, C.A., 2019. Environmental life cycle assessment for potable water production—a case study of seawater desalination and mine-water reclamation in South Africa. Water SA 45, 700–709.

Hancock, N.T., Black, N.D., Cath, T.Y., 2012. A comparative life cycle assessment of hybrid osmotic dilution desalination and established seawater desalination and wastewater reclamation processes. Water Res. 46, 1145–1154.

Hsien, C., Choong Low, J.S., Chan Fuchen, S., Han, T.W., 2019. Life cycle assessment of water supply in Singapore—a water-scarce urban city with multiple water sources. Resour. Conserv. Recycl. 151, 104476.

Jijakli, K., Arafat, H., Kennedy, S., Mande, P., Theeyattuparampil, V.V., 2012. How green solar desalination really is? Environmental assessment using life-cycle analysis (LCA) approach. Desalination 287, 123–131.

Karami, S., Karami, E., Zand-Parsa, S., 2017. Environmental and economic appraisal of agricultural water desalination use in South Iran: a comparative study of tomato production. Journal of Applied Water Engineering and Research 5 (2), 91–102.

Kim, J.E., Phuntsho, S., Chekli, L., Hong, S., Ghaffour, N., Leiknes, T.O., Choi, J.Y., Shon, H.K., 2017. Environmental and economic impacts of fertilizer drawn forward osmosis and nanofiltration hybrid system. Desalination 416, 76–85.

Lane, J.L., de Haas, D.W., Lant, P.A., 2015. The diverse environmental burden of city-scale urban water systems. Water Res. 81, 398–415.

Li, Y., Xiong, W., Zhang, W., Wang, C., Wang, P., 2016. Life cycle assessment of water supply alternatives in water-receiving areas of the south-to-north water diversion project in China. Water Res. 89, 9–19.

Linares, R.V., Li, Z., Yangali-Quintanilla, V., Ghaffour, N., Amy, G., Leiknes, T., Vrouwenvelder, J.S., 2016. Life cycle cost of a hybrid forward osmosis–low pressure reverse osmosis system for seawater desalination and wastewater recovery. Water Res. 88, 225–234.

Lundie, S., Peters, G.M., Beavis, P.C., 2004. Life cycle assessment for sustainable metropolitan water systems planning. Environ. Sci. Technol. 38 (13), 3465–3473.

Lyons, E., Zhang, P., Benn, T., Sharif, F., Li, K., Crittenden, J., Costanza, M., Chen, Y., 2009. Life cycle assessment of three water supply systems: importation, reclamation and desalination. Water Sci. Technol. Water Supply 9, 439–448.

Mannan, M., Alhaj, M., Mabrouk, A.N., Al-Ghamdi, S.G., 2019. Examining the life-cycle environmental impacts of desalination: a case study in the State of Qatar. Desalination 452, 238–246.

Marín, D., Juncà, S., Massagué, A., Cortina, J.L., Fonseca, I., Valero, F., 2012. Impacts on climate change of three drinking water treatment plants supplying Barcelona Metropolitan Area. In: World Congress on Water, Climate and Energy, pp. 1–8.

Mekonnen, M.M., Gerbens-Leenes, W., 2020. The water footprint of global food production. Water 12 (10), 2696.

Meneses, M., Pasqualino, J.C., Céspedes-Sánchez, R., Castells, F., 2010. Alternatives for reducing the environmental impact of the main residue from a desalination plant. J. Ind. Ecol. 14 (3), 512–527.

Meron, N., Blass, V., Thoma, G., 2020. A national-level LCA of a water supply system in a Mediterranean semi-arid climate—Israel as a case study. Int. J. Life Cycle Assess. 25, 1133–1144.

Munoz, I., Fernández-Alba, A.R., 2008. Reducing the environmental impacts of reverse osmosis desalination by using brackish groundwater resources. Water Res. 42, 801–811.

Munoz, I., Rodriguez, A., Rosal, R., Fernandez-Alba, A.R., 2009. Life cycle assessment of urban wastewater reuse with ozonation as tertiary treatment: a focus on toxicity-related impacts. Sci. Total Environ. 407 (4), 1245–1256.

Norwood, Z., Kammen, D., 2012. Life cycle analysis of distributed concentrating solar combined heat and power: economics, global warming potential and water. Environ. Res. Lett. 7 (4), 044016.

Opher, T., Friedler, E., Shapira, A., 2019. Comparative life cycle sustainability assessment of urban water reuse at various centralization scales. Int. J. Life Cycle Assess. 24 (7), 1319–1332.

Pasqualino, J.C., Meneses, M., Castells, F., 2011. Life cycle assessment of urban wastewater reclamation and reuse alternatives. J. Ind. Ecol. 15 (1), 49–63.

Proskynitopoulou, V., Katosoyiannis, I.A., 2018. Review of recent desalination developments for more. New Material, Compounds and Applications 2, 179–195.

Raluy, R.G., Serra, L., Uche, J., Valero, A., 2004. Life-cycle assessment of desalination technologies integrated with energy production systems. Desalination 167, 445–458.

Raluy, R.G., Serra, L., Uche, J., 2005a. Life cycle assessment of desalination technologies integrated with renewable energies. Desalination 183, 81–93.

Raluy, R.G., Serra, L., Uche, J., 2005b. Life cycle assessment of water production technologies—Part 1 Life cycle assessment of different commercial desalination technologies. Int. J. Life Cycle Assess. 10, 285–293.

Raluy, R.G., Serra, L., Uche, J., Valero, A., 2005c. Life cycle assessment of water production technologies—Part 2: reverse osmosis desalination versus the Ebro river water transfer (9 pp). Int. J. Life Cycle Assess. 10 (5), 346–354.

Raluy, G., Serra, L., Uche, J., 2006. Life cycle assessment of MSF, MED and RO desalination technologies. Energy 31 (13), 2361–2372.

A. General on membrane, materials and application

Ronquim, F.M., Sakamoto, H.M., Mierzwa, J.C., Kulay, L., Seckler, M.M., 2020. Eco-efficiency analysis of desalination by precipitation integrated with reverse osmosis for zero liquid discharge in oil refineries. J. Clean. Prod. 250, 119547.

Salcedo, R., Antipova, E., Boer, D., Jiménez, L., Guillén-Gosálbez, G., 2012. Multi-objective optimization of solar Rankine cycles coupled with reverse osmosis desalination considering economic and life cycle environmental concerns. Desalination 286, 358–371.

Shahabi, M.P., McHugh, A., Anda, M., Ho, G., 2015a. Comparative economic and environmental assessments of centralised and decentralised seawater desalination options. Desalination 376, 25–34.

Shahabi, M.P., McHugh, A., Ho, G., 2015b. Environmental and economic assessment of beach well intake versus open intake for seawater reverse osmosis desalination. Desalination 357, 259–266.

Shahabi, M.P., Anda, M., Ho, G., 2015c. Influence of site-specific parameters on environmental impacts of desalination. Desalin. Water Treat. 55 (9), 2357–2363.

Shahabi, M.P., McHugh, A., Anda, M., Ho, G., 2017. A framework for planning sustainable seawater desalination water supply. Sci. Total Environ. 575, 826–835.

Sohani, A., Hoseinzadeh, S., Berenjkar, K., 2021. Experimental analysis of innovative designs for solar still desalination technologies: an in-depth technical and economic assessment. Journal of Energy Storage 33, 101862.

Stokes, J., Horvath, A., 2006. Life cycle energy assessment of alternative water supply systems. Int. J. Life Cycle Assess. 11, 335–343.

Stokes, J.R., Horvath, A., 2009. Energy and air emission effects of water supply. Environ. Sci. Technol. 43 (8), 2680–2687.

Tahir, F., Mabrouk, A., Al-Ghamdi, S.G., Krupa, I., Sedlacek, T., Abdala, A., Koc, M., 2021. Sustainability assessment and techno-economic analysis of thermally enhanced polymer tube for multi-effect distillation (med) technology. Polymers 13, 1–20.

Tarnacki, K.M., Melin, T., Jansen, A.E., Van Medevoort, J., 2011. Comparison of environmental impact and energy efficiency of desalination processes by LCA. Water Sci. Technol. Water Supply 11 (2), 246–251.

Tarnacki, K., Meneses, M., Melin, T., van Medevoort, J., Jansen, A., 2012. Environmental assessment of desalination processes: reverse osmosis and Memstill®. Desalination 296, 69–80.

Tarpani, R.R.Z., Miralles-Cuevas, S., Gallego-Schmid, A., Cabrera-Reina, A., Cornejo-Ponce, L., 2019. Environmental assessment of sustainable energy options for multi-effect distillation of brackish water in isolated communities. J. Clean. Prod. 213, 1371–1379.

Uche, J., Martínez, A., Castellano, C., Subiela, V., 2013. Life cycle analysis of urban water cycle in two Spanish areas: inland city and island area. Desalin. Water Treat. 51, 280–291.

Vince, F., Aoustin, E., Bréant, P., Marechal, F., 2008. LCA tool for the environmental evaluation of potable water production. Desalination 220 (1–3), 37–56.

Yu, T.H., Shiu, H.Y., Lee, M., Chiueh, P.T., Hou, C.H., 2016. Life cycle assessment of environmental impacts and energy demand for capacitive deionization technology. Desalination 399, 53–60.

Zhang, J., Yuan, H., Deng, Y., Zha, Y., Abu-Reesh, I.M., He, Z., Yuan, C., 2018. Life cycle assessment of a microbial desalination cell for sustainable wastewater treatment and saline water desalination. J. Clean. Prod. 200, 900–910.

Zhang, J., Yuan, H., Abu-Reesh, I.M., He, Z., Yuan, C., 2019. Life cycle environmental impact comparison of bioelectrochemical systems for wastewater treatment. Procedia Cirp 80, 382–388.

Zhou, J., Chang, V.W.C., Fane, A.G., 2011. Environmental life cycle assessment of brackish water reverse osmosis desalination for different electricity production models. Energ. Environ. Sci. 4 (6), 2267–2278.

Zhou, J., Chang, V.W.C., Fane, A.G., 2013. An improved life cycle impact assessment (LCIA) approach for assessing aquatic eco-toxic impact of brine disposal from seawater desalination plants. Desalination 308, 233–241.

Applications of membrane technology for water and wastewater treatment

Aerobic and anaerobic membrane bioreactors for seafood processing wastewater treatment

Nguyen Nhat Thoai[a], Tran Thi Thai Hang[a], La Vinh Trung[a], Tran Hung Thuan[b], Nguyen Van Tuyen[b], Chu Xuan Quang[b], Alberto Figoli[c], Francesco Galiano[c], and Tran Le Luu[a]

[a]Master Program in Water Technology, Reuse & Management, Vietnamese German University, Binh Duong, Vietnam [b]Center for Advanced Materials Technology, National Center for Technological Progress, Ha Noi, Vietnam [c]Institute on Membrane Technology, CNR, Rende, Italy

1. Introduction

The seafood processing industry discharges a tremendous amount of wastewater containing very high dissolved and suspended organic substances, resulting in high biochemical oxygen demand (BOD) and chemical oxygen demand (COD). Nutrients mainly exist in this type of wastewater. High total nitrogen (TN) value is probably a result of the high protein content of fish and invertebrates (Chowdhury et al., 2010). Besides, total phosphorus (TP) was discovered from the fish and during the processing and cleaning agents (Chowdhury et al., 2010). Along with specific characteristics, seafood processing wastewater probably contains a high concentration of sodium chloride (NaCl) (Pham and Nguyen, 2020). Measured values of COD, BOD, total solid (TS), total suspended solids (TSS), TN, ammonium (NH_4^+-N), TP, and fat, oil, grease (FOG), salinity from different sources of seafood industries are summarized in Table 1.

Membrane bioreactors (MBRs) have been increasingly used for wastewater treatment and reuse. It is known as a compact system comprising biological degradation and physical filtration (Tan et al., 2019). The technologies can be applied in treating wastewater from the dairy industry, winery industry, field crop, leachate, brewery, textile industries,

89

Copyright © 2023 Elsevier Inc. All rights reserved.

TABLE 1 The characteristics of seafood processing wastewater.

	COD (mg/L)	BOD (mg/L)	TS (mg/L)	TSS (mg/L)	TN (mg/L)	NH_4^+-N (mg/L)	TP (mg/L)	FOG (mg/L)	Salinity (mg/L)	References
Canned seafood industry	10,400	–	9370	–	870	–	53.6	130	–	Panpong et al. (2014)
Shrimp Processing	1442	964	2211.5	191.5	–	36.2	–	0.9	875	Thomas et al. (2015)
Fish filleting	1882	1342	2366.6	125.6	–	29.1	–	5.1	838	Thomas et al. (2015)
Cephalopod preprocessing	2570	2061	3035.9	355.9	–	32.6	–	2.9	809	Thomas et al. (2015)
Shrimp preprocessing	2700	2250	3779.9	680.8	–	36.1	–	2.9	800	Thomas et al. (2015)
Seafood processing	1220.8	–	–	–	121.07	–	57.32	–	–	Gao et al. (2018)
Centralized seafood WWTP	–	–	–	465.5	107.25	–	11.04	–	–	Nguyen et al. (2019a)
Seafood industry	306.5	206.2	–	–	92.7	–	9.5	–	–	Nguyen et al. (2019b)
Seafood industry	5600	–	–	–	–	–	–	–	69,500	Jamal et al. (2020)
Seafood processing	5600	–	–	–	–	–	–	–	39,700	Pugazhendi et al. (2020)
Seafood processing	2000	–	–	–	150	90	50	–	30,000	Pham and Nguyen (2020)

pharmaceutical industries, etc. (Sari Erkan et al., 2018; Maaz et al., 2019). This review points out two main variants of MBR aerobic MBR and anaerobic membrane bioreactor (AnMBR) in seafood wastewater treatment. The mechanisms of their biodegradability and membrane characteristics are mentioned. Besides, the influence of salinity on such processes is discussed to have a general correlation of different factors to membrane performance and their fouling behaviors.

2. Membrane bioreactor technologies

In this section, two main categories of the membrane bioreactor technologies are described: aerobic membrane bioreactors (MBR) and anaerobic ones. In addition, two types of membrane showed their treatment performance with added salinity, which is one of the factors impacting operation efficiency and membrane fouling.

2.1 Aerobic membrane bioreactor (AMBR)

Basically, AMBR operates in the presence of oxygen and membrane modules. In other words, it is a modification of aerobic technology with membrane systems which are set up to replace the secondary sedimentation systems.

2.1.1 Mechanism of aerobic process

Generally, MBR is a treatment technology of the biological process; hence, the activity of microorganisms is the main role of treatment function. The microbes signify their removal of dissolved and particulate carbonaceous BOD and the stabilization of organic matter existing in wastewater. They have the ability to oxidize the dissolved and particulate organic matter into simple-end products and biomass. In the aerobic process, there is an equation representing the oxidation of organic substances by the bacteria (Tchobanoglous et al., 2014).

$$v_1 (\text{organic material}) + v_2 O_2 + v_3 NH_4^+ - N + v_4 PO_4^{3-}$$

$$-P \xrightarrow{\text{microorganism}} v_3 (\text{new cells}) + v_6 CO_2 + v_7 H_2 O$$

where v_1, v_2, ..., v_7 are stoichiometric coefficients.

The main fractions of nutrients are represented by ammonia (NH_4^+-N) and phosphate (PO_4^{3-}-P), which are needed for the conversion of organic matter into basic products. The microbes must exist to commence the oxidation process. The term new cells means that the biomass is generated due to the oxidation of organic matter. In terms of TP removal, the process is able to encourage the development of microorganisms with the function of storing inorganic phosphorus amounts (Tchobanoglous et al., 2014).

There are some specific bacteria that are utilized to convert (NH_4^+-N) into nitrite (NO_2^-) and nitrate (NO_3^-) (nitrification). In the nitrification process, the typical bacteria, NH_4^+-N oxidizing bacteria and NO_2^- oxidizing bacteria, are presented in the liquor (Qiang et al., 2020). These bacteria are vital to reduce the amount of (NH_4^+-N) present in wastewater. The microorganisms involved in these processes are autotrophs for which (CO_2) is the major source of carbon; the oxidation of inorganic matter (NH_4^+-N) to mineralized forms generates the amount of energy (Chen et al., 2020):

$$2NH_4^+ - N + 3O_2 \xrightarrow{\text{nitrosonomas}} 2NO_2^- + 4H_+ + 2H_2O$$

$$2NO_2^- + O_2 \xrightarrow{\text{nitrobacter}} 2NO_3^-$$

2.1.2 Technical characteristics of MBR

MBR is known as a new technology for wastewater treatment and reuses (Sari Erkan et al., 2018). According to Goswami et al. (2018), this type of technology is an innovation of conventional biological activated sludge process coupling membrane filtration. The solid separation reflects the typical membranes with pore sizes ranging from 0.01 to 0.1 µm. There are two methods of MBR including internal and external installation. In 1969, the external one was introduced and replaced the function of the sedimentation tank after the activated sludge process with the installation of an ultrafiltration membrane outside the bioreactor (Tchobanoglous et al., 2014). On the other hand, the latter method is the membrane submerged in the mixed liquor. It was admitted that submerged MBRs have numerous advantages rather than external MBRs in terms of the easier cleaning process, lower investment cost, and lower operating flux (Sari Erkan et al., 2018). In particular, the application of two types of methods is able to be operated under aerobic and anaerobic conditions.

According to the membrane material, there are two common types of materials: polymeric and ceramic (Judd, 2016). Likewise, there are various modifications from polymeric materials consisting of polyvinylidene difluoride (PVDF), polyethersulfone (PES), polyethylene (PE), polypropylene (PP), polysulfone (PS), etc. (Iorhemen et al., 2016). Additionally, three types of the membrane module are flat sheet (FS), hollow fiber (HF), and multitube (MT) (Judd, 2016), while HF is commonly used in both schematic and large-scale MBR systems (Park et al., 2015). Along with polymeric types, ceramic membranes are applied with their advanced resistance, stability, and safety in terms of fouling and chemical impacts. PS and PES are commonly used for the fabrication of ultrafiltration membranes (Tiron et al., 2017).

Membrane fouling is known as a major factor that impacts negatively on MBR processes. It is the result of the interaction between activated sludge substances and membrane pores (Sari Erkan et al., 2018). The specific point of fouling is the decreasing flux or increasing constant flux mode in operation (Park et al., 2015). Several types of factors contributing to fouling include wastewater characteristics, activated sludge properties, membrane characteristics, and operating conditions (Iorhemen et al., 2016). Additionally, the presence of fouling is more on hydrophobic membranes than on hydrophilic ones (Attiogbe, 2013). Therefore, these types of membranes are coated with hydrophilic materials to battle the fouling (Iorhemen et al., 2016).

2.1.3 Effects of salinity on MBR

Along with the high strength of organic matters within seafood processing wastewater, salinity is one of the components present in this type of sewage. According to Luo et al. (2021), as the shock threshold is high, MBR systems can handle saline wastewater. The salinity itself, especially high concentration, has severe effects on the microbe community structure which are vital for biological performance (García-Ruiz et al., 2018). Therefore, it is implied to influence membrane performance.

2.1.4 Physicochemical properties

Oxygen is essential for respiratory microorganisms in an aerobic reactor, especially aerobic MBR. It is utilized for metabolism purposes and contaminant oxidation. Biomass is a heterogeneous mixture of particles, microbes, colloids, organic polymers, and cations, which all vary in shapes, sizes and densities, and influence oxygen transfer factor (α). In other words, the increase in biomass leads to the decrease of the α-factor which has an interrelationship to the oxygen transfer rate (Tchobanoglous et al., 2014). It is also verified that the volume airflow rate, mixed liquor suspended solid (MLSS) (mixed liquor volatile suspended solid—MLVSS & viscosity), soluble microbial product (SMP), and extracellular polymeric substance (EPS) have an impact on the α-factor. In relation to the presence of high salinity directly impacting on viscosity and the coalescence of air bubbles (Lay et al., 2010), the α-factor is significantly influenced by salinity. The β-factor is known to determine oxygen transfer rate for different oxygen solubility due to the presence of salts, particulates, and surface-active substances (Tchobanoglous et al., 2014). Once the salt content is expected to be high, the value of the β-factor is probably altered remarkably (Lay et al., 2010). The interconnection is expressed in the following equation:

$$OTR_f = SOTR * \left[\frac{\tau * \beta * C^*_{\infty 20} - C}{C^*_{\infty 20}} \right] * \theta^{t-20} * \alpha * F$$

where OTR_f—filled oxygen transfer rate; SOTR—oxygen transfer rate under standard condition (20°C, 1 atm, C = 0 mg/L); τ—temperature correction factor; C^*_∞—steady-state dissolved oxygen (DO) saturation concentration; C—average dissolved oxygen concentration; θ—empirical temperature correction factor; α—relative oxygen transfer rate; β—relative DO saturation factor; and F—fouling factor.

Other factors including density, turbidity, and viscosity of suspension could be impacted by the elevated salt content (Lay et al., 2010). The study verified that the increase of salt content results in a higher density of mixed liquor. The link between these factors is presented in the following equation:

$$\rho_{ML} = \frac{1}{\left(\dfrac{1 - \sum_i w_i}{\rho_L} \right)} + \sum_i \left(\frac{w_i}{\rho_i} \right)$$

where ρ_{ML}—density of mixed liquor; ρ_L—density of water; ρ_i—density of the constituents; w_i—weight fraction of constituents. The equation clearly describes that the higher the concentration of salt content (w_i), the higher the increase of mixed liquor. Therefore, it results in greater resistance to decantation via higher buoyant forces (Jang et al., 2013).

The rapid alteration of NaCl causes poor sludge settling due to the inhibition of filamentous bulking (Ng et al., 2005). In other words, the lack of filamentous organisms results in turbidity. Likewise, the presence of salt contents causes the collapse of protozoa and rotifers helping to reduce turbidity. As a result, their cellular fractions are released in the solution causing the increase of turbidity. Additionally, the rise of NaCl, higher stress on microbial cells which cannot lyse probably reduces the interaction surface, resulting in decreasing coagulation and flocculation and risen turbidity. To strengthen it, Kara et al. (2008) verified that as the concentration of NaCl increases, the negatively charged carbohydrate increases relatively. Also, the zeta potential decreases due to the positive surface charge of activated sludge by protein polymers and the presence of salinity. As a result, sodium ions cause deterioration of bacteria flocculation (Jang et al., 2013). According to Reid et al. (2006), EPS turbidity has a correlation to the EPS protein. The term salting—in is used to reflect the presence of salt contents leading to the increasing solubility of protein due to the ions carrying a large amount of hydration water into the vicinity of the protein (Zhou and Pang, 2018). It performs salt ions binding electrostatically to the protein molecules that results in the increase in solubility (Sun et al., 2010).

In terms of viscosity suspension, Judd (2011) verified that viscosity has a negative influence on the oxygen transfer coefficient. The correlation between the α-factor and viscosity (η) is expressed in the following equation:

$$\alpha = \eta^{-x.}$$

where x is an experimentally determined value. According to Bird et al. (2002), there is a possible link between salinity and viscosity. In other words, the anticipation of salinity could lead to the higher viscosity of a liquid due to the rise of dissolved solid (DS) content. The state is predicted by Einstein's equation:

$$\frac{\mu_{eff}}{\mu_0} = 1 + \frac{5}{2}\phi$$

where μ_{eff}—effective viscosity; μ_0—viscosity of the suspended medium; ϕ—volume fraction of the spheres. In the combination, the increase of salinity or called the volume fraction of spheres from Einstein's equation influences the α-factor, and therefore it has a correlation to the oxygen transfer rate.

2.1.4.1 Microorganism properties

The presence of salinity in wastewater is able to cause effects on the activity of nitrifying bacteria and characteristics of floc (Bassin et al., 2012). It could create high osmotic pressure on bacteria cells, resulting in plasmolysis, dehydration, and disintegration of cells and the absence of cell activity (Ng et al., 2005; Moussa et al., 2006; Luo et al., 2016). Higher osmotic pressure can damage the outer cells of microbes that are responsible for the release of SMP and EPS (Johir et al., 2013). According to Reid et al. (2006), both of the organic cellular components were palpably impacted by salinity shocking which had no influence on the distribution of particle size. The correlation of salinity and specific growth rate of bacteria are described in the Ghose and Tyagi model equation:

$$\mu = \mu_m\left(1 - \frac{I}{K_I}\right)\left(\frac{S}{S + K_S}\right) = \mu_{obs}\frac{S}{S + K_S}$$

where μ—specific growth rate; μ_m—maximum specific growth rate in salt-free solution; I—salt content; S—concentration of substrate; K_I—critical salt content above which reaction stops; μ_{obs}—observed maximum growth rate; K_S—half velocity constant. By contrast, Xie et al. (2014) argued that the presence of a smaller particle size leads to a higher membrane fouling tendency. The term is expressed by the Carman–Kozeny equation

$$\alpha = \frac{180 * (1 - \varepsilon)}{\rho_p * d_p^2 * \varepsilon^3}$$

where α—specific cake resistance; ε—cake porosity; dp—particle size; ρ_p—particle density. Needless to say, based on the equation, a smaller particle size is responsible for greater specific resistance due to the acceleration of particles on the membrane surface. For more information, salinity stimulates microbes to need more oxygen for cell growth that means a saline environment probably exerts selective pressure boosting the predominance of species which have a higher rate of respiratory to synthesis (Lefebvre and Moletta, 2006). The intense of biomass activity was obtained with the absence of saline concentration; likewise, the respiration rate decreased profoundly due to the presence of salinity (Di Bella et al., 2013).

With a shock load of saline concentration, the removal efficiency of COD and NH_4^+-N was affected profoundly (Reid et al., 2006). A study performed by Xie et al. (2014) revealed the removal efficiency of total organic compound (TOC) and NH_4^+-N and membrane fouling with the added saline concentration. The experiment was run in batch mode with a 2-h aerobic phase and 1-h anoxic phase. Hollow fiber PVDF membranes were operated at a flux of 5 LMH and were submerged in five reactors, where the saline concentration ranged from 0 to 35 g/L with every 5 g/L. The removal efficiency of TOC and NH_4^+-N decreased from 95% to 83% and 98% to 70%, the saline concentration ranged from 0 to 35 g/L. The attention of decreasing TOC and NH_4^+-N removal were paid at 15 and 20 g/L. They also revealed that the increasing concentration of SMP and EPS is due to the elevated salt content. An adverse impact was observed at a saline concentration of 35 g/L. As a result, membrane pore blocking is related to the appearance of SMP and EPS under conditions of high salinity, which contributes to low microbial activity. Likewise, Di Bella et al. (2013) admitted that with a remarkable increase of salinity, the removal efficiency and biomass activity are affected significantly.

The nitrification process was impacted negatively due to the increase of salinity (Jang et al., 2013). The MBR system was installed with an HDPE hollow fiber membrane. The ratio of constituents was COD:N:P = 100:10:2 with the influent COD of 1200 mg/L. The experiment was conducted at an HRT of 18 h, SRT of 50 days, constant flux of 3.5 LMH. Removal rate of NH_4^+-N decreased from 87% to 46% with the rise in salinity (NaCl) from 0 to 20 g/L. The poor removal efficiency of NH_4^+-N indicates that NH_4^+-N and NO_2^- oxidizers are sensitive to a high concentration of salt.

2.1.4.2 Membrane properties

Several studies have been conducted to evaluate the effect of saline wastewater on treatment performance and fouling behaviors. Fouling in MBR systems is the most considerable manner in operation works. Specifically, scaling is one of the main issues leading to fouling, which occurs by the precipitation of dissolved metal salt in the feed water on the membrane surface (Baker, 2004). Seawater is typical for causing scaling due to the presence of high concentrations of SO_4^{2-}, Ca^{2+}, Mg^{2+}, and other multivalent ions enabling precipitation fouling. Precipitation fouling mitigates the quality and flux of permeate and also damages membrane systems; therefore, it faces difficulty in scale removal and tackling irreversible membrane pore clogging (Schäfer et al., 2004). It implies that the elevated salt milieu probably intensifies scaling and fouling. Furthermore, a palpable flux decline was obtained with the elevated growth of crystal accumulates on the membrane surface (Chan et al., 2005).

A study performed by Jang et al. (2013) verified that transmembrane pressure (TMP) was obtained at a critical flux with a salt concentration of 20 g/L within 60 h of operation that results in accelerating fouling in a high saline milieu. The reduced fraction of TMP is related to the resistance caused by a cake layer, while the rest proportion of TMP signifies the resistance caused by pore blocking. TMP has a correlation with permeate flux that is described by Darcy's law equation:

$$J_V = \frac{dV}{Adt} = \frac{\Delta P}{\mu R_t} = \frac{\Delta P}{\mu (R_m + R_c + R_f)}$$

where J_V—permeate flux; V—total permeate volume; A—membrane area; R_t—total resistance; R_m—intrinsic membrane resistance; R_c—cake layer resistance; R_f—fouling resistance. Once the shock load of salinity added significantly, membrane permeate quality can be deteriorated (Sun et al., 2010). On the other hand, the development of TMP did not alter remarkably due to the elevating concentration of salinity (Johir et al., 2013). They identified that the higher TMP development with a high saline concentration resulted in a high amount of organic production.

Sludge flocs are known as a major factor resulting in the formation of cake layer on the membrane surface (Le-Clech et al., 2006). They comprise microorganisms, colloids, EPS, and cations. Di Bella et al. (2013) confirmed that cake resistance is the major component in total resistance. In addition, irreversible cake deposition is known as the predominant fouling mechanism. Furthermore, EPS has a strong impact on the hydrophobicity of microbial flocs which accumulate easier on the membrane surface (Tian and Su, 2012). A study conducted by Pendashteh et al. (2012) revealed that EPS concentration was obtained at 10 mg/g of the fouling gel layer and concentration of protein was higher than that of polysaccharides in the cake EPS. They stated that the secretion of polymeric compounds from the microbes results in the decrease in membrane permeability.

SMP was found to be accelerated around the membrane due to the elevation of salinity (Sun et al., 2010). They confirmed that the performance of the membrane has a binding relation to the saline milieu due to the impact of higher salinity connecting to a higher SMP concentration, resulting in a higher membrane fouling potential. According to Xie et al. (2014), carbohydrate, a major component of SMP, is an important factor in membrane fouling. Furthermore, they verified that the protein secreted in mixed liquor elevated with a dropping of microbial activity coincided with the increase of saline concentration and the carbohydrate as the available carbon declined.

2.1.5 Application of aerobic MBR in seafood processing wastewater treatment

Huang et al. (2000) studied the organic removal by using submerged MBR in artificial wastewater comprised of COD (300–600 mg/L), TOC (110–125 mg/L), BOD (150–300 mg/L), NH_4^+-N (36–72 mg/L), and pH (6.5). Hollow fiber (PP) membrane was used under conditions of hydraulic retention time (HRT) 8.3 h, sludge retention time (SRT)

20 days, DO 5 mg/L, and MLSS 3000–4000 mg/L. The study obtained 90%, 94%, 95% in terms of COD, TOC, and BOD removal efficiency, respectively. However, the removal of NH_4^+-N did not examine in the experiment. Another study stated that the nitrification performance between the submerged MBR and conventional activated sludge process (CAS) has a nearly similar efficiency, while the completed conversion of NH_4^+-N to NO_3^- of MBR is much faster than (CAS) performs (Gao et al., 2004).

In a study of treatment wastewater from surimi products conducted in Thailand, the flat sheet (PES) membrane was examined in a lab-scale pilot. The wastewater contained (150 mg/L) SS, 1700 (mg/L) COD, (1200 mg/L) BOD, (95 mg/L) TKN, (19 mg/L) P, and (5.67) pH. The membrane module was operated with a permeation flux of 5 L/m^2 h. The operation conditions comprised a flow rate of 5/7 L/d, HRT of 6.1 days, DO of (4–5 mg/L), airflow rate of (15–20 L/min), and organic loading rate (OLR) of (0.28 kg COD/m^3 d). The experiment strongly showed a result of 99% in BOD removal, while the COD and TOC removal rate comprised of 85%. The efficiency of TKN also examined with the removal of NH_4^+-N was 94%. The study revealed that a clear permeate achieved a turbidity of 5 NTU (Sridang et al., 2006). Subsequently, a study performed by Sridang et al. (2008) revealed a flat sheet (PES) MBR filterability in surimi wastewater. The characteristic of wastewater is similar to that in their previous study. They confirmed that macromolecular substances existing in soluble compounds have a significant fouling potential leading to the filterability.

Treatment of wastewater originating in the fish meal industry located in Chile by aerobic MBR was also investigated (Afonso and Bórquez, 2002). A mono-tubular (UF) membrane coated with a thin layer of ZrO_2-TiO_2 was tested in the study. The ideal hydraulic permeability of the membrane was 55.6 L/m^2 h bar. Under the operational conditions comprising a pressure of 4 bar, flow of 4 m/s, the permeability dropped to 36.5 L/m^2 h bar even after performing a backwash at 1 bar and 4 m/s for 2 h. The reduction rates of TS and VS were obtained at 78.86% (from 42.1 to 8.9 g/L) and 86.62% (from 26.9 to 3.6 g/L), respectively, while the figures for FOG and total proteins were 98.4% (from 20.5 to 0.32 g/L) and 61.73% (from 8.1 to 3.1 g/L), respectively. Besides, the study revealed that total recirculation of protein rejection is low at 14% with a volume reduction factor VRF = 1 and then slightly increased to 39% at VRF = 1.5.

Hybrid MBR is known as one of the modified membrane technologies and was applied in a study of treatment of tuna cooking wastewater in Spain. The experiment was performed by Artiga et al. (2008). The module consisted of three chambers including an anoxic chamber, an aerobic chamber with a suspended biomass and plastic carriers, and a membrane chamber. Three types of membranes were used in the study. The (UF) hollow fiber membrane (0.04 µm) was used in the first stage (0–98 days) where wastewater was collected during tuna cooking with saline concentration ranging from 73 to 83 g/L. During the experiment, the COD removal efficiency dropped from 95% to 65% with the OLR of 0.3 and 0.6 kg COD/m^3. On the other hand, the COD efficiency slowly rose from 77% to 92% at the OLR of 1.4 kg COD/m^3. Due to the high saline concentration, nitrification did not exist during the first stage. In terms of membrane characteristics, the permeability severely decreased from 200 LMH to 50 LMH. Scaling also existed on the membrane surface due to the precipitation of $Ca_3(PO_4)_2$. The latter stage (day 100–day 225) witnessed the examination of a (UF) hollow fiber membrane (0.04 µm), a (UF) hollow fiber membrane (0.4 µm), and an external tubular UF membrane (0.03 µm) in a steam tuna cooking stream where the salinity varied from 2 to 15 g/L. They revealed that the COD removal efficiency ranged from 50% to 99% with the maximum OLR of 16,000 mg/L and maximum OLR of 4 kgCOD/m^3. The concentration of TN was lower than 100 mg/L compared to the initial concentration of 2.5–4 g/L thanks to the hybrid system. Regarding filterability, the (UF) hollow fiber module had a diminishing permeability from 106 to 10 L/m^2.bar in 12 days, while the external module had an initial permeability of 200 L/m^2.bar from day 117 to 176 and dropped to 100–145 L/m^2.bar during days 218–225.

Other advanced modifications of MBR processes were used in different studies. For instance, the combination of a heterogeneous Fenton fluidized bed and a membrane photobioreactor (MPBR) is utilized to deal with the high presence of salinity in seafood processing wastewater (C. Li et al., 2019). It was stated that modified MBR has a better biodegradation and antifouling ability than conventional MBR at a salinity ranging from 10 to 100 g/L (Tan et al., 2019). The pretreatment process commenced initially with the inoculation of mixture of H_2O_2 and wastewater into a fluidized bed from the bottom of the column, and heterogeneous Fenton catalysts were pumped into the reaction column to enhance antifouling in the MBR process (Qin et al., 2018). Membranes were fabricated from graphene oxide (GO) and polyvinylidene fluoride (PVDF) and applied in the MPBR system with a high critical flux (48–50 L/m^2 h) (Qin et al., 2018). The highest rate of COD reduction reached around 54% following the pretreatment process. After the MPBR process, COD and NH_4^+-N reduction rates were around 95% and 98%, respectively, with a high biomass production of 105 mg/L.d.

2.2 Anaerobic membrane bioreactor (AnMBR)

Along with the aerobic MBR technology, the membrane can be modified to be suitable for working in anaerobic condition. The anaerobic mechanism is still the center of the reactor; membrane modules are added to increase the treatment efficiency.

2.2.1 Mechanism of the anaerobic process

The anaerobic process refers to the decomposition of organic substances by a wide range of microorganisms in the absence of oxygen. The microbes convert complex organic substances and mineralize them into methane (CH_4), CO_2, (NH_4^+-N), hydrogen sulfide (H_2S), and water (van Lier et al., 2008).

The process comprises three main stages including hydrolysis, acidogenesis and acetogenesis, and methanogenesis. Initially, the particulate material is transformed into soluble compounds which then can be hydrolyzed to simple monomers that are utilized by bacteria to perform fermentation. The stage conducts with extracellular enzymes which are produced by various anaerobes. Subsequently, bacteria enable to break down products of the hydrolysis stage (monosaccharides, amino acid, and fatty acid) into volatile fatty acids (VFAs), CO_2, and hydrogen (H_2) (von Sperling, 2007). In addition to acetogenesis, it refers to further fermentation by bacteria to convert products of the previous stage (propionate and butyrate) to acetate, CO_2, and H_2, which are the main components of CH_4 formation. Finally, methanogenesis happens by a group of *Archaea* organisms which convert acetate into CH_4 and CO_2 (Tchobanoglous et al., 2014). The simplest equation of carbonaceous matter conversion under an anaerobic milieu is mentioned below:

$$C_6H_{12}O_6 \,(\text{organic matter}) \rightarrow 3CH_4 + 3CO_2 + \text{energy}$$

On the other hand, the equation can be described in a generic way for an organic compound

$$C_xH_yO_z + \frac{4x - y - 2z}{4}H_2O \rightarrow \frac{4x + y - 2z}{8}CH_4 + \frac{4x - y + 2z}{8}CO_2$$

Basically, the presence of nitrogen in wastewater does not enable to be treated by the anaerobic process due to the absence of oxygen to convert the NH_4^+-N into NO_2^- and NO_3^- in terms of nitrification. Additionally, there is no denitrification existing in the anaerobic process unless the influent consists of nitrate (van Lier et al., 2008). However, anaerobic ammonium oxidation (Anammox) is able to handle, but the anammox bacteria faces a long doubling time, as a result of a long start-up for a batch reactor and difficulties in enrichment (Zhang et al., 2019). For some cases, it is an alternative nitrogen removal process which oxidizes ammonium into nitrogen by using nitrite as an electron acceptor. The reaction is shown below

$$NH_4^+ - N + 1.32NO_2^- + 0.066HCO_3^- + 0.13H^+ \rightarrow 1.02N_2 + 0.26NO_3^- + 0.066CH_2O_{0.5}N_{0.15}$$
$$+ 2.03H_2O$$

In terms of sulfate (SO_4^{2-}) removal, SO_4^{2-} reducing bacteria can convert SO_4^{2-} into H_2S. Along with direct utilization of methanogenic substrates, these bacteria enable to use propionate, butyrate, higher and branched fatty acids, lactate, ethanol, etc. (van Lier et al., 2008). The study also points out the function of SO_4^{2-} reducing bacteria in the following equation:

$$H_2 + 0.25SO_4^{2-} + 0.25H^+ \rightarrow 0.25HS^- + H_2O$$

$$CH_3COO^- + SO_4^{2-} \rightarrow HS^- + 2HCO_3^-$$

2.2.2 Technical characteristics of AnMBR

AnMBR is a hybrid system consisting of membranes submerged in the anaerobic reactor or outside the reactor (Maaz et al., 2019). The process is able to combat different types of wastewaters containing a high number of toxicants, inorganic or organic compounds from leachate, brewery, textile industries, pharmaceutical industries, etc. Likewise, AnMBR technology is suitable for treatment of wastewater from the food processing industry and municipality (Lin et al., 2013). Similar to MBR technology, AnMBR is also applied with a flat sheet, hollow fiber, and tubular form. The flat and hollow forms are more appropriate and used in submerged reactors, while tubular membranes are preferred to external cross-flow ones (Shahid et al., 2020; Ozgun et al., 2013). In terms of membrane material, the membranes are fabricated with modified material including PVDF, PES, PE, PP, PSF, ceramic, and metal (Lin et al., 2013). The presence of fouling in AnMBR is unavoidable. There are two types of fouling including internal fouling and external fouling. Internal fouling represents the presence of particles which are larger than the pore size, while external fouling signifies fine particles, solutes, and undissolved matter running inside the membrane pores (Maaz et al., 2019). In order to mitigate fouling, AnMBR is coupled with the forward osmosis (FO) system as a pretreatment stage which enhances low fouling potential and easy cleaning (Schneider et al., 2019). Likewise, different physical and chemical methods are able to be used for clean membranes (Shahid et al., 2020).

Along with similar points to the MBR process, AnMBR has its own advantages in terms of biogas production, high treatment efficiency, and energy recovery (Stuckey, 2012). According to Kong et al. (2021), the conversion of methane comprised 60%–64% of COD reduction, and energy saving reached approximately 80%. A large pilot-scale process achieved high removal rates of COD and BOD over 90% and 95%, respectively. Another study performed by Robles et al. (2020) highlighted the efficiency of COD removal as approximately 92%. In terms of methane production, it was above 70% at an HRT of 40 h and slightly fluctuated due to the influence of the temperature.

Furthermore, AnMBR has a higher treatment efficiency than MBR; generally, 90% or higher COD removal is achieved by the AnMBR process. In addition, CH_4 production ranges from 0.25 to 0.35 m^3 CH_4/kg COD and the OLR of 15–30 g/L or higher. HRT varies from 1 to 25 days and the membrane filtration flux from 5 to 10 LMH (Chang, 2014). In Istanbul, the AnMBR process applied in the treatment of wastewater from the confectionery industry revealed a COD removal rate up to 99% (Balcıoğlu et al., 2020).

2.2.3 Effects of salinity on AnMBR

In terms of AnMBR, a saline environment also has some influence on the treatment process. Previous studies revealed how a significant salt content can damage aerobic microbes as well as anaerobic microorganism activity. High salinity has a remarkable impact on anaerobic digesters than activated sludge processes (Chowdhury et al., 2010). Also, the presence of fouling on membranes in an anaerobic milieu leads to low performance and other related issues.

2.2.3.1 Microorganism properties

Along with the inhibition of salinity to aerobic microbes, anaerobic microbes are vulnerable to salt concentration (Dan et al., 2003), especially with the high salinity (Lefebvre and Moletta, 2006). According to Lefebvre et al. (2007) and Sierra et al. (2018), the increase of NaCl concentration indicates the slowdown of microbial activities, leading to the low rate of biogas production, which is impacted by the inhibition of methanogenesis. Also, their experience showed a significant decrease in COD removal efficiency, specific loading rate (SLR), OLR, and specific methanogenic activity (SMA) as well. Further studies indicated the effects of high salt concentration on anaerobic biodegradation and a decrease in biomass production (Foglia et al., 2020). Sierra et al. (2017) verified that the addition of 23 g NaCl leads to 50% inhibition of methanogenic activity and totally completed methanogenesis inhibition at 34 g NaCl. Additionally, a decrease in the COD removal rate was observed at the increased salinity above 10 g/L (Song et al., 2016). Also, the study revealed the increase of MLSS concentration and the decrease of MLVSS concentration with the salinity ranging from 16 to 22 g/L. Thus, they concluded that the buildup salinity has negative effects on microbial activity in anaerobic reactors. Yogalakshmi and Joseph (2010) explained that the decrease in organic removal efficiency occurs due to the inhibition of salt content to microbial community resulting in loss of metabolic function and plasmolysis leading to the release of intracellular constituents and SMPs. A study performed by Vyrides and Stuckey (2009) indicated that the presence of higher salinity causes the higher accumulation of SMPs with higher production of molar weight compounds. In addition, methane production performs low at a high salt concentration due to the generation of compatible solutes and extracellular polysaccharides by anaerobic microbes using substrate to under high osmotic pressure.

Furthermore, salt concentration effects on size of granulation presenting in the anaerobic reactor. A study performed by Sudmalis et al. (2018) showed the increase of particle size resulting in the increasing salinity concentration. Specifically, the strength of microbial granular was not affected by sodium concentration. By contrast, Ismail et al. (2008) argued that a striking decrease in granule strength occurs under conditions of long-term exposure to high sodium concentration. Gagliano et al. (2020) verified that the accumulation of salinity concentration can be an advantage for biomass aggregation. Also, the study pointed out that the thicker EPS gel layer is probably an adaptive response of microbes to the presence of high salinity. Ismail et al. (2009) revealed that the binding of EPS can be lost by the presence of an excessive amount of sodium cation and leads to the swollen granule structure in terms of repulsive electrostatic forces. In other words, the bigger the granular size, the weaker it becomes with high salinity. Therefore, the fraction of EPS and microbes are probably the factors leading to the large effluent COD content (Sudmalis et al., 2018).

2.2.3.2 Membrane properties

Similar to the MBR in aerobic condition, AnMBR also faces fouling issues created from suspension properties which affect the filtration performance of the membrane. Also, particle size plays an important role in the accumulation of solids on the membrane surface (Jeison, 2007). Besides, most significant foulants including inorganic precipitation and deposition of organic forming specific cake resistance (Choo et al., 2000).

The presence of a high concentration of sodium causes the deterioration of the membrane filtration performance which is attributed to the decrease in biomass particle size (Sierra et al., 2018; Foglia et al., 2020). They revealed that the reduction of particle size and biomass properties affected by salinity are the components leading to fluctuating membrane filtration resistance. Also, a study performed by Lin et al. (2011) confirmed that cake sludge has a smaller particle distribution causing a higher specific filtration rate. The presence of a fine particle size ranging from 0.15 to 0.4 μm causes membrane fouling (Hu and Stuckey, 2006). In addition, Zhou et al. (2019) found that the microparticles ranging from 5 to 10 μm containing of filamentous bacteria which was also mentioned in Section 2.1.4.2 contributes the accumulation of cake layer on membrane surface and biofilm formation. Their previous study showed that a subvisible particle in the range of 0.45–10 μm, which comprises 90% of total organics (proteins and polysaccharides), probably plays an important role in cake resistance in submerged AnMBR (Zhou et al., 2016).

Similarly, for aerobic MBR fouling, SMP and EPS are the main components causing gel fouling in the AnMBR process with the same concept due to the presence of salinity concentration. It was confirmed that raw sludge contributes a lower fouling resistance than colloid and solute fraction (Christensen et al., 2018). According to Rosenberger et al. (2006), the polysaccharide fraction (polysaccharides, proteins, and organic colloids), which consists of a microbiological origin (soluble EPS and SMP), plays an important role in the alteration of membrane performance. A high concentration of polysaccharide causes a high fouling rate. Nevertheless, Aquino and Stuckey (2008) verified that the production of EPS is only a part of SMP, leading to the argument of membrane fouling resulting in SMP or EPS. Vyrides and Stuckey (2011) agreed with the term of cell lysis and substrate intermediates are probably part of SMPs and contributes to membrane fouling. For more information, EPS is attached on a membrane surface due to the filter function of the reactor contents including SMPs by the membrane that is considered increase in cake resistance. For more detail, sludge with a high ratio of proteins to polysaccharides (PN/PS) in bound EPS has a tendency of stickiness and prefers to cake formation (Lin et al., 2011). A study performed by Liu et al. (2019) revealing quorum quenching for fouling control verified that the decrease of EPS is probably the main solution for membrane biofouling mitigation.

With the possible presence of salinity in seafood processing wastewater, a lab scale of AnMBR with an external tubular ceramic membrane module was installed by Hemmelmann et al. (2012). The module was fed with synthetic wastewater containing an NaCl concentration of 25 g/L. Air sparging was used by recirculated biogas. A flux of 5 LMH was applied during the operation time. They identified that the deposition of the cake layer on the membrane surface was the main cause inhibiting the applicable flux.

Another study investigated the AnMBR with a PTFE hollow fiber which was set up outside the reactor (Li et al., 2018). Synthetic wastewater was used with COD of 20,000 mg/L, NH_4^+-N of 195 mg/L, and TP of 470 mg/L. The module was operated at an HRT of 5 days and SRT of 226 days. The experiment was divided into four stages including the gradual addition of saline concentration (NaCl) of 11 g/L (I), 19 g/L (II), 27 g/L (III), and 35.5 g/L (IV). The study revealed that the ratio of internal residual fouling resistance and external fouling resistance gradually rose at 1.98, 2.21, 2.73, and 3.18 in four phases. Consequently, the internal fouling had an interconnection with salinity. Furthermore, the size of the sludge particle was affected by the increased saline concentration. The initial mean size was 126.02 μm, then reduced to 76.95, 56.46, 36.80, and 25.19 μm from phase (I) to phase (IV). The study concluded that the acceleration of membrane fouling was caused by the reduced particle size, which was influenced by the rise of salinity.

2.2.3.3 Biogas production

A study performed by Rodriguez-Sanchez et al. (2019) investigated the impact of salinity on the performance of biogas production. Two lab-scale AnMBR were installed with the flat sheet (PVDF) membranes submerged in the reactors where the latter one had an addition of salt. The operational configurations included an HRT of 6 h, SRT of 120 days, air scouring from biogas, and a flux of 15 LMH. Synthetic wastewater was fed with the accumulation of saline concentration of 5, 10, 20, 40 g/L. The amount of biogas produced by first module is of 0.35–0.49 L/g COD removed. By contrast, the latter one witnessed a significant decrease of biogas production of 0.26, 0.11, 0.02 L/g COD removed at a saline concentration (NaCl) of 5, 10, and 20 g/L, respectively. Likewise, with the gradual addition of salt concentration from 10 to 20 and 40 g/L, the efficiency of COD reduction dropped from 91.4% to 86.7% and 77.7%, respectively.

In Italy, another experiment performed by Foglia et al. (2020) revealed the performance of a pilot-scale AnMBR in handling urban wastewater with gradual addition of NaCl concentration from 200 to 1500 mg/L. The sewage had constituents of (550 mg/L) COD, (30 mg/L) NH_4-N. Biogas production was examined with remarkable results. At the initial stage, 1.2–1.3 L/d of the biogas was produced at an NaCl concentration of 200 mg/L before dropping to 0.13–0.57 L/d at a saline concentration of 500 mg/L. Biogas production was only obtained at 0.08 L/d at a salinity of 2200 mg/L where the failure of the system occurred.

2.2.4 Application of AnMBR in seafood processing wastewater treatment

A study performed by Chaiprapat et al. (2016) highlighted the treatment efficiency in seafood processing wastewater by using the AnMBR process with PVDF hollow membrane fibers submerged in a transparent PVC tube as a reactor. The experiment was performed with three variant recirculation streams including permeate (I), permeate and granular activated sludge (GAC) within reactor (II), and permeate, GAC, biogas for air scouring (III). The reactor was fed with UASB effluent which comprised (106–408 mg/L) COD, (108–260 mg/L) TN, (65.8–188 mg/L) NH_4^+-N, (67.7–158.9 mg/L) SO_4^{2-}. They found that the removal of NH_4^+-N was obtained at 18%–35% due to biomass synthesis. The study revealed the highest removal efficiency of COD was obtained at 90.5% at an HRT of 4 h in the (II) module, while the figure for SO_4^{2-} was 32.9% at an HRT of 8 h in the (I) module. In three variances, the latter had the best efficiency of CH_4 formation (51.3%–54.5%). In terms of TMP, they concluded the result that biogas scouring (III) could effectively resist membrane fouling. The membrane fouling rate of (III) was recorded at the lowest value of 0.1, 0.9, and 6 mbar/d at an HRT of 8, 6, and 4 h, respectively.

2.2.5 Energy recovery

AnMBR technology plays an important role in product recovery. It was examined by a study performed by Galib et al. (2016). The AnMBR was installed with a (UF) hollow fiber membrane submerged in the reactor. The sewage stream including COD of 4398 mg/L, NH_4^+-N of 77 mg/L, and PO_4^{3-}-P of 101 mg/L was fed from the meat processing factory. The module was operated at an HRT of 5, 2, and 1 days, corresponding to an OLR of 0.4, 1.3, and 3.1 kg COD/m^3.d, respectively. SRT was 50 days, and the flux was maintained at 1.17, 3.13, and 6.4 LMH for each HRT condition. Biogas was used as air flushing for reduction of membrane fouling. They figured out that the energy consumption was approximately 0.39 kWh/m^3 for each phase. The study revealed that the module provided a net energy benefit of 0.13–5.1 kWh/m^3. The highest rate of energy benefit was achieved at the highest OLR which witnessed the adverse irreversible fouling. Therefore, the flux of 6.4 LMH was set to obtain a sustainable energy benefit and no significant appearance of membrane fouling.

To strengthen the energy production from AnMBR, a study performed by Balcıoğlu et al. (2020), showed the performance of AnMBR in the treatment of the confectionery industry wastewater. In the study, the commercial flat sheet (PES) membrane was applied outside the reactor equipped with a mixing device. The lab-scale module was run at a constant flux of 4.5 LMH, OLR of 1.1, 2.2, 4.4, 6.6, and 7.9 kg COD/m^3.d. The fed wastewater characteristics contained COD of 18,000–18,900 mg/L, TSS of 1600–3900 mg/L, TKN of 125–170 mg/L. The biogas production was achieved at 1.5, 3, 5.8, 8.3, and 9.7 L/d at each OLR. In addition, the CH_4 ratio of biogas was 79%, 81%, and 80% at an OLR of 1.1, 2.2, 4.4 kg COD/m^3.d, while the figure for an OLR of 6.6 and 7.9 kg COD/m^3.d dropped to 70%. Therefore, a high OLR was able to adversely influence biogas production.

Another study investigated by Gong et al. (2019) figured out the energy balance using a hollow fiber (PVDF) membrane submerged in a reactor. Municipal wastewater was used with COD of 231 mg/L, TN of 34.4 mg/L, NH_4^+-N of 29.3 mg/L, TP of 2.9 mg/L. The experiment of energy balance was examined in two conditions of ideal operation and severe fouling. Pumping energy was set at 0.006 kWh/m^3, and energy of air scouring was 0.02 kWh/m^3. In the latter situation, energy consumption from pumps and air flushing were 0.027 and

$0.02\ kWh/m^3$, respectively. They indicated that the energy production was achieved at 0.076 kWh/m^3 thanks to the high organic recovery and anaerobic biodegradability. Finally, the study revealed that the net energy production was obtained at 0.05 and 0.029 kWh/m^3 for each scenario.

2.3 Advantages and disadvantages of MBR and AnMBR

According to Oreopoulou and Russ (2007), the anaerobic process refers to the high-polluted wastewater. It has the advantage of decomposing a high number of organic compounds with the absence of oxygen. As the specificity of anaerobic bacteria, biogas is generated due to the degradation of the organic substances. Additionally, the biomass is much more sustainable and more active without feeding than the aerobic process. The volume of excess sludge yield in anaerobic treatment is generally lower as compared to the aerobic system. Due to a slow rate of growth, a longer period of time is required to commence the process, but they can fully grow without being washed out from the reactor (Guo et al., 2016) and give a chance to strengthen the salt-tolerant biomass (Sierra et al., 2018).

As little energy is required for methane yield from microorganisms, their rate of growth is slower than under the aerobic process and only a small amount of waste is converted to new cells. As a result, the problem of disposal of excess sludge is remarkably minimized. In addition to the absence of oxygen, it reduces the power requirement for treatment process and provides a valuable source of CH_4 which is utilized for running engines or generating electricity. Guo et al. (2016) verified that AnMBR can be utilized to achieve renewable sources of energy with no extra energy consumption. Lin et al. (2013) verified a benefit value of 341,640 US$ annually from energy recovery from a full-scale AnMBR.

In spite of the high quality of treated effluent in terms of organic and nutrients removal, MBR only achieved fully nitrification and denitrification with the addition of an anoxic tank prior to a circular aeration tank (Guo et al., 2016). The high removal efficiency of nutrients such as TN and TP shows the poor performance of AnMBR due to the absence of anoxic and aerobic zones.

In terms of operating parameters, the anaerobic process is operated in a high temperature ranging from 85°F to 95°F (30–35°C), which seems to be a disadvantage of the technology. Besides good function in high temperature at mesophilic and thermophilic (50–60°C), methanogenic process is impacted in psychrophilic (<30°C) (Guo et al., 2016). In contrast, aerobic MBR obtained good performance at 25°C (Kris and Ghawi, 2008).

Along with the pros and cons of both technologies in terms of removal efficiency, energy recovery, effects of operating parameters, biomass growth, the comparison of fouling control is also considerable. It was verified that SRT has a negative effect on both of the membrane technologies, resulting in cake layer resistance (Wang et al., 2018). The SMPs are a result of membrane fouling. In the processes, SMPs accumulate to fill in the gap of the microbial flocs, leading to the decrease in permeability (Du et al., 2020). It was concluded that the permeability in AnMBR seems to be lower than for aerobic MBR in spite of the lower presence of SMPs in AnMBR than aerobic MBR (Judd, 2011). In terms of membrane cleaning, both the physical and chemical methods have similar effectiveness on aerobic MBR and AnMBR (Wang et al., 2018).

TABLE 2 Comparison of aerobic MBR and AnMBR.

Categories	Aerobic MBR	AnMBR
Organic removal efficiency	High	High
Nutrient removal	High	Low
Operational manner	Simple	Complicated
Start-up time	Short	Long
Energy requirement	High	Low
Fouling accumulation	Higher	Lower
Salinity tolerance	Low	High
Sludge yield	High	Low
Energy recovery	No	Yes

In summary, the AnMBR technology has a more significant advantage than the aerobic one. Despite the temperature sensitivity and challenges in operation, the advantages of AnMBR technology are outweighed. The pros and cons summary of the two membrane methods is described in Table 2

3. Conclusions and perspectives

MBR is known as the major method to deal with different types of wastewaters, especially in the fishery industry. Likewise, AnMBR, the modified technology from its family, exposed efficient treatment for high organic compounds waste stream and the ability of biogas production and energy recovery. However, the application from this kind of technology is currently limited in seafood processing wastewater treatment, but it still has been used to adapt with saline wastewater. Further research of AnMBR is necessary for the seafood industry wastewater. Besides, AnMBR technology is still a promising method to adapt the current situation of environmental issues due to some remarks mentioned below:

- Wide range of application in different types of wastewaters containing a high organic loading rate including: pulp and paper industries, food processing industries, pharmaceutical industries, seafood processing industries, etc.
- High adaptation in saline wastewater.
- Along with the complexity of operational parameter control, it can be conducted at low investment cost, require less space, and reduce operation cost.
- The ability of biogas production can be utilized for energy supply and heat. The method is called a future green bioprocess.

Acknowledgments

This research is funded by the Vietnam Ministry of Science and Technology (MOST) under the Vietnam Italia bilateral project with grant number NĐT/IT/21/22.

References

Afonso, M.D., Bórquez, R., 2002. Review of the treatment of seafood processing wastewaters and recovery of proteins therein by membrane separation processes—prospects of the ultrafiltration of wastewaters from the fish meal industry. Desalination 142 (1), 29–45. https://doi.org/10.1016/S0011-9164(01)00423-4.

Aquino, S.F., Stuckey, D.C., 2008. Integrated model of the production of soluble microbial products (SMP) and extracellular polymeric substances (EPS) in anaerobic chemostats during transient conditions. Biochem. Eng. J. 38 (2), 138–146. https://doi.org/10.1016/j.bej.2007.06.010.

Artiga, P., García-Toriello, G., Méndez, R., Garrido, J.M., 2008. Use of a hybrid membrane bioreactor for the treatment of saline wastewater from a fish canning factory. Desalination 221 (1–3), 518–525. https://doi.org/10.1016/j.desal.2007.01.112.

Attiogbe, F., 2013. Comparison of membrane bioreactor technology and conventional activated sludge system for treating bleached Kraft mill effluent. African J. Environ. Sci. Technol. Full 7 (5), 292–306. https://doi.org/10.5897/AJEST2013.1429.

Baker, R., 2004. Membrane Technology and Application, second ed. John Wiley & Sons, Ltd, https://doi.org/10.1016/C2009-0-19129-8.

Balcıoğlu, G., Yilmaz, G., Gönder, Z.B., 2020. Evaluation of anaerobic membrane bioreactor (AnMBR) treating confectionery wastewater at long-term operation under different organic loading rates: performance and Gökhan Balcıoğlu, Gulsum Yilmaz, Z. Beril Gönder. Chem. Eng. J., 126261. https://doi.org/10.1016/j.cej.2020.126261.

Bassin, J.P., Kleerebezem, R., Muyzer, G., Rosado, A.S., Van Loosdrecht, M.C.M., Dezotti, M., 2012. Effect of different salt adaptation strategies on the microbial diversity, activity, and settling of nitrifying sludge in sequencing batch reactors. Appl. Microbiol. Biotechnol. 93 (3), 1281–1294. https://doi.org/10.1007/s00253-011-3428-7.

Bird, B.R., Stewart, W.E., & Lightfoot, E.N. (2002). Transport phenomena. In Advances in Contraceptive Delivery Systems (Second, vol. 1, Issue 2). John Wiley & Son, Inc.

Chaiprapat, S., Thongsai, A., Charnnok, B., Khongnakorn, W., Bae, J., 2016. Influences of liquid, solid, and gas media circulation in anaerobic membrane bioreactor (AnMBR) as a post treatment alternative of aerobic system in seafood industry. J. Membr. Sci. 509, 116–124. https://doi.org/10.1016/j.memsci.2016.02.029.

Chan, M.T., Fane, A.G., Matheickal, J.T., Sheikholeslami, R., 2005. Membrane distillation crystallization of concentrated salts - flux and crystal formation. J. Membr. Sci. 257 (1–2), 144–155. https://doi.org/10.1016/j.memsci.2004.09.051.

Chang, S., 2014. Anaerobic membrane bioreactors (AnMBR) for wastewater treatment. Adv. Chem. Eng. Sci. 04 (01), 56–61. https://doi.org/10.4236/aces.2014.41008.

Chen, G., van Loodrecht, M.C., Ekama, G.A., & Brdjanovic, D. (2020). Biological Wastewater Treatment: Principles, Modeling and Design (second ed.). IWA Publishing. https://books.google.com.vn/books?hl=en&lr=&id=0Wn2DwAAQBAJ&oi=fnd&pg=PP1&dq=Biological+Wastewater+Treatment:+2nd+edition&ots=fIn0qdhWJs&sig=xrf5Hhk9VF-5XOtlphx1BPx1zx4&redir_esc=y#v=onepage&q=Biological wastewater treatment%3A 2nd edition&f=false.

Sridang, P.C., Kaiman, J., Pottier, A., Wisniewski, C., 2006. Benefits of MBR in seafood wastewater treatment and water reuse: study case in southern part of Thailand. Desalination 200 (1–3), 712–714. https://doi.org/10.1016/j.desal.2006.03.509.

Choo, K.H., Kang, I.J., Yoon, S.H., Park, H., Kim, J.H., Adiya, S., Lee, C.H., 2000. Approaches to membrane fouling control in anaerobic membrane bioreactors. Water Sci. Technol. 41 (10 – 11), 363–371. https://doi.org/10.2166/wst.2000.0681.

Chowdhury, P., Viraraghavan, T., Srinivasan, A., 2010. Biological treatment processes for fish processing wastewater—A review. Bioresour. Technol. 101 (2), 439–449. https://doi.org/10.1016/j.biortech.2009.08.065.

Christensen, M.L., Niessen, W., Sørensen, N.B., Hansen, S.H., Jørgensen, M.K., Nielsen, P.H., 2018. Sludge fractionation as a method to study and predict fouling in MBR systems. Sep. Purif. Technol. 194, 329–337. https://doi.org/10.1016/j.seppur.2017.11.055.

Dan, N.P., Visvanathan, C., Basu, B., 2003. Comparative evaluation of yeast and bacterial treatment of high salinity wastewater based on biokinetic coefficients. Bioresour. Technol. 87 (1), 51–56. https://doi.org/10.1016/S0960-8524(02)00204-3.

Di Bella, G., Di Trapani, D., Torregrossa, M., Viviani, G., 2013. Performance of a MBR pilot plant treating high strength wastewater subject to salinity increase: analysis of biomass activity and fouling behaviour. Bioresour. Technol. 147, 614–618. https://doi.org/10.1016/j.biortech.2013.08.025.

Du, X., Shi, Y., Jegatheesan, V., Ul Haq, I., 2020. A review on the mechanism, impacts and control methods of membrane fouling in MBR system. In. Membranes 10 (2). https://doi.org/10.3390/membranes10020024.

Foglia, A., Akyol, Ç., Frison, N., Katsou, E., Eusebi, A.L., Fatone, F., 2020. Long-term operation of a pilot-scale anaerobic membrane bioreactor (AnMBR) treating high salinity low loaded municipal wastewater in real environment. Sep. Purif. Technol. 236 (October), 116279. https://doi.org/10.1016/j.seppur.2019.116279.

Gagliano, M.C., Sudmalis, D., Pei, R., Temmink, H., Plugge, C.M., 2020. Microbial community drivers in anaerobic granulation at high salinity. Front. Microbiol. 11 (February), 1–15. https://doi.org/10.3389/fmicb.2020.00235.

Galib, M., Elbeshbishy, E., Reid, R., Hussain, A., Lee, H.S., 2016. Energy-positive food wastewater treatment using an anaerobic membrane bioreactor (AnMBR). J. Environ. Manage. 182, 477–485. https://doi.org/10.1016/j.jenvman.2016.07.098.

Gao, M., Yang, M., Li, H., Yang, Q., Zhang, Y., 2004. Comparison between a submerged membrane bioreactor and a conventional activated sludge system on treating ammonia-bearing inorganic wastewater. J. Biotechnol. 108 (3), 265–269. https://doi.org/10.1016/j.jbiotec.2003.12.002.

Gao, F., Peng, Y.Y., Li, C., Yang, G.J., Deng, Y.B., Xue, B., Guo, Y.M., 2018. Simultaneous nutrient removal and biomass/lipid production by chlorella sp. in seafood processing wastewater. Sci. Total Environ. 640–641, 943–953. https://doi.org/10.1016/j.scitotenv.2018.05.380.

García-Ruiz, M.J., Castellano-Hinojosa, A., González-López, J., Osorio, F., 2018. Effects of salinity on the nitrogen removal efficiency and bacterial community structure in fixed-bed biofilm CANON bioreactors. Chem. Eng. J. 347, 156–164. https://doi.org/10.1016/j.cej.2018.04.067.

Gong, H., Jin, Z., Xu, H., Yuan, Q., Zuo, J., Wu, J., Wang, K., 2019. Enhanced membrane-based pre-concentration improves wastewater organic matter recovery: pilot-scale performance and membrane fouling. J. Clean. Prod. 206, 307–314. https://doi.org/10.1016/j.jclepro.2018.09.209.

Goswami, L., Vinoth Kumar, R., Borah, S.N., Arul Manikandan, N., Pakshirajan, K., Pugazhenthi, G., 2018. Membrane bioreactor and integrated membrane bioreactor systems for micropollutant removal from wastewater: A review. J. Water Process Eng. 26 (October), 314–328. https://doi.org/10.1016/j.jwpe.2018.10.024.

Guo, W., Ngo, H.H., Chen, C., Pandey, A., Tung, K.L., Lee, D.J., 2016. Anaerobic membrane bioreactors for future green bioprocesses. In: Green Technologies for Sustainable Water Management., https://doi.org/10.1061/9780784414422.ch25.

Hemmelmann, A., Torres, A., Vergara, C., Azocar, L., Jeison, D., 2012. Application of anaerobic membrane bioreactors for the treatment of protein-containing wastewaters under saline conditions. J. Chem. Technol. Biotechnol. https://doi.org/10.1002/jctb.3882.

Hu, A.Y., Stuckey, D.C., 2006. Treatment of dilute wastewaters using a novel submerged anaerobic membrane bioreactor. J. Environ. Eng. 132 (2), 190–198. https://doi.org/10.1061/(asce)0733-9372(2006)132:2(190).

Huang, X., Liu, R., Qian, Y., 2000. Behaviour of soluble microbial products in a membrane bioreactor. Process Biochem. 36 (5), 401–406. https://doi.org/10.1016/S0032-9592(00)00206-5.

Iorhemen, O.T., Hamza, R.A., Tay, J.H., 2016. Membrane bioreactor (Mbr) technology for wastewater treatment and reclamation: membrane fouling. Membranes 6 (2), 13–16. https://doi.org/10.3390/membranes6020033.

Ismail, S.B., Gonzalez, P., Jeison, D., Van Lier, J.B., 2008. Effects of high salinity wastewater on methanogenic sludge bed systems. Water Sci. Technol. 58 (10), 1963–1970. https://doi.org/10.2166/wst.2008.528.

Ismail, S.B., de La Parra, C.J., Temmink, H., van Lier, J.B., 2009. Extracellular polymeric substances (EPS) in upflow anaerobic sludge blanket (UASB) reactors operated under high salinity conditions. Water Res. 44 (6), 1909–1917. https://doi.org/10.1016/j.watres.2009.11.039.

Jamal, M.T., Pugazhendi, A., Jeyakumar, R.B., 2020. Application of halophiles in air cathode MFC for seafood industrial wastewater treatment and energy production under high saline condition. Environ. Technol. Innov. 20, 101119. https://doi.org/10.1016/j.eti.2020.101119.

Jang, D., Hwang, Y., Shin, H., Lee, W., 2013. Effects of salinity on the characteristics of biomass and membrane fouling in membrane bioreactors. Bioresour. Technol. 141, 50–56. https://doi.org/10.1016/j.biortech.2013.02.062.

B. Applications of membrane technology for water and wastewater treatment

Jeison, D., 2007. Anaerobic Membrane Bioreactors for Wastewater Treatment: Feasibility and Potential Application. PhD thesis, Wageningen University, Wageningen, The Netherlands.

Johir, M.A.H., Vigneswaran, S., Kandasamy, J., BenAim, R., Grasmick, A., 2013. Effect of salt concentration on membrane bioreactor (MBR) performances: detailed organic characterization. Desalination 322, 13–20. https://doi.org/10.1016/j.desal.2013.04.025.

Judd, S., 2011. Once again, for Oliver and Samuel. And also for our family e Ivor and Margaret, Lorna, Ciss, Robert and Jane, Daisy and Heyes, John and Patricia, Lucy, Cameron and Dynamite. www.elsevier.com.

Judd, S.J., 2016. The status of industrial and municipal effluent treatment with membrane bioreactor technology. Chem. Eng. J. 305, 37–45. https://doi.org/10.1016/j.cej.2015.08.141.

Kara, F., Gurakan, G.C., Sanin, F.D., 2008. Monovalent cations and their influence on activated sludge floc chemistry, structure, and physical characteristics. Biotechnol. Bioeng. 100 (2), 231–239. https://doi.org/10.1002/bit.21755.

Kong, Z., Wu, J., Rong, C., Wang, T., Li, L., Luo, Z., Ji, J., Hanaoka, T., Sakemi, S., Ito, M., Kobayashi, S., Kobayashi, M., Qin, Y., Li, Y.Y., 2021. Large pilot-scale submerged anaerobic membrane bioreactor for the treatment of municipal wastewater and biogas production at 25°C. Bioresour. Technol. 319 (September 2020). https://doi.org/10.1016/j.biortech.2020.124123.

Kris, J., Ghawi, A.H., 2008. Study the Effect of Temperature on the Performance of Hollow. (April).

Lay, W.C.L., Liu, Y., Fane, A.G., 2010. Impacts of salinity on the performance of high retention membrane bioreactors for water reclamation: A review. Water Res. 44 (1), 21–40. https://doi.org/10.1016/j.watres.2009.09.026.

Le-Clech, P., Chen, V., Fane, T.A.G., 2006. Fouling in membrane bioreactors used in wastewater treatment. J. Membr. Sci. 284 (1–2), 17–53. https://doi.org/10.1016/j.memsci.2006.08.019.

Lefebvre, O., Moletta, R., 2006. Treatment of organic pollution in industrial saline wastewater: A literature review. Water Res. 40 (20), 3671–3682. https://doi.org/10.1016/j.watres.2006.08.027.

Lefebvre, O., Quentin, S., Torrijos, M., Godon, J.J., Delgenès, J.P., Moletta, R., 2007. Impact of increasing NaCl concentrations on the performance and community composition of two anaerobic reactors. Appl. Microbiol. Biotechnol. 75 (1), 61–69. https://doi.org/10.1007/s00253-006-0799-2.

Li, J., Jiang, C., Shi, W., Song, F., He, D., Miao, H., Wang, T., Deng, J., Ruan, W., 2018. Polytetrafluoroethylene (PTFE) hollow fiber AnMBR performance in the treatment of organic wastewater with varying salinity and membrane cleaning behavior. Bioresour. Technol. 267 (July), 363–370. https://doi.org/10.1016/j.biortech.2018.07.063.

Li, C., Li, X., Qin, L., Wu, W., Meng, Q., Shen, C., Zhang, G., 2019. Membrane photo-bioreactor coupled with heterogeneous Fenton fluidized bed for high salinity wastewater treatment: pollutant removal, photosynthetic bacteria harvest and membrane anti-fouling analysis. Sci. Total Environ. 696, 133953. https://doi.org/10.1016/j.scitotenv.2019.133953.

Lin, H., Liao, B.Q., Chen, J., Gao, W., Wang, L., Wang, F., Lu, X., 2011. New insights into membrane fouling in a submerged anaerobic membrane bioreactor based on characterization of cake sludge and bulk sludge. Bioresour. Technol. 102 (3), 2373–2379. https://doi.org/10.1016/j.biortech.2010.10.103.

Lin, H., Peng, W., Zhang, M., Chen, J., Hong, H., Zhang, Y., 2013. A review on anaerobic membrane bioreactors: applications, membrane fouling and future perspectives. Desalination 314, 169–188. https://doi.org/10.1016/j.desal.2013.01.019.

Liu, J., Eng, C.Y., Ho, J.S., Chong, T.H., Wang, L., Zhang, P., Zhou, Y., 2019. Quorum quenching in anaerobic membrane bioreactor for fouling control. Water Res. 156, 159–167. https://doi.org/10.1016/j.watres.2019.03.029.

Luo, W., Phan, H.V., Hai, F.I., Price, W.E., Guo, W., Ngo, H.H., Yamamoto, K., Nghiem, L.D., 2016. Effects of salinity build-up on the performance and bacterial community structure of a membrane bioreactor. Bioresour. Technol. 200, 305–310. https://doi.org/10.1016/j.biortech.2015.10.043.

Luo, L., Zhou, W., Yuan, Y., Zhong, H., Zhong, C., 2021. Effects of salinity shock on simultaneous nitrification and denitrification by a membrane bioreactor: performance, sludge activity, and functional microflora. Sci. Total Environ. 801. https://doi.org/10.1016/j.scitotenv.2021.149748.

Maaz, M., Yasin, M., Aslam, M., Kumar, G., Atabani, A.E., Idrees, M., Anjum, F., Jamil, F., Ahmad, R., Khan, A.L., Lesage, G., Heran, M., Kim, J., 2019. Anaerobic membrane bioreactors for wastewater treatment: novel configurations, fouling control and energy considerations. Bioresour. Technol. 283, 358–372. https://doi.org/10.1016/j.biortech.2019.03.061.

Moussa, M.S., Sumanasekera, D.U., Ibrahim, S.H., Lubberding, H.J., Hooijmans, C.M., Gijzen, H.J., Van Loosdrecht, M.C.M., 2006. Long term effects of salt on activity, population structure and floc characteristics in enriched bacterial cultures of nitrifiers. Water Res. 40 (7), 1377–1388. https://doi.org/10.1016/j.watres.2006.01.029.

Ng, H.Y., Ong, S.L., Ng, W.J., 2005. Effects of sodium chloride on the performance of a sequencing batch reactor. J. Environ. Eng. 131 (11), 1557–1564. https://doi.org/10.1061/(asce)0733-9372(2005)131:11(1557).

Nguyen, T.D.P., Nguyen, D.H., Lim, J.W., Chang, C.K., Leong, H.Y., Tran, T.N.T., Vu, T.B.H., Nguyen, T.T.C., Show, P.L., 2019a. Investigation of the relationship between bacteria growth and lipid production cultivating of microalgae *Chlorella vulgaris* in seafood wastewater. Energies 12 (12). https://doi.org/10.3390/en12122282.

Nguyen, T.D.P., Tran, T.N.T., Le, T.V.A., Nguyen Phan, T.X., Show, P.L., Chia, S.R., 2019b. Auto-flocculation through cultivation of *Chlorella vulgaris* in seafood wastewater discharge: influence of culture conditions on microalgae growth and nutrient removal. J. Biosci. Bioeng. 127 (4), 492–498. https://doi.org/10.1016/j.jbiosc.2018.09.004.

Oreopoulou, V., Russ, W., 2007. Utilization of by-products and treatment of waste in the food industry. In: Utilization of By-Products and Treatment of Waste in the Food Industry, pp. 1–316, https://doi.org/10.1007/978-0-387-35766-9. November.

Ozgun, H., Dereli, R.K., Ersahin, M.E., Kinaci, C., Spanjers, H., Van Lier, J.B., 2013. A review of anaerobic membrane bioreactors for municipal wastewater treatment: integration options, limitations and expectations. Sep. Purif. Technol. 118, 89–104. https://doi.org/10.1016/j.seppur.2013.06.036.

Panpong, K., Srisuwan, G., O-Thong, S., Kongjan, P., 2014. Anaerobic co-digestion of canned seafood wastewater with glycerol waste for enhanced biogas production. Energy Procedia 52, 328–336. https://doi.org/10.1016/j.egypro.2014.07.084.

Park, H.-D., Chang, I.-S., Lee, K.-J., 2015. Principles of Membrane Bioreactors for Wastewater Treatment., https://doi.org/10.1201/b18368.

Pendashteh, A.R., Abdullah, L.C., Fakhru'L-Razi, A., Madaeni, S.S., Zainal Abidin, Z., Awang Biak, D.R., 2012. Evaluation of membrane bioreactor for hypersaline oily wastewater treatment. Process Saf. Environ. Prot. 90 (1), 45–55. https://doi.org/10.1016/j.psep.2011.07.006.

Pham, T.T.H., Nguyen, T.M.H., 2020. A study to use activated sludge anaerobic combining aerobic for treatment of high salt seafood processing wastewater. Curr. Chem. Lett. 9 (2), 79–88. https://doi.org/10.5267/j.ccl.2019.8.002.

Pugazhendi, A., Al-Mutairi, A.E., Jamal, M.T., Jeyakumar, R.B., Palanisamy, K., 2020. Treatment of seafood industrial wastewater coupled with electricity production using air cathode microbial fuel cell under saline condition. Int. J. Energy Res. 44 (15), 12535–12545. https://doi.org/10.1002/er.5774.

Qiang, J., Zhou, Z., Wang, K., Qiu, Z., Zhi, H., Yuan, Y., Zhang, Y., Jiang, Y., Zhao, X., Wang, Z., Wang, Q., 2020. Coupling ammonia nitrogen adsorption and regeneration unit with a high-load anoxic/aerobic process to achieve rapid and efficient pollutants removal for wastewater treatment. Water Res. 170, 115280. https://doi.org/10.1016/j.watres.2019.115280.

Qin, L., Zhang, Y., Xu, Z., Zhang, G., 2018. Advanced membrane bioreactors systems: new materials and hybrid process design. Bioresour. Technol. 269, 476–488. https://doi.org/10.1016/j.biortech.2018.08.062.

Reid, E., Liu, X., Judd, S.J., 2006. Effect of high salinity on activated sludge characteristics and membrane permeability in an immersed membrane bioreactor. J. Membr. Sci. 283 (1–2), 164–171. https://doi.org/10.1016/j.memsci.2006.06.021.

Robles, Á., Durán, F., Giménez, J.B., Jiménez, E., Ribes, J., Serralta, J., Seco, A., Ferrer, J., Rogalla, F., 2020. Anaerobic membrane bioreactors (AnMBR) treating urban wastewater in mild climates. Bioresour. Technol. 314 (July), 123763. https://doi.org/10.1016/j.biortech.2020.123763.

Rodriguez-Sanchez, A., Leyva-Diaz, J.C., Poyatos, J.M., Gonzalez-Lopez, J., 2019. Influent salinity conditions affect the bacterial communities of biofouling in hybrid MBBR-MBR systems. J. Water Process Eng. 30 (March 2018), 100650. https://doi.org/10.1016/j.jwpe.2018.07.001.

Rosenberger, S., Laabs, C., Lesjean, B., Gnirss, R., Amy, G., Jekel, M., Schrotter, J.C., 2006. Impact of colloidal and soluble organic material on membrane performance in membrane bioreactors for municipal wastewater treatment. Water Res. 40 (4), 710–720. https://doi.org/10.1016/j.watres.2005.11.028.

Sari Erkan, H., Bakaraki Turan, N., Önkal Engin, G., 2018. Membrane bioreactors for wastewater treatment. Compr. Anal. Chem. 81, 151–200. https://doi.org/10.1016/bs.coac.2018.02.002.

Schäfer, A.I., Andritsos, N., Anastasios, J., Hoek, E.M.V., Schneider, R., 2004. Chapter 8 Fouling in nanofiltration. In: Nanofiltration—Principles and applications. Elsevier, pp. 169–239.

Schneider, C., Sathyadev, R., Zarebska, A., Tsapekos, P., Hélix-nielsen, C., 2019. Science of the Total environment treating anaerobic ef fl uents using forward osmosis for combined water purification and biogas production. Sci. Total Environ. 647, 1021–1030. https://doi.org/10.1016/j.scitotenv.2018.08.036.

Shahid, M.K., Kashif, A., Rout, P.R., Aslam, M., Fuwad, A., Choi, Y., Banu, J., Park, J.H., Kumar, G., 2020. A brief review of anaerobic membrane bioreactors emphasizing recent advancements, fouling issues and future perspectives. J. Environ. Manage. 270 (June), 110909. https://doi.org/10.1016/j.jenvman.2020.110909.

B. Applications of membrane technology for water and wastewater treatment

Sierra, M., David, J., Lafita, C., Gabaldón, C., Spanjers, H., van Lier, J.B., 2017. Trace metals supplementation in anaerobic membrane bioreactors treating highly saline phenolic wastewater. Bioresour. Technol. 234, 106–114. https://doi.org/10.1016/j.biortech.2017.03.032.

Sierra, M., Julian, D., Oosterkamp, M.J., Wang, W., Spanjers, H., van Lier, J.B., 2018. Impact of long-term salinity exposure in anaerobic membrane bioreactors treating phenolic wastewater: performance robustness and endured microbial community. Water Res. 141, 172–184. https://doi.org/10.1016/j.watres.2018.05.006.

Song, X., McDonald, J., Price, W.E., Khan, S.J., Hai, F.I., Ngo, H.H., Guo, W., Nghiem, L.D., 2016. Effects of salinity build-up on the performance of an anaerobic membrane bioreactor regarding basic water quality parameters and removal of trace organic contaminants. Bioresour. Technol. 216, 399–405. https://doi.org/10.1016/j.biortech.2016.05.075.

Sridang, P.C., Pottier, A., Wisniewski, C., Grasmick, A., 2008. Performance and microbial surveying in submerged membrane bioreactor for seafood processing wastewater treatment. J. Membr. Sci. 317 (1–2), 43–49. https://doi.org/10.1016/j.memsci.2007.11.011.

Stuckey, D.C., 2012. Recent developments in anaerobic membrane reactors. Bioresour. Technol. 122, 137–148. https://doi.org/10.1016/j.biortech.2012.05.138.

Sudmalis, D., Gagliano, M.C., Pei, R., Grolle, K., Plugge, C.M., Rijnaarts, H.H.M., Zeeman, G., Temmink, H., 2018. Fast anaerobic sludge granulation at elevated salinity. Water Res. 128, 293–303. https://doi.org/10.1016/j.watres.2017.10.038.

Sun, C., Leiknes, T.O., Weitzenböck, J., Thorstensen, B., 2010. Salinity effect on a biofilm-MBR process for shipboard wastewater treatment. Sep. Purif. Technol. 72 (3), 380–387. https://doi.org/10.1016/j.seppur.2010.03.010.

Tan, X., Acquah, I., Liu, H., Li, W., Tan, S., 2019. A critical review on saline wastewater treatment by membrane bioreactor (MBR) from a microbial perspective. Chemosphere 220, 1150–1162. https://doi.org/10.1016/j.chemosphere.2019.01.027.

Tchobanoglous, G., Stensel, H.D., Tsuchihashi, R., & Burton, F. (2014). Metcalf and Eddy, AECOM—wastewater engineering_ treatment and Resource recovery (fifth ed.).Pdf. McGraw-Hill Education, 2 Penn Plaza, New York, NY 10121. Inc. Metcalf & Eddy. Wastewater Engineering: treatment and reuse (p. iv). McGraw-Hill Education. Kindle Edition.

Thomas, S., Harindranathan Nair, M., Singh, I., 2015. Physicochemical analysis of seafood processing effluents in Aroor Gramapanchayath, Kerala. IOSR J. Environ. Sci. Ver. III 9 (6), 2319–2399. https://doi.org/10.9790/2402-09633844.

Tian, Y., Su, X., 2012. Relation between the stability of activated sludge flocs and membrane fouling in MBR: under different SRTs. Bioresour. Technol. 118, 477–482. https://doi.org/10.1016/j.biortech.2012.05.072.

Tiron, L.G., Pintilie, C., Vlad, M., Birsan, I.G., Baltă., 2017. Characterization of Polysulfone membranes prepared with thermally induced phase separation technique. IOP Conf. Ser.: Mater. Sci. Eng. 209 (1). https://doi.org/10.1088/1757-899X/209/1/012013.

van Lier, J.B., Mahmoud, N., Zeeman, G., 2008. Anaerobic wastewater treatment. IWA 151 (4), 402–442. https://doi.org/10.4064/aa151-4-5.

von Sperling, M., 2007. Basic Priniples of Wastewater Treatment. vol. 2 IWA.

Vyrides, I., Stuckey, D.C., 2009. Effect of fluctuations in salinity on anaerobic biomass and production of soluble microbial products (SMPs). Biodegradation 20 (2), 165–175. https://doi.org/10.1007/s10532-008-9210-6.

Vyrides, I., Stuckey, D.C., 2011. Fouling cake layer in a submerged anaerobic membrane bioreactor treating saline wastewaters: curse or a blessing? Water Sci. Technol. 63 (12), 2902–2908. https://doi.org/10.2166/wst.2011.461.

Wang, K., Garcia, N.M., Soares, A., Jefferson, B., McAdam, E.J., 2018. Comparison of fouling between aerobic and anaerobic MBR treating municipal wastewater. H2Open J. 1 (2), 131–159. https://doi.org/10.2166/h2oj.2018.109.

Xie, K., Xia, S., Song, J., Li, J., Qiu, L., Wang, J., Zhang, S., 2014. The effect of salinity on membrane fouling characteristics in an intermittently aerated membrane bioreactor. J. Chem. 2014. https://doi.org/10.1155/2014/765971.

Yogalakshmi, K.N., Joseph, K., 2010. Effect of transient sodium chloride shock loads on the performance of submerged membrane bioreactor. Bioresour. Technol. 101 (18), 7054–7061. https://doi.org/10.1016/j.biortech.2010.03.135.

Zhang, L., Lv, W., Li, S., Geng, Z., Yao, H., 2019. Nitrogen removal characteristics and comparison of the microbial community structure in different anaerobic ammonia oxidation reactors. Water (Switzerland) 11 (2). https://doi.org/10.3390/w11020230.

Zhou, H.X., Pang, X., 2018. Electrostatic interactions in protein structure, folding, binding, and condensation. Chem. Rev. 118 (4), 1691–1741. https://doi.org/10.1021/acs.chemrev.7b00305.

Zhou, Z., Tan, Y., Xiao, Y., Stuckey, D.C., 2016. Characterization and significance of sub-visible particles and colloids in a submerged anaerobic membrane bioreactor (sanmbr). Environ. Sci. Technol. 50 (23), 12750–12758. https://doi.org/10.1021/acs.est.6b03581.

Zhou, Z., Tao, Y., Zhang, S., Xiao, Y., Meng, F., Stuckey, D.C., 2019. Size-dependent microbial diversity of sub-visible particles in a submerged anaerobic membrane bioreactor (SAnMBR): implications for membrane fouling. Water Res. 159, 20–29. https://doi.org/10.1016/j.watres.2019.04.050.

6

Ultralow pressure membrane filtration for water and wastewater treatment

Chew Lee Leong[a], Muhammad Roil Bilad[b], Norazanita Shamsuddin[b], Hazwani Suhaimi[b], Nasrul Arahman[c], Adewale Giwa[d], and Ahmed Yusuf[d]

[a]Department of Chemical Engineering, Universiti Teknologi PETRONAS, Bandar Seri Iskandar, Malaysia [b]Faculty of Integrated Technologies, Universiti Brunei Darussalam, Gadong, Brunei [c]Department of Chemical Engineering, Engineering Faculty, Syiah Kuala University, Banda Aceh, Indonesia [d]Department of Chemical Engineering, Khalifa University of Science and Technology, Abu Dhabi, United Arab Emirates

1. Introduction

Ultrafiltration (UF) is an established technology for drinking water treatment. Initially, it was used for turbidity and pathogen removals. The cost of the membrane is seen as the main constraint in operating UF; as such, it operates at a high flux and high pressure (>1 bar) to make it economical and by incorporating means for limiting membrane fouling. The system is thus complex to accommodate physical and chemical cleanings. The use of high pressure driven by a pump produces high flux yet promotes fouling. High flux drags the foulant, allows it to accumulate, consolidate, and restrict the permeate flow. Membrane fouling requires intensive cleaning using chemicals or mechanical disinfection. Biofilm formation on the membrane surface is the main culprit in membrane fouling and becomes a water filtration limitation.

The significant drop in membrane price has changed the operation of UF. The flux can be slightly lowered to simplify the process (i.e., less chemical cleaning). The pumping (energy) cost is lowered by lowering the transmembrane pressure (ΔP) of 0.2–1.0 bar to achieve acceptable fluxes of 50–100 $L/m^2 h$ (LMH) (Crittenden et al., 2012). Under this operation, the control of membrane fouling is still indispensable to achieve a sustainable operation. Peter-Varbanets

Copyright © 2023 Elsevier Inc. All rights reserved.

et al. (2010) unraveled the phenomenon of biofilm-controlled flux stabilization. Their findings have shifted the paradigm in operating a dead-end UF. The filtration is run under operating pressures of 0.04–0.1 bar which offers a prolonged stable flux (of 2–10 LMH) without the need of any cleaning (physical or chemical). Since the typical time required to reach a stable flux is about 30 days, the prolonged operation is defined for at least 30 days. The biofilm formed on the membrane surface was attributed to dictates the hydraulic performance over a prolonged operation (Peter-Varbanets et al., 2017). The hydrostatic pressure provided the low pressure. The process was thus labeled as "gravity-driven membrane filtration" or "biofilm-controlled ultrafiltration" due to the crucial roles of biofilm in dictating the performance. However, because of the possibility of applying other sources of pressure (i.e., from pump work) and applying the system without biofilm, a more general term of ultralow pressure membrane filtration (ULPMF) is used in this chapter.

A concept of ULPMF has proven to offer a sustained filtration process without any membrane cleaning during the prolonged operation. It achieves a stable flux thanks to the biofilm's mutual role, which is naturally being a culprit causing the membrane fouling (Peter-Varbanets et al., 2010). Treated wastewater also experiences additional treatment by the biofilm leading to enhanced permeate quality. Several reports emphasized that the transmembrane pressure (hereafter termed to as hydrostatic pressure driven by gravity) affected membrane permeability and is thus considered an essential parameter (Li et al., 2020; Wu et al., 2019a).

Applying ULPMF driven exclusively by gravity may offer stable flux when treating surface water. However, the low flux can be compensated for with a high membrane area. The biofilm on the membrane surface is treated as a secondary biodegradation process. Instead of constantly removing the biofilm for membrane fouling control, it can grow to emulate the traditionally attached growth system. It is known that the biofilm is formed on the membrane surface (Waqas et al., 2020).

Extensive research has been done to explore and understand ULPMF better. Recently, Pronk et al. (2019) comprehensively reviewed the progress on ULPMF research. This chapter builds from the earlier review to establish the fundamental knowledge and process and integrates recent findings in ULPMF. A few important factors were discussed in great detail. They included the dynamic ecosystem of the biofilm and its mechanism in stabilizing the flux, the parameters influencing the process, the roles of feed properties and feed pretreatment through process integration, and the new potential applications of ULPMF. Furthermore, the sustainability aspect of ULPMF and important recommendations are detailed for further research.

2. Terminology and applications of ultralow pressure membrane filtration (ULPMF)

2.1 Ultralow pressure

The term "ultralow pressure" in the ULPMF system is defined as transmembrane pressure of less than 0.1 bar. The use of the new terminology is important to distinguish it from the common microfiltration and ultrafiltration process. Such a low pressure is normally obtained from the feed hydrostatic pressure from the water head of <1.00 m. This definition excludes the submerged membrane bioreactors (MBRs) that typically work under low ΔP, some even driven by gravity (Ding et al., 2017a). Most research has been done using pressure driven by gravity to treat water or wastewater feeds. Therefore, the term gravity-driven membrane (GDM) filtration is mostly used. Due to the expansion of the concepts to other applications, and the possibility to impose pressure by different means, the term ultralow pressure is used in this chapter. The ultralow pressure is defined as the transmembrane pressure of less than 0.1 bar, corresponding to a feedwater head of <1.0 m.

2.2 ULPMF system and operation

Fig. 1 illustrates the basic system of ULPMF. It consists of a feed tank in which the membrane can be submerged and linked externally. The system operates under a constant hydrostatic pressure that can be adjusted by changing the level of feed. The membrane can be either a hollow fiber or a flat-sheet configuration. In many cases, the performance can be enhanced by dosing a small amount of coagulant/flocculant, adsorbent, and prefilter. The pressure can also be generated by pumping the feed to generate positive pressure or suction from the permeate side to generate negative pressure to drive the filtration.

The main advantage offered by the ULPMF system is the possibility of achieving a stable flux without membrane cleaning over a prolonged operation. The stable flux decreased significantly due to increased filtration resistance from the foulant layer accumulated atop the membrane surface, including biofilm growth. After a certain operation time, the flux reaches a stable value where the filtration resistance is constant. The permeate pump in Fig. 1B is used to evacuate the permeate and not to drive the filtration. Under these circumstances, the system still qualifies for the constant-pressure system.

The level of stable flux is closely related to the organic content in the feed and the biofilm's properties: morphology, living ecosystem and composition of organics/inorganics (i.e., exopolymer substances, EPS). The biofilm's properties are affected by the abundance of organic content in the feed and the presence of predators, as discussed in Section 3. More

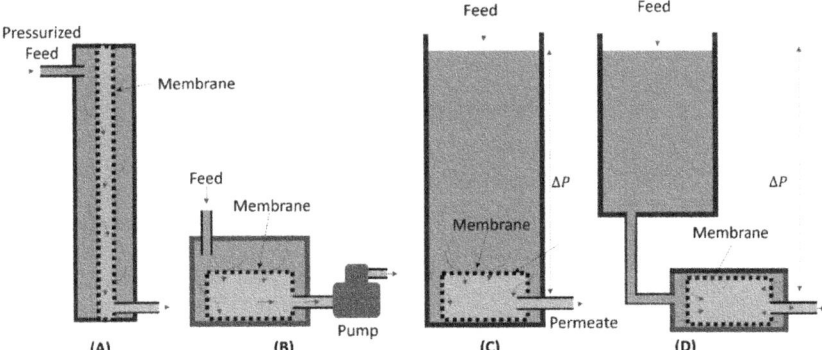

FIG. 1 Basic illustrations of the ultralow pressure membrane filtration (ULPMF) system driven by mechanical and hydrostatic pressure showing (A) prepressured feed, (B) vacuum system involving a permeate pump and gravity-driven with (C) internal and (D) external membrane placements.

importantly, biofouling enhances the permeate quality by improving the removals of organics [i.e., humic acid, algal toxin, biopolymers, assimilated organic matter (AOC)] through biodegradation. The value of stable flux can be altered by implementing a standstill period in an intermittent filtration or by incorporating feed pretreatment (i.e., coagulation/flocculation, enhancing biodegradation, or adsorption).

2.3 Applications

Different types of feedwater have been treated with ULPMF. They include river and pond water, gray water, diluted wastewater, and seawater (Table 1). The simple system and operation of ULPMF make it very attractive for small-scale applications in decentralized water

TABLE 1   Summary of the study of membrane filtration performance with different types of water sources.

Type of wastewater	Stable flux (L/m² h)	Membrane and pore size	Type of membrane module	Transmembrane pressure	Ref.
River water	10				Peter-Varbanets et al. (2010)
Diluted wastewater 12.5 mg TOC/L	4	PES 100 kDa	Flat sheet	65 mbar	
Pretreated river water by sand filtration	11.1	PES 100 kDa	Flat sheet	65 mbar	Peter-Varbanets et al. (2011)
Mixture of pond and tap water					
No fouling control	**			**	Oka et al. (2017)
Air sparging	**	PVDF 40 nm	Hollow fiber	**	
Backwashing	**			**	
Diluted wastewater					
7.5 mg COD/L	4	PES 20 kDa	Flat sheet	45 mbar	Wang et al. (2017)
15 mg COD/L	2				
Graywater	1.0	PVDF 0.2 μm	Hollow fiber	30 mbar	Jabornig and Podmirseg (2015)
Synthetic graywater	2.0	PES 150 kDa	Flat sheet	50 mbar	Ding et al. (2016)
Rainwater	6.0	PES 150 kDa	Flat sheet	50 mbar	Ding et al. (2017a)
Seawater	18.6	PVDF 80 nm	Flat sheet	40 mbar	Wu et al. (2017b)
Rainwater (with addition of nematodes)	18–20	HPS 100 kDa	Flat sheet	61.5 mbar	Klein et al. (2016)
Synthetic produced water	37.8	PVDF/PEG 9.34 ± 1.57 μm	Flat sheet	200 mbar	Mat Nawi et al. (2020)

TABLE 1 Summary of the study of membrane filtration performance with different types of water sources—cont'd

Type of wastewater	Stable flux (L/m² h)	Membrane and pore size	Type of membrane module	Transmembrane pressure	Ref.
Primary effluent wastewater (aeration rate of 0.17 L/min)	3.5	PVDF 150 kDa	Hollow fiber	30 mbar	Lee et al. (2019b)
Sewage wastewater	5	PVDF 0.08 μm	Hollow fiber	40 mbar	Liu et al. (2020)
Municipal wastewater	6.1	PP, 0.2 μm	Flat sheet	30 mbar	Hey et al. (2018)
Synthetic and real fermentation broth	100–600	SiC ceramic, 450 Da	Tubular monochannel	0.4–1.4 MPa	Woźniak and Prochaska (2014)
Municipal wastewater	41.7	α-Al₂O₃, 100 nm	Flat sheet	**	Zhao et al. (2020)
Graywater (from laundry) Tilting angle of 20° Aeration rate of 1.5 L/min	22.14 ± 10.8 19.96 ± 21.6	PVDF, 300 kDa	Flat sheet	0.1 bar	Barambu et al. (2020)
Landry wastewater The fouling was controlled by imposing coarse air bubbles	15–19	PSF, 0.01 μm	Flat sheet	0.05–0.1 bar	Bilad et al. (2020)
Algae-laden water treatment	7.1	PVDF, 0.45 μm	Flat sheet	0.06 bar	Jiang et al. (2021)
Tapioca Wastewater (in membrane fixed film bioreactor) Biological aeration was provided, but there was no fouling control	1–3	PAN, 0.01 μm	Hollow fiber	0.022–0.1 bar	Zainuddin et al. (2021)
Microalgae broth (*C. vulgaris*) Membrane fouling was controlled by imposing coarse bubble aeration.	5.7 = 50.0	PVDF, 0.14 μm	Flat sheet	0.025–0.19 bar	Wan Osman et al. (2021)
Microalgae broth (*Spirulina* sp.) Membrane fouling was controlled by aeration and panel tilting	10–38	PVDF, 0.42 μm and PSF, 0.04 μm	Flat sheet	0.1 bar	Ismail et al. (2021)
Activated sludge Membrane fouling was controlled by employing Patterned membrane on tilted and aerated module	44–84	PVDF, 0.5 (flat) and 0.78 (corrugated) μm	Flat sheet	0.2 bar	Osman et al. (2020)

Continued

B. Applications of membrane technology for water and wastewater treatment

TABLE 1   Summary of the study of membrane filtration performance with different types of water sources—cont'd

Type of wastewater	Stable flux (L/m² h)	Membrane and pore size	Type of membrane module	Transmembrane pressure	Ref.
Filtration of shale gas flowback and produced water	0.65–0.82	PVDF, <0.01 µm (MWCO 100 kDa)	Hollow fiber	0.04–0.16 bar	Chang et al. (2019)
Decentralized wastewater Membrane fouling was controlled by internal recirculation and membrane packing density optimization and addition of granular activated carbon	3.1–4.5	PVDF; 150 kDa	Hollow fiber	0.03 bar	Lee et al. (2021)
Lake water The flux was enhanced by a combination of membrane relaxation and shear stress	2.5–4.0	PES, 150 kDa	Flat sheet	0.055 bar	Shi et al. (2020)
River water The flux was enhanced by intermittent shear via cross-flow	12–15	PES, 150 kDa	Flat sheet	0.1 bar	Derlon et al. (2022)
Simulated micropolluted water Flux was enhanced by a biomimetic membrane system involving activated carbon, graphene oxide and carbon nanotube	Up to 70	PVDF, 0.3 µm (as base material)	Flat-sheet	0.012 bar	Zhu et al. (2020)
Roofing water The system was enhanced by incorporating electrochemical oxidation disinfection	8.0–12.0	PVDF, 0.45 µm	Flat sheet	0.04–0.06 bar	Du et al. (2022b)
Roofing rainwater Standard GDM without flux enhancement	2.4–4.0	PVDF, 0.45 µm	Flat sheet	0.04–0.06 bar	Du et al. (2019)
Riverwater Stable flux was enhanced by integrating granular activated carbon	7–12	PES, 150 kDa	Flat sheet	0.05 bar	Tang et al. (2021)
Riverwater Stable flux was enhanced by integrating biocarrier and intermittent aeration	0.9–3.5	PVDF, 150 kDa	Hollow fiber	0.03 bar	Lee et al. (2019b)

TABLE 1 Summary of the study of membrane filtration performance with different types of water sources—cont'd

Type of wastewater	Stable flux (L/m² h)	Membrane and pore size	Type of membrane module	Transmembrane pressure	Ref.
Wastewater Algae-polluted surface water had significantly lower filterability	1.45–9.42	PS, 100 kDa and PVDF, 0.1 μm	Flat sheet	0.075 bar	Truttmann et al. (2020)
Riverwater Stable flux was enhanced by addition of powdered activated carbon/zeolite layer	5.8–6.8 (PAC) 3.8–6.1 (zeolite)	PES, 150 kDa	Flat sheet	0.07 bar	Ding et al. (2021)
Roofing rainwater GDM was integrated with electrocoagulation	17.0–24.6	PVDF, <0.1 μm	Flat sheet	4–6 kPa	Xu et al. (2021a)
Model feed (0.2 g/L BSA) The stable flux was obtained by PVDF membrane development blending with amphiphilic multiarms polymer PPG-Si-PEG	4.6–12.1	PVDF, 38–59 nm	Flat sheet	4 kPa	Jiang et al. (2019b)
Simulated wastewater The stable flux was obtained by employing multifunctional microporous membranes	30 (control) 110 (modified)	PVDF, 0.45–1.2 μm	Flat sheet	2.3 kPa	Xu et al. (2021b)
Sinthetic raw water The stable flux was obtained by combination with electrocoagulation	22.7	Ceramic (MF)	Flat sheet	5–10 kPa	Du et al. (2021)
Tap water The stable flux was obtained by Packing granular activated carbon	4.3–52	PVDF, 150 kDa (30–50 nm)	Flat sheet	3.25 kPa	Schumann et al. (2020)
River water (regulated-biofilms by in situ coagulation combined with activated alumina filtration)	From 2 to 3.1–8.3	PVDF, 100 kDa	Hollow fiber	10 kPa	Du et al. (2022b)
Model feed (0.2 g/L BSA) (PVDF membrane development via in-situ formed rigid pore structure)	4.3–19.7	PVDF, 0.48–1.28 μm	Flat sheet	6 kPa	Jiang et al. (2022)

Continued

B. Applications of membrane technology for water and wastewater treatment

TABLE 1 Summary of the study of membrane filtration performance with different types of water sources—cont'd

Type of wastewater	Stable flux (L/m² h)	Membrane and pore size	Type of membrane module	Transmembrane pressure	Ref.
Synthetic groundwater (PVDF membrane development via grafted quaternary ammonium moieties)	32.2	PVDF, N/A	Flat sheet (electrospun)	3.13 kPa	Wan et al. (2022)
Diluted domestic sewage (improvement of predator intensity at higher temperature)	3.5–12.2	PES, 150 kDa	Flat sheet	6 kPa	Chen et al. (2021)
Primary wastewater effluent (periodical NaOCl cleaning)	Up to 10	PES, 100 kDa	Flat sheet	5.75 kPa	Guðjónsdóttir et al. (2022)
River water (space reduction for biofilm growth)	8.0 ± 0.8 (large) and 3.9 ± 1.2 (low)	PES, 0.02 μm	Hollow fiber	10.1 kPa	Stoffel et al. (2022)
Surface water surface water contaminated by Mn^{2+} and $NH^{3-}N$ (integration of iron-manganese co-oxide (FMO)	6	PVDF, 150 kDa	Flat sheet	6 kPa	Li et al. (2022)
Synthetic river water (novel housing design of spent RO membrane)	0.27	NF- or UF like from Spent RO	Flat sheet	16 kPa	García-Pacheco et al. (2021)
Primary wastewater (relaxation and air bubble scouring in membrane bioreactor mode)	≈ 1	PS, 20 kDa	Flat sheet	4 kPa	Fortunato et al. (2020)

Abbreviations: **, membrane systems operated at constant flux instead of constant pressure; BSA, bovine serum albumin; HPS, hydrophylized polysulfone; PAN, polyacrylonitrile; PEG, polyethlene glycol; PES, polyethersulfone; PP, polypropylene; PS, polysulfone; PVDF, polyvinylidene fluoride; SiC, silicon carbide; VIPS, vapor-induced phase separation; α-Al₂O₃, alpha-aluminium oxide.

and wastewater treatments. Due to its lower flux compared to traditional membrane filtration, the membrane material becomes the highest cost factor. However, the operation, maintenance, and axillary equipment costs are very low, making it highly attractive in remote areas even without electricity supply. ULPMF systems are low in energy footprint. Practically, there is no energy input in a household-scale system. However, larger-scale systems for drinking water production require pumping for feedwater transport up to the inlet of the system, which was very low at 0.006–0.04 kWh/m^3 (Boulestreau et al., 2012; Frechen et al., 2011).

Comprehensive information on the fundamentals of ULPMF has been reported. The system has been implemented mainly for drinking water treatment in decentralized applications, i.e., to serve the needs of households and small communities in rural areas with poor access to clean drinking water supply. Recently, the concept of ULPM that initially did not incorporate any cleaning has been expanded. Many reports demonstrated substantial improvement in stable flux value by employing periodic membrane cleaning or by developing membrane materials customized for ULPM.

2.3.1 Decentralized potable water treatment treating surface or rainwater

Simplification of membrane-based filtration systems has long been pursued, but few systems have reached the implementation stage. A portable, low-cost, gravity-driven ULPMF has been implemented to treat microbiological water serving households and communities. The system was equipped with a hollow fiber UF membrane and operated with periodical backwashing for daily production of 0.216 m^3 of clean water at a cost of ~US$1 (Naranjo et al., 2009). A small-scale ULPMF custom made from a membrane module with an area of 6 and 11 m^2 was developed for emergency relief by treating the surface water with a turbidity of 70 NTU (Frechen et al., 2011). It achieved a stable flux of around 5 LMH without cleaning by operating under hydrostatic pressure. Ultralow pressure can also be provided mechanically, which was the basis for developing the IGW Emergency Ultrafiltration system (www.gdpfilter.co.id). The systems were initially developed as a disaster emergency relief but have recently been commercialized for household uses. A much simpler gravity-driven system has also been explored for commercialization that was developed by Eawag (www.eawag.ch).

A unique hollow fiber module design that allows a simple mechanical cleaning has long been commercialized to serve small communities in rural areas (https://skyjuice.org.au/). However, the implementation of this system still follows the traditional view on having a high flux by facilitating a simple cleaning system through an innovative membrane module. The operation and the cleaning of the system were driven by hydrostatic pressure and manual cleaning, respectively.

A community-scale ULPMF with a 5 m^3/day capacity was developed to treat river water for drinking (Boulestreau et al., 2012). The system was equipped with a 40 m^2 membrane and operated without cleaning and regular blowdown to remove debris. A stable flux of 5–7 LMH was achieved when the feedwater turbidity was <160 NTU and deteriorated to 2–4 LMH when the feedwater turbidity was >600 NTU. A larger scale of the gravity-driven ULPMF system with a membrane area of 75 m^2 was run over a period of 1 year by treating lake water (Peter-Varbanets et al., 2017). Stable fluxes of 5.2–11.6 LMH were achieved under a hydrostatic pressure of 75–100 mbar. A rather low flux of 0.47 LMH was achieved by ULPMF combined with GAC for treating rainwater. It was equipped with a hollow fiber polysulfone membrane with a pore size of 0.1 μm operated for 60 days (Kus et al., 2013a, b).

2.3.2 Pretreatment of seawater desalination using reverse osmosis

A pilot scale gravity-driven ULPMF system reached a rather high steady-state flux of 18–20 LMH. Such performance was achieved during the pretreatment of seawater for reverse osmosis desalination by employing a transmemnrane pressure of 40 mbar (Wu et al., 2017b). The high flux was almost comparable to conventional UF, which was attributed to the low organic content in the feed and the abundance of predatory eukaryotes favorable to enhancing the biofilm porosity.

2.3.3 Filtration of wastewater and gray water

Table 1 summarizes research on the application of ULPMF for treatment of wastewater. Unlike for the surface water (rain/rivel), the achievable stable flux for these feeds was substantially low ($<5 \, \text{L/m}^2 \, \text{h}$). Therefore, attempts have been made to enhance the system throughput by incorporating membrane fouling control. They were done by implementing aeration with various modifications (module tilting, relaxation, etc.) (Barambu et al., 2020; Bilad et al., 2020; Ismail et al., 2021; Lee et al., 2019a, 2021; Mat Nawi et al., 2020; Shi et al., 2020; Wan Osman et al., 2021). Other works also developed membrane materials customized for the ULPMF systems by incorporation of biomimetic and hydrophilic additives (Jiang et al., 2019a; Wan et al., 2022; Zhu et al., 2020). A more holistic approach to deal with the severity of membrane fouling and low stable flux was to increase biological activity in the form of membrane bioreactors (Fortunato et al., 2020; Osman et al., 2020; Zainuddin et al., 2021).

3. Characteristics of ULPMF processes

3.1 The stable flux of ULPMF

The possibility of achieving a constant flux can be seen as the equilibrium process in the thermodynamics. The concept of stable flux in ULPMF is seen as the "threshold flux" in which the membrane fouling rate is considered to be low and thus can be accepted for long-term operation (Field and Pearce, 2011). In a classic definition, a stable flux is achieved when foulant buildup is balanced by the back transport of foulant promoted by the tangential shear on the feed side due to crossflow, bubbling, and other means of membrane fouling controls. In the context of ULPMF, an equilibrium condition can be achieved when the rate of foulant input equals the rate of biodegradation by the biofilm on the membrane surface, where the system reaches a quasi-equilibrium stage in which microscopic changes still occur microscopically within the system.

3.2 Enhanced organic removal

The dynamic layer of the foulant and biofilm atop the membrane surface in the ULPMF system is expected to enhance membrane rejection and perform biodegradation of organics, as demonstrated in earlier reports and summarized in Fig. 2. Sufficient research findings support this theory. Humic acid ($<1500 \, \text{kDa}$) could be removed by a factor of up to 40% by using a ULPMF system equipped with a UF membrane with $>>1 \, \text{kDa}$ MWCO for treatment of river water. Peter-Varbanets et al. (2010) suggested the active role of biofilm in the biodegradation

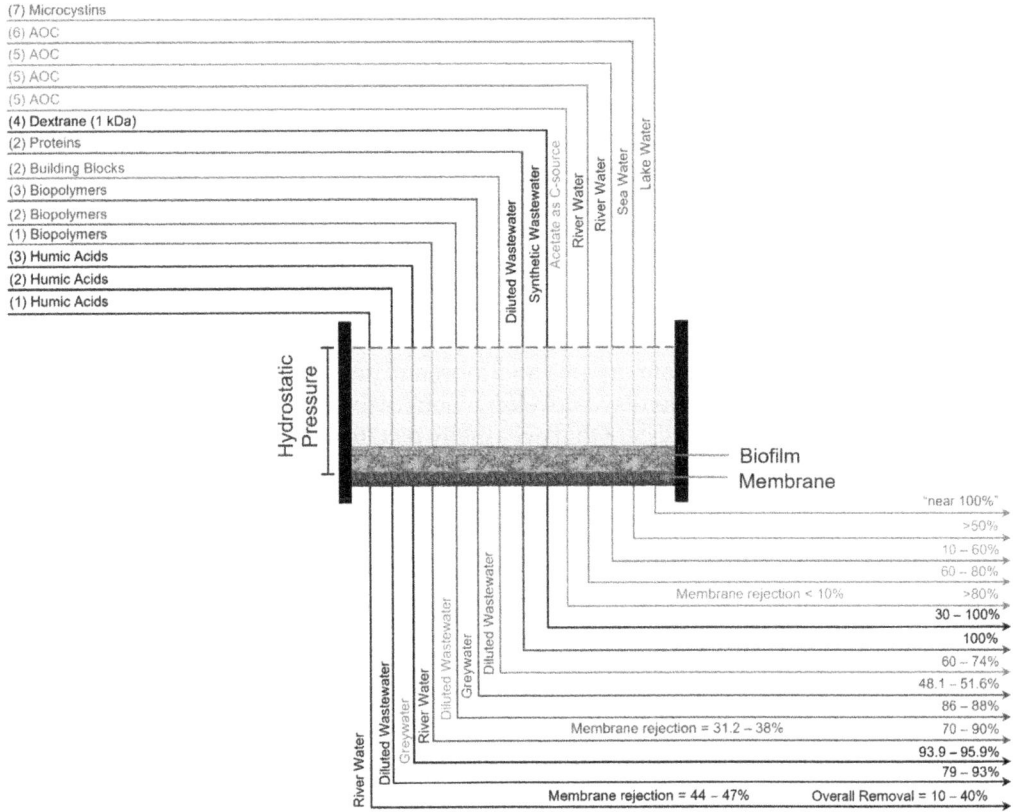

FIG. 2 Graphical summary of reports showing active roles of biofilm atop the membrane surface to enhance permeate quality. (1) Peter-Varbanets et al. (2010), (2) Wang et al. (2017), (3) Ding et al. (2017a), (4) Chomiak et al. (2015), (5) Derlon et al. (2014)), (6) Wu et al. (2017a), (7) Kohler et al. (2014).

of humic acid. Another report compared UF filtration and ULPMF in natural organic matter (NOM) and humic acid retention (Ding et al., 2017a). The former retained biopolymer and humic acid at 44%–48% and 31%–38%, respectively, while the latter did so at 94%–96% and 48%–52%, respectively. Substantial reductions of dissolved organic carbon (DOC) were observed for biopolymer (88%), humics (93%), and building blocks (97%) (Wang et al., 2017). The removal of dextran with molecular weight (10 kDa) caused a spike in the feedwater and gradually reached complete degradation on day 7. The membrane retained the higher molecular weight (150–2000 kDa) via size exclusion to allow a longer retention time for biodegradation by the biofilm (Chomiak et al., 2015).

More evidence on the active roles of biofilm in removing organics can be seen from the assimilable organic carbon (AOC) degradation. AOC is a good indicator due to its high biodegradability despite undergoing a complex process. Derlon et al. (2014) reported a >80% AOC biodegradation by the biofilm dosed in the form of acetate, which was much higher than the <10% removal by the virgin membrane. However, fluctuations of AOC removals were observed when treating river water. The removals were 60%–80% during the first 2 months and dropped to 10%–60% later. The AOC removal fluctuations were attributed to the interference of organic matter degradation that produces AOC-like by-products. High AOC removals of >50% were also observed during seawater filtration leading to improvement of the subsequent RO performances (Wu et al., 2017b). Apart from the common organics, cyanobacterial toxins from *Microcystis aeruginosa*, which may present problems in drinking water safety, could also be removed by biofilm in a ULPMF system (Kohler et al., 2014). The concentration was reduced to below the acceptable limit of 1 ppb.

3.3 Biofilm ecosystem

During ULPMF, all materials in the feed retained by the membrane (i.e., microorganisms, organic aggregates, and particulate organic and inorganic materials) accumulate on the membrane surface. The biofilm layer tends to develop when treating surface water. The formed biofilm can be considered as an ecosystem that eventually dictates the overall system performance. The knowledge of the biofilm system can help in understanding its positive rules in ULPMF.

3.3.1 Morphology

Flux stabilization associates well with the biofilm morphology. During operation, biofilm structures (i.e., roughness and porosity) experienced dynamic changes (Fig. 3). Matured biofilm is heterogeneous as the operation progresses; it becomes thicker and more porous, composed with voids between patches observed from OCT and confirmed using CLSM. There are spatial densities across the biofilm depth, denser in the base to contact the membrane surface (Derlon et al., 2012). Other methods have also been implemented to unravel the complex morphology of biofilm in ULPMF that led to a similar conclusion (Pronk et al., 2019).

3.3.2 Composition

Biofilm is a dynamic ecosystem that evolves in response to the changes in the environment. Biofilm in ULPMF showed high viability thanks to the biofilm's heterogeneity. Such heterogeneity enriches the structure with cavities and channels (Peter-Varbanets et al., 2010). Without the presence of shear, the biofilm can be thicker than normal. It can further grow if the substrate can

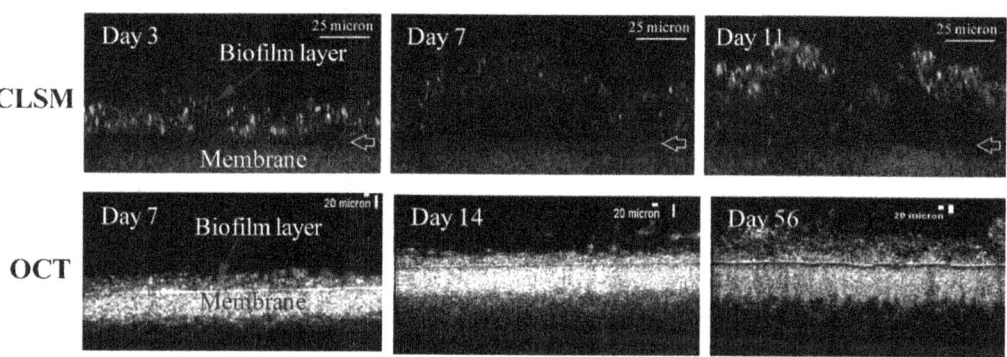

FIG. 3   The evolution of biofilm morphology atop the membrane surface in ULPMF systems as observed using a confocal laser scanning microscopy (CLSM) and an optical coherence tomography (OCT) (Pronk et al., 2019).

diffuse down to the base layer. The convective flow of the permeate from the top to the base layer of the biofilm toward the pore mouth seems to play a crucial role in supplying nutrients to the base layer. In order to allow more supply of nutrition to the base, cavities and channels are formed within the biofilm that lower the density and eventually the filtration resistance.

The dynamics of the biofilm are affected by the biological parameters. They typically include the viability of the cells, the composition of the microbial community that has consistently been reported to be highly diversified (Akhondi et al., 2015), biopolymeric matter, etc. High diversity of the bacterial community within the biofilm was observed. Moreover, the presence and behaviors of predations are highly important in determining the biofilm's characteristics. Wu et al. (2017b) implied that the prokaryotic and the eukaryotic composition and dynamics could explain the stable flux value in ULPMF. Some predators alter the community distribution and the morphology of the biofilm. The spiking of metazoans substantially improved the stable flux value in a ULPMF system treating river water (Klein et al., 2016). Compared with the control with fluxes of 5.7–8.8 LMH at $\Delta P = 61.5$ mbar, oligochaetes and nematodes increased the stable fluxes to 9.2–12.5 and 14.0–20.8 LMH, respectively. The metazoan activities, particularly nematodes, alter the biofilm to be more heterogeneous and porous and sweep the base structure of the biofilm completely. Similarly, the activity of eukaryotes in a ULPMF system pretreating seawater resulted in a long-term (>1 year) stable flux of about 20 LMH (at $\Delta P = 40$ mbar) (Wu et al., 2017b).

3.3.3 *Component and composition of organic/inorganic substances*

The component and composition of organic and/or inorganic substances in the feedwater are the main culprits of the flux decline. They accumulate on the membrane surface and act as the nutrients for microorganisms in the biofilm. The composition and abundance of the organic/inorganic substances thus affect the biofilm ecosystem. The amount dictates the total biovolume of the biofilm by defining the food to microorganism (F/M) ratio, while the composition affects the heterogeneity of the biofilm. A high quantity of organics is typically not required as it lowers the stable flux value, most likely by inducing a thick biofilm that increases the filtration resistance. Microorganisms in the biofilm also secrete metabolite biopolymeric microbial products, like protein and polysaccharides, that may be rejected by the membrane through size exclusion and accumulate in the biofilm to increase its density and eventually reduce the stable flux value.

Persistent or inert substances rejected by the membrane through size exclusion also accumulate in the (bio)fouling, eventually lowering the stable flux value. They can affect the structure of the biofilm by increasing its density. The accumulation of 3.6 µm kaolin particles present in the feedwater made the fouling layer more compact and lowered the stable flux value (Chomiak et al., 2015).

4. Factors influencing ULPMF performance

4.1 Feed

ULPMF systems have been explored to produce drinking water from river and rainwater, gray water treatment, and seawater filtration for reverse osmosis desalination. The stable flux is related to the feed. Organic substances in the feed can be excessive, beyond microbial utilization capability, and thus accumulate in the matrix. It could be worsened under depleted dissolved oxygen unfavorable to the growth and predation activities. Feedwater with high organics formed biofilm with higher filtration resistance, which eventually lowered the stable flux value, resulting in the flux in the following order/sequence: rainwater > river water/ seawater > (diluted) wastewater/gray water. The rank order was supported by a decrease in stable flux when river water was added with wastewater (Peter-Varbanets et al., 2011) and a substantial decrease of stable flux value from ~5 to ~2 LMH when lake water was affected by algal bloom exerting algogenic matters during ULPMF operations (Lee et al., 2019a). In addition, the treatment of seawater with low organic content resulted in 18–20 LMH stable flux values (Wu et al., 2017b).

4.2 Dissolved oxygen and temperature

In the ULPMF system, high DO levels, and high temperature benefited the performance, as proven from earlier reports. They promote the activity of the aerobic microorganisms in the biofilm that digests the organics. Oxygen is used as the acceptor of the electron in the metabolism process through an enzymatic reaction in which the rate is faster at higher temperatures. The optimum temperatures of the microorganisms for metabolism are around 30C–35°C. Filtration of wastewater saturated with oxygen at a concentration of 7.9 ppm in ULPMF reached a stable flux of 8.9 LMH. In contrast, the ones with an oxygen concentration of 0.1 achieved a stable flux of merely 0.8 LMH (Peter-Varbanets et al., 2010). Without sufficient oxygen, the organic fraction in the feed accumulates on the biofilm and makes it denser, leading to higher filtration resistance. Similarly, feed gray water with low DO concentration (0.4–0.6 mg/L) showed a higher cake resistance than with a higher DO of 6.0–6.5 mg/L (Ding et al., 2017a). This phenomenon could be attributed to the lower activity, thicker, smooth biofilm and higher EPS concentration under low DO conditions.

Temperature influences ULPMF by affecting the permeate viscosity and microbial activity in the biofilm. Increasing temperature from 21°C to 29°C increased the stable flux by 25% at ΔP of 40 mbar and by 21% at ΔP of 100 mbar (Akhondi et al., 2015). It was attributed to higher predation, reduced the biovolume, increased the biofilm porosity at a higher temperature, and increased the biochemical activities of cell metabolism at a higher temperature.

4.3 Membrane material and type and properties

Membrane material and intrinsic membrane resistance showed less contribution to stable flux value and organics rejection in ULPMF. In ULPMF, the overall filtration resistance is dictated by the cake or biofilm resistance. Membranes with a very low intrinsic resistance may benefit from a high initial flux, but as the operation progresses, the cake resistance increases and eventually controls the hydraulic performance.

Comparison of the membrane with a pore size of 0.1 and 0.01 μm resulted in a similar stable flux of around 4–5 LMH for river water treatment using the ULPMF system (Frechen et al., 2011). As expected, a membrane with a larger pore size had a higher initial flux and lost the advantage as the operation progressed. Wu et al. (2017a) compared several UF and MF membranes with variable clean water flux of 27–795 LMH using a dead-end cell and submerged system in ULPMF for seawater pretreatment. Irrespective of the membrane material, the stable flux values for the dead-end and submerged ULPMF systems were 2.7–8.4 and 16.3–18.6 LMH, respectively. Lee et al. (2019b) compared three UF membranes (with MWCO of 100–120 kDa) made from PSF and PVDF and one MF of a 0.3 μm membrane for lake water filtration. Almost similar trends of the fluxes were observed during the ULPMF systems operation, irrespective of the membrane materials. Despite showing a slightly higher stable flux, Pronk et al. (2019) suggested that the hollow fiber membrane offers greater productivity per footprint. It is less vulnerable to cake layer fouling because of the effective detachment of the foulant from the membrane surface. A hollow fiber membrane offers a great advantage when space is constricted due to its high membrane packing density and greater productivity per footprint.

The module types between the hollow fiber and flat-sheet membranes showed no significant difference in stable flux for river water filtration in ULPMF systems. After a month of operation, the flux stabilized at around 2 LMH for the flat-sheet membrane showed a similar value for the hollow fiber membrane despite the latter still seeming to decline (Chawla et al., 2017). The basic requirement of membrane materials for a ULPMF seems similar to the traditional membrane UF of MF for water filtration. They include high permeability, low membrane fouling propensity, low irreversible fouling, high packing density, etc. Hollow fiber with a high loose fiber will allow a high surface area for biofilm growth and permeation.

4.4 Operation pressure

Conventionally, higher ΔP under a certain range proportionally increases the filtration flux in the absence of membrane fouling. However, the variation of ΔP in ULPMF only slightly affected the stable flux. It means that higher ΔP somehow also increases the filtration resistance. The fouling layer's compaction explained an absurd phenomenon that lowered the biofilm porosity (Derlon et al., 2016). Filtration of reservoir water under variable ΔPs resulted in stable fluxes of 8.6 and 6.6 LMH for a ΔP of 200 and 60 mbar, respectively (Tang et al., 2016). The stable flux values were independent of the ΔP in the range of 15–200 mbar, favoring a low ΔP of 40–60 mbar (Kus et al., 2013a).

4.5 Continuous vs intermittent operation

Applications of ULPMF for water treatment in household or community scales are mostly made in discontinuous mode. They can be used on demand in the household or by supplying a water reservoir tank for the community scale. Therefore, research has been done on the impact of the discontinuous operation on the performance of ULPMF. The standstill period helped expand the fouling layer, thus increasing the initial flux when the filtration started (Derlon et al., 2012). Flushing after a standstill period further intensifies this process through the formation of larger foulant aggregates. Flux decline and recovery during intermittent operation with or without flushing were reversible.

Applying short cleaning (a few minutes a day) via air bubbling with an intermittent idle phase (a few hours per day) could also increase the stable flux value when a ULPMF was used to treat a mixture of pond and tap water (Oka et al., 2017). An increase in the initial flux by 40% was achieved, almost twice higher than without air sparging. Intermittent relaxation also increased the stable flux by 40%–60%, depending on the relaxation frequencies (a few hours per day). The implementation of intermittent filtration operation can lower the overall productivity if run under suboptimum conditions. For instance, relaxation could temporarily increase the flux level when in operation, but if done for too long, it may decrease the system's daily production. For instance, a lower stable flux in continuous operation may result in a higher total permeate volume than a system with a higher flux but operated intermittently (Tang et al., 2016).

4.6 Shear conditions

Shear forces can either increase or decrease the stable flux value depending on its implication on the biofilm characteristics. Imposing shear to the biofilm increased the flux (Peter-Varbanets et al., 2011). For example, a 50% flux recovery was achieved through gentle shaking after a standstill period. The standstill period allowed the expansion of the biofilm and the shaking action sloughing off the biofilm. On the other hand, Ding et al. (2016) obtained a lower stable flux value of 0.5 LMH for the sheared system compared to the nonsheared system of 2 LMH for gray water treatment. In this case, an imposing shear made the biofilm thinner and denser. In the long term, constant shearing allows survival of only the strongly binding microorganisms that produce high EPS; they need to obtain the binding capacity, which eventually increases the filtration resistance.

4.7 Biotechnology and bioengineering

In most of the past reports on ULPMF, the long-term stability of the process was attributed to the dynamic of the biofilm formed on the membrane surface. The filtration resistance imposed by the biofilm is far greater than the membrane resistance, as such it is almost exclusively responsible for the hydraulic performance. All of the factors mentioned in Sections 4.1–4.6 can be closely associated with the dynamic of the biofilm.

In the traditional membrane filtration, biofilm is constantly revolved through series of cleaning to maximize the flux. Such cleanings are not recommended in the ULPMF, but emphasis is given on how to exploit the biofilm in favor of the operational objective. One of the inherent benefits by enhancing the permeate quality has been widely reported and the issue associated with a low stable flux remains a challenge. In recent works, spreading the biofilm over a larger area either on the membrane surface or other solid surface seems effective. By simply altering the flow direction configuration that allows a higher membrane surface area (i.e., acting as the substrate of the biofilm), the flux can be significantly enhanced (Stoffel et al., 2022). Similarly, the addition of activated carbon provides an additional surface for biofilm to be formed, hence also increasing the stable flux (Schumann et al., 2020; Tang et al., 2021). Alternatively, when a simple mechanical washing can be implemented, some sort of biofilm control also be imposed to enhance the operational flux of the system. For instance, physical washing by using a spacer that can be turned mechanically, as well as providing a platform for biofilm growth (Waqas et al., 2021) and a module that can be shaken or vibrated (Wu et al., 2019b), would be effective.

5. Process integration

The performance of ULPMF in terms of flux and organic removal can be enhanced through combination with other processes through process intensification. Other processes that have been explored include a biofilm reactor, adsorption using granular or powdered activated carbon, and precoagulation. The added process helped in reducing organic substances through degradation or physical removal. Wu et al. (2016) integrated a biofilm reactor (BR) with ULPMF into the ULPMF-BR system. The BR promoted more biofilm for organic degradation, hence the permeate quality. The hybrid system showed a stable flux of ~5.7–8.6 LMH and a standalone ULPMF of ~4.5 LMH.

The addition of a granular (GAC) and powdered activated carbon (PAC) or sand layers as the prefilter or pretreatment improved organic removal efficiencies by 20%–25% thanks to adsorption of the fluorescent aromatic organic compounds by the GAC or the PAC layers. However, the GAC or the PAC layers reduced the stable fluxes from 4.5 LMH to 3.0–3.2 LMH due to the denser and EPS-rich biofilm formation. However, prefiltration with the sand layer did not improve organic removal and lowered the stable flux (Ding et al., 2018a, b). Pretreatment of feed by dosing the coagulant/flocculent in synthetic sewage increased the stable flux of the ULPMF-BR system by $>2 \times$ and $>1 \times$. The coagulants limited the dissolution of microbial metabolites to restrict membrane pore blocking (Ding et al., 2017b).

6. Economic assessment

6.1 Cost factors

The operation of the membrane filtration process incurs costs for both capital (CAPEX) and operating expenditures (OPEX). The CAPEX of the membrane filtration system constitutes membrane material, auxiliary equipment, and initial installation costs. OPEX consists of labor, energy, chemicals, membrane replacement, and others (i.e., water supply and wastewater handling). OPEX also includes the depreciation of auxiliary equipment. Table 2 compares the characteristics of CAPEX and OPEX in ULPMF and the traditional UF.

The stable fluxes reported in laboratory experiments were in the range of 1–2 LMH for wastewater treatment, much lower than those applied in MBRs of 8–40 LMH (Bilad, 2017). Higher fluxes were achieved in drinking water treatment of 4–20 LMH but still did not match the fluxes of traditional microfiltration/ultrafiltration of 50–100 LMH (Baker, 2013). On the contrary, the initial capital cost of membrane materials is much higher in the ULPMF system than in the traditional micro-/ultrafiltration. Nevertheless, there is no data yet on the life span of the membrane materials in the ULPMF system. When operating without chemical cleaning, the membrane is expected to last longer than 3–5 years, typically found in traditional

TABLE 2   the characteristics of CAPEX and OPEX in the ULPMF and the traditional UF.

Items	ULPMF	Traditional UF
Auxiliary equipment	Minimum auxiliary equipment is used due to the simple system	Auxiliary equipment is necessary to support the operation and the complex cleaning protocols
Membrane installation	Since the system operates under low flux, a higher membrane area is required contributing to high initial investment cost	The required membrane area is 5–20 times less, providing the main operation gain under high fluxes
Labor	No skilled labor is required due to very simple operation	Complex operation necessitates skilled labor, which contributes to operational costs
Energy	The very low energy input for feed pumping	Energy is consumed for pumping of feed at high velocity and pressure or under vacuum in the permeate. Energy can also be consumed for membrane fouling control (i.e., air bubbling, dynamic membrane system)
Chemicals	Use almost no or minimum chemicals	Chemicals are used for maintenance and intensive cleanings
Membrane replacement	Due to the absence of chemicals, a longer membrane lifespan (which can reach >10 years) is expected	The use of chemical cleaning contributed to shortening the membrane lifespan (3–11 years), increasing membrane replacement costs

micro-/ultrafiltration. Intensive chemical cleaning normally leads to material degradation and shortens the membrane life span (Cote et al., 2012).

6.2 Cost comparison of ULPMF with the conventional MF/UF or MBR

Since a limited report is available on the true cost of a ULPMF, a phenomenological approach has been provided elsewhere (Pronk et al., 2019) to compare the economy of ULPMF and traditional microfiltration/ultrafiltration processes, as illustrated in Fig. 4. In order to allow counting for CAPEX depreciation, the cost was presented in terms of m^3 water produced.

Due to the nature of the cost factors listed in Table 2, the scale highly affects the cost balance between ULPMF and the traditional UF. The cost of membrane materials is almost proportional to the treatment capacity since they are assembled into modules. This way, the membrane investment cost of ULPMF is higher than that of the traditional UF. The relative investment cost can be roughly predicted from the flux data. The typical ULPMF fluxes are 5–10 LMH and 1–2 LMH, while the traditional UF/MBR flux is 30–50 LMH and 10–30 LMH for surface water and gray water treatment, respectively. The relative membrane costs of ULPMF are 3–10 times and 5–30 times of UF/MBR for treatment of surface and gray water, respectively.

A qualitative presentation of CAPEX for ULPMF and traditional UF/MBR for surface water and gray water treatment is shown in Fig. 4. The costs of auxiliary equipment are much higher for the traditional UF/MBR systems compared to ULPMF. The cost increases

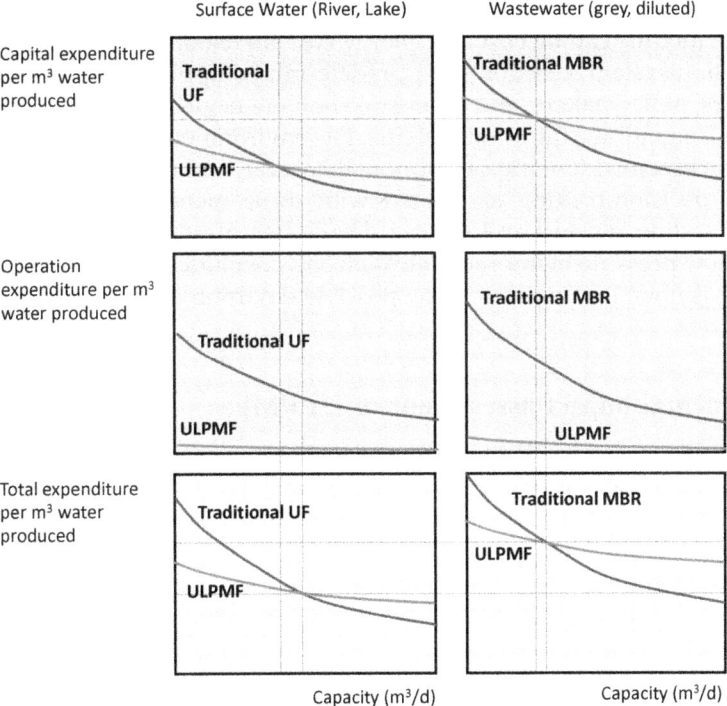

FIG. 4 Qualitative comparison of capital (CAPEX) and operation expenditures (OPEX) of ultralow pressure membrane filtration and traditional ultrafiltration or membrane bioreactor treating surface and gray water. The vertical dashed lines highlight the contribution of OPEX in enhancing the competitive capacity of ULPMF, while the horizontal ones compare the expenses between surface water and wastewater.

exponentially as a function of the treatment capacity on the order of 0.6 (Westney, 1997). Meanwhile, the OPEX of the ULPMF system is almost negligible. Based on the total costs, ULPMF is more economical than the traditional UF/MBR under low treatment capacities and vice versa. The cost balance will indeed depend on many factors and vary from site to site. However, it can be projected that ULPMF is more economically attractive for household scale with a capacity of 20–50 L/day and water or gray water treatment for small communities with a capacity of 1–10 m^3/day. On the other hand, a full-scale system with more than 100 m^3/day is better to run under the traditional UF/MBR.

7. Environmental impact and sustainability assessments

The contribution or influence that membrane filtration technologies wield in water and wastewater treatment applications cannot be overemphasized. The ease with which they can be adopted and retrofitted into any system make them attractive as a water treatment technology. When a membrane filtration technology is going to be selected for any application, membrane module capital cost and energy cost are usually the major inputs to reach an optimum filtration system selection for a predetermined application. Oftentimes, manufacturers, engineers, policymakers, and other stakeholders neglect the environmental cost associated with these membrane filtration systems. The environmental impact assessment of these technologies has become more important now than ever, and environmental factors must be factored in all decision-making associated with these membrane filtration technologies. Therefore, this section reviews and gives a detailed account of the possible environmental impact of ultralow pressure membrane filtration and concludes by assessing how sustainable this technology is now; it also projects the possible environmental effects in the near future.

7.1 Environmental impact assessment of ULPMF

The most popular framework used for environmental impact assessment is the life cycle assessment (LCA). LCA can give detailed analyses on the environmental burden technologies imposes from cradle to grave. These environmental burdens usually include how any technology depletes natural resources, consumes energy, and possible emissions to the environment (that is air, water, soil, etc.) (Lawler et al., 2015). Environmental impact assessment is usually carried out within a consistent LCA framework which includes: (i) an explicit definition of the goal and scope, (ii) life cycle inventory analysis, (iii) life cycle impact assessment, and (iv) life cycle improvement analysis and interpretation (Tangsubkul et al., 2006). The strength of an LCA lies in its system boundary definition; the more the number of processes included within the boundary, the more robust the result of the analyses. For membrane filtration systems, the commonly employed environmental impacts include global warming potential (GWP), eutrophication potential (EP), photo-oxidant formation potential (POCP), human toxicity potential (HTP), freshwater, marine and terrestrial ecotoxicity potentials (FAETP, MAETP, and TETP) (Ho et al., 2021).

A notable LCA of an ULPMF was first reported some half a decade ago (Tangsubkul et al., 2006). This LCA study was performed with the GaBi4 software, an engineering design tool that can perform life cycle balance with process visualization. This software is designed in accordance with the ISO 14040 series. Their LCA study investigated different operational scenarios which lie within the ULPMF systems. They tested maximum transmembrane pressure (TMPmax) in the range of (0.2–0.5 bar), imposed flux in the range of 10–100 LMH, and required total modules required in the range of 33–283. These operating conditions were varied and 20 distinct scenarios in two groups of A (TMP$_{max}$ = 0.2 bar) and B (TMP$_{max}$ = 0.2 bar) were generated. The MF process consists of membrane modules with each having 15 m^2 of area, a nominal pore size of 0.2 μm, and the secondary effluent was used as the feed to this MF system. The MF was operated at a controlled flux with high pressure (6 bar) backwashing with air. Backwashing is only initiated when the transmembrane pressure (TMP) reaches a certain predetermined set point. The system boundary (see Fig. 5) considered the production of energy and material involved in the manufacturing of the microfiltration membrane (MF), transportation of membrane modules and cleaning chemicals, and the MF operation. The after-use phase (basically demolition) was excluded from the study because it was assumed to be the same for all the scenarios that were assessed. The materials considered in the

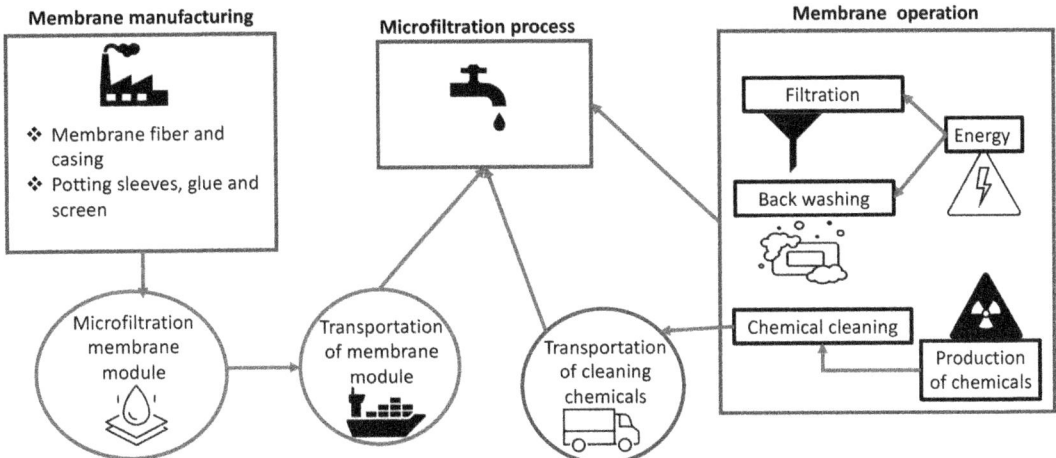

FIG. 5 System boundary definition for LCA study. The graphic was developed from a concept proposed earlier (Tangsubkul et al., 2006).

membrane manufacturing process include polypropylene (PP) for the production of membrane fiber, acrylonitrile-butadiene-styrene (ABS) for producing membrane casing and potting sleeves, polyethylene (PE) for membrane protective mesh and polyurethane for potting of the membrane module. NaOH was assumed to be the cleaning chemical in the MF process operation. Processes considered for MF operation include backwashing pumping of filtrate and chemical cleaning of the membranes.

Some assumptions were made whene defining the system boundary. Additional assumptions involved the scope definition. They include: (1) a plant size of 1 ML/day capacity, (2) plant lifetime of 20 years, (3) treated effluents in all scenarios are the same, (4) membrane lifetime was assumed to be 5 years, (5) all ancillary units are the same for all scenarios, (6) the chemical is used for cleaning in all scenarios, and membrane modules are housed in the same building size for all scenarios. The cleaning interval was based on the imposed flux; at 10 LMH cleaning is required every 6 months, for 50 LMH it is 10 days and for 100 LMH it is every 2 days.

The best conditions for optimal environmental protection were found to be at low flux and high TMPmax for all the environmental impact categories considered. For instance, at 10 LMH and 0.5 bar, the least GWP was 29 kg CO_2 equivalent emission of greenhouse gas for all the chemical cleaning scenarios considered. The POCP impacts result from the emissions of organic compounds released into the environment via transportation, electricity generation, and chemical production. At 30 LMH and 0.5 bar, the lowest POCP value was 0.00294 kg ethane, which was equivalent to all chemical cleaning options. MF operations at a mid-range flux of 50–60 LMH and TMP_{max} 0.5 bar yielded the lowest EP value of 2.0 g O_2 equivalent for all chemical cleaning options. For HTP, at 10 LMH and 0.5 bar, the lowest values of 0.15, 0.14, 0.13 kg DCB equivalent for chemical cleaning scenarios 1, 2, and 3 were obtained, respectively. Besides, the lowest MAETP, FAETP, and TETP were obtained when the MF was operated at 10 LMH and 0.5 bar. Finally, the authors concluded that, at a low flux, the choice of chemical cleaning frequency can significantly influence the overall environmental performance of the MF process. At low flow flux (10–30 LMH), the MF process performs better, but the trade-off is that more membrane module is required, thus increasing the environmental cost. However, the extra cost from more membrane modules was estimated and it was found to be negligible compared to the energy cost during the MF operation. The findings from the LCA reports also reveal that with a cleaner source of energy, the MF operation becomes more attractive at a high flux.

7.2 Overall sustainability assessment of ULPMF

The attractiveness and strength of ULPMF lie in its ability to reduce the environmental footprints associated with the use of other high-pressure membrane technology, advanced oxidation processes, adsorption technologies, etc. ULPMF has also attracted interest in recycling and reuse of end-of-life (EOL) membranes from other high-pressure membrane filtration processes such as Reverse osmosis. In fact, ULPMF has been used to reduce the environmental footprint of other processes by integrating it with other water treatment technologies as either a pretreatment or posttreatment step. Although ULPMF requires an occasional chemical cleaning, the environmental footprint is significantly negligible compared to other high-pressure membrane filtration processes.

7.2.1 Gravity-driven membrane filtration as a sustainable ULPMF

The Gravity-driven membrane (GDM) filtration system offers lower energy consumption, lower chemical consumption, lower membrane replacement cost, and reduced labor cost compared to conventional filtration systems such as UF or membrane bioreactors (MBR). GDM relies solely on the hydrostatic pressure of the feedwater to produce clean water for use; it requires almost no external energy at all to push the feed through the membrane. GDM can achieve stable flux at a very low TMP (<0.1 bar), thanks to the synergistic effect of the porous biofilm layer formed on the membrane. GDM can reduce the environmental footprint of other filtration technologies making it more sustainable. In fact, it can achieve a stable flux all year round even if there are intermittent changes in the TMP. Its clean water production is independent of the TMP; if the TMP is increased significantly, it does not increase the stable flux drastically. Because GDM is highly sustainable, it is more attractive to be used in a household water filtration technology, especially in a water-stressed area without a reliable source of energy. GDM, a type of ULPMF, is considered sustainable because of the factors detailed in Table 2.

7.2.2 ULPMF as a pre- or posttreatment technique for other technologies

The integration of ULPMF with, for instance, an RO system will ultimately contribute to the reduction in energy consumption (i.e., reduced RO TMP) and increase in the time interval for chemical cleaning of the membranes or backwashing of the RO process. This means that incorporating ULPMF into a system with a low sustainability assessment index can eventually elevate the parent system sustainability ranking. For instance, a GDM water system was employed to pretreat seawater for an RO system (Wu et al., 2017b). At a gravitational pressure of about 0.04 bar and for a period of 1 year, a stable flux of c. 18.6 LMH can be maintained. The findings also revealed that the addition of GDM system had a beneficial effect on RO compared to a conventional UF (i.e., GDM/RO > conventional UF/RO). The feeding of the GDM pretreated seawater to the RO system did not increase the TMP over time, but a significant increase of the RO TMP was observed when UF pretreated seawater was fed. The GDM system was also employed as a posttreatment technology (Wang et al., 2017). They employed the GDM system as an ultralow energy MBR to filter secondary effluent from a WWTP. The GDM system can achieve a stabilized flux after 30 days of operation; this was impossible for a conventional MBR system. The permeate quality was of a high standard and this was a credit to the synergistic effect of the biofilm layer; this cake can degrade some unwanted organic loading. The ease with which the cake layer on the membrane can be physically removed, when needed, reinforces the potential of the GDM system for long-term operation for secondary effluent treatment. The combination of a submerged GDM system with a biofilm reactor as a seawater pretreatment technology for RO has been shown to improve flux compared to filtration cells (Wu et al., 2016). Elsewhere, ULPMF was combined with aeration for low energy *Cholera vulgaris* broth filtration; the objective was to lower the energy requirement while maximizing permeate flux (Wan Osman et al., 2021). Combining low TMP (c. 0.019 bar) with aeration resulted in low energy input of about $0.0044 \, kWh \, m^{-3}$. Their work provides a new perspective toward energy saving and a fouling-free MF membrane for microalgae harvesting. The use of aeration is obviously an extra energy cost, but the duration and interval of application will be much lower and subsequently have a negligible effect.

7.2.3 *ULPMF as a sink for EOL membranes*

The sustainability of a ULPMF filtration goes beyond lowering energy and chemical consumption; it can be used as a sink for recycling or reusing condemned or EOL membranes from the RO process. To recycle EOL membranes, it will require applying a low amount of energy and chemical for refurbishment before being transferred for use in a ULPMF system. This kind of approach is still desirable as the time to the landfill of such membranes will be prolonged, thereby reducing the environmental impact. EOL membranes from the RO system were reused in a ULPMF, microfiltration system operating at a fixed TMP (1 bar) (Khoo et al., 2021). Prior to the reuse of the EOL membranes, the membranes were treated with sonication to remove all foulant. They compared the use of sonication and sonication/KMnO$_4$ or refurbishment of the EOL membrane. They found that sonication was efficient for converting the discarded RO membrane to an MF membrane within a really short time (15 min). The ultrasonicated membrane was able to achieve about 24 times the permeate water flux of an EOL at a fixed TMP of 1 bar, but with a low salt rejection of about 11%. They were able to increase the salt rejection to 82% and maintained the permeate water flux in the range of 8–12 LMH/bar by treating the EOL membrane with ultrasonication plus 5000 ppm of an oxidizing agent (KMnO$_4$). Their findings demonstrated that is possible to transform EOL membranes for use in other processes such as MF within a short time compared to other chemical methods.

Overall, without any doubt, it is evident that ULPMF is sustainable either as a stand-alone technology or when integrated with other water treatment technologies. It is obvious that the use of ULPMF always results in a better sustainability index which can be easily translated to a lower environmental impact. Table 3 summarizes and compares qualitatively the

TABLE 3 Qualitative comparison of environmental and sustainability assessment of ULPMF and RO on different scales.

	Pilot scale			Community scale			Large scale		
Items	ULPMF	ULMPF/RO	RO	ULPMF	ULPMF	RO	ULPMF	ULMPF/RO	RO
Energy cost	UL	M	H	M	H	UH	M	H	UH
CAPEX	UL	M	H	M	H	UH	M	H	UH
Membrane replacement cost	M	M	H	H	H	UH	H	H	UH
OPEX	UL	M	H	M	H	UH	M	H	UH
Environmental impact and sustainability	UL	M	H	L	H	UH	L	H	UH

H, *high*; L, *low*; M, *medium*; U, *ultra*.

environmental and sustainability assessments of ULPMF with RO systems. There is a consensus that ULPMF reduces the environmental impact of RO and thereby increases its sustainability.

8. Conclusions and perspectives

The key factor determining the application potential of ULPMF is the stable flux value that vastly determines the treatment cost. Increasing the stable flux remains the main obstacle in the widespread acceptance of ULPMF in large-scale implementations. The high stable flux values (10–20 LMH) are attributed to the low organic content and predator activity that enhances the biofilm porosity. Conversely, the feed contains substances with high membrane fouling propensity (i.e., gray water) leading to low, stable flux value (<5 LMH). Deliberately introducing and growing metazoans in the biofilm has been considered, particularly for large-scale applications. However, it is questionable if this measure is economically and technically feasible. Process intensification involving pretreatment of adsorption, coagulation/flocculation, and biodegradation has shown a viable option to increase the stable flux value for river and grey water treatments.

The generic approach seems to improve the biofilm properties favorable for low filtration resistance (i.e., heterogeneous and low density) or minimize biofilm through pretreatment. The former seems challenging and hard to control when operating small-scale and possibly unsupervised systems. The latter seems more promising, especially via the extended volume of biofilm in the attached growth system.

It is generally accepted that the presence of biofilm atop the membrane surface determines the stable flux value and enhances the permeate quality through the rejection and biodegradation of organics, including humic acid and polysaccharide proteins AOC and microcystins.

For the AOC removal that is generally converted into the cell building blocks, regular removal of the biofilm is probably required to minimize leaching from more mature biofilms as a result of the hydrolysis and slaughtering (also known as the seeding phenomenon) of biomass from the biofilm itself. Effective system design and operation of ULPMF require further investigations. Its application from decentralized water treatment seems highly attractive.

The most important consideration for ULPMF implementation is the total cost (CAPEX + OPEX). It is more favorable than the traditional UF for installation with low or medium scales. Despite the few available costs data available in the literature, it highly depends on the site condition. Therefore, individual assessment per site is required to judge the economy of the process. The most established application of ULPMF is for household-scale and community-scale water treatment systems. The "water backpack" has been commercialized for disaster relief supporting aids, and community-scale water kiosks have also been reported. Under this scale, ULPMF can demonstrate its true benefit of simple operation and maintenance, minimum supervision without process control, and running without electricity supply.

References

Akhondi, E., Wu, B., Sun, S., Marxer, B., Lim, W., Gu, J., Liu, L., Burkhardt, M., McDougald, D., Pronk, W., Fane, A.G., 2015. Gravity-driven membrane filtration as pretreatment for seawater reverse osmosis: linking biofouling layer morphology with flux stabilization. Water Res. 70, 158–173. https://doi.org/10.1016/j.watres.2014.12.001.

Baker, R., 2013. Membrane Technology and Applications. Wiley, Hoboken, NJ.

Barambu, N.U., Peter, D., Yusoff, M.H.M., Bilad, M.R., Shamsuddin, N., Marbelia, L., Nordin, N.A.H., Jaafar, J., 2020. Detergent and water recovery from laundry wastewater using tilted panel membrane filtration system. Membranes 10, 260. https://doi.org/10.3390/membranes10100260.

Bilad, M.R., 2017. Membrane bioreactor for domestic wastewater treatment: principles, challanges and future research directions. Indones. J. Sci. Technol. 2, 97. https://doi.org/10.17509/ijost.v2i1.5993.

Bilad, M.R., Mat Nawi, N.I., Subramaniam, D.D., Shamsuddin, N., Khan, A.L., Jaafar, J., Nandiyanto, A.B.D., 2020. Low-pressure submerged membrane filtration for potential reuse of detergent and water from laundry wastewater. J. Water Process Eng., 36. https://doi.org/10.1016/j.jwpe.2020.101264.

Boulestreau, M., Hoa, E., Peter-Verbanets, M., Pronk, W., Rajagopaul, R., Lesjean, B., 2012. Operation of gravity-driven ultrafiltration prototype for decentralised water supply. Desalination Water Treat. 42, 125–130. https://doi.org/10.1080/19443994.2012.683073.

Chang, H., Liu, B., Wang, H., Zhang, S.-Y., Chen, S., Tiraferri, A., Tang, Y.-Q., 2019. Evaluating the performance of gravity-driven membrane filtration as desalination pretreatment of shale gas flowback and produced water. J. Membr. Sci. 587, 117187. https://doi.org/10.1016/j.memsci.2019.117187.

Chawla, C., Zwijnenburg, A., Kemperman, A.J.B., Nijmeijer, K., 2017. Fouling in gravity driven point-of-use drinking water treatment systems. Chem. Eng. J. 319, 89–97. https://doi.org/10.1016/j.cej.2017.02.120.

Chen, R., Liang, H., Wang, J., Lin, D., Zhang, H., Cheng, X., Tang, X., 2021. Effects of predator movement patterns on the biofouling layer during gravity-driven membrane filtration in treating surface water. Sci. Total Environ. 771, 145372. https://doi.org/10.1016/j.scitotenv.2021.145372.

Chomiak, A., Traber, J., Morgenroth, E., Derlon, N., 2015. Biofilm increases permeate quality by organic carbon degradation in low pressure ultrafiltration. Water Res. 85, 512–520. https://doi.org/10.1016/j.watres.2015.08.009.

Cote, P., Alam, Z., Penny, J., 2012. Hollow fiber membrane life in membrane bioreactors (MBR). Desalination 288, 145–151. https://doi.org/10.1016/j.desal.2011.12.026.

Crittenden, J.C., Borchardt, J.H., Crittenden, J.C., Harza, M.W. (Eds.), 2012. MWH's Water Treatment: Principles and Design, 3rd. John Wiley & Sons, Hoboken, N.J.

Derlon, N., Peter-Varbanets, M., Scheidegger, A., Pronk, W., Morgenroth, E., 2012. Predation influences the structure of biofilm developed on ultrafiltration membranes. Water Res. 46, 3323–3333. https://doi.org/10.1016/j.watres.2012.03.031.

Derlon, N., Mimoso, J., Klein, T., Koetzsch, S., Morgenroth, E., 2014. Presence of biofilms on ultrafiltration membrane surfaces increases the quality of permeate produced during ultra-low pressure gravity-driven membrane filtration. Water Res. 60, 164–173. https://doi.org/10.1016/j.watres.2014.04.045.

Derlon, N., Grütter, A., Brandenberger, F., Sutter, A., Kuhlicke, U., Neu, T.R., Morgenroth, E., 2016. The composition and compression of biofilms developed on ultrafiltration membranes determine hydraulic biofilm resistance. Water Res. 102, 63–72. https://doi.org/10.1016/j.watres.2016.06.019.

Derlon, N., Desmond, P., Rühs, P.A., Morgenroth, E., 2022. Cross flow frequency determines the physical structure and cohesion of membrane biofilms developed during gravity-driven membrane ultrafiltration of river water: implication for hydraulic resistance. J. Membr. Sci. 643, 120079. https://doi.org/10.1016/j.memsci.2021.120079.

Ding, A., Liang, H., Li, G., Derlon, N., Szivak, I., Morgenroth, E., Pronk, W., 2016. Impact of aeration shear stress on permeate flux and fouling layer properties in a low pressure membrane bioreactor for the treatment of grey water. J. Membr. Sci. 510, 382–390. https://doi.org/10.1016/j.memsci.2016.03.025.

Ding, A., Liang, H., Li, G., Szivak, I., Traber, J., Pronk, W., 2017a. A low energy gravity-driven membrane bioreactor system for grey water treatment: permeability and removal performance of organics. J. Membr. Sci. 542, 408–417. https://doi.org/10.1016/j.memsci.2017.08.037.

Ding, A., Wang, J., Lin, D., Tang, X., Cheng, X., Li, G., Ren, N., Liang, H., 2017b. In situ coagulation versus pre-coagulation for gravity-driven membrane bioreactor during decentralized sewage treatment: permeability stabilization, fouling layer formation and biological activity. Water Res. 126, 197–207. https://doi.org/10.1016/j.watres.2017.09.027.

Ding, A., Wang, J., Lin, D., Cheng, X., Wang, H., Bai, L., Ren, N., Li, G., Liang, H., 2018a. Effect of PAC particle layer on the performance of gravity-driven membrane filtration (GDM) system during rainwater treatment. Environ. Sci. Water Res. Technol. 4, 48–57. https://doi.org/10.1039/C7EW00298J.

Ding, A., Wang, J., Lin, D., Zeng, R., Yu, S., Gan, Z., Ren, N., Li, G., Liang, H., 2018b. Effects of GAC layer on the performance of gravity-driven membrane filtration (GDM) system for rainwater recycling. Chemosphere 191, 253–261. https://doi.org/10.1016/j.chemosphere.2017.10.034.

Ding, A., Song, R., Cui, H., Cao, H., Ngo, H.H., Chang, H., Nan, J., Li, G., Ma, J., 2021. Presence of powdered activated carbon/zeolite layer on the performances of gravity-driven membrane (GDM) system for drinking water treatment: ammonia removal and flux stabilization. Sci. Total Environ. 799, 149415. https://doi.org/10.1016/j.scitotenv.2021.149415.

Du, P., Li, X., Yang, Y., Zhou, Z., Fan, X., Chang, H., Liang, H., 2022a. Regulated-biofilms enhance the permeate flux and quality of gravity-driven membrane (GDM) by in situ coagulation combined with activated alumina filtration. Water Res. 209, 117947. https://doi.org/10.1016/j.watres.2021.117947.

Du, X., Wang, Z., Liu, Y., Ma, R., Lu, S., Lu, X., Liu, L., Liang, H., 2022b. Gravity-driven membrane bioreactor coupled with electrochemical oxidation disinfection (GDMBR-EO) to treat roofing rainwater. Chem. Eng. J. 427, 131714. https://doi.org/10.1016/j.cej.2021.131714.

Du, X., Xu, J., Mo, Z., Luo, Y., Su, J., Nie, J., Wang, Z., Liu, L., Liang, H., 2019. The performance of gravity-driven membrane (GDM) filtration for roofing rainwater reuse: implications of roofing rainwater energy and rainwater purification. Sci. Total Environ. 697, 134187. https://doi.org/10.1016/j.scitotenv.2019.134187.

Du, X., Zhao, W., Wang, Z., Ma, R., Luo, Y., Wang, Z., Sun, Q., Liang, H., 2021. Rural drinking water treatment system combining solar-powered electrocoagulation and a gravity-driven ceramic membrane bioreactor. Sep. Purif. Technol. 276, 119383. https://doi.org/10.1016/j.seppur.2021.119383.

Field, R.W., Pearce, G.K., 2011. Critical, sustainable and threshold fluxes for membrane filtration with water industry applications. Adv. Colloid Interface Sci. Membr. Separ. Colloid Sci. 164, 38–44. https://doi.org/10.1016/j.cis.2010.12.008.

Fortunato, L., Ranieri, L., Naddeo, V., Leiknes, T., 2020. Fouling control in a gravity-driven membrane (GDM) bioreactor treating primary wastewater by using relaxation and/or air scouring. J. Membr. Sci. 610, 118261. https://doi.org/10.1016/j.memsci.2020.118261.

Frechen, F.-B., Exler, H., Romaker, J., Schier, W., 2011. Long-term behaviour of a gravity-driven dead end membrane filtration unit for potable water supply in cases of disasters. Water Supply 11, 39–44. https://doi.org/10.2166/ws.2011.006.

García-Pacheco, R., Li, Q., Comas, J., Taylor, R.A., Le-Clech, P., 2021. Novel housing designs for nanofiltration and ultrafiltration gravity-driven recycled membrane-based systems. Sci. Total Environ. 767, 144181. https://doi.org/10.1016/j.scitotenv.2020.144181.

Guðjónsdóttir, S., Ge, L., Zhao, K., Lisak, G., Wu, B., 2022. Gravity-driven membrane filtration of primary wastewater effluent for edible plant cultivations: membrane performance and health risk assessment. J. Environ. Chem. Eng. 10, 107046. https://doi.org/10.1016/j.jece.2021.107046.

Hey, T., Bajraktari, N., Davidsson, Å., Vogel, J., Madsen, H.T., Hélix-Nielsen, C., Jansen, J.l.C., Jönsson, K., 2018. Evaluation of direct membrane filtration and direct forward osmosis as concepts for compact and energy-positive municipal wastewater treatment. Environ. Technol. 39, 264–276. https://doi.org/10.1080/09593330.2017.1298677.

Ho, K.C., Teoh, Y.X., Teow, Y.H., Mohammad, A.W., 2021. Life cycle assessment (LCA) of electrically-enhanced POME filtration: environmental impacts of conductive-membrane formulation and process operating parameters. J. Environ. Manag. 277, 111434. https://doi.org/10.1016/j.jenvman.2020.111434.

Ismail, I., Kurnia, K.A., Samsuri, S., Bilad, M.R., Marbelia, L., Ismail, N.M., Khan, A.L., Budiman, A., Susilawati, S., 2021. Energy efficient harvesting of *Spirulina* sp. from the growth medium using a tilted panel membrane filtration. Bioresour. Technol. Rep. 15, 100697. https://doi.org/10.1016/j.biteb.2021.100697.

Jabornig, S., Podmirseg, S.M., 2015. A novel fixed fibre biofilm membrane process for on-site greywater reclamation requiring no fouling control: novel fixed fibre biofilm membrane process. Biotechnol. Bioeng. 112, 484–493. https://doi.org/10.1002/bit.25449.

Jiang, H., Zhao, Q., Wang, P., Ma, J., Zhai, X., 2019a. Improved separation and antifouling properties of PVDF gravity-driven membranes by blending with amphiphilic multi-arms polymer PPG-Si-PEG. J. Membr. Sci. 588, 117148. https://doi.org/10.1016/j.memsci.2019.05.072.

Jiang, H., Zhao, Q., Wang, P., Ma, J., Zhai, X., 2019b. Improved separation and antifouling properties of PVDF gravity-driven membranes by blending with amphiphilic multi-arms polymer PPG-Si-PEG. J. Membr. Sci. 588, 117148. https://doi.org/10.1016/j.memsci.2019.05.072.

Jiang, H., Zhao, Q., Wang, P., Chen, M., Wang, Z., Ma, J., 2021. Inhibition of algae-induced membrane fouling by in-situ formed hydrophilic micropillars on ultrafiltration membrane surface. J. Membr. Sci. 638, 119648. https://doi.org/10.1016/j.memsci.2021.119648.

Jiang, H., Wang, P., Zhao, Q., Wang, Z., Sun, X., Chen, M., Han, B., Ma, J., 2022. Enhanced water permeance and antifouling performance of gravity-driven ultrafiltration membrane with in-situ formed rigid pore structure. J. Membr. Sci. 644, 120154. https://doi.org/10.1016/j.memsci.2021.120154.

Khoo, Y.S., Lau, W.J., Hasan, S.W., Salleh, W.N.W., Ismail, A.F., 2021. New approach of recycling end-of-life reverse osmosis membranes via sonication for microfiltration process. J. Environ. Chem. Eng. 9, 106731. https://doi.org/10.1016/j.jece.2021.106731.

Klein, T., Zihlmann, D., Derlon, N., Isaacson, C., Szivak, I., Weissbrodt, D.G., Pronk, W., 2016. Biological control of biofilms on membranes by metazoans. Water Res. 88, 20–29. https://doi.org/10.1016/j.watres.2015.09.050.

Kohler, E., Villiger, J., Posch, T., Derlon, N., Shabarova, T., Morgenroth, E., Pernthaler, J., Blom, J.F., 2014. Biodegradation of microcystins during gravity-driven membrane (GDM) ultrafiltration. PLoS ONE 9, e111794. https://doi.org/10.1371/journal.pone.0111794.

Kus, B., Kandasamy, J., Vigneswaran, S., Shon, H.K., Moody, G., 2013a. Gravity driven membrane filtration system to improve the water quality in rainwater tanks. Water Supply 13, 479–485. https://doi.org/10.2166/ws.2013.046.

Kus, B., Kandasamy, J., Vigneswaran, S., Shon, H.K., Moody, G., 2013b. Household rainwater harvesting system—pilot scale gravity driven membrane-based filtration system. Water Supply 13, 790–797. https://doi.org/10.2166/ws.2013.067.

Lawler, W., Alvarez-Gaitan, J., Leslie, G., Le-Clech, P., 2015. Comparative life cycle assessment of end-of-life options for reverse osmosis membranes. Desalination 357, 45–54. https://doi.org/10.1016/j.desal.2014.10.013.

Lee, D., Lee, Y., Choi, S.S., Lee, S.-H., Kim, K.-W., Lee, Y., 2019a. Effect of membrane property and feed water organic matter quality on long-term performance of the gravity-driven membrane filtration process. Environ. Sci. Pollut. Res. 26, 1152–1162. https://doi.org/10.1007/s11356-017-9627-8.

Lee, S., Sutter, M., Burkhardt, M., Wu, B., Chong, T.H., 2019b. Biocarriers facilitated gravity-driven membrane (GDM) reactor for wastewater reclamation: effect of intermittent aeration cycle. Sci. Total Environ. 694, 133719. https://doi.org/10.1016/j.scitotenv.2019.133719.

Lee, S., Badoux, G.O., Wu, B., Chong, T.H., 2021. Enhancing performance of biocarriers facilitated gravity-driven membrane (GDM) reactor for decentralized wastewater treatment: effect of internal recirculation and membrane packing density. Sci. Total Environ. 762, 144104. https://doi.org/10.1016/j.scitotenv.2020.144104.

Li, J., Chang, H., Tang, P., Shang, W., He, Q., Liu, B., 2020. Effects of membrane property and hydrostatic pressure on the performance of gravity-driven membrane for shale gas flowback and produced water treatment. J. Water Process Eng. 33, 101117. https://doi.org/10.1016/j.jwpe.2019.101117.

B. Applications of membrane technology for water and wastewater treatment

Li, K., Xu, W., Han, M., Cheng, Y., Wen, G., Huang, T., 2022. Integration of iron-manganese co-oxide (FMO) with gravity-driven membrane (GDM) for efficient treatment of surface water containing manganese and ammonium. Sep. Purif. Technol. 282, 119977. https://doi.org/10.1016/j.seppur.2021.119977.

Liu, C., Song, D., Zhang, W., He, Q., Huangfu, X., Sun, S., Sun, Z., Cheng, W., Ma, J., 2020. Constructing zwitterionic polymer brush layer to enhance gravity-driven membrane performance by governing biofilm formation. Water Res. 168, 115181. https://doi.org/10.1016/j.watres.2019.115181.

Mat Nawi, N.I., Chean, H.M., Shamsuddin, N., Bilad, M.R., Narrkun, T., Faungnawakij, K., Khan, A.L., 2020. Development of hydrophilic PVDF membrane using vapour induced phase separation method for produced water treatment. Membranes 10, 121. https://doi.org/10.3390/membranes10060121.

Naranjo, J., Clasen, T., Gerba, C., Frauchiger, D., 2009. Laboratory assessment of a gravity-fed ultrafiltration water treatment device designed for household use in low-income settings. Am. J. Trop. Med. Hyg. 80, 819–823. https://doi.org/10.4269/ajtmh.2009.80.819.

Oka, P.A., Khadem, N., Bérubé, P.R., 2017. Operation of passive membrane systems for drinking water treatment. Water Res. 115, 287–296. https://doi.org/10.1016/j.watres.2017.02.065.

Osman, A., Mat Nawi, N.I., Samsuri, S., Bilad, M.R., Shamsuddin, N., Khan, A.L., Jaafar, J., Nordin, N.A.H., 2020. Patterned membrane in an energy-efficient tilted panel filtration system for fouling control in activated sludge filtration. Polymers 12, 432. https://doi.org/10.3390/polym12020432.

Peter-Varbanets, M., Hammes, F., Vital, M., Pronk, W., 2010. Stabilization of flux during dead-end ultra-low pressure ultrafiltration. Water Res. 44, 3607–3616. https://doi.org/10.1016/j.watres.2010.04.020.

Peter-Varbanets, M., Margot, J., Traber, J., Pronk, W., 2011. Mechanisms of membrane fouling during ultra-low pressure ultrafiltration. J. Membr. Sci. 377, 42–53. https://doi.org/10.1016/j.memsci.2011.03.029.

Peter-Varbanets, M., Dreyer, K., McFadden, N., Ouma, H., Wanyama, K., Etenu, C., Meierhofer, R., 2017. Evaluating Novel Gravity-Driven Membrane (GDM) Water Kiosks in Schools. Shaw, R.J. (ed). Local action with international cooperation to improve and sustain water, sanitation and hygiene (WASH) services: Proceedings of the 40th WEDC International Conference, Loughborough, UK. WEDC, Loughborough University.

Pronk, W., Ding, A., Morgenroth, E., Derlon, N., Desmond, P., Burkhardt, M., Wu, B., Fane, A.G., 2019. Gravity-driven membrane filtration for water and wastewater treatment: a review. Water Res. 149, 553–565. https://doi.org/10.1016/j.watres.2018.11.062.

Schumann, P., Ordóñez Andrade, J.A., Jekel, M., Ruhl, A.S., 2020. Packing granular activated carbon into a submerged gravity-driven flat sheet membrane module for decentralized water treatment. J. Water Process Eng. 38, 101517. https://doi.org/10.1016/j.jwpe.2020.101517.

Shi, D., Liu, Y., Fu, W., Li, J., Fang, Z., Shao, S., 2020. A combination of membrane relaxation and shear stress significantly improve the flux of gravity-driven membrane system. Water Res. 175, 115694. https://doi.org/10.1016/j.watres.2020.115694.

Stoffel, D., Rigo, E., Derlon, N., Staaks, C., Heijnen, M., Morgenroth, E., Jacquin, C., 2022. Low maintenance gravity-driven membrane filtration using hollow fibers: effect of reducing space for biofilm growth and control strategies on permeate flux. Sci. Total Environ. 811, 152307. https://doi.org/10.1016/j.scitotenv.2021.152307.

Tang, X., Ding, A., Qu, F., Jia, R., Chang, H., Cheng, X., Liu, B., Li, G., Liang, H., 2016. Effect of operation parameters on the flux stabilization of gravity-driven membrane (GDM) filtration system for decentralized water supply. Environ. Sci. Pollut. Res. 23, 16771–16780. https://doi.org/10.1007/s11356-016-6857-0.

Tang, X., Pronk, W., Traber, J., Liang, H., Li, G., Morgenroth, E., 2021. Integrating granular activated carbon (GAC) to gravity-driven membrane (GDM) to improve its flux stabilization: respective roles of adsorption and biodegradation by GAC. Sci. Total Environ. 768, 144758. https://doi.org/10.1016/j.scitotenv.2020.144758.

Tangsubkul, N., Parameshwaran, K., Lundie, S., Fane, A.G., Waite, T.D., 2006. Environmental life cycle assessment of the microfiltration process. J. Membr. Sci. 284, 214–226. https://doi.org/10.1016/j.memsci.2006.07.047.

Truttmann, L., Su, Y., Lee, S., Burkhardt, M., Brynjólfsson, S., Chong, T.H., Wu, B., 2020. Gravity-driven membrane (GDM) filtration of algae-polluted surface water. J. Water Process Eng. 36, 101257. https://doi.org/10.1016/j.jwpe.2020.101257.

Wan Osman, W.N.A., Mat Nawi, N.I., Samsuri, S., Bilad, M.R., Khan, A.L., Hunaepi, H., Jaafar, J., Lam, M.K., 2021. Ultra low-pressure filtration system for energy efficient microalgae filtration. Heliyon 7. https://doi.org/10.1016/j.heliyon.2021.e07367.

Wan, H., Mills, R., Wang, Y., Wang, K., Xu, S., Bhattacharyya, D., Xu, Z., 2022. Gravity-driven electrospun membranes for effective removal of perfluoro-organics from synthetic groundwater. J. Membr. Sci. 644, 120180. https://doi.org/10.1016/j.memsci.2021.120180.

Wang, Y., Fortunato, L., Jeong, S., Leiknes, T., 2017. Gravity-driven membrane system for secondary wastewater effluent treatment: filtration performance and fouling characterization. Sep. Purif. Technol. 184, 26–33. https://doi.org/10.1016/j.seppur.2017.04.027.

Waqas, S., Bilad, M.R., Man, Z., Wibisono, Y., Jaafar, J., Indra Mahlia, T.M., Khan, A.L., Aslam, M., 2020. Recent progress in integrated fixed-film activated sludge process for wastewater treatment: a review. J. Environ. Manag. 268, 110718. https://doi.org/10.1016/j.jenvman.2020.110718.

Waqas, S., Bilad, M.R., Man, Z.B., Suleman, H., Hadi Nordin, N.A., Jaafar, J., Dzarfan Othman, M.H., Elma, M., 2021. An energy-efficient membrane rotating biological contactor for wastewater treatment. J. Clean. Prod. 282, 124544. https://doi.org/10.1016/j.jclepro.2020.124544.

Westney, R.E., 1997. The Engineer's Cost Handbook: Tools for Managing Project Costs, 0 ed. CRC Press, https://doi.org/10.1201/9780203910016.

Woźniak, M.J., Prochaska, K., 2014. Fumaric acid separation from fermentation broth using nanofiltration (NF) and bipolar electrodialysis (EDBM). Sep. Purif. Technol. 125, 179–186. https://doi.org/10.1016/j.seppur.2014.01.051.

Wu, B., Hochstrasser, F., Akhondi, E., Ambauen, N., Tschirren, L., Burkhardt, M., Fane, A.G., Pronk, W., 2016. Optimization of gravity-driven membrane (GDM) filtration process for seawater pretreatment. Water Res. 93, 133–140. https://doi.org/10.1016/j.watres.2016.02.021.

Wu, B., Christen, T., Tan, H.S., Hochstrasser, F., Suwarno, S.R., Liu, X., Chong, T.H., Burkhardt, M., Pronk, W., Fane, A.G., 2017a. Improved performance of gravity-driven membrane filtration for seawater pretreatment: implications of membrane module configuration. Water Res. 114, 59–68. https://doi.org/10.1016/j.watres.2017.02.022.

Wu, B., Suwarno, S.R., Tan, H.S., Kim, L.H., Hochstrasser, F., Chong, T.H., Burkhardt, M., Pronk, W., Fane, A.G., 2017b. Gravity-driven microfiltration pretreatment for reverse osmosis (RO) seawater desalination: microbial community characterization and RO performance. Desalination 418, 1–8. https://doi.org/10.1016/j.desal.2017.05.024.

Wu, B., Soon, G.Q.Y., Chong, T.H., 2019a. Recycling rainwater by submerged gravity-driven membrane (GDM) reactors: effect of hydraulic retention time and periodic backwash. Sci. Total Environ. 654, 10–18. https://doi.org/10.1016/j.scitotenv.2018.11.068.

Wu, B., Zhang, Y., Mao, Z., Tan, W.S., Tan, Y.Z., Chew, J.W., Chong, T.H., Fane, A.G., 2019b. Spacer vibration for fouling control of submerged flat sheet membranes. Sep. Purif. Technol. 210, 719–728. https://doi.org/10.1016/j.seppur.2018.08.062.

Xu, J., Du, X., Zhao, W., Wang, Z., Lu, X., Zhu, L., Wang, Z., Liang, H., 2021a. Roofing rainwater cleaner production using pilot-scale electrocoagulation coupled with a gravity-driven membrane bioreactor (EC-GDMBR): water treatment and energy efficiency. J. Clean. Prod. 314, 128055. https://doi.org/10.1016/j.jclepro.2021.128055.

Xu, S., Lu, D., Qi, J., Wang, P., Zhao, Y., Zhang, H., Ma, J., 2021b. Gravity-driven multifunctional microporous membranes for household water treatment: simultaneous pathogenic disinfection, metal recycling, and biofouling mitigation. Chem. Eng. J. 410, 128289. https://doi.org/10.1016/j.cej.2020.128289.

Zainuddin, N.I., Bilad, M.R., Marbelia, L., Budhijanto, W., Arahman, N., Fahrina, A., Shamsuddin, N., Zaki, Z.I., El-Bahy, Z.M., Nandiyanto, A.B.D., Gunawan, P., 2021. Sequencing batch integrated fixed-film activated sludge membrane process for treatment of tapioca processing wastewater. Membranes 11, 875. https://doi.org/10.3390/membranes11110875.

Zhao, Y., Li, P., Li, R., Li, X., 2020. Characterization and mitigation of the fouling of flat-sheet ceramic membranes for direct filtration of the coagulated domestic wastewater. J. Hazard. Mater. 385, 121557. https://doi.org/10.1016/j.jhazmat.2019.121557.

Zhu, Z., Chen, Z., Luo, X., Zhang, W., Meng, S., 2020. Gravity-driven biomimetic membrane (GDBM): an ecological water treatment technology for water purification in the open natural water system. Chem. Eng. J. 399, 125650. https://doi.org/10.1016/j.cej.2020.125650.

Commercial scale membrane-based produced water treatment plant

Utjok W.R. Siagian[a], L. Lustiyani[b], K. Khoiruddin[b],
I.N. Widiasa[c], Tjandra Setiadi[d,e], and I.G. Wenten[b,e]

[a]Department of Petroleum Engineering, Bandung Institute of Technology, Bandung, Indonesia
[b]Department of Chemical Engineering, Bandung Institute of Technology, Bandung, Indonesia
[c]Chemical Engineering Department, Diponegoro University, Semarang, Indonesia [d]Center for Environmental Studies, Bandung Institute of Technology, Bandung, Indonesia [e]Research Center for Bioscience and Biotechnology, Bandung Institute of Technology, Bandung, Indonesia

1. Introduction

The oil and gas extraction processes generate a large volume of wastewater known as produced water, which has a complex composition and can cause environmental problems (Bagheri et al., 2018; Chen et al., 2014; Liang et al., 2019; Varjani et al., 2017). The properties and pollutants in produced water may vary, depending on the location and geological conditions (Fillo and Evans, 1990; Johnson et al., 2008; Veil et al., 2004). Typical contaminants in produced water are dispersed oil or grease, dissolved organics, suspended solids, dissolved gas, salts, and heavy metals (Arthur et al., 2005; Siagian et al., 2018a). These contaminants will be the target of produced water treatment, and their removal will depend on the final purpose of the produced water treatment. Produced water treatment usually requires a series of separation steps to remove the contaminants since a single separation unit would not be adequate (Al-Ghouti et al., 2019).

Produced water can be reinjected, reused, or discharged after an appropriate treatment to minimize environmental problems (Nasiri and Jafari, 2017). In reinjection scheme, produced water is injected into a formation well. It sometimes helps to enhance oil recovery in a formation well. Produced water contains a large volume of water that can be reused for other purposes, including process water or land application. Reusing produced water may save the water resource. For the discharge objective, treated produced water should meet the

Copyright © 2023 Elsevier Inc. All rights reserved.

regulation for onshore or offshore discharge (Arthur et al., 2005). The treatment should ensure the treated produced water is not harmful to the receiving environment.

Various separation methods are available for produced water treatment. Typical conventional methods, which are typically used for produced-water treatment are coagulation-flocculation, hydro-cyclone, adsorption, and gas flotation (Halim et al., 2021). However, conventional methods usually do not adequately satisfy petroleum industry requirements of discharge and reuse standards (Alzahrani and Mohammad, 2014; Fakhru'l-Razi et al., 2009). They are usually limited to removing oil/water emulsion or small oil droplets, especially those with sizes of less than 10 μm (Halim et al., 2021). In addition, conventional methods usually require chemical consumption additions, high energy requirement, and large footprint (Samuel et al., 2022).

Membrane technology has been considered as an interesting alternative to conventional processes in the petroleum industry since the early 20th century (Winston Ho and Sirkar, 1992). Membranes have been applied in various processes commercially, including wastewater treatment and water recycling (Bernardo and Drioli, 2010; Buonomenna, 2013). Membranes offer several attractive features in the separation process. Membrane-based processes use semipermeable membranes with fine pores, where separation can take place at the molecular level. Hence, membrane-based processes can attain high separation performance. It has been reported that membranes have greatly improved the separation of different components in oilfield produced water (Ji, 2015; Nazirah Wan Ikhsan et al., 2017; Qiao et al., 2008; Scott, 1995; Yu et al., 2017). In addition, membrane-based separation processes do not involve phase changes. Therefore, they require less energy consumption than conventional separation processes, such as distillation or evaporation (Ashaghi et al., 2008; Cui et al., 2008; Ebrahimi et al., 2010; Hua et al., 2007). Another attractive feature of membranes is theirs high packing density or large surface area per volume, which lowers the footprint. Membranes are also modular, making them easy to scale up (Ashaghi et al., 2008; Owen et al., 1995; Padaki et al., 2015; Tawalbeh et al., 2018). These advantages have attracted researchers from academia and industry to explore more membrane applications in produced water treatment.

Numerous studies related to produced water treatment and membranes have been reported. One of the major disadvantages of membrane operation is the fouling phenomenon. Fouling occurs due to contaminant accumulation on the membrane surfaces and pores. This accumulation forms a new layer on the membrane surface or plug in the membrane pores. As a result, membrane productivity decreases over time. To keep the membrane flux at the desired value, the transmembrane pressure is increased. However, this action results in increasing energy requirement. Therefore, periodic cleaning is needed to regenerate the membrane. In some cases, membrane cleaning may require chemical agents. Thus, it will add more operational costs. To address the fouling issue and related problems, many efforts have been devoted to find fouling control strategies (Nawaz et al., 2021; Silalahi and Leiknes, 2011; Virga et al., 2021; Xia et al., 2018; Xue et al., 2016). As shown in Fig. 1, membrane fouling is one of the important topics in produced-water treatment using membranes.

This chapter summarizes the application of membranes in produced-water treatment. The chapter starts with a discussion of various technologies for produced-water treatment, membrane processes, and integrated membrane systems. Membrane processes are generally classified based on the driving force involved in the separation process. There is also a particular

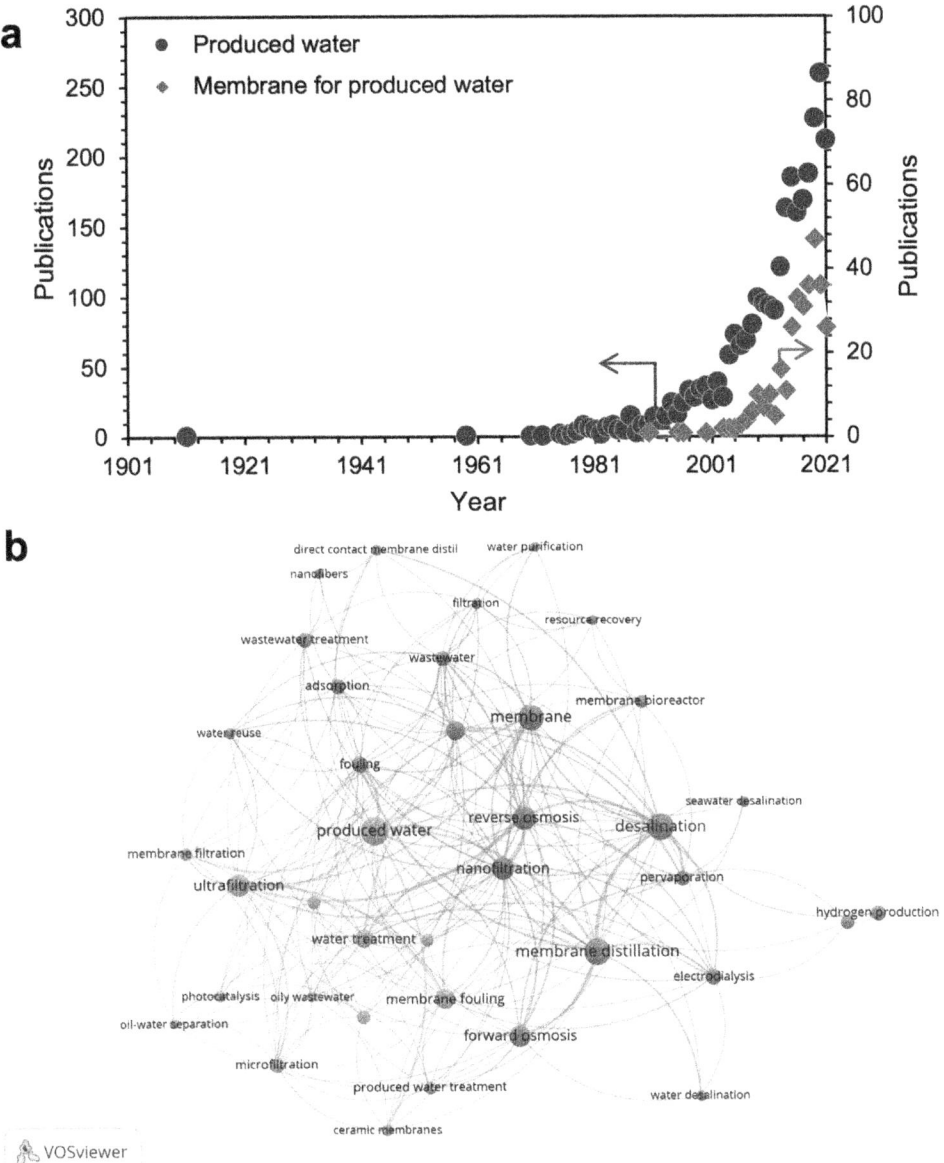

FIG. 1 The result of bibliographic analysis by using VOSviewer. Analysis is based on co-occurrence of keywords related to membranes for produced-water treatment.

class of membrane processes, known as hybrid processes, such as membrane bioreactors (MBRs). Even though most membrane-based water treatments employ membrane filtration, other membrane processes are also discussed here. Special attention is given to commercial-scale membrane technologies applied in produced-water treatment plants.

B. Applications of membrane technology for water and wastewater treatment

2. Produced-water treatment steps and technologies

Produced-water treatment is typically aimed to remove oil, dissolved organics, suspended solids, dissolved gas, salinity, hardness, and naturally occurring radioactive materials (Arthur et al., 2005). Produced-water treatment is divided into the primary, secondary, and tertiary processes to achieve these goals (Jafarinejad and Jiang, 2019). Oilfield produced-water treatment usually applies primary and secondary oil-water separation followed by biological and tertiary treatments or polishing (Al-Ghouti et al., 2019; Alzahrani and Mohammad, 2014; Jafarinejad and Jiang, 2019; Munirasu et al., 2016). These steps comprise several technologies, such as conventional technologies summarized in Table 1.

The primary oil- and water-separation step involves the removal of oil droplets and particles by using common separators such as American Petroleum Institute (API) separator, corrugated plate interceptor (CPI), parallel plate interceptor (PPI), tilted plate interceptor (TPI), hydro-cyclone separators, and buffer and/or equalization tanks (Al-Ghouti et al., 2019; Jafarinejad and Jiang, 2019). Meanwhile, gas flotation is usually employed in secondary oil and water separation to remove much smaller oil droplets and particles (Al-Ghouti et al., 2019). The most common gas-flotation technologies used in this step are dissolved air flotation (DAF), dissolved gas flotation (DGF), induced air flotation (IAF), and induced gas flotation (IGF) (Jafarinejad and Jiang, 2019). Akvola Technologies introduced another alternative system. Akvola Technologies has developed a flotation-filtration system called as AkvoFloat, which combines air flotation in the first stage and filtration with submerged ceramic membranes in the next stage. This technology can effectively achieve suspended solids and oil removal from produced water (Beery et al., 2015).

Conventional technologies, such as API gravity separator, hydro-cyclone, and gas flotation have their associated disadvantages in produced-water treatment, particularly in oil separation. A gravity separator is usually effective only for free oil, while hydro-cyclone is less effective for removing oil droplets that have particle sizes of less than 10 μm (Drioli et al., 2016). Air flotation needs huge volume of air and long retention time (or large footprint) (Drioli et al., 2016). In this case, membrane filtration (microfiltration (MF) and ultrafiltration (UF)) could be a promising alternative.

Biological treatments consisting of aerobic and anaerobic processes are usually considered as a secondary treatment to remove remaining dissolved oil and organic compounds (Cerqueira et al., 2011). Activated sludge, sequencing batch reactors, continuous stirred-tank bioreactor, and MBR are available biological treatments that can be used to remove organic substances from produced water (Jafarinejad and Jiang, 2019). Sometimes, biological processes are combined with physical processes. A typical example of combined physical-biological processes is operated by Pertamina Indonesia (unit IV) (Edwan Kardena, 2015). This treatment facility has a capacity of $4000\,\text{m}^3$ produced water/day. The primary treatment is to remove oil and organics compounds associated with oil by using physical processes consisting of API, CPI, and DAF. In the next step, an activated sludge system serves as the tertiary treatment that removes the remaining organic component and production chemicals within produced water. The system is able to achieve 88.4%, 90.2%, 97.8%, 96%, 98%, and 99.4% removal efficiencies for chemical oxygen demand (COD), biochemical oxygen demand (BOD), oil, ammonia, sulfide, and phenol, respectively (Edwan Kardena, 2015).

TABLE 1 Conventional produced water treatment technologies (Fakhru'l-Razi et al., 2009; Igunnu and Chen, 2014; Rezakazemi et al., 2017; Saththasivam et al., 2016).

Technology	Description	Pre- and post-treatment	Energy	Chemical use	Advantages	Disadvantages
Media filtration	Media: walnut, sand, and anthracite	None	For backwashing filters	Regenerant and Coagulants	Oil and grease removal rate: >90% For final polishing	Chemical regeneration of media Inefficient for <50mg/L oil content Disposal of used media filter
Gas flotation	Gas bubbles facilitates the flotation of oil particles	Pretreatment: coagulation (optional) Posttreatment: none	For aeration	Coagulants may be required to remove target contaminants	~100% water recovery No moving parts Simple operation Robust and durable	Sludge handling is required Energy consumption is relatively high compared to other technologies Generation of large volume of air Long retention time Large skim volume
Evaporation pond	Applicable to any kind of produced water	Pretreatment: none Posttreatment: depending on product water quality	None, except for pumping water to/from the pond	None	Chemicals and energy are not required	Water loss by evaporation Waste or sludge disposal
Ion exchange	Ion exchange between bulk solution and resins	Pre- and post-treatments are required for high efficiency	0.07 kWh/bbl (for 200gpm flow rate and 5m pumping head)	Regenerant solutions: H_2SO_4, NaOH, HCl, NaCl or Na_2CO_3, H_2O_2 or NaOCl	Suitable for ions and salt removal The life cycles of anion-exchange resins and cation-exchange resins are 4–8years and 10–15years, respectively	Large scale application may be costly (especially for high concentration)

Continued

TABLE 1 Conventional produced water treatment technologies (Fakhru'l-Razi et al., 2009; Igunnu and Chen, 2014; Rezakazemi et al., 2017; Saththasivam et al., 2016)—cont'd

Technology	Description	Pre- and post-treatment	Energy	Chemical use	Advantages	Disadvantages
Biological aerated filter (BAF)	Active biomass attaches on medium Biological process	Pretreatment: sedimentation may be required Posttreatment: none	1–4 kWh	None	Water recovery is almost 100%	Sludge disposal cost is almost 40% of the overall cost
Chemical oxidation	Degradation of organic compounds by chemical oxidants	None	Energy consumption: 18% of the total operation and maintenance	Oxidants: chlorine, chlorine dioxide, permanganate, oxygen, and ozone	Simple equipment No waste generation Almost 100% water recovery	Chemical consumption depends on pollutant concentration and produced water volume The chemical dosing pump requires regular calibration and maintenance

The higher environmental standard and economic constraints require more advanced methods, but some of these methods are still developed in small-scale applications (Ali et al., 2018; Jafarinejad and Jiang, 2019). Typical examples are chemical treatment technologies, such as chemical precipitation (Fakhru'l-Razi et al., 2009; Frankiewicz and Gerlach, 2000), electrochemical processes (Ma and Wang, 2006), room temperature ionic liquids (McFarlane et al., 2005), and demulsifiers (Holloway, 1977). Other technologies such as chemical oxidation can remove hydrocarbons, acid, base and neutral organics, volatile and nonvolatile organics, hydrogen sulfide (H_2S), and particulates (Ashaghi et al., 2008). They rely on oxidation/reduction reactions (Guerra et al., 2011). Furthermore, advanced oxidation processes (AOPs) such as electrochemical oxidation, Fenton reaction, ozone treatment, and photocatalytic oxidation can remove oils, fats, and metals as well as 90% of BOD/COD from produced water (Coday et al., 2014; dos Santos et al., 2014; Neyens and Baeyens, 2003). Chemical oxidation requires chemical agents, including the handling, storage, and byproduct treatment, while ozonation usually needs high energy consumption.

Adsorption can be another alternative technology for produced-water treatment. A variety of materials, including organoclays (Masooleh et al., 2010), zeolites (Feng et al., 2015), chitosan, and activated carbon (Fouladi Tajar et al., 2009), can be adsorbents of pollutants in produced-water treatment. Adsorption mostly removes oil, grease, organic substances, and heavy metals (Allen, 2008; Fakhru'l-Razi et al., 2009; Guerra et al., 2011). In practice, the adsorption process is utilized together with other separation units rather than as a stand-alone process (Saleh and Gupta, 2014). It is usually applied in the tertiary treatment step. However, chemical agents will be needed to regenerate the medium after its active sites are fully occupied (Apul and Karanfil, 2015). Moreover, the used material needs disposal handling. Adsorption also works at a high retention time, requiring ample space (Drioli et al., 2016).

Total oil remediation and recovery (TORR) is a commercial technology that The EARTH Canada Corporation develops. TORR combines multistage adsorption and another separation system to effectively remove dispersed oil in water (Plebon, 2007). Another commercial technology that shows excellent suspended solids and hydrocarbon removal from produced water is Power Clean Nutshell filters from Veolia Water Technologies (Veolia Water Technologies, 2014). Both TORR and Power Clean are equally effective in removing suspended solids, oils, and metallic hydroxides from wastewater (Veolia Water Technologies, 2014). Besides the technologies mentioned above, other technologies that can remove suspended solids and oil effectively are membranes. Membrane technologies for produced-water treatment will be discussed in the next section.

3. Membrane technologies in produced-water treatment

3.1 Membrane processes

The application of membrane technology in the tertiary treatment of produced water is rapidly increasing (Alzahrani et al., 2013; Munirasu et al., 2016). Membranes are applied for recycling produced water to overcome water supply problems while complying with environmental regulations regarding produced-water discharge (Alzahrani and Mohammad,

2014). Membranes are generally classified based on the driving force of the process, namely, pressure-driven, concentration-driven, thermal-driven, and electrical-driven membrane processes. Microfiltration (MF), ultrafiltration (UF), nanofiltration (NF), and reverse osmosis (RO) are classified as pressure-driven membranes that utilize pressure differences to drive the transport of species through the membrane. The membrane has a pore size that follows the following order: $MF > UF > NF > RO$ (Siagian et al., 2021; Wenten et al., 2017b). The RO membrane has the smallest pore size, allowing only water to pass through it. Hence, it typically needs a high operating pressure.

The concentration-driven process includes dialysis, membrane gas separation, and pervaporation, wherein concentration differences across the membrane drive species separation. Pervaporation is frequently used in the dehydration of organic solvents, such as ethanol dehydration (Wenten et al., 2017a, 2020c). In this application, water is removed by permeation and evaporation processes through the membrane simultaneously. There are also many efforts to employ pervaporation in the desalination process, including produced water-desalination (Almarzooqi et al., 2021; Huth et al., 2014). However, the flux of this process is usually low.

Membrane distillation (MD) is a thermal-driven membrane that employs a hydrophobic porous membrane to separate the feed and distillate phase. Water vapor from the feed phase passes through the membrane and condenses in the distillate phase. This process is suitable for the desalination of high salinity solution (Salehi and Rostamani, 2013). Relatively pure water can be obtained from MD. It is also suitable for treating produced water with at high temperatures, such as those coming from steam-assisted gravity drainage (Drioli et al., 2016). To treat such produced water, MD will require no additional heating processes. It was found in a study that the highest water flux of $195 \, kg/m^2.h$ is obtained by direct contact MD when it is used to treat produced water containing $10,000 \, mg/L$ NaCl at $128°C$ (Singh and Sirkar, 2012). Meanwhile, electrodialysis (ED), electrodialysis reversal (EDR), electrodeionization (EDI), and membrane capacitive deionization (MDCI) are membrane processes that utilize electric potential difference as the driving force of species transport and charged membranes as separator (Hakim et al., 2019; Handojo et al., 2019; Wenten et al., 2020b). These processes are only effective for charged species.

In addition, there is also a hybrid membrane, which combines membrane and other processes in a single system, such as membrane reactor (MR) and MBR. MBR has been widely applied to wastewater treatment. Combining biological reactors with membrane separation provides advantages compared to conventional biological processes (Steven et al., 2022; Subagjo et al., 2015; Wenten et al., 2020a). Therefore, the application of MBR in wastewater treatment keeps increasing.

Membranes are made of polymer, inorganics, liquid, or mixed matrix (a mixture of organic and inorganic). Polymeric membranes are usually less costly, but their application is limited under nonharsh conditions. Inorganic membranes, such as ceramic, are mostly thermally and chemically resistant, but they have high cost. Sometimes, combining polymeric material and inorganic nanoparticles is preferable to obtain both materials' synergetic effect. Membrane properties, including pore size, hydrophilicity, chemical properties, and surface roughness, determine the separation performance of membrane-based processes. Permeability and selectivity are two most important parameters of membrane performance. A higher permeability is favorable for producing the same capacity but with lower energy and smaller surface area.

Meanwhile, a higher selectivity is needed to achieve a high separation level. A materials approach is one of the effective ways to improve membrane separation properties, such as permeability, selectivity, and fouling resistivity (Lee et al., 2011; Li et al., 2016; Ong et al., 2016; Werber et al., 2016).

Table 2 shows the comparison of produced-water treatment with membrane technologies. MF, UF, NF, and RO have been applied in produced-water treatment for removing suspended solids, turbidity, oil, organics, and dissolved solids. MF can remove suspended solids and fine particulates effectively, while UF can attain high removals of colloids and emulsified oil (Saththasivam et al., 2016). MF is able to remove suspended solids from produced water. A ceramic MF membrane with 0.1 μm pore size can remove dispersed oil, but the removal rate is ~60% (Ebrahimi et al., 2010). An α-alumina membrane with 0.8, 0.1 μm pore size shows 99% removal of oil from a synthetic produced water containing 250 mg/L of hydrocarbon (Mueller et al., 1997). An UF membrane is usually employed to remove colloids, viruses, and other macromolecules from water. A UF membrane with a lower pore size (0.05 μm) than MF, can achieve ~99% dispersed-oil removal (Ebrahimi et al., 2010). A study showed that a PVDF membrane modified by nano-alumina could achieve 98% rejection of TOC and 90% rejection of COD (Li et al., 2006). However, both MF and NF cannot be used to remove salts from a high salinity produced water. Membrane with smaller pore size, such as NF and RO, are required to remove the smaller molecular weight molecules, including dissolved solids, salts, etc. NF is suitable for removing bivalent and multivalent salts, while the rejection of monovalent salts is relatively low. RO is efficient in eliminating total dissolved solids (TDS), COD, BOD_5, and color (Salahi et al., 2010, 2012). RO is usually used for the desalination of produced water, but it typically uses high operating pressure. For instance, to reduce TDS of produced water from 18,900 to 192 mg/L, an RO is operated at 45 bars (Guo et al., 2018). For higher TDS concentration, RO would need more considerable operating pressure. Hence, it would be impractical to desalinate produced water with high salinity due to osmotic-pressure limitations.

ED and EDR are electro-membrane processes often used for desalination or deionization purposes (Drewes et al., 2009). Separation in ED is based on the Donnan exclusion mechanism. ED uses cation- and anion-exchange membranes, respectively, which are permselective

TABLE 2 Produced water treatment by membrane technologies (Drewes et al., 2009; Igunnu and Chen, 2014; Interstate Oil and Gas Compact Commission (IOGCC), 2006).

Technology	Description	Pre- and post-treatment	Energy use	Chemical use
MF	Micro-particles removal Suspended solids removal Use low hydraulic pressure Use porous membrane Operated at <2 bar	Pretreatment: cartridge filtration and coagulation Posttreatment: depending on the product water	No data available	Coagulants for pretreatment Cleaning chemicals

Continued

TABLE 2   Produced water treatment by membrane technologies (Drewes et al., 2009; Igunnu and Chen, 2014; Interstate Oil and Gas Compact Commission (IOGCC), 2006)—cont'd

Technology	Description	Pre- and post-treatment	Energy use	Chemical use
UF	Ultra-particles removal Turbidity and colloids removal Macromolecule removal Use low hydraulic pressure Use porous membrane Operated at <10 bar	Pretreatment: cartridge filtration and coagulation Posttreatment: if product water quality is unsatisfied	No data available	Coagulants for pretreatment Cleaning chemicals
NF	Hardness removal Multivalent ion removal Metal removal Organic removal Use high hydraulic pressure Use dense membrane Operated at 5–25 bar	Pretreatment: antiscaling injection Posttreatment: product water may require remineralization	0.08 kWh/bbl	Antiscaling agent Cleaning chemicals
RO	Desalination Use high hydraulic pressure Use dense membrane Operated at 10–60 bar	Pretreatment: fouling and scaling prevention step Posttreatment: product water remineralization or pH stabilization	Seawater RO: 0.46–0.67 kWh/bbl (with energy recovery device) Brackish water RO: 0.02–0.13 kWh/bbl	Antiscaling agent Cleaning chemicals
FO	Removal of all particulate matter and almost all dissolved constituents Utilize osmotic pressure difference Use dense membrane Use draw solution	Pretreatment: prefilter for large debris removal; antiscaling agent Draw solution regeneration system	Energy for circulation of draw solution and feed or wastewater through FO module	Antiscaling agent Cleaning chemicals
ED/EDR	Desalination Salt recovery Use electrical potential difference Use charged membrane	Pretreatment: particulates removal, pH adjustment, and antiscaling injection Posttreatment: product water remineralization and disinfection	0.14–0.20 kWh/lb NaCl equivalent	Scale inhibitor Cleaning chemicals
MD	Desalination Use porous hydrophobic membrane Use thermal	Pretreatment: removal of components in feed that can wet the membrane Posttreatment: product water remineralization	For heater	Scale inhibitor Cleaning chemicals

for cations and anions. When a solution containing ions is transferred to the ED unit and electric potential is supplied to the ED electrode, ions migrate from the feed to the concentrate compartment through the charge membranes. Two streams are produced from ED, i.e., diluate stream, which has less ion concentrations and the concentrate stream, which contains high ion concentrations. EDR is almost the same as ED, except that it is operated with periodic polarity reversal. The electrode is reversed periodically to detach the pollutant from the membrane surface. The periodic-polarity reversal is effective in controlling fouling (Hansima et al., 2021; Merkel and Ashrafi, 2019). Produced-water treatment using ED and EDR mainly removes salinity or desalination and valuable salt recovery (Finklea et al., 2022; Jing et al., 2011; Sirivedhin et al., 2004; Sosa-Fernandez et al., 2019, 2021). ED can achieve higher salt concentrations in concentrate solution and more significant water recovery than RO. The higher concentration and recovery are because ED and EDR are not limited by osmotic pressure. Despite the advantages of high water recovery compared to RO, these processes usually need high energy consumption and high cost (Sirivedhin et al., 2004).

MBR combines biological reactions for pollutant degradation with membrane separation for extracting clean water from the bioreactor. The membrane also retains the suspended solids in a bioreactor, resulting in mixed high-liquor suspended solids (MLSS). A high MLSS makes the biodegradation of organic compounds in MBR highly effective. MBR has been applied in industrial and domestic wastewater treatment at megaliter-per-day capacities (thembrsite.com, 2021). MBR technology has also been employed for produced-water treatment (Fulazzaky et al., 2020). MBR displays high separation efficiency (more than 90%) for COD, oil, grease, and NH_3, confirming its potential application in produced-water treatment. A more detailed discussion regarding the application of MBR in produced-water treatment can be found elsewhere (Asante-Sackey et al., 2022). It is worth noting that the biological processes of MBR are significantly affected by salt content. It should be considered in designing wastewater-treatment facilities of produced water with high salinity. Moreover, when working with real produced water, MBR suffers from fouling phenomena (Asante-Sackey et al., 2022; Drioli et al., 2016). Therefore, an appropriate fouling mitigation approach is needed.

Membranes are potential processes for produced-water treatment with advantages and features, especially in tertiary treatment or polishing (Alzahrani and Mohammad, 2014). Moreover, regarding their benefits, it is predicted that membranes applied in case of zero discharge should be implemented in produced-water treatment (Jepsen et al., 2018).

3.2 Membrane fouling

Despite their excellent separation performance, membrane operations are usually hampered by fouling. Fouling is the most important limitation for applying membrane technology in industrial water and wastewater treatment (Dickhout et al., 2017; Madaeni, 1999; Siagian et al., 2018b; Tawalbeh et al., 2018). Membrane fouling is indicated by flux decline over time, caused by solute accumulation in membrane pores and on the surface. The mechanism can be different, depending on the membrane structure and pollutant properties (Fig. 2A and B). In pressure-driven membranes, fouling increases transmembrane pressure (TMP) over time to overcome the loss of membrane productivity (Badrnezhad and Beni, 2013; Khor, 2017). The

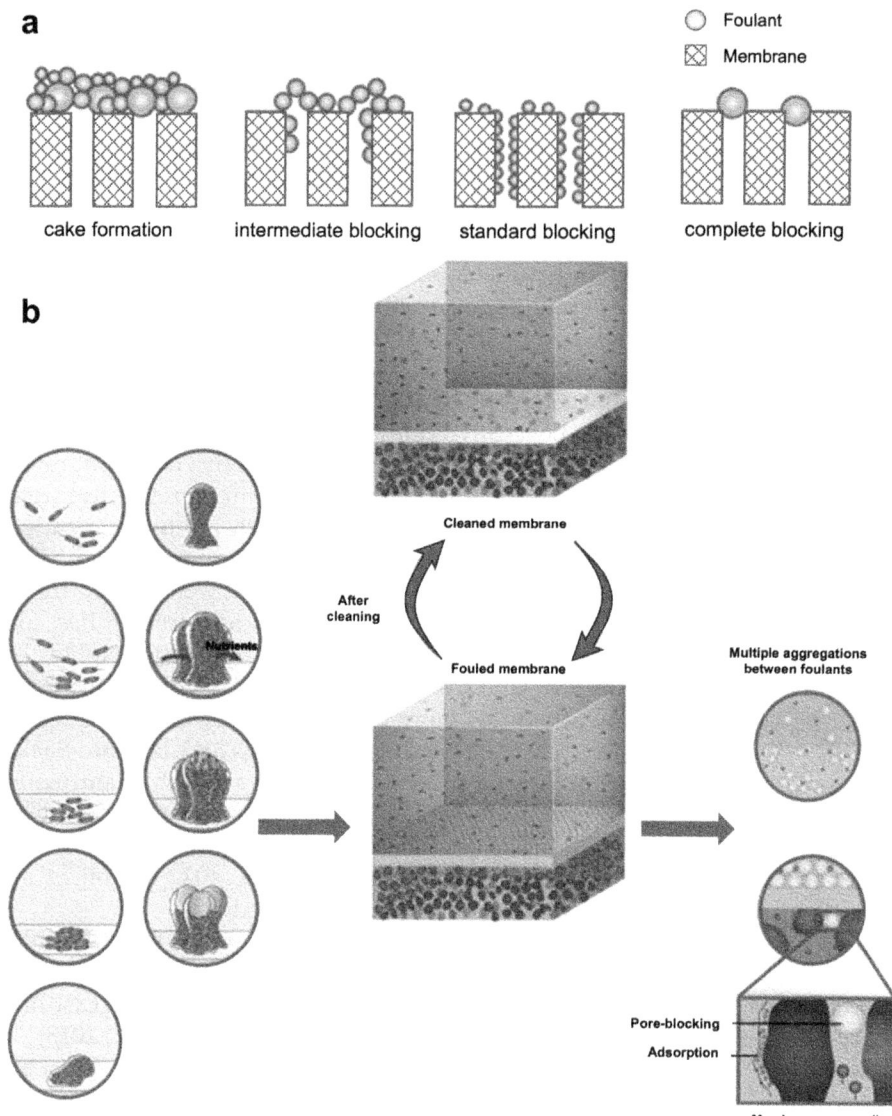

FIG. 2 Membrane fouling. (A) Fouling formed on porous membrane. (B) Various fouling observed on NF and RO during produced water treatment. *(A) Reproduced with permission from Wenten, I.G., Victoria, A.V, Tanukusuma, G., Khoiruddin, K., Zunita, M., 2019. Simultaneous clarification and dehydration of crude palm oil using superhydrophobic polypropylene membrane. J. Food Eng. 248, 23–27. https://doi.org/10.1016/j.jfoodeng.2018.12.010, copyright © 2019, Elsevier. (B) Reproduced with permission from Alzahrani, S., Mohammad, A.W., Hilal, N., Abdullah, P., Jaafar, O., 2013. Identification of foulants, fouling mechanisms and cleaning efficiency for NF and RO treatment of produced water. Sep. Purif. Technol. 118, 324–341. https://doi.org/10.1016/j.seppur.2013.07.016, copyright © 2013, Elsevier.*

productivity loss will eventually increase operational and maintenance costs (Jafarinejad, 2020; Xue et al., 2016). The produced water exhibits high fouling and scaling potential upon concentration due to complex contaminants.

Some fouling control strategies have been proposed by many studies reported in the literature. In general, membrane fouling control includes:

- Selection of appropriate pretreatment processes.
- Development of an antifouling membrane.
- Improvement of hydrodynamic conditions.
- Optimization of operating conditions.
- Regeneration of the membrane with an optimized cleaning procedure.

Membrane property is one of the crucial parameters that affect fouling phenomena. Fig. 3A shows how the pore size of the MF membrane affects the membrane fouling rate. There is increasing research in developing antifouling membranes (Karkooti et al., 2020; Liang et al., 2021; Mahmodi et al., 2022; Shamaei et al., 2020; Yang et al., 2015; Zhang et al., 2017). Membrane properties are engineered for minimizing foulant attachment or adsorption on the membrane surface. Membrane hydrophilization is usually selected for decreasing foulant attachment (Khoiruddin et al., 2021; Wardani et al., 2019, 2020; Wenten et al., 2020d). Since contaminants in produced water, such as oil and grease, and other organic compounds, are primarily hydrophobic, these contaminants are hard to attach on a highly hydrophilic membrane. In addition, foulant removal by cleaning procedures will be more efficient in highly hydrophilic surfaces.

In San Ardo's produced water-reclamation plant, organics fouling and silica scaling are observed, and they can be controlled by operating the membrane process at high pH conditions (Webb et al., 2009). Pretreatment is also provided before RO systems such as chemical softening and ion-exchange softening technologies to reduce fouling (Webb et al., 2009). Fig. 3B shows how pretreatment, such as flocculation, improves membrane flux stability. Another approach for fouling control is improving the hydrodynamic conditions by a turbulence promotor. Regarding turbulence promotors, promoting turbulence often gives better fouling control, but at the cost of membrane integrity. Fouling by oil is more difficult to remove than fouling by particulates, which makes cleaning efficiency low (Tanudjaja et al., 2019). In addition, chemical cleaning with harsh conditions may damage the membrane (Mansouri et al., 2010). Therefore, optimum fouling control is crucial to maintain membrane performance and decrease cleaning frequencies (Tanudjaja et al., 2019). Physical cleaning, such as backflushing, can minimize flux loss. A typical example is shown in Fig. 3C. This backflushing was operated on a ceramic membrane. Backflushing is usually effective in removing foulants from the membrane pore. However, for irreversible fouling, chemical cleaning is inevitable.

4. Integrated membrane system

Most of the technologies described previously are suitable for removing specific components but unsuitable for others. In many cases, produced-water treatment facility is a combination of different processes or technologies. This combination or integration aims to achieve

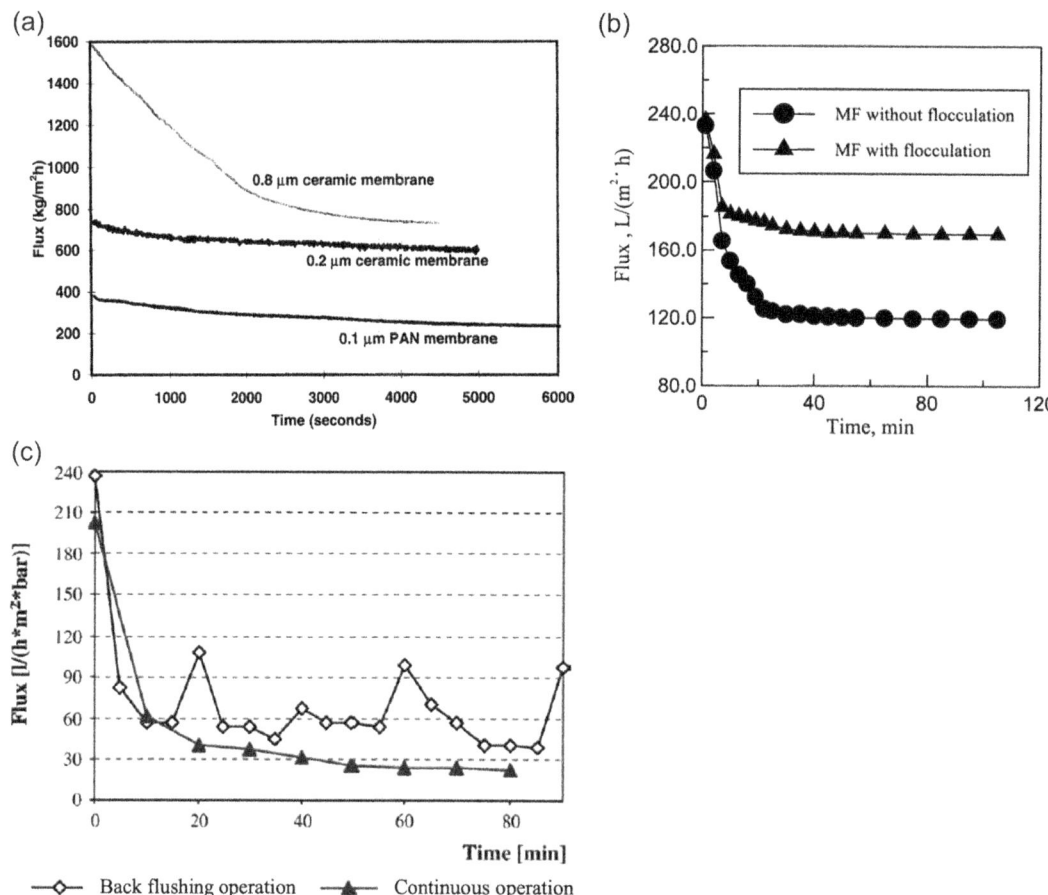

FIG. 3 Flux decline caused by membrane fouling during the treatment of oily wastewater including produced water. (A) Flux decline on different membrane types. (B) Effect of flocculation on MF flux profile. (C) Effect of back flushing on membrane flux. *(A) Reprinted with permission from Mueller, J., Cen, Y., Davis, R.H., 1997. Crossflow microfiltration of oily water. J. Membr. Sci. 129, 221–235. https://doi.org/10.1016/S0376-7388(96)00344-4, copyright © 1997, Elsevier. (B) Reprinted with permission from Zhong, J., Sun, X., Wang, C., 2003. Treatment of oily wastewater produced from refinery processes using flocculation and ceramic membrane filtration. Sep. Purif. Technol. 32, 93–98. https://doi.org/10.1016/S1383-5866(03)00067-4, copyright © 2003, Elsevier. (C) Reprinted with permission from Ebrahimi, M., Willershausen, D., Shams, K., Engel, L., Placido, L., Mund, P., Bolduan, P., Czermak, P., 2010. Investigations on the use of different ceramic membranes for efficient oil-field produced water treatment. Desalination 250, 991–996. https://doi.org/10.1016/j.desal.2009.09.088, copyright © 2010, Elsevier.*

a higher separation efficiency or produce higher quality effluent. For instance, a standalone membrane unit will experience severe fouling from various pollutants. Thus, other processes are installed before the membrane to remove contaminants and minimize fouling formation (Qiao et al., 2008).

A combination of membrane and conventional processes has been implemented in many membrane installations worldwide (Cheryan and Rajagopalan, 1998). Integrating several

TABLE 3 Configuration of membrane system classified on treatment purposes.

Purpose of treatment	Configuration	Ref.
Reinjection into low permeability oil reservoirs	Gravity sedimentation, coagulation, sand filtration, disinfection, and UF (hydrophilized PVDF membrane)	Xu et al. (2016)
Reinjection into low permeability oil reservoirs	Scheme 1: MBR and NF; Scheme 2: MBR and RO	Kose Mutlu et al. (2019)
Discharging to surface water or injection into low-osmosis oilfield	Aeration, DAF, sand filtration, and UF	Qiao et al. (2008)
Discharging into surface water	Coarse filtration and combined air-lift reactor (biological treatment)-MF	Campos et al. (2002)
Reuse for drinking water	Media filter, chemical softening, fixed-bed biological organics oxidation, pressure filtration, ion-exchange softening, and high-pH RO	Doran et al. (1997) and Wong and Hung (2006)
Reuse for drinking water	DAF, clarifier, softening processes (using lime, zeolite, and weak acid), RO, and water reconditioner	Tao et al. (1993)
Reuse for irrigation	Chemical oxidation, electrocoagulation, DAF, UF, and activated carbon	Pica et al. (2017)
Reuse for agriculture	Warm softening, filtration (pressure filtration and cartridge filtration), RO, and disinfection	Funston et al. (2002)
Reuse for industrial process water	Warm softening, equalization storage, and pH adjustment	Funston et al. (2002)
Reuse for industrial process water	Clarification, media filtration, UF, and RO	Madwar and Tarazi (2003)
Reuse for industrial process water (cooling tower)	Gravity separation (API separator), evaporation (basin), chemical treatment, flotation, bioprocess, clarification (clarifier), sedimentation, granular activated carbon, and RO	Manouchehr and Mishana (2008)

membrane processes or traditional techniques is usually called an integrated membrane system. An integrated system can attain maximum separation performance in produced water treatment, as reported in Han et al. (2016). In an integrated system, membrane technology acts as a complement rather than a replacement of conventional methods in produced-water treatment. Hence, integration of membrane and other processes offers more advantages while overcoming the drawbacks of each (Tanudjaja et al., 2019). Unfortunately, no standard systems are available for produced-water treatment according to the final purpose (Alzahrani and Mohammad, 2014). Table 3 shows different configurations of produced-water treatment, while examples of the membrane process schematic in produced-water treatment are shown in Fig. 4.

In oily wastewater treatment, it is suggested to employ an aeration tank, air flotation, and sand filter as the pretreatment of UF membranes (Qiao et al., 2008). This configuration has been evaluated in a pilot-scale test. The reported results show that this configuration

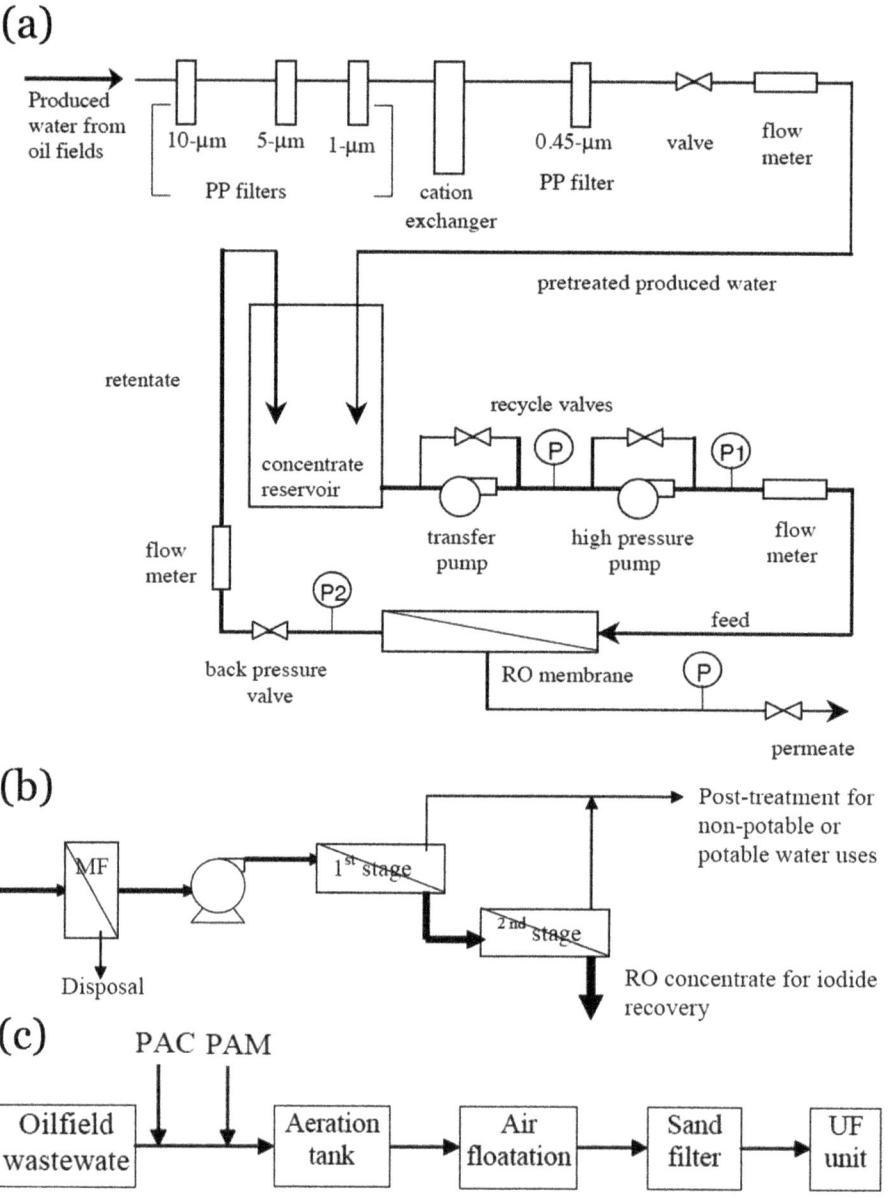

FIG. 4 Schematics of membranes for produced-water treatment. (A) Filter, ion exchange, and RO. (B) MR and double-stage RO. (C) Coagulation, flotation, sand filter, and UF. *(A) Reprinted with permission from Murray-Gulde, C., Heatley, J.E., Karanfil, T., Rodgers Jr., J.H., Myers, J.E., 2003. Performance of a hybrid reverse osmosis-constructed wetland treatment system for brackish oil field produced water. Water Res. 37, 705–713. https://doi.org/10.1016/S0043-1354(02)00353-6, copyright © 2003, Elsevier. (B) Reprinted with permission from Xu, P., Drewes, J.E., Heil, D., 2008. Beneficial use of co-produced water through membrane treatment: technical-economic assessment. Desalination 225, 139–155. https://doi.org/10.1016/j.desal.2007.04.093, copyright © 2008, Elsevier. (C) Reprinted with permission from Qiao, X., Zhang, Z., Yu, J., Ye, X., 2008. Performance characteristics of a hybrid membrane pilot-scale plant for oilfield-produced wastewater. Desalination 225, 113–122. https://doi.org/10.1016/j.desal.2007.04.092, copyright © 2008, Elsevier.*

effectively removes oil and suspended solids. The removal rates of this system for oil and suspended solids are 0.5 and 1.0 mg/L, respectively (Qiao et al., 2008). This result can meet effluent specifications for produced-water discharge or reinjection (Qiao et al., 2008).

A pilot study was conducted for produced-water treatment at Placerita Canyon Oil Field, Los Angeles County, California (Alzahrani and Mohammad, 2014; Funston et al., 2002). The researchers evaluated process conditions and schemes to achieve beneficial reuse options of produced water, including industrial, irrigation, and potable water use. Unit operations used in the pilot plant were walnut shell filter, warm lime clarifier, fin fan cooling, trickling filter, pressure filter, ion exchange, and RO. The hardness removal obtained by warm softening was ~95%, while TDS removal by RO was more than 95%. At a pH of 10.5 or above, the removal of boron was effective at about ~90% removal rate. Meanwhile, ammonia removal was relatively effective at 80% at a pH of 8.7 or below (Alzahrani and Mohammad, 2014; Funston et al., 2002). Furthermore, different membrane processes can be integrated to create a whole-membrane system. An integrated whole-membrane system would show less energy consumption, lower raw material consumption, and lower waste while satisfying environmental standards, decreasing overall cost, and ensuring quality (Fakhru'l-Razi et al., 2009; Iulianelli and Drioli, 2020). Membrane manufacturers such as Osmonics, Koch, Alcoa, Veolia, and Filtration Solution Inc. have commercialized specific membranes for oil-water separation (Tanudjaja et al., 2019). Table 4 shows several integrated membrane technologies applied in wastewater treatment.

Pemex's Caderayta Refinery (Mexico) uses a process consisting of clarification, dissolved air flotation, biological process, immersed UF, and RO for produced-water treatment with a treatment capacity of 6.8 mgd (Peeters and Theodoulou, 2005). This process had an overall water recycle efficiency of 70%. The final product water coming from the RO permeate was reused for cooling tower makeup and low-pressure boiler feed (Peeters and Theodoulou, 2005). Similarly, Marathon Ashland Refinery (Kentucky, United States) also uses immersed

TABLE 4 Integrated membrane technology applied in refinery wastewater.

Company/ location	Wastewater	Wastewater treatment	Purpose	Results	Ref.
PEMEX's refinery, Mexico	Produced water, 173,200 bpd	Clarification, DAF, biological process, UF, and RO	RO permeate: cooling tower make-up; low-pressure boilers feed RO reject: discharge	RO permeate: 0.2 mg/L oil 0.5 mg/L TSS 36.4 mg/L COD 2.4 mg/L BOD	Jafarinejad and Jiang (2019) and Peeters and Theodoulou (2005)
Marathon Ashland refinery	Produced water, 50,000 gpd. Raw water characteristics: 775 mg/L BOD; 1000 mg/L COD; 66 mg/L TSS; 10 mg/L BTEX; 165 mg/L free oil and grease	Screening, equalization, DAF, MBR, UF, and RO	Discharge	97% oil removal 89% TSS removal 99.7% BOD removal	Peeters and Theodoulou (2005) and Peteers (2005)

Continued

B. Applications of membrane technology for water and wastewater treatment

TABLE 4   Integrated membrane technology applied in refinery wastewater.—cont'd

Company/ location	Wastewater	Wastewater treatment	Purpose	Results	Ref.
Tehran refinery, Iran	Oily wastewater; Raw water characteristics: 60 mg/L BOD; 2028 mg/L TDS; 78 mg/L oil and grease; 124 mg/L COD; 81 mg/L TOC; 53 NTU turbidity	API separator, gravity separator and skimmer, DAF, coagulation and flocculation, aeration, RO	Reuse for cooling tower	87% TDS removal 95% COD removal 95% BOD_5 removal 90% TOC removal 82% turbidity removal 87% oil and grease removal Complete removal of color, free oil and TSS	Salahi et al. (2010, 2012)
San Ardo reclamation facility	Produced water, 50,000 bpd. Raw water characteristics: 7000 mg/L TDS; 75 mg/L TOC; 325 mg/L hardness (as $CaCO_3$)	Degasification, softener (chemical and ion exchange), multimedia filter, cartridge filter, double-pass RO, pH adjustment, and remineralization	Discharged to the aquifer via recharge basins	RO permeate: 180 mg/L TDS 0.1 mg/L boron	Alzahrani and Mohammad (2014), Dahm and Chapman (2014), and Webb et al. (2009)
Sinopec YanShan refinery	Refinery wastewater; Raw water characteristics: ≤1000 mg/L TDS; ≤5 mg/L TSS; ≤40 mg/L COD; ≤2 mg/L oil	Biological treatment, media filtration, UF, RO	Reuse	RO permeate: 3.4 mg/L NO_3^- 1.2 mg/L SO_4^{2-} 1.1 mg/L TOC	Alzahrani and Mohammad (2014), Dow Chemical Company (2012), and Wang et al. (2011)
Riyadh refinery	Refinery wastewater; Raw water characteristics: 29 mg/L TSS; 34 mg/L BOD; 48 mg/L COD; 14 mg/L NH_{4-N}	Lime clarification, dual media filtration, RO, ion exchange demineralization	Reuse	RO permeate: 0 mg/L TSS 17 mg/L TDS	Madwar and Tarazi (2003)

Bbl, barrel; *bpd*, barrel per day; *BTEX*, benzene, toluene, ethylbenzene, and xylene; *gpd*, gallon per day; *TOC*, total organic carbon.

UF and RO membranes for produced-water treatment. The process used MBR to treat oily wastewater with a treatment capacity of 50,000 gpd. The results show that oil, total suspended solids (TSS), and BOD can be removed by 97%, 89%, and 99.7%, respectively (Peeters and Theodoulou, 2005). Moreover, industrial oily wastewater treatment in a Tehran refinery has been performed with a combined biological process/RO process. An experiment was conducted by using an original oily wastewater composition. By using the original oily wastewater composition, the membrane showed an excellent performance with rejections of 87%, 95%, 95%, 90%, 82%, and 87% for TDS, COD, BOD_5, TOC, turbidity, and oil and grease content, respectively, as well as complete removal of color, free oil, and TSS (Salahi et al., 2010, 2012). The membrane also achieved a relatively high flux of $50 L/m^2 h$.

Another produced water reuse plant is the membrane system at the Sinopec Beijing Yanshan Refinery China (Alzahrani and Mohammad, 2014; Dow Chemical Company, 2012; Wang et al., 2011). It was operated successfully for 4 years. Processes in the wastewater treatment system are biological process, media filter, UF, and RO. In this plant, UF decreased the wastewater turbidity to 0.3 NTU from its initial value of 5 NTU, before being transferred to RO units. Meanwhile, the remaining pollutants were successfully removed by RO, and then a portion of the RO permeate was reused as boiler feed water in the refinery (Alzahrani and Mohammad, 2014; Dow Chemical Company, 2012; Wang et al., 2011). However, membrane filtration often generates a concentrated stream containing rejected components from the feed with a higher concentration than the initial. The concentrate stream still has a large volume of water; thus, water recovery of the membrane system should be increased to minimize waste generation. Methods to increase water recovery usually integrate membrane with others, such as dual RO-chemical precipitation, dual RO-high efficiency RO (HERO), slurry precipitation and recycling RO (SPARRO), and FO/RO systems (Drewes et al., 2009), as shown in Table 5.

Dual RO-chemical precipitation couples conventional technologies such as lime-soda softening and two-stage RO (Drewes et al., 2009). For a dual RO-HERO system, the processes are softener, first-stage RO, ion exchange, degasification, and pH adjustment of the first RO concentrate. Then, after pH adjustment, the first-stage RO concentrate goes to the secondary RO of HERO system (Guerra et al., 2011). The pretreatment helps the second RO recover the remaining water, which results in higher overall water recovery. HERO developed by GE Ionics is expected to provide higher water recovery with better product quality, more significant flux, and fewer costs than conventional RO (Shenvi et al., 2015; Venzke et al., 2018). The most crucial stage of HERO is the pretreatment step, which elevates the pH of the feed water for attaining higher efficiency (Hayter et al., 2010; Interstate Oil and Gas Compact Commission (IOGCC), 2006). The SPARRO process uses a single-stage RO membrane unit combined with precipitation. Precipitation is conducted by using a seeded crystalline slurry. Subsequently, crystal separation occurs in a cyclone separator, and the remaining water is recirculated back to the RO (Guerra et al., 2011). As a result, high recovery is obtained.

In a hybrid FO/RO system, the feed solution is concentrated by a draw solution. After this process, the draw solution becomes diluted and needs reconcentration before reuse in the next cycle. The RO system is used for a draw solution regeneration of reconcentration. This option is viable since the draw solution has no sparingly soluble salts such as multivalent ions or other foulants (Drewes et al., 2009). Therefore, this combination can also improve water recovery of the membrane system.

5. Commercial membrane technologies for produced-water treatment

Several commercial membrane technologies are specially developed for produced-water treatment. The first is CDM produced-water technology commercialized by CDM Smith

TABLE 5   Multiple stages of membrane-based treatment processes (Drewes et al., 2009; Guerra et al., 2011; Interstate Oil and Gas Compact Commission (IOGCC), 2006).

Parameters	Dual RO-chemical precipitation	Dual RO with HERO	SPARRO	FO/RO
Application	Pilot scale at municipal desalination plants	Lab-scale	Pilot scale	Pilot scale
Pretreatment	The second stage RO: chemical precipitation and filtration	Coagulation and filtration Other pretreatment: antiscaling and acid addition The second stage RO: chemical precipitation and filtration	Coagulation and prefiltration prior to the slurry reaction chamber Other pretreatment options: antiscalant and acid addition The second stage RO: chemical precipitation and filtration	Pretreatment options: antiscaling and acid addition
Posttreatment	pH stabilization or remineralization	pH stabilization or remineralization	pH stabilization or remineralization	pH stabilization or remineralization RO concentrate: posttreatment or disposal consideration
Feed water	TDS range: 1000–35,000 mg/L	TDS range: 500–10,000 mg/L	TDS range: 500–10,000 mg/L	TDS range: 500–35,000 mg/L
TDS removal and water recovery	TDS removal = 94% (pilot scale) Water recovery >90%	TDS removal = 94% (lab scale) Water recovery >90%	TDS removal = 94% (pilot scale) Water recovery >94%	TDS removal >94% (pilot scale) Water recovery >96%
Energy	No data available	11–19 kwh/m^3	18.2 kWh/kgal	5.68–11.36 kWh/kgal
Chemical use	Lime or caustic soda Cleaning chemicals: NaOH, Na$_4$EDTA, or HCl	Cleaning chemicals: NaOH, Na$_4$EDTA, or HCl. IX process chemical regeneration: strong acid such as H$_2$SO$_4$ or HCl	Cleaning chemicals: NaOH, Na$_4$EDTA, or HCl	Cleaning chemicals: NaOH, Na$_4$EDTA, or HCl
Overall cost	No data available	Operation and management costs = $3.5/kgal ($0.14/bbl)	Operation and management costs = currently unknown	No data available

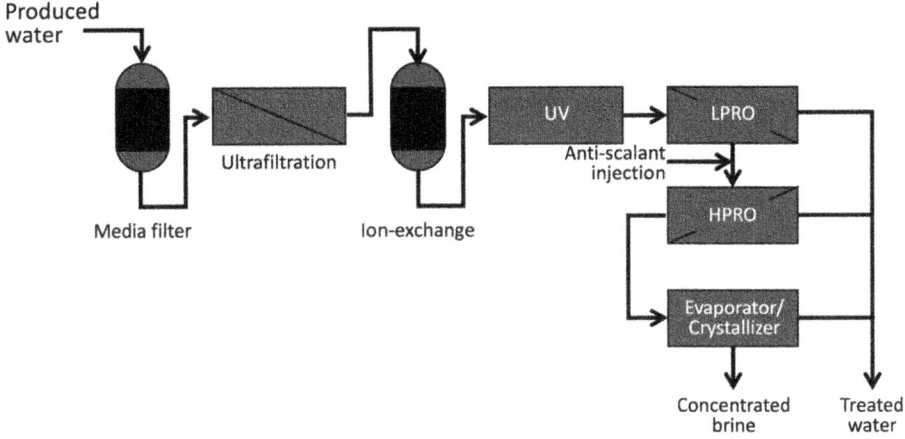

FIG. 5 Schematic CDM technology for produced water treatment.

Inc. (Guerra et al., 2011; Kimball and Locke, 2013). CDM technology comprises of a series of processes, including advanced filtration, ion-exchange units, UV irradiation, low-pressure RO, high-pressure RO, and evaporation or crystallization (Fig. 5). Advanced filtration consists of a media filter and UF. The main task of advanced filtration is to remove particulates, oil, and microbes. The product of advanced filtration is then sent to the ion-exchange unit. The ion-exchange unit contains a weak acid cation exchanger for a softening process. Hardness and metal ions are removed in this unit. Afterward, the soft water is transferred to UV irradiation for the disinfection process. The disinfected water then goes to low- and high-pressure RO units for desalination. The low-pressure RO has 85% water recovery. The reject stream of the low-pressure RO is then treated in high-pressure RO after being injected with an antiscaling agent. The antiscaling injection allows 80% water recovery in the high-pressure unit. For achieving zero liquid discharge, evaporation and crystallization units can be added after high-pressure RO. These units produce concentrated brine and distillate, increasing the overall water recovery. This technology is applicable for produced water containing total dissolved solids up to 20 g/L (Guerra et al., 2011; Kimball and Locke, 2013). In addition, an overall water recovery of 97% is achievable by this technology (depending on the feed condition) (Guerra et al., 2011; Kimball and Locke, 2013).

Another commercial technology applied commercially is OPUS technology or Optimized Pretreatment and Unique Separation developed by Veolia Water Technologies (Tanudjaja et al., 2019). OPUS technology also comprises a series of treatment processes, i.e., degassifier, softener, filter, double-pass RO, pH adjustment, and remineralization (Webb et al., 2009). This technology has been applied to treat produced water with a capacity of 50,000 bpd (330 m^3/h) in San Ardo, California (Webb et al., 2009). The aims of the produced-water treatment are (1) discharge to a recharge basin and (2) production of makeup water for once through steam generator (Dahm and Chapman, 2014). The description of OPUS technology used in San Ardo is as follows (Webb et al., 2009). The first step of produced-water treatment is induced gas flotation followed by walnut-shell filtration. The goal of this first step is free oil removal with a final concentration of less than 1 ppm. The second process is degasification, which removes carbon dioxide and hydrogen sulfide. Dissolved gas removal comprises acid injection and excess alkali addition. The next step is a softening process consisted of chemical softening, ion-exchange, and cartridge filter. Hardness, metal, and particulate concentrations are significantly reduced in this step. The product of this step is sent to a double-pass RO operated at elevated pH. An elevated pH allows a high boron removal rate. The double-pass RO has a relatively high recovery, i.e., 80.5% and 90% for the first and second pass, respectively. Permeate of the second pass RO contained 76 mg/L TDS from its initial value of 7000 mg/L. In the final part, acid and carbon dioxide are used for pH neutralization followed by constructed wetland as a buffer between the desalination plant and the basin. During the operation, organic fouling was observed in RO (Webb et al., 2009).

A GeoPure desalination process developed by GeoPure Water Technologies, LLC is a combination of pretreatment, UF, and RO dedicated to treating produced water (Jafarinejad, 2017). The pretreatment can be coagulation/flocculation processes, depending on the produced-water composition. TDS removal is within the range of 98%−99% with 50% water recovery (Jafarinejad, 2017). The GeoPure desalination process can generate clean water for discharge and reuse (Jafarinejad, 2017).

Another example of a commercial-scale membrane plant operated for produced water treatment is located at Tambun Field, Indonesia (PT Artha Envirotama, 2022). The membrane plant has been operational since 2012. The plant is used to treat produced water for reinjection purposes. The plant has a capacity of 33,000 barrels of water per day (bwpd). It consists of dissolved air flotation, media filtration, and UF membrane. Suspended solids and oils are removed by dissolved air flotation and a media filter. The UF membrane does further clarification and turbidity removal. The product of the UF membrane then undergoes to chemical treatment and deaeration before being reinjected.

6. Conclusions and perspectives

If not correctly handled, produced water has toxic contaminants that can negatively impact the environment. Therefore, several works have been dedicated to find promising treatment technology that can satisfy both technological and economic points. Membranes have been viewed as one of the promising technologies in produced-water treatment. Commercially available membrane technologies have been increasingly used to separate oil and water in the past few years, with ceramic tubular membranes and flat-sheet polymeric membranes being the most popular. Typical examples of available commercial membrane-based technology are CDM, OPUS, and GeoPure water technology. Membranes are usually integrated with other methods to harness the advantages and circumvent the shortcomings of each. In produced water treatment, membrane technology acts as a complement rather than a replacement of conventional methods. Hence, the process becomes more efficient with high-quality water.

Despite the advantages, the application of membranes is still hampered by fouling, which causes deterioration of membrane performance and durability. Several approaches have been proposed to mitigate fouling, such as membrane and module modification, pretreatment optimization, operating condition adjustment, and membrane cleaning protocol selection.

Numerous studies regarding these approaches have been reported in the literature. In commercial applications, membranes are usually integrated with other methods. The other methods act as pretreatments to partially remove some of the pollutants that potentially harm or cause fouling to the membrane. As a result, fouling can be minimized.

In general, membranes have shown excellent performance in produced-water treatment regarding the produced effluent quality. This is evidenced especially in abundant lab-scale studies. However, reports on commercial-scale performance are still limited. In the future, more studies may be needed on the application of membrane-based processes in produced-water treatment at a larger scale with real influent.

References

Al-Ghouti, M.A., Al-Kaabi, M.A., Ashfaq, M.Y., Da'na, D.A., 2019. Produced water characteristics, treatment and reuse: a review. J. Water Process Eng. 28, 222–239. https://doi.org/10.1016/j.jwpe.2019.02.001.

Ali, A., Quist-Jensen, C.A., Drioli, E., Macedonio, F., 2018. Evaluation of integrated microfiltration and membrane distillation/crystallization processes for produced water treatment. Desalination 434, 161–168. https://doi.org/10.1016/j.desal.2017.11.035.

Allen, E.W., 2008. Process water treatment in Canada's oil sands industry: II. A review of emerging technologies. J. Environ. Eng. Sci. 7, 499–524. https://doi.org/10.1139/S08-020.

Almarzooqi, K., Ashrafi, M., Kanthan, T., Elkamel, A., Pope, M.A., 2021. Graphene oxide membranes for high salinity, produced water separation by pervaporation. Membranes (Basel) 11, 475. https://doi.org/10.3390/membranes11070475.

Alzahrani, S., Mohammad, A.W., 2014. Challenges and trends in membrane technology implementation for produced water treatment: a review. J. Water Process Eng. 4, 107–133. https://doi.org/10.1016/j.jwpe.2014.09.007.

Alzahrani, S., Mohammad, A.W., Abdullah, P., Jaafar, O., 2013. Potential tertiary treatment of produced water using highly hydrophilic nanofiltration and reverse osmosis membranes. J. Environ. Chem. Eng. 1, 1341–1349. https://doi.org/10.1016/j.jece.2013.10.002.

Apul, O.G., Karanfil, T., 2015. Adsorption of synthetic organic contaminants by carbon nanotubes: a critical review. Water Res. 68, 34–55. https://doi.org/10.1016/j.watres.2014.09.032.

Arthur, J., Langhus, B., Patel, C., 2005. Technical summary of oil & gas produced water treatment technologies. ALL, Tulsa, OK, pp. 1–53, https://doi.org/10.2166/aqua.2010.004.

Asante-Sackey, D., Rathilal, S., Tetteh, E.K., Armah, E.K., 2022. Membrane bioreactors for produced water treatment: a mini-review. Membranes. https://doi.org/10.3390/membranes12030275.

Ashaghi, K.S., Ebrahimi, M., Czermak, P., 2008. Ceramic ultra- and nanofiltration membranes for oilfield produced water treatment: a mini review. Open Environ. Sci. 1, 1–8. https://doi.org/10.2174/1876325100701010001.

Badrnezhad, R., Beni, A.H., 2013. Ultrafiltration membrane process for produced water treatment: experimental and modeling. J. Water Reuse Desalin. 3, 249. https://doi.org/10.2166/wrd.2013.092.

Bagheri, M., Roshandel, R., Shayegan, J., 2018. Optimal selection of an integrated produced water treatment system in the upstream of oil industry. Process. Saf. Environ. Prot. 117, 67–81. https://doi.org/10.1016/j.psep.2018.04.010.

Beery, M., Ludwig, J., León, L., 2015. Hybrid flotation-filtration process for oil water separation based on ceramic membranes. In: Chem. Technol. June, 20–23.

Bernardo, P., Drioli, E., 2010. Membrane gas separation progresses for process intensification strategy in the petrochemical industry. Pet. Chem. 50, 271–282. https://doi.org/10.1134/S0965544110040043.

Buonomenna, M.G., 2013. Membrane processes for a sustainable industrial growth. RSC Adv. 3, 5694–5740. https://doi.org/10.1039/c2ra22580h.

Campos, J.C., Borges, R.M.H., Oliveira Filho, A.M., Nobrega, R., Sant'Anna, G.L., 2002. Oilfield wastewater treatment by combined microfiltration and biological processes. Water Res. 36, 95–104. https://doi.org/10.1016/S0043-1354(01)00203-2.

Cerqueira, V.S., Hollenbach, E.B., Maboni, F., Vainstein, M.H., Camargo, F.A.O., Peralba, M.d.C.R., Bento, F.M., 2011. Biodegradation potential of oily sludge by pure and mixed bacterial cultures. Bioresour. Technol. 102, 11003–11010. https://doi.org/10.1016/j.biortech.2011.09.074.

Chen, Y.M., Lin, W.Y., Chan, C.C., 2014. The impact of petrochemical industrialisation on life expectancy and per capita income in Taiwan: an 11-year longitudinal study. BMC Public Health 14, 1–8. https://doi.org/10.1186/1471-2458-14-247.

Cheryan, M., Rajagopalan, N., 1998. Membrane processing of oily streams. Wastewater treatment and waste reduction. J. Membr. Sci. 151, 13–28. https://doi.org/10.1016/S0376-7388(98)00190-2.

Coday, B.D., Xu, P., Beaudry, E.G., Herron, J., Lampi, K., Hancock, N.T., Cath, T.Y., 2014. The sweet spot of forward osmosis: treatment of produced water, drilling wastewater, and other complex and difficult liquid streams. Desalination 333, 23–35. https://doi.org/10.1016/j.desal.2013.11.014.

Cui, J., Zhang, X., Liu, H., Liu, S., Yeung, K.L., 2008. Preparation and application of zeolite/ceramic microfiltration membranes for treatment of oil contaminated water. J. Membr. Sci. 325, 420–426. https://doi.org/10.1016/j.memsci.2008.08.015.

Dahm, K., Chapman, M., 2014. Produced Water Treatment Primer: Case Studies of Treatment Applications.

Dickhout, J.M., Moreno, J., Biesheuvel, P.M., Boels, L., Lammertink, R.G.H., de Vos, W.M., 2017. Produced water treatment by membranes: a review from a colloidal perspective. J. Colloid Interface Sci. 487, 523–534. https://doi.org/10.1016/j.jcis.2016.10.013.

Doran, G.F., Carini, F.H., Fruth, D.A., Drago, J.A., Leong, L.Y.C., 1997. Evaluation of technologies to treat oil field produced water to drinking water or reuse quality. In: Proceedings—SPE Annual Western Regional Meeting. ARCO Western Energy, United States, pp. 811–822.

dos Santos, E.V., Bezerra Rocha, J.H., de Araújo, D.M., de Moura, D.C., Martínez-Huitle, C.A., 2014. Decontamination of produced water containing petroleum hydrocarbons by electrochemical methods: a minireview. Environ. Sci. Pollut. Res. 21, 8432–8441. https://doi.org/10.1007/s11356-014-2780-4.

Dow Chemical Company, 2012. Ultrafiltration Membranes Meet Challenge of High COD, Oil-Contaminated Waste Water.

Drewes, J., Cath, T., Debroux, J., Veil, J., 2009. An integrated framework for treatment and management of produced water—Technical assessment of produced water treatment technologies.

Drioli, E., Ali, A., Lee, Y.M., Al-Sharif, S.F., Al-Beirutty, M., Macedonio, F., 2016. Membrane operations for produced water treatment. Desalin. Water Treat. 57, 14317–14335. https://doi.org/10.1080/19443994.2015.1072585.

Ebrahimi, M., Willershausen, D., Shams, K., Engel, L., Placido, L., Mund, P., Bolduan, P., Czermak, P., 2010. Investigations on the use of different ceramic membranes for efficient oil- field produced water treatment. Desalination 250, 991–996. https://doi.org/10.1016/j.desal.2009.09.088.

Edwan Kardena, Q.H., 2015. Petroleum oil and gas industry waste treatment; common practice in Indonesia. J. Pet. Environ. Biotechnol. 06. https://doi.org/10.4172/2157-7463.1000241.

Fakhru'l-Razi, A., Pendashteh, A., Abdullah, L.C., Biak, D.R.A., Madaeni, S.S., Abidin, Z.Z., 2009. Review of technologies for oil and gas produced water treatment. J. Hazard. Mater. 170, 530–551. https://doi.org/10.1016/j.jhazmat.2009.05.044.

Feng, C., Khulbe, K.C., Matsuura, T., Farnood, R., Ismail, A.F., 2015. Recent progress in zeolite/zeotype membranes. J. Membr. Sci. Res. 1, 49–72.

Fillo, J.P., Evans, J.M., 1990. Characterization and management of produced waters from underground natural gas storage reservoirs. Am. Gas Assoc. Oper. Sect. Proc. 448459.

Finklea, H., Lin, L.-S., Khajouei, G., 2022. Electrodialysis of softened produced water from shale gas development. J. Water Process Eng. 45, 102486. https://doi.org/10.1016/j.jwpe.2021.102486.

Fouladi Tajar, A., Kaghazchi, T., Soleimani, M., 2009. Adsorption of cadmium from aqueous solutions on sulfurized activated carbon prepared from nut shells. J. Hazard. Mater. 165, 1159–1164. https://doi.org/10.1016/j.jhazmat.2008.10.131.

Frankiewicz, T.C., Gerlach, J., 2000. Removal of hydrocarbons, mercury and arsenic from oil-field produced water. US006117333A.

Fulazzaky, M., Setiadi, T., Fulazzaky, M.A., 2020. An evaluation of the oilfield-produced water treatment by the membrane bioreactor. J. Environ. Chem. Eng. 8, 104417. https://doi.org/10.1016/j.jece.2020.104417.

Funston, R., Ganesh, R., Leong, L.Y.C., Consultants, K.J., Road, N.S., 2002. Evaluation of Technical and Economic Feasibility of Treating Oilfield Produced Water to Create a "New" Water Resource 1–14.

Guerra, K., Dahm, K., Dundorf, S., 2011. Oil and gas produced water management and beneficial use in the Western United States. Sci. Technol. Progr. Rep. 157. U.S. Department of the Interior Bureau of Reclamation, pp. 1–113. https://www.usbr.gov/research/dwpr/reportpdfs/report157.pdf.

Guo, C., Chang, H., Liu, B., He, Q., Xiong, B., Kumar, M., Zydney, A.L., 2018. A combined ultrafiltration-reverse osmosis process for external reuse of Weiyuan shale gas flowback and produced water. Environ. Sci. Water Res. Technol. 4, 942–955. https://doi.org/10.1039/c8ew00036k.

Hakim, A.N., Khoiruddin, K., Ariono, D., Wenten, I.G., 2019. Ionic separation in electrodeionization system: mass transfer mechanism and factor affecting separation performance. Sep. Purif. Rev. 1–23. https://doi.org/10.1080/15422119.2019.1608562.

Halim, N.S.A., Wirzal, M.D.H., Hizam, S.M., Bilad, M.R., Nordin, N.A.H.M., Sambudi, N.S., Putra, Z.A., Yusoff, A.R.M., 2021. Recent development on electrospun nanofiber membrane for produced water treatment: a review. J. Environ. Chem. Eng. 9. https://doi.org/10.1016/j.jece.2020.104613.

Han, R., Zeng, J., Han, L., Wang, Y., Chang, Q., 2016. Application of integrated membrane technology in purification of chlorogenic acid 3994., https://doi.org/10.1080/19443994.2014.937755.

Handojo, L., Wardani, A.K., Regina, D., Bella, C., Kresnowati, M.T.A.P., Wenten, I.G., 2019. Electro-membrane processes for organic acid recovery. RSC Adv. 9, 7854–7869. https://doi.org/10.1039/C8RA09227C.

Hansima, M.A.C.K., Makehelwala, M., Jinadasa, K.B.S.N., Wei, Y., Nanayakkara, K.G.N., Herath, A.C., Weerasooriya, R., 2021. Fouling of ion exchange membranes used in the electrodialysis reversal advanced water treatment: a review. Chemosphere 263, 127951. https://doi.org/10.1016/j.chemosphere.2020.127951.

Hayter, S., Urbatsch, E., Tanner, S., Zuboy, J., 2010. Saving energy, water, and money with efficient water treatment technologies. U.S. Dep. Energy, Energy Effic. Renew. Energy 132.

Holloway, F.H., 1977. The chemical treatment production. In: Proceedings of the Offshore Europe Conference.

Hua, F.L., Tsang, Y.F., Wang, Y.J., Chan, S.Y., Chua, H., Sin, S.N., 2007. Performance study of ceramic microfiltration membrane for oily wastewater treatment. Chem. Eng. J. 128, 169–175. https://doi.org/10.1016/j.cej.2006.10.017.

Huth, E., Muthu, S., Ruff, L., Brant, J.A., 2014. Feasibility assessment of pervaporation for desalinating high-salinity brines. J. Water Reuse Desalin. 4, 109–124. https://doi.org/10.2166/wrd.2014.038.

Igunnu, E.T., Chen, G.Z., 2014. Produced water treatment technologies. Int. J. Low-Carbon Technol. 9, 157–177. https://doi.org/10.1093/ijlct/cts049.

Interstate Oil and Gas Compact Commission (IOGCC), 2006. A guide to practical management of produced water from onshore oil and gas operations in the United States. Rep. No. DE-PS26-04NT15460-02, Technical Report prepared for U.S. Department of Energy.

Iulianelli, A., Drioli, E., 2020. Membrane engineering: latest advancements in gas separation and pre-treatment processes, petrochemical industry and refinery, and future perspectives in emerging applications. Fuel Process. Technol. 206, 106464. https://doi.org/10.1016/j.fuproc.2020.106464.

Jafarinejad, S., 2017. A comprehensive study on the application of reverse osmosis (RO) Technology for the petroleum industry wastewater treatment. J. Water Environ. Nanotechnol. 2, 243–264. https://doi.org/10.22090/jwent.2017.04.003.

Jafarinejad, S., 2020. Forward osmosis membrane technology for nutrient removal/recovery from wastewater: recent advances, proposed designs, and future directions. Chemosphere. https://doi.org/10.1016/j.scitotenv.2019.135577.

Jafarinejad, S., Jiang, S.C., 2019. Current technologies and future directions for treating petroleum refineries and petrochemical plants (PRPP) wastewaters. J. Environ. Chem. Eng. 7, 103326. https://doi.org/10.1016/j.jece.2019.103326.

Jepsen, K.L., Bram, M.V., Pedersen, S., Yang, Z., 2018. Membrane fouling for produced water treatment: a review study from a process control perspective. Water (Switzerland) 10. https://doi.org/10.3390/w10070847.

Ji, Y., 2015. Membrane technologies for water treatment and reuse in the gas and petrochemical industries. In: Advances in Membrane Technologies for Water Treatment: Materials, Processes and Applications, pp. 519–536, https://doi.org/10.1016/B978-1-78242-121-4.00016-2.

Jing, G., Xing, L., Li, S., Han, C., 2011. Reclaiming polymer-flooding produced water for beneficial use: salt removal via electrodialysis. Desalin. Water Treat. 25, 71–77. https://doi.org/10.5004/dwt.2011.1766.

Johnson, B.M., Kanagy, L.E., Rodgers, J.H., Castle, J.W., 2008. Chemical, physical, and risk characterization of natural gas storage produced waters. Water Air Soil Pollut. 191, 33–54.

Karkooti, A., Rastgar, M., Nazemifard, N., Sadrzadeh, M., 2020. Graphene-based electro-conductive anti-fouling membranes for the treatment of oil sands produced water. Sci. Total Environ. 704. https://doi.org/10.1016/j.scitotenv.2019.135365.

B. Applications of membrane technology for water and wastewater treatment

Khoiruddin, K., Ariono, D., Subagjo, S., Wenten, I.G., 2021. Improved anti-organic fouling of polyvinyl chloride-based heterogeneous anion-exchange membrane modified by hydrophilic additives. J. Water Process Eng. 41, 102007. https://doi.org/10.1016/j.jwpe.2021.102007.

Khor, E.H., 2017. Review on produced water treatment technology.

Kimball, R.J., Locke, A.L., 2013. Innovations in sustainable shale gas water managment high recovery reverse osmosis case study. In: Eur. HSE Conf. Exhib. SPE 164964., https://doi.org/10.2118/164964-MS.

Kose Mutlu, B., Ozgun, H., Ersahin, M.E., Kaya, R., Eliduzgun, S., Altinbas, M., Kinaci, C., Koyuncu, I., 2019. Impact of salinity on the population dynamics of microorganisms in a membrane bioreactor treating produced water. Sci. Total Environ. 646, 1080–1089. https://doi.org/10.1016/j.scitotenv.2018.07.386.

Lee, K.P., Arnot, T.C., Mattia, D., 2011. A review of reverse osmosis membrane materials for desalination-development to date and future potential. J. Membr. Sci. 370, 1–22. https://doi.org/10.1016/j.memsci.2010.12.036.

Li, Y.S., Yan, L., Xiang, C.B., Hong, L.J., 2006. Treatment of oily wastewater by organic-inorganic composite tubular ultrafiltration (UF) membranes. Desalination 196, 76–83. https://doi.org/10.1016/j.desal.2005.11.021.

Li, D., Yan, Y., Wang, H., 2016. Recent advances in polymer and polymer composite membranes for reverse and forward osmosis processes. Prog. Polym. Sci. 61, 104–155. https://doi.org/10.1016/j.progpolymsci.2016.03.003.

Liang, J., Mai, W., Tang, J., Wei, Y., 2019. Highly effective treatment of petrochemical wastewater by a super-sized industrial scale plant with expanded granular sludge bed bioreactor and aerobic activated sludge. Chem. Eng. J. 360, 15–23. https://doi.org/10.1016/j.cej.2018.11.167.

Liang, H., Zou, C., Tang, W., 2021. Development of novel polyether sulfone mixed matrix membranes to enhance antifouling and sustainability: treatment of oil sands produced water (OSPW). J. Taiwan Inst. Chem. Eng. 118, 215–222. https://doi.org/10.1016/j.jtice.2020.12.022.

Ma, H., Wang, B., 2006. Electrochemical pilot-scale plant for oil field produced wastewater by M/C/Fe electrodes for injection. J. Hazard. Mater. 132, 237–243. https://doi.org/10.1016/j.jhazmat.2005.09.043.

Madaeni, S.S., 1999. The application of membrane technology for water disinfection. Water Res. 33, 301–308. https://doi.org/10.1016/S0043-1354(98)00212-7.

Madwar, K., Tarazi, H., 2003. Desalination techniques for industrial wastewater reuse. Desalination 152, 325–332. https://doi.org/10.1016/S0011-9164(02)01080-9.

Mahmodi, G., Ronte, A., Dangwal, S., Wagle, P., Echeverria, E., Sengupta, B., Vatanpour, V., Mcllroy, D.N., Ramsey, J.D., Kim, S.-J., 2022. Improving antifouling property of alumina microfiltration membranes by using atomic layer deposition technique for produced water treatment. Desalination 523. https://doi.org/10.1016/j.desal.2021.115400.

Manouchehr, N., Mishana, J., 2008. Reuse of refinery treated wastewater in cooling towers. Iran. J. Chem. Chem. Eng. 27, 1–7.

Mansouri, J., Harrisson, S., Chen, V., 2010. Strategies for controlling biofouling in membrane filtration systems: challenges and opportunities. J. Mater. Chem. 20, 4567–4586. https://doi.org/10.1039/B926440J.

Masooleh, M.S., Bazgir, S., Tamizifar, M., Nemati, A., 2010. Adsorption of petroleum hydrocarbons on organoclay archive of SID. J. Appl. Chem. Res. 4, 19–23.

McFarlane, J., Ridenour, W.B., Luo, H., Hunt, R.D., DePaoli, D.W., Ren, R.X., 2005. Room temperature ionic liquids for separating organics from produced water. Sep. Sci. Technol. 40, 1245–1265. https://doi.org/10.1081/SS-200052807.

Merkel, A., Ashrafi, A.M., 2019. An investigation on the application of pulsed electrodialysis reversal in whey desalination. Int. J. Mol. Sci. 20, 1918. https://doi.org/10.3390/ijms20081918.

Mueller, J., Cen, Y., Davis, R.H., 1997. Crossflow microfiltration of oily water. J. Membr. Sci. 129, 221–235. https://doi.org/10.1016/S0376-7388(96)00344-4.

Munirasu, S., Haija, M.A., Banat, F., 2016. Use of membrane technology for oil field and refinery produced water treatment—a review. Process. Saf. Environ. Prot. 100, 183–202. https://doi.org/10.1016/j.psep.2016.01.010.

Nasiri, M., Jafari, I., 2017. Produced water from oil-gas plants: a short review on challenges and opportunities. Period. Polytech. Chem. Eng. 61, 73–81. https://doi.org/10.3311/PPch.8786.

Nawaz, M.S., Son, H.S., Jin, Y., Kim, Y., Soukane, S., Al-Hajji, M.A., Abu-Ghdaib, M., Ghaffour, N., 2021. Investigation of flux stability and fouling mechanism during simultaneous treatment of different produced water streams using forward osmosis and membrane distillation. Water Res. 198. https://doi.org/10.1016/j.watres.2021.117157.

Nazirah Wan Ikhsan, S., Yusof, N., Aziz, F., Misdan, N., 2017. A review of oilfield wastewater treatment using membrane filtration over conventional technology. Malaysian J. Anal. Sci. 21, 643–658. https://doi.org/10.17576/mjas-2017-2103-14.

Neyens, E., Baeyens, J., 2003. A review of classic Fenton's peroxidation as an advanced oxidation technique. J. Hazard. Mater. 98, 33–50. https://doi.org/10.1016/S0304-3894(02)00282-0.

Ong, C.S., Goh, P.S., Lau, W.J., Misdan, N., Ismail, A.F., 2016. Nanomaterials for biofouling and scaling mitigation of thin film composite membrane: a review. Desalination 393, 2–15. https://doi.org/10.1016/j.desal.2016.01.007.

Owen, G., Bandi, M., Howell, J.A., Churchouse, S.J., 1995. Economic assessment of membrane processes for water and waste water treatment. J. Membr. Sci. 102, 77–91. https://doi.org/10.1016/0376-7388(94)00261-V.

Padaki, M., Surya Murali, R., Abdullah, M.S., Misdan, N., Moslehyani, A., Kassim, M.A., Hilal, N., Ismail, A.F., 2015. Membrane technology enhancement in oil-water separation. A review. Desalination 357, 197–207. https://doi.org/10.1016/j.desal.2014.11.023.

Peeters, J., Theodoulou, S., 2005. Membrane technology treating oily wastewater for reuse. In: CORROSION 2005. OnePetro, Houston, Texas, April 2005., p. NACE-05534.

Peteers, J., 2005. Membrane technology treating oily wastewater for reuse. [WWW Document]. SAWEA Work URL http://www.sawea.org/pdf/2005/MainSession/Nov29/MembraneTechTreatingOilyWWReuse.pdf. (Accessed 1 March 2022).

Pica, N.E., Carlson, K., Steiner, J.J., Waskom, R., 2017. Produced water reuse for irrigation of non-food biofuel crops: effects on switchgrass and rapeseed germination, physiology and biomass yield. Ind. Crop. Prod. 100, 65–76. https://doi.org/10.1016/j.indcrop.2017.02.011.

Plebon, M.J., 2007. Further advances in produced water oiling utilizing a recently developed technology which removes recovers dispersed oil in produced water 2 microns and larger 3. pp. 147–158.

PT. Artha Envirotama, 2022. Oil & Gas. [WWW Document]. URL https://arthaenvirotama.com/id/our_project/oil-gas-wwtp-id/. (Accessed 1 Febrauary 2022).

Qiao, X., Zhang, Z., Yu, J., Ye, X., 2008. Performance characteristics of a hybrid membrane pilot-scale plant for oilfield-produced wastewater. Desalination 225, 113–122. https://doi.org/10.1016/j.desal.2007.04.092.

Rezakazemi, M., Khajeh, A., Mesbah, M., 2017. Membrane filtration of wastewater from gas and oil production. Environ. Chem. Lett. 16, 367–388. https://doi.org/10.1007/s10311-017-0693-4.

Salahi, A., Mohammadi, T., Rekabdar, F., Mahdavi, H., Science, P., Division, T., 2010. Reverse osmosis of refinery oily wastewater effluents. J. Environ. 7, 413–422.

Salahi, A., Mohammadi, T., Nikbakht, M., Golshenas, M., Noshadi, I., 2012. Purification of biologically treated Tehran refinery oily wastewater using reverse osmosis. Desalin. Water Treat. 48, 27–37. https://doi.org/10.1080/19443994.2012.698752.

Saleh, T.A., Gupta, V.K., 2014. Processing methods, characteristics and adsorption behavior of tire derived carbons: a review. Adv. Colloid Interf. Sci. 211, 93–101. https://doi.org/10.1016/j.cis.2014.06.006.

Salehi, M.A., Rostamani, R., 2013. Review of membrane distillation for the production of fresh water from saline water. J. Novel Appl. Sci., 1072–1075.

Samuel, O., Othman, M.H.D., Kamaludin, R., Sinsamphanh, O., Abdullah, H., Puteh, M.H., Kurniawan, T.A., Li, T., Ismail, A.F., Rahman, M.A., Jaafar, J., El-badawy, T., Chinedu Mamah, S., 2022. Oilfield-produced water treatment using conventional and membrane-based technologies for beneficial reuse: a critical review. J. Environ. Manag. 308, 114556. https://doi.org/10.1016/j.jenvman.2022.114556.

Saththasivam, J., Loganathan, K., Sarp, S., 2016. An overview of oil-water separation using gas flotation systems. Chemosphere 144, 671–680. https://doi.org/10.1016/j.chemosphere.2015.08.087.

Scott, K., 1995. Handbook of Industrial Membranes. Elsevier, Oxford.

Shamaei, L., Khorshidi, B., Islam, M.A., Sadrzadeh, M., 2020. Development of antifouling membranes using agro-industrial waste lignin for the treatment of Canada's oil sands produced water. J. Membr. Sci. 611. https://doi.org/10.1016/j.memsci.2020.118326.

Shenvi, S.S., Isloor, A.M., Ismail, A.F., 2015. A review on RO membrane technology: developments and challenges. Desalination 368, 10–26. https://doi.org/10.1016/j.desal.2014.12.042.

Siagian, U.W.R., Dwipramana, A.S., Perwira, S.B., Khoiruddin, Wenten, I.G., 2018a. Ceramic membrane ozonator for soluble organics removal from produced water. IOP Conf. Ser. Mater. Sci. Eng. 285, 012012. https://doi.org/10.1088/1757-899X/285/1/012012.

Siagian, U.W.R., Widodo, S., Wardani, A.K., Wenten, I.G., 2018b. Oilfield produced membrane water reuse and reinjection with membrane. MATEC Web Conf., 1–10. https://doi.org/10.1051/matecconf/201815608005. 08005.

Siagian, U.W.R., Khoiruddin, K., Wardani, A.K., Aryanti, P.T.P., Widiasa, I.N., Qiu, G., Ting, Y.P., Wenten, I.G., 2021. High-performance ultrafiltration membrane: recent progress and its application for wastewater treatment. Curr. Pollut. Rep. 7, 448–462. https://doi.org/10.1007/s40726-021-00204-5.

B. Applications of membrane technology for water and wastewater treatment

Silalahi, S.H.D., Leiknes, T., 2011. High frequency back-pulsing for fouling development control in ceramic microfiltration for treatment of produced water. Desalin. Water Treat. 28, 137–152. https://doi.org/10.5004/dwt.2011.2482.

Singh, D., Sirkar, K.K., 2012. Desalination of brine and produced water by direct contact membrane distillation at high temperatures and pressures. J. Membr. Sci. 389, 380–388. https://doi.org/10.1016/j.memsci.2011.11.003.

Sirivedhin, T., McCue, J., Dallbauman, L., 2004. Reclaiming produced water for beneficial use: salt removal by electrodialysis. J. Membr. Sci. 243, 335–343. https://doi.org/10.1016/j.memsci.2004.06.038.

Sosa-Fernandez, P.A., Post, J.W., Leermakers, F.A.M., Rijnaarts, H.H.M., Bruning, H., 2019. Removal of divalent ions from viscous polymer-flooding produced water and seawater via electrodialysis. J. Membr. Sci. 589. https://doi.org/10.1016/j.memsci.2019.117251.

Sosa-Fernandez, P.A., Miedema, S.J., Bruning, H., Leermakers, F.A.M., Post, J.W., Rijnaarts, H.H.M., 2021. Effects of feed composition on the fouling on cation-exchange membranes desalinating polymer-flooding produced water. J. Colloid Interface Sci. 584, 634–646. https://doi.org/10.1016/j.jcis.2020.10.077.

Steven, S., Friatnasary, D.L., Wardani, A.K., Khoiruddin, K., Suantika, G., Wenten, I.G., 2022. High cell density submerged membrane photobioreactor (SMPBR) for microalgae cultivation. IOP Conf. Ser. Earth Environ. Sci. 963, 12034. https://doi.org/10.1088/1755-1315/963/1/012034.

Subagjo, S., Prasetya, N., Wenten, I.G., 2015. Hollow fiber membrane bioreactor for COD biodegradation of tapioca wastewater. J. Membr. Sci. Res. 1, 79–84. https://doi.org/10.22079/jmsr.2015.13533.

Tanudjaja, H.J., Hejase, C.A., Tarabara, V.V., Fane, A.G., Chew, J.W., 2019. Membrane-based separation for oily wastewater: a practical perspective. Water Res. 156, 347–365. https://doi.org/10.1016/j.watres.2019.03.021.

Tao, F.T., Curtice, S., Hobbs, R.D., Sides, J.L., Wieser, J.D., Dyke, C.A., Tuohey, D., Pilger, P.F., 1993. Conversion of oilfield produced water into an irrigation/drinking quality water. In: SPE 26003, pp. 571–580.

Tawalbeh, M., Al Mojjly, A., Al-Othman, A., Hilal, N., 2018. Membrane separation as a pre-treatment process for oily saline water. Desalination, 1–21. https://doi.org/10.1016/j.desal.2018.07.029.

thembrsite.com, 2021. Largest MBR Plants (Over 100 MLD)—Worldwide. [WWW Document]. URL https://www.thembrsite.com/largest-mbr-plants/largest-membrane-bioreactor-plants-worldwide/. (Accessed 27 December 2021).

Varjani, S.J., Gnansounou, E., Pandey, A., 2017. Comprehensive review on toxicity of persistent organic pollutants from petroleum refinery waste and their degradation by microorganisms. Chemosphere 188, 280–291. https://doi.org/10.1016/j.chemosphere.2017.09.005.

Veil, J.A., Puder, M.G., Elcock, D., Redweik Jr., R.J., 2004. A White Paper Describing Produced Water From Production of Crude Oil, Natural Gas, and Coal Bed Methane. Argonne National Lab, Lemont, IL.

Venzke, C.D., Giacobbo, A., Ferreira, J.Z., Bernardes, A.M., Rodrigues, M.A.S., 2018. Increasing water recovery rate of membrane hybrid process on the petrochemical wastewater treatment. Process. Saf. Environ. Prot. 117, 152–158. https://doi.org/10.1016/j.psep.2018.04.023.

Veolia Water Technologies, 2014. Power Clean Nutshell Filters. [WWW Document]. URL https://www.veoliawatertechnologies.com/sites/g/files/dvc2476/files/document/2019/03/PowerClean_5-25-12.pdf.

Virga, E., Parra, M.A., de Vos, W.M., 2021. Fouling of polyelectrolyte multilayer based nanofiltration membranes during produced water treatment: The role of surfactant size and chemistry. J. Colloid Interface Sci. 594, 9–19. https://doi.org/10.1016/j.jcis.2021.02.119.

Wang, D., Tong, F., Aerts, P., 2011. Application of the combined ultrafiltration and reverse osmosis for refinery wastewater reuse in Sinopec Yanshan Plant. Desalin. Water Treat. 25, 133–142. https://doi.org/10.5004/dwt.2011.1137.

Wardani, A.K., Ariono, D., Subagjo, Wenten, I.G., 2019. Hydrophilic modification of polypropylene ultrafiltration membrane by air-assisted polydopamine coating. Polym. Adv. Technol. 30, 1148–1155. https://doi.org/10.1002/pat.4549.

Wardani, A.K., Ariono, D., Subagjo, S., Wenten, I.G., 2020. Fouling tendency of PDA/PVP surface modified PP membrane. Surf. Interfaces 19, 100464. https://doi.org/10.1016/j.surfin.2020.100464.

Webb, C., North, C., Exploration, A., Nagghappan, L., Water, V.N., 2009. Desalination of oilfield-produced water at the San Ardo Water Reclamation Facility, CA. Soc. Pet. Eng., 1–21. https://doi.org/10.2118/121520-MS.

Wenten, I.G., Dharmawijaya, P.T., Aryanti, P.T.P., Mukti, R.R., Khoiruddin, K., 2017a. LTA zeolite membranes: current progress and challenges in pervaporation. RSC Adv. 7, 29520–29539. https://doi.org/10.1039/C7RA03341A.

Wenten, I.G., Khoiruddin, K., Hakim, A.N., Himma, N.F., 2017b. The bubble gas transport method. In: Hilal, N., Ismail, A.F., Matsuura, T., Oatley-Radcliffe, D. (Eds.), Membrane Characterization, pp. 199–218, https://doi.org/10.1016/B978-0-444-63776-5.00011-5.

Wenten, I.G., Friatnasary, D.L., Khoiruddin, K., Setiadi, T., Boopathy, R., 2020a. Extractive membrane bioreactor (EMBR): recent advances and applications. Bioresour. Technol. 297, 122424. https://doi.org/10.1016/j.biortech.2019.122424.

Wenten, I.G., Khoiruddin, K., Alkhadra, M.A., Tian, H., Bazant, M.Z., 2020b. Novel ionic separation mechanisms in electrically driven membrane processes. Adv. Colloid Interf. Sci. 284, 102269. https://doi.org/10.1016/j.cis.2020.102269.

Wenten, I.G., Khoiruddin, K., Kadja, G.T.M., Mukti, R.R., Sutrisna, P.D., 2020c. Modified zeolite-based polymer nanocomposite membranes for pervaporation. In: Thomas, S., George, S.C., Jose, T. (Eds.), Polymer Nanocomposite Membranes for Pervaporation. Elsevier, pp. 263–300, https://doi.org/10.1016/B978-0-12-816785-4.00011-2.

Wenten, I.G., Khoiruddin, K., Wardani, A.K., Aryanti, P.T.P., Astuti, D.I., Komaladewi, A.A.I.A.S., 2020d. Preparation of antifouling polypropylene/ZnO composite hollow fiber membrane by dip-coating method for peat water treatment. J. Water Process Eng. 34, 101158. https://doi.org/10.1016/j.jwpe.2020.101158.

Werber, J.R., Osuji, C.O., Elimelech, M., 2016. Materials for next-generation desalination and water purification membranes. Nat. Rev. Mater. 1. https://doi.org/10.1038/natrevmats.2016.18.

Winston Ho, W.S., Sirkar, K.K., 1992. Membrane Handbook. Van Nostrand Reinhold, https://doi.org/10.1016/0376-7388(95)90055-1.

Wong, J.M., Hung, Y., 2006. Treatment of oilfield and refinery wastes. In: Waste Treat. Process Ind., https://doi.org/10.1016/j.jhazmat.2006.05.112. 138, 235.

Xia, Q., Guo, H., Ye, Y., Yu, S., Li, L., Li, Q., Zhang, R., 2018. Study on the fouling mechanism and cleaning method in the treatment of polymer flooding produced water with ion exchange membranes. RSC Adv. 8, 29947–29957. https://doi.org/10.1039/c8ra05575k.

Xu, J., Ma, C., Cao, B., Bao, J., Sun, Y., Shi, W., Yu, S., 2016. Pilot study on hydrophilized PVDF membrane treating produced water from polymer flooding for reuse. Process. Saf. Environ. Prot. 104, 564–570. https://doi.org/10.1016/j.psep.2016.06.020.

Xue, W., Yamamoto, K., Tobino, T., 2016. Membrane fouling and long-term performance of seawater-driven forward osmosis for enrichment of nutrients in treated municipal wastewater. J. Membr. Sci. 499, 555–562. https://doi.org/10.1016/j.memsci.2015.11.009.

Yang, R., Goktekin, E., Gleason, K.K., 2015. Zwitterionic antifouling coatings for the purification of high-salinity shale gas produced water. Langmuir 31, 11895–11903. https://doi.org/10.1021/acs.langmuir.5b02795.

Yu, L., Han, M., He, F., 2017. A review of treating oily wastewater. Arab. J. Chem. 10, S1913–S1922. https://doi.org/10.1016/j.arabjc.2013.07.020.

Zhang, R., Yu, S., Shi, W., Wang, X., Cheng, J., Zhang, Z., Li, L., Bao, X., Zhang, B., 2017. Surface modification of piperazine-based nanofiltration membranes with serinol for enhanced antifouling properties in polymer flooding produced water treatment. RSC Adv. 7, 48904–48912. https://doi.org/10.1039/c7ra09496e.

Membrane bioreactor for wastewater treatment: Fouling and abatement strategies

Shamas Tabraiz[a,b], Muhammad Zeeshan[c,d], Muhammad Bilal Asif[e], Uchenna Egwu[a], Sidra Iftekhar[f], and Paul Sallis[a]

[a]School of Engineering, Newcastle University, Newcastle, United Kingdom [b]Natural and Applied Sciences Section, School of Psychology and Life Sciences, Canterbury Christ Church University, Canterbury, United Kingdom [c]German Environment Agency, Section II 3.3, Berlin, Germany [d]Department of Water Quality Control, Technical University of Berlin, Berlin, Germany [e]Advanced Membranes and Porous Material Center, Physical Sciences and Engineering, King Abdullah University of Science and Technology (KAUST), Thuwal, Saudi Arabia [f]Department of Applied Physics, University of Eastern Finland, Kuopio, Finland

1. Introduction

The idea of combining a fine filter with membrane as a separation stage during sludge digestion was first proposed by Dorr Oliver in the mid-1960s (Sutherland, 2010). A flat sheet membrane was suggested as a sidestream filter unit. The submerged membrane bioreactor concept, introducing the membrane directly into the activated sludge, was first presented in 1989 (Yamamoto et al., 1988). The idea behind integrating the membrane within the activated sludge tank, creating a membrane bioreactor (MBR), was to remove the need for a separate secondary clarifier, as shown in Fig. 1. Polymeric membranes with pore sizes ranging from 0.003 to 0.01 μm were used (Paul and Ebra-Lima, 1970). However, it was difficult to recommend membrane technology for commercial treatment technologies due to some problems, such as membrane cost, the low economic value of the product, and poor performance due to membrane fouling (Enegess et al., 2003). Nevertheless, over time, the progress in the material technology of the membranes reduced the cost significantly and made way for wider use. In addition, subsequent improvements to our understanding of the relationship

Copyright © 2023 Elsevier Inc. All rights reserved.

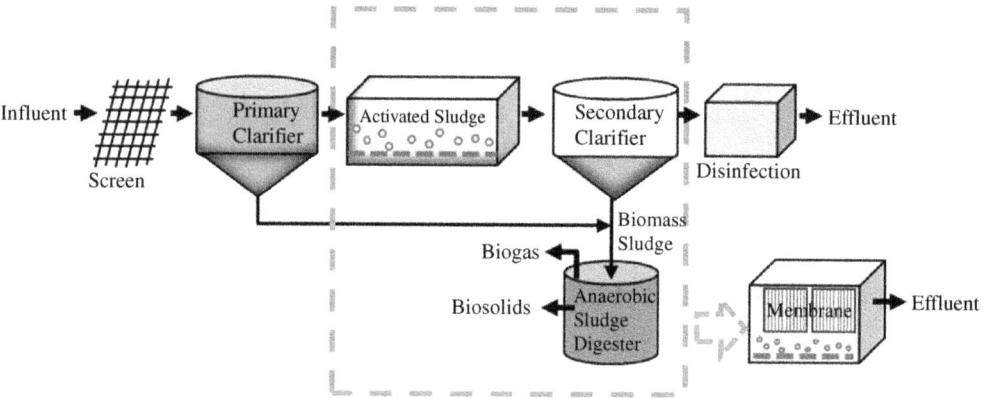

FIG. 1 Schematic diagram of activated sludge process and showing the equivalent role of MBR in wastewater treatment.

between the operational parameters, fouling, and changes in the reactor operation improved MBR performance substantially (Ahmed et al., 2007; Belfort et al., 1994).

Furthermore, higher removal efficiencies, low production of sludge, the smaller size of the reactor, shorter hydraulic retention time (HRT), elimination of the secondary clarifier, and simultaneous nitrification and denitrification due to high solid retention time (SRT) make MBR a promising technology for high-quality treatment of sewage and industrial wastewater compared to conventional activated sludge (CAS) (Melin et al., 2006).

2. Configuration and types of membrane bioreactors

Commercially, the most common trend has been toward MBRs with submerged membranes, although MBRs with external membrane modules (side streams) are also available in the market. A schematic diagram of the submerged and sidestream MBR systems is shown in Fig. 2, and typically, both types of MBR can employ flat sheet membranes or hollow fiber membranes (Chan and Chen, 2004). The economics and efficiency of the membrane unit depend on the operationally achievable permeate flux, which can be controlled by keeping fouling rates below critical flux (Enegess et al., 2003). The membrane bioreactor can be aerobic and anaerobic. The anaerobic membrane bioreactor can be energy neutral as the anaerobic digestion (AD) process converts the carbonaceous compounds into methane.

3. Aerobic and anaerobic MBR

3.1 Aerobic membrane bioreactor

Membrane filtration technology has been widely adopted for wastewater treatment to permit reuse. These membranes were developed primarily to replace secondary clarifiers in conventional activated sludge systems (CAS), reducing the reactor size by retaining high concentrations of biomass (i.e., high mixed liquor suspended solids, MLSS concentrations).

a)

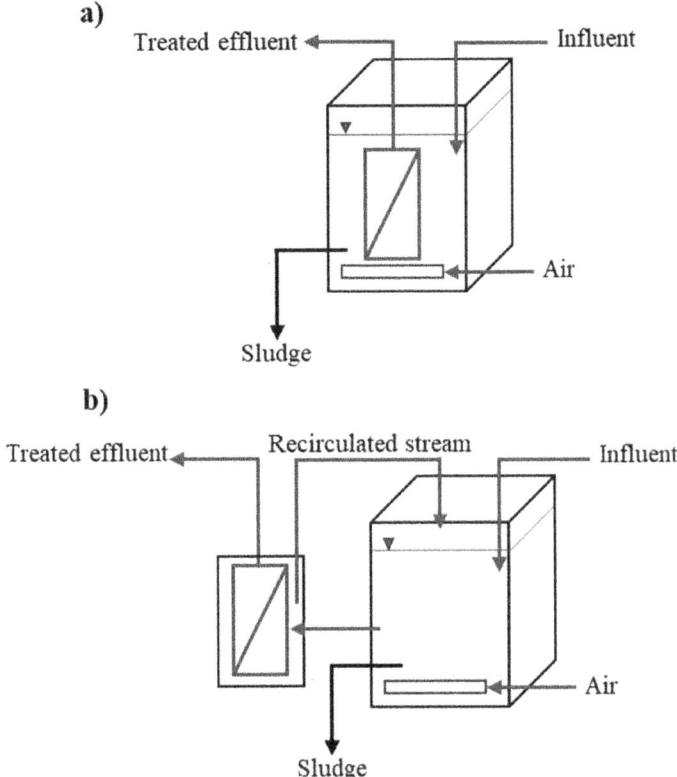

FIG. 2 (A) Submerged (integrated) membrane bioreactor, (B) external sidestream membrane bioreactor.

The growth in the membrane bioreactor (MBR) technology is driven by the need for high-quality treated effluent that satisfies stricter regulations, and MBR systems can usually provide this more reliably than other biological wastewater treatment technologies (Fane, 1996; Melin et al., 2006; Oron et al., 2008; Tabraiz et al., 2016, 2018). Recently, due to novel developments in membranes, the capital and operational cost of MBR has decreased over the last two decades (Li et al., 2019). It has been successfully employed for both industrial wastewater and sewage treatment. The chemical oxygen demand (COD) removal efficiency of MBR can reach 94%–99%, while nitrogen and phosphorus removal are possible up to $\sim$90% and $\sim$92%, respectively (Monclús et al., 2010). Moreover, 90%–100% removal efficiencies have been reported in MBR for pharmaceutical emerging pollutants, e.g., Sulfamethazine, Ibuprofen (Gu et al., 2018). For the highest degree of treatment, i.e., to achieve potable water quality, MBR has been coupled with reverse osmosis, forward osmosis, and membrane distillation (Luo et al., 2017; Nguyen et al., 2016; Song et al., 2018; Wang et al., 2017b).

3.2 Anaerobic membrane bioreactor

A paradigm shift has been observed in wastewater treatment from end-of-pipe treatment to resource recovery (Sheik et al., 2014), and wastewater is now considered a valuable source of water, nutrients, and energy rather than just a pollution source (Guest et al., 2009; Puyol et al., 2017). To recover many of these valuable resources from wastewater, the anaerobic digestion (AD) is considered more sustainable treatment option compared to the conventional aerobic activated sludge process (CAS) (McCarty et al., 2011; Wan et al., 2016), mainly because it does not require costly aeration when converting organic carbon to CO_2 and methane (Egwu et al., 2022; Petropoulos et al., 2020; Shamurad et al., 2019, 2020a,b,c,d). In contrast, CAS is a considerably more energy-intensive process. AD offers several other advantages, such as the production of methane and low sludge production (Lei et al., 2018; Verstraete and Vlaeminck, 2011). Different types of reactor configurations have been employed to treat wastewater. However, upflow anaerobic sludge blanket (UASB) and expanded granular sludge bed reactor (ESB) systems provide higher removal efficiencies and higher production rates than earlier systems (Tauseef et al., 2013; Van Lier et al., 2015). The competitive benefits of these high-rate reactors are due to the retention of valuable biomass (by avoiding washout), which increases the solid retention time (SRT). Full-scale UASB reactors have been employed for sewage treatment in warm climate countries like India, Brazil, and the Middle East. However, in many cases, the UASB treatment efficiencies (i.e., COD removal) are low ($\sim$60%) (Chernicharo et al., 2015; Heffernan et al., 2011; Van Lier et al., 2010). This results in a low-quality effluent with biodegradable carbonaceous compounds remaining in the effluent. Often, this poor performance has been attributed to the inappropriate design of the reactors and poor maintenance and operational procedures (Batstone et al., 2015; Vinardell et al., 2020).

AnMBR can be a better option than UASB reactors as they overcome some limitations of the latter (Foglia et al., 2019; Giménez et al., 2011; Ozgun et al., 2015). A micro- or ultrafiltration membrane is placed inside the reactor and effluent is filtered through the membrane using a pump while retaining virtually all the biomass inside, which yields a high-quality effluent (Petropoulos et al., 2021, 2019). However, fouling of this membrane and the associated cleaning and maintenance costs are the main drawbacks that hinder the wider uptake of AnMBR for treating wastewater.

4. Aerobic versus anaerobic treatment and AnMBR

Aerobic treatment is the most widely applied wastewater treatment process. Although it produces a very good quality effluent, it is an energy-intensive practice that also produces a large amount of sludge compared to anaerobic treatment. The sludge produced requires handling, treatment, and disposal, and many technologies fail to recover the valuable resources from the wastewater, i.e., nutrients and energy (Martinez-Sosa et al., 2012; Smith et al., 2012). The sludge produced by aerobic treatment can be digested anaerobically, which allows the recovery of only a fraction of the energy associated with the original dissolved and suspended organic waste components in the wastewater. Furthermore, the overall energy produced by sludge digestion is only a fraction of the total energy required to operate the whole conventional aerobic treatment process, so an external energy input is required to run CAS treatment plants (McCarty et al., 2011). To overcome these limitations, anaerobic treatment is an attractive alternative with the advantages of lower sludge production and energy recovery through biogas (Bae et al., 2014; Chong et al., 2012b). A few studies have compared anaerobic wastewater treatment with the CAS process for domestic wastewater treatment having a typical influent COD value of $500\,mg\,L^{-1}$. The results indicated that anaerobic treatment gave twice the biogas production compared to CAS treatment with sludge digestion, and the energy produced exceeded the energy required for plant operation (McCarty et al., 2011).

On the other hand, aerobic systems have higher removal efficiencies for soluble biodegradable organic material than anaerobic wastewater treatment processes. Furthermore, the biomass produced by aerobic systems has better settling properties which result in less biomass in the effluent quality than in anaerobic systems (based on flocculent sludge). In general, the effluent quality of anaerobic systems is low and contains a high concentration of suspended solids, pathogens, organic matter, and nutrients. However, the effluent from both processes often requires additional downstream treatment (tertiary treatment) to meet the stringent effluent discharge standards (Chan et al., 2009; Chernicharo et al., 2015; Kim et al., 2011). The lower removal efficiencies of AD systems are due to the low metabolic capacity of anaerobic bacteria; thus, anaerobic processes require a longer hydraulic retention time (HRT) (Van Haandel et al., 2006), which implies larger, more costly reactors.

The demerits of longer HRT and shorter SRT in anaerobic systems can be overcome by employing high-rate anaerobic reactors such as UASB, anaerobic filters (AF), expanded granular sludge bed (EGSB), and anaerobic baffled reactors (ABR) (Stazi and Tomei, 2018) as these all intensify the biological process by retaining more active biomass in the same size reactor. Consequently, to retain the biomass completely and achieve higher internal biomass concentrations and higher effluent quality (compared to the other high-rate anaerobic reactors such as UASB reactors), the AnMBR system should be employed. The very high quality of the effluent is a distinctive advantage of the AnMBR design over other high-rate AD systems. In addition, the AnMBR effluent is essentially free from biomass and pathogens (Batstone and Virdis, 2014; Smith et al., 2012; Stuckey, 2012). Many studies have demonstrated that AnMBR can be an energy-neutral or even energy-positive wastewater treatment technology due to the efficient production of methane. The zero- or low-energy requirement of AnMBR is a distinctive benefit over aerobic MBR. However, a few limitations hinder the application of the AnMBR system on a wider scale, the main limitation being the biofouling of the membrane, which increases its operational and maintenance costs (Cogert et al., 2019; Evans et al., 2019; Pretel et al., 2015; Smith et al., 2014).

5. Fouling, the main hindrance in the widespread use of MBR

For all types of membrane filtration systems, membrane fouling, especially biofouling, impairs filtration performance, and consequently, this affects MBR performance adversely over extended periods of operation. Fouling of membranes increases the energy required to maintain the flux through the fouled membrane and requires frequent backwashing of the membrane, which eventually deteriorates the surface and ultimately necessitates the replacement of the membrane. Consequently, membrane fouling in MBR is one of the main factors contributing to high operational and maintenance costs, limiting their widespread use for wastewater treatment (Judd, 2017; Luo et al., 2018; Teng et al., 2019).

5.1 Types of fouling

Fouling can be classified based on the cleaning method required to reduce it. Reversible fouling is the form that can be removed with the physical cleaning of the membrane, such as backwashing, vibration, and shaking in water. It is comprised chiefly of bio-cake/biofilm. Irremovable fouling is a form that cannot be removed by physical treatment but can be removed with chemical or surfactant washing (Meng et al., 2017). Only a few colloids and solute particles possess sufficiently strong bonding to the surface of the membrane that they cannot be removed with any type of treatment, and such fouling is termed irreversible fouling (Tsuyuhara et al., 2010), as shown in Fig. 3.

Based on the type of foulant, fouling can be divided into biofouling, organic fouling, and inorganic fouling. Biofouling is due to the growth of microorganisms and excretion of extracellular polymeric substances (EPS) comprising proteins, polysaccharides, humic substances, and other biopolymers (Meng et al., 2007, 2009, 2010). Organic fouling is due to the attachment of the organic chemicals on the membrane surface originating from the wastewater, such as grease, oil, hydrophobic humic organic matter, and hydrophilic carbohydrate-based organic matter. A proportion of organic fouling is reversible and can be removed from the membrane surface with chemical treatment (Gao et al., 2010; Iorhemen et al., 2016; Tian et al., 2013).

Inorganic fouling is caused by inorganic compound deposition on the membrane surface and pores. The typical inorganic species which cause inorganic fouling consist of cationic and anionic species such as PO_4^{3-}, Ca^{2+}, Mg^{2+}, Al^{3+}, SO_4^{2-}, OH^-, and CO_3^{2-} (Lin et al., 2014). The most common inorganic foulant compound in membrane-based wastewater treatment systems is Struvite $MgNH_4PO_4 \cdot 6H_2O$. Struvite is formed during the anaerobic decomposition of organics due to the release of ammonium ions and phosphate. In addition, $CaCO_3$ and $K_2NH_4PO_4$ can also contribute to the inorganic fouling of MBR (Meng et al., 2010; Stuckey, 2010). $CaCO_3$ deposition on the membrane surface happens if there is high alkalinity in

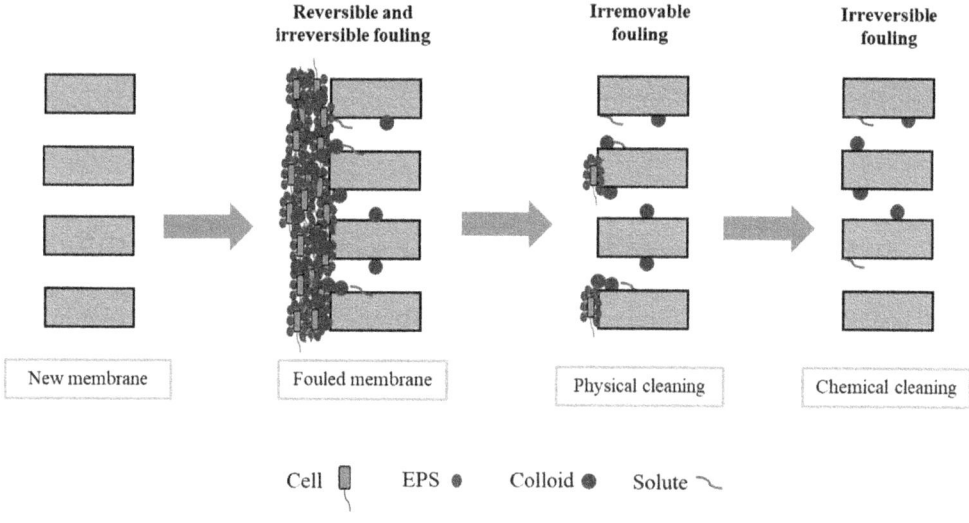

FIG. 3   Types of membrane fouling and membrane cleaning requirements.

the sludge, and Ca^{2+} and PO_4^{3-} precipitation on the membrane surface is more problematic under higher pH conditions. Inorganic fouling contributes to the irremovable fouling that can be removed with chemical treatment Fig. 3. There are two mechanisms of inorganic fouling: crystallization and particulate, the latter being the major contributor to inorganic fouling. In crystallization, the ions get precipitated and deposited on the membrane surface; in particulate fouling, the colloidal particulate matter present in the wastewater deposits on the surface of the membrane during the process of filtration (Iorhemen et al., 2016). Among all three types of fouling, biofouling and the organics originating from the microorganism (EPS) are the main sources of operational fouling problems that severely hamper the filtration capacity of membranes.

5.2 Fouling mechanism in membrane bioreactors

Based on transmembrane pressure (TMP), the membrane fouling profile can be divided into three phases: initial phase, maturation phase, and jump phase. In the initial phase, membrane fouling is caused by the pore blocking/clogging and narrowing due to the adsorption or binding of solute and colloids on the membrane surface. In addition, dead or live cell adhesion on the membrane surface or pores also contributes to the initial phase pore blockage. In the maturation phase, EPS from the sludge promotes adhesion and makes a gel-like layer on the membrane surface, which further helps cells/microorganisms and colloids adhere to the membrane's surface (Lin et al., 2014; Wang et al., 2008). Consequently, more pore blockage occurs and causes a further rise in the TMP. The process of cell adhesion on the gel-like layer on the membrane surface continues until it reaches a specific cell density. At this point, quorum sensing (QS) processes are activated (Yeon et al., 2009). Consequently, higher concentrations of EPS formed as a result of QS further help cell adhesion and form a thick fouling layer (bio-cake/biofilm) (Fig. 4). Enhanced EPS production and biofilm formation block the pores severely and a sudden jump in the membrane TMP then occurs, known as the jump phase (Hong et al., 2014). After the jump phase, membranes have to be detached from the treatment train and cleaned; otherwise, the reversible initial fouling shifts to an irreversible form of fouling.

6. Factors affecting fouling in MBR

Different factors can affect membrane fouling in the MBR and are divided into three categories: membrane characteristics, feed/influent characteristics, and reactor operating conditions, as shown in Fig. 5.

6.1 Membrane characteristics

6.1.1 Membrane material
Membranes can be divided into three types depending on the construction material, namely: organic, ceramic, and metal membranes. Organic membranes are widely used due

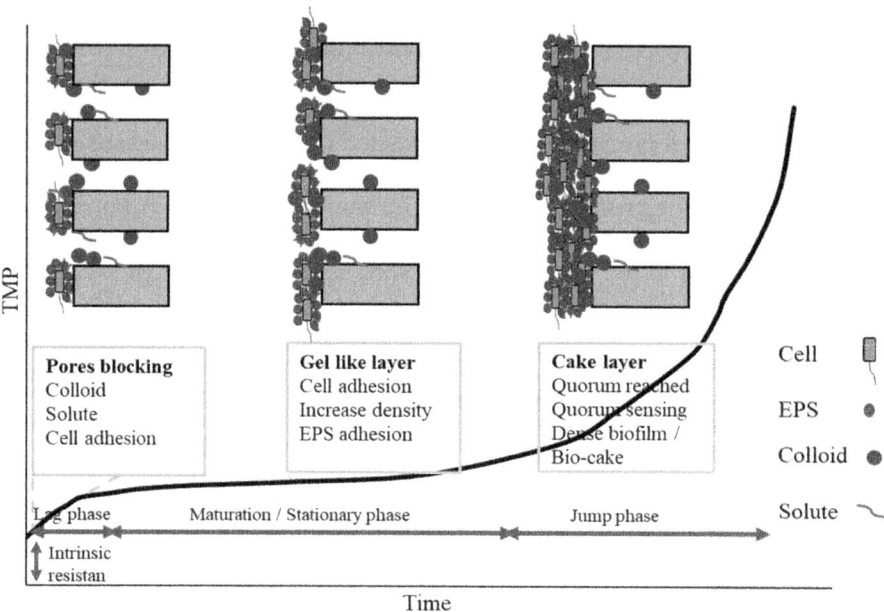

FIG. 4   Membrane fouling mechanism, foulants, and fouling pattern in different phases of fouling.

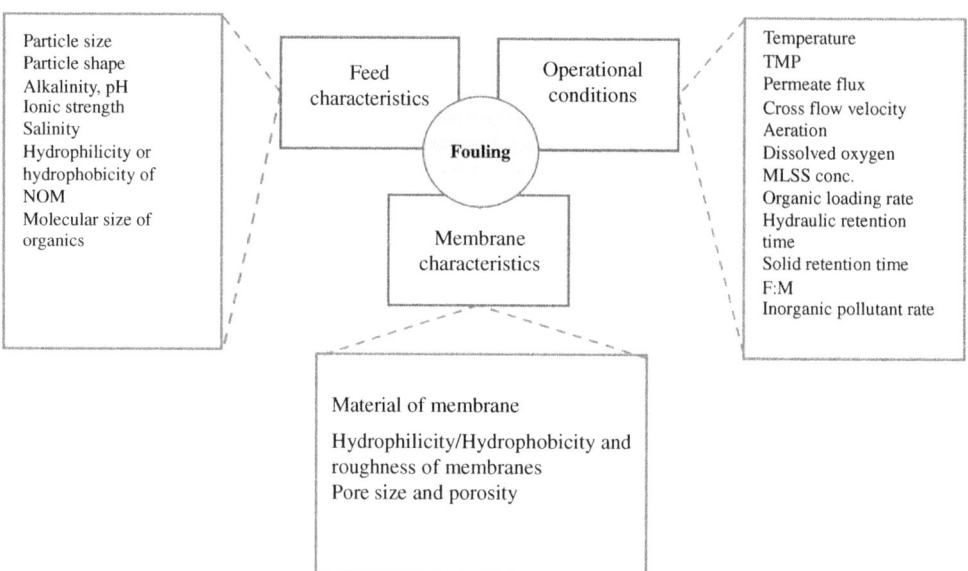

FIG. 5   Factors affecting membrane fouling in MBR. *F:M*, food to microorganism ratio; *MLSS*, mixed liquor suspended solids; *NOM*, natural organic molecules; *TMP*, transmembrane pressure.

to their low cost and easy manufacturing process. However, organic membranes are lower in mechanical strength and have higher fouling rates than the other two types.

The common organic polymers used in membrane manufacture are polyethersulfone (PES), polyethylene (PE), polysulfone (PS), polyacrylonitrile (PAN), and polyvinylidene fluoride (PVDF). PVDF is generally accepted as a better option compared to the other materials due to its good mechanical strength, high thermal stability, good chemical resistance, and membrane forming properties. PVDF membrane performance for fouling retardation has been improved after modification by blending other chemicals (Kang and Cao, 2014; Liu et al., 2014; Zhao and Chen, 2015). Recently, novel materials, e.g., graphene oxide, boron nitride, and MXene, were used to make membranes with better stability (Asif et al., 2021).

6.1.2 *Hydrophobicity, hydrophilicity, and roughness of the membrane*

Hydrophobic membranes are more prone to fouling than hydrophilic membranes due to enhanced interface interaction between the foulant and membrane surface. The lack of hydrogen bonding sites on the surface of the membrane causes hydrophobicity. A spontaneous process of entropy increase of hydrophobic solutes happens while approaching the hydrophobic membrane surface, which leads to adsorption of these solutes on the membrane surface and causes blockage of the membrane (Sun et al., 2013). Ceramic membranes are hydrophilic and experience lower fouling rates compared to organic membranes. It has also been reported that the membrane surface's higher roughness (texture) imparts a greater tendency to bind organic matter/foulants and stimulates increased fouling compared to a smoother surface (Myat et al., 2014). The membrane's cleaning efficiencies are also affected by hydrophobicity/hydrophilicity. The hydrophilic membrane has higher cleaning efficiencies and lower irreversible foulant absorption (Fan et al., 2018).

6.1.3 *Pore size and porosity*

Pore size and size distribution play a key role in determining the flux of membranes. A wider pore size usually helps to maintain a higher flux at the start. However, it allows larger particles to entrap in the larger pores, resulting in blockage of the membrane (Judd, 2010). The large size particle entrapment in the membrane causes rapid flux reduction and irreversible fouling (Shang et al., 2015). So smaller pore size (0.01–0.4 μm) membranes are recommended in MBR to control fouling from larger particles (Meng et al., 2009). The membrane fouling rates can be reduced further by increasing the porosity, which also helps to maintain the flux (Kang et al., 2006).

6.2 Feed characteristics

Properties of feed (wastewater influent) also significantly affect membrane fouling. Particles in liquid feed cause fouling by pore blocking and pore narrowing, dependent on the particle size. Some of the important properties of feed wastewater are particle concentration, particle properties, ionic strength and pH, hydrophilicity/hydrophobicity of the natural organic molecules (NOM), and affect of the fouling of the membrane in MBR.

6.2.1 Particle size and concentration

Particle size distribution has a significant effect on the membrane fouling rate and membrane cleaning. Large particles in the feed coalesce and form flocs that span across pores and inhibit the trapping of small particles within the pores. The presence of both fine and coarse particles in the feed causes lower porosity of the cake, as fine particles slide in between coarse particles, thus filling empty voids between coarse particles, restricting flow, and enhancing the fouling. Similarly, increases in the concentration of the feed particles also increase the membrane fouling rates (Vyas et al., 2000).

6.2.2 Ionic strength, pH, and salinity

The ionic strength and the electric charge on particles in feed water significantly affect membrane fouling. The pH and ionic strength of the feed solution affect the charge on the membrane surface and the charge on particles, resulting in the adhesiveness of particles and the size of the cake. The ionic strength in the range of 0.005–$0.05 \, mol \, L^{-1}$ does not affect the fouling significantly (Wang et al., 2014a). However, ionic strength up to $0.5 \, mol \, L^{-1}$ increased the fouling rates. Higher ionic strength increases the fouling of the membrane by reducing the floc size, production of colloidal particles and polymers out of the sludge, and the subsequent production of a more compact fouling layer (Christensen et al., 2021). Salinity shocks have been reported for reduced permeability and increased fouling rates. Salinity affects the surface charges on the particles present in wastewater and the sludge EPS production, which clog the pores of the membrane and lead to higher fouling rates (Jang et al., 2013; Sun et al., 2010). It has been observed that a reduction in pH can decrease the molecular size of NOM, which increases the adsorption on the membrane and thus results in high fouling (Abdelrasoul et al., 2013). Moreover, changes in pH also change the community structure and the EPS structure, which increases the membrane fouling. Aliphatic and aromatic components present in EPS contain phenolic alcohols (with the OH group) and carboxylic acids (with COOH). The EPS adsorption on the negative surface increases as the pH decreases. The extreme pH increase also induces changes in the EPS structure which can be either coiled (dense) or linear (Sanguanpak et al., 2015; Sweity et al., 2011).

6.2.3 Hydrophilicity/hydrophobicity of NOM

The nature of the fouling material within the wastewater and its interaction with the membrane material are important factors in determining the initial fouling phase of the membrane.

Specifically, the hydrophilicity or hydrophilicity of NOM plays an important role in membrane fouling. Hydrophilic particles dominate at the start of the operation and have been reported to affect the fouling rates severely (Carroll et al., 2000). The relative order of importance for four different particle types in terms of flux reduction/fouling is neutral hydrophilic fraction > hydrophobic acids fraction > transphilic acids fraction > polar hydrophilic fraction (Fan et al., 2001). Gray found that polar hydrophilic organics and neutral hydrophilic organics could form a gel-like layer on microfiltration membranes, and this caused a rapid flux decline, while hydrophobic fractions caused a slow flux decline (Gray et al., 2007). Chen et al. (2007) found that hydrophilic organics of the influent water caused a slow flux decline in ultrafiltration membranes while hydrophobic organic macromolecules caused a rapid flux decline. Wastewater contains a variety of NOM, which include both hydrophobic and hydrophilic

NOM. The wastewater with a nominally moderate fraction of hydrophobic/hydrophilic NOM (between 5 and 50) had a higher fouling propensity (Mu et al., 2019). Another recent study demonstrated that removing hydrophobic NOM from the wastewater with ozonation reduced the fouling (Huang et al., 2019b).

6.2.4 Molecular size of organics

The molecular size of organics has a great influence on the performance of membranes. The organic matter having a molecular size greater than the pore size of the membrane can clog it, which results in a flux decline. The organic matter having a molecular size smaller than the membrane pore size can enter inside the membrane, which can also cause a flux decline (Sun et al., 2013). It has also been reported that organic materials of high molecular weight, such as polysaccharides and long-chain aliphatic molecules, can be retained on the membrane, while aromatic compounds of medium molecular weight, such as lignin and tannic acid, can enter inside the membrane (Lankes et al., 2008). Other studies have reported that small organic matter also causes severe types of fouling in membrane systems (Li and Chen, 2004).

6.3 Operational conditions

6.3.1 MLSS concentrations

The activated sludge mainly consists of dissolved organic matter (DOM), nutrients, microbial cells, and EPS. In general, a higher organic loading rate (OLR), a lower hydraulic retention time (HRT), and a long solids retention time (SRT) can lead to higher concentrations of mixed liquor suspended solids (MLSS). Membrane fouling increases as the MLSS concentration increases due to increased viscosity (Pan et al., 2010). Yigit et al. (2008) studied the impact of MLSS concentration on the fouling of the submerged MBR and reported that increases in the MLSS concentration increased the concentration of EPS (proteins and polysaccharides), which then increased the fouling rates of the MBR. A few studies have reported decreases in the fouling rates at MLSS concentrations above $15 \, \text{g} \, \text{L}^{-1}$, which can plausibly be due to the decrease in the food-to-microorganism (F:M) ratio (Rosenberger et al., 2005).

6.3.2 OLR, SRT, HRT, F:M

SRT is one of the most important parameters influencing membrane fouling. SRT controls the biomass characteristics in membrane systems, and increasing SRT from 5 to 15 days has been shown to reduce fouling rates (Mirzavandi et al., 2019). Another recent study reported increased pore blockage in the integrated fixed-film MBR as the SRT was decreased from 30 to 15 days (Mannina et al., 2019). An SRT less than 20 days increased the concentration of bound SMP, which increased the cake resistance and resulted in higher TMP/fouling. However, at an SRT greater than 60 days, the EPS content was lower, but the fouling rates were higher. This observation was attributed to inorganic fractions of dead cells (Ahmed et al., 2007). Equally, a reduction in the EPS at a higher SRT could have happened due to a change in the community structure or the social behavior of the microorganisms. A recent study investigated this aspect and found that increases in the SRT (4, 10, and 4 days) increased the quorum quenching (QQ) bacterial group proportion in the activated sludge of an MBR.

The higher abundance of QQ bacteria reduced the QS signal molecules (AHL), which resulted in lower levels of EPS and reduced the fouling rates (Yu et al., 2016).

Generally, a higher OLR increases the fouling rates of MBR due to the hydrophilic substances in the EPS of the MBR sludge. In addition, the biopolymeric-type substances, humic acid, low molecular weight neutral and acidic compounds were found in the foulant material of membranes at a higher OLR when increased from 0.5 to $3.0\,kg\,COD\,m^{-3}\,day^{-1}$ (Johir et al., 2012; Sharghi et al., 2020). Another study reported higher fouling rates at a higher OLR and changes in the microbial communities of the biofilm and sludge (Xia et al., 2010). However, further research is needed to understand the role of microbial communities in the biofilm/ sludge and how the fouling mechanism is affected by operational conditions.

A decrease in the HRT from 5 to 3 days increased the fouling rates in an AnMBR as the average SMP and EPS per unit of biomass increased (Santos et al., 2017). In an aerobic MBR, a decrease in HRT (from 6.67 to 5.33 and 4.00 h) increased the concentration of filamentous bacteria in the reactor, leading to higher EPS levels and hence higher fouling rates (Deng et al., 2016). The EPS foulant layer developed was loose, soft, and had high fluidity at a lower HRT, which was the reason for the higher fouling rates (Wang et al., 2017a). In addition, increased viscosity of SMP of the sludge at a lower HRT (decreased from 24 to 8 h) has also been reported in the membrane sequencing reactors (Shariati et al., 2011).

F:M can be varied by varying the OLR. A higher F:M leads to higher substrate utilization rates, resulting in more inorganic fraction and increased passage of untreated organics through the membrane, which accelerates fouling. Moreover, higher concentrations of EPS were reported at a higher F:M ($3.8\,g\,COD\,g^{-1}\,MLSS\,day^{-1}$) compared to a lower F:M ($0.1\,g\,COD\,g^{-1}\,MLSS\,day^{-1}$), which are the main component of biofouling in MBR (Liu et al., 2012). In addition to the elevated concentration of EPS, the nature of the EPS (a higher protein concentration) also changes at a higher F:M, which plays a key role in fouling (Iorhemen et al., 2016; Kimura et al., 2005).

6.3.3 Temperature

Temperature is a key parameter for biological treatment and affects the microorganisms' metabolism and many other parameters in MBR (Ma et al., 2013). Generally, lower temperatures increase the fouling rates. Four phenomena are linked to higher fouling rates in MBR at lower temperatures: (1) reduced levels of the treatment efficacy in the MBR, hence affecting the permeate quality, which elevates the fouling rates (van den Brink et al., 2011), (2) deflocculation of sludge flocs which causes the release of EPS (Morgan-Sagastume and Allen, 2005), (3) an increase in the EPS concentration in the MBR sludge which increases the viscosity of the sludge and reduces the shear of the aeration in the reactor, and (4) a sudden decrease in temperature affects the biomass and increases the EPS concentration with increased levels of proteins, which results in increased fouling rates (Drews et al., 2007). However, further studies are needed to investigate the reason for the release of EPS at low temperatures, which caused this fouling in MBR. A greater understanding in this area would help to devise strategies that reduce the level of EPS at a lower temperature to sustain the longer operation of MBR under such conditions. A recent study has reported that a low temperature induces a high stress on the biofilm community of an AnMBR, triggering higher quorums sensing activity, resulting in a higher level of EPS and subsequently higher fouling rates (Tabraiz et al., 2021a).

6.3.4 COD:N

A high COD:N ratio in the influent increases the concentration of EPS and affects its properties (change in MW distribution) and proportion (increased carbohydrates fraction). The increased concentration and changed properties of EPS can then increase the fouling rates in submerged MBR. Moreover, higher concentrations of NH_4^+ would occur at a low COD:N, and the higher concentration of NH_4^+ replaced the multivalent cations in the extracted EPS, which led to lower fouling rates (Feng et al., 2012). Similarly, another study reported increased concentrations of humic acid, carbohydrates, and proteins in EPS, as the COD:N ratio increased. Overall, a change in composition at high COD:N ratios increased the adsorption capacity of EPS (Han et al., 2015). COD:N affects the microbial community composition in the reactor, with an appropriate ratio of COD:N promoting the growth of nitrifying bacteria. Nitrifying bacteria produce fewer EPS compared to heterotrophs. Hence, increases in the proportion of nitrifying bacteria reduced the fouling rates of MBR (Sepehri and Sarrafzadeh, 2018).

6.3.5 *Operating mode/transmembrane pressure (TMP)*

Membranes can be operated on constant TMP, reducing flux over time or constant flux, increasing TMP over time. The latter is a common mode of operation as it keeps the HRT of the reactor constant (Guglielmi and Andreottola, 2010). The fouling is usually measured by TMP increase at a constant flux operation mode. The flux above which the deposition of the solids on the membrane becomes evident is called critical flux. The membrane in MBR is continuously operated below the critical flux to prolong the operation of the MBR (Bacchin et al., 2006; van der Marel et al., 2009). Operational flux should provide only a small and sustainable increase in the TMP. However, since increases in the flux increase the fouling rates, keeping the operating flux lower than the critical flux will lead to slower fouling rates in MBR (Kimura et al., 2015).

6.3.6 *Aeration*

In submerged MBR systems, aeration has three major roles, i.e., it provides oxygen to the biomass, keeps activated sludge in suspension, and mitigates fouling by the constant scouring of the membrane surface (Dufresne et al., 1997). The proximity of the membrane next to the airstream and the size and velocity of the bubbles from the aeration are important parameters that affect the biofilm layer and/or the size of flocs. A tilted angle (20 degrees) between the membrane module and the aeration stream reduced fouling 2.7 fold by eroding the biofilm effectively with shear (Eliseus et al., 2017). Bubbles near the membrane surface induce local shear transients and liquid flow fluctuations, which lead to back transport phenomena, and tangential shear on the membrane surface prevents the deposition of large particles on its surface, but tangential shear depends on the bubble diameter (Choo and Lee, 1998). Aeration optimization studies through computational models have been conducted to get the maximum shear near the membrane surface through water velocity. A critical water velocity of $0.232\,\mathrm{m\,s^{-1}}$ is reported to remove the fouling effectively (Wang et al., 2021). Increased aeration velocity reduced the fouling rates, but energy consumption was also increased. In addition, the bubble size is very important and affects the sludge characteristics, fouling removal, and oxygen transfer rates. A fine bubble size produces foaming, which tends to reduce the fouling and high oxygen transfer rates. So an optimized size of the bubble must be provided to optimize all these factors (Zhang et al., 2019). For this purpose, artificial intelligence was implied using a harmony search algorithm to optimize the aeration to mitigate fouling and reduce energy requirements (Nam et al., 2021).

6.3.7 Dissolved oxygen

The growth of microorganisms highly depends on the concentration of dissolved oxygen (DO) as it is key to bacterial metabolic function in aerobic systems. In MBR, DO levels may increase the growth of some species but decrease others, but a small alteration in DO may not change the composition of microbial communities. However, by decreasing DO to $0.5\,mg\,L^{-1}$, a significant change in the microbial composition and diversity was observed (Gao et al., 2011; Ma et al., 2006; Tocchi et al., 2012). Low DO affects the biofilm microbial community more than the mixed liquor, which contributed to the membrane fouling. Two dominant genera, *Paracoccus* and *Rhizobiales*, only appeared under moderate and low DO levels $(4, 2\,mg\,L^{-1})$ in biofilms. The DO level also influenced the generation, composition, and release of EPS. Low levels of DO increased the EPS concentration in the MLSS and their accumulation on the membrane surface, which induced higher membrane fouling (Gao et al., 2011). Low DO levels in MBR decrease the size of microbial flocs, leading to a compacted and dense cake layer formation on the membrane surface and elevating the fouling rates. Low DO levels $(0.25\,mg\,L^{-1})$ induced higher biopolymers, a small floc size, poor filterability, and low critical fluxes. However, an increase in the DO level to $0.38\,mg\,L^{-1}$ improved the aforementioned sludge properties (Díaz et al., 2017). In addition, decreasing DO concentrations to $0.5\,mg\,L^{-1}$ or below resulted in a decrease in COD removal efficiencies, which enhanced the fouling rate due to the deposition of unhydrolyzed organics on the membrane (Komesli et al., 2007).

7. Fouling abatement strategies

Different types of physical, chemical, and biological techniques are used to reduce the fouling in MBR (Iorhemen et al., 2016; Zeeshan et al., 2017). The physical techniques include relaxation of a continuous mode of operation, backwashing with permeate, air sparging, sonication, and scouring the biofilm from the surface of the membrane with an adsorbent or granular medium (Chang et al., 2019; Yusuf et al., 2016). Chemical techniques involved in situ backflushing of chemicals and coagulation while biological techniques involved reducing the QS activity in the reactor by quorum quenching.

7.1 Physiochemical strategies

7.1.1 Relaxation and backwashing with permeate

Relaxation is a technique in which the membrane is relaxed for some time between two consecutive operational modes. The TMP reduces, and reversibly attached foulants diffuse away from the membrane surface due to concentration gradients during relaxation. The efficiency of this process can be improved further by air scouring during the relaxation period (Chua et al., 2002; Hong et al., 2002). A recent study has reported that relaxation optimization times may vary from 0.2 to 4 min, depending on the type of the sludge, and can vary depending upon the sludge characteristics within the reactor (Christensen et al., 2016).

Backwashing (backflushing) is a reversed filtration process in which the permeate is flushed back through the membrane. When a backward flow is applied in porous membranes, the pores are flushed from inside outward. The pressure applied to the permeate side of the membrane during backwash helps pores to be cleaned. Backwashing is considered better than relaxation for removing the reversible fouling due to pore blocking. It dislodges loosely attached sludge cake from the membrane surface and transports it back into the bioreactor (Wu et al., 2008). The key parameters on which the efficiency of backwashing depends are the duration and frequency of the backwashing cycle. Less frequent and longer backwashing (600 s filtration/45 s backwashing) has been observed to be most efficient and better than more frequent backwashing (200 s filtration/20 s backwashing) (Jiang et al., 2005). However, keeping the backwashing time and flow constant, the short cycle time was more effective than the longer one. In addition, permeate backwashing has been a better option than relaxation as it reduces fouling more effectively. A plausible reason is the efficient removal of reversible foulants due to the shear provided by permeate backwashing (Tabraiz et al., 2017).

7.1.2 Chemicals enhanced backwash

Sodium hydroxide (NaOH) (Lee et al., 2012) and sodium hypochlorite (NaOCl) (Cai et al., 2017; Lee et al., 2016) solution has been used for in situ intermittent backwashing in MBR to reduce biofouling. Backwashing with NaOH detached the foulant from the membrane, overall reducing 50%–69% of the total fouling rates and 40%–50% of the irreversible fouling rates, and long-term exposure to the NaOH solution (0.01 N) did not change the chemical structure of the membrane (Zhou et al., 2014). NaOCl reduced the fouling effectively by hydrolyzing the proteins in EPS while the polysaccharides in EPS were more resistant (Lee et al., 2016). Wang has reported that NaOCl backwashing did not change the surface properties of the membrane. In addition, their study also reported that NaOCl backwashing reduced the abundance of filamentous bacteria, especially *Thiothrix eikelboomii*, in the biofilm, which could be the one reason for lower EPS levels and reduced fouling rates (Wang et al., 2014b). An NaOCl backwash solution concentration of $\sim 300\,\mathrm{mg\,L^{-1}}$ was found optimal for fouling reduction. Beyond that concentration, there was no significant improvement in the fouling reduction (Jiang et al., 2019). However, optimal backwash concentrations could vary from case to case.

7.1.3 Air sparging

Aeration with increased velocity near the membrane surface helps to slough off the biofilm with shear, prolonging the membrane run. Air sparging is most effective and economical in the sidestream configuration of MBR. From the membrane top and bottom, simultaneous downward and upward sparging is more effective than upward sparging only. The aeration rate for sparging in the membrane chamber ($1.18\,\mathrm{m^3\,min^{-1}}$) was fourfold higher than the normal aeration rate applied in the mixed liquor tank ($0.3\,\mathrm{m^3\,min^{-1}}$). The fouling rates were twofold lower in simultaneous upward and downward sparging mode compared to upward sparging only (Park et al., 2010). Another study optimized the air injection ratio (air velocity/(air velocity + water velocity)) for air sparging in a sidestream MBR. The results showed that air injection ratios between 0.4 and 0.5 were optimal and reduced fouling by 30%.

Moreover, the study has reported that turbulent flow conditions below Reynolds number 400 were sufficient to slough off the biofilm, but above that, no significant improvement in the fouling reduction was observed (Psoch and Schiewer, 2005). A comparison between the external and internal air sparging in ceramic membrane treating agro-industrial wastewater revealed that internal air sparging increased the permeability 2.5 times better than external air sparging (Son et al., 2021).

7.1.4 Sonication

Ultrasonication can be used in situ or ex situ in the MBR. For biofilm detachment, ultrasonic treatment is used within the MBR. The biofilm removal can be attributed to physical phenomena caused by the ultrasonic waves in the heterogeneous solid-liquid system. Sonication mainly forms microstreaming, acoustic streaming, microjets, and shock waves, which detach the foulant from the membrane surface (Kyllönen et al., 2005; Qasim et al., 2018). Contraction and expansion of the bubble size generated by sonication create turbulence in the liquid flow called microstreaming (Fig. 6A and B). Without producing bubble cavitation, the absorption of ultrasonic energy by water molecules generates a liquid stream called an acoustic stream (Johns, 2002; Lamminen et al., 2004). The sudden stop of the cavitation bubble reflects the water moving with it and generates an energy wave or shock wave (Fig. 6C). Macrojets are produced due to the collapse of the cavitation bubbles (Fig. 6D). Compression of the bubbles near the membrane surface pulls the foulant from the membrane, while expansion pushes the liquid toward the membrane surface. These rapid changes near the membrane surface produce shear and dragging forces that lead to the removal of the foulants

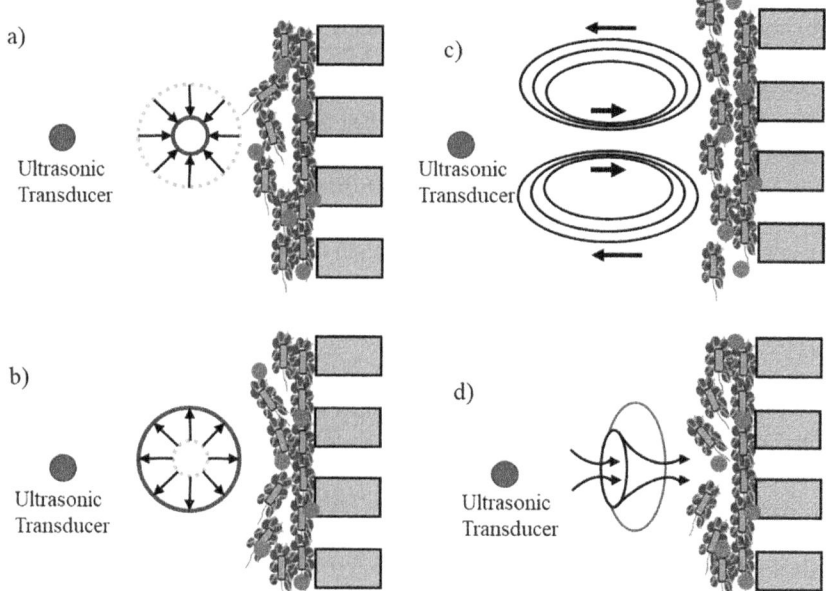

FIG. 6   Schematic diagram of contraction (A) and expansion (B) of microstreaming, acoustic streaming (C), and microjet (D).

(Arefi-Oskoui et al., 2019; Naji et al., 2021). In addition, ultrasonication can be used to pretreat the wastewater to control biofouling. It converts the contaminant compounds to biodegradable intermediates that can be hydrolyzed easily by microorganisms, which leads to reduced membrane surface attachments (Hakata et al., 2011).

7.1.5 Adsorbent and granular media addition

Adsorbent material supplements, such as activated carbon and reactive minerals, provide a large surface area for the adsorption of organics, inorganics, colloids, and the EPS, hence reducing the fouling propensity of sludge of MBR. In addition to the fouling reduction, adsorbents have been reported to improve the removal efficiency of COD in the MBR (Skouteris et al., 2015). Activated carbon (powdered and granular), zeolite (clinoptilolite) (Rezaei and Mehrnia, 2014; Sajadian et al., 2020), macroporous adsorption resins (Chen et al., 2017), diatomite (Yang et al., 2010), and sludge-based adsorbents (Pan et al., 2016) have been used to reduce the fouling of the MBR. The granular activated carbon (GAC) size was optimized for the fouling reduction in MBR; a size between 0.3 and 0.6 mm was optimal and reduced 60% of the membrane resistance (Johir et al., 2013). Another study has reported an optimal size of 2–3 mm to reduce the fouling in an AnMBR. A GAC size of 0.18–0.5 mm increased the fouling by fostering the development of a cake layer on the membrane (Charfi et al., 2017). Limited information is available on the reactivation and life span of these adsorbents in MBR. In addition, the economics of adsorbent addition in bench- or full-scale MBR needs to be evaluated.

7.1.6 Coagulant addition

The addition of coagulants to MBR enlarges the size of sludge flocs and fine organic particles. The deposition of large flocs on the biofilm of MBR increases its porosity and helps to maintain the filtration rates for longer periods. Both ferric chloride and alum addition to the MLSS have been reported to control biofouling by increasing the filterability of the MBR biofilm layer (Lee et al., 2001; Zhang et al., 2004). Polymeric coagulants: polymeric aluminum chloride and polymeric ferric sulfate have also been reported to control biofouling better than the monomer coagulants (alum and ferric chloride) and polymeric aluminum ferric chloride. Polymeric coagulants effectively reduced the gel layer formation. Among all polymeric coagulants, polymeric persulfate was the most effective at a dose of 1.05 mM Fe (Wu et al., 2006). Wu also investigated the basis for the polymeric persulfate effect on MLSS and biofilm in MBR and concluded that it provided a positive charge to organics and neutralized negative ones, reducing the charged particles deposition on the membrane and reduced fouling. It also removed organics with a high molecular weight and created large suspended flocs, which led to the reduced fouling rates (Wu and Huang, 2008). Another study compared ferric chloride and aluminum sulfate and reported that ferric chloride was a better option, with the addition of 2.28 g L^{-1} of ferric chloride increasing operational time (by two times) by retarding biofilm formation (Mishima and Nakajima, 2009). A recent study used the ferric hydroxide in MBR treating pharmaceutical wastewater and reported a 35% fouling reduction due to the reduced EPS levels and dissolved organic matter (Huang et al., 2019a). However, the addition of coagulants changes the pH of the reactor, affecting the activity of MBR sludge. In addition to this, coagulants may deposit on the surface of the membrane and can affect long-term operation. Moreover, the cost of coagulants needs to be compared with other cleaning chemicals to assess the overall benefits of coagulant addition.

7.2 Biological strategies: Quorum sensing abatement through quorum quenching

7.2.1 Quorum sensing

Attachment and growth of microbial cells to the membrane surface leads to biofilm formation, which is the main cause of biofouling in MBR (Le-Clech et al., 2006; Meng et al., 2009, 2017). The biofilm buildup is usually attributed to a phenomenon called quorum sensing (QS). QS is a mechanism of communication between bacteria that is dependent on population density and external environmental factors. In this mechanism, bacteria regulate gene expression and synchronize their communal behavior in the environment. This controls processes such as biofilm formation, excretion of toxins, virulence, exoenzyme production, granulation, etc. (Chong et al., 2012a; Dong et al., 2001; Ochiai et al., 2013; Tan et al., 2014), and involves the excretion of low molecular weight chemicals known as autoinducers (AI), QS molecules, or signal molecules. Briefly, the QS circuit consists of four main steps: (i) generation of QS signal molecules, (ii) release of QS signal molecule into the environment, (iii) recognition of QS signal molecule by receptor proteins on the same or another cell, and (iv) activation of regulatory genes in that cell after binding with the signal-receptor protein complex (Lade et al., 2014; Roy et al., 2011).

A group of chemical molecules, known as acyl-homoserine lactones (AHL), is mostly used as the signal molecule for communication between Gram-negative *Proteobacteria*. These chemicals contain an acyl chain length between C4 and C18 attached with a homoserine lactone (HSL) ring. The acyl chain comes with or without the "oxo" or "hydroxy" group at the C3 position (Krick et al., 2007; Milton et al., 1997, 2001; Schaefer et al., 1999), as shown in Fig. 7A–C.

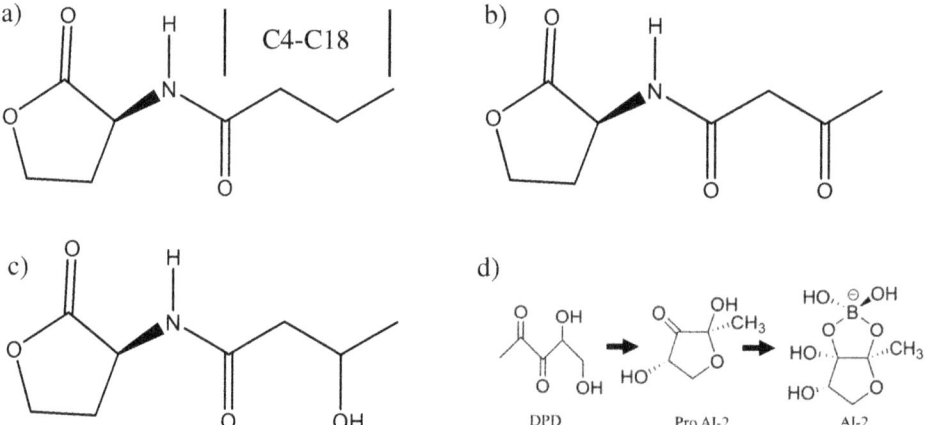

FIG. 7 Acyl homoserine lactone (AHL) autoinducers/quorum sensing molecule molecular structure: (A) butanoyl homoserine lactone *N*-butanoyl-L-homoserine lactone (C4-HSL), (B) *N*-3-oxo-butanoyl-L-homoserine lactone (3-oxo-C4-HSL), (C) *N*-3-hydroxy-butanoyl-L-homoserine lactone (3-hydroxy-C4-HSL), (D) 4,5-dihydroxy-2,3-pentadione (DPD), pro AI-2 and AI-2, *AI*, autoinducer.

AHL is one of the main types of AI used by Gram-negative bacteria to orchestrate and adopt social behavior (i.e., biofilm formation). A higher abundance of Gram-negative bacteria, compared to Gram-positive bacteria, has been reported previously in MBR (Aryal et al., 2016; Yavuztürk Gül et al., 2018). Hence, AHL has been reported as one of the main signal molecules responsible for biofilm formation (biofouling) in the MBR. Similarly, *Enterobacter* and *Aeromonas* species are reported to use a wide range of AHL to control the expression of biofilm in MBR (Lade et al., 2014). Yeon et al. reported three different AHL in the bio-cake of MBR and linked them to biofouling: N-3-oxo-octonoyl-L-homoserine lactone (C8-oxo-HSL), N-octonoyl-L-homoserine lactone (C8-HSL), N-hexanoyl-L-homoserine lactone (C6-HSL) (Yeon et al., 2009). Therefore, many studies have targeted AHL for the retardation of biofouling by QQ bacteria in aerobic MBR (Kim et al., 2014). In addition, AHL is found in an AnMBR and linked to biofouling in the AnMBR (Tabraiz et al., 2020, 2021a).

Moreover, another molecule, 4,5-dihydroxy-2,3-pentadione (DPD), is used as a signal molecule and one of its derivatives is referred to as AI-2 (Fig. 7D). It is a universal signal molecule used by both Gram-negative and Gram-positive bacteria. In addition to the AHL, DPD's role in biofilm development in MBRs has also been reported (Lee et al., 2018).

7.2.2 *Quorum quenching*

Interfering with the QS communication process between bacteria to reduce/prevent the expression of synchronized behavior is termed quorum quenching (QQ). The interference to quench the communication could disturb any bacterial social behaviors. For example, a chemical with the same properties and shape (analog) as the precursor of AHL could interfere with signal molecule generation or result in different chemical production, either of which would stop AHL production (Rasmussen and Givskov, 2006).

Similarly, another strategy can be used to quench the sensing by blocking the signal receptor site in the receptor proteins with an analog of AHL. Some of these AHL analog chemicals shown to act in this way have been obtained from nature or were synthetic (Castang et al., 2004; Persson et al., 2005; Smith et al., 2003). Other methods of QQ include oxidizing the AHL signal molecule by chemical reaction, for example, with oxidized halogens (Borchardt et al., 2001), but such chemicals might lead to the production of unwanted by-products in bioreactors, e.g., trihalomethanes. Similarly, some enzymes produced by bacteria have been reported to reduce the QS activity by degrading the AHL signal molecules (Dong and Zhang, 2005). These enzymes are known as QQ enzymes, and bacteria that produce these enzymes are termed QQ bacteria. The QQ enzymes are classified into three categories according to their enzymatic mechanism: AHL acylase, AHL lactonase, and AHL oxidase and reductase (Lade et al., 2014; Riaz et al., 2008). AHL acylase cleaves the amide bond in the AHL and results in fatty acid and HSL (Lin et al., 2003), while AHL lactonase opens the HSL ring of the AHL molecule. In the case of AHL oxidoreductase enzymes, the acyl side chain is targeted, and the structure of AHL is modified to prohibit QS (Dong et al., 2001; Uroz et al., 2005). QQ enzymes have been applied in MBR to successfully quench the QS and reduce biofouling. However, the low stability of enzymes in bioreactors and high cost make it an unsustainable option for the full-scale MBR treating actual wastewater, although efforts were made to immobilize the enzymes on the magnetic carrier to make them more stable and durable in the bioreactor. However, still a frequent change in magnetic carriers is required, which shall be an expensive alternate for full-scale MBR, and hence not viable. QQ bacterial applications in the bioreactors are a more sustainable and feasible option and have been tested on lab scale and pilot scale. QQ bacteria have been applied in the MBR with and without immobilization (Tabraiz et al., 2021b). Different materials (polyvinyl alcohol-alginate, alginate-polysulfone, alginate-polysulfone-activated carbon) have been evaluated for immobilization including the micron membrane (Kim et al., 2015; Ouyang et al., 2020; Syafiuddin et al., 2021). However, the sustainability of the addition of exogenous QQ bacteria in MBR and its effect on the community for longer periods need to be evaluated further for its application in full-scale applications.

8. Conclusions and perspectives

Wastewater treatment to a level of reusable or recyclable quality by sustainable wastewater treatment technologies is an ideal solution to reduce the stress on freshwater resources. The membrane bioreactor technology could be a feasible solution due to its high-quality effluent. An AnMBR can be an energy-positive or at least an energy-neutral wastewater treatment option. However, fouling of the membrane is the main hindrance to achieving sustainable and widespread use of this technology. QS has been reported as the main biological phenomenon that bacteria use during biofilm formation through diffusible QS chemicals. QQ is a process that could attenuate the QS mechanism of biofilm formation. Consequently, prohibit or reduce the biofilm formation rate in MBR by reducing the concentration of QS molecules. In addition, operational conditions can affect QS in MBR and need to be investigated to understand and address the biofouling problem in such systems.

The feed characteristics and operational parameters, directly or indirectly, can influence the bacterial communities in an MBR, affecting the fouling of the MBR membrane. However, despite the key role that the microbial community structure has on the fouling process, very few studies have investigated any changes in the microbial diversity of the sludge and biofilm in an AnMBR and linked them to changes in these parameters. Investigation of the effects of these parameters on the community structure in the sludge and biofilm of MBR, and the QS activity together could provide critical information that could help devise better strategies to reduce biofouling in the AnMBR.

Fouling of the membrane in MBR originates from the biological and nonbiological phenomenon. So the combination of physical, chemical, and biological approaches might help prolong the operation of MBR by reducing fouling effectively and making the technology more sustainable. The physical and chemical methods used for biofouling reduction must be investigated in detail for their effect on the membrane surface and life. The reaction between backwash chemicals and different compounds in biofilm/wastewater must be probed, and their effect on membrane permeability should be evaluated. QQ bacteria immobilization in beads could be a promising technology. However, the beads material is not very resilient to the wastewater environment. Hence, it requires frequent replacement very often, making it an expensive option. New durable material should be evaluated to immobilize the QQ bacteria and applied in MBR treating real wastewater.

References

Abdelrasoul, A., Doan, H., Lohi, A., 2013. Fouling in membrane filtration and remediation methods. In: Mass Transfer-Advances in Sustainable Energy and Environment Oriented Numerical Modeling. IntechOpen, p. 195.

Ahmed, Z., Cho, J., Lim, B.-R., Song, K.-G., Ahn, K.-H., 2007. Effects of sludge retention time on membrane fouling and microbial community structure in a membrane bioreactor. J. Membr. Sci. 287 (2), 211–218.

Arefi-Oskoui, S., Khataee, A., Safarpour, M., Orooji, Y., Vatanpour, V., 2019. A review on the applications of ultrasonic technology in membrane bioreactors. Ultrason. Sonochem. 58, 104633.

Aryal, R., Yadav, M., Hussain, S., Beecham, S., Diprose, D., 2016. Tracking changes in fluorescent organic composition in leachates using excitation emission matrix-parallel factor analysis. Process Saf. Environ. Prot. 104, 507–516.

Asif, M.B., Iftekhar, S., Maqbool, T., Paramanik, B.K., Tabraiz, S., Sillanpää, M., Zhang, Z., 2021. Two-dimensional nanoporous and lamellar membranes for water purification: reality or a myth? Chem. Eng. J., 134335.

Bacchin, P., Aimar, P., Field, R.W., 2006. Critical and sustainable fluxes: theory, experiments and applications. J. Membr. Sci. 281 (1–2), 42–69.

Bae, J., Shin, C., Lee, E., Kim, J., McCarty, P.L., 2014. Anaerobic treatment of low-strength wastewater: a comparison between single and staged anaerobic fluidized bed membrane bioreactors. Bioresour. Technol. 165, 75–80.

Batstone, D.J., Virdis, B., 2014. The role of anaerobic digestion in the emerging energy economy. Curr. Opin. Biotechnol. 27, 142–149.

Batstone, D., Hülsen, T., Mehta, C., Keller, J., 2015. Platforms for energy and nutrient recovery from domestic wastewater: a review. Chemosphere 140, 2–11.

Belfort, G., Davis, R.H., Zydney, A.L., 1994. The behavior of suspensions and macromolecular solutions in crossflow microfiltration. J. Membr. Sci. 96 (1–2), 1–58.

Borchardt, S., Allain, E., Michels, J., Stearns, G., Kelly, R., McCoy, W., 2001. Reaction of acylated homoserine lactone bacterial signaling molecules with oxidized halogen antimicrobials. Appl. Environ. Microbiol. 67 (7), 3174–3179.

Cai, W., Liu, J., Zhu, X., Zhang, X., Liu, Y., 2017. Fate of dissolved organic matter and byproducts generated from on-line chemical cleaning with sodium hypochlorite in MBR. Chem. Eng. J. 323, 233–242.

Carroll, T., King, S., Gray, S., Bolto, B.A., Booker, N., 2000. The fouling of microfiltration membranes by NOM after coagulation treatment. Water Res. 34 (11), 2861–2868.

Castang, S., Chantegrel, B., Deshayes, C., Dolmazon, R., Gouet, P., Haser, R., Reverchon, S., Nasser, W., Hugouvieux-Cotte-Pattat, N., Doutheau, A., 2004. N-Sulfonyl homoserine lactones as antagonists of bacterial quorum sensing. Bioorg. Med. Chem. Lett. 14 (20), 5145–5149.

Chan, R., Chen, V., 2004. Characterization of protein fouling on membranes: opportunities and challenges. J. Membr. Sci. 242 (1–2), 169–188.

Chan, Y.J., Chong, M.F., Law, C.L., Hassell, D., 2009. A review on anaerobic–aerobic treatment of industrial and municipal wastewater. Chem. Eng. J. 155 (1–2), 1–18.

Chang, Y.-R., Lee, Y.-J., Lee, D.-J., 2019. Membrane fouling during water or wastewater treatments: current research updated. J. Taiwan Inst. Chem. Eng. 94, 88–96.

Charfi, A., Aslam, M., Lesage, G., Heran, M., Kim, J., 2017. Macroscopic approach to develop fouling model under GAC fluidization in anaerobic fluidized bed membrane bioreactor. J. Ind. Eng. Chem. 49, 219–229.

Chen, Y., Dong, B., Gao, N., Fan, J., 2007. Effect of coagulation pretreatment on fouling of an ultrafiltration membrane. Desalination 204 (1–3), 181–188.

Chen, W., Luo, J., Cao, R., Li, Y., Liu, J., 2017. Effect of macroporous adsorption resin-membrane bioreactor hybrid system against fouling for municipal wastewater treatment. Bioresour. Technol. 224, 112–117.

Chernicharo, C., Van Lier, J., Noyola, A., Ribeiro, T.B., 2015. Anaerobic sewage treatment: state of the art, constraints and challenges. Rev. Environ. Sci. Biotechnol. 14 (4), 649–679.

Chong, G., Kimyon, O., Rice, S.A., Kjelleberg, S., Manefield, M., 2012a. The presence and role of bacterial quorum sensing in activated sludge. Microb. Biotechnol. 5 (5), 621–633.

Chong, S., Sen, T.K., Kayaalp, A., Ang, H.M., 2012b. The performance enhancements of upflow anaerobic sludge blanket (UASB) reactors for domestic sludge treatment—a state-of-the-art review. Water Res. 46 (11), 3434–3470.

Choo, K.-H., Lee, C.-H., 1998. Hydrodynamic behavior of anaerobic biosolids during crossflow filtration in the membrane anaerobic bioreactor. Water Res. 32 (11), 3387–3397.

Christensen, M.L., Bugge, T.V., Hede, B.H., Nierychlo, M., Larsen, P., Jørgensen, M.K., 2016. Effects of relaxation time on fouling propensity in membrane bioreactors. J. Membr. Sci. 504, 176–184.

Christensen, M.L., Jørgensen, M.K., Van De Staey, G., De Cock, L., Smets, I., 2021. Hydraulic resistance and osmotic pressure effects in fouling layers during MBR operations. J. Membr. Sci. 627, 119213.

Chua, H., Arnot, T., Howell, J., 2002. Controlling fouling in membrane bioreactors operated with a variable throughput. Desalination 149 (1–3), 225–229.

Cogert, K.I., Ziels, R.M., Winkler, M.K., 2019. Reducing cost and environmental impact of wastewater treatment with denitrifying methanotrophs, anammox, and mainstream anaerobic treatment. Environ. Sci. Technol. 53 (21), 12935–12944.

Deng, L., Guo, W., Ngo, H.H., Du, B., Wei, Q., Tran, N.H., Nguyen, N.C., Chen, S.-S., Li, J., 2016. Effects of hydraulic retention time and bioflocculant addition on membrane fouling in a sponge-submerged membrane bioreactor. Bioresour. Technol. 210, 11–17.

Díaz, O., González, E., Vera, L., Macías-Hernández, J.J., Rodríguez-Sevilla, J., 2017. Fouling analysis and mitigation in a tertiary MBR operated under restricted aeration. J. Membr. Sci. 525, 368–377.

Dong, Y.-H., Zhang, L.-H., 2005. Quorum sensing and quorum-quenching enzymes. J. Microbiol. 43 (1), 101–109.

Dong, Y.-H., Wang, L.-H., Xu, J.-L., Zhang, H.-B., Zhang, X.-F., Zhang, L.-H., 2001. Quenching quorum-sensing-dependent bacterial infection by an N-acyl homoserine lactonase. Nature 411 (6839), 813–817.

Drews, A., Mante, J., Iversen, V., Vocks, M., Lesjean, B., Kraume, M., 2007. Impact of ambient conditions on SMP elimination and rejection in MBRs. Water Res. 41 (17), 3850–3858.

Dufresne, R., Lebrun, R., Lavallee, H., 1997. Comparative-study on fluxes and performances during paper-mill wastewater treatment with membrane bioreactor. Can. J. Chem. Eng. 75 (1), 95–103.

Egwu, U., Onyelowe, K., Tabraiz, S., Johnson, E., Mutshow, A.D., 2022. Investigation of the effect of equal and unequal feeding time intervals on process stability and methane yield during anaerobic digestion grass silage. Renew. Sust. Energ. Rev. 158, 112092.

Eliseus, A., Bilad, M., Nordin, N., Putra, Z.A., Wirzal, M., 2017. Tilted membrane panel: a new module concept to maximize the impact of air bubbles for membrane fouling control in microalgae harvesting. Bioresour. Technol. 241, 661–668.

Enegess, D., Togna, A., Sutton, P., 2003. Membrane separation applications to biosystems for wastewater treatment. Filtr. Sep. 40 (1), 14–17.

Evans, P.J., Parameswaran, P., Lim, K., Bae, J., Shin, C., Ho, J., McCarty, P.L., 2019. A comparative pilot-scale evaluation of gas-sparged and granular activated carbon-fluidized anaerobic membrane bioreactors for domestic wastewater treatment. Bioresour. Technol. 288, 120949.

Fan, L., Harris, J.L., Roddick, F.A., Booker, N.A., 2001. Influence of the characteristics of natural organic matter on the fouling of microfiltration membranes. Water Res. 35 (18), 4455–4463.

Fan, H., Xiao, K., Mu, S., Zhou, Y., Ma, J., Wang, X., Huang, X., 2018. Impact of membrane pore morphology on multicycle fouling and cleaning of hydrophobic and hydrophilic membranes during MBR operation. J. Membr. Sci. 556, 312–320.

Fane, A., 1996. Membranes for water production and wastewater reuse. Desalination 106 (1–3), 1–9.

Feng, S., Zhang, N., Liu, H., Du, X., Liu, Y., Lin, H., 2012. The effect of COD/N ratio on process performance and membrane fouling in a submerged bioreactor. Desalination 285, 232–238.

Foglia, A., Cipoletta, G., Frison, N., Sabbatini, S., Gorbi, S., Eusebi, A.L., Fatone, F., 2019. Anaerobic membrane bioreactor for urban wastewater valorization: operative strategies and fertigation reuse. Chem. Eng. Trans. 74, 247–252.

Gao, W.J., Lin, H., Leung, K., Liao, B., 2010. Influence of elevated pH shocks on the performance of a submerged anaerobic membrane bioreactor. Process Biochem. 45 (8), 1279–1287.

Gao, D.-w., Fu, Y., Tao, Y., Li, X.-x., Xing, M., Gao, X.-h., Ren, N.-q., 2011. Linking microbial community structure to membrane biofouling associated with varying dissolved oxygen concentrations. Bioresour. Technol. 102 (10), 5626–5633.

Giménez, J., Robles, A., Carretero, L., Durán, F., Ruano, M., Gatti, M.N., Ribes, J., Ferrer, J., Seco, A., 2011. Experimental study of the anaerobic urban wastewater treatment in a submerged hollow-fibre membrane bioreactor at pilot scale. Bioresour. Technol. 102 (19), 8799–8806.

Gray, S.R., Ritchie, C., Tran, T., Bolto, B., 2007. Effect of NOM characteristics and membrane type on microfiltration performance. Water Res. 41 (17), 3833–3841.

Gu, Y., Huang, J., Zeng, G., Shi, L., Shi, Y., Yi, K., 2018. Fate of pharmaceuticals during membrane bioreactor treatment: status and perspectives. Bioresour. Technol. 268, 733–748.

Guest, J.S., Skerlos, S.J., Barnard, J.L., Beck, M.B., Daigger, G.T., Hilger, H., Jackson, S.J., Karvazy, K., Kelly, L., Macpherson, L., 2009. A New Planning and Design Paradigm to Achieve Sustainable Resource Recovery from Wastewater. ACS Publications.

Guglielmi, G., Andreottola, G., 2010. Selection and design of membrane bioreactors in environmental bioengineering. In: Environmental Biotechnology. Springer, pp. 439–516.

Hakata, Y., Roddick, F., Fan, L., 2011. Impact of ultrasonic pre-treatment on the microfiltration of a biologically treated municipal effluent. Desalination 283, 75–79.

Han, X., Wang, Z., Ma, J., Zhu, C., Li, Y., Wu, Z., 2015. Membrane bioreactors fed with different COD/N ratio wastewater: impacts on microbial community, microbial products, and membrane fouling. Environ. Sci. Pollut. Res. 22 (15), 11436–11445.

Heffernan, B., Van Lier, J., Van Der Lubbe, J., 2011. Performance review of large scale up-flow anaerobic sludge blanket sewage treatment plants. Water Sci. Technol. 63 (1), 100–107.

Hong, S., Bae, T.-H., Tak, T., Hong, S., Randall, A., 2002. Fouling control in activated sludge submerged hollow fiber membrane bioreactors. Desalination 143 (3), 219–228.

Hong, H., Zhang, M., He, Y., Chen, J., Lin, H., 2014. Fouling mechanisms of gel layer in a submerged membrane bioreactor. Bioresour. Technol. 166, 295–302.

Huang, S., Shi, X., Bi, X., Lee, L.Y., Ng, H.Y., 2019a. Effect of ferric hydroxide on membrane fouling in membrane bioreactor treating pharmaceutical wastewater. Bioresour. Technol. 292, 121852.

Huang, W., Lv, W., Zhou, W., Hu, M., Dong, B., 2019b. Investigation of the fouling behaviors correlating to water characteristics during the ultrafiltration with ozone treatment. Sci. Total Environ. 676, 53–61.

Iorhemen, O.T., Hamza, R.A., Tay, J.H., 2016. Membrane bioreactor (MBR) technology for wastewater treatment and reclamation: membrane fouling. Membranes 6 (2), 33.

Jang, D., Hwang, Y., Shin, H., Lee, W., 2013. Effects of salinity on the characteristics of biomass and membrane fouling in membrane bioreactors. Bioresour. Technol. 141, 50–56.

Jiang, T., Kennedy, M.D., Guinzbourg, B., Vanrolleghem, P.A., Schippers, J., 2005. Optimising the operation of a MBR pilot plant by quantitative analysis of the membrane fouling mechanism. Water Sci. Technol. 51 (6–7), 19–25.

Jiang, C.-K., Tang, X., Tan, H., Feng, F., Xu, Z.-M., Mahmood, Q., Zeng, W., Min, X.-B., Tang, C.-J., 2019. Effect of scrubbing by NaClO backwashing on membrane fouling in anammox MBR. Sci. Total Environ. 670, 149–157.

Johir, M.A., Vigneswaran, S., Sathasivan, A., Kandasamy, J., Chang, C., 2012. Effect of organic loading rate on organic matter and foulant characteristics in membrane bio-reactor. Bioresour. Technol. 113, 154–160.

Johir, M., Shanmuganathan, S., Vigneswaran, S., Kandasamy, J., 2013. Performance of submerged membrane bioreactor (SMBR) with and without the addition of the different particle sizes of GAC as suspended medium. Bioresour. Technol. 141, 13–18.

Johns, L.D., 2002. Nonthermal effects of therapeutic ultrasound: the frequency resonance hypothesis. J. Athl. Train. 37 (3), 293.

Judd, S., 2010. The MBR Book: Principles and Applications of Membrane Bioreactors for Water and Wastewater Treatment. Elsevier.

Judd, S.J., 2017. Membrane technology costs and me. Water Res. 122, 1–9.

Kang, G.-d., Cao, Y.-m., 2014. Application and modification of poly (vinylidene fluoride)(PVDF) membranes—a review. J. Membr. Sci. 463, 145–165.

Kang, S., Hoek, E.M., Choi, H., Shin, H., 2006. Effect of membrane surface properties during the fast evaluation of cell attachment. Sep. Sci. Technol. 41 (7), 1475–1487.

Kim, J., Kim, K., Ye, H., Lee, E., Shin, C., McCarty, P.L., Bae, J., 2011. Anaerobic fluidized bed membrane bioreactor for wastewater treatment. Environ. Sci. Technol. 45 (2), 576–581.

Kim, A.L., Park, S.-Y., Lee, C.-H., Lee, C.-H., Lee, J.-K., 2014. Quorum quenching bacteria isolated from the sludge of a wastewater treatment plant and their application for controlling biofilm formation. J. Microbiol. Biotechnol. 24 (11), 1574–1582.

Kim, S.-R., Lee, K.-B., Kim, J.-E., Won, Y.-J., Yeon, K.-M., Lee, C.-H., Lim, D.-J., 2015. Macroencapsulation of quorum quenching bacteria by polymeric membrane layer and its application to MBR for biofouling control. J. Membr. Sci. 473, 109–117.

Kimura, K., Yamato, N., Yamamura, H., Watanabe, Y., 2005. Membrane fouling in pilot-scale membrane bioreactors (MBRs) treating municipal wastewater. Environ. Sci. Technol. 39 (16), 6293–6299.

B. Applications of membrane technology for water and wastewater treatment

Kimura, K., Ogyu, R., Miyoshi, T., Watanabe, Y., 2015. Transition of major components in irreversible fouling of MBRs treating municipal wastewater. Sep. Purif. Technol. 142, 326–331.

Komesli, O.T., Teschner, K., Hegemann, W., Gokcay, C.F., 2007. Vacuum membrane applications in domestic wastewater reuse. Desalination 215 (1–3), 22–28.

Krick, A., Kehraus, S., Eberl, L., Riedel, K., Anke, H., Kaesler, I., Graeber, I., Szewzyk, U., König, G.M., 2007. A marine *Mesorhizobium* sp. produces structurally novel long-chain N-acyl-L-homoserine lactones. Appl. Environ. Microbiol. 73 (11), 3587–3594.

Kyllönen, H., Pirkonen, P., Nyström, M., 2005. Membrane filtration enhanced by ultrasound: a review. Desalination 181 (1–3), 319–335.

Lade, H., Paul, D., Kweon, J.H., 2014. N-Acyl homoserine lactone-mediated quorum sensing with special reference to use of quorum quenching bacteria in membrane biofouling control. BioMed Res. Int. 2014.

Lamminen, M.O., Walker, H.W., Weavers, L.K., 2004. Mechanisms and factors influencing the ultrasonic cleaning of particle-fouled ceramic membranes. J. Membr. Sci. 237 (1–2), 213–223.

Lankes, U., Lüdemann, H.-D., Frimmel, F.H., 2008. Search for basic relationships between "molecular size" and "chemical structure" of aquatic natural organic matter—answers from 13C and 15N CPMAS NMR spectroscopy. Water Res. 42 (4–5), 1051–1060.

Le-Clech, P., Chen, V., Fane, T.A., 2006. Fouling in membrane bioreactors used in wastewater treatment. J. Membr. Sci. 284 (1–2), 17–53.

Lee, J., Kim, J., Kang, I., Cho, M., Park, P., Lee, C., 2001. Potential and limitations of alum or zeolite addition to improve the performance of a submerged membrane bioreactor. Water Sci. Technol. 43 (11), 59–66.

Lee, E.-J., Kim, K.-Y., Lee, Y.-S., Nam, J.-W., Lee, Y.-S., Kim, H.-S., Jang, A., 2012. A study on the high-flux MBR system using PTFE flat sheet membranes with chemical backwashing. Desalination 306, 35–40.

Lee, E.-J., An, A.K., Hadi, P., Yan, D.Y., Kim, H.-S., 2016. A mechanistic study of in situ chemical cleaning-in-place for a PTFE flat sheet membrane: fouling mitigation and membrane characterization. Biofouling 32 (3), 301–312.

Lee, K., Kim, Y.-W., Lee, S., Lee, S.H., Nahm, C.H., Kwon, H., Park, P.-K., Choo, K.-H., Koyuncu, I., Drews, A., 2018. Stopping autoinducer-2 chatter by means of an indigenous bacterium (*Acinetobacter* sp. DKY-1): a new antibiofouling strategy in a membrane bioreactor for wastewater treatment. Environ. Sci. Technol. 52 (11), 6237–6245.

Lei, Z., Yang, S., Li, Y.-y., Wen, W., Wang, X.C., Chen, R., 2018. Application of anaerobic membrane bioreactors to municipal wastewater treatment at ambient temperature: a review of achievements, challenges, and perspectives. Bioresour. Technol. 267, 756–768.

Li, C.-W., Chen, Y.-S., 2004. Fouling of UF membrane by humic substance: effects of molecular weight and powder-activated carbon (PAC) pre-treatment. Desalination 170 (1), 59–67.

Li, P., Liu, L., Wu, J., Cheng, R., Shi, L., Zheng, X., Zhang, Z., 2019. Identify driving forces of MBR applications in China. Sci. Total Environ. 647, 627–638.

Lin, Y.H., Xu, J.L., Hu, J., Wang, L.H., Ong, S.L., Leadbetter, J.R., Zhang, L.H., 2003. Acyl-homoserine lactone acylase from *Ralstonia* strain XJ12B represents a novel and potent class of quorum-quenching enzymes. Mol. Microbiol. 47 (3), 849–860.

Lin, H., Zhang, M., Wang, F., Meng, F., Liao, B.-Q., Hong, H., Chen, J., Gao, W., 2014. A critical review of extracellular polymeric substances (EPSs) in membrane bioreactors: characteristics, roles in membrane fouling and control strategies. J. Membr. Sci. 460, 110–125.

Liu, Y., Liu, H., Cui, L., Zhang, K., 2012. The ratio of food-to-microorganism (F/M) on membrane fouling of anaerobic membrane bioreactors treating low-strength wastewater. Desalination 297, 97–103.

Liu, T.-Y., Zhang, R.-X., Li, Q., Van der Bruggen, B., Wang, X.-L., 2014. Fabrication of a novel dual-layer (PES/PVDF) hollow fiber ultrafiltration membrane for wastewater treatment. J. Membr. Sci. 472, 119–132.

Luo, W., Phan, H.V., Li, G., Hai, F.I., Price, W.E., Elimelech, M., Nghiem, L.D., 2017. An osmotic membrane bioreactor–membrane distillation system for simultaneous wastewater reuse and seawater desalination: performance and implications. Environ. Sci. Technol. 51 (24), 14311–14320.

Luo, W., Arhatari, B., Gray, S.R., Xie, M., 2018. Seeing is believing: insights from synchrotron infrared mapping for membrane fouling in osmotic membrane bioreactors. Water Res. 137, 355–361.

Ma, B.-C., Lee, Y.-N., Park, J.-S., Lee, C.-H., Lee, S.-H., Chang, I.-S., Ahn, T.-S., 2006. Correlation between dissolved oxygen concentration, microbial community and membrane permeability in a membrane bioreactor. Process Biochem. 41 (5), 1165–1172.

Ma, C., Yu, S., Shi, W., Heijman, S., Rietveld, L., 2013. Effect of different temperatures on performance and membrane fouling in high concentration PAC–MBR system treating micro-polluted surface water. Bioresour. Technol. 141, 19–24.

Mannina, G., Capodici, M., Cosenza, A., Di Trapani, D., Viviani, G., 2019. The influence of solid retention time on IFAS-MBR systems: analysis of system behavior. Environ. Technol. 40 (14), 1840–1852.

Martinez-Sosa, D., Helmreich, B., Horn, H., 2012. Anaerobic submerged membrane bioreactor (AnSMBR) treating low-strength wastewater under psychrophilic temperature conditions. Process Biochem. 47 (5), 792–798.

McCarty, P.L., Bae, J., Kim, J., 2011. Domestic wastewater treatment as a net energy producer—can this be achieved? Environ. Sci. Technol. 45 (17), 7100–7106.

Melin, T., Jefferson, B., Bixio, D., Thoeye, C., De Wilde, W., De Koning, J., Van der Graaf, J., Wintgens, T., 2006. Membrane bioreactor technology for wastewater treatment and reuse. Desalination 187 (1–3), 271–282.

Meng, F., Zhang, H., Yang, F., Liu, L., 2007. Characterization of cake layer in submerged membrane bioreactor. Environ. Sci. Technol. 41 (11), 4065–4070.

Meng, F., Chae, S.-R., Drews, A., Kraume, M., Shin, H.-S., Yang, F., 2009. Recent advances in membrane bioreactors (MBRs): membrane fouling and membrane material. Water Res. 43 (6), 1489–1512.

Meng, F., Liao, B., Liang, S., Yang, F., Zhang, H., Song, L., 2010. Morphological visualization, componential characterization and microbiological identification of membrane fouling in membrane bioreactors (MBRs). J. Membr. Sci. 361 (1–2), 1–14.

Meng, F., Zhang, S., Oh, Y., Zhou, Z., Shin, H.-S., Chae, S.-R., 2017. Fouling in membrane bioreactors: an updated review. Water Res. 114, 151–180.

Milton, D.L., Hardman, A., Camara, M., Chhabra, S.R., Bycroft, B.W., Stewart, G., Williams, P., 1997. Quorum sensing in *Vibrio anguillarum*: characterization of the vanI/vanR locus and identification of the autoinducer N-(3-oxodecanoyl)-L-homoserine lactone. J. Bacteriol. 179 (9), 3004–3012.

Milton, D.L., Chalker, V.J., Kirke, D., Hardman, A., Cámara, M., Williams, P., 2001. The LuxM homologue VanM from *Vibrio anguillarum* directs the synthesis of N-(3-hydroxyhexanoyl) homoserine lactone and N-hexanoylhomoserine lactone. J. Bacteriol. 183 (12), 3537–3547.

Mirzavandi, A., Hazrati, H., Ebrahimi, S., 2019. Investigation of influence of temperature and solid retention time on membrane fouling in MBR. Membr. Water Treat. 10 (2), 179–189.

Mishima, I., Nakajima, J., 2009. Control of membrane fouling in membrane bioreactor process by coagulant addition. Water Sci. Technol. 59 (7), 1255–1262.

Monclús, H., Sipma, J., Ferrero, G., Rodriguez-Roda, I., Comas, J., 2010. Biological nutrient removal in an MBR treating municipal wastewater with special focus on biological phosphorus removal. Bioresour. Technol. 101 (11), 3984–3991.

Morgan-Sagastume, F., Allen, D.G., 2005. Activated sludge deflocculation under temperature upshifts from 30 to 45 C. Water Res. 39 (6), 1061–1074.

Mu, S., Wang, S., Liang, S., Xiao, K., Fan, H., Han, B., Liu, C., Wang, X., Huang, X., 2019. Effect of the relative degree of foulant "hydrophobicity" on membrane fouling. J. Membr. Sci. 570, 1–8.

Myat, D.T., Mergen, M., Zhao, O., Stewart, M.B., Orbell, J.D., Merle, T., Croué, J.-P., Gray, S.R., 2014. Membrane fouling mechanism transition in relation to feed water composition. J. Membr. Sci. 471, 265–273.

Naji, O., Al-juboori, R.A., Khan, A., Yadav, S., Altaee, A., Alpatova, A., Soukane, S., Ghaffour, N., 2021. Ultrasound-assisted membrane technologies for fouling control and performance improvement: a review. J. Water Process Eng. 43, 102268.

Nam, K., Heo, S., Rhee, G., Kim, M., Yoo, C., 2021. Dual-objective optimization for energy-saving and fouling mitigation in MBR plants using AI-based influent prediction and an integrated biological-physical model. J. Membr. Sci. 626, 119208.

Nguyen, N.C., Nguyen, H.T., Chen, S.-S., Ngo, H.H., Guo, W., Chan, W.H., Ray, S.S., Li, C.-W., Hsu, H.-T., 2016. A novel osmosis membrane bioreactor-membrane distillation hybrid system for wastewater treatment and reuse. Bioresour. Technol. 209, 8–15.

Ochiai, S., Morohoshi, T., Kurabeishi, A., Shinozaki, M., Fujita, H., Sawada, I., Ikeda, T., 2013. Production and degradation of N-acylhomoserine lactone quorum sensing signal molecules in bacteria isolated from activated sludge. Biosci. Biotechnol. Biochem. 77 (12), 2436–2440.

Oron, G., Gillerman, L., Bick, A., Manor, Y., Buriakovsky, N., Hagin, J., 2008. Membrane technology for sustainable treated wastewater reuse: agricultural, environmental and hydrological considerations. Water Sci. Technol. 57 (9), 1383–1388.

Ouyang, Y., Hu, Y., Huang, J., Gu, Y., Shi, Y., Yi, K., Yang, Y., 2020. Effects of exogenous quorum quenching on microbial community dynamics and biofouling propensity of activated sludge in MBRs. Biochem. Eng. J. 157, 107534.

Ozgun, H., Gimenez, J.B., Ersahin, M.E., Tao, Y., Spanjers, H., Van Lier, J.B., 2015. Impact of membrane addition for effluent extraction on the performance and sludge characteristics of upflow anaerobic sludge blanket reactors treating municipal wastewater. J. Membr. Sci. 479, 95–104.

Pan, J.R., Su, Y., Huang, C., 2010. Characteristics of soluble microbial products in membrane bioreactor and its effect on membrane fouling. Desalination 250 (2), 778–780.

Pan, Z., Zhang, C., Huang, B., 2016. Using adsorbent made from sewage sludge to enhance wastewater treatment and control fouling in a membrane bioreactor. Desalin. Water Treat. 57 (20), 9070–9081.

Park, H.-D., Lee, Y.H., Kim, H.-B., Moon, J., Ahn, C.-H., Kim, K.-T., Kang, M.-S., 2010. Reduction of membrane fouling by simultaneous upward and downward air sparging in a pilot-scale submerged membrane bioreactor treating municipal wastewater. Desalination 251 (1–3), 75–82.

Paul, D., Ebra-Lima, O., 1970. Pressure-induced diffusion of organic liquids through highly swollen polymer membranes. J. Appl. Polym. Sci. 14 (9), 2201–2224.

Persson, T., Hansen, T.H., Rasmussen, T.B., Skindersø, M.E., Givskov, M., Nielsen, J., 2005. Rational design and synthesis of new quorum-sensing inhibitors derived from acylated homoserine lactones and natural products from garlic. Org. Biomol. Chem. 3 (2), 253–262.

Petropoulos, E., Yu, Y., Tabraiz, S., Yakubu, A., Curtis, T.P., Dolfing, J., 2019. High rate domestic wastewater treatment at 15°C using anaerobic reactors inoculated with cold-adapted sediments/soils–shaping robust methanogenic communities. Environ. Sci. Water Res. Technol. 5 (1), 70–82.

Petropoulos, E., Shamurad, B., Acharya, K., Tabraiz, S., 2020. Domestic wastewater hydrolysis and lipolysis during start-up in anaerobic digesters and microbial fuel cells at moderate temperatures. Int. J. Environ. Sci. Technol. 17 (1), 27–38.

Petropoulos, E., Shamurad, B., Tabraiz, S., Yu, Y., Davenport, R., Curtis, T.P., Dolfing, J., 2021. Sewage treatment at 4°C in anaerobic upflow reactors with and without a membrane–performance, function and microbial diversity. Environ. Sci. Water Res. Technol. 7 (1), 156–171.

Pretel, R., Shoener, B., Ferrer, J., Guest, J., 2015. Navigating environmental, economic, and technological trade-offs in the design and operation of submerged anaerobic membrane bioreactors (AnMBRs). Water Res. 87, 531–541.

Psoch, C., Schiewer, S., 2005. Long-term study of an intermittent air sparged MBR for synthetic wastewater treatment. J. Membr. Sci. 260 (1–2), 56–65.

Puyol, D., Batstone, D.J., Hülsen, T., Astals, S., Peces, M., Krömer, J.O., 2017. Resource recovery from wastewater by biological technologies: opportunities, challenges, and prospects. Front. Microbiol. 7, 2106.

Qasim, M., Darwish, N.N., Mhiyo, S., Darwish, N.A., Hilal, N., 2018. The use of ultrasound to mitigate membrane fouling in desalination and water treatment. Desalination 443, 143–164.

Rasmussen, T.B., Givskov, M., 2006. Quorum-sensing inhibitors as anti-pathogenic drugs. Int. J. Med. Microbiol. 296 (2), 149–161.

Rezaei, M., Mehrnia, M., 2014. The influence of zeolite (clinoptilolite) on the performance of a hybrid membrane bioreactor. Bioresour. Technol. 158, 25–31.

Riaz, K., Elmerich, C., Moreira, D., Raffoux, A., Dessaux, Y., Faure, D., 2008. A metagenomic analysis of soil bacteria extends the diversity of quorum-quenching lactonases. Environ. Microbiol. 10 (3), 560–570.

Rosenberger, S., Evenblij, H., Te Poele, S., Wintgens, T., Laabs, C., 2005. The importance of liquid phase analyses to understand fouling in membrane assisted activated sludge processes—six case studies of different European research groups. J. Membr. Sci. 263 (1–2), 113–126.

Roy, V., Adams, B.L., Bentley, W.E., 2011. Developing next generation antimicrobials by intercepting AI-2 mediated quorum sensing. Enzym. Microb. Technol. 49 (2), 113–123.

Sajadian, Z.S., Hazrati, H., Rostamizadeh, M., 2020. Investigation of influence of nano H-ZSM-5 and NH 4-ZSM-5 zeolites on membrane fouling in semi batch MBR. Adv. Nano Res. 8 (2), 183–190.

Sanguanpak, S., Chiemchaisri, C., Chiemchaisri, W., Yamamoto, K., 2015. Influence of operating pH on biodegradation performance and fouling propensity in membrane bioreactors for landfill leachate treatment. Int. Biodeterior. Biodegradation 102, 64–72.

Santos, F.S., Ricci, B.C., Neta, L.S.F., Amaral, M.C., 2017. Sugarcane vinasse treatment by two-stage anaerobic membrane bioreactor: effect of hydraulic retention time on changes in efficiency, biogas production and membrane fouling. Bioresour. Technol. 245, 342–350.

Schaefer, A.L., Hanzelka, B.L., Parsek, M.R., Greenberg, E.P., 1999. Detection, purification, and structural elucidation of the acylhomoserine lactone inducer of *Vibrio fischeri* luminescence and other related molecules. Methods Enzymol. 305, 288–301.

Sepehri, A., Sarrafzadeh, M.-H., 2018. Effect of nitrifiers community on fouling mitigation and nitrification efficiency in a membrane bioreactor. Chem. Eng. Process. Process Intensif. 128, 10–18.

Shamurad, B., Gray, N., Petropoulos, E., Tabraiz, S., Acharya, K., Quintela-Baluja, M., Sallis, P., 2019. Co-digestion of organic and mineral wastes for enhanced biogas production: reactor performance and evolution of microbial community and function. Waste Manag. 87, 313–325.

Shamurad, B., Gray, N., Petropoulos, E., Dolfing, J., Quintela-Baluja, M., Bashiri, R., Tabraiz, S., Sallis, P., 2020a. Low-temperature pretreatment of organic feedstocks with selected mineral wastes sustains anaerobic digestion stability through trace metal release. Environ. Sci. Technol. 54 (14), 9095–9105.

Shamurad, B., Gray, N., Petropoulos, E., Tabraiz, S., Membere, E., Sallis, P., 2020b. Predicting the effects of integrating mineral wastes in anaerobic digestion of OFMSW using first-order and Gompertz models from biomethane potential assays. Renew. Energy 152, 308–319.

Shamurad, B., Gray, N., Petropoulos, E., Tabraiz, S., Sallis, P., 2020c. Improving the methane productivity of anaerobic digestion using aqueous extracts from municipal solid waste incinerator ash. J. Environ. Manag. 260, 110160.

Shamurad, B., Sallis, P., Petropoulos, E., Tabraiz, S., Ospina, C., Leary, P., Dolfing, J., Gray, N., 2020d. Stable biogas production from single-stage anaerobic digestion of food waste. Appl. Energy 263, 114609.

Shang, R., Vuong, F., Hu, J., Li, S., Kemperman, A.J., Nijmeijer, K., Cornelissen, E.R., Heijman, S.G., Rietveld, L.C., 2015. Hydraulically irreversible fouling on ceramic MF/UF membranes: comparison of fouling indices, foulant composition and irreversible pore narrowing. Sep. Purif. Technol. 147, 303–310.

Sharghi, E.A., Shourgashti, A., Bonakdarpour, B., 2020. Considering a membrane bioreactor for the treatment of vegetable oil refinery wastewaters at industrially relevant organic loading rates. Bioprocess Biosyst. Eng., 1–15.

Shariati, S.R.P., Bonakdarpour, B., Zare, N., Ashtiani, F.Z., 2011. The effect of hydraulic retention time on the performance and fouling characteristics of membrane sequencing batch reactors used for the treatment of synthetic petroleum refinery wastewater. Bioresour. Technol. 102 (17), 7692–7699.

Sheik, A.R., Muller, E.E., Wilmes, P., 2014. A hundred years of activated sludge: time for a rethink. Front. Microbiol. 5, 47.

Skouteris, G., Saroj, D., Melidis, P., Hai, F.I., Ouki, S., 2015. The effect of activated carbon addition on membrane bioreactor processes for wastewater treatment and reclamation—a critical review. Bioresour. Technol. 185, 399–410.

Smith, K.M., Bu, Y., Suga, H., 2003. Induction and inhibition of *Pseudomonas aeruginosa* quorum sensing by synthetic autoinducer analogs. Chem. Biol. 10 (1), 81–89.

Smith, A.L., Stadler, L.B., Love, N.G., Skerlos, S.J., Raskin, L., 2012. Perspectives on anaerobic membrane bioreactor treatment of domestic wastewater: a critical review. Bioresour. Technol. 122, 149–159.

Smith, A.L., Stadler, L.B., Cao, L., Love, N.G., Raskin, L., Skerlos, S.J., 2014. Navigating wastewater energy recovery strategies: a life cycle comparison of anaerobic membrane bioreactor and conventional treatment systems with anaerobic digestion. Environ. Sci. Technol. 48 (10), 5972–5981.

Son, D.-J., Kim, D.-G., Kim, W.-Y., Hong, K.-H., 2021. Anti-fouling effect by internal air injection in plate-type ceramic membrane fabricated for the treatment of agro-industrial wastewater. J. Water Process Eng. 41, 102021.

Song, X., Luo, W., McDonald, J., Khan, S.J., Hai, F.I., Price, W.E., Nghiem, L.D., 2018. An anaerobic membrane bioreactor–membrane distillation hybrid system for energy recovery and water reuse: removal performance of organic carbon, nutrients, and trace organic contaminants. Sci. Total Environ. 628, 358–365.

Stazi, V., Tomei, M.C., 2018. Enhancing anaerobic treatment of domestic wastewater: state of the art, innovative technologies and future perspectives. Sci. Total Environ. 635, 78–91.

Stuckey, D.C., 2010. Anaerobic membrane reactors. In: Environmental Anaerobic Technology: Applications and New Developments. World Scientific, pp. 137–161.

Stuckey, D.C., 2012. Recent developments in anaerobic membrane reactors. Bioresour. Technol. 122, 137–148.

Sun, C., Leiknes, T., Weitzenböck, J., Thorstensen, B., 2010. Salinity effect on a biofilm-MBR process for shipboard wastewater treatment. Sep. Purif. Technol. 72 (3), 380–387.

Sun, W., Liu, J., Chu, H., Dong, B., 2013. Pretreatment and membrane hydrophilic modification to reduce membrane fouling. Membranes 3 (3), 226–241.

Sutherland, K., 2010. The rise of membrane bioreactors. Filtr. Sep. 47 (5), 14–16.

Sweity, A., Ying, W., Belfer, S., Oron, G., Herzberg, M., 2011. pH effects on the adherence and fouling propensity of extracellular polymeric substances in a membrane bioreactor. J. Membr. Sci. 378 (1–2), 186–193.

Syafiuddin, A., Boopathy, R., Mehmood, M.A., 2021. Recent advances on bacterial quorum quenching as an effective strategy to control biofouling in membrane bioreactors. Bioresour. Technol. Rep. 15, 100745.

Tabraiz, S., Haydar, S., Hussain, G., 2016. Evaluation of a cost-effective and energy-efficient disc material for rotating biological contactors (RBC), and performance evaluation under varying condition of RPM and submergence. Desalin. Water Treat. 57 (43), 20439–20446.

Tabraiz, S., Haydar, S., Sallis, P., Nasreen, S., Mahmood, Q., Awais, M., Acharya, K., 2017. Effect of cycle run time of backwash and relaxation on membrane fouling removal in submerged membrane bioreactor treating sewage at higher flux. Water Sci. Technol. 76 (4), 963–975.

Tabraiz, S., Hassan, S., Abbas, A., Nasreen, S., Zeeshan, M., Fida, S., Shamurad, B.A., Acharya, K., Petropoulos, E., 2018. Effect of effluent and sludge recirculation ratios on integrated fixed films A2O system nutrients removal efficiency treating sewage. Desalin. Water Treat. 114, 120–127.

Tabraiz, S., Shamurad, B., Petropoulos, E., Charlton, A., Mohiudin, O., Danish Khan, M., Ekwenna, E., Sallis, P., 2020. Diversity of acyl homoserine lactone molecules in anaerobic membrane bioreactors treating sewage at psychrophilic temperatures. Membranes 10 (11), 320.

Tabraiz, S., Petropoulos, E., Shamurad, B., Quintela-Baluja, M., Mohapatra, S., Acharya, K., Charlton, A., Davenport, R.J., Dolfing, J., Sallis, P.J., 2021a. Temperature and immigration effects on quorum sensing in the biofilms of anaerobic membrane bioreactors. J. Environ. Manag. 293, 112947.

Tabraiz, S., Shamurad, B., Petropoulos, E., Quintela-Baluja, M., Charlton, A., Dolfing, J., Sallis, P.J., 2021b. Mitigation of membrane biofouling in membrane bioreactor treating sewage by novel quorum quenching strain of *Acinetobacter* originating from a full-scale membrane bioreactor. Bioresour. Technol. 334, 125242.

Tan, C.H., Koh, K.S., Xie, C., Tay, M., Zhou, Y., Williams, R., Ng, W.J., Rice, S.A., Kjelleberg, S., 2014. The role of quorum sensing signalling in EPS production and the assembly of a sludge community into aerobic granules. ISME J. 8 (6), 1186–1197.

Tauseef, S., Abbasi, T., Abbasi, S., 2013. Energy recovery from wastewaters with high-rate anaerobic digesters. Renew. Sust. Energ. Rev. 19, 704–741.

Teng, J., Zhang, M., Leung, K.-T., Chen, J., Hong, H., Lin, H., Liao, B.-Q., 2019. A unified thermodynamic mechanism underlying fouling behaviors of soluble microbial products (SMPs) in a membrane bioreactor. Water Res. 149, 477–487.

Tian, J.-y., Ernst, M., Cui, F., Jekel, M., 2013. Effect of particle size and concentration on the synergistic UF membrane fouling by particles and NOM fractions. J. Membr. Sci. 446, 1–9.

Tocchi, C., Federici, E., Fidati, L., Manzi, R., Vincigurerra, V., Petruccioli, M., 2012. Aerobic treatment of dairy wastewater in an industrial three-reactor plant: effect of aeration regime on performances and on protozoan and bacterial communities. Water Res. 46 (10), 3334–3344.

Tsuyuhara, T., Hanamoto, Y., Miyoshi, T., Kimura, K., Watanabe, Y., 2010. Influence of membrane properties on physically reversible and irreversible fouling in membrane bioreactors. Water Sci. Technol. 61 (9), 2235–2240.

Uroz, S., Chhabra, S.R., Camara, M., Williams, P., Oger, P., Dessaux, Y., 2005. N-Acylhomoserine lactone quorum-sensing molecules are modified and degraded by *Rhodococcus erythropolis* W2 by both amidolytic and novel oxidoreductase activities. Microbiology 151 (10), 3313–3322.

van den Brink, P., Satpradit, O.-A., van Bentem, A., Zwijnenburg, A., Temmink, H., van Loosdrecht, M., 2011. Effect of temperature shocks on membrane fouling in membrane bioreactors. Water Res. 45 (15), 4491–4500.

van der Marel, P., Zwijnenburg, A., Kemperman, A., Wessling, M., Temmink, H., van der Meer, W., 2009. An improved flux-step method to determine the critical flux and the critical flux for irreversibility in a membrane bioreactor. J. Membr. Sci. 332 (1–2), 24–29.

Van Haandel, A., Kato, M.T., Cavalcanti, P.F., Florencio, L., 2006. Anaerobic reactor design concepts for the treatment of domestic wastewater. Rev. Environ. Sci. Biotechnol. 5 (1), 21–38.

Van Lier, J.B., Vashi, A., Van Der Lubbe, J., Heffernan, B., 2010. Anaerobic sewage treatment using UASB reactors: engineering and operational aspects. In: Environmental Anaerobic Technology: Applications and New Developments. World Scientific, pp. 59–89.

Van Lier, J., Van der Zee, F., Frijters, C., Ersahin, M., 2015. Celebrating 40 years anaerobic sludge bed reactors for industrial wastewater treatment. Rev. Environ. Sci. Biotechnol. 14 (4), 681–702.

Verstraete, W., Vlaeminck, S.E., 2011. ZeroWasteWater: short-cycling of wastewater resources for sustainable cities of the future. Int. J. Sustain. Dev. World Ecol. 18 (3), 253–264.

Vinardell, S., Astals, S., Peces, M., Cardete, M., Fernández, I., Mata-Alvarez, J., Dosta, J., 2020. Advances in anaerobic membrane bioreactor technology for municipal wastewater treatment: a 2020 updated review. Renew. Sust. Energ. Rev. 130, 109936.

Vyas, H.K., Bennett, R., Marshall, A., 2000. Influence of feed properties on membrane fouling in crossflow microfiltration of particulate suspensions. Int. Dairy J. 10 (12), 855–861.

Wan, J., Gu, J., Zhao, Q., Liu, Y., 2016. COD capture: a feasible option towards energy self-sufficient domestic wastewater treatment. Sci. Rep. 6 (1), 1–9.

Wang, Z., Wu, Z., Yin, X., Tian, L., 2008. Membrane fouling in a submerged membrane bioreactor (MBR) under subcritical flux operation: membrane foulant and gel layer characterization. J. Membr. Sci. 325 (1), 238–244.

Wang, F., Zhang, M., Peng, W., He, Y., Lin, H., Chen, J., Hong, H., Wang, A., Yu, H., 2014a. Effects of ionic strength on membrane fouling in a membrane bioreactor. Bioresour. Technol. 156, 35–41.

Wang, Z., Meng, F., He, X., Zhou, Z., Huang, L.-N., Liang, S., 2014b. Optimisation and performance of NaClO-assisted maintenance cleaning for fouling control in membrane bioreactors. Water Res. 53, 1–11.

Wang, X., Cheng, B., Ji, C., Zhou, M., Wang, L., 2017a. Effects of hydraulic retention time on adsorption behaviours of EPS in an A/O-MBR: biofouling study with QCM-D. Sci. Rep. 7 (1), 1–9.

Wang, X., Wang, C., Tang, C.Y., Hu, T., Li, X., Ren, Y., 2017b. Development of a novel anaerobic membrane bioreactor simultaneously integrating microfiltration and forward osmosis membranes for low-strength wastewater treatment. J. Membr. Sci. 527, 1–7.

Wang, J., Lv, M., Huang, Y., Huang, L., Ying, X., Xu, Y., Shen, D., Feng, H., Zhang, X., 2021. Numerical simulation and optimization of a cold model of a flat membrane bioreactor air scouring for membrane fouling control. J. Membr. Sci. 640, 119814.

Wu, J., Huang, X., 2008. Effect of dosing polymeric ferric sulfate on fouling characteristics, mixed liquor properties and performance in a long-term running membrane bioreactor. Sep. Purif. Technol. 63 (1), 45–52.

Wu, J., Chen, F., Huang, X., Geng, W., Wen, X., 2006. Using inorganic coagulants to control membrane fouling in a submerged membrane bioreactor. Desalination 197 (1–3), 124–136.

Wu, J., Le-Clech, P., Stuetz, R.M., Fane, A.G., Chen, V., 2008. Effects of relaxation and backwashing conditions on fouling in membrane bioreactor. J. Membr. Sci. 324 (1–2), 26–32.

Xia, S., Li, J., He, S., Xie, K., Wang, X., Zhang, Y., Duan, L., Zhang, Z., 2010. The effect of organic loading on bacterial community composition of membrane biofilms in a submerged polyvinyl chloride membrane bioreactor. Bioresour. Technol. 101 (17), 6601–6609.

Yamamoto, K., Hiasa, M., Mahmood, T., Matsuo, T., 1988. Direct solid-liquid separation using hollow fiber membrane in an activated sludge aeration tank. In: Water Pollution Research and Control Brighton. Elsevier, pp. 43–54.

Yang, X.-L., Song, H.-L., Lu, J.-L., Fu, D.-F., Cheng, B., 2010. Influence of diatomite addition on membrane fouling and performance in a submerged membrane bioreactor. Bioresour. Technol. 101 (23), 9178–9184.

Yavuztürk Gül, B., Imer, D.Y., Park, P.-K., Koyuncu, I., 2018. Evaluation of a novel anti-biofouling microorganism (Bacillus sp. T5) for control of membrane biofouling and its effect on bacterial community structure in membrane bioreactors. Water Sci. Technol. 77 (4), 971–978.

Yeon, K.-M., Cheong, W.-S., Oh, H.-S., Lee, W.-N., Hwang, B.-K., Lee, C.-H., Beyenal, H., Lewandowski, Z., 2009. Quorum sensing: a new biofouling control paradigm in a membrane bioreactor for advanced wastewater treatment. Environ. Sci. Technol. 43 (2), 380–385.

Yigit, N., Harman, I., Civelekoglu, G., Koseoglu, H., Cicek, N., Kitis, M., 2008. Membrane fouling in a pilot-scale submerged membrane bioreactor operated under various conditions. Desalination 231 (1–3), 124–132.

Yu, H., Xu, G., Qu, F., Li, G., Liang, H., 2016. Effect of solid retention time on membrane fouling in membrane bioreactor: from the perspective of quorum sensing and quorum quenching. Appl. Microbiol. Biotechnol. 100 (18), 7887–7897.

Yusuf, Z., Abdul Wahab, N., Sahlan, S., 2016. Fouling control strategy for submerged membrane bioreactor filtration processes using aeration airflow, backwash, and relaxation: a review. Desalin. Water Treat. 57 (38), 17683–17695.

Zeeshan, M., Haydar, S., Tabraiz, S., 2017. Effect of fixed media surface area on biofouling and nutrients removal in fixed film membrane bioreactor treating sewage at medium and high fluxes. Water Air Soil Pollut. 228 (9), 1–10.

Zhang, Y., Bu, D., Liu, C., Luo, X., Gu, P., 2004. Study on retarding membrane fouling by ferric salts dosing in membrane bioreactors. In: WEMT2004, IWA Specialty Conference.

B. Applications of membrane technology for water and wastewater treatment

Zhang, M., Leung, K.-T., Lin, H., Liao, B.-Q., 2019. Characterization of foaming and non-foaming sludge relating to aeration and the implications for membrane fouling control in submerged membrane bioreactors. J. Water Process Eng. 28, 250–259.

Zhao, G., Chen, W.-N., 2015. Enhanced PVDF membrane performance via surface modification by functional polymer poly (N-isopropylacrylamide) to control protein adsorption and bacterial adhesion. React. Funct. Polym. 97, 19–29.

Zhou, Z., Meng, F., Lu, H., Li, Y., Jia, X., He, X., 2014. Simultaneous alkali supplementation and fouling mitigation in membrane bioreactors by on-line NaOH backwashing. J. Membr. Sci. 457, 120–127.

Membrane and filtration processes for microplastic removal

Linh-Thy Le[a,b], *Xuan-Bui Bui*[b,c], *Cong-Sac Tran*[b,c],
Chart Chiemchaisri[d], *and Ashok Pandey*[e,f]

[a]Faculty of Public Health, University of Medicine and Pharmacy at Ho Chi Minh City (UMP), Ho Chi Minh City, Vietnam [b]Key Laboratory of Advanced Waste Treatment Technology, Ho Chi Minh City University of Technology (HCMUT), Ho Chi Minh City, Vietnam [c]Vietnam National University Ho Chi Minh City (VNU-HCM), Ho Chi Minh City, Vietnam [d]Department of Environmental Engineering, Faculty of Engineering, Kasetsart University, Bangkok, Thailand [e]Centre for Innovation and Translational Research, CSIR-Indian Institute of Toxicology Research, Lucknow, Uttar Pradesh, India [f]Sustainability Cluster, School of Engineering, University of Petroleum and Energy Studies, Dehradun, India

1. Introduction

Nowadays, the population is increasing rapidly, and with the development of industrialization, anthropogenic activities have made the problem of microplastic pollution more and more serious. The world's plastic production constantly grew from approximately 330 million tons to 360 million tons in 2016 (Azizi et al., 2021; Yang et al., 2021). It could be seen that the amount of plastic produced every year in the world is tremendous; so, the amount of plastic waste is also increasing. It was estimated that estuaries released between 4.8 and 12.7 tons of plastic waste into the oceans (Razeghi et al., 2021). The annual amount of plastics released on land is approximately 23 times the amount of plastics released into the ocean (Horton et al., 2017). According to Uddin et al. (2020), the density of microplastics floating in the ocean ranges from 15 to 51 million particles. According to Statista (2019), Asia generated about half of the total global plastic waste, followed by Canada, the United States, and Mexico, with about 20% of the total, Europe accounted for 14%, and the remaining areas were all less than

Copyright © 2023 Elsevier Inc. All rights reserved.

10%. The dramatic increase in the use and production of plastics worldwide caused growing concerns about microplastic pollution to biological populations and humans. With different sampling methods, scientists have shown that microplastics are present everywhere, from lakes and rivers to oceans and in municipal wastewater (van Emmerik et al., 2018; Park and Park, 2021). Many studies have shown the harmful effects of microplastics on organisms, especially on human health (Waring et al., 2018; Rahman et al., 2020). Macroplastics affect large animals (e.g., large fish, reptiles, birds, and mammals), while microplastics affect tiny organisms (e.g., plankton, worms, corals, crustaceans, mollusks, and small fish) (van Emmerik et al., 2018). MPs are pretty durable, so they take longer to decompose. In addition, they are resistant to corrosion and lightweight, so they can be easily moved over long distances and can become carriers of harmful pollutants and pathogens as they float on the water surface. Studies have estimated that about 80% of microplastics (MPs) released into the aquatic environment originates from land (Rochman, 2018), mainly from inadequately managed waste and wastewater treatment.

Various studies have shown that less than 20% of municipal wastewater was treated by wastewater treatment plants (WWTPs), (especially in developing countries. Although the MPs' treatment efficiency of wastewater treatment plants was relatively high, the water flow rate of WWTPs was quite large, resulting in a large amount of MP that could be discharged into the aquatic environment. The number of pollutants discharged into the environment depends on the water processes. Commonly, microplastics are removed by wastewater treatment plants through primary and secondary treatment (biological treatment). Various studies reported that about 60%–90% of MPs could be removed by primary treatment (Sun et al., 2019). Secondary treatment could further reduce the number of MPs by 7%–20% (Talvitie et al., 2017; Ziajahromi et al., 2017; Gies et al., 2018). So far, not much research has been published reporting the application of membrane processes in tertiary treatment to remove microplastics. Therefore, this chapter focuses on microplastic removal through filtration and membrane processes. The removal efficiency and factors affecting the MPs' removal processing will be discussed. In addition, the negative impacts of MPs on the environment and human health are also mentioned in this section.

2. Microplastic generation and pollution

2.1 Microplastics

Microplastics (MPs) are plastic particles less than 5 mm in size (Bilgin et al., 2020; Sun et al., 2019; Bui et al., 2020). Today, microplastics have been found everywhere, such as in the atmosphere (Abbasi et al., 2019), soil (Guo et al., 2020), tap water (Anagnosti et al., 2021), freshwater sediments (Yang et al., 2021), beach (Razeghi et al., 2021), and deep water areas of the ocean (Liu et al., 2019). Besides, microplastics could be divided into two main sources in the environment, primary microplastics and secondary microplastics (Fig. 1) (Pirsaheb et al., 2020). Primary sources of microplastics in the environment were generated from specific needs, such as microparticles in facial cleansers and shower gels (Bilgin et al., 2020; He et al., 2021; Razeghi et al., 2021), glitter beads in makeup and craft materials (Bilgin et al., 2020), abrasive cleaning beads (He et al., 2021), preproduction resin beads (He et al., 2021),

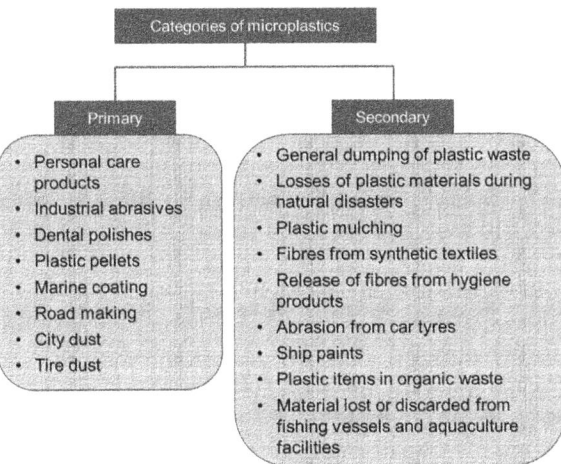

FIG. 1   Sources for primary and secondary microplastics present in the environment.

plastic beads, microfibers in textile production (Bilgin et al., 2020), fibers, films, granules, and powders used in cosmetics (Pirsaheb et al., 2020). Secondary sources of microplastics in the environment are large pieces of plastic that were severed and broken down into small pieces during physical processes (e.g., solar radiation, wind, and water waves), chemical processes (e.g., acidic, oxidizing, heat, and radiation conditions), and biological processes (microbiological erosion) (Bilgin et al., 2020; Anagnosti et al., 2021).

2.2 Classification of microplastics

In various studies, microplastics are often classified based on their morphology, including fragments, fibers, foams, films, sheets, and pellets, also known as granules (Bui et al., 2020; He et al., 2021). It has been found that MPs as fiber type occurred commonly in wastewater. Gündoğdu et al. (2018) showed that most MPs in the secondary effluent of a wastewater treatment plant in Turkey were approximately 60% as microfibers. Gies et al. (2018) reported the apparent presence of microfibers with more than 65%, followed by fragments with 27%, the remainder being pellets, granules, foams, and sheets accounting for insignificant amounts in all wastewater samples, although they make no difference between particles, flakes, and pellets. In the study of Long et al. (2019), granules were most identified, present in both influent (48%) and effluent (35%), followed by fragments (28%) in the influent and fibers (29%) in the posttreatment. Raju et al. (2020) determined the number of microplastics presented in a WWTPs located in Australia, including plastic fragments and fibers in the influent accounting for 38% and 36%, respectively; fibers and fragments in the UV-treated wastewater accounted for 58% and 25%, respectively.

Microplastics could be made up of different types of materials, including polyethylene (PE), polypropylene (PP), polyvinyl chloride (PVC), polyethylene terephthalate (PET), polybutylene terephthalate (PBT), polyamide (Nylon6, 66) (PA), polycarbonate (PC), polyoxymethylene (POM), polyphenylene sulfide (PPS), polystyrene (PS) (Bui et al., 2020; He et al., 2021).

2.3 Biodegradation of microplastics

In recent years, microorganism applications for treating microplastics have drawn attention worldwide. Some research results showed that from 2013, microplastic-degrading microorganisms had been detected in different environments with different environmental conditions such as living conditions, decomposition mechanism, and degradation efficiency (Zurier and Goddard, 2020). Microorganisms carry out the mechanism of microplastic decomposition through the following stages: first, microorganisms attach to the polymer surface, forming biofilms. Especially in the process of microplastic treatment in wastewater treatment plants, biofilm formation is considered an essential step for biodegradation. In the second step, extracellular enzymes are released to interact with the polymer. During this phase, enzymes bind to the polymer, then begin to cleave to form oligomers, dimers, and monomers. Microorganisms use oligomers as substrate and release mineralized substances (CO_2 and H_2O) and energy to the environment. These products then return to natural biogeochemical cycles, completing the transformation from harmful starting materials into valuable products.

2.4 Transport of microplastics in the environment

Up to now, microplastic pollution has arisen from human activities. Microplastics have usually been discharged from agricultural, industrial, and urban areas. Wastewater in agricultural areas might contain plastic fragments that were broken under the effects of ultraviolet (UV) radiation, physical wear, and food wrappers. Agricultural land also accumulates a significant amount of microplastics (Zeng, 2018). It was estimated that 130–900 tons of MPs/(million people. year) were discharged into agricultural land in Europe (Horton and Dixon, 2018). In the industrial sector, soil and sediment samples also contained a large number of microplastics (Koutnik et al., 2021). Runoff containing MPs could arise from soil leaching from landfills and direct discharge of plastic debris from industrial land use to the environment (Lutz et al., 2021).

MPs found in municipal wastewater are often the result of daily anthropogenic activities. During laundry in households, MPs from garment materials derived from artificial fibers (e.g., polyesters and polyamides) have commonly been shed from clothes due to the action of the washing-drying process (Napper and Thompson, 2016). In addition, the packaging of personal care products such as toothpaste, face wash, and shower gel that we use every day also emits a small number of MPs that go into WWTPs (Magni et al., 2018) due to the decomposition of plastic waste. Untreated MPs were often released from WWTPs, entered receiving sources, and eventually accumulated in the environment along the food chain (Carr et al., 2016). These wastewater treatment systems could not capture tiny microplastics in size, partly because the primary purpose of WWTPs was not designed to remove microplastics. In addition, large amounts of MPs were also found in stormwater drainage systems. Lutz et al. (2021) found approximately 1–3500 microplastic particles per kilogram of dry sediment, with an average concentration of 664 particles per kilogram occurring in the drainage network. In addition, MPs were also detected in the atmosphere; they were transported by wind from one place to another. In China, the atmospheric deposition fluxes of MPs ranged from 51 to about 200 particles/(m^2 day) (Huang et al., 2021).

Although microplastics originate in cities and densely populated areas, air currents and winds can lead particles to be transported farther from the source (Horton and Dixon, 2018).

It can be concluded that MPs can be transported into the environment in three ways:

- MPs are accumulated in the soil along with sludge as fertilizer.
- MPs are transported into the ocean by surface currents.
- MPs are widely dispersed in the atmosphere or even into the ocean.

Liu et al. (2019) demonstrated that about 1.21 tons of atmospheric MPs are thought to be transported from primary sources to the marine environment each year. In 2010, the amount of plastic waste from land to the ocean in Iran was 0.05–0.3 million tons (Razeghi et al., 2021). Every year, about 3.8 million tons of plastic debris enter the oceans through rivers. It is estimated that, by 2025, about 250 million tons of MPs will be deposited in the ocean (Xu et al., 2020). According to Harris (2020), sediment was the final destination of MPs; they were ubiquitous in sediments in the marine. With many MPs in the ocean, they threaten the life of microorganisms living in the sea, even humans (Fig. 2).

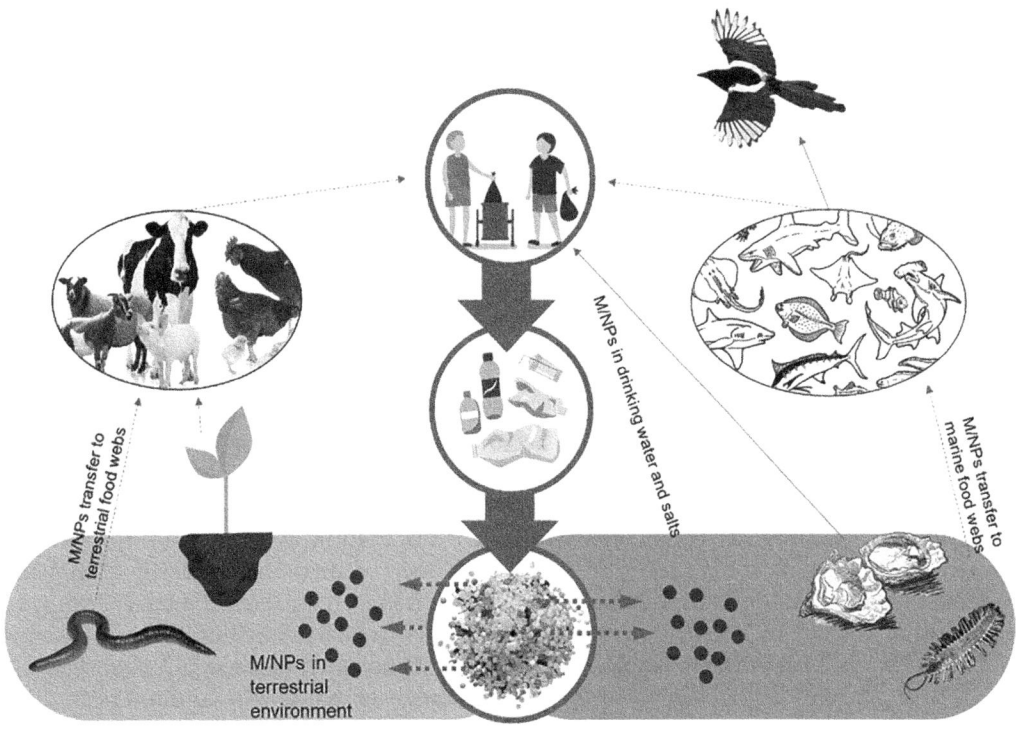

FIG. 2 Classification, origins, and impact of MPs.

B. Applications of membrane technology for water and wastewater treatment

3. Effect of synthetic microplastics pollution on the environment and human health

In recent years, microplastics gained the attention of scientists worldwide due to their presence and existence in all environments (soil, water, and air) and the effects of MPs on organisms and humans. MPs were considered a vector for disease transmission to aquatic organisms. Thirty types of bacteria were found on the surface of microplastics collected in the North Adriatic (Viršek et al., 2017). Through adsorption, heavy metals (e.g., Hg, Cu, Zn, and Cd) and pathogens could be trapped in the biofilm that forms on the surface of microplastics. Some tiny aquatic organisms consume microplastics because they are mistaken for food. Through the food chain, larger organisms such as large fish, reptiles, birds, and mammals consuming smaller organisms (e.g., plankton, worms, corals, crustaceans, animals, mollusks, and small fish) would be affected by microplastics. When marine animals ingest MPs, MPs could damage their digestive systems (Zhang et al., 2020; Wu et al., 2020). After entering the fish body, MPs, especially microfibers, would be accumulated in their intestines (Qiao et al., 2019) or trapped in the gills. In particular, the sharp edges of the MPs have caused intestinal damage, obstruction, and growth inhibition in small aquatic animals. According to some studies, because plastic is hydrophobic, materials derived from PS, PE, and PP can absorb hydrophobic organic substances (i.e., PCBs, PBDEs, and PAHs) or heavy metals on surface MPs (Pirsaheb et al., 2020; Joo et al., 2021). After entering the fish's body, a part of the smaller particles will travel throughout the fish's body, spreading toxins into fish tissues and causing harm. Some of the deleterious effects on living organisms that have been studied include neurotoxicity and genotoxicity and reduced feeding, survival, and reproduction (Amelia et al., 2021).

In order to improve the physical and chemical properties of plastics, many harmful additives such as bisphenol A, phthalates, and flame retardants have been added to the plastic manufacturing process. Bisphenol A could cause cancer and endocrine disorders and adversely affect fertility in humans and animals. Phthalates harm children's brain development, behavioral development, and muscle coordination, affecting thyroid hormone levels necessary for fetal and infant brain development and causing a decrease in testosterone (male sex hormone). Flame retardants are primarily chemical substances that mimic estrogen's effects and are unsuitable for thyroid function; they can disrupt hormones and promote cell proliferation and tumor growth. Most of this poison will spread out and pollute the ecological environment. The remaining toxins in MPs, however small, still could potentially be detrimental to natural and human health after accumulating in food webs.

Not only harmful to aquatic animals, but MPs can also harm human health when we consume foods (e.g., seafood) or drink bottled water (Mason et al., 2018; Jiang et al., 2020). When we ingest food, granular MPs could enter the body and then be absorbed into the intestines. Alternatively, the smallest MPs could move into the body by drinking water; then, they move into the circulatory system. After entering the bloodstream, the clear MPs move around the body to other organs such as the liver (Harvey and Watts, 2018). Scientists have identified about 4–18 MPs in 1 liter of disposable bottled water. Thus, one person could absorb approximately 1950 particles in a week. Various evidence suggests that plastic containers could also introduce microplastics into the human body. Cox et al. (2019) estimated that humans consume between 4×10^3 and 5×10^3 MPs per year. MPs containing 243–684 particles per litter were also detected in drinking water (Eerkes-Medrano et al., 2019). In addition, microplastics were also present in the sugars and salts we eat daily (Lee et al., 2019; Peixoto et al., 2019; Selvam et al., 2020; Zhang et al., 2020). Peixoto et al. (2019) found that sea salt contains up to 20 particles of MPs per gram of salt. In short, MPs are present everywhere and enter our bodies in many ways. However, the scientific research community's interest in the organisms that consume MPs and the environmental pollution from packaging and plastic containers is disproportionate (Rist et al., 2020). Few studies have been done to estimate human exposure to MPs and their effects on human health.

4. Microplastic removal in wastewater treatment plants

Various treatment technologies applied in the WWTPs included primary treatment processes (primary settling treatment, grit, and grease treatment), secondary treatment processes (A2O, biofilters, and other bioreactors), and tertiary treatment processes (UV, O_3, chlorination, biologically active filters (BAFs), disc filters (DFs), and rapid sand filters (RSFs)) (Fig. 3). Many reports have shown that microplastics could be effectively removed in WWTPs

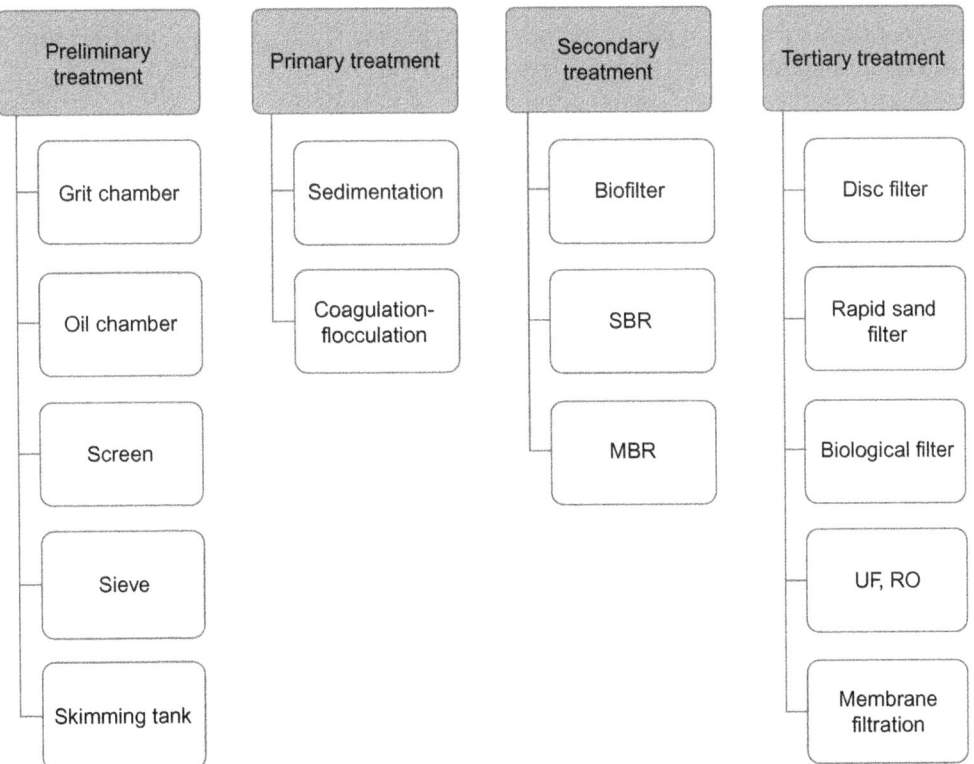

FIG. 3   Various techniques for MP removal in WWTPs.

through preliminary and primary treatment (e.g., screening, skimming, and sedimentation). The pretreatment and primary treatment could effectively remove approximately 65%–80% of microplastics (Hidayaturrahman and Lee, 2019; Iyare et al., 2020; Ziajahromi et al., 2021). Indeed, Gies et al. (2018) found 92.8% fibers and 88.7% fragments of MP retention by the primary treatment. Ziajahromi et al. (2017) investigated microfibers as the dominant type of microplastic detected in all wastewater effluent samples. The concentration of microfibers in the primary effluent of about two fibers per liter (the capacity of WWTPs of 310 mL per day) was discharged into deep water ocean outfall. According to the mesh size, the screen is divided into a coarse screen (mesh size 50–100 mm), medium screen (mesh size 10–40 mm), and fine screen (mesh size 2.5–10 mm) (Zhang et al., 2020). This is because MPs were defined as plastics less than 5 mm in size (small MPs of 0.3–1 mm and prominent MPs of 1–5 mm) (Bui et al., 2020). Therefore, only a fine screen could remove microplastic particles larger than 2.5 mm, and other sizes of plastic particles are expected to be removed during preliminary treatment (Liu et al., 2019; Iyare et al., 2020). Besides, the density of MPs is lower than that of water, so that MPs could easily float on the water surface. Therefore, floating MPs could be removed effectively with a removal rate of approximately 65%–80% through the skimming process, while the others were trapped in sludge flocs in secondary processes (Talvitie et al., 2017; Long et al., 2019; Magni et al., 2018; Bayo et al., 2020a).

Secondary treatment operations were typically biological processes that further reduce the residual suspended and dissolved solids remaining in wastewater after primary treatment. So far, secondary treatment usually includes an activated sludge process, biofiltration, trickling filters, and solid contact tanks (Lares et al., 2018; Alvim et al., 2020; He et al., 2021). According to the microscopic observation results of Ziajahromi et al. (2017), the number of microplastics reduced by approximately 70% after secondary treatment (i.e., from 1.5 MPs per liter to 0.3 MPs per liter). Based on the result of He et al. (2021), the total amount of MPs in wastewater can reach 3160 particles/L, and most of them are found to be transferred to sludge during the wastewater process. According to He et al. (2021), the mechanism to remove microplastics in influents is to convert them from the liquid phase to the solid phase, which can be understood as the MPs being enriched in sludge rather than in final removal. Furthermore, according to Alvim et al. (2020), microplastics which correspond to plastic fragments smaller than 5 mm, are primarily retained in the activated sludge. Therefore, the removal of microplastics in sludge is necessary. The activated sludge process has been observed to remove MPs in the range of 17%–98% (Lee et al., 2019; Liu et al., 2019). Specifically, Liu et al. (2019) evaluated the MPs' removal with an anaerobic-anoxic-oxic (A2O) process and observed that 17 particles/L were removed using biological treatment. Talvitie et al. (2017) also describe similar results; in the secondary wastewater treatment in Finland, around 87% of MPs were removed by the activated sludge process. Lares et al. (2018) also investigated the performance of the conventional activated sludge (CAS) process at the municipal wastewater treatment plant in Finland. After passing this stage, the microplastic removal efficiency was approximately 98%.

Tertiary treatment [e.g., UV, O₃, chlorination, biologically active filters (BAFs), disc filters (DFs), and rapid sand filters (RSFs)] is the last stage of a WWTP, used to treat specific pollutants, and disinfection. Tertiary treatment processes can be highly effective for further removing MPs from wastewater. After treatment, the concentration of MPs is reduced on average to between 0.1% and 7.8% of the influent concentration (Xu et al., 2020). After tertiary treatment processes, the microplastic abundance further decreased in most investigated WWTPs. Sun et al. (2019) found that after tertiary treatment, MPs were mainly found in fibers and microbeads. The minor sizes' fraction (20–190 μm) MPs were found to be the most abundant.

However, a significant amount of microplastics (maximum approximately 50%) from wastewater treatment plants still escapes through the treated effluent and enters various water receiving bodies (Liu et al., 2021a,b). The concentration of MPs in treated wastewater will be affected by different wastewater treatment technologies applied. Besides, the number of treatment processes used for the influent could affect MPs' release into the receiving reservoirs. Table 1 briefly describes the results of the MPs' removal efficiency of various wastewater treatment techniques.

4.1 Microplastic removal by filter-based technologies

According to the information mentioned above, filter-based technologies [e.g., biological filtration, ultrafiltration (UF), and rapid sand filtration (RSF)] microplastics were usually applied in the polishing step (third-order processing) to remove MPs thoroughly (Talvitie et al., 2017). Typically, disc filtration (DF), rapid sand filtration (RSF), and activated carbon

TABLE 1 Microplatics removal by various wastewater treatment techniques.

Wastewater	Technologies/ operating conditions	Influent (MPs/L)	Effluent (MPs/L)	Removal (%)	Comments	References
Municipal	Sedimentation (Primary, secondary)	Fibers: 180 Particles: 430	Fibers: 4.9 Particles: 8.6	95–98	Easy to install and operate, effective with most of MPs	Talvitie et al. (2015)
Synthetic	Flotation	1.03–1.6 g/cm^3	0	100	Used as tertiary treatment. Not efficient in removing particles <20 μm	Zhang et al. (2021)
Municipal	Aerated grit chamber	110	84	60	• Could remove particles of 1–5 mm and fibers >1 mm. • Need adding gravity settling devices and coagulant to improve MPs removal	Bilgin et al. (2020)
Municipal	Aerated grit chamber	–	–	59		Yang et al. (2021)
Municipal	Grit chamber	57.6	0.6	98.5		Lares et al. (2018)
Municipal	A$_2$O	28,400	12.8	>90	MPs of 25 μm could be released through effluent.	Edo et al. (2020)

Continued

TABLE 1   Microplatics removal by various wastewater treatment techniques—cont'd

Wastewater	Technologies/ operating conditions	Influent (MPs/L)	Effluent (MPs/L)	Removal (%)	Comments	References
Municipal	Conventional activated sludge	12.2	2	83	• Simple operation • moderate energy cost • need combine with another techniques	Hongprasith et al. (2020)
Municipal	MBR	57.6	0.4	98.5	• Membrane fouling • high operating cost • slightly high energy consumption	Lares et al. (2018)
Municipal	MBR	6.9	0.005	>98		Talvitie et al. (2017)
Municipal	MBR	0.28	0.05	82		Lv et al. (2019)
Urban	MBR	4.4	0.92	78		Bayo et al. (2020a)
Municipal	RSF	4.4	1.08	76	Easy to install and operate, low cost	Bayo et al. (2020b)
Domestic and industrial	Coagulation, RSF	5840	2080	64.3	The combinations could enhance MPs removal	Hidayaturrahman and Lee (2019)
Domestic and industrial	Coagulation, O_3	4200	33	90		Hidayaturrahman and Lee (2019)
Domestic and industrial	Coagulation, DF	31400	297	81.2		Hidayaturrahman and Lee (2019)
Municipal	Coagulation			76–95		Rajala et al. (2020)
Municipal	DF (10-µm pore size filters)	0.5	0.3	40	• Effective in removing small MPs • Simple operation • Low energy consumption	Talvitie et al. (2017)
Municipal	DF (20-µm pore size filters)	2	0.03	> 98		Talvitie et al. (2017)
Households and industrial	DF			89.7	• Required large amount of chemicals • Large volume of sludge generated • Low energy consumption • Filter clogging	Simon et al. (2019)

filtration (ACF) were commonly applied in wastewater treatment plants because they require no special equipment and simple practicability. Magni et al. (2018) investigated MPs' removal capacity of one of the biggest WWTPs in Italy (flow rate of about 4×10^8 L per day). They recognized that sand filter treatment had a significant MPs' removal ability with the efficiencies of about 85%. In a study at a WWTP in Spain, treating domestic and industrial wastewater with a capacity of $12,000\,m^3$/day was investigated for 3 years; rapid gravity sand filters (RSF) of the tertiary stage had an MP retention of approximately 76%. The average effluent MP concentration of the RSF process was approximately 1.2 MPs/L. The prominent type of MPs in the effluent was microfiber (Bayo et al., 2020a,b). The mechanism of sand filtration was mainly based on the size of microplastic particles and filter material particles. Microplastic particles were placed individually in the spaces between the filter particles that are smaller than the particle size. Hidayaturrahman and Lee (2019) found that MP removal rates of DF and RSF were insignificantly different. A sand filter with 6.8 m depth and used sand grain size of approximately 0.8–1.2 mm could remove about 80% MPs. In contrast, another WWTP had a percentage of microplastics removed of 75% after an RSF step. Talvitie et al. (2017) also evaluated the effectiveness of removing MPs by RSF containing 1 m of gravel and 0.5 m of quartz at the sewage treatment plant in Finland. A high MP removal efficiency of approximately 96% (from 0.7 ± 0.1 MPs/L down to 0.02 ± 0.007 MPs/L) was achieved. The lower MP removal efficiency result of Hidayaturrahman and Lee (2019) than that of Talvitie et al. (2017) might be caused by the different sizes of particles. Some particles having a dimension smaller than the diameter of filter materials (e.g., sand) could pass through the rapid sand filtration process.

In addition, a commonly used filtration-based microplastic removal method is the use of disc filtration (DF). DF consists of support plates and multiple layers of circular screens on top of each other and is often applied in the polishing step (tertiary treatment) to remove small impurities in wastewater. The material for the screens could be polypropylene, polyester, or polyamide with pore sizes of 10–40 μm (Padervand et al., 2020; Xu et al., 2020). The DF method could help to remove 89% of microplastic particles (Simon et al., 2019). Removing microplastics by DF was based on physical capture by the filter and the formation of sludge cake on the filter plates. The sludge cake was periodically removed by high-pressure backflushing (Talvitie et al., 2017; Simon et al., 2019; Xu et al., 2020; Zhang et al., 2021).

Talvitie et al. (2017) investigated the secondary effluent filtration process and showed that the disc filter removed MPs ranging from 45% to over 97% during treatment. The study applied an inside-out pilot disc filter consisting of two discs, each of 24 filter plates. However, the results obtained in the study were not as expected. Typically, a small pore size disc filter would be expected to remove more MPs than a large pore size disc. In this case, the DF with a 10 μm pore size only reduced the concentration of MPs from 0.7 to 0.4, while the DF with 20 μm pore size could achieve MP concentration of 0.03 after treatment. It was explained by inappropriate polymers, leading to clogged filters. During backwashing at high pressure, part of MPs might have passed through the filter cloth to the final waste stream, reducing the removal efficiency.

In short, the main disadvantage of the filtration process is that the biofilm formation process due to the adhesion of impurities on the microplastic surface will clog the membrane surface. In addition, some nanometer-sized microplastics will also create blockages. The filter pressure increases, the filter diaphragm becomes more prone to clogging, and the need for replacement occurs more often.

4.2 Microplastic removal by membrane-based technologies

So far, membranes were defined as a barrier that selectively allows certain substances (e.g., molecules, ions, or particles as MPs) to pass through and retain others (Fig. 4). Categorized by pore size, membranes were divided into different types, including microfiltration (MF), ultrafiltration (UF), nanofiltration (NF), and reverse osmosis (RO) membranes. MF, UF, and NF membranes could effectively remove particles of size in the range 0.08–2 μm, 0.005–0.02 μm, and 0.002 μm, respectively, while RO membranes were generally applied in desalination works and reuse of seawater or brackish water. Besides, the membrane bioreactor (MBR) was considered as a combination of biological treatment and membrane filtration.

Ultrafiltration (UF) has a pore size of 0.01 μm, which can remove particles and macromolecules such as proteins, fatty acids, bacteria, protozoa, viruses, and suspended solids. Many municipal WWTPs have been gradually using UF filtration to replace traditional processes such as sedimentation or sand filtration. Besides, using UF filtration in water treatment could reduce dissolved organic matter (DOM) present in surface water by approximately 30%–40%, significantly reducing turbidity and achieving pathogen removal of about 90%–100% (Bui et al., 2020; Tang and Hadibarata, 2021). However, the UF membrane was not specifically designed to retain MPs because they could be clogged quickly. Typically, UF was combined with coagulation and could be used as a secondary, tertiary treatment, or pre-treatment in reverse osmosis plants to preserve RO membranes. Various reports have shown that MPs presented in water could not be removed effectively by UF combined with coagulation (Ma et al., 2019; Ziajahromi et al., 2017). It depended on whether the used coagulant kinds were appropriate. Ma et al. (2019) investigated that ultrafiltration and coagulation could remove polyethylene (PE) by adding Fe-based coagulants. However, PE removal efficiency was only about 20%. It was thought that PE is a common microplastic, having a density of 0.92–0.97 g/cm^3, very close to that of water, making it easy to pass through filter holes. After

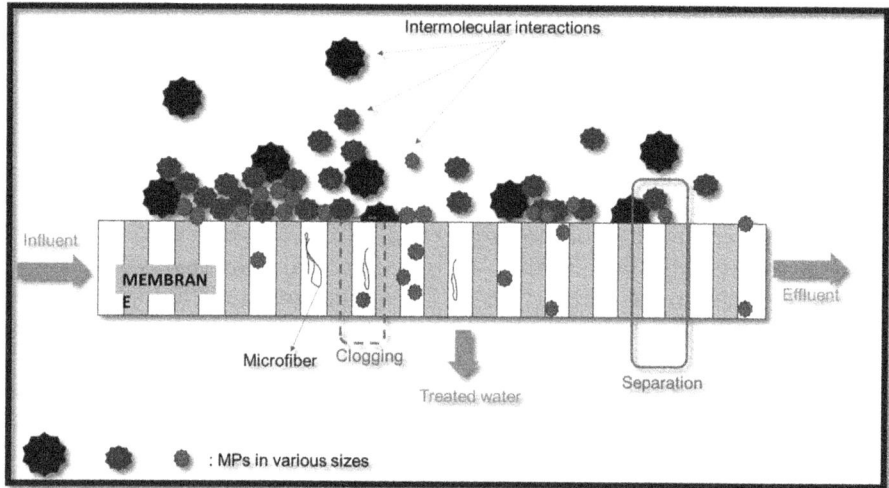

FIG. 4   Schematic diagram of membrane filtration for MP removal.

that, polyacrylamide (PAM) was added to enhance the coagulation efficiency, and the ability to significantly remove small-sized polyethylene particles ($d < 0.5$ mm) was up to 90% (Ma et al., 2019).

In addition to using UF membranes, some other studies have used reverse osmosis membranes to remove microplastics in the aquatic environment. According to the information mentioned above, filter-based technologies [e.g., biological filtration, ultrafiltration (UF), and rapid sand filtration (RSF)] microplastics were usually applied in the polishing step (third-order processing) to remove MPs thoroughly (Talvitie et al., 2017). Typically, disc filtration (DF), rapid sand filtration (RSF), and activated carbon filtration (ACF) were commonly applied in wastewater treatment plants because they require no special equipment and simple practicability. Magni et al. (2018) investigated MP removal capacity of one of the biggest WWTP in Italy (flow rate of approximately 4×10^8 L per day). They recognized that sand filter treatment had a significant MP removal ability with the efficiencies was about 85%. In a study at a WWTP in Spain, treating domestic and industrial wastewater with a capacity of 12,000 m^3/day was investigated for 3 years; rapid gravity sand filters (RSFs) of the tertiary stage had an MP retention of approximately 76%. The average effluent MP concentration of the RSF process was approximately 1.2 MPs/L. The prominent type of MPs in the effluent was microfiber (Bayo et al., 2020a,b). The mechanism of sand filtration was mainly based on the size of microplastic particles and filter material particles. Microplastic particles were placed individually in the spaces between the filter particles that are smaller than the particle size. Hidayaturrahman and Lee (2019) found that MP removal rates of DF and RSF were insignificantly different. A sand filter with 6.8 m depth and used sand grain size approximately 0.8–1.2 mm could remove about 80% of MPs. In contrast, another WWTP had a percentage of microplastics removed of 75% after an RSF step. Talvitie et al. (2017) also evaluated the effectiveness of removing MPs by RSF containing 1 m of gravel and 0.5 m of quartz at the sewage treatment plant in Finland. A high MP removal efficiency of approximately 96% (from 0.7 ± 0.1 MPs/L to 0.02 ± 0.007 MPs/L) was achieved. The lower MP removal efficiency result of Hidayaturrahman and Lee (2019) than that of Talvitie et al. (2017) might be caused by the different sizes of particles. Some particles having a dimension smaller than the diameter of filter materials (e.g., sand) could pass through the rapid sand filtration process.

Membrane bioreactor (MBR) is being considered as a particular form of tertiary biological treatment. This technology has been and widely applied worldwide because of several advantages such as less footprint area requirement, good build-up area, high biomass retention time, ease to upgrade from the existing conventional treatment system (Al-Asheh et al., 2021). MBR technology is integrated between the activated sludge process and the membrane (e.g., MF, UF, and NF) submerged or placed outside the bioreactor. The primary mechanism of microplastic removal by MBR is mainly the adsorption of MPs into the activated sludge, and the MPs are retained in the sludge recirculation of the membrane filtration process (Fig. 1). Some studies (Talvitie et al., 2017; Lares et al., 2018) show that the efficiency of MPs removal from wastewater by biofilm filtration is approximately 5%–10% higher than that of other filtration processes. (e.g., DF, RSF). More than 98% of microplastics were retained by applying the MBR process (Tang and Hadibarata, 2021). However, various studies have suggested that MBR can remove MPs from wastewater less efficiently than traditional processes (Bayo et al., 2020a,b). They show that the MBR only retains approximately 75% of the microplastics as fiber type, which is lower than the conventional activated sludge process with above 89% (Bayo et al., 2020a,b). In contrast, the particle removal was more effective with above 98% in the same MBR system. It was thought that a low microfiber removal efficiency in Bayo's study might be due to applied high pressure in the MBR system, making some tiny microfibers go through the larger membrane pores. From the information mentioned above, the significant variation in MP removal efficiency in the above studies could depend on MPs' size and material differences. A new issue that needs attention is the membrane material. Some membranes are made from synthetic polymers such as polyvinylidene fluoride (PVDF), polysulfone (PSF), and polyacrylonitrile (PAN). After a period of use, the filter material can be aged under the influence of chemicals (e.g., chemicals used to wash the membrane) and thus can be a source of artificial MP emissions into the aquatic environment. Furthermore, when the MBR process operates at high pressures up to 800 psi, it is possible to push some microplastics smaller than the longitudinal pore size of the system into the receiving reservoirs (Talvitie et al., 2017). Besides, the microparticle removal process by MBR is also affected by several other factors such as film thickness, inlet wastewater properties, membrane pressure, and formation of cake residue on the membrane surface. In summary, the above information shows that biofilm filtration is one of the practical solutions to remove microplastics in water, especially when combined with RO pretreatment and polishing processes, it will be highly effective.

5. Conclusions and perspectives

The speed of industrialization has increased, leading to an increase in human activities related to the production and consumption of microplastics. Since then, studies related to microplastics in wastewater systems have been increasingly carried out by scientists worldwide to find the most sustainable methods of minimizing microplastic emissions and reducing negative environmental impacts on human health. WWTPs have been considered a barrier to reducing microplastic discharge into the environment for a long time. The primary mechanism for MP removal is through adsorption into the activated sludge, and the MPs are disposed of with the sludge. Not only that, but WWTPs played an essential role in releasing MPs to the receiving reservoirs. Indeed, municipal WWTPs treated massive amounts of wastewater so that the total discharges of the microplastics from WWTPs were still considered significant. Various studies indicated that tiny MPs (<20 micros net) could leak out from the effluent of WWTPs into the aquatic environment. The WWTPs could remove some microplastics depending on the treatment processes applied. Some WWTPs with tertiary treatment processes had a lower amount of MPs in the effluent than those with primary or secondary treatment processes. Among the processes used as tertiary treatments in WWTPs, filtration-based processes (RSF, DF, and membrane filtration) could retain MPs as high as 78% to over 95%. In particular, when applying a combination of pre-treatment, membrane bioreactor (MBR) process, and RO for the polishing step before being discharged into the aquatic environment, more than 98% of MPs could be retained. However, the biggest challenge of these technologies is that the fouling process and membrane materials need to be studied and discussed more in the future. Besides, various studies showed that many MPs come from household discharges such as laundry activities, and using plastic products. However, the biggest challenge of these technologies is the fouling process and membrane materials that need to be studied and discussed more in the future. Besides, various studies show that a large number of MPs come from household activities such as laundry activities and the use of plastic products. Therefore, in addition to applying wastewater treatment technologies to remove microplastics, an important issue that needs to be paid more attention to is the management and minimization of discharging microplastic at their sources. For example, each family can use filter devices attached to the washing machine to increase the retaining efficiency of fiber-like microplastics generated from household laundry wastewater. In addition, limiting the use of garment products derived from artificial fibers (such as PE and Nylon) is also a solution to reduce the generation of microplastics that needs more profound attention.

Acknowledgments

This research was funded by Vietnam National University Ho Chi Minh City (VNU-HCM) under grant number NCM2021-20-01. We acknowledge the support of time and facilities from Ho Chi Minh City University of Technology (HCMUT), VNU-HCM for this study. We gratefully acknowledge the technical support of the CARE-RESCIF initiative within the International Joint Laboratory LECZ-CARE project for conducting the research.

References

Abbasi, S., Keshavarzi, B., Moore, F., Turner, A., Kelly, F.J., Dominguez, A.O., Jaafarzadeh, N., 2019. Distribution and potential health impacts of microplastics and microrubbers in air and street dusts from Asaluyeh County, Iran. Environ. Pollution. 244, 153–164. https://doi.org/10.1016/j.envpol.2018.10.039.

Al-Asheh, S., Bagheri, M., Aidan, A., 2021. Membrane bioreactor for wastewater treatment: a review. Case Studies Chem. Environ. Eng. 4, 100109. https://doi.org/10.1016/j.cscee.2021.100109.

Alvim, B.C., Mendoza-Roca, J.A., Bes-Piá, A., 2020. Wastewater treatment plant as microplastics release source—quantification and identification techniques. J. Environ. Manag. 255, 109739. https://doi.org/10.1016/j.jenvman.2019.109.

Amelia, T.S.M., Khalik, W.M.A., Ong, M.C., Shao, Y.T., Pan, H.-J., Bhubalan, K., 2021. Marine microplastics as vectors of major ocean pollutants and its hazards to the marine ecosystem and humans. Progress in Earth and Planetary. Science 8 (1). https://doi.org/10.1186/s40645-020-00405-4.

Anagnosti, L., Varvaresou, A., Pavlou, P., Protopapa, E., Carayanni, V., 2021. Worldwide actions against plastic pollution from microbeads and microplastics in cosmetics focusing on European policies. Has the issue been handled effectively? Mar. Pollut. Bull. 162, 111883. https://doi.org/10.1016/j.marpolbul.2020.111883.

Azizi, N., Khoshnamvand, N., Nasseri, S., 2021. The quantity and quality assessment of microplastics in the freshwater fishes: a systematic review and meta-analysis. Reg. Stud. Mar. Sci. 47, 101955. https://doi.org/10.1016/j.rsma.2021.101955.

Bayo, J., López-Castellanos, J., Olmos, S., 2020a. Membrane bioreactor and rapid sand filtration for the removal of microplastics in an urban wastewater treatment plant. Mar. Pollut. Bull. 156, 111211. https://doi.org/10.1016/j.marpolbul.2020.111211.

Bayo, J., Olmos, S., López-Castellanos, J., 2020b. Microplastics in an urban wastewater treatment plant: the influence of physicochemical parameters and environmental factors. Chemosphere, 124593. https://doi.org/10.1016/j.chemosphere.2019.124593.

Bilgin, M., Yurtsever, M., Karadagli, F., 2020. Microplastics removal by aerated grit chambers versus settling tanks of a municipal wastewater treatment plant. J. Water Process Eng. 38, 101604. https://doi.org/10.1016/j.jwpe.2020.101604.

Bui, X.-T., Vo, T.-D.-H., Nguyen, P.-T., Nguyen, V.-T., Dao, T.-S., Nguyen, P.-D., 2020. Microplastics pollution in wastewater: characteristics, occurrence and removal technologies. Environ. Technol. Innov. 101013. https://doi.org/10.1016/j.eti.2020.101013.

Carr, S.A., Liu, J., Tesoro, A.G., 2016. Transport and fate of microplastic particles in wastewater treatment plants. Water Res. 91, 174–182. https://doi.org/10.1016/j.watres.2016.01.002.

Cox, K.D., Covernton, G.A., Davies, H.L., Dower, J.F., Juanes, F., Dudas, S.E., 2019. Human consumption of microplastics. Environ. Sci. Technol. 53 (12), 7068–7074. https://doi.org/10.1021/acs.est.9b01517.

Edo, C., González-Pleiter, M., Leganés, F., Fernández-Piñas, F., Rosal, R., 2020. Fate of microplastics in wastewater treatment plants and their environmental dispersion with effluent and sludge. Environ. Pollut. 259, 113837.

Eerkes-Medrano, D., Thompson, R.C., Aldridge, D.C., 2019. Microplastics in freshwater systems: a review of the emerging threats, identification of knowledge gaps and prioritisation of research needs. Water Res. 75, 63–82. https://doi.org/10.1016/j.watres.2015.02.012.

van Emmerik, T., Kieu-Le, T.C., Loozen, M., van Oeveren, K., Strady, E., Bui, X.T., Egger, M., Gasperi, J., Lebreton, L., Nguyen, P.D., Schwarz, A., Slat, B., Tassin, B., 2018. A methodology to characterize macroplastic dynamics in rivers. Front. Mar. Sci. 1-11. https://doi.org/10.3389/fmars.2018.00372.

Gies, E.A., LeNoble, J.L., Noël, M., Etemadifar, A., Bishay, F., Hall, E.R., Ross, P.S., 2018. Retention of microplastics in a major secondary wastewater treatment plant in Vancouver, Canada. Mar. Pollut. Bull. 133, 553–561.

Gündoğdu, S., Çevik, C., Güzel, E., Kilercioğlu, S., 2018. Microplastics in municipal wastewater treatment plants in Turkey: a comparison of the influent and secondary effluent concentrations. Environ. Monit. Assess. 190 (11). https://doi.org/10.1007/s10661-018-7010-y.

Guo, J., Huang, X., Xiang, L., Wang, Y., Li, Y., Cai, Q., Mo, C., Wong, M., 2020. Source, migration and toxicology of microplastics in soil. Environ. Int. 137, 105263. https://doi.org/10.1016/j.envint.2019.105263.

Harris, P.T., 2020. The fate of microplastic in marine sedimentary environments: a review and synthesis. Mar. Pollut. Bull. 158, 111398. https://doi.org/10.1016/j.marpolbul.2020.111398.

Harvey, F., Watts, J., 2018. Microplastics found in human stools for the first time. The Guardian.

He, Z.-W., Yang, W.-J., Ren, Y.-X., Jin, H.-Y., Tang, C.-C., Liu, W.-Z., Wang, A.-J., 2021. Occurrence, effect, and fate of residual microplastics in anaerobic digestion of waste activated sludge: a state-of-the-art review. Bioresour. Technol. 331, 125035. https://doi.org/10.1016/j.biortech.2021.125035.

Hidayaturrahman, H., Lee, T.-G., 2019. A study on characteristics of microplastic in wastewater of South Korea: identification, quantification, and fate of microplastics during treatment process. Mar. Pollut. Bull. 146, 696–702. https://doi.org/10.1016/j.marpolbul.2019.06.0.

Hongprasith, N., Kittimethawong, C., Lertluksanaporn, R., Eamchotchawalit, T., Kittipongvises, S., Lohwacharin, J., 2020. IR microspectroscopic identification of microplastics in municipal wastewater treatment plants. Environ. Sci. Pollut. Res. 27 (15), 18557–18564.

Horton, A., Dixon, S., 2018. Microplastics: an introduction to environmental transport processes. Wiley Interdiscip. Rev. https://doi.org/10.1002/wat2.1268.

Horton, A., Walton, A., Spurgeon, D., Lahive, E., Svendsen, C., 2017. Microplastics in freshwater and terrestrial environments: evaluating the current understanding to identify the knowledge gaps and future research priorities. Sci. Total Environ. 586, 127–141. https://doi.org/10.1016/j.scitotenv.2017.01.190.

Huang, Y., He, T., Yan, M., Yang, L., Gong, H., Wang, W., Wang, J., 2021. Atmospheric transport and deposition of microplastics in a subtropical urban environment. J. Hazard. Mater. 416, 126168. https://doi.org/10.1016/j.jhazmat.2021.126168.

Iyare, P.U., Ouki, S.K., Bond, T., 2020. Microplastics removal in wastewater treatment plants: a critical review. Environ. Sci.: Water Res. Technol. 6 (10), 2664–2675. https://doi.org/10.1039/d0ew00397b.

Jiang, B., Kauffman, A.E., Li, L., McFee, W., Cai, B., Weinstein, J., Xiao, S., 2020. Health impacts of environmental contamination of micro- and nanoplastics: a review. Environ. Health Prev. Med. 25 (1). https://doi.org/10.1186/s12199-020-00870-9.

Joo, S.H., Liang, Y., Kim, M., Byun, J., Choi, H., 2021. Microplastics with adsorbed contaminants: mechanisms and treatment. Environ. Challenges 3, 100042. https://doi.org/10.1016/j.envc.2021.100042.

Koutnik, V.S., Leonard, J., Alkidim, S., DePrima, F.J., Ravi, S., Hoek, E.M., Mohanty, S.K., 2021. Distribution of microplastics in soil and freshwater environments: global analysis and framework for transport modelling. Environ. Pollut. https://doi.org/10.1016/j.envpol.2021.116552.

Lares, M., Ncibi, M.C., Sillanpää, M., Sillanpää, M., 2018. Occurrence, identification and removal of microplastic particles and fibers in conventional activated sludge process and advanced MBR technology. Water Res. 133, 236–246.

Lee, H., Kunz, A., Shim, W.J., Walther, B.A., 2019. Microplastic contamination of table salts from Taiwan, including a global review. Sci. Rep. 9, 10145.

Liu, W., Zhang, J., Liu, H., Guo, X., Zhang, X., Yao, X., Zhang, T., 2021a. A review of the removal of microplastics in global wastewater treatment plants: characteristics and mechanisms. Environ. Int. 146, 106277. https://doi.org/10.1016/j.envint.2020.106277.

Liu, K., Zhang, F., Song, Z., Zong, C., Wei, N., Li, D., 2019. A novel method enabling the accurate quantification of microplastics in the water column of deep ocean. Mar. Pollut. Bull. 146, 462–465. https://doi.org/10.1016/j.marpolbul.2019.07.008.

Liu, J., Zhao, M., Duan, C., Yue, P., Li, T., 2021b. Removal characteristics of dissolved organic matter and membrane fouling in ultrafiltration and reverse osmosis membrane combined processes treating the secondary effluent of wastewater treatment plant. Water Sci. Technol. 83 (3), 689–700. https://doi.org/10.2166/wst.2020.589.

Long, Z., Pan, Z., Wang, W., Ren, J., Yu, X., Lin, L., Jin, X., 2019. Microplastic abundance, characteristics, and removal in wastewater treatment plants in a Coastal City of China. Water Res. https://doi.org/10.1016/j.watres.2019.02.028.

Lutz, N., Fogarty, J., Rate, A., 2021. Accumulation and potential for transport of microplastics in stormwater drains into marine environments, Perth region, Western Australia. Mar. Pollut. Bull. https://doi.org/10.1016/j.marpolbul.2021.112362.

Lv, X., Dong, Q., Zuo, Z., Liu, Y., Huang, X., Wu, W.M., 2019. Microplastics in a municipal wastewater treatment plant: fate, dynamic distribution, removal efficiencies, and control strategies. J. Clean. Prod. 225, 579–586.

Ma, B., Xue, W., Ding, Y., Hu, C., Liu, H., Qu, J., 2019. Removal characteristics of microplastics by Fe-based coagulants during drinking water treatment. J. Environ. Sci. https://doi.org/10.1016/j.jes.2018.10.006.

Magni, S., Binelli, A., Pittura, L., Avio, C.G., Della Torre, C., Parenti, C.C., Regoli, F., 2018. The fate of microplastics in an Italian Wastewater Treatment Plant. Sci. Total Environ. https://doi.org/10.1016/j.scitotenv.2018.10.2.

Mason, S.A., Welch, V.G., Neratko, J., 2018. Synthetic polymer contamination in bottled water. Front. Chem. 6. https://doi.org/10.3389/fchem.2018.00407.

Napper, I.E., Thompson, R.C., 2016. Release of synthetic microplastic plastic fibres from domestic washing machines: effects of fabric type and washing conditions. Mar. Pollut. Bull. 112 (1-2), 39–45. https://doi.org/10.1016/j.marpolbul.2016.09.025.

Padervand, M., Lichtfouse, E., Robert, D., Wang, C., 2020. Removal of microplastics from the environment. A review. Environ. Chem. Lett. https://doi.org/10.1007/s10311-020-00983-1.

Park, H., Park, B., 2021. Review of microplastic distribution, toxicity, analysis methods, and removal technologies. Water 13, 2736. https://doi.org/10.3390/w13192736.

Peixoto, D., Pinheiro, C., Amorim, J., Oliva-Teles, L., Guilhermino, L., Vieira, M.N., 2019. Microplastic pollution in commercial salt for human consumption: a review. Estuar. Coast. Shelf Sci. 219, 161–168.

Pirsaheb, M., Hossini, H., Makhdoumi, P., 2020. Review of microplastic occurrence and toxicological effects in marine environment: experimental evidence of inflammation. Process Saf. Environ. Prot. https://doi.org/10.1016/j.psep.2020.05.050.

Qiao, R., Deng, Y., Zhang, S., Wolosker, M.B., Zhu, Q., Ren, H., Zhang, Y., 2019. Accumulation of different shapes of microplastics initiates intestinal injury and gut microbiota dysbiosis in the gut of zebrafish. Chemosphere 236, 124334. https://doi.org/10.1016/j.chemosphere.2019.07.065.

Rahman, A., Sarkar, A., Yadav, O.P., Achari, G., Slobodnik, J., 2020. Potential human health risks due to environmental exposure to microplastics and knowledge gaps: a scoping review. Sci. Total Environ., 143872. https://doi.org/10.1016/j.scitotenv.2020.143872.

Rajala, K., Grönfors, O., Hesampour, M., Mikola, A., 2020. Removal of microplastics from secondary wastewater treatment plant effluent by coagulation/flocculation with iron, aluminium and polyamine-based chemicals. Water Res. 183, 116045. https://doi.org/10.1016/j.watres.2020.116045.

Raju, S., Carbery, M., Kuttykattil, A., Senthirajah, K., Lundmark, A., Rogers, Z., Palanisami, T., 2020. Improved methodology to determine the fate and transport of microplastics in a secondary wastewater treatment plant. Water Res., 115549. https://doi.org/10.1016/j.watres.2020.115549.

Razeghi, N., Hamidian, A.H., Wu, C., Zhang, Y., Yang, M., 2021. Scientific studies on microplastics pollution in Iran: an in-depth review of the published articles. Mar. Pollut. Bull. 162, 111901. https://doi.org/10.1016/j.marpolbul.2020.111901.

Rist, S., Vianello, A., Winding, M.H.S., Nielsen, T.G., Almeda, R., Torres, R.R., Vollertsen, J., 2020. Quantification of plankton-sized microplastics in a productive coastal arctic marine ecosystem. Environ. Pollut. 115248. https://doi.org/10.1016/j.envpol.2020.115248.

Rochman, C.M., 2018. Microplastics research—from sink to source. Science 360 (6384), 28–29. https://doi.org/10.1126/science.aar7734.

Selvam, S., Manisha, A., Venkatramanan, S., Chung, S.Y., Paramasivam, C.R., Singaraja, C., 2020. Microplastic presence in commercial marine sea salts: a baseline study along Tuticorin Coastal salt pan stations, Gulf of Mannar, South India. Mar. Pollut. Bull. 150, 110675.

Simon, M., Vianello, A., Vollertsen, J., 2019. Removal of >10 μm microplastic particles from treated wastewater by a disc filter. Water 11, 1935. https://doi.org/10.3390/w11091935.

B. Applications of membrane technology for water and wastewater treatment

Statista, 2019. Distribution of Global Plastic Materials Production in 2020, by Region. https://www.statista.com/statistics/281126/global-plastics-production-share-of-various-countries-and-regions/.

Sun, J., Dai, X., Wang, Q., Loodrecht, M., Ni, B., 2019. Microplastics in wastewater treatment plants: detection, occurrences and removal. Water Res. 152, 21–37. https://doi.org/10.1016/j.watres.2018.12.050.

Talvitie, J., Heinonen, M., Pääkkönen, J.P., Vahtera, E., Mikola, A., Setälä, O., Vahala, R., 2015. Do wastewater treatment plants act as a potential point source of microplastics? Preliminary study in the coastal Gulf of Finland, Baltic Sea. Water Sci. Technol. 72 (9), 1495–1504.

Talvitie, J., Mikola, A., Koistinen, A., Setälä, O., 2017. Solutions to microplastic pollution—removal of microplastics from wastewater effluent with advanced wastewater treatment technologies. Water Res. 123, 401–407. https://doi.org/10.1016/j.watres.2017.07.005.

Tang, K.H.D., Hadibarata, T., 2021. Microplastics removal through water treatment plants: its feasibility, efficiency, future prospects and enhancement by proper waste management. Environ. Challenges 5, 100264. https://doi.org/10.1016/j.envc.2021.100264.

Uddin, S., Fowler, S.W., Behbehani, M., 2020. An assessment of microplastics inputs into the aquatic environment from wastewater stream. Mar. Pollut. Bull. 160, 111538. https://doi.org/10.1016/j.marpolbul.2020.111538.

Viršek, M.K., Lovšin, M.N., Koren, Š., Kržan, A., Peterlin, M., 2017. Microplastics as a vector for the transport of the bacterial fish pathogen species Aeromonas salmonicida. Mar. Pollut. Bull. 125 (1-2), 301–309. https://doi.org/10.1016/j.marpolbul.2017.08.024.

Waring, R.H., Harris, R.M., Mitchell, S.C., 2018. Plastic contamination of the food chain: a threat to human health? Maturitas 115, 64–68. https://doi.org/10.1016/j.maturitas.2018.0.

Wu, J., Lai, M., Zhang, Y., Li, J., Zhou, H., Jiang, R., Zhang, C., 2020. Microplastics in the digestive tracts of commercial fish from the marine ranching in east China sea, China. Case Studies Chem. Environ. Eng. 2, 100066. https://doi.org/10.1016/j.cscee.2020.100066.

Xu, S., Ma, J., Ji, R., Pan, K., Miao, A.-J., 2020. Microplastics in aquatic environments: occurrence, accumulation, and biological effects. Sci. Total Environ. 703, 134699. https://doi.org/10.1016/j.scitotenv.2019.1.

Yang, L., Zhang, Y., Kang, S., Wang, Z., Wu, C., 2021. Microplastics in freshwater sediment: a review on methods, occurrence, and source. Sci. Total Environ. 754, 141948. https://doi.org/10.1016/j.scitotenv.2020.141948.

Zeng, E., 2018. Microplastic Contamination in Aquatic Environments: An Emerging Matter of Environmental Urgency. Elsevier, ISBN: 978-0-12-813747-5, https://doi.org/10.1016/C2016-0-04784-8.

Zhang, Y., Jiang, H., Bian, K., Wang, H., Wang, C., 2021. Is froth flotation a potential scheme for microplastics removal? Analysis on flotation kinetics and surface characteristics. Sci. Total Environ. 792, 148345.

Zhang, Y., Wolosker, M.B., Zhao, Y., Ren, H., Lemos, B., 2020. Exposure to microplastics cause gut damage, locomotor dysfunction, epigenetic silencing, and aggravate cadmium (Cd) toxicity in Drosophila. Sci. Total Environ. 744, 140979. https://doi.org/10.1016/j.scitotenv.2020.140979.

Ziajahromi, S., Neale, P.A., Rintoul, L., Leusch, F.D.L., 2017. Wastewater treatment plants as a pathway for microplastics: development of a new approach to sample wastewater-based microplastics. Water Res. 112, 93–99. https://doi.org/10.1016/j.watres.2017.01.042.

Ziajahromi, S., Neale, P.A., Silveira, I.T., Chua, A., Leusch, F.D.L., 2021. An audit of microplastic abundance throughout three Australian wastewater treatment plants. Chemosphere, 128294.

Zurier, H.S., Goddard, J.M., 2020. Biodegradation of microplastics in food and agriculture. Curr. Opin. Food Sci. https://doi.org/10.1016/j.cofs.2020.09.001.

10

Membrane bioreactor for municipal solid waste leachate treatment and organic micropollutant removals

Chart Chiemchaisri, Wilai Chiemchaisri, Varinthorn Boonyaroj, Anekpracha Kaewmanee, Chayanid Witthayaphirom, and Ngech Horng Heang

Department of Environmental Engineering, Faculty of Engineering, Kasetsart University, Bangkok, Thailand

1. Introduction

Municipal solid wastes (MSW) are commonly disposed of in land either at open dumps or landfills where solid waste leachate is produced from excess water drained out from their disposal area. This leachate generally contains organic matter, inorganic macrocomponents such as ammonium nitrogen, toxic compounds such as heavy metals, xenobiotic organic compounds such as chlorinated organics, and plasticizers (Kjeldsen et al., 2002). Its composition can be widely varied depending on several factors such as waste compositions, age of wastes, climate conditions, landfill designs, and operational practices. Treatment of solid waste leachate is an obligation to control pollution prior to its discharge to public waterways because it can pose high risk in polluting surface and groundwater sources (Derviševic et al., 2016). In order to meet effluent standards for direct discharge, several treatment technologies are being proposed for removing biochemical oxygen demand (BOD), chemical oxygen demand (COD), and ammonia nitrogen (NH_3), as well as other treatment technologies such as biological treatment, physical/chemical treatment or a combination of them (Gao et al., 2015). Among them, conventional processes such as activated sludge have been applied to the treatment of landfill leachate but their constraints in solid–liquid separation using gravity sedimentation as well as long treatment time requirement

Copyright © 2023 Elsevier Inc. All rights reserved.

and large excess sludge production were experienced and therefore require more robust treatment technologies (Renou et al., 2008). One of the emerging novel technologies applied to landfill leachate treatment is the membrane bioreactor (MBR) in which biological treatment is integrated with membrane filtration (Torretta et al., 2017). The MBR has several advantages over conventional suspended growth biological processes including more compact reactor size with high biomass concentration, producing high-quality effluent, applicable to landfill leachate of different degrees of stabilization, and less excess sludge production (Ahmed and Lan, 2012). Recently, the full-scale application of MBR to leachate treatment is increasing, especially in China (Zhang et al., 2020). From the technological viewpoint, the use of membrane separation instead of gravity sedimentation allows MBRs to operate with independent control of solid retention time (SRT) and hydraulic retention time (HRT), regardless of sludge settleability. Therefore, MBRs operated with prolonged SRT could reduce the amount of excess sludge to be wasted from the system. This unique feature would benefit the application of MBR to landfill leachate containing high organic substances as sludge discharge from the system would be seldom required during its operation as a core treatment technology at remote landfill sites.

2. Solid waste leachate characteristics

Solid waste leachate can be produced at several stages along the MSW management pathway. Fresh solid waste leachate is produced at solid waste collection vehicles as well as solid waste transfer stations where it serves as temporary storage facilities after MSW is collected from waste sources and prior to its transport to final disposal site. At those facilities, MSW is compressed to reduce its volume and excess liquid is drained as fresh leachate from the waste. After the MSW is delivered to the final disposal site, it is placed into the landfill where further compaction takes place. After its final placing, solid waste leachate is still produced from its excess moisture and intrusion of rainwater into the waste layer. The produced leachate from the landfill has its quantity and characteristics varied depending on the amount of infiltrated water and waste decomposition stage, which can be considered as young, moderate, or partially stabilized, and old or well-stabilized leachate.

Table 1 shows some characteristics of fresh and partially stabilized leachate. The leachate produced from fresh MSW is highly contaminated with organic compounds, mainly organic acids, thus leading to rapid microbial decomposition activities. It is also containing high suspended solids, nitrogen, inorganic salts, and other chemicals constituted in the original waste. The characteristics of fresh MSW can vary widely depending on several factors such as waste composition and its original moisture, degree of waste compaction, and local climate. They are presented as acidic liquid containing high suspended solids (SS), dark color, and foul odor. Fresh MSW leachate has a high degree of organic biodegradability indicated in terms of BOD/COD as it contains a high percentage of low molecular weight (MW) volatile fatty acids (VFA). Their characteristics make it biologically effective for the treatment processes preceded by anaerobic units and followed by aerobic units.

For partially stabilized leachate, its organic concentrations are weaker than those of fresh MSW as it composes leaching products in the transition stage of organic biodegradation between acidogenesis and methanogenesis. As the degree of its stabilization increases, the

TABLE 1 Characteristics of fresh and partially stabilized MSW leachate.

Parameter (unit)	Fresh leachate						Partially stabilized leachate			
	Ye et al. (2011)	Liu et al. (2011)	Zhao et al. (2013)	Zhang et al. (2015)	Kaewmanee et al. (2016)	Nuansawan et al. (2016)	Chiemchaisri et al. (2011a)	Chiemchaisri et al. (2011b)	Zin et al. (2014)	Heang et al. (2020)
pH (−)	3.9–6.4	3.8–6.3	2.7–5.0	6.5–6.7	5.5±0.6	2.9–3.7	8.0–9.4	6.0–8.6	7.9–8.5	6.9±0.9
BOD (g/L)	39.3–46.5	–	–	–	12.2±1.7	41.3–72.5	0.9–3.6	0.05–1.19	0.05–0.28	1.48±0.45
COD (g/L)	70.4–75.5	51.2–71.3	12.5–157.2	60.1–64.7	20.7±5.2	72.4–91.2	2.0–7.3	1.16–1.75	0.47–1.26	2.87±0.91
SS (g/L)	–	15.9–33.7*	0.8–11.3	56.3–59.1*	1.2±1.0	9.3–13.0	0.3–1.2	0.01–0.02	0.2–1.62	0.89±0.45
NH$_3$-N (g/L)	0.64–0.90	0.23–0.80	0.04–0.78	2.16–2.42	0.20±0.06	2.3–3.7	0.06–1.75	0.10–0.18	0.3–1.62	0.05±0.01
EC (dS/m)	–	–	0.8–13.8	–	–	–	14.0–30.6	–	–	–
VFA (g/L)	4.8–15.9	–	–	25.2–25.7	–	–	–	–	–	1.3±0.5

–: Data not available *: measured as total solids (TS).

stabilized leachate become more alkaline and contains more complex organic compounds with a low degree of biodegradability. Nevertheless, some remaining biodegradable organic compounds in partially stabilized leachate still make it possible to treat by biological processes (Chiemchaisri et al., 2011a,b; Bernat et al., 2021).

3. Emerging organic micropollutants in leachate

Due to ineffective upstream segregation of waste components in many countries, the collected MSW transferred to intermediate storage facilities and landfill disposal is commonly contaminated with various toxic organic pollutants. Typical groups of xenobiotic organic compounds (XOCs) found in MSW and its leachates are aromatic hydrocarbons, halogenated hydrocarbons, phenols, alkylphenols, pesticides, phthalates, aromatic sulfonates, phosphonates, and other miscellaneous compounds (Slack et al., 2005). XOCs originating from household or industrial chemicals usually present in relatively low concentrations (usually <1 mg/L of individual compounds). The more frequently detected compounds are mono-aromatic hydrocarbons (benzene, toluene, ethylbenzene, and xylenes) and halogenated hydrocarbons such as tetrachloroethylene and trichloroethylene. Among them, phenol and phthalate compounds are considered of pollutants high priority of concern and normally their removals in conventional landfill leachate treatment processes are limited. The major phenols and phthalates observed in landfill leachate are bisphenol A (BPA), 2,6-di-tert-butylphenol (BHT), di-(2-hexylethyl)-phthalate (DEHP), di-ethyl-phthalate (DEP), and di-butyl-phthalate (DBP). Additionally, contamination of large number of pharmaceuticals and personal care products (PPCPs) including antibiotic compounds is also reported (Peng et al., 2014; Sui et al., 2017). These PPCPs are mainly originated from unused pharmaceuticals codisposed with household wastes. They enter into the natural environment mainly through leachate formed at solid waste transfer stations, landfills, and wastewater treatment plants. Among them, antihistamines, antiepileptics, hormones, and antibiotics were the most common pharmaceuticals found at landfill sites (Bound and Voulvoulis, 2005). They can transport through seepage and leakage, and are detected even at the base of landfill liners (Schwarzbauer et al., 2002; Slack et al., 2007). Table 2 presents the reported concentrations of selected XOCs detected in solid waste leachate.

TABLE 2   Xenobiotic organic compounds detected in solid waste leachate.

Compounds	Abbreviation	Country	Concentration (µg/L)	Reference
Phenols				
Bisphenol A	BPA	UK	200–240	Slack et al. (2005)
		Norway	0.7–200	Morin et al. (2015)
		Japan	26–8400	Urase and Miyashita (2003)
		Thailand	75.4 ± 7.2	Boonyaroj et al. (2012)
		Sweden	4–136	Paxéus (2000)

TABLE 2 Xenobiotic organic compounds detected in solid waste leachate—cont'd

Compounds	Abbreviation	Country	Concentration (µg/L)	Reference
2,6-di-tert-butylphenol	BHT	Sweden	2–3	Paxéus (2000)
		Thailand	173 ± 158 10,750 ± 608	Boonnorat et al. (2014a) Boonyaroj et al. (2012)
Phthalates				
Di-ethyl-phthalate	DEP	UK	0.1–660	Slack et al. (2005)
		Sweden	ND-33	Jonsson et al. (2003)
		Thailand	12.5 ± 1.0 130 ± 121	Boonyaroj et al. (2012) Boonnorat et al. (2014a)
Di-butyl-phthalate	DBP	UK	0.1–70	Slack et al. (2005)
		Thailand	275 ± 230	Boonnorat et al. (2014a)
Di-(2-ethylhexyl)-phthalate	DEHP	UK	0.6–236	Slack et al. (2005)
		Sweden	ND-460	Jonsson et al. (2003)
		Thailand	276 ± 293	Boonnorat et al. (2014a)
PPCPs (antibiotics)				
Trimethoprim	TMP	China	ND-8.1	Sui et al. (2017)
		Thailand	0.4–80	Kaewmanee et al. (2019)
Sulfamethazine	SMZ	China	0.73–2.39	Sui et al. (2017)
		Singapore	0.06–0.44	Yi et al. (2017)
Sulfamethoxazole	SMX	China	ND-2.33	Sui et al. (2017)
		Thailand	0.01–0.20	Kaewmanee et al. (2019)
Carbamazepine	CBZ	China	2.12–6.27	Sui et al. (2017)
Ciprofloxacin	CIP	Thailand	1.2–4.0	Kaewmanee et al. (2019)
Norfloxacin	NOR	Thailand	0.08–2.0	Kaewmanee et al. (2019)
Lincomycin	LIN	Thailand	0.10–1.0	Kaewmanee et al. (2019)
		Singapore	ND-0.02	Yi et al. (2017)

ND: Not detected.

4. Application of MBRs for MSW leachate treatment

Solid waste leachate constitutes a very complex mixture and may contain a variety of toxic organic contaminants. There is an urgent need to treat solid waste leachate in terms of major pollutants including organic substances and nitrogen to minimize its impact on the natural environment upon its discharge. For fresh leachate, biological techniques can yield a reasonable treatment performance with respect to BOD, COD, SS, and NH_3. On the other hand, a combination of biological and physicochemical treatment was found to be a suitable treatment method for partially stabilized (less biodegradable) leachate, in order to remove organic refractory substances. In recent years, with more stringent discharge standards adopted in many countries and aging of landfill sites with more stabilized leachates, conventional biological or simple physicochemical treatments are not sufficient to reach the level of purification needed to fully mitigate the negative impact of solid waste leachate on the environment.

Earlier development of the MBR process for landfill leachate treatment (Lubbecke et al., 1995) has demonstrated successful application of MBR operated under high biomass concentrations to reduce COD in feeding leachate (2700–4300 mg/L) by 65%–80% at a HRT operation of 15–25 h. The operating pressures of membranes were 2.5–4.5 bars at a permeate flux of $15 L/m^2 h$ for nanofiltration (NF) and $40 L/m^2 h$ for ultrafiltration (UF) membranes. A later trial (Jensen et al., 2001) has confirmed the feasibility of its operation in the treatment of landfill leachate at HRT of 2.7 days achieving more than 90% COD removal efficiency.

Results from previous studies indicated that MBR processes have great potential with respect to biomass retention and their treatment efficiency. The incorporation of membranes in the system retains active microbial population and produces a high-quality effluent. The MBR processes have potential capacities to handle the treatment under higher organic loading rates and have elevated their maximum treatment capacities. This was made possible with good control of microbial population in the reactor provided that there was negligible biomass loss through the effluent during membrane filtration. When the MBRs were applied to leachate treatment, organic removal efficiencies were found to be between 80% and 90% for fresh leachate with COD > 10,000 mg/L, 60%–65% for partially stabilized leachate with COD ranging from 5000 to 7000 mg/L and 20%–50% in the case of stabilized leachate with COD <2500 mg/L. Nevertheless, the application of MBRs as single-stage treatment unit may not produce the effluent which meet directly regulatory discharge limits as their capabilities were still limited in terms of removals of chlorides, sulfate, ammonia nitrogen, and refractory organic compounds.

The removals of organic micropollutants in MBRs applied to the treatment of landfill leachate were also investigated. High effectiveness of MBRs in removing xenobiotic substances (nonylphenol, BPA) in landfill leachate of more than 80% was reported (Wintgens et al., 2003). The combination of anaerobic MBR technology was later examined for its removals of organic micropollutants in the treatment of landfill leachate (Xu et al., 2008) and high removals of BOD (99%) and COD (89%) whereas 94% of organochlorine (OCP) > 77% of 4-nonylphenol (4-NP) > 59% of polycyclic aromatic hydrocarbon (PAH) removals were simultaneously achieved. Another study (Matosić et al., 2008) reported a wide range of pharmaceutical compound removals from landfill leachate in the MBR. Less removals (16%) for propylphenazone, moderate for vitamin C synthesis reaching 30% for diacetone sorbose (DAS), and higher (69%) for diacetone alpha-keto-gulonic acid (DAG) were observed.

As the removal of emerging PPCPs is important for the prevention of aquatic toxicity and for enabling safe water reuse, anoxic-aerobic MBR possesses characteristics that were found beneficial for PPCP removals. The process is principally designed for nitrogen removal and recent studies have demonstrated a close relationship between nitrogen and PPCP removals (Fernandez-Fontaina et al., 2012). The biodegradation of the recalcitrant PPCP and phenolic compounds has been demonstrated to occur only under stable nitrifying conditions (Helbling et al., 2012; Boonyaroj et al., 2017) and better performance of PPCP removal by aerobic

TABLE 3   PPCP removals in MBR operated under anaerobic/anoxic and aerobic conditions.

Process condition	PPCPs	Removal performance	Reference
An/A/O-MBR	Estrogens	>90% removal	Joss et al. (2004)
A/O-MBR	8 PPCPs	0 to >90% removal	Clara et al. (2005)
An/A/O-MBR	PPCPs	60%–99% removal	Zuehlke et al. (2006)
An/A/O-MBR	Sulfonamide, macrolides and trimethoprim	80% for sulfonamide 50% for macrolide and trimethoprim up to 90% for trimethoprim at longer SRT	Göbel et al. (2007)
A/O-MBR	EDCs	0–91% removal	Hu et al. (2007)
(An/A)/O-MBR	Bisphenol A	74%–78% removal	Kim et al. (2009)
An/A/O-MBR	19 EDCs/ PPCPs	EDCs >70% most of PPCPs: 50%–100%	Xue et al. (2010)

Notes: An/A/O: anaerobic/anoxic/oxic condition, EDCs: Endocrine disrupting compounds.

nitrifying bioreactor was achieved. Meanwhile, the combination of anoxic and aerobic bioreactors creates multiple microniches of different redox conditions (i.e., different DO levels). Better biodegradation of recalcitrant PPCPs under low DO condition or under alternate exposure to different redox conditions was also noticed (Stadler et al., 2015). For example, carbamazepine and sulfamethoxazole underwent better biodegradation under a range of redox conditions, particularly under low dissolved oxygen (DO) condition (Hai et al., 2011). Table 3 summarizes reported performance of MBR operated under alternate anaerobic/anoxic and aerobic conditions in terms of PPCP removals.

Furthermore, the SRT condition of the MBR also affects PPCP removals as longer SRT helped in enhancing their removals (Göbel et al., 2007). Thus, an operation of MBR under long SRT condition or minimization of excess sludge wastage could improve its capabilities in removing recalcitrant pollutants.

5. Development of two-stage MBR for MSW leachate treatment

In order to approach a no-excess sludge discharge condition, the concept of sludge allocation between an anoxic reactor equipped with an inclined plate and an aerobic membrane reactor proposed by Xing et al. (2006) was proposed. The two-stage MBR (Fig. 1) comprising of an anaerobic or anoxic reactor in which an inclined tube module (0.15 m square tube, 60° inclination, 0.45 m depth) was installed for sludge separation followed by an aerobic reactor with a submerged hollow-fiber membrane module for solid–liquid separation. Internal recirculation of sludge from the aerobic reactor back to its preceding reactor allowed the system to keep the majority of the produced sludge in the storage zone below the tube settler of

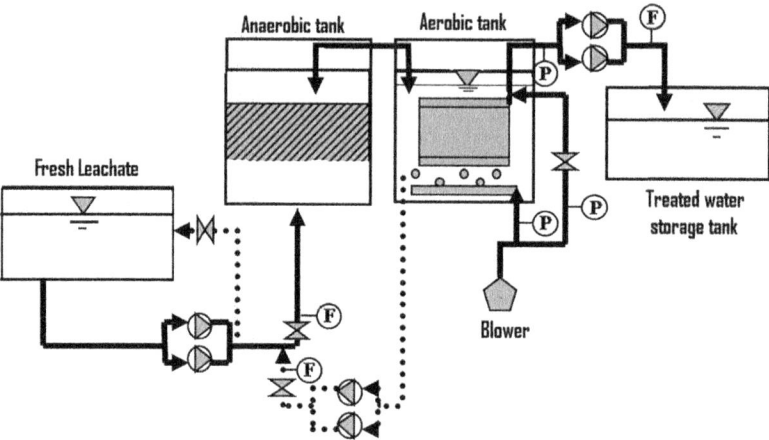

FIG. 1   Schematic of two-stage MBR.

the anaerobic tank while maintaining a lower mixed liquor suspended solid concentration in the aerobic reactor to limit membrane fouling. The MBR was applied to the treatment of leachate containing different organic concentrations in terms of BOD, COD, NH_3, and total Kjeldahl nitrogen (TKN) while having different degrees of stabilization in terms of BOD/COD, demonstrating its capabilities to sustain its long-term operation without sludge wastage (Chiemchaisri et al., 2011a; Kaewmanee et al., 2016; Nuansawan et al., 2016; Heang et al., 2020).

When the two-stage MBR was applied to the treatment of medium strength or partially stabilized leachate having BOD and COD concentrations of 1.0–2.5 g/L and 2.0–5.0 g/L [9], the HRTs in anoxic and aerobic reactors were maintained at 15 and 12 h, yielding food (BOD basis) to microorganisms (F/M) ratio of 0.1–0.2 and 0.3–0.45 day^{-1}, respectively, in those reactors. As a result, high BOD and COD removals of >99% and 60%–70% in which 30%–40% took place in the anoxic reactor could be achieved. High organic removals were explained by the fact that hydrophilic compounds presented in landfill leachate were efficiently removed with the biological part of the MBR process whereas partial removals of recalcitrant humic substances were achieved through adsorption to sludge particles and membrane retention (Sanguanpak et al., 2013). During the operation, biomass in the aerobic reactor in terms of mixed liquor suspended solids (MLSS) was kept lower than 10 g/L so that severe membrane fouling was not observed. The transmembrane pressure (TMP) of the membrane modules was maintained relatively constant during long-term operation of almost 300 days at a permeate flux operation of 0.08 m^3/m^2/d. The majority of produced sludge in the aerobic reactor was recirculated to the anoxic reactor at a rate of 100% of influent flow rate where the inclined tube helped in separating 95% of solid concentrations prior to their overflow back to the aerobic reactor. The stored biomass in the anoxic reactor was highly compacted and was partly digested; therefore, the storage capacity of 25% of the anoxic reactor volume was sufficient to sustain the system operation without sludge wastage during long-term operation.

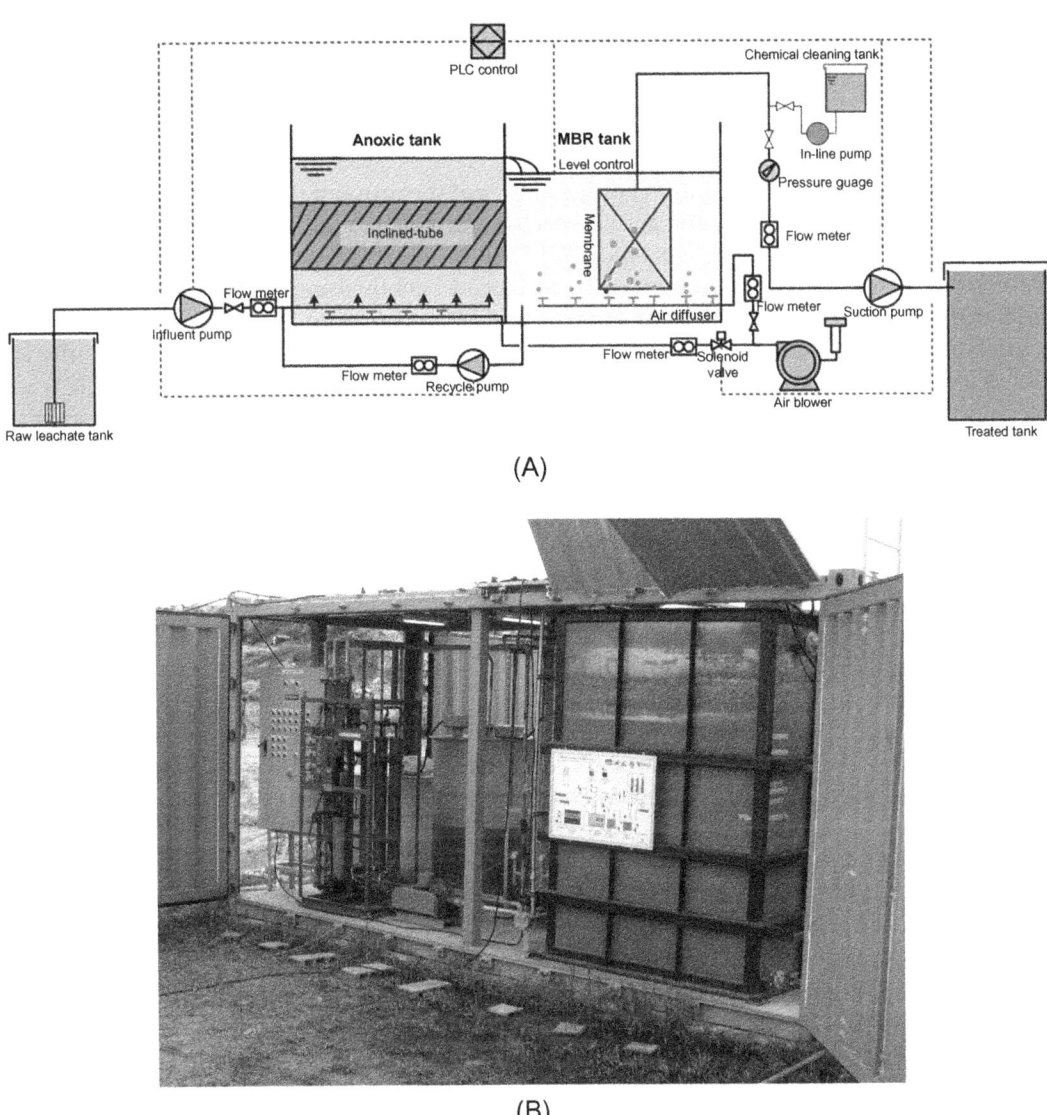

(A)

(B)

FIG. 2 Configuration (A) and image (B) of two-stage MBR installed at the MSW disposal site.

A field-scale MBR unit (Fig. 2) was further developed for the treatment of fresh solid waste leachate containing high BOD and COD concentrations up to 15 and 30 g/L, respectively. It comprises a 4.0 m^3 of the anoxic tank with an effective volume of 3.0 m^3 connected to an equal effective volume of an aerobic tank. An inclined tube module (0.15 m tube size, 60° inclination) of 0.45 m depth was installed inside the anoxic tank. In an aerobic tank, two hollow fiber membrane modules with a total surface area of 12 m^2 and an average pore size of 0.4 μm were installed for solid–liquid separation. The MBR was operated at a HRT of 12–20 h in each reactor; so, the organic loading rates (OLR) were varied between 4 and 30 kg BOD/m^3/d over 550 days of operation (Kaewmanee et al., 2016). High BOD (97%) and COD (94%) removals were achieved during which optimal OLR was determined as 18 kgBOD/m^3/d. The organic removals in the anoxic reactor accounted for 30%–50% in the total removals. Under this high organic loading operation, MLSS in the aerobic reactor was maintained mostly between 10 and 15 g/L while the recirculation sludge at 67% of the influent flow rate allowed the solids to accumulate at the bottom of the anoxic reactor and reached almost 25 g/L during its long-term operation without sludge wastage. The TMP of the membrane module could be effectively kept below 40 kPa using chemically enhanced backflushing by 0.05% NaOCl and 2% citric acid at every 7–10 days to limit organic fouling and scaling of the membrane.

Effective nitrogen removals were also achieved in two-stage MBR applied to partially stabilized and fresh leachate. Their removal efficiencies were observed at more than 80% and 90% for partially stabilized and fresh leachate, respectively, as the influent nitrogen concentrations were varied between 200 and 2000 mg/L, being higher for partially stabilized leachate. The removals of nitrogenous compounds in the system could be mainly associated with ammonia volatilization, biomass assimilation, and biological nitrification and denitrification reactions. During the treatment of fresh leachate, nitrogen uptake for biomass assimilation would predominate whereas biological reactions were responsible for nitrogen removals during partially stabilized leachate treatment (Chiemchaisri et al., 2011a; Kaewmanee et al., 2016). It was anticipated that nitrifying and denitrifying microorganisms could be proliferated in the reactors operated under sludge recirculation and long SRT condition.

The nitrogen removals in the aerobic MBR treating landfill leachate were also found affected by influent characteristics. With the COD/N of 10–30 found in partially stabilized and fresh leachate, high nitrogen removals could be achieved; but, inhibition of biological activities was observed when stabilized leachate with COD/N of 1 was applied (Chiemchaisri et al., 2011b).

Microbial community analyses in two-stage MBR sludge revealed the presence of *Pseudomonas* sp., *Burkholderia* sp., *Nitrosomonas* sp., *Enterobacter* sp., *Sphingobacterium* sp., and *Rhodopseudomonas* sp. during its long-term operation treating solid waste leachate (Boonnorat et al., 2014b). Gradual development of a microbial consortium capable of biodegrading slowly biodegradable compounds such as phenols and phthalates, which were presented in feeding leachate were observed. For example, they included *D. aromatica* that could biodegrade organic compounds with aromatic ring structures, *B. subtilis* that had been reported to biodegrade phthalate, and *K. pneumoniae* that could degrade phenols as well as phthalates. The study has shown that MBR operated without intentional sludge wastage resulting in the development of unique environmental condition for microbial growth, which enhanced its capacities in treating recalcitrant compounds present in solid waste leachate (Boonnorat et al., 2014a).

6. Enhancement of two-stage MBR performance through bioaugmentation

An operation of a two-stage MBR for the treatment of highly concentrated leachate (BOD of 20 g/L, COD of 40 g/L, and TKN of 2.5 g/L) could also yield significant greenhouse gas emissions in terms of methane (CH$_4$) and nitrous oxide (N$_2$O) if longer HRT conditions (2–5 days) are employed (Nuansawan et al., 2016; Chiemchaisri et al., 2020). High CH$_4$ emission would occur from the anaerobic reactor especially when sludge recirculation was not practiced. Meanwhile, N_2O emissions were also observed in both anaerobic and aerobic reactors. The introduction of sludge recirculation helped in reducing the emissions of both gases through the increase in hydraulic flow and decrease in HRT in the reactor as well as reduction in the abundance of CH_4- and N_2O-producing microorganisms in microbial sludge.

The challenging task in developing a highly compact leachate treatment system for high-strength leachate is to shorten the treatment time (or HRT) without compromising its treatment efficiencies. In terms of organic removals, an MBR system operated without excess sludge wastage could handle high input of biodegradable organic substances by allowing an increase in biomass (or MLSS) concentrations in response to the OLR and thus maintain its organic removal efficiencies. Nevertheless, an increase in MLSS concentrations in the aerobic MBR would increase sludge viscosity and reduce oxygen transfer to microorganisms, especially autotrophic nitrifying organisms. It was found that the nitrification reaction was significantly inhibited when the MLSS concentration in the aerobic MBR increased to 18 g/L (Prasertkulsak et al., 2016). To overcome such a limitation, integration of heterotrophic nitrifiers into the treatment system has been proposed.

Some bacterial species, e.g., *Agrobacterium* sp. LAD9, *Achromobacter* sp. GAD3, and *Comamonas* sp. GAD4 have been found capable of removing ammonia through heterotrophic nitrification-aerobic denitrification reactions during landfill leachate treatment (Chen and Ni, 2011). Meanwhile, *Alcaligenes faecalis* no. 4 has also been reported to improve ammonium contained in wastewater at high concentrations under mixed culture condition (Joo et al., 2005a, 2007). This microorganism was therefore introduced into two-stage MBR to enhance nitrogen removals during the treatment of partially stabilized leachate containing high ammonium concentrations. *A. faecalis* no. 4 is capable of converting 50% of removed ammonium nitrogen to intracellular nitrogen and 40%–50% of removed ammonium to denitrified products whereas 90% of denitrified outcome was N_2 with small N_2O production (Joo et al., 2005b). These specific bacteria provided several advantages during their application in the system such as (1) operational simplicity, where nitrification and denitrification can take place simultaneously in the aerobic MBR stage, (2) shorter acclimatization period due to its higher growth rate, and (3) lesser buffer quantity needed due to the alkalinity generated during denitrification can partly compensate for the alkalinity consumption in nitrification. They have also been demonstrated to remove more than 2000 mg/L of ammonium nitrogen in wastewaters within 24 h (Joo et al., 2005a).

The introduction of *A. faecalis* no. 4 in the two-stage MBR was investigated during the treatment of fresh BOD = 22 g/L, COD = 29 g/L, and TKN = 0.4 g/L and mixture of fresh and partially stabilized leachate BOD = 12 g/L, COD = 24 g/L, and TKN = 0.3 g/L. High organic (>95%) and nitrogen (90%) removal efficiencies were achieved when the system was supplied with partially stabilized leachate mixture even under higher loading conditions (HRT of 1 day) when *A. faecalis* no. 4 was bio-augmented into the aerobic MBR. Their pollutant removal efficiencies were maintained at the same level as those during the operation without *A. faecalis* no. 4 addition even though leachate loading rate was increased by 2.5 times (Nuansawan et al., 2016).

Under the presence of *A. faecalis* no. 4, the operation of two-stage MBR with sludge recirculation could reduce CH_4 and N_2O emissions by 50% as compared to its operation without sludge recirculation. An operation under short HRT of 1 day also yielded high organic carbon and nitrogen removals while emitting lower CH_4 and N_2O emission when operated with sludge recirculation (Nuansawan et al., 2018). Introduction of *A. faecalis* no. 4 reduced CH_4 and N_2O emissions in both reactors as it became the predominant microorganism under an elevated pH condition (pH = 8.9–9.2). *A. faecalis* no. 4 could suppress methanogenic activities in the anaerobic reactor during sludge recirculation and converted a majority of nitrogen into its cell mass, thus reducing N_2O production through a heterotrophic nitrification and aerobic denitrification pathway. The presence of high volatile fatty acids (VFA) in fresh leachate was the key factor promoting the growth and sustaining *A. faecalis* no. 4 population in the system.

7. Overall performance of two-stage MBR treating solid waste leachate

The aforementioned studies have demonstrated successful application of two-stage MBR to the treatment of landfill and solid waste leachate having different organic concentrations and stabilization degrees. Table 4 summarized the treatment performance in terms of organic and nitrogen removals in two-stage MBR and its optimum operating conditions.

TABLE 4   Two-stage MBR performance treating landfill and fresh solid waste leachate.

Leachate	HRT (d)	Operating condition	Removal (%)				Reference
			BOD	COD	NH_3	TKN	
Fresh	2.5	w/o recirculation	91	87	84	85	Heang et al. (2020)
	2.5	With recirculation	88	86	82	85	
Fresh	2.5	With recirculation	91	85	71	74	Nuansawan et al. (2018)
	2.5	With recirculation + *A. faecalis* no. 4	98	90	100	94	
	1	With recirculation + *A. faecalis* no. 4	93	76	86	87	
Fresh	1	With recirculation	97	94	96	95	Kaewmanee et al. (2016)
Mixed (fresh + partially stabilized)	1	With recirculation	86	79	81	80	Heang et al. (2020)
	1	With recirculation + *A. faecalis* no. 4	91	87	90	94	
Partially stabilized	4	With recirculation	>99	97	99	94	Chiemchaisri et al. (2020)
Partially stabilized	1	With recirculation	>99	72	82	83	Chiemchaisri et al. (2011a)
Partially stabilized	1	With recirculation	>99	95	95	75	Boonnorat et al. (2014a)

TABLE 5 Phenolic compound and phthalates removals during two-stage MBR treatment.

Leachate	HRT (d)	Operating condition	Removal (%)				Reference
			BHT	DEP	DBP	DEHP	
Partially stabilized	1	With recirculation	83	81	87	96	Boonyaroj et al. (2012)
	1	With recirculation	>99	98	97	96	Boonnorat et al. (2014a)
Mixed (fresh + partially stabilized)	2.5	w/o recirculation	81	83	82	80	
	2.5	With recirculation	77	78	76	74	
	1	With recirculation	66	70	68	63	
	1	With recirculation & A. faecalis no. 4	71	74	73	70	

ND: not determined.

For organic micropollutants, high removals (80%–90%) of most phenolic compounds and phthalates were achieved in two-stage MBR operated under long SRT condition (Boonyaroj et al., 2012). Those compounds were removed mostly through microbial degradation with the exception of hydrophobic compounds, i.e., DEHP, which was removed through adsorption onto sludge particles and subsequently retained by a microfiltration membrane. The long sludge age (SRT) condition maintained in two-stage MBR helped the development of a phenolic and phthalate-degrading bacterial consortium to improve the biodegradation of those compounds (Boonnorat et al., 2014b). Furthermore, their removals in the aerobic treatment stage were also improved in the mixed liquor containing enriched nitrifying sludge cultivated under long sludge age condition through cometabolism pathways including heterotrophic and nitrifying activities (Boonyaroj et al., 2017). Moreover, high MLSS concentrations maintained in the MBR increased the initial adsorption of moderate recalcitrant hydrophobic compounds to acclimatized sludge, yielding longer retention in the mixed liquor for their subsequent biodegradative removals.

Table 5 shows the removals of selected micropollutants in the two-stage MBR operated under different conditions. Comparing their removal efficiencies, an increase in HRT and introduction of sludge recirculation and Alcaligenes faecalis no. 4 improved the removals of most compounds. High removal efficiencies during the treatment of partially stabilized leachate were also noticed. The biodegradative removals of compounds by microbial sludge involving heterotrophic and ammonia-oxidizing bacteria were improved under appropriate organic and nitrogen loadings, i.e., carbon (C) to nitrogen (N) ratio of 6 and long SRT (90 days) condition (Boonnorat et al., 2016).

For the removals of antibiotics in two-stage MBR, laboratory investigation has demonstrated that up to 77% of NOR and 50%–57% of TMP and SMX removals could be achieved when it was operated at a HRT of 12 h in each stage receiving solid waste leachate contaminated mixed antibiotic with a dose of 1 mg/L each. An extension of HRT to 24 h in each stage did not affect antibiotic removals in the system. While high antibiotic dose applied to the system did not have significant impact on organic removals, it had a negative effect on nitrite transformation during a denitrification process and yielded lower antibiotic removal. Nevertheless, supplementation of nitrate solution in an anoxic reactor helped recovering antibiotic removal efficiencies in the system (Kaewmanee et al., 2019).

8. Conclusions and perspectives

Two-stage MBR specifically developed for operation at municipal solid waste sites has demonstrated its capability to remove macro- (BOD, COD, and TKN) and micropollutants (antibiotics, phenolic, PAEs) presented in solid waste and landfill leachates with different degrees of stabilization. The allocation of sludge biomass in anaerobic/anoxic and aerobic reactors of two-stage MBR enabled the system to operate without excess sludge discharge. Long sludge age condition maintained in the system promoted the development of a strong consortium in microbial sludge capable of biodegrading slowly biodegradable organic compounds such as phenolics, phthalates, and antibiotics. The two-stage MBR could therefore be applied remove organic, nitrogenous compounds, and recalcitrant organic micropollutants effectively. Other operating conditions such as sludge recirculation and bioaugmentation of specific heterotrophic nitrifiers (*A. faecalis* no. 4) also helped in improving the treatment efficiencies especially when the system is operated at short hydraulic retention time.

In addition to its implementation as a stand-alone solid waste leachate treatment unit at municipal solid waste transfer or disposal facilities, further application of two-stage MBR can be considered at other medium- to high-strength organic wastewater sources at which their wastewaters contain complex organic pollutants such as hospital or industrial wastewaters. Flexible operation of the system in terms of hydraulic retention time and sludge recirculation allows the treatment unit to adjust its operation to variable hydraulic and organic loading rates of wastewater through its maintenance of appropriate biomass distributions between the reactors. Moreover, its operation under long sludge age condition would minimize excess sludge disposal requirement, which make the system attractive for the wastewater sources facing difficulties in sludge handling.

Despite several anticipated beneficial features of two-stage MBR, there is still some room for improvement to attract wider ranges of applications. Its integration with anaerobic pretreatment including biogas recovery and utilization to reach carbon neutral condition as well as posttreatment to polish the effluent for reuse purposes can be further explored.

References

Ahmed, F.N., Lan, C.Q., 2012. Treatment of landfill leachate using membrane bioreactors: a review. Desalination 287, 41–54.

Bernat, K., Zaborowska, M., Zielińska, M., Wojnowska-Baryla, I., Ignalewski, W., 2021. Biological treatment of leachate from stabilization of biodegradable municipal solid waste in a sequencing batch biofilm reactor. Int. J. Environ. Sci. Technol. 18, 1047–1060.

Boonnorat, J., Chiemchaisri, C., Chiemchaisri, W., Yamamoto, K., 2014a. Removals of phenolic compounds and phthalic acid esters in landfill leachate by microbial sludge of two-stage membrane bioreactor. J. Hazard. Mater. 277, 93–101.

Boonnorat, J., Chiemchaisri, C., Chiemchaisri, W., Yamamoto, K., 2014b. Microbial adaptation to biodegrade toxic organic micro-pollutants in membrane bioreactor using different sludge sources. Bioresour. Technol. 165, 50–59.

Boonnorat, J., Chiemchaisri, C., Chiemchaisri, W., Yamamoto, K., 2016. Kinetics of phenolic and phthalic acid esters biodegradation in membrane bioreactor (MBR) treating municipal landfill leachate. Chemosphere 150, 639–649.

Boonyaroj, V., Chiemchaisri, C., Chiemchaisri, W., Theepharaksapan, S., Yamamoto, K., 2012. Toxic organic micropollutants removal mechanisms in long-term operated membrane bioreactor treating municipal solid waste leachate. Bioresour. Technol. 113, 174–180.

Boonyaroj, V., Chiemchaisri, C., Chiemchaisri, W., Yamamoto, K., 2017. Enhanced biodegradation of phenolic compounds in landfill leachate by enriched nitrifying membrane bioreactor sludge. J. Hazard. Mater. 323, 311–318.

Bound, J.P., Voulvoulis, N., 2005. Household disposal of pharmaceuticals as a pathway for aquatic contamination in the United Kingdom. Environ. Health Perspect. 113, 1705–1711.

Chen, Q., Ni, J., 2011. Heterotrophic nitrification-aerobic denitrification by novel isolated bacteria. J. Ind. Microbiol. Biotechnol. 38, 1305–1310.

Chiemchaisri, C., Chiemchaisri, W., Nindee, P., Chang, C.Y., Yamamoto, K., 2011a. Treatment performance and microbial characteristics in two-stage membrane bioreactor applied to partially stabilized leachate. Water Sci. Technol. 64 (5), 1064–1072.

Chiemchaisri, C., Putthiwara, S., Chanpeng, A., 2011b. Application of biofilm membrane bioreactor and reverse osmosis membrane to the treatment of fresh and partially stabilised leachate. Int. J. Environ. Eng. 3 (3–4), 210–220.

Chiemchaisri, C., Chiemchaisri, W., Manochai, N., 2020. Methane and nitrous oxide emissions from a two-stage membrane bioreactor applied in municipal landfill leachate treatment. J. Mater. Cycles Waste Manage. 22, 365–374.

Clara, M., Strenn, B., Gans, O., Martinez, E., Kreuzinger, N., Kroiss, H., 2005. Removal of selected pharmaceuticals, fragrances and endocrine disrupting compounds in a membrane bioreactor and conventional wastewater treatment plants. Water Res. 39, 4797–4807.

Derviševič, I., Đokić, J., Elezović, N., Milentijević, G., Ćosivić, V., Derviševič, A., 2016. The impact of leachate on the quality of surface and groundwater and proposal of measures for pollution remediation. J. Environ. Prot. 7, 745–759.

Fernandez-Fontaina, E., Omil, F., Lema, J.M., Carballa, M., 2012. Influence of nitrifying conditions on the biodegradation and sorption of emerging micropollutants. Water Res. 46, 5434–5444.

Gao, J., Oloibiri, V., Chys, M., Audenaert, W., Decostere, B., He, Y., Van Langenhove, H., Demeestrere, K., Van Hulle, S.W.H., 2015. The present status of landfill leachate treatment and its development trend from a technological point of view. Rev. Environ. Sci. Biotechnol. 14, 93–122.

Göbel, A., McArdell, C.S., Joss, A., Siegrist, H., Giger, W., 2007. Fate of sulfonamides, macrolides, and trimethoprim in different wastewater treatment technologies. Sci. Total Environ. 372, 361–371.

Hai, F.I., Li, X., Price, W.E., Nghiem, L.D., 2011. Removal of carbamazepine and sulfamethoxazole by MBR under anoxic and aerobic conditions. Bioresour. Technol. 102, 10386–10390.

Heang, N.H., Chiemchaisri, C., Chiemchaisri, W., Shoda, M., 2020. Treatment of municipal landfill leachate at different stabilization states in two-stage membrane bioreactor bioaugmented with *Alcaligenes faecalis* no. 4. Biores. Technol. Rep. 11, 100528.

Helbling, D.E., Johnson, D.R., Honti, M., Fenner, K., 2012. Micropollutant biotransformation kinetics associate with WWTP process parameters and microbial community characteristics. Environ. Sci. Technol. 46, 10579–10588.

Hu, J.Y., Chen, X., Tao, G., Kekred, K., 2007. Fate of endocrine disrupting compounds in membrane bioreactor systems. Environ. Sci. Technol. 41, 4097–4102.

Jensen, J.C., Cameron, D.H., Penny, J.P., 2001. Operating experience with innovative high efficiency membrane treatment of landfill leachate. In: Proceedings of the 6th Annual Symposium of Solid Waste Association of North America (SWANA).

Jonsson, S.J., Ejlertsson, J., Svensson, B.H., 2003. Transformation of phthalates in young landfill cells. Waste Manag. 23, 641–651.

Joo, H.S., Hirai, M., Shoda, M., 2005a. Nitrification and denitrification in high-strength ammonium by *Alcaligenes faecalis*. Biotechnol. Lett. 27, 773–778.

Joo, H.S., Hirai, M., Shoda, M., 2005b. Characteristics of ammonium removal by heterotrophic nitrification-aerobic denitrification by *Alcaligenes faecalis* no. 4. J. Biosci. Bioeng. 100, 184–191.

Joo, H.S., Hirai, M., Shoda, M., 2007. Improvement in ammonium removal efficiency in wastewater treatment by mixed culture of *Alcaligenes faecalis* no. 4 and L1. J. Biosci. Bioeng. 103, 66–73.

Joss, A., Andersen, H., Ternes, T., Richle, P.R., Siegrist, H., 2004. Removal of estrogens in municipal wastewater treatment under aerobic and anaerobic conditions: consequences for plant optimization. Environ. Sci. Technol. 38, 3047–3055.

Kaewmanee, A., Chiemchaisri, W., Chiemchaisri, C., Yamamoto, K., 2016. Treatment performance and membrane fouling characteristics of inclined tube anoxic/aerobic membrane bioreactor applied to municipal solid waste leachate. Desalin. Water Treat. 57, 29201–29211.

Kaewmanee, A., Chiemchaisri, W., Chiemchaisri, C., 2019. Influence of high doses of antibiotics on anoxic-aerobic membrane bioreactor in treating solid waste leachate. Int. Biodeter. Biodegr. 138, 15–22.

Kim, K.P., Ahmed, Z., Ahn, K.H., Paeng, K.J., 2009. Biodegradation of two model estrogenic compounds in a preanoxic/anaerobic nutrient removing membrane bioreactor. Desalination 243, 265–272.

Kjeldsen, P., Barlaz, M.A., Rooker, A.P., Baun, A., Ledin, A., Christensen, T.H., 2002. Present and long-term composition of MSW landfill leachate: a review. Crit. Rev. Environ. Sci. Technol. 32, 297–336.

Liu, J., Hu, J., Zhong, J., Luo, J., Zhao, A., Liu, F., Hong, R., Qian, G., Xu, Z.P., 2011. The effect of calcium on the treatment of fresh leachate in an expanded granular sludge bed bioreactor. Bioresour. Technol. 102, 5466–5472.

Lubbecke, S., Vogelpohl, A., Dewjanin, W., 1995. Wastewater treatment in a biological high-performance system with high biomass concentration. Water Res. 29, 793–802.

Matosić, M., Terzic, S., Jakopović, H.K., Mijatović, I., Ahel, M., 2008. Treatment of a landfill leachate containing compounds of pharmaceutical origin. Water Sci. Technol. 58 (3), 597–602.

Morin, N., Arp, H.P.H., Hale, S.E., 2015. Bisphenol a in solid waste materials, leachate water, and air particles from Norwegian waste-handling facilities: presence and partitioning behavior. Environ. Sci. Technol. 49, 7675–7683.

Nuansawan, N., Boonnorat, J., Chiemchaisri, W., Chiemchaisri, C., 2016. Effect of hydraulic retention time and sludge recirculation on greenhouse gas emission and related microbial communities in two-stage membrane bioreactor treating solid waste leachate. Bioresour. Technol. 210, 35–42.

Nuansawan, N., Chiemchaisri, C., Chiemchaisri, W., Shoda, M., 2018. Treatment of concentrated leachate with low greenhouse gas emission in two-stage membrane bioreactor bio-augmented with *Alcaligenes faecalis* no. 4. J. Air Waste Manage. Assoc. 68, 1378–1390.

Paxéus, N., 2000. Organic compounds in municipal landfill leachates. Water Sci. Technol. 42 (7), 323–332.

Peng, X., Ou, W., Wang, C., Wang, Z., Huang, Q., Jin, J., Tan, J., 2014. Occurrence and ecological potential of pharmaceuticals and personal care products in groundwater and reservoirs in the vicinity of municipal landfills in China. Sci. Total Environ. 490, 889–898.

Prasertkulsak, S., Chiemchaisri, C., Chiemchaisri, W., Itonaga, T., Yamamoto, K., 2016. Removal of pharmaceutical compounds from hospital wastewater in membrane bioreactor operated under short hydraulic retention time. Chemosphere 150, 624–631.

Renou, S., Givaudan, J.G., Poulain, S., Dirassouyan, F., Moulin, P., 2008. Landfill leachate treatment: review and opportunity. J. Hazard. Mater. 150, 468–498.

Sanguanpak, S., Chiemchaisri, C., Chiemchaisri, W., Yamamoto, K., 2013. Removal and transformation of dissolved organic matter (DOM) during the treatment of partially stabilized leachate in membrane bioreactor. Water Sci. Technol. 68 (5), 1091–1099.

Schwarzbauer, J., Sabine, H., Sabine, B., Ralf, L., 2002. Occurrence and alteration of organic contaminants in seepage and leakage water from a waste deposit landfill. Water Res. 36, 2275–2287.

Slack, R.J., Gronow, J.R., Voulvoulis, N., 2005. Household hazardous waste in municipal landfills: contaminants in leachate. Sci. Total Environ. 337, 119–137.

Slack, R.J., Gronow, J.R., Hall, D.H., Voulvoulis, N., 2007. Household hazardous waste disposal to landfill: using LandSim to model leachate migration. Environ. Pollut. 146, 501–509.

Stadler, L.B., Su, L., Moline, C.J., Ernstoff, A.S., Aga, D.S., Love, N.G., 2015. Effect of redox conditions on pharmaceutical loss during biological wastewater treatment using sequencing batch reactors. J. Hazard. Mater. 282, 106–115.

Sui, Q., Zhao, W., Cao, X., Lu, S., Qiu, Z., Gu, X., Yu, G., 2017. Pharmaceuticals and personal care products in the leachates from a typical landfill reservoir of municipal solid waste in Shanghai, China: occurrence and removal by a full-scale membrane bioreactor. J. Hazard. Mater. 323, 99–108.

Torretta, V., Ferronato, N., Katsoyiannis, I.A., Tolkou, A.K., Airoldi, M., 2017. Novel and conventional technologies for landfill leachate treatment: a review. Sustainability 9, 9.

Urase, T., Miyashita, K., 2003. Factors affecting the concentration of bisphenol a in leachates from solid waste disposal sites and its fate in treatment processes. J. Mater. Cycles Waste Manage. 5, 77–82.

Wintgens, T., Gallenkemper, M., Melin, T., 2003. Occurrence and removal of endocrine disrupters in landfill leachate treatment plants. Water Sci. Technol. 48 (3), 127–134.

Xing, C.H., Yamamoto, K., Fukushi, K., 2006. Performance of an inclined-plate membrane bioreactor at zero excess sludge discharge. J. Membr. Sci. 275, 175–186.

Xu, Y., Zhou, Y., Wang, D., Chen, S., Liu, J., Wang, Z., 2008. Occurrence and removal of organic micropollutants in the treatment of landfill leachate by combined anaerobic-membrane bioreactor technology. J. Environ. Sci. (China) 20, 1281–1287.

Xue, W., Wu, C., Xiao, K., Huang, X., Zhou, H., Tsuno, H., Tanaka, H., 2010. Elimination and fate of selected micro-organic pollutants in a full-scale anaerobic—anoxic—aerobic process combined with membrane bioreactor for municipal wastewater reclamation. Water Res. 44, 5999–6010.

Ye, J., Mu, Y., Cheng, X., Sun, D., 2011. Treatment of fresh leachate with high-strength organics and calcium from municipal solid waste incineration plant using UASB reactor. Bioresour. Technol. 102, 5498–5503.

Yi, X., Tran, N.H., Yin, T., He, Y., Gin, K.Y., 2017. Removal of selected PPCPs, EDCs, and antibiotic resistance genes in landfill leachate by a full-scale constructed wetlands system. Water Res. 121, 46–60.

Zhang, W., Zhang, L., Li, A., 2015. Anaerobic co-digestion of food waste with MSW incineration plant fresh leachate: process performance and synergistic effects. Chem. Eng. J. 259, 795–805.

Zhang, J., Xiao, K., Huang, X., 2020. Full-scale MBR applications for leachate treatment in China: practical, technical and economic features. J. Hazard. Mater. 389, 122318.

Zhao, J., Lu, X.Q., Luo, J.H., Liu, J.Y., Xu, Y.F., Zhao, A.H., Liu, F., Tai, J., Qian, G.R., Peng, B., 2013. Characterization of fresh leachate from a refuse transfer station under different seasons. Int. Biodeter. Biodegr. 85, 631–637.

Zin, N.S.M., Aziz, H.A., Adlan, M.N., Ariffin, A., Yusoff, M.S., Dahlan, I., 2014. Treatability study of partially stabilized leachate by composite coagulant (prehydrolyzed iron and tapioca flour). Int. J. Scient. Res. Knowl. 2 (7), 313–319.

Zuehlke, S., Duennbier, U., Lesjean, B., Gnirss, R., Buisson, H., 2006. Long-term comparison of trace organics of removal performances between conventional and membrane activated sludge processes. Water Environ. Res. 78, 2480–2486.

B. Applications of membrane technology for water and wastewater treatment

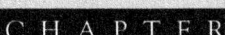

CHAPTER

11

Hybrid membrane bioreactors for wastewater treatment

Shamas Tabraiz[a,b], Muhammad Zeeshan[c,d],
Muhammad Bilal Asif[e], Sidra Iftekhar[f], and Zohaib Abbas[g]

[a]School of Engineering, Newcastle University, Newcastle, United Kingdom [b]Natural and Applied Sciences Section, School of Psychology and Life Sciences, Canterbury Christ Church University, Canterbury, United Kingdom [c]Department of Water Quality Control, Technical University of Berlin, Berlin, Germany [d]German Environment Agency, Section II 3.3, Berlin, Germany [e]Institute of Environmental Engineering & Nano-Technology, Tsinghua-Shenzhen International Graduate School, Tsinghua University, Shenzhen, China [f]Department of Applied Physics, University of Eastern Finland, Kuopio, Finland [g]Department of Environmental Sciences and Engineering, Government College University, Lahore, Pakistan

1. Introduction

Membrane bioreactor (MBR) has emerged as a better alternative to the conventional activated sludge process for wastewater treatment. It is a combination of biological treatment (activated sludge process) and physical separation through the membrane (Meng et al., 2009; Wu et al., 2021). It has overcome many of the demerits of the activated sludge process and offers many merits, including less sludge production, a smaller area requirement by eliminating the secondary clarifier and shorter hydraulic retention time (HRT), higher volumetric loading rates, higher removal efficiencies of organics and recalcitrant, higher solid retention time (SRT), and the potential of higher nitrogen removal efficiencies through simultaneous nitrification and denitrification at longer SRTs (Iorhemen et al., 2016; Strathmann et al., 2011).

MBR technology was introduced in the 1960s. In the earlier days, only ultrafiltration and microfiltration membranes were available on a commercial scale (Schenkelberg and Horton, 2009). However, the commercial use of membranes was limited because of the low membrane flux, low permeability, limited life of the membrane, and its higher cost. In the early 1990s, a

Copyright © 2023 Elsevier Inc. All rights reserved.

new group of membranes evolved that considerably overcame many of the earlier limitations. This led to the commercial use of membranes for wastewater treatment. Although the MBR system outperformed the conventional activated sludge process, fouling is one of the main hindrances in the widespread use of MBR technology (Petropoulos et al., 2021). Detailed mechanisms of fouling have been reported and different strategies have been adopted accordingly to reduce the fouling in MBR (Tabraiz et al., 2017, 2020, 2021a,b). Moreover, there is a lot of room to improve the nutrients and recalcitrance chemical removals in the conventional MBR technology. So different types of hybrid-MBR have been suggested for improvement in this area to make the technology sustainable.

The use of hybrid-MBR has been reported for reduced membrane biofouling, higher biomass concentrations, higher effluent quality, and lower footprints (Leiknes and Ødegaard, 2001; Odegaard, 2000; Zhang et al., 2019c). Many researchers had worked with different biofilm growth media like polyurethane cubes, polystyrene beads, polyethylene carriers (Kaldnes), activated carbon (granular and powdered), polymers like MPE50, and sponge in MBR (Artiga et al., 2008; Belli et al., 2021; Lee et al., 2006; Yang et al., 2006; Ying and Ping, 2006; Yoon and Collins, 2006; Zhang et al., 2019c). This chapter discusses recent advancements in different types of hybrid-MBRs, their working principles, and benefits in the fouling reduction and removal efficiencies.

2. Attached growth membrane bioreactors

In the attached growth MBR (AG-MBR), the attached media is being provided in the bioreactor of the MBR. Media in the reactor provides a surface where microorganisms attach, grow, and make biofilms and ultimately sustain a higher biomass. The advantages of providing attached media in conventional MBR are lower operational complexity, higher biomass concentrations, high resistance to loading rates and toxic compounds, higher removal efficiencies of emerging pollutants, shorter hydraulic retention times (HRTs), and higher removal efficiencies of carbonaceous compounds and nutrients (Odegaard, 2000; Skouteris et al., 2020; Wilderer et al., 2000). The dissolved oxygen, moisture, and nutrients make a gradient from the inner to the outer layers of the biofilms attached to the surface of attached media which results in diverse micro-ecosystems (aerobic, anoxic, and anaerobic zones) within the biofilms. These micro-ecosystems help to do simultaneous nitrification and denitrification (SND) in a single aerated bioreactor. As a result, the total nitrogen removal can be enhanced (Salcedo Moyano et al., 2021). The anoxic zones within the biofilm allow the heterotrophic denitrifying microorganisms to grow. At present, SND phenomena have become an attractive technology for nitrogen removal (Guo et al., 2005; Saidulu et al., 2021; Salcedo Moyano et al., 2021; Tabraiz et al., 2018; Third et al., 2005). Higher biomass concentrations are achieved due to the high specific surface area present for biofilm growth which results in higher removal efficacies of nutrients (Xu et al., 2021). Interestingly, the provision of attached media reduces the membrane fouling rates in AG-MBR compared to the conventional MBR. The attached growth media is preferred to having a low density close to water (sponge or plastic carriers), a high specific surface area, a good holding capacity for biomass, and a good diffusion of nutrients and oxygen into the deeper layers of the culture in the biofilm (Lessel, 1991; Morgan-Sagastume, 2018). Moreover, the AG-MBR has a more diverse community compared to conventional MBR. Besides the nitrifying and denitrifying communities, more diverse and high concentrations of protozoa, metazoan, and other microbial predators have been reported, which keeps a check on the biofilm biomass and hence control sludge production (Zhang et al., 2017c). AG-MBR can be categorized into fixed bed MBR (FBMBR) and moving bed MBR (MBMBR). The details of these two types of MBR are given below.

2.1 Fixed bed MBR

In this type of AG-MBR, the attached media is kept static in the MBR tank. The media could be thin plates, a sponge, or a honeycomb-like structure. Different types of shapes and materials have been used as static attached media in different studies.

The FB-MBR has been reported for the higher removal efficiencies of nutrients, i.e., total nitrogen (TN) and total phosphorus (TP), even at a higher organic loading rate (OLR). In addition, lower fouling rates were reported and linked with the high stability of biofilm of fixed bed attached media which improved the degradation of the organic pollutants. As a result, lower pore blockage and low membrane fouling rates were observed (Zhang et al., 2017c). A study has reported that the majority of the particle sizes in the soluble microbial product (SMP) of the biofilm of FB-MBR were less compared to the conventional MBR sludge which resulted in low fouling rates. In addition, a low MLSS could be one of the reasons for the low fouling rates (Ng et al., 2011).

A comparison between conventional MBR and FB-MBR treating domestic wastewater showed that FB-MBR has higher nitrogen, phosphorus, and carbon removal efficiencies and lower fouling rates (1.41 times) (Khastoo et al., 2021b). Similarly, enhanced simultaneous nitrification and denitrification were reported in the FB-MBR (Sun et al., 2020). FB-MBR were also investigated and proved for their suitability to treat the industrial wastewater for high nutrient removal efficiencies (Izadi et al., 2019).

Although the majority of FB studies reported a decrease in fouling as the fixed bed attached media was introduced in the MBR, the optimal density needs to be investigated in terms of fouling and nutrients removal. An increase in the surface area of attached media can lead to an increase in the fouling rates, though it increases the nutrients removal as well (Zeeshan et al., 2017).

2.2 Moving bed MBR

Moving bed MBR (MB-MBR) is another type of AG-MBR, in which suspended media is added and the biofilm is devolved on the suspended media. The media kept moving with the help of aeration or an agitator. Due to the shear between the liquid and surface, a thin but active biofilm is formed. In addition, moving media indeed provides physical cleaning of the membrane by scrubbing its surface (Lee et al., 2006). The MB-MBR are often favored because they are compact and volume-efficient and can tolerate high particulate loadings without having problems with the clogging of the system (Ivanovic and Leiknes, 2012).

Filtration performance of suspended growth systems in a submerged membrane bioreactor (SMBR) using looped core media (BioMatrix) with a total surface area of $4.37\,m^2$ operated at the constant flux of $25\,L\,m^{-2}\,d^{-1}$ has been reported (Lee et al., 2001). The treatment efficiencies of both reactors at 8 h HRT were greater than 98% and 95% for COD and $NH4^+$-N concentrations, respectively. Better filtration performance with moving attached media was explained by the formation of dynamic biofilm on it. Most of the biomass is attached to media which reduces the mixed liquor suspended solids. It subsequently helps to reduce the cake formation on the membrane surface and specific cake resistance due to the loose cake layer (Ghaffour and Qamar, 2020).

The size of the suspended media can affect the MB-MBR efficiency. Sponges of different sizes $(1 \times 1 \times 1\,cm, 2 \times 2 \times 2\,cm, 3 \times 3 \times 3\,cm)$ were evaluated and it was found that the $2 \times 2 \times 2\,cm$ sponge had the best removal efficiency (Nguyen et al., 2010). Under aerobic conditions, the COD, TN, and TP removal efficiencies were up to 70%, 45%, and 55%, respectively. The effect of high- and low-density polyester-urethane sponges, with a sponge volume fraction of 10% and MLSS of $10\,g\,L^{-1}$, was studied (Ngo et al., 2006). The study concluded that the sponge resulted in increasing the permeate flux, better nutrient removal, and lowered the TMP development. The sponge has been considered one of the best-attached growth media because it can act as a mobile carrier for active biomass, retain microorganisms by incorporating a hybrid growth system (both attached and suspended growth), and reduce fouling of the membrane (Gzar et al., 2021). The performance of a sponge-submerged membrane bioreactor (SSMBR) with SMBR alone in terms of critical flux was observed. The sustainable flux of the sponge-SMBR system increased 2 times with the addition of a 10% volume fraction of sponge in the bioreactor, thus reducing its fouling (Ngo et al., 2006). Similarly, the use of polyurethane with a density of $30\,kg\,m^{-3}$ and 90% porosity reduced the fouling rate effectively (Yang et al., 2006).

The combined effect of suspended and attached growth on nitrogen removal via SND by comparing a conventional suspended growth MBR (CMBR) with an MB-MBR was observed in a study. A nonwoven carrier with a density of $0.27\,g\,cm^{-3}$ as a biofilm carrier in the MB-MBR was used and has reported high nitrogen removal (70% and 80%) at different COD/N ratios (22.1 to 8.9) (Yang et al., 2006). The COD/N ratio of the influent is one of the most critical parameters for the wastewater nitrogen removal process because it directly affects the functional microorganism population. It has been reported that by decreasing the COD/N ratio, the removal efficiencies of nitrogen and phosphorus also decrease in conventional MBR (Fu et al., 2009). MB-MBR reduced the nitrogen footprint of the plant by 50% compared to the conventional activated sludge process (Güneş et al., 2019). Sponge with a dense layer of polyurethane sponge was used and it was reported that the membrane fouling was reduced and the operation time was increased by 154%. In addition, the nitrogen and phosphorus removal efficiencies were also increased (Almusawy et al., 2021).

According to the available literature, it is recommended that the filling fraction of carriers in the reactor should not exceed 70% (Rusten et al., 2006) so that the carriers can move freely in the suspension. The only drawback of the moving attached media (carriers) in MBR is associated with its high cost (Ivanovic and Leiknes, 2012).

More examples of AG-MBR research studies are given in Table 1.

TABLE 1 Attached growth MBR (AG-MBR) studies and key findings.

Reactor type, membrane, wastewater, attached media	Operational parameters	Results/key findings	Reference
MB-MBR/CMBR: with and without membrane scouring, 12.8 L, aeration = 4 L min^{-1}, domestic wastewater **Membrane:** flat-sheet ceramic ultrafiltration (UF), a surface area of 0.08 m^2 **Bio-carrier** = polypropylene, 20% of reactor volume, surface area = 500 m^2 m^{-3}, density = 573.3 kg m^{-3}	HRT: 6 h SRT: 20 days OLR: 1.25 kg COD m^{-3} day^{-1} Flux: 20 L m^{-2} h^{-1}	COD, TN, and TP removal efficiencies were significantly higher in MB-MBR than in CMBR. MB-MBR had 1.5 times less fouling than CMBR. MB-MBR scouring had 8 times less fouling than CMBR. MB-MBR had low carbohydrates, biopolymers, and low molecular weight compounds in MLSS than CMBR. MB-MBR scouring had less TSS, organic and inorganic matter in the cake than CMBR	Chen et al. (2016)
MB-MBR: 12 L, DO = 5.0–6.0 mg L^{-1}, synthetic wastewater **Membrane:** hollow fiber, polyethylene (0.1-μm pore size), surface area of 0.195 m^2 **Bio-carrier** = plastic carrier with and without sponge, 20% of reactor volume, surface area = 500 m^2 m^{-3}, density = 950 kg m^{-3} polyester-polyurethane sponge: 28–30 kg m^{-3}	HRT: 6–12 h SRT: infinite OLR: 1.38 kg COD m^{-3} day^{-1} Flux: 10.26 L m^{-2} h^{-1}	MB-MBR R.E: COD (95.89 %), TN (71.8 %), TP (70. 2%) Sponge-MB-MBR R.E: COD (97.52 %), TN (86.7 %), TP (84.5%) Low SMP and EPS in MLSS of sponge- MB-MBR than MB-MBR which decreased the fouling in sponge-MB-MBR fouling 1.5 times	Deng et al. (2016)
MB-MBR/CMBR: 5 L, DO = 3.0 mg L^{-1}, synthetic textile wastewater **Membrane:** ceramic flat sheet membrane, surface area = 0.057 m^2, pore size = 0.1 μm **Bio-carrier: bio-carrier** = plastic carrier with and without sponge, 20% of reactor volume, surface area = 1036 m^2 m^{-3}, density = 170 kg m^{-3}	HRT: 16.9 h SRT: 30 days OLR: 1.17 kg COD m^{-3} day^{-1} Flux: 5–5.8 L m^{-2} h^{-1}	MB-MBR R.E: COD (98.5 %), color (89.5%) CMBR R.E: COD (93.1%), color (87.1%) Low polysaccharides in EPS, low floc size, the higher zeta potential of sludge, and low fouling were observed in MB-MBR than in MBR	Erkan et al. (2020)
MB-MBR/CMBR: 36 L, aeration = 6.6 L min^{-1}, DO = 3.0 mg L^{-1}, graywater **Membrane:** PVDF membrane, surface area = 32 m^2, pore size = 0.1 μm **Bio-carrier:** plastic carrier with and without sponge, 10%–40% of reactor volume	HRT: 8 h SRT: 40 days OLR: 3.21 kg COD m^{-3} day^{-1} Flux: 3.4 L m^{-2} h^{-1}	MB-MBR R.E: COD (91%), BOD$_5$ (97%) TN (66 %), NH3-N (93%), TP (90%) CMBR R.E: COD (91%), BOD$_5$ (97%) TN (58 %), NH3-N (88%), TP (90%)	Palmarin and Young (2019)
MB-MBR/CMBR: 40 L, DO = 3.0 mg L^{-1}, graywater	HRT: 6 h SRT: infinite days	8 times low fouling and significantly higher removal efficiencies of	Luo et al. (2015)

Continued

TABLE 1 Attached growth MBR (AG-MBR) studies and key findings—cont'd

Reactor type, membrane, wastewater, attached media	Operational parameters	Results/key findings	Reference
Membrane: PVDF hollow fiber, membrane, surface area = 0.2 m^2, pore size = 0.2 μm **Bio-carrier:** polyurethane sponge size 2 × 2 × 2 cm, 20% of reactor volume, contained biomass of 0.41 g g^{-1} of sponge	OLR: 0.36 kg COD m^{-3} day^{-1} Flux: 8.34 L m^{-2} h^{-1}	22 micropollutants in MB-MBR than n CMBR. 5 times higher SMP in CMBR than MB-MBR	Khastoo et al. (2021a)
FB-MBR/CMBR: 140 L, aeration = 6.6 L min^{-1}, DO in aerobic parts = 4.0 mg L^{-1}, DO anoxic part < 1 mg L^{-1}, Municipal wastewater **Membrane:** PVDF flat sheet membrane, surface area = 1.4 m^2, pore size = 0.2–0.8 μm **Attached media** = PVC, density = 26 kg m^{-3}, surface area = 4400 m^2 m^{-3}, nominal area = 200 m^2	HRT: 6–24 h SRT (aerobic): 75 days SRT (anoxic): 75 days OLR: 0.58, 0.71, 1.55 kg COD m^{-3} day^{-1}	FB-MBR RE: COD (>95%), TN (74%), TP (86%) CMBR R.E: COD (>95%), TN (39%), TP (66%) Low fouling rates in FB-MBR	Sun et al. (2015)
FB-MBR/CMBR: 4 L, synthetic wastewater **Membrane:** PVDF hollow fiber membrane, surface area = 0.2 m^2, pore size = 0.4 μm **Attached media =** polyethylene square meshes, density = 950 kg m^{-3}, dimensions = 1.5 × 1.5 × 0.11 cm, surface area = 210 m^2 m^{-3}, nominal area = 3.6 m^2	HRT: 12 h OLR: 0.67 kg COD m^{-3} day^{-1} Flux: 18.75 L m^{-2} h^{-1}	Low fouling rates in FB-MBR than CMBR An increase in attached media surface area decreased the fouling rates EPS decreased as the surface area of the attached media increased Removal efficiencies are comparable in C-MBR and FB-MBR, slightly higher in FB-MBR. It can be speculated that the attached media provided was very less. Further increase in the attached media would have increased the R.E significantly	
FB-MBR/CMBR: 1.8 m^3, aeration = 8–9 m^3 h^{-1}, municipal wastewater **Membrane:** PVDF hollow fiber membrane, surface area = 2 m^2, pore size = 0.4 μm **Attached media =** Polyethylene square meshes, density = 950 kg m^{-3}, dimensions = 1.5 × 1.5 × 0.11 cm, surface area = 119 m^2 m^{-3}, 50% of reactor volume	HRT: 9–12 h SRT: 37–140 days OLR: 0.36–3.87 kg COD m^{-3} day^{-1} Flux: 10 L m^{-2} h^{-1}	FB-MBR RE: COD (85%), BOD$_5$ (98%), TN (75%), TP (42%) CMBR R.E: COD (80%), BOD$_5$ (96%), TN (38%), TP (37%) Halved the fouling rates in FB-MBR due to low concentration of colloidal biopolymers in sludge	Rodríguez-Hernández et al. (2014)

3. Adsorbent-based MBR

In adsorbent-based MBR, an adsorbent is added to the activated sludge tank of the MBR. Adsorbent addition helps to improve the effluent quality by adsorption of many unwanted chemicals. Moreover, due to the provision of porosity and a large surface area by adsorbents, a unique microenvironment develops which helps to grow diverse microorganism communities in the form of thin motile biofilms (Alvarino et al., 2017). It ultimately helps to improve the nutrients, recalcitrant chemicals, and carbonaceous pollutants removal (Sohn et al., 2021). Activated carbon (AC) has been widely used in MBR because of its high adsorption capacity, enhancement of biodegradation, and subsequent removal of recalcitrant pollutants. Additionally, it mitigates membrane fouling efficiently (Zhang et al., 2019a, 2017a). Powdered activated carbon (PAC) and granular activated carbon (GAC) are the two types of AC. AC has a high porosity and surface area which can result in a high adsorption capacity and remove recalcitrant chemicals, odor, and color (Zhang et al., 2019a).

GAC has a larger size than PAC, which allows it to be more easily retained in the reactor and is more economical when used continuously because it can be regenerated thermally. GAC has stronger physical interactions with the membrane surface due to its larger size (Zamani et al., 2017).

3.1 Granular activated carbon MBR

The addition of GAC in MBR adds an unsteady-state shear on the membrane surface. Many other methods, such as bubbling and vibration, have been used to reduce the shear on the surface of the membrane to mitigate membrane fouling in MBR. Particle fluidization has recently been proposed as an alternative to bubbling because it may have the same effect on membrane fouling reduction while requiring at least ten times less energy than bubbling. Fluidization of GAC, in particular, has received a lot of attention, because larger GAC scours the cake layer on the surface of the membrane; hence, it is more effective in the long run of MBR (Wang et al., 2016b,c). Numerous studies have reported that GAC fluidization significantly reduces membrane fouling in anaerobic fluidized MBR (Aslam et al., 2018; Gao et al., 2014; Shin et al., 2014; Wu et al., 2015). In addition to fouling reduction, GAC also provides the surface to grow the biofilms. The efficiency of the activated carbon carrier is attributed to its excellent microbial attachment properties and high adsorption capacity for different compounds. It was found that the addition of granular activated carbon to MBR causes a linear rise in TMP as compared to conventional MBR (Thuy, 2003). Flux permeability increased due to the addition of powdered activated carbon to MBR (Kim et al., 1998). Biomass grows on GAC as biofilm/attached growth, which is more resilient to temperature changes fluctuation in the influent composition, and shock loadings. A comparative pilot-scale study between sparging anaerobic MBR (AnMBR) and GAC AnMBR demonstrated that GAC AnMBR achieved the equivalent removal efficiencies of the COD in less (65%) hydraulic retention time. The GAC addition increased the life of the membrane and decreased the fouling significantly compared to gas sparging (Evans et al., 2019). The membrane fouling reduction is attributed to the adsorption of proteins in the AnMBR reactor. A decrease in protein concentration in the cake layer of the biofilm of AnMBR has been reported due to the

GAC addition which improved the membrane filterability. Increased concentration of GAC has been reported to be beneficial in fouling reduction, and increased size of GAC required higher energy in AnMBR (Aslam et al., 2014). However, a complete and detailed cost-benefit analysis is required to reveal the beneficial size and concentration in AnMBR.

A study used $50\,g\,L^{-1}$ of GAC in MBR and it showed a remarkable reduction of COD and soluble microbial products (SMP) concentration. The cake layer resistance was also decreased by 53.5% (Deng et al., 2014). Coated activated carbon on polyurethane cubes (surface area $35,000\,m^2\,m^{-3}$) was also used in MBR and has significantly reduced membrane fouling (Lee et al., 2006). In another study, only $2\,g\,L^{-1}$ of GAC was effective for COD and SMP removal (Wang et al., 2018b). GAC has an adsorption capacity to capture compounds like SMP and reduce the membrane fouling significantly (Johir et al., 2011). Complete removal of pharmaceutical contaminants (Sulfamethoxazole, Ibuprofen, and Diclofenac) in a short HRT of 8.7h has been reported due to the GAC addition to AnMBR (Lim et al., 2019). A pilot-scale study of AnMBR treating domestic wastewater demonstrated that GAC addition can help to operate the AnMBR for a long period (487 days) without chemical cleaning (Shin et al., 2014). Furthermore, GAC addition also reduced the biofouling by 2.5 times and enhanced the nitrate removal by 3 times in nitrate-dependent anaerobic methane oxidation in AnMBR (Lu et al., 2020).

However, the adsorption capacity of GAC may reduce over time and was a major limitation. Thus, the exhausted GAC needs to be replaced or regenerated by the thermal process to recover the adsorption capacity (Korotta-Gamage and Sathasivan, 2017).

3.2 Powdered activated carbon MBR

Powdered activated carbon (PAC) gets attached to the sludge flocs and alters their physio-mechanical properties, reduces the concentration of polysaccharides and proteins in the soluble microbial product (SMP) (Iorhemen et al., 2016), and adsorbs the fraction of organics coming with wastewater to reduce the irreversible fouling (Campinas and Rosa, 2010). PAC has more surface area compared to granular activated carbon (GAC), hence providing better adsorption and more microbial concentration in the MBR. However, PAC forms a cake layer around the membrane which is termed reversible fouling (Shao et al., 2017). An increase in the PAC concentration increased the cake layer formation and subsequently fouling. At a certain PAC concentration, the PAC cake layer fouling shall be offset by its fouling reduction due to organic and SMP adsorption (Stoquart et al., 2012). To find the optimal concentration, the effect of PAC concentrations was investigated and concluded that a low dose of PAC, i.e., $0.5\,g\,L^{-1}$, and an increase in the solids retention time (SRT) up to 50 days favor a long-term filtration interval (Remy et al., 2009). However, a concentration of $0.75\,g\,L^{-1}$ of PAC was also reported for the effective fouling reduction in MBR (Ying and Ping, 2006). Similarly, $4\,g\,L^{-1}$ of PAC was reported for significant fouling reduction by effectively decreasing the concentration of protein and polysaccharides in extracellular polymeric substances (EPS) (Khan et al., 2012). To combine the benefits of PAC and GAC, both were being employed in the water treatment coupled with ultrafiltration for better fouling control (Yu et al., 2022). Combining both types of AC reduced the biofouling, increased the life span of the membrane, and reduced the overall energy requirement. However, such a combination needs to be investigated in the MBR treating wastewater. In addition, the size distribution of the PAC and membrane pore size are the main factors that need to be optimized for effective fouling abatement of MBR. Though an overall reduction in the EPS has been reported due to the

addition of PAC in MBR, the types of EPS reduced and change in the EPS physiochemical properties need to be investigated for a better understanding of the mechanism involved in the PAC-sludge interaction.

The PAC-MBR has been reported for high removal efficiencies of micropollutants by simultaneous adsorption and biodegradation (Alvarino et al., 2017; Jiang et al., 2020; Serrano et al., 2011). Other benefits of the addition of PAC to MBR are: improved sludge flocs size, improved specific oxygen uptake rates (Hu et al., 2015), and reduced effect of the toxic chemicals shock loadings on nitrification (Widjaja et al., 2004). This can be attributed to the development of microenvironments in the pores of the PAC which are suitable for the growth of a variety of microbes and the increase in the diversity of a microbial community (Zhang et al., 2019b). An increase in the diversity and richness of the nitrifying bacteria has been reported due to the addition of PAC in MBR. Moreover, 20 particular genera growth was favored in PAC-MBR which resulted in increased removal of nitrogen, phosphorus, and micropollutants (Asif et al., 2020).

PAC produced from the sludge/biosolid has also been reported to control the irreversible fouling in MBR and better performance (Pan et al., 2016; Villamil et al., 2016). Sulfur-based PAC was also used to enhance the denitrification of MBR. However, the fouling retardation propensity of the sulfur-PAC was not evaluated. Similarly, magnetic-PAC was prepared and used for the polishing of the MBR effluent. It effectively reduced the dissolved organic matter. However, such a composite of PAC must be evaluated in the bioreactor of MBR for its benefit of micropollutant, nutrient, and phosphorus removal as well as biofouling reduction (Ittisupornrat et al., 2019).

3.3 Zeolite MBR

Zeolite is a porous, crystalline substance that mainly consists of aluminum, oxygen, and metals such as titanium, tin, and zinc, while natural zeolite can be found in rocks near volcanoes. Its working principle in the MBR is like GAC and PAC, i.e., adsorption of organics, proteins, and polysaccharides from SMP, providing the surface for attached growth, and increasing the size of the floc (if used as powdered). Hence, its addition results in the reduced membrane fouling in the MBR (Nguyen et al., 2013). Due to zeolite's high porosity, large surface area, and active groups, it provides suitable sites for the attachment of bacteria and enhances the settleability of the sludge, improves the nutrients removal, and results in fouling reduction (Wei et al., 2013). The application of zeolites facilitates the performance of MBR in terms of COD, suspended solids (SS) removal, and fouling reduction (Chen et al., 2019a; Li et al., 2020b). A comparative study between the GAC-MBR and granular zeolite showed that both reduced the membrane fouling significantly, with GAC showing a slightly higher fouling reduction. Both have approximately the same effect on the reduction of EPS and SMP; however, zeolite showed a slightly higher adsorption of polysaccharides (Mohamadi et al., 2020).

Zeolite addition in the AnMBR improved the system efficiency as a porous attached media and ion exchanger. Its property of ion exchange helps to remove the ammonia from the anaerobic digester, which is reported as an inhibitor in the anaerobic digestion process. The zeolite was modified with sodium chloride which further enhanced the ammonium removal rate by increasing the mass transfer through the biofilm by modifying the surface morphology (Lin et al., 2013). The addition of zeolite has also helped to improve the methane yield and COD removal (Dutta et al., 2014; Lin et al., 2013; Wang et al., 2019). This improvement in anaerobic digestion can be attributed to the ion exchange of cations, ammonia, and long-chain fatty acids (Nordell et al., 2013). The addition of zeolite has a significant effect on the community of the reactor as well. A significant increase in the growth of the *Methanobacterium* and *Methanosarcrina* has been reported due to the addition of the zeolite (Poirier et al., 2017).

Zeolite has been reported to form bigger and rigid flocs which improved the filterability of the membrane and reduced the fouling in AnMBR. Moreover, significant AnMBR enhanced the COD and suspended solids (SS) which improved membrane performance and reduced the fouling (Chen et al., 2019a; Wei et al., 2013). A significant increase (22%) in COD removal efficiencies and sludge settleability has been reported due to the addition of zeolite which helped to grow the anaerobic microorganism on the surface. No fouling was observed due to low SS and COD concentrations (Chen et al., 2019a).

More examples of adsorbent-based MBR research studies are given in Table 2.

4. Coagulation-MBR

Coagulants are commonly used for the removal of colloidal and soluble organic matters from wastewater during the wastewater treatment process. The addition of a coagulant modifies the mixed liquor sludge properties in the MBR, such as an increase in the particle size of the floc, resulting in the formation of cake more permeable than that formed from small and compacted flocs (Ji et al., 2010). It was reported that in situ coagulation significantly increased permeability, compared to noncoagulated or precoagulated rectors (Ding et al., 2017). Many studies have reported that the use of alum in MBR increased the permeate flow rate and phosphate was removed effectively, as the alum concentration increased (Song et al., 2008). Effects of three coagulants viz. polymeric ferric sulfate (PFS), polymeric aluminum ferric chloride (PAFC), and poly aluminum chloride (PACl) were studied for membrane fouling mitigation. PFS was found to be the best among these three for fouling reduction with an optimal concentration of 1.05 mM as Fe. The addition of these coagulants reduced the initial transmembrane pressure (TMP) and rate of TMP which reduced the overall fouling of the membrane in MBR. This was attributed to the inhibition of the gel layer formation on the surface of the membrane and delayed the cake layer formation. The delayed gel layer formation was attributed to charge neutralization of the SMP (Wu et al., 2006). The suitability of PFS for the long-time MBR operation was also evaluated (Wu and Huang, 2008).

Modified cationic polymer (MPE50TM) in MBR increased the overall oxygen transfer rate by 10%–20% and fouling rates were almost constant at the flux that was moderately higher than the critical flux (Yoon and Collins, 2006). Another modified cationic polymer (MPE 30TM) was tested in MBR, and it increased the permeability up to 400% and reduced the TMP constantly by more than 75%. A significant reduction in the chemical cleaning frequency was also observed (Wozniak, 2010).

TABLE 2 Adsorbent-based MBR studies and key findings.

Reactor type, membrane, wastewater, PAC/GAC/Zeolite	Operational parameters	Results/key findings	Reference
AnMBR: 3.25L, textile wastewater **Membrane**: PVDF hollow fiber membrane, surface area $=0.8\,m^2$, pore size $=0.4\,\mu m$ **Adsorbent media**: PAC $(4\,g\,L^{-1})$	HRT: 24h OLR: $0.67\text{–}4.2\,kg\,COD\,m^{-3}\,day^{-1}$ Flux: $0.17\,L\,m^{-2}\,h^{-1}$	PAC-AnMBR RE: COD (88.6%), AnMBR RE: COD (78.3%) Increased stability of PAC-AnMBR, PAC adsorbed SMP and toxic compounds, aromatic, amines, and VFA	Baêta et al. (2016), Baeta et al. (2013)
AnMBR: 3L, saline sewage **Membrane**: polyethylene membrane, surface area $=0.1\,m^2$, pore size $=0.4\,\mu m$ **Adsorbent media**: PAC $(1.7\,g\,L^{-1})$, surface area $=1300\,m^2\,g^{-1}$	HRT: 8h SRT: 20 days OLR: $2\,kg\,COD\,m^{-3}\,day^{-1}$ Flux: $5\text{–}8\,L\,m^{-2}\,h^{-1}$	The addition of PAC increased DOC and high molecular weight chemical removal by 30% and 70%, respectively. It decreased the SMP and large flocs	Vyrides and Stuckey (2009)
AnMBR: 15L, palm oil wastewater **Membrane**: polyethylene membrane, surface area $=0.1\,m^2$, pore size $=0.4\,\mu m$ **Adsorbent media**: PAC $(1\text{–}5\,g\,L^{-1})$, surface area $=1300\,m^2\,g^{-1}$	HRT: 6h SRT: 30 days OLR: $0.8\,kg\,COD\,m^{-3}\,day^{-1}$ Flux: $25\text{–}30\,L\,m^{-2}\,h^{-1}$	Higher COD removal efficiencies, increased floc size, reduction in SMP, and low fouling rates were observed at higher PAC dose	Chong (2015)
Anoxic-aerobic MBR: anoxic (2L), aerobic (3L), membrane tank (2L) anoxic DO $<0.5\,mg\,L^{-1}$, aerobic tank DO $=1.5\text{–}2.5\,mg\,L^{-1}$ **Membrane**: ceramic membrane, surface area $=0.0425\,m^2$, pore size $=0.1\,\mu m$ **Adsorbent media**: PAC $(20\,g\,L^{-1})$	HRT: 11h SRT: 30 days OLR: $1.25\,kg\,COD\,m^{-3}\,day^{-1}$ Flux: $15\,L\,m^{-2}\,h^{-1}$	Increase in the diversity and richness of the nitrifying bacteria 20 particular genera growth was favored in PAC-MBR Increased removal of nitrogen, phosphorus, and micropollutants	Asif et al. (2020)
MBR: 1.5L, diluted sewage, DO $=6\,mg\,L^{-1}$ **Membrane**: hollow-fiber membrane, surface area $=0.1\,m^2$, pore size $=0.01\,\mu m$ **Adsorbent media**: PAC $(3\,g\,L^{-1})$	HRT: 1h OLR: $0.1\,kg\,COD\,m^{-3}\,day^{-1}$ Flux: $15\,L\,m^{-2}\,h^{-1}$	PAC addition prevented polypeptide adhesion in membrane pores Reduced hydrophilic molecules adhesion on the membrane surface Reduced 60% membrane resistance	Zhang et al. (2021)
Fluidized AnMBR: domestic wastewater **Membrane**: hollow-fiber membrane, surface area $=0.1\,m^2$, pore size $=0.01\,\mu m$ **Adsorbent media**: GAC $(30\,g\,L^{-1})$	HRT: 2.3h OLR: $1\text{–}3.5\,kg\,COD\,m^{-3}\,day^{-1}$ Flux: $6\text{–}11\,L\,m^{-2}\,h^{-1}$	GAC addition decreased energy requirement and fouling rates High removal efficiencies (COD $=84\%$, BOD$_5$ $=92\%$) were achieved at very low HRT	Yoo et al. (2012)

Continued

TABLE 2 Adsorbent-based MBR studies and key findings—cont'd

Reactor type, membrane, wastewater, PAC/GAC/Zeolite	Operational parameters	Results/key findings	Reference
MBR: with a GAC-packed zone, 0.85L, textile wastewater **Membrane**: polyethylene hollow-fiber membrane, surface area $=0.256\,cm^2$, pore size $=0.4\,\mu m$ **Adsorbent media**: GAC (0–160 g L^{-1}, size $=1.1\,mm$)	HRT: 24h OLR: 0.1–1 kg COD m^{-3} day^{-1} Flux: 0.14 L m^{-2} h^{-1}	High and stable removal of color, Total organic carbon, and TN were achieved under high dye-loading conditions due to the addition of an anaerobic GAC zone in aerobic MBR	Hai et al. (2011)
Methane oxidation-AnMBR: 2.3L, synthetic wastewater **Membrane**: hollow-fiber membrane, surface area $=452\,cm^2$, pore size $=0.1\,\mu m$ **Adsorbent media**: GAC (3.25 g L^{-1}, size $=0.5$–2 mm)	HRT: 3 days Flux: 12.95 L m^{-2} h^{-1}	The addition of GAC in anaerobic methane oxidation-MBR increased the nitrate removal rate from 10 to 31 mg NL^{-1} d^{-1} and decreased membrane flux decay rates 3 times	Lu et al. (2020)
MBR: volume 9.3L, DO $=3$–4 mg L^{-1}, synthetic wastewater, **Membrane**: surface area $=0.11\,m^2$ and pore size 0.1 μm **Adsorbent**: zeolite	HRT: 15h SRT: 35 days OLR: 2.24 kg COD m^{-3} day^{-1} Flux: 4.24 L m^{-2} h^{-1}	The addition of zeolite reduced fouling by 66% by reducing the cake formation and pore blockage (84% reduction). Moreover, EEM analysis clearly showed that zeolite addition decreased the level of proteins and fulvic acids in SMP	Hazrati et al. (2018)
MBR: synthetic municipal wastewater, DO 2–3 mg L^{-1} **Membrane**: surface area of 0.11 m^2 and pore size of 0.4 μm **Adsorbent media**: zeolite powder	SRT: 10 days Flux: 17 L m^{-2} h^{-1}	The addition of zeolite contributed mainly to the alleviation of irreversible fouling. Significant SMP removal was observed at 5 g L^{-1}	Gkotsis and Zouboulis (2019)

Several MBR studies showed that ferrous and aluminum-based salts have been used as coagulation reagents (Lefebvre et al., 2019). It was reported that aluminum sulfate effectively removed 40% of organic substances. Moreover, higher removals of chemical oxygen demand (COD), nitrogen, and phosphorus were observed from domestic wastewater by dosing $40\,mg\,L^{-1}$ of polyaluminum chloride (Ji et al., 2015). Polyaluminum chloride was used as a coagulant in MBR and reported for increased removal efficiencies of personal care products. Moreover, polyaluminum chloride addition increased the oxygen uptake rates of sludge and nitrification rates and decreased the fouling rates of MBR (Park et al., 2018).

Three different types of polyelectrolyte coagulants: polyacrylamides, polydiallyldimethylammonium chloride (PDADMAC), and polyamine, were used in MBR, and sludge was characterized. The polyacrylamide performed better than the other two and its application significantly increased the elasticity, viscosity, and Bingham yield stress of the sludge. It produced more compacted sludge with a low water content compared to polyamine and PDADMAC (Yousefi et al., 2020).

An excessive dosage of the coagulant decreases the pH which can affect the activity of microorganisms in activated sludge. So a sustainable dosage is required to avoid such problems.

More examples of Coagulation-MBR research studies are given in Table 3.

5. Electro-MBR/electrocoagulation-MBR

Electric field application to MBR can neutralize positively and negatively charged particles and suppress fouling, and this type of MBR is called electro-MBR. Four different mechanisms work due to the electrochemical applications in MBR: electrocoagulation, electrophoresis, electroosmosis, and electrochemical quorum quenching. Altogether, these four processes can neutralize the foulant, degrade the high fouling substances, and control the foulant placement on the surface of the membrane which result in low fouling and less sludge production (Ensano et al., 2016; Khan et al., 2019). In electrocoagulation, the fouling compounds like biopolymers and coagulants are removed with the help of coagulant cations generated from sacrificial anodes, i.e., iron and aluminum. In the case of electrophoresis, the electric field is applied to take away the charged particles including EPS from the membrane (Akamatsu et al., 2010; Chen et al., 2007). In the case of electrochemical quorum quenching, reactive species of chloride or oxygen are being generated by the electric field which degrades the bacterial signal molecules. It alleviates quorum sensing and helps to reduce biofouling (Borea et al., 2018). Sludge properties of sludge such as floc size, settleability, surface charge, etc., change due to the application of electric field and cause an overall fouling reduction in the MBR (Giwa et al., 2015; Hua et al., 2015; Lin et al., 2016; Tafti et al., 2015). A range of DC intensities ($0-20\,v\,cm^{-1}$) was investigated for its effects on membrane biofouling. The results showed an up to 80% reduction in the specific cake reduction at higher DC intensities. The significant fouling reduction was due to electrophoresis (Giwa et al., 2015). Another study investigated the effect of the intermittent electric field during the crossflow filtration. The intermittent electric field produced the electric repulsive force which helped to detach the foulant from the membrane surface and reduced the fouling significantly (Akamatsu et al., 2010).

TABLE 3 Coagulation-MBR studies and key findings.

Reactor type, membrane, wastewater, coagulant	Operational parameters	Results/key findings	Reference
Batch and continuous MBR: 10 L, synthetic wastewater, 1.5 h aerobic, 0.5 h anaerobic, DO 2–3 mg L^{-1} **Membrane**: chlorinated polyethylene flat sheet, surface area = 0.04 m^2, pore size = 0.4 μm **Coagulant**: Ferric chloride solutions 2260 and 4520 mg L^{-1} (dose = 200 mL d^{-1})	HRT: 24 h OLR: 1 kg BOD m^{-3} day^{-1} Flux: 10.4 L m^{-2} h^{-1}	The addition of ferric chloride increased the removal efficiencies of BOD, COD, and TP, decreased 50% SMP, and increased flocs size which resulted in 2–4 times. TP removal was increased from 33.4% to 99.5%. Fe coagulants were more effective than aluminum coagulants	Mishima and Nakajima (2009)
MBR: synthetic wastewater, DO 2–3 mg L^{-1} **Membrane**: PVDF hollow fiber, microfiltration membrane, surface area = 0.75 m^2, pore size = 0.1 μm **Coagulants**: 28 inorganic, organic polyelectrolytes, AL-based and Fe-based	HRT: 15.6, 10 h SRT: 20, 30 days OLR: 0.87, 3.1 kg COD m^{-3} day^{-1} Flux: 30 L m^{-2} h^{-1}	Cationic polyelectrolyte FO 4350 SSH was the optimal coagulant for membrane fouling mitigation at a concentration of 0.16 mg L^{-1}. The coagulant reduced reversible fouling	Gkotsis et al. (2017)
MBR: anoxic (2.5 L), aerobic (2.5 L), municipal wastewater **Membrane**: PVDF hollow fiber, microfiltration membrane, surface area = 0.04 m^2, pore size = 0.4 μm **Coagulants**: propylammonium chloride (PACl) and chitosan	HRT: 9 h SRT: 25 days OLR: 0.32 kg COD m^{-3} day^{-1} Flux: 10.41 L m^{-2} h^{-1}	Optimal dosage addition of PACl and chitosan in MBR improved the removal of pharmaceutical and personal care products (PPCPs), and reduced the fouling rates. The PPCP increased removal was linked to increased degradability in coagulation-MBR. Moreover, increased oxygen uptake rates and nitrification rates were observed coagulation-MBR	Park et al. (2018)
MBR: 5.0 L, biological filter-treated sewage **Membrane**: PVDF hollow-fiber membrane, surface area = 0.3 m^2, pore size = 0.2 μm **Coagulants**: ferric chloride and aluminum sulfate	HRT: 1.6 h OLR: 0.72–1 kg COD m^{-3} day^{-1} Flux: 10–15 L m^{-2} h^{-1}	The optimal dose of ferric chloride was 20 mg L^{-1} and aluminum sulfate 20 mg L^{-1} for the removal of total phosphorus (>80%). Both coagulants mitigated fouling by decreasing polysaccharides in SMP	Li et al. (2017a)
Fluidized AnMBR: 0.48 L, domestic wastewater, **Membrane**: PVDF hollow-fiber membrane, surface area = 0.013 m^2, pore size = 0.1 μm **Coagulant**: FeCl$_3$ (10, 20 mg L^{-1})	HRT: 5.5 h OLR: 1.8 kg COD m^{-3} day^{-1} Flux: 7 L m^{-2} h^{-1}	Sulfur removal increased from 59% to 95% when FeCl$_3$ was added FeCl$_3$ addition enhanced the fouling rate due to the deposition of the inorganic compounds in pores	Lee et al. (2016)

MBR: 5.9L, aeration$=2$ L min^{-1}, Pharmaceutical wastewater **Membrane:** flat sheet ceramic membrane, surface area$=0.08$ m^2, pore size$=0.1$ μm **Coagulant:** ferric hydroxide	HRT: 9h SRT: 10 days OLR: 3.7 kg COD m^{-3} day^{-1} Flux: 8.2 L m^{-2} h^{-1}	35% fouling reduction was observed in coagulant-MBR. Ferric hydroxide addition increased sludge flocs and decreased organics (carbohydrates, biopolymers, low molecular weight compounds) in sludge	Huang et al. (2019)
MBR:, 30L, aeration$=5$ L min^{-1}, synthetic wastewater **Membrane:** PVDF hollow-fiber membrane, surface area$=0.4$ m^2, pore size$=0.2$ μm **Coagulant:** ferric chloride (Fe(III)) and ferrous sulfate (Fe(II))	HRT: 10h SRT: 30 days OLR: 2.0 kg COD m^{-3} day^{-1} Flux: 16 L m^{-2} h^{-1}	Phosphorus removal was significantly improved using ferric chloride and ferrous sulfate in MBR. Fe/P ratio of 4 increased the fouling rates. Fe/P ratio of 2 or less was found to be beneficial in terms of fouling mitigation and phosphorus removal	Zhang et al. (2015b)

Different materials of cathode have been tested and compared in the electric-MBR. Titanium and iron cathode were compared in Electro-MBR with a stainless steel anode. The results showed that the iron cathode was better in controlling biofouling as the iron cathode facilitated an iron-based coagulant which resulted in lower SMP concentrations and subsequently led to a loose cake layer (Zhang et al., 2015a). Few studies have used a conductive membrane as a cathode or placed the cathode inside the membrane. Such a membrane keeps away the microorganisms from the surface of the membrane by electrophoretic forces. Moreover, under constant aeration, hydrogen peroxide could be electro-generated and has contributed to SMP and EPS oxidation which thereby reduced fouling (Khalid et al., 2018; Li et al., 2014b; Liu et al., 2013a,b; Malaeb et al., 2013). Copper wires have been used in the flat sheet membrane as a cathode and in the stainless steel mesh as an anode (Liu et al., 2012). In addition to cathodic membrane, few studies have reported the performance of nanocomposite-based anodic membranes that can generate the reactive species for pollutant degradation. A PVDF membrane was coated with multiwalled carbon nanotubes as an anode. During the filtration, the application of an electric field resulted in the production of reactive chloridespecies which oxidized the foulants on the surface of the membrane or within the pores, which reduced membrane fouling (Hashaikeh et al., 2014; Yang et al., 2019a). Similarly, the anodic membrane of carbon fiber cloth coated with a PVDF/PVP composite demonstrated electrooxidation to control in situ membrane fouling. It reduced up to 70% of the fouling for long-term operation.

However, accumulation of the coagulants (such as aluminum and iron ions) leached from the electrode of the electro-MBR could impede the biological activity of the microorganism, thereby reducing pollutant removal efficiency (Bani-Melhem and Smith, 2012; Brillas and Martínez-Huitle, 2015; Khalid et al., 2018; Liu et al., 2013a). The iron precipitates have been reported to hinder the nutrient diffusion to the microbial cell which resulted in reduced nitrification rates (Bani-Melhem and Elektorowicz, 2010). Similarly, few studies have reported the change in the microbial community structure due to the use of different types of electrodes and electric fields in electro-MBR (Chen et al., 2019b; Song et al., 2020; Sun et al., 2019; Yang et al., 2019b). So better anode and cathode materials need to be designed for lower toxic effects on the microbial community and pollutant removal.

Table 4 contains more examples of electro-MBR research studies with details of reactor, operational conditions, and key findings.

6. Microbial fuel cell—MBR

A promising hybrid process was developed by integrating microbial fuel cells (MFC) with MBR to achieve a high-quality effluent and harvest the energy contained in wastewater (Kim et al., 2016). Normally, three different configurations of MBR and MFC have been used to acquire a higher energy density and pollutants removal. In the first configuration, MFC served as pretreatment of MBR; in the second, MFC served as posttreatment of the MBR effluent, and in the third configuration, MFC is immersed into MBR directly. Normally, the first and second integrations are termed external integrations and the third is known as internal integration (Li et al., 2020a). Internal integration is more widely used because of its compact and cost-

TABLE 4 Electro-MBR studies and key findings.

Reactor type, membrane, wastewater	Operational parameters	Results/key findings	Reference
electro-MBR: pair of aluminum (Al) flat-plate electrodes ($18\,cm^2$), Electric current density 10 to $40\,A\,m^{-2}$, volume 9 L, DO > 5, synthetic wastewater **Membrane**: PTFE flat sheet membrane and pore size of $0.4\,\mu m$	HRT: 25 h SRT: 60 days OLR: $0.786\,kg\,COD\,m^{-3}\,day^{-1}$ Flux: $27\,L\,m^{-2}\,h^{-1}$	The average fouling rate was substantially reduced by 7.8-fold resulting in no chemical cleaning requirements during the entire operation of the electro-MBR Effective removal of DOC and ammonia was consistently maintained, with an average efficiency rate of more than 98%. Charge neutralization and adsorption are ascribed to mitigate soluble foulants	Hua et al. (2015)
electro-MBR: 22.5 L, aluminum as anode and stainless steel as cathode, voltage gradient $1.18\,V\,cm^{-1}$, synthetic wastewater **Membrane**: (MF) flat sheet membrane surface area $0.11\,m^2$ and pore size $0.4\,\mu m$	HRT: 13.5 h SRT: 10 days OLR: $0.2\,kg\,COD\,m^{-3}\,day^{-1}$ Flux: $15.2\,L\,m^{-2}\,h^{-1}$	Sludge filterability was enhanced with a 70% reduction in time-to-filter (TTF) 57% decrease in conductivity and 100% removal of phosphates were achieved in electro-MBR	Giwa et al. (2015)
electro-MBR: 25.8 L, synthetic wastewater, voltage 1, 3, 5 V, iron anode **Membrane**: hollow fiber, surface area = $0.3\,m^2$, and pore size = $0.1\,\mu m$	HRT: 15.3 h OLR: $0.32\,kg\,COD\,m^{-3}\,day^{-1}$ Flux: $6.52\,L\,m^{-2}\,h^{-1}$	When electrocoagulation was carried out in MBR, the operating cycle was extended from 4–7 days to 14–20 days with 0.15 A constant current Application of 1 and 3 V increased the filamentous bacteria which caused serious membrane fouling. However, under 5 V/0.15 A filamentous bacteria number decreased and fouling was reduced. The removal of ammonia nitrogen, phosphate, and TOC increased significantly in electro-MBR	Liu et al. (2019)
electro-MBR: 5 L, DC Power (3 A, 35 V), Ti or Fe were used as electrodes, the current density of $2\,A\,m^{-2}$, synthetic domestic sewage **Membrane**: PVDF hollow fiber, pore size of 30 nm	HRT: 9 h	Electro-MBR and CMBR energy consumption was 45 and $5\,kWh\,m^{-3}$, respectively. $4\,V\,cm^{-1}$ had the highest EPSs removal rate (90.6%) and membrane flux (65.2%) Microbial community diversity was enriched in electro-MBR total phosphorus (TP) removal of ~99% and COD removal of ~85% were achieved Fouling was reduced by 1.3 times in e-MBR	Su et al. (2020)

effective design, better treatment, and low membrane fouling. Compared to the conventional MBR, the MFC-MBR performance improved effectively in terms of pollutant removal efficiencies, energy recovery, biofouling, and sludge production (Liu et al., 2013a). Few studies have investigated the effect of electrochemical treatment on fouling and sludge characteristics (Borea et al., 2017). The current generation was beneficial in terms of fouling reduction as the negative electric charge collected at the cathode prevented the cake attachment on the membrane surface. MFC-MBR sludge has less loosely bound EPS, less soluble microbial product, larger particle size, and decreased zeta potential which effectively enhanced the sludge and membrane filterability and reduced the fouling compared to conventional MBR (Zhou et al., 2015). The maximum power efficiencies of MFC-MBR varied greatly. In one study, it was $6\,W\,m^{-3}$ (Malaeb et al., 2013) and the other reported $0.2–300\,W\,m^{-3}$ (Janicek et al., 2014). The lower electricity generation might be because of the reactor integration, electrode materials, separation membrane, and influent substrate (Logan et al., 2015). Integration of MFC-MBR was reported to be an effective way to reduce membrane fouling in sewage treatment and the bioelectricity process (Li et al., 2020a).

A 90% COD removal has been reported in MFC-MBR using a hollow fiber membrane. The same system was then modified and converted to anaerobic-aerobic conditions for enhancing the nitrification and 69% of total inorganic nitrogen was removed. Energy consumption in this system was lower than the conventional MBR (Ge et al., 2013; Li et al., 2014a). For better performance to remove recalcitrant chemicals and high-power generation, the coated cathode membrane was investigated in MFC-MBR which demonstrated better performance compared to the noncoated cathode membranes in MFC-MBR (Li et al., 2017c). Different types of materials including graphene, graphitic carbon nitride, carbon nanotubes, and their composite were being employed on the cathode/anode of MFC to enhance its performance (Zhang et al., 2017b). A novel membrane was made by coating $CoFe_2O_4$-rGO/PVDF on the carbon fiber cloth and used in the photocatalysis-assisted MFC-MBR. The modified coated cathode membrane greatly improved the filterability and conductivity and reduced the membrane fouling significantly (Li et al., 2017b). Integrated MFC-MBR systems are energy-saving and have low footprints compared to the other electrochemical wastewater treatment systems in use (Wang et al., 2018a).

Table 5 contains more examples of MFC-MBR research studies with details of reactor, operational conditions, and key findings.

7. Conclusions and perspectives

Hybrid MBRs show significant improvement in the removal of nutrients and recalcitrant pollutant as well as can achieve enhanced fouling reduction. However, lifecycle assessment of hybrid MBR and comparison with the conventional MBR and other biological treatment technologies is required to completely understand their benefits and sustainability. For example, a life cycle assessment of the PAC/GAC-MBR along with the conventional MBR/AnMBR needs to be done to have a complete idea of the full benefits of the addition of PAC/GAC. Moreover, cost-benefit analysis and comparison between different types of adsorbents and attached media can reveal the best option.

TABLE 5 MFC-MBR studies and key findings.

Reactor type, membrane, wastewater	Operational parameters	Results/key findings	Reference
MFC-MBR: anaerobic anode chamber (0.76L), aerobic cathode MBR (0.83L), simulated phenol wastewater **Membrane**: PVDF hollow fiber surface area $=0.02\,m^2$ and pore size $=0.2\,\mu m$	HRT: 36h OLR: $0.13\,kgCOD\,m^{-3}\,day^{-1}$ Flux: $20\,L\,m^{-2}\,h^{-1}$	A high aeration rate reduces the electric field to mitigation of membrane fouling Less SMP and LB-EPS, and high PN/PS of LB-EPS were observed in MFC-MBR The mass of the foulants of the membrane surface in MFC-MBR were less than the conventional MBR	Hou et al. (2022)
Two-stage MFC-MBR: fabricated anode and cathode, petroleum refinery wastewater **Membrane**: PVDF hollow fiber, surface area $=0.3\,m^2$ and mean pore size $=0.03\,\mu m$	HRT: 8h OLR: $17.35\,kgCOD\,m^{-3}\,day^{-1}$ Flux: $7\,L\,m^{-2}\,h^{-1}$	Removal efficiencies of COD, ammonium nitrogen (NH4+N), and total nitrogen (TN) were 96.3%, 92.4%, and 86.6%, respectively, and significantly higher than control system The maximum energy recovery of the system ($0.00258\,kWh\,m^{-3}$) was five times higher than that of the control system. A 50% reduction in membrane fouling was observed	Zhao et al. (2022)
MFC-MBR: anodic chamber (240mL), aerobic cathode (12L), DO 2–4mgL^{-1}, synthetic domestic wastewater **Membrane**: PVDF hollow fiber surface area $0.2\,m^2$, $0.03\,\mu m$	HRT in anodic chamber: 0.13h HRT aeration tank: 6.4h Flux: $1.87\,L\,m^{-2}\,h^{-1}$	The SMP in MFC-MBR induced a lower membrane flux decline than that in C-MBR The sludge floc in MFC-MBR showed lower surface hydrophobicity and negative charge The energy barrier for the SMPs in MFC-MBR was 58.9% higher than that for SMPs in Conventional MBR	Li et al. (2021)
MFC-MBR: anoxic tank and oxic tank were 3.5L and 6.5L, respectively, DO 2–4mgL^{-1}, synthetic wastewater **Membrane**: surface area $=0.1\,m^2$ and pore size $=0.22\,\mu m$	HRT: 6.3h Flux: $7.5\,L\,m^{-2}\,h^{-1}$	The electric field intensity could reach $0.114\,V\,cm^{-1}$ Low dissolved oxygen feed water increased the performance of membrane-less MFC 0.5% of the feed COD was translated into electricity	Wang et al. (2016a)
MFC-MBR: Aeration rate $0.25\,m^3\,h^{-1}$, synthetic wastewater, carbon anode electrode with coatings **Membrane**: PVDF hollow fiber, surface area $=0.10\,m^2$ and pore size $=0.02\,\mu m$	HRT: 6.3h OLR: 0.5, $2.0\,kgCOD\,m^{-3}\,day^{-1}$ Flux: $5.0\,L\,m^{-2}\,h^{-1}$	At low-loading, the biofouling increased in MFC-MBR compared to CMBR At high-loading stage, high MLSS showed an active effect on organics degradation, and electric field could be a positive factor with sufficient energy in MFC-MBR Reduction in irreversible fouling mainly resulted from the effect of electric field on the adhesions of the SMP and LB-EPS to membrane surface of MFC-MBR	Wang et al. (2018a)

Composite adsorbents made with different chemicals/nanoparticles and AC has shown promising results. However, only few studies have reported on the efficacy of composite adsorbants. Further research in this area can help develop better adsorbents which can make the MBR technology more sustainable. Moreover, the synergetic effect of GAC and PAC in MBR/AnMBR can further reduce the footprint of the technology in terms of fouling and removal efficiencies.

A detailed analysis of the pollutants adsorbed on AC/zeolite and their chemistry could give us a better understanding which can be used for the modification of the AC/zeolite to improve their performance. Extra energy requirement in electro-MBR needs to be taken into account and assessed for a long period and should be compared with the energy saved through fouling reduction. A comparison of the energy consumed due to the provision of electric current and energy saved due to the fouling reduction shall give an idea about the sustainability of the process.

References

Akamatsu, K., Lu, W., Sugawara, T., Nakao, S.-I., 2010. Development of a novel fouling suppression system in membrane bioreactors using an intermittent electric field. Water Res. 44 (3), 825–830.

Almusawy, A., Al-Anbari, R., Alsalhy, Q., 2021. Mitigation of membrane fouling in waste water treatment plants by using MBBR & sponge membrane bioreactor (sponge-MBR). In: IOP Conference Series: Materials Science and Engineering. IOP Publishing, p. 012108.

Alvarino, T., Torregrosa, N., Omil, F., Lema, J., Suarez, S., 2017. Assessing the feasibility of two hybrid MBR systems using PAC for removing macro and micropollutants. J. Environ. Manag. 203, 831–837.

Artiga, P., García-Toriello, G., Méndez, R., Garrido, J., 2008. Use of a hybrid membrane bioreactor for the treatment of saline wastewater from a fish canning factory. Desalination 221 (1), 518–525.

Asif, M.B., Ren, B., Li, C., Maqbool, T., Zhang, X., Zhang, Z., 2020. Powdered activated carbon-membrane bioreactor (PAC-MBR): impacts of high PAC concentration on micropollutant removal and microbial communities. Sci. Total Environ. 745, 141090.

Aslam, M., McCarty, P.L., Bae, J., Kim, J., 2014. The effect of fluidized media characteristics on membrane fouling and energy consumption in anaerobic fluidized membrane bioreactors. Sep. Purif. Technol. 132, 10–15.

Aslam, M., Yang, P., Lee, P.-H., Kim, J., 2018. Novel staged anaerobic fluidized bed ceramic membrane bioreactor: energy reduction, fouling control and microbial characterization. J. Membr. Sci. 553, 200–208.

Baêta, B., Lima, D., Silva, S.Q., Aquino, S., 2016. Influence of the applied organic load (OLR) on textile wastewater treatment using submerged anaerobic membrane bioreactors (SAMBR) in the presence of redox mediator and powdered activated carbon (PAC). Braz. J. Chem. Eng. 33, 817–825.

Baeta, B.E.L., Luna, H., Sanson, A.L., Silva, S.d.Q., Aquino, S.F.d., 2013. Degradation of a model azo dye in submerged anaerobic membrane bioreactor (SAMBR) operated with powdered activated carbon (PAC). J. Environ. Manag. 128, 462–470.

Bani-Melhem, K., Elektorowicz, M., 2010. Development of a novel submerged membrane electro-bioreactor (SMEBR): performance for fouling reduction. Environ. Sci. Technol. 44 (9), 3298–3304.

Bani-Melhem, K., Smith, E., 2012. Grey water treatment by a continuous process of an electrocoagulation unit and a submerged membrane bioreactor system. Chem. Eng. J. 198, 201–210.

Belli, T.J., Bassin, J.P., Costa, R.E., Akaboci, T.R., Battistelli, A.A., Lobo-Recio, M.A., Lapolli, F.R., 2021. Evaluating the effect of air flow rate on hybrid and conventional membrane bioreactors: implications on performance, microbial activity and membrane fouling. Sci. Total Environ. 755, 142563.

Borea, L., Naddeo, V., Belgiorno, V., Choo, K.-H., 2018. Control of quorum sensing signals and emerging contaminants in electrochemical membrane bioreactors. Bioresour. Technol. 269, 89–95.

Borea, L., Puig, S., Monclús, H., Naddeo, V., Colprim, J., Belgiorno, V., 2017. Microbial fuel cell technology as a downstream process of a membrane bioreactor for sludge reduction. Chem. Eng. J. 326, 222–230.

Brillas, E., Martínez-Huitle, C.A., 2015. Decontamination of wastewaters containing synthetic organic dyes by electrochemical methods. An updated review. Appl. Catal. B Environ. 166, 603–643.

Campinas, M., Rosa, M.J., 2010. Assessing PAC contribution to the NOM fouling control in PAC/UF systems. Water Res. 44 (5), 1636–1644.

Chen, F., Bi, X., Ng, H.Y., 2016. Effects of bio-carriers on membrane fouling mitigation in moving bed membrane bioreactor. J. Membr. Sci. 499, 134–142.

Chen, W.-H., Tsai, C.-Y., Chen, S.-Y., Sung, S., Lin, J.-G., 2019a. Treatment of campus domestic wastewater using ambient-temperature anaerobic fluidized membrane bioreactors with zeolites as carriers. Int. Biodeterior. Biodegradation 136, 49–54.

Chen, M., Xu, J., Dai, R., Wu, Z., Liu, M., Wang, Z., 2019b. Development of a moving-bed electrochemical membrane bioreactor to enhance removal of low-concentration antibiotic from wastewater. Bioresour. Technol. 293, 122022.

Chen, J.-P., Yang, C.-Z., Zhou, J.-H., Wang, X.-Y., 2007. Study of the influence of the electric field on membrane flux of a new type of membrane bioreactor. Chem. Eng. J. 128 (2-3), 177–180.

Chong, C.T., 2015. Performance of Anaerobic Membrane Bioreactors (AnMBRs) With Different Dosages of Powdered Activated Carbon (PAC) at Mesophilic Regime in Membrane Fouling Control. UTAR.

Deng, L., Guo, W., Ngo, H.H., Zhang, J., Liang, S., Xia, S., Zhang, Z., Li, J., 2014. A comparison study on membrane fouling in a sponge-submerged membrane bioreactor and a conventional membrane bioreactor. Bioresour. Technol. 165, 69–74.

Deng, L., Guo, W., Ngo, H.H., Zhang, X., Wang, X.C., Zhang, Q., Chen, R., 2016. New functional biocarriers for enhancing the performance of a hybrid moving bed biofilm reactor-membrane bioreactor system. Bioresour. Technol. 208, 87–93.

Ding, A., Wang, J., Lin, D., Tang, X., Cheng, X., Li, G., Ren, N., Liang, H., 2017. In situ coagulation versus pre-coagulation for gravity-driven membrane bioreactor during decentralized sewage treatment: permeability stabilization, fouling layer formation and biological activity. Water Res. 126, 197–207.

Dutta, K., Tsai, C.-Y., Chen, W.-H., Lin, J.-G., 2014. Effect of carriers on the performance of anaerobic sequencing batch biofilm reactor treating synthetic municipal wastewater. Int. Biodeterior. Biodegradation 95, 84–88.

Ensano, B., Borea, L., Naddeo, V., Belgiorno, V., De Luna, M.D., Ballesteros Jr., F.C., 2016. Combination of electrochemical processes with membrane bioreactors for wastewater treatment and fouling control: a review. Front. Environ. Sci. 57.

Erkan, H.S., Çağlak, A., Soysaloglu, A., Takatas, B., Engin, G.O., 2020. Performance evaluation of conventional membrane bioreactor and moving bed membrane bioreactor for synthetic textile wastewater treatment. J. Water Process Eng. 38, 101631.

Evans, P.J., Parameswaran, P., Lim, K., Bae, J., Shin, C., Ho, J., McCarty, P.L., 2019. A comparative pilot-scale evaluation of gas-sparged and granular activated carbon-fluidized anaerobic membrane bioreactors for domestic wastewater treatment. Bioresour. Technol. 288, 120949.

Fu, Z., Yang, F., Zhou, F., Xue, Y., 2009. Control of COD/N ratio for nutrient removal in a modified membrane bioreactor (MBR) treating high strength wastewater. Bioresour. Technol. 100 (1), 136–141.

Gao, D.-W., Hu, Q., Yao, C., Ren, N.-Q., 2014. Treatment of domestic wastewater by an integrated anaerobic fluidized-bed membrane bioreactor under moderate to low temperature conditions. Bioresour. Technol. 159, 193–198.

Ge, Z., Ping, Q., He, Z., 2013. Hollow-fiber membrane bioelectrochemical reactor for domestic wastewater treatment. J. Chem. Technol. Biotechnol. 88 (8), 1584–1590.

Ghaffour, N., Qamar, A., 2020. Membrane fouling quantification by specific cake resistance and flux enhancement using helical cleaners. Sep. Purif. Technol. 239, 116587.

Giwa, A., Ahmed, I., Hasan, S.W., 2015. Enhanced sludge properties and distribution study of sludge components in electrically-enhanced membrane bioreactor. J. Environ. Manag. 159, 78–85.

Gkotsis, P.K., Mitrakas, M.M., Tolkou, A.K., Zouboulis, A.I., 2017. Batch and continuous dosing of conventional and composite coagulation agents for fouling control in a pilot-scale MBR. Chem. Eng. J. 311, 255–264.

Gkotsis, P., Zouboulis, A., 2019. The use of bio-carriers and zeolite in a lab-scale MBR for membrane fouling mitigation. Glob. NEST J. 21, 58–63.

Güneş, G., Hallaç, E., Özgan, M., Ertürk, A., Taş, D.O., Çokgor, E., Güven, D., Takacs, I., Erdinçler, A., Insel, G., 2019. Enhancement of nutrient removal performance of activated sludge with a novel hybrid biofilm process. Bioprocess Biosyst. Eng. 42 (3), 379–390.

Guo, H., Zhou, J., Su, J., Zhang, Z., 2005. Integration of nitrification and denitrification in airlift bioreactor. Biochem. Eng. J. 23 (1), 57–62.

Gzar, H.A., Al-Rekabi, W.S., Shuhaieb, Z.K., 2021. Applicaion of moving bed biofilm reactor (MBBR) for treatment of industrial wastewater: a mini review. J. Phys. Conf. Ser., 012024.

B. Applications of membrane technology for water and wastewater treatment

Hai, F.I., Yamamoto, K., Nakajima, F., Fukushi, K., 2011. Bioaugmented membrane bioreactor (MBR) with a GAC-packed zone for high rate textile wastewater treatment. Water Res. 45 (6), 2199–2206.

Hashaikeh, R., Lalia, B.S., Kochkodan, V., Hilal, N., 2014. A novel in situ membrane cleaning method using periodic electrolysis. J. Membr. Sci. 471, 149–154.

Hazrati, H., Jahanbakhshi, N., Rostamizadeh, M., 2018. Fouling reduction in the membrane bioreactor using synthesized zeolite nano-adsorbents. J. Membr. Sci. 555, 455–462.

Hou, B., Liu, X., Zhang, R., Li, Y., Liu, P., Lu, J., 2022. Investigation and evaluation of membrane fouling in a microbial fuel cell-membrane bioreactor systems (MFC-MBR). Sci. Total Environ. 814, 152569.

Hu, Q.-Y., Li, M., Wang, C., Ji, M., 2015. Influence of powdered activated carbon addition on water quality, sludge properties, and microbial characteristics in the biological treatment of commingled industrial wastewater. J. Hazard. Mater. 295, 1–8.

Hua, L.-C., Huang, C., Su, Y.-C., Chen, P.-C., 2015. Effects of electro-coagulation on fouling mitigation and sludge characteristics in a coagulation-assisted membrane bioreactor. J. Membr. Sci. 495, 29–36.

Huang, S., Shi, X., Bi, X., Lee, L.Y., Ng, H.Y., 2019. Effect of ferric hydroxide on membrane fouling in membrane bioreactor treating pharmaceutical wastewater. Bioresour. Technol. 292, 121852.

Iorhemen, O.T., Hamza, R.A., Tay, J.H., 2016. Membrane bioreactor (MBR) technology for wastewater treatment and reclamation: membrane fouling. Membranes 6 (2), 33.

Ittisupornrat, S., Phihusut, D., Kitkaew, D., Sangkarak, S., Phetrak, A., 2019. Performance of dissolved organic matter removal from membrane bioreactor effluent by magnetic powdered activated carbon. J. Environ. Manag. 248, 109314.

Ivanovic, I., Leiknes, T., 2012. The biofilm membrane bioreactor (BF-MBR)—a review. Desalin. Water Treat. 37 (1-3), 288–295.

Izadi, A., Hosseini, M., Darzi, G.N., Bidhendi, G.N., Shariati, F.P., 2019. Performance of an integrated fixed bed membrane bioreactor (FBMBR) applied to pollutant removal from paper-recycling wastewater. Water Resourc. Ind. 21, 100111.

Janicek, A., Fan, Y., Liu, H., 2014. Design of microbial fuel cells for practical application: a review and analysis of scale-up studies. Biofuels 5, 79–92.

Ji, J., Qiu, J., Wai, N., Wong, F.-S., Li, Y., 2010. Influence of organic and inorganic flocculants on physical-chemical properties of biomass and membrane-fouling rate. Water Res. 44 (5), 1627–1635.

Ji, B., Yang, K., Wang, H., 2015. Impacts of poly-aluminum chloride addition on activated sludge and the treatment efficiency of SBR. Desalin. Water Treat. 54 (9), 2376–2381.

Jiang, Y., Liu, Y., Shi, D., Fu, W., Sun, P.-F., Li, J., Shao, S., 2020. Membrane fouling in a powdered activated carbon-membrane bioreactor (PAC-MBR) for micro-polluted water purification: fouling characteristics and the roles of PAC. J. Clean. Prod. 277, 122341.

Johir, M., Aryal, R., Vigneswaran, S., Kandasamy, J., Grasmick, A., 2011. Influence of supporting media in suspension on membrane fouling reduction in submerged membrane bioreactor (SMBR). J. Membr. Sci. 374 (1-2), 121–128.

Khalid, A., Abdel-Karim, A., Atieh, M.A., Javed, S., McKay, G., 2018. PEG-CNTs nanocomposite PSU membranes for wastewater treatment by membrane bioreactor. Sep. Purif. Technol. 190, 165–176.

Khan, M., Khan, S.J., Hasan, S.W., 2019. Quorum sensing control and wastewater treatment in quorum quenching/submerged membrane electro-bioreactor (SMEBR (QQ)) hybrid system. Biomass Bioenergy 128, 105329.

Khan, S.J., Visvanathan, C., Jegatheesan, V., 2012. Effect of powdered activated carbon (PAC) and cationic polymer on biofouling mitigation in hybrid MBRs. Bioresour. Technol. 113, 165–168.

Khastoo, H., Hassani, A.H., Mafigholami, R., Mahmoudkhani, R., 2021. Comparing the performance of the conventional and fixed-bed membrane bioreactors for treating municipal wastewater. J. Environ. Health Sci. Eng. 19 (1), 997–1004.

Kim, J.-S., Lee, C.-H., Chun, H.-D., 1998. Comparison of ultrafiltration characteristics between activated sludge and BAC sludge. Water Res. 32 (11), 3443–3451.

Kim, K.-Y., Yang, W., Ye, Y., LaBarge, N., Logan, B.E., 2016. Performance of anaerobic fluidized membrane bioreactors using effluents of microbial fuel cells treating domestic wastewater. Bioresour. Technol. 208, 58–63.

Korotta-Gamage, S.M., Sathasivan, A., 2017. A review: potential and challenges of biologically activated carbon to remove natural organic matter in drinking water purification process. Chemosphere 167, 120–138.

Lee, J., Ahn, W.-Y., Lee, C.-H., 2001. Comparison of the filtration characteristics between attached and suspended growth microorganisms in submerged membrane bioreactor. Water Res. 35 (10), 2435–2445.

Lee, W.-N., Kang, I.-J., Lee, C.-H., 2006. Factors affecting filtration characteristics in membrane-coupled moving bed biofilm reactor. Water Res. 40 (9), 1827–1835.

Lee, E., McCarty, P.L., Kim, J., Bae, J., 2016. Effects of FeCl3 addition on the operation of a staged anaerobic fluidized membrane bioreactor (SAF-MBR). Water Sci. Technol. 74 (1), 130–137.

Lefebvre, O., Ng, K.K., Tang, K.Y., Ng, T.C.A., Ng, H.Y., 2019. Hybrid Processes, New Generation Membranes and Novel MBR Designs. IWA Publishing.

Leiknes, T., Ødegaard, H., 2001. Moving bed biofilm membrane reactor (MBB-MR): characteristics and potentials of a hybrid process design for compact wastewater treatment plants. In: Proceedings of Engineering With Membranes, pp. 52–57.

Lessel, T., 1991. First practical experiences with submerged rope-type bio-film reactors for upgrading and nitrification. Water Sci. Technol. 23 (4-6), 825–834.

Li, T., Cai, Y., Yang, X.-L., Wu, Y., Yang, Y.-L., Song, H.-L., 2020a. Microbial fuel cell-membrane bioreactor integrated system for wastewater treatment and bioelectricity production: overview. J. Environ. Eng. 146 (1), 04019092.

Li, J., Ge, Z., He, Z., 2014a. Advancing membrane bioelectrochemical reactor (MBER) with hollow-fiber membranes installed in the cathode compartment. J. Chem. Technol. Biotechnol. 89 (9), 1330–1336.

Li, X., Liu, Y., Liu, F., Liu, A., Feng, Q., 2017a. Comparison of ferric chloride and aluminum sulfate on phosphorus removal and membrane fouling in MBR treating BAF effluent of municipal wastewater. J. Water Reuse Desal. 7 (4), 442–448.

Li, N., Liu, L., Yang, F., 2014b. Power generation enhanced by a polyaniline-phytic acid modified filter electrode integrating microbial fuel cell with membrane bioreactor. Sep. Purif. Technol. 132, 213–217.

Li, Y., Liu, L., Yang, F., 2017c. Destruction of tetracycline hydrochloride antibiotics by FeOOH/TiO2 granular activated carbon as expanded cathode in low-cost MBR/MFC coupled system. J. Membr. Sci. 525, 202–209.

Li, Y., Sim, L.N., Ho, J.S., Chong, T.H., Wu, B., Liu, Y., 2020b. Integration of an anaerobic fluidized-bed membrane bioreactor (MBR) with zeolite adsorption and reverse osmosis (RO) for municipal wastewater reclamation: comparison with an anoxic-aerobic MBR coupled with RO. Chemosphere 245, 125569.

Li, Y., Sun, J., Liu, L., Yang, F., 2017b. A composite cathode membrane with CoFe2O4–rGO/PVDF on carbon fiber cloth: synthesis and performance in a photocatalysis-assisted MFC-MBR system. Environ. Sci.: Nano 4 (2), 335–345.

Li, H., Xing, Y., Cao, T., Dong, J., Liang, S., 2021. Evaluation of the fouling potential of sludge in a membrane bioreactor integrated with microbial fuel cell. Chemosphere 262, 128405.

Lim, M., Ahmad, R., Guo, J., Tibi, F., Kim, M., Kim, J., 2019. Removals of micropollutants in staged anaerobic fluidized bed membrane bioreactor for low-strength wastewater treatment. Process Saf. Environ. Prot. 127, 162–170.

Lin, J.-L., Hua, L.-C., Wu, Y., Huang, C., 2016. Pretreatment of algae-laden and manganese-containing waters by oxidation-assisted coagulation: effects of oxidation on algal cell viability and manganese precipitation. Water Res. 89, 261–269.

Lin, L., Lei, Z., Wang, L., Liu, X., Zhang, Y., Wan, C., Lee, D.-J., Tay, J.H., 2013. Adsorption mechanisms of high-levels of ammonium onto natural and NaCl-modified zeolites. Sep. Purif. Technol. 103, 15–20.

Liu, J., Ju, X., Gao, B., Wang, L., 2019. Effect of electrocoagulation on MBR under different power supply conditions. Biochem. Eng. J. 152, 107371.

Liu, L., Liu, J., Gao, B., Yang, F., 2012. Minute electric field reduced membrane fouling and improved performance of membrane bioreactor. Sep. Purif. Technol. 86, 106–112.

Liu, J., Liu, L., Gao, B., Yang, F., 2013a. Integration of bio-electrochemical cell in membrane bioreactor for membrane cathode fouling reduction through electricity generation. J. Membr. Sci. 430, 196–202.

Liu, L., Zhao, F., Liu, J., Yang, F., 2013b. Preparation of highly conductive cathodic membrane with graphene (oxide)/PPy and the membrane antifouling property in filtrating yeast suspensions in EMBR. J. Membr. Sci. 437, 99–107.

Logan, B.E., Wallack, M.J., Kim, K.-Y., He, W., Feng, Y., Saikaly, P.E., 2015. Assessment of microbial fuel cell configurations and power densities. Environ. Sci. Technol. Lett. 2 (8), 206–214.

Lu, P., Wang, X., Tang, Y., Ding, A., Yang, H., Guo, J., Cui, Y., Ling, C., 2020. Granular activated carbon assisted nitrate-dependent anaerobic methane oxidation-membrane bioreactor: strengthening effect and mechanisms. Environ. Int. 138, 105675.

Luo, Y., Jiang, Q., Ngo, H.H., Nghiem, L.D., Hai, F.I., Price, W.E., Wang, J., Guo, W., 2015. Evaluation of micropollutant removal and fouling reduction in a hybrid moving bed biofilm reactor-membrane bioreactor system. Bioresour. Technol. 191, 355–359.

B. Applications of membrane technology for water and wastewater treatment

Malaeb, L., Katuri, K.P., Logan, B.E., Maab, H., Nunes, S.P., Saikaly, P.E., 2013. A hybrid microbial fuel cell membrane bioreactor with a conductive ultrafiltration membrane biocathode for wastewater treatment. Environ. Sci. Technol. 47 (20), 11821–11828.

Meng, F., Chae, S.-R., Drews, A., Kraume, M., Shin, H.-S., Yang, F., 2009. Recent advances in membrane bioreactors (MBRs): membrane fouling and membrane material. Water Res. 43 (6), 1489–1512.

Mishima, I., Nakajima, J., 2009. Control of membrane fouling in membrane bioreactor process by coagulant addition. Water Sci. Technol. 59 (7), 1255–1262.

Mohamadi, S., Hazrati, H., Shayegan, J., 2020. Influence of a new method of applying adsorbents on membrane fouling in MBR systems. Water Environ. J. 34, 355–366.

Morgan-Sagastume, F., 2018. Biofilm development, activity and the modification of carrier material surface properties in moving-bed biofilm reactors (MBBRs) for wastewater treatment. Crit. Rev. Environ. Sci. Technol. 48 (5), 439–470.

Ng, K.-K., Lin, C.-F., Panchangam, S.C., Hong, P.-K.A., Yang, P.-Y., 2011. Reduced membrane fouling in a novel bio-entrapped membrane reactor for treatment of food and beverage processing wastewater. Water Res. 45 (14), 4269–4278.

Ngo, H., Nguyen, M., Sangvikar, N., Hoang, T., Guo, W., 2006. Simple approaches towards the design of an attached-growth sponge bioreactor (AGSB) for wastewater treatment and reuse. Water Sci. Technol. 54 (11-12), 191–197.

Nguyen, T.T., Ngo, H.H., Guo, W., 2013. Pilot scale study on a new membrane bioreactor hybrid system in municipal wastewater treatment. Bioresour. Technol. 141, 8–12.

Nguyen, T.T., Ngo, H.H., Guo, W., Johnston, A., Listowski, A., 2010. Effects of sponge size and type on the performance of an up-flow sponge bioreactor in primary treated sewage effluent treatment. Bioresour. Technol. 101 (5), 1416–1420.

Nordell, E., Hansson, A.B., Karlsson, M., 2013. Zeolites relieves inhibitory stress from high concentrations of long chain fatty acids. Waste Manag. 33 (12), 2659–2663.

Odegaard, H., 2000. Advanced compact wastewater treatment based on coagulation and moving bed biofilm processes. Water Sci. Technol. 42 (12), 33–48.

Palmarin, M.J., Young, S., 2019. Comparison of the treatment performance of a hybrid and conventional membrane bioreactor for greywater reclamation. J. Water Process Eng. 28, 54–59.

Pan, Z., Zhang, C., Huang, B., 2016. Using adsorbent made from sewage sludge to enhance wastewater treatment and control fouling in a membrane bioreactor. Desalin. Water Treat. 57 (20), 9070–9081.

Park, J., Yamashita, N., Tanaka, H., 2018. Membrane fouling control and enhanced removal of pharmaceuticals and personal care products by coagulation-MBR. Chemosphere 197, 467–476.

Petropoulos, E., Shamurad, B., Tabraiz, S., Yu, Y., Davenport, R., Curtis, T.P., Dolfing, J., 2021. Sewage treatment at 4° C in anaerobic upflow reactors with and without a membrane—performance, function and microbial diversity. Environ. Sci.: Water Res. Technol. 7 (1), 156–171.

Poirier, S., Madigou, C., Bouchez, T., Chapleur, O., 2017. Improving anaerobic digestion with support media: mitigation of ammonia inhibition and effect on microbial communities. Bioresour. Technol. 235, 229–239.

Remy, M., van der Marel, P., Zwijnenburg, A., Rulkens, W., Temmink, H., 2009. Low dose powdered activated carbon addition at high sludge retention times to reduce fouling in membrane bioreactors. Water Res. 43 (2), 345–350.

Rodríguez-Hernández, L., Esteban-García, A., Tejero, I., 2014. Comparison between a fixed bed hybrid membrane bioreactor and a conventional membrane bioreactor for municipal wastewater treatment: a pilot-scale study. Bioresour. Technol. 152, 212–219.

Rusten, B., Eikebrokk, B., Ulgenes, Y., Lygren, E., 2006. Design and operations of the Kaldnes moving bed biofilm reactors. Aquac. Eng. 34 (3), 322–331.

Saidulu, D., Majumder, A., Gupta, A.K., 2021. A systematic review of moving bed biofilm reactor, membrane bioreactor, and moving bed membrane bioreactor for wastewater treatment: comparison of research trends, removal mechanisms, and performance. J. Environ. Chem. Eng., 106112.

Salcedo Moyano, A.J., Delforno, T.P., Subtil, E.L., 2021. Simultaneous nitrification-denitrification (SND) using a thermoplastic gel as support: pollutants removal and microbial community in a pilot-scale biofilm membrane bioreactor. Environ. Technol., 1–15.

Schenkelberg, K., Horton, B., 2009. Application of membrane bioreactor technology to meet stringent load limits on Virginia's. Proc. Water Environ. Fed. 2009 (15), 2203–2215.

Serrano, D., Suárez, S., Lema, J., Omil, F., 2011. Removal of persistent pharmaceutical micropollutants from sewage by addition of PAC in a sequential membrane bioreactor. Water Res. 45 (16), 5323–5333.

Shao, S., Cai, L., Li, K., Li, J., Du, X., Li, G., Liang, H., 2017. Deposition of powdered activated carbon (PAC) on ultrafiltration (UF) membrane surface: influencing factors and mechanisms. J. Membr. Sci. 530, 104–111.

Shin, C., McCarty, P.L., Kim, J., Bae, J., 2014. Pilot-scale temperate-climate treatment of domestic wastewater with a staged anaerobic fluidized membrane bioreactor (SAF-MBR). Bioresour. Technol. 159, 95–103.

Skouteris, G., Rodriguez-Garcia, G., Reinecke, S., Hampel, U., 2020. The use of pure oxygen for aeration in aerobic wastewater treatment: a review of its potential and limitations. Bioresour. Technol. 312, 123595.

Sohn, W., Guo, W., Ngo, H.H., Deng, L., Cheng, D., Zhang, X., 2021. A review on membrane fouling control in anaerobic membrane bioreactors by adding performance enhancers. J. Water Process Eng. 40, 101867.

Song, K.-G., Kim, Y., Ahn, K.-H., 2008. Effect of coagulant addition on membrane fouling and nutrient removal in a submerged membrane bioreactor. Desalination 221 (1-3), 467–474.

Song, J., Yin, Y., Li, Y., Gao, Y., Liu, Y., 2020. In-situ membrane fouling control by electrooxidation and microbial community in membrane electro-bioreactor treating aquaculture seawater. Bioresour. Technol. 314, 123701.

Stoquart, C., Servais, P., Bérubé, P.R., Barbeau, B., 2012. Hybrid membrane processes using activated carbon treatment for drinking water: a review. J. Membr. Sci. 411, 1–12.

Strathmann, H., Giorno, L., Drioli, E., 2011. Introduction to Membrane Science and Technology. Wiley-VCH Verlag & Company.

Su, F., Liang, Y., Liu, G., Mota Filho, C.R., Hu, C., Qu, J., 2020. Enhancement of anti-fouling and contaminant removal in an electro-membrane bioreactor: significance of electrocoagulation and electric field. Sep. Purif. Technol. 248, 117077.

Sun, F., Li, P., Li, J., Li, H., Ou, Q., Sun, T., Dong, Z., 2015. Hybrid biofilm-membrane bioreactor (Bf-MBR) for minimization of bulk liquid-phase organic substances and its positive effect on membrane permeability. Bioresour. Technol. 198, 772–780.

Sun, F., Wu, D., Chua, F.D., Zhu, W., Zhou, Y., 2019. Free nitrous acid (FNA) induced transformation of sulfamethoxazole in the enriched nitrifying culture. Water Res. 149, 432–439.

Sun, H., Yang, Z., Yang, F., Wu, W., Wang, J., 2020. Enhanced simultaneous nitrification and denitrification performance in a fixed-bed system packed with PHBV/PLA blends. Int. Biodeterior. Biodegradation 146, 104810.

Tabraiz, S., Hassan, S., Abbas, A., Nasreen, S., Zeeshan, M., Fida, S., Shamurad, B.A., Acharya, K., Petropoulos, E., 2018. Effect of effluent and sludge recirculation ratios on integrated fixed films A2O system nutrients removal efficiency treating sewage. Desalin. Water Treat. 114, 120–127.

Tabraiz, S., Haydar, S., Sallis, P., Nasreen, S., Mahmood, Q., Awais, M., Acharya, K., 2017. Effect of cycle run time of backwash and relaxation on membrane fouling removal in submerged membrane bioreactor treating sewage at higher flux. Water Sci. Technol. 76 (4), 963–975.

Tabraiz, S., Petropoulos, E., Shamurad, B., Quintela-Baluja, M., Mohapatra, S., Acharya, K., Charlton, A., Davenport, R.J., Dolfing, J., Sallis, P.J., 2021a. Temperature and immigration effects on quorum sensing in the biofilms of anaerobic membrane bioreactors. J. Environ. Manag. 293, 112947.

Tabraiz, S., Shamurad, B., Petropoulos, E., Charlton, A., Mohiudin, O., Danish Khan, M., Ekwenna, E., Sallis, P., 2020. Diversity of acyl homoserine lactone molecules in anaerobic membrane bioreactors treating sewage at psychrophilic temperatures. Membranes 10 (11), 320.

Tabraiz, S., Shamurad, B., Petropoulos, E., Quintela-Baluja, M., Charlton, A., Dolfing, J., Sallis, P.J., 2021b. Mitigation of membrane biofouling in membrane bioreactor treating sewage by novel quorum quenching strain of Acinetobacter originating from a full-scale membrane bioreactor. Bioresour. Technol. 334, 125242.

Tafti, A.D., Mirzaii, S.M.S., Andalibi, M.R., Vossoughi, M., 2015. Optimized coupling of an intermittent DC electric field with a membrane bioreactor for enhanced effluent quality and hindered membrane fouling. Sep. Purif. Technol. 152, 7–13.

Third, K., Gibbs, B., Newland, M., Cord-Ruwisch, R., 2005. Long-term aeration management for improved N-removal via SND in a sequencing batch reactor. Water Res. 39 (15), 3523–3530.

Thuy, Q.T.T., 2003. Treatment of Inhibitory Phenolic Compounds by Membrane Bioreactor. Asian Institute of Technology.

Villamil, J., Monsalvo, V., Lopez, J., Mohedano, A., Rodriguez, J., 2016. Fouling control in membrane bioreactors with sewage-sludge based adsorbents. Water Res. 105, 65–75.

Vyrides, I., Stuckey, D., 2009. Saline sewage treatment using a submerged anaerobic membrane reactor (SAMBR): effects of activated carbon addition and biogas-sparging time. Water Res. 43 (4), 933–942.

Wang, J., Bi, F., Ngo, H.-H., Guo, W., Jia, H., Zhang, H., Zhang, X., 2016a. Evaluation of energy-distribution of a hybrid microbial fuel cell-membrane bioreactor (MFC-MBR) for cost-effective wastewater treatment. Bioresour. Technol. 200, 420–425.

Wang, Y., Jia, H., Wang, J., Cheng, B., Yang, G., Gao, F., 2018a. Impacts of energy distribution and electric field on membrane fouling control in microbial fuel cell-membrane bioreactor (MFC-MBR) coupling system. Bioresour. Technol. 269, 339–345.

Wang, J., Wu, B., Liu, Y., Fane, A.G., Chew, J.W., 2018b. Monitoring local membrane fouling mitigation by fluidized GAC in lab-scale and pilot-scale AnFMBRs. Sep. Purif. Technol. 199, 331–345.

Wang, J., Wu, B., Yang, S., Liu, Y., Fane, A.G., Chew, J.W., 2016b. Characterizing the scouring efficiency of Granular Activated Carbon (GAC) particles in membrane fouling mitigation via wavelet decomposition of accelerometer signals. J. Membr. Sci. 498, 105–115.

Wang, H., Xu, J., Sheng, L., Liu, X., Zong, M., Yao, D., 2019. Anaerobic digestion technology for methane production using deer manure under different experimental conditions. Energies 12 (9), 1819.

Wang, J., Zamani, F., Cahyadi, A., Toh, J.Y., Yang, S., Wu, B., Liu, Y., Fane, A.G., Chew, J.W., 2016c. Correlating the hydrodynamics of fluidized granular activated carbon (GAC) with membrane-fouling mitigation. J. Membr. Sci. 510, 38–49.

Wei, D., Xue, X., Chen, S., Zhang, Y., Yan, L., Wei, Q., Du, B., 2013. Enhanced aerobic granulation and nitrogen removal by the addition of zeolite powder in a sequencing batch reactor. Appl. Microbiol. Biotechnol. 97 (20), 9235–9243.

Widjaja, T., Miyata, T., Nakano, Y., Nishijima, W., Okada, M., 2004. Adsorption capacity of powdered activated carbon for 3, 5-dichlorophenol in activated sludge. Chemosphere 57 (9), 1219–1224.

Wilderer, P., Arnz, P., Arnold, E., 2000. Application of biofilms and biofilm support materials as a temporary sink and source. Water Air Soil Pollut. 123 (1-4), 147–158.

Wozniak, T., 2010. MBR design and operation using MPE-technology (Membrane Performance Enhancer). Desalination 250 (2), 723–728.

Wu, J., Chen, F., Huang, X., Geng, W., Wen, X., 2006. Using inorganic coagulants to control membrane fouling in a submerged membrane bioreactor. Desalination 197 (1-3), 124–136.

Wu, J., Huang, X., 2008. Effect of dosing polymeric ferric sulfate on fouling characteristics, mixed liquor properties and performance in a long-term running membrane bioreactor. Sep. Purif. Technol. 63 (1), 45–52.

Wu, B., Wong, P.C.Y., Fane, A.G., 2015. The potential roles of granular activated carbon in anaerobic fluidized membrane bioreactors: effect on membrane fouling and membrane integrity. Desalin. Water Treat. 53 (6), 1450–1459.

Wu, B., Zamani, F., Lim, W., Liao, D., Wang, Y., Liu, Y., Chew, J.W., Fane, A.G., 2017. Effect of mechanical scouring by granular activated carbon (GAC) on membrane fouling mitigation. Desalination 403, 80–87.

Wu, J., Zhang, Y., Wang, J., Zheng, X., Chen, Y., 2021. Municipal wastewater reclamation and reuse using membrane-based technologies: a review. Desalin. Water Treat. 224, 65–82.

Xu, X., Liu, G.-H., Li, Q., Wang, H., Sun, X., Shao, Y., Zhang, J., Liu, S., Luo, F., Wei, Q., 2021. Optimization nutrient removal at different volume ratio of anoxic-to-aerobic zone in integrated fixed-film activated sludge (IFAS) system. Sci. Total Environ. 795, 148824.

Yang, Q., Chen, J., Zhang, F., 2006. Membrane fouling control in a submerged membrane bioreactor with porous, flexible suspended carriers. Desalination 189 (1), 292–302.

Yang, Y., Qiao, S., Jin, R., Zhou, J., Quan, X., 2019a. A novel aerobic electrochemical membrane bioreactor with CNTs hollow fiber membrane by electrochemical oxidation to improve water quality and mitigate membrane fouling. Water Res. 151, 54–63.

Yang, C., Qiu, C., He, C., Hu, Z., Wang, W., 2019b. Influence of aluminium accumulation on biological nitrification and phosphorus removal in an anoxic-oxic membrane bioreactor. Environ. Sci. Pollut. Res. 26 (27), 28127–28134.

Ying, Z., Ping, G., 2006. Effect of powdered activated carbon dosage on retarding membrane fouling in MBR. Sep. Purif. Technol. 52 (1), 154–160.

Yoo, R., Kim, J., McCarty, P.L., Bae, J., 2012. Anaerobic treatment of municipal wastewater with a staged anaerobic fluidized membrane bioreactor (SAF-MBR) system. Bioresour. Technol. 120, 133–139.

Yoon, S.-H., Collins, J.H., 2006. A novel flux enhancing method for membrane bioreactor (MBR) process using polymer. Desalination 191 (1), 52–61.

Yousefi, S.A., Nasser, M.S., Hussein, I.A., Benamor, A., El-Naas, M.H., 2020. Influence of polyelectrolyte structure and type on the degree of flocculation and rheological behavior of industrial MBR sludge. Sep. Purif. Technol. 233, 116001.

Yu, S., Wang, J., Zhao, Z., Cai, W., 2022. Simultaneous coupling of fluidized granular activated carbon (GAC) and powdered activated carbon (PAC) with ultrafiltration process: a promising synergistic alternative for water treatment. Sep. Purif. Technol. 282, 120085.

Zeeshan, M., Haydar, S., Tabraiz, S., 2017. Effect of fixed media surface area on biofouling and nutrients removal in fixed film membrane bioreactor treating sewage at medium and high fluxes. Water Air Soil Pollut. 228 (9), 1–10.

Zhang, S., Chang, J., Lin, C., Pan, Y., Cui, K., Zhang, X., Liang, P., Huang, X., 2017a. Enhancement of methanogenesis via direct interspecies electron transfer between Geobacteraceae and Methanosaetaceae conducted by granular activated carbon. Bioresour. Technol. 245, 132–137.

Zhang, J., Chua, Q.W., Mao, F., Zhang, L., He, Y., Tong, Y.W., Loh, K.-C., 2019a. Effects of activated carbon on anaerobic digestion-methanogenic metabolism, mechanisms of antibiotics and antibiotic resistance genes removal. Bioresource Technol. Rep. 5, 113–120.

Zhang, Y., Liu, L., Van der Bruggen, B., Yang, F., 2017b. Nanocarbon based composite electrodes and their application in microbial fuel cells. J. Mater. Chem. A 5 (25), 12673–12698.

Zhang, J., Satti, A., Chen, X., Xiao, K., Sun, J., Yan, X., Liang, P., Zhang, X., Huang, X., 2015a. Low-voltage electric field applied into MBR for fouling suppression: performance and mechanisms. Chem. Eng. J. 273, 223–230.

Zhang, W., Tang, B., Bin, L., 2017c. Research progress in biofilm-membrane bioreactor: a critical review. Ind. Eng. Chem. Res. 56 (24), 6900–6909.

Zhang, Z., Wang, Y., Leslie, G.L., Waite, T.D., 2015b. Effect of ferric and ferrous iron addition on phosphorus removal and fouling in submerged membrane bioreactors. Water Res. 69, 210–222.

Zhang, Y., Wang, X., Ye, H., Zhou, L., Zhao, Z., 2021. Effect and mechanism of reduced membrane bioreactor fouling by powdered activated carbon. Water Sci. Technol. 83 (5), 1005–1016.

Zhang, S., Xiong, J., Zuo, X., Liao, W., Ma, C., He, J., Chen, Z., 2019b. Characteristics of the sludge filterability and microbial composition in PAC hybrid MBR: effect of PAC replenishment ratio. Biochem. Eng. J. 145, 10–17.

Zhang, S., Zuo, X., Xiong, J., Ma, C., Hu, B., 2019c. Effect of powdered activated carbon dosage on sludge properties and membrane bioreactor performance in a hybrid MBR-PAC system. Environ. Technol. 40 (9), 1156–1165.

Zhao, S., Yun, H., Khan, A., Salama, E.-S., Redina, M.M., Liu, P., Li, X., 2022. Two-stage microbial fuel cell (MFC) and membrane bioreactor (MBR) system for enhancing wastewater treatment and resource recovery based on MFC as a biosensor. Environ. Res. 204, 112089.

Zhou, G., Zhou, Y., Zhou, G., Lu, L., Wan, X., Shi, H., 2015. Assessment of a novel overflow-type electrochemical membrane bioreactor (EMBR) for wastewater treatment, energy recovery and membrane fouling mitigation. Bioresour. Technol. 196, 648–655.

Engineered membrane processes for nutrient removal and microalgae harvesting

Lijuan Deng, Huu Hao Ngo, Bing-Jie Ni, Wei Wei, Qilin Wang, and Wenshan Guo

Centre for Technology in Water and Wastewater, School of Civil and Environmental Engineering, University of Technology Sydney, Sydney, NSW, Australia

1. Introduction

Membrane technology, especially membrane bioreactor (MBR), has been widely used in wastewater treatment due to its small footprint, solid–liquid separation, high effluent quality, and removal of protozoans, bacteria, and viruses. Nevertheless, to improve nutrient removal, it requires additional anoxic and/or anaerobic conditions, which increase operational complexity, operational and maintenance cost, and land space demand (Kimura et al., 2008; Luo et al., 2017). Microalgae-based wastewater treatment is also applied as a cost-effective and environmentally friendly method for nutrient removal and recovery from wastewater. Microalgae consume inorganic carbon (i.e., CO_2) with light by photosynthesis and/or organic carbon under dark condition, nitrogen and phosphorus from wastewater for their growth (Babaei et al., 2016; Vu et al., 2022). Consequently, this technology not only generates high-quality treated water, but also reduces greenhouse gas emission (Chong et al., 2019). However, the wide application of microalgal biotechnology is limited by the high production cost of the microalgal biomass. This is mainly determined by microalgae harvesting, which is a process that increases microalgal cells concentration to an economically advantageous level attractive for downstream steps (i.e., extraction, drying, etc.) through separating microalgae from culture solution (Cho et al., 2020; Bilad et al., 2014; Hafiz et al., 2020). Development of efficient and cost-saving harvesting methods confronts some challenges related to the inherent properties of microalgae, including small particle size (2–30 μm), similar density with

Copyright © 2023 Elsevier Inc. All rights reserved.

water, low concentration of cultured suspension ($\sim$99.9% water), negative surface charge, and various culture conditions. These properties induce high difficulty in collecting and recovering microalgae. Considerable energy input is required in the harvesting and dewatering steps during the microalgal production process (Min et al., 2022; Uduman et al., 2010). The harvesting process is the most challenging obstacle in the production process and accounts for 50% of the total specific energy consumption of the process (Hafiz et al., 2020). Moreover, microalgae release organic matters (i.e., microalgal extracellular organic matters (EOM)) during growth and storage (Wang et al., 2019; Cho et al., 2020; Zhao et al., 2021a). To address these issues, conventional microalgal harvesting methods have been employed, e.g., centrifugation, gravity sedimentation, and chemical coagulation/flocculation. Nevertheless, these methods are environmentally unfriendly, energy-consuming, time-consuming, and labor-intensive. The cell structure may be destroyed during the centrifugation process by the high gravitational force and shear stresses. Gravity sedimentation is an unreliable and ineffective process owing to its low slurry output and requirement of further thickening. Chemical coagulation/flocculation is pH-sensitive and not eco-friendly due to the contamination of final products by metal salts. Hence, metal salts need to be separated from the algal biomass (Bamba et al., 2020; Li et al., 2020; Yin et al., 2020).

Membrane filtration has been considered as a promising alternative to traditional microalgae harvesting approaches owing to its high microalgal biomass retention, easy operation and upscaling, and reduced application of chemicals (Razak et al., 2020). This technology can also generate stable and purified filtrate without the yield of harvested biomass impurities, as well as the recycle of culture medium by retaining residual nutrients for cost-effective microalgae harvesting and nutrient removal (Wang et al., 2019). However, membrane fouling occurs during the filtration process and is caused by the accumulation of microsized microalgal cells and EOM on the membrane surface and/or inside membrane pores. It results in permeate flux decline, increases the operational and maintenance costs, and shortens the membrane life span (Cho et al., 2020; Razak et al., 2020). Hence, membrane fouling control strategies should be developed and applied during the microalgae harvesting process.

This chapter discusses the performance of membrane-based systems (membrane bioreactor-based systems and membrane-based hybrid systems) in terms of microalgae harvesting and nutrient removal. Furthermore, challenges on the application of membrane-based systems are demonstrated with a major focus on membrane fouling, fouling control approaches, and comparison with other systems. It also provides future research directions on the development and applications of novel carriers, immobilized beads, and advanced in situ physical cleaning methods as well as techno-economic analyses for membrane-based systems.

2. Performance of membrane-based systems

During the membrane filtration process, microalgal cells are separated from the culture medium using microfiltration (MF), ultrafiltration (UF), macrofiltration, dead-end filtration, tangential flow filtration (TFF), vacuum filtration, or pressure filtration for microalgae harvesting (Singh and Patidar, 2018; Yin et al., 2020).

2.1 Membrane bioreactor-based systems

2.1.1 *Microalgae membrane photobioreactor (MPBR)*

The microalgae membrane photobioreactor (MPBR) (or microalgae-based membrane bioreactor (MMBR)) consists of an enclosed cultivation process [photobioreactor (PBR)] and one submerged or sidestream pressure-driven membrane filtration process (microfiltration or ultrafiltration) (Fig. 1A and B). MPBR can completely retain microalgal cells by using a membrane, which avoids biomass washout, enables complete separation of microalgal cells from effluent, and decouples control of solids retention time (SRT) and hydraulic retention time

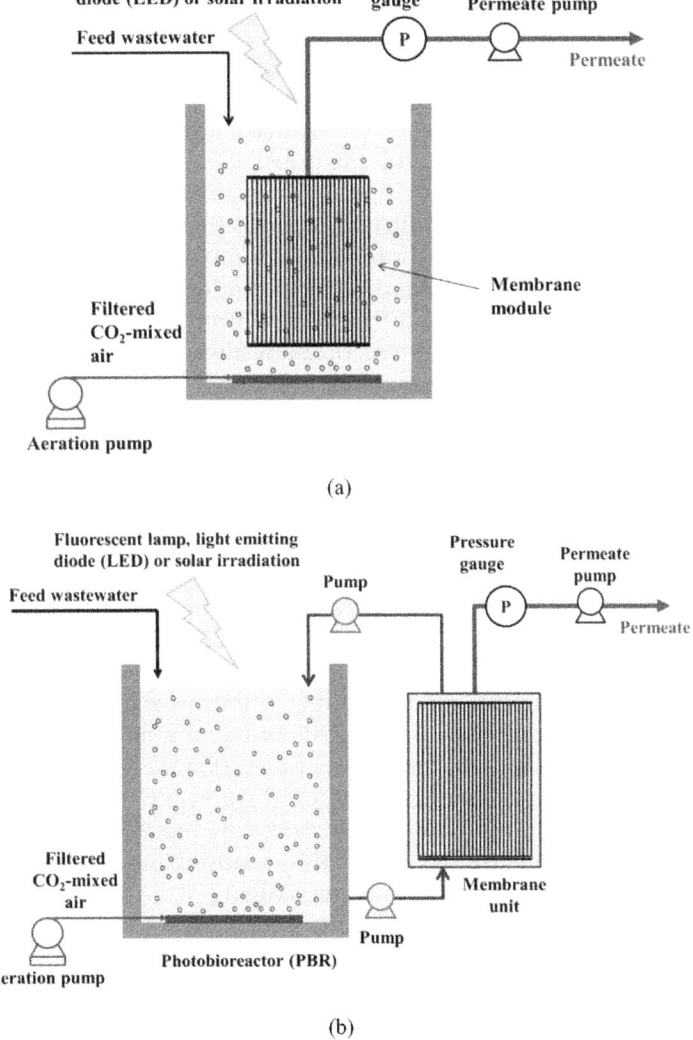

(a)

(b)

FIG. 1 Schematic diagrams of microalgae membrane photobioreactor (MPBR) with (A) a submerged membrane filtration process and (B) a sidestream membrane filtration process.

(HRT). It prompts greater microalgal harvesting, increases biomass productivity and nutrient removal, and reduces footprint compared to PBR (Luo et al., 2017). Under mixotrophic cultivation condition, microalgae (*Chlorella vulgaris*) could simultaneously take up and assimilate nutrients, inorganic carbon (CO_2 supplied by air) and organic carbon (glucose supplied by feed wastewater) into biomass when treating high strength ammonium wastewater by MMBR. This encouraged microalgal cultivation, resulting in significantly higher mixotrophic microalgal biomass productivity (1.8–3.5 times higher) than corresponding productivities in heterotrophic culture (utilization of organic carbon under darkness) and photoautotrophic culture (utilization of inorganic carbon under light) with NH_4^+-N as the nitrogen source. Moreover, the highest nutrient removal was also obtained under mixotrophic cultivation condition, reaching nitrogen removal rates of 23.64–28.80 mg/L · d and PO_4^{3-}-P removal rates of 4.38–5.08 mg/L · d with NH_4^+-N or NO_3^--N as the nitrogen source (Babaei et al., 2016).

In the MPBR, the pH value is generally kept in the range of 6–8 (close to neutral pH), which effectively stimulates the production of microalgae and avoids loss of nutrients. Under the neutral pH condition, nutrient removal is mainly accomplished by microalgae assimilation, while chemical processes commonly involving nutrient removal in microalgae-based wastewater treatment processes (ammonia stripping via volatilization and chemical precipitation of phosphates) have minor effects on nutrient removal (Babaei et al., 2016; Peng et al., 2020; Gao et al., 2021). During the operational process, CO_2-mixed air is supplied from the bottom of the reactor, which mixes microalgae, directly provides inorganic carbon for microalgal growth and scours membrane surface for fouling control. In lab-scale experiments, the air is needed to be filtered by some filters (i.e., 0.22-μm filter, 0.45-μm cartridge filter) prior to air supply, in order to avoid airborne contamination (Lee et al., 2018; Peng et al., 2020).

Aeration and concentration of CO_2 in air as important factors should be optimized to encourage cell growth and enable microalgae to more easily contact light and nutrients. Relatively high temperature (25–30°C) is favorable for microalgae growth (Luo et al., 2017). Control of SRT or biomass retention time (BRT) encourages the microalgae production and nutrient removal. Compared to no biomass extraction or longer SRT, the daily biomass extraction or reduced SRT accelerates microalgae growth in the bioreactor through moderating limitations of light and carbon source, which gives rise to higher biomass viability and better nutrient removal via microalgal assimilation (Nguyen et al., 2020; Luo et al., 2018). However, biomass harvesting will be not satisfactory at low biomass concentrations when operating at lower SRT. Moreover, SRT control requires direct withdrawal of large volumes of biomass from the MPBR, which compromises the benefits of membrane filtration and increases the demand from the extra downstream treatment system (Parakh et al., 2020).

The shortened HRT improves the microalgae production and nutrient removal rate. This is ascribed to the fact that the volume of effluent withdrawing from the photobioreactor and supply of feed wastewater increase at a lower HRT, resulting in a higher nutrient supply rate of the photobioreactor and greater growth of microalgal cells. Additionally, the complete retention of microalgal cells by membrane inhibits the loss of microalgae within effluent (Gao et al., 2016, 2018). A greater volumetric microalgae production rate was achieved at a shorter HRT (0.118 g/L · d at an HRT of 1 d) than at an extended HRT (0.064 g/L · d at an HRT of 3 d) when treating a secondary treatment effluent. Moreover, the reduction of HRT also increased total nitrogen (TN) and total phosphorus (TP) removal rates (5.55 and 0.40 mg/L · d at an HRT of 1 d vs 2.86 mg/L·d and 0.17 mg/L·d at an HRT of 3 d, respectively) (Solmaz and Işık, 2020). During tertiary treatment of livestock wastewater, the declined HRT from 5 to 3 d increased nitrogen and phosphorus removals by 19%–36%, reaching up to 96% of TN removal and 85% of TP removal (Lee et al., 2018). However, when treating wastewater containing very high nitrogen and phosphorus concentrations (i.e., TN of 1012 mg/L, TP of 318 mg/L), the longer HRT favored nutrient removal (30% higher at an HRT of 5 d than at an HRT of 3 d) through prolonging contact time for nutrient assimilation by microalgae and increasing pH for ammonia volatilization (Vu et al., 2022).

The critical value of nutrient supply exists concerning microalgae growth. When nutrient supply is higher than the critical value, microalgae growth is affected by other factors (i.e., light intensity, light wavelength, and light/dark cycle). At an HRT in the range of 6–2 d, the declined HRT improved microalgae generation (from 26.69 mg/L·d at 6 d to 48.78 mg/L·d at 2 d) by elevating the nutrient supply rate for microalgal growth and completely retaining microalgae by membrane. However, the shortened HRT within the range of 2–1 d did not affect the microalgal growth, i.e., 48.50 mg/L d at an HRT of 1 d, 48.78 mg/L d at an HRT of 2 d (Gao et al., 2018; Luo et al., 2017). Besides, it should select HRT and SRT based on the performance requirements (i.e., biomass productivity and harvesting potential, discharge quality). Short HRT (1 d) and SRT (9 d) gave rise to higher microalgal cell viability and productivity (up to 0.25×10^7 cells/mL·d) owing to the increased nutrient loading and less nutrient requirement by faster-growing algal cells. On the other hand, at longer HRT (4 d) and SRT (30 d), the microalgal biomass flocculation was elevated due to the declined surface charge and release of higher amounts of algal organic matters (AOM). Moreover, NO_3^--N and PO_4^{3-}-P removals also increased (up to 79% and 78%, respectively) owing to the increased nutrient uptake for new cell growth at lower nutrient loadings (Luo et al., 2018).

Apart from light intensity, wavelength, light/dark cycle, and illumination (Luo et al., 2017), the light path also affects microalgae biomass productivity and nutrient removal. The narrower light path increased biomass concentration (0.920 gVSS/L in the 10-cm MPBR plant vs 0.288 gVSS/L in the 25-cm light path MPBR plant) and areal biomass productivity (20.0 g VSS/m^2·d at 10 cm vs 15.7 g VSS/m^2·d at 25 cm) since light could be more efficiently utilized for photosynthesis. Moreover, the nitrogen and phosphorus recovery rates were significantly higher with a narrower light path (150% and 103% higher) than with a wider light path (González-Camejo et al., 2020).

2.1.2 Forward osmosis-based osmotic membrane photobioreactor (OMPBR)

Forward osmosis (FO) has high potential in concentrating microalgae as it employs an osmotic pressure gradient to allow pure water to move from concentrated feed solution (FS) across the hydrophilic and porous membrane to a diluted draw solution (DS). Compared to pressure-driven membranes (i.e., microfiltration, ultrafiltration), FO shows less fouling propensity, better fouling reversibility, greater separation efficiency, and higher recovery of microalgal cells (Jiang et al., 2020; Nawi et al., 2020). Fig. 2A and B show two different configurations of OMPBR. The FO membrane process demonstrates excellent nutrient rejection and efficiently retains nutrients in the photobioreactor (Jafarinejad, 2021). Therefore, OMPBR showed better nutrient removal (NH_4^+-N and NO_3^--N removals, 86%–99%; PO_4^{3-}-P removal, 100%) than MPBR with microfiltration (NH_4^+-N and NO_3^--N removals, 48%–97%; PO_4^{3-}-P removal, 46%) and good microalgae harvesting ability (total microalgae

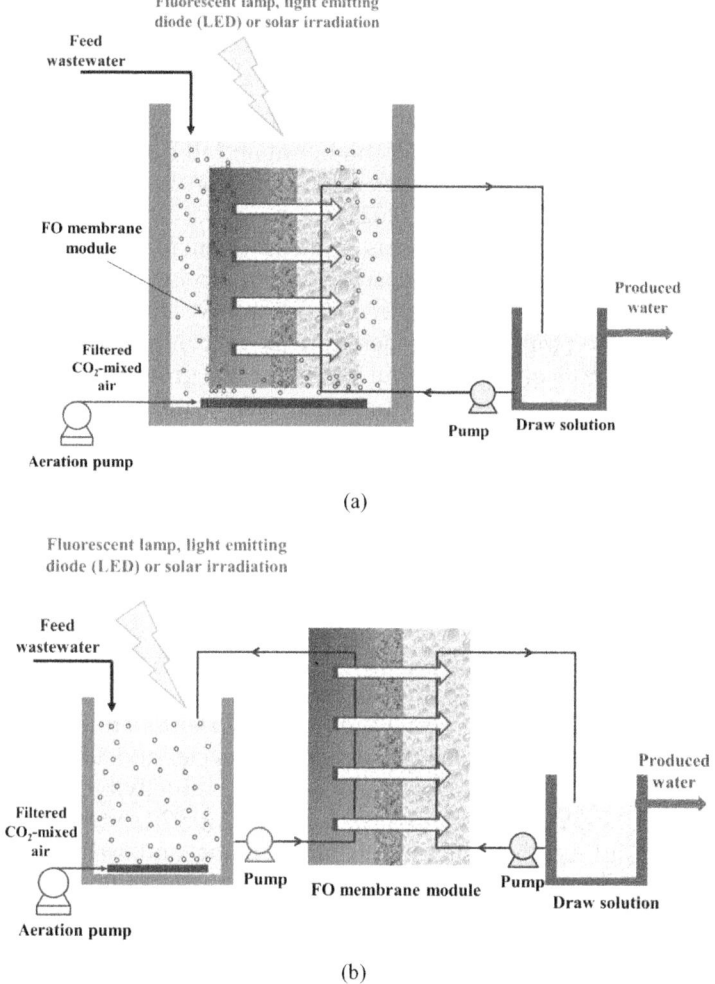

FIG. 2   Forward osmosis-based osmotic membrane photobioreactor (OMPBR) with (A) a submerged FO membrane module and (B) a sidestream FO membrane module.

biomass >2 g/L) when operating at HRTs of 1–2 d (Praveen et al., 2016). At the extended HRT of 3.33 d, OMPBR could resist NH_4^+-N overloading from feed wastewater (45 mg/L NH_4Cl in mimicked tertiary wastewater) and effectively eliminate NH_4^+-N and PO_4^{3-}-P ($\sim$ 2.0 mg/L and <0.6 mg/L in the effluent, respectively). This was accomplished via microalgal uptake, the presence of dead volume in the FO hollow fiber chamber, and the formation of ammonium and phosphate-based precipitates. When operating at a much longer HRT, microalgal cells were prone to endogenous decay at a lower nutrient loading, reducing the percentage of microalgae in the harvested biomass (from >85% at an HRT of 3.33 d to around 65% at an HRT of 6.67 d). The prolonged SRT increased the biomass density and the corresponding concentration factor, i.e., 2.03-fold at 9.41 d, and 2.93-fold at an SRT of 17.82 d. However, a too high SRT (25.26 d) decreased the activity and harvesting rate of microalgae due to the accumulation of soluble algal product. During the operational process, periodical supplementation of necessary elements (i.e., Ca^{2+}, Mg^{2+}, SO_4^{2-}) was required to compensate the depletion of essential cations for microalgal growth induced by the forward ion exchange (Na^+ and K^+) and chemical precipitation (Mg^{2+} and Ca^{2+}). This reduced the accumulation of NH_4^+-N in OMPBR by approximately 27% and increased microalgal density by almost 60% through enhancing photosynthetic activity for dissolved CO_2 consumption (Wang et al., 2020).

The operational cost of OMPBR was 32%–45% higher (\$ 0.29–0.32/d) than that of MPBR ($<$ \$ 0.22/d) when operating at relatively long HRTs (i.e., 1–2 d). The difference in the cost was mainly ascribed to the operation of a peristaltic pump for recirculation of DS in the membrane module, stirring plate and feedback control system in OMPBR. The feedback control system was employed to maintain DS concentration in the DS reservoir with a magnetic stirrer according to conductivity measurements. To keep the DS concentration in the DS reservoir at a constant level, another peristaltic pump was required to add a concentrated DS stock solution. Besides, the filtration cost for OMPBR was also considerably higher (3.5–4.6 times higher) than that for MPBR due to the high cost of DS recirculation from the diluted effluent and energy-intensive feedback control system. The operational cost of OMPBR could be reduced when operating at shorter HRTs (i.e., 6–8 h), prolonging operational duration, or consistently supplying DS from seawater or wasted brine solution (Praveen et al., 2016).

2.2 Membrane-based hybrid systems

2.2.1 MMBR with immobilized algal cells or solid carriers

Immobilized algal cells are obtained by immobilization technology which entraps microalgae within a porous polymeric matrix (i.e., alginate matrix). The immobilized beads are generally prepared as small balls with a diameter of 3–5 mm. Suspended algal cells commonly suffer from a light shading effect, while substrates and light more easily contact the algal cells on the surface of alginate-based beads. Moreover, the immobilized algal cells have higher robust capacities in resisting the high concentrations of nutrient and organic matters from wastewater, show uniform distribution in beads and long-lasting cell viability within the matrix beads, and reduce cell leakage and the related products [extracellular polymeric substances (EPS), soluble microbial products (SMP)]. Thus, the immobilized microalgal cells feature higher activity and better physiological functions. This accelerates the growth of microalgal cells around the surface of the beads and further achieves a good biomass harvest (Xie et al., 2018; Wu et al., 2021). In an immobilized MMBR containing alginate-immobilized microalgae beads, nutrient removal was enhanced compared to the suspended microalgae due to microalgae uptake of nutrient (PO_4^{3-}-P, NH_4^+-N and/or NO_3^--N), chemical precipitation of PO_4^{3-}-P with calcium released from the beads, and adsorption of PO_4^{3-}-P by the beads. Furthermore, the immobilized MMBR under intermittent filtration mode showed better NO_3^--N and PO_4^{3-}-P removals (63% and 58%, respectively) than that under continuous filtration mode (corresponding removals of around 19%) because of the better microalgae harvesting (total biomass concentration, 0.14 g/L under intermittent filtration vs 0.4 g/L under continuous filtration) (Wu et al., 2021). After adding powdered activated carbon (PAC) into the immobilized MMBR, the light intensity transmissivity increased as some chromophore from wastewater (anaerobic digestion effluent) was adsorbed by PAC, which improved algal cell photosynthesis and biomass accumulation (biomass concentration,

2.33 g/L with PAC vs 2.24 g/L without PAC). This led to the higher proportion of living algal cells with PAC (91% vs 89.9% without PAC) (Xie et al., 2018).

During the long-term operation, immobilized beads have to be regularly replaced by fresh ones. The increased density of microalgal cells at the outer layer of the immobilized beads over time led to the formation of a shading layer and further aggravated the light shading effect, limiting nutrient removal and microalgae growth. Besides, the initial cell density of microalgae inside the immobilized beads, quantity of immobilized beads, and cell size should be taken into account. Moderate initial cell density (i.e., 7.4×10^5 cell/bead) was more preferable compared to lower or higher density. A low initial cell density not only deteriorated the nutrient removal during the growth process of microalgae in beads, but also increased the consumption of the immobilized matrix (alginate) for NH_4^+-N/PO_4^{3-}-P removal. The higher initial cell density adversely affected the specific growth rates of microalgae, significantly prompted the formation of the shading layer on the surface of the immobilized beads, and reduced the light available for the inner layer of beads (self-shading). The small cells with high initial cell density might aggravate the shading layer formation. Moreover, the microalgae cells inside the beads might be released over time caused by the gradual decomposition of the immobilized beads. Hence, harvesting duration of beads and the balance between initial microalgal cell density and alginate concentration during bead preparation also need to be considered (Cao et al., 2020).

To more easily concentrate and harvest the microalgal biomass, an attached microalgal culture system is employed to enable the formation of attached microalgal cells on solid carriers in MMBR or MPBR. This not only eliminates time-consuming and expensive downstream processes (i.e., flocculation, centrifugation) for concentrating the harvested microalgae, but also saves harvesting costs as microalgae embedded in the attached algal biofilm reduces the harvesting frequency (Gao et al., 2015; Peng et al., 2020). The presence of fixed solid carriers (i.e., flexible fiber bundles submerged in the reactor) in the microalgae biofilm membrane photobioreactor (BMPBR) increased the volumetric microalgae production by 44% compared to MMBR. Most of the increased microalgal cells were presented as an attached algal biofilm on solid carriers (around 72% of the total microalgae yield coming from the attached microalgae). The higher algal biomass productivity in BMPBR gave rise to a higher total inorganic nitrogen (TIN) removal rate (6.19 mg/L · d) than that in MPBR (4.87 mg/L · d) (Gao et al., 2015). When employing moving solid carriers (i.e., fiber balls) in BMPBR during the long-term cultivation process (70 d), the biomass productivity of the attached microalgae was higher (up to 14.00 mg/L · d) than that of the suspended microalgae (7.31–8.03 mg/L · d). After the concentration of the suspended microalgae reached maximum value (50 d later), it experienced a descending trend, which was ascribed to two reasons: firstly, the gradual decline in the growth rate of the suspended microalgal cells at the high concentration of microalgae induced by limited light; and secondly, the gradual transfer of suspended cells to the media surface, which encouraged the formation of microalgal biofilm on the media surface. Moreover, dissolved inorganic nitrogen (DIN) and dissolved inorganic phosphorus (DIP) removals reached 91.0%–99.6% and 92.1%–98.4%, respectively (Peng et al., 2020).

2.2.2 MBR combined with MPBR

The hybrid systems combining MBR with MPBR can (1) treat more types of wastewaters; (2) effectively remove nitrogen and phosphorus from the single MBR effluent via nutrient uptake by microalgae to generate reclaimed water; (3) reduce the carbon footprint of wastewater treatment through biofixing carbon dioxide as the carbon source by microalgae and harvesting solar energy into the algal biomass; (4) improve the effluent quality without requirement of multiple biological processes (i.e., combination of aerobic, anoxic, and anaerobic stages for nitrogen removal, chemical precipitation process for phosphorus removal), which reduces the energy demand and amounts of equipment instruments; (5) generate valuable microalgal biomass due to the nutrient from the MBR effluent available for microalgal growth; and (6) provide a promising technology to realize sustainable wastewater treatment (Marbelia et al., 2014; Viruela et al., 2018; Nguyen et al., 2020; Vu et al., 2020; Gao et al., 2021). However, the hybrid system confronts two issues. One is the deterioration of microalgae culture after the long-term operation. Another is the large reactor volume required for the formation of the favored microalgal culture conditions (i.e., aeration, light intensity, mass transfer, etc.), which compromises the small footprint design of MBR and wide application of wastewater treatment plants (Vu et al., 2020).

When treating high-strength nitrogen wastewater (i.e., anaerobic digestate), an aerobic MBR with heterotrophic microorganisms was employed to improve the availability of the digestate (10% (v/v) mixed with municipal wastewater) for microalgae cultivation and nutrient removal. The MBR converted NH_4^+-N to NO_3^--N and increased the phosphorus to nitrogen (P/N) ratio through nitrification, thereby alleviating ammonia toxicity on microalgae cultivation and providing extra PO_4^{3-}-P for microalgae growth. Moreover, the high removal of organics by MBR (COD removal of 85%) minimized the competition for the organic carbon between heterotrophic microorganisms and autotrophic microalgae in the subsequent MPBR. This encouraged the microalgal growth using inorganic carbon provided by CO_2-enriched air. As a result, the MPBR not only improved NH_4^+-N removal (about 97%), but also eliminated NO_3^--N as an extra nitrogen source for microalgae, resulting in total nitrogen removal of around 75%. After increasing the light intensity (doubling the light intensity) in MPBR, biomass concentration sharply increased from 0.49 to 4.92 g/L and the PO_4^{3-}-P level reduced from 10 mg/L to less than 1 mg/L (removal of 99%) (Praveen et al., 2018).

In anaerobic MBR (AnMBR), some complex organic compounds from municipal wastewater could be converted to simple organics (micromolecular organic matters) via hydrolysis, which facilitated the cultivation of mixotrophic microalgae (*Chlorella pyrenoidosa* FACHB-5) in MPBR. The mixotrophic microalgae consumed organic matters via heterotrophic metabolism as well as an inorganic carbon source (CO_2) provided by both aeration and the carbonate-bicarbonate buffer system via photoautotrophic metabolism for their growth. Moreover, the biomass production rate in the MPBR of the hybrid system was higher (91.10 mg/L d) than that in the case of single MPBR (60.13 mg/L · d). Excellent nutrient removal was obtained in the MPBR (TN removal of 96.7%, TP removal of 98.0%), mainly through biological assimilation, along with a stable microalgal biomass concentration of 1.65–1.86 g/L (Gao et al., 2021). The subsequent MPBR could also effectively utilize NH_4^+-N and PO_4^{3-}-P from the nutrient-rich AnMBR effluent for microalgae cultivation (biomass production up to 700 mg/L) and nutrient removal (around 75%) during the short-term operation (6 d). However, the long-term operation of the microalgae-based bioreactor deteriorated the microalgae culture when the microalgae growth reached the lag phase, leading to a decline in the biomass production and nutrient removal (Vu et al., 2020).

2.2.3 *Other hybrid systems*

To enhance microalgal biomass harvesting, MPBR was combined with a gravity settler, namely the MPBR-settler system, in which microalgal cell suspension was circulated between the MPBR and the settler. This reduced the suspended biomass in MPBR, encouraging light penetration and alleviating the light-shading effect. Consequently, the suspended biomass could be effectively concentrated and harvested in the settler. The increased biomass recirculation rates (BRRs, 0.15–0.6 L/d) allowed a higher steady-state biomass inflow from the MPBR to the settler (volume of 250 mL), resulting in a less suspended biomass (from 3.31 to 1.75 g/L) and higher biomass productivity (from 0.13 to 0.22 g/L d) in the MPBR due to greater light availability. However, extremely high BRRs (1.2 and 2.4 L/d) reduced the biomass harvest and their productivity during steady state as the biomass settling ability was deteriorated by the higher overflow velocity and shorter HRT in the settler. These problems could be addressed by enlarging the settler volume (i.e., MPBR with 1000 mL of settler ($MPBR_{1000}$)) as the larger cross-sectional area of the settler prolonged the HRT and decreased the overflow velocity. In $MPBR_{1000}$, biomass productivity and harvest biomass concentration could reach high levels up to 0.26 g/L · d and 31 g/L at BRR of 0.6 L/d, respectively (Parakh et al., 2020).

Flocculation is also employed to improve microalgal biomass harvesting. Flocculants addition (i.e., chitosan, cationic polyacrylamide polymers) in MPBR or MMBR could enlarge the floc size of microalgae and prompt the better settling ability of microalgae via adsorption and bridging effect or charge neutralization on the surface of microalgae. The harvesting efficiency of the microalgal biomass taken from MMBR reached almost 100% at optimal dosages of flocculants, i.e., 75.5%–100% with the addition of cationic polyacrylamide at 36 mg/g dry weight (Vu et al., 2020), 98% at 24 mg/g dry weight of cationic polyacrylamide (Nguyen et al., 2020), close to 100% with chitosan at 5–10 mg/L (Zhao et al., 2020a).

A hybrid forward osmosis-reverse osmosis (FO-RO) system was developed, which acted as a dual barrier to almost completely reject organic matters, nutrients, and suspended solids to further extract fresh water (portable water production) from the diluted DS. An algal photobioreactor for the mixotrophic algal process combined with the hybrid FO-RO system increased the algae concentration by more than 9 times compared to that in the algae effluent and completely removed nutrients N and P (NH_3-N and PO_4^{3-}-P). The majority of organic matters and multivalent cations (i.e., Ca^{2+}, Mg^{2+}), and anions (i.e., SO_4^{3-}) from the algal effluent could be rejected by the FO membrane, limiting the RO membrane fouling and scaling (Jiang et al., 2020).

3. Challenges on the application of membrane-based systems

3.1 Membrane fouling

During the filtration process, attachment, adsorption, deposition, and/or accumulation of foulants (i.e., microalgae cell debris, inorganic colloidal particles, natural organic matters (NOM), EOM, AOM, SMP, etc.) on the membrane surface, on pore surfaces, and/or inside membrane pores result in four different types of membrane fouling in MPBR or MMBR, namely pore blocking, adsorption, formation of a cake layer and a gel layer. Pore blocking is mainly caused by deposition of foulants into membrane pore surfaces and membrane pores. Adsorption is defined as adsorption of biopolymers (SMP and EPS) released from microalgal cells into membrane surfaces and pore surfaces, as well as accumulation of polymers on the membrane surface and inside pores. The cake layer is formed due to the deposition and accumulation of mixed foulants (i.e., microalgal cells, cell debris, biopolymers, inorganic colloidal particles, etc.) on the membrane surface. Accumulation of biopolymers and organic matters from feed wastewater on the membrane surface and pore surfaces leads to the formation of a gel layer (Liao et al., 2018).

In OMPBR, membrane fouling is mainly ascribed to microalgal biomass clogging as some microalgal cells could be trapped and accumulated in the dead zone and in the lumen chamber inlet of the enclosed hollow fiber FO membranes. This elevates flow resistance and encourages polarization concentration which reduces the transport driving force. Moreover, the reverse diffusion of draw solutes from the draw solution (i.e., magnesium and calcium-based draw solution) to the feed solution caused by the concentration gradient leads to accumulation of undesirable compounds in the feed solution and declined osmotic pressure difference. The reverse diffusion of solutes also increases the cake-enhanced osmotic pressure by the buildup of salt on the membrane surface. Divalent ions (i.e., Mg^{2+}, Ca^{2+}) interact with microalgal cells and the released extracellular polysaccharides through charge neutralization or the cation bridge effect, which aggravates FO fouling (Wang et al., 2020; Zou et al., 2011).

3.2 Fouling control strategies

Membrane fouling could be controlled by optimizing operational parameters (transmembrane pressure (TMP), aeration, cross-flow velocity (CFV), SRT, HRT, pH, and temperature). The increase in TMP to a certain level (i.e., 125 kPa) exerts compressive force on the fouling layer on the membrane surface. This limits the fouling thickness and prompts more permeate penetration, which in turn reduces the filtration resistance. The relatively high CFV (i.e., up to 4 m/s) induces greater turbulence to detach the biomass particles, NOM and microalgal cells from the membrane surface, which alleviates membrane fouling and enhances the permeate flux. Air and/or CO_2 enhanced aeration generates shear stress to remove microalgal particulates from the membrane surface. However, the excessive aeration damages the microalgal cells. A relatively short SRT should be selected as a higher SRT elevates microalgal biomass concentration and further aggravates membrane fouling. HRT could be extended to reduce loading rates and further prompt the formation of larger particles in homogeneous distribution, giving rise to porous layer formation and lower fouling rates. Higher temperatures (25–35°C) limit the release of EOM by reducing microalgal biomass concentration and decrease water viscosity, which ameliorates membrane fouling. pH can be kept at a relatively high level (i.e., range of 7–8) to hinder the interaction between EOM and membrane as well as reduce electrostatic attraction, alleviating membrane fouling (Liao et al., 2018; Novoa et al., 2020).

Membrane properties need to be considered when considering fouling control approaches (i.e., pore size, surface charge, hydrophilicity, surface roughness, etc.). Compared to tubular membranes, multichannel membranes (19 channels per tube) moderated permeate flux decline due to their low fouling tendency. Membranes with smaller pores (i.e., 0.2 μm) limited the deposition of EOM into membrane pores, thereby curtailing irreversible resistance (Bamba et al., 2020). Enhanced hydrophilicity of membrane by surface modification is one of the preferred strategies for fouling control (antifouling property) (Liao et al., 2018). The Ag/graphene oxide (GO) nanohybrids incorporating the polyvinylidene difluoride (PVDF) membrane improved hydrophilicity and increased the negative charge of the membrane, which could reject transphilic and hydrophilic carbohydrate. The repulsion force between the membranes and microalgae with the negative charges facilitated the detachment of microalgal cells from the membrane surface (Chong et al., 2019). Incorporating nano-TiO_2 into the membrane could also increase the hydrophilicity of membrane, which decreased the cake layer formation on the membrane surface (Hu et al., 2015).

Chemical cleaning methods using different chemical reagents (i.e., NaOH + NaClO, NaOH, NaOH + citric acid + NaClO, etc.) have been also used for fouling control, especially for irreversible fouling reduction (Liao et al., 2018). Physical cleaning is more widely applied for in situ reversible fouling mitigation. Both backwash and relaxation reduced the fouling rate by 50%, while nitrogen gas scouring caused a decline in the fouling rate of 60%. Microalgal biomass was temporarily detached from the membrane surface via backwash (i.e., 20-min filtration and 10-min backwash), which led to deposition of more particles with smaller size on the membrane surface. Relaxation (i.e., 20-min filtration and 10-min relaxation) induced the deposition of a large amount of microalgae on the membrane surface and the formation of looser and more irregular fouling layer with high thickness (180 μm). Nitrogen gas scouring (i.e., 20-min filtration and 10-min nitrogen gas scouring) ameliorated the deposition of biomass on the membrane surface, which generated a thin microalgal film layer (5–10 μm). However, the smaller flocs and higher microalgal concentrations generated by the high shear forces from gas bubbles aggravated membrane fouling (Fortunato et al., 2020).

Membrane modification and in situ physical cleaning methods can enhance membrane permeability/filtration performance and increase the filtration flux (Table 1). As a result, the specific energy consumption declines compared to the conventional membrane module, although the extra electricity energy required for pumping feed into the membrane module increases to some extent. After combining the patterned membrane (patterns located at the active side of the membrane) and flocculation, formation of larger microalgal flocs and reduction of EOM by flocculation encouraged detachment of flocs from the membrane surface and deposition of the flocs above valleys of the patterned membrane, which left more area for water passage than that of the commonly used membranes with a flat surface. Thus, high membrane permeability (110 L/m^2 · h) could be obtained at a very low cross-flow velocity (CFV, 0.0025 m/s) during cross-flow filtration. This significantly reduced energy consumption (0.28 kWh/kg) when compared with other studies under cross flow or dead-end flat membrane filtration (0.59–1.27 kWh/kg). Synergy between membrane patterning and flocculation saved the total cost (0.16 €/kg) compared to the cross-flow system (0.26 €/kg) and membrane filtration combined with chitosan flocculation (0.297 €/kg) (Zhao et al., 2021b).

When operating with a turbulent jet-assisted membrane (TJ) module which possessed a perforated cylinder containing 40 holes, the specific energy requirement was remarkably lower (8 Wh/L) than that with the conventional membrane module (one small inlet and one small outlet at different sides of the module) (13 Wh/L). When compared with the conventional module, the extra electricity energy required for pumping only increased by

TABLE 1 Membrane fouling control strategies during microalgae harvesting.

Methods[a]	Membrane properties[b]	Effects on fouling control[c]	
In situ physical cleaning approaches			
Membrane filtration with orifice plate • Installing an orifice plate on the frontal area (inlet) of membrane module • Fitting orifice holes with a diameter of 0.5 mm into the flow channel of membrane module • Horizontally uniform distribution of holes in the inlet area of membrane	→ Ultrafiltration membrane → Effective filtration area of 24 cm^2 (3 cm width, 8 cm length)	√ Vigorous and evenly turbulence (shear stress) on membrane surface, which removed fouling layer and limit accumulation of microalgae cells or EPS on membrane surface √ Higher shear stress obtained with less orifice holes (5.63 Pa with 24 holes, 31.24 Pa with 16 holes and 52.3 Pa with 8 holes) √ Enhanced filtration performance (reduction of total resistance by 36–63%) compared to the control one	Cho et al. (2020)
Tilting the membrane panel	→ Phase inverted PVDF membrane → Thickness, 0.18 mm → Maximum pore size, 0.19 mm → Mean pore, 0.16 mm → Contact angle, 75 ± 5° → Surface pore size, 0.1 mm → Surface pore density, 19 pores/μm^2 → Surface porosity, 11%	√ Encouraging air bubbles to go upward and consistently contact membrane surface to scour off the foulants on membrane surface through drag force √ Higher permeability at higher tilting angles (up to 2.7 times at angles <20°) than the vertical panel √ Better performance when using one-side panel, which could increase panel permeability (21% higher) and halve total membrane surface compared to the two-sided one (two sides of panel were aerated by switching)	Eliseus et al. (2017)
Housing the finned spacer to mount the membrane panel	→ Phase inverted PVDF membrane → Thickness, 0.18 mm → Maximum pore size, 0.19 mm → Mean pore, 0.16 mm → Contact angle, 75 ± 5° → Surface pore size, 0.1 mm → Surface pore density, 19 pores/μm^2	√ Encouraging air bubbles to move toward membrane surface by fins, which improved drag forces and shear stress rate to drive foulants away from membrane surface √ Effective fouling mitigation at narrow fins gaps (2 cm), higher aeration rate	Razak et al. (2020)

Continued

TABLE 1 Membrane fouling control strategies during microalgae harvesting—cont'd

Methods[a]	Membrane properties[b]	Effects on fouling control[c]	
	→ Surface porosity, 11%	(1.5 L/min) and short fin switching period (<5 min), leading to the highest clean water permeability of 870 L/(m² hbar)	Kim et al. (2015)
Adding a perforated disk to a dynamic microfiltration system • Installing the rotating disk on the rotating shaft at the center of the module which was connected to a rotor	→ Hydrophilic PVDF membranes → Pore size, 0.2 μm	√ Inducing a high fluid velocity near the membrane surface (two times higher) compared to that by unperforated disk, which reduced fouling and increased maximum permeate flux	Kim et al. (2019)
Turbulent jet-assisted microfiltration containing a perforated cylinder with uniform distribution of holes located at the central area of the membrane module • Implementing turbulent jet into membrane module with complicated and limited inner space	→ Hollow fiber polyvinylidene floride membrane module → Pore size, 0.2 μm → Length and diameter of the module, 0.32 and 0.05 m → Inner and the outer diameter of a fiber, 0.9 and 1.2 mm → Effective filtration area, 0.25 m²	√ Generating turbulent jets to perpendicularly push the feed to the membrane surface through the holes (radical flow direction) √ Creating wall jet by the turbulent jet along membrane surface, which induced higher shear stress and local fluid velocity √ Enhancing permeate flux (by 126% compared to conventional-type membrane module), while maintaining overall feed flowrate	
Axial vibratory membrane • Fixing membrane module on a cassette which was oscillated by a servo motor • Operating at the frequency of 10 Hz and amplitude of 1 cm	→ Flat PVDF membrane → Total area, 0.02 m² → Nominal pore size, 0.1 μm	√ High fluid velocity by vibration close to membrane, which alleviated membrane fouling √ Low attractive interaction between microalgae cells and membrane √ Consistently rejecting proteins and polysaccharides (main components of EOM)	Zhao et al. (2019)

AF mode with backwashing which changed flow direction every 3min and backwashed membrane every 15min	→ PVC hollow fiber UF membrane modules → Filtration area, 0.13 m² → MWCO, 50 kDa → Water contact angle, 67 ±2° (hydrophilic membrane) → Inner diameter of the fiber, 1.0mm → Outer diameter of the fiber, 1.66mm	√ Increase in average flux by 28% compared to the BF mode with backwashing √ High shear rates on the top or bottom regions of membrane fibers √ Dynamic shear stress across the membrane surface, which removed microalgal cells and organic foulants (proteins and polysaccharides) on membrane surface, resulting in higher flux	Yang et al. (2019)

Membrane modification

Fe_2O_3 nanoparticles incorporated PVC membranes (1% Fe_2O_3)		√ Increase in hydrophilicity, leading to less deposition of protein substances on membrane surface √ Considerably greater average filtration flux (138 L/m² · h) than unmodified PVC membrane (83 L/m² · h)	Liu et al. (2019)
Silver nanowires-based highly electro-conductive membrane (C-AgNWs) • Coating AgNWs on PES membrane via dead-end vacuum filtration • Constructing an electroplating reactor as an electrolytic cell for Ag-based electroplating, consisting of an anode and conductive AgNWs film as cathode (placing horizontally and perpendicularly facing each other)	→ Enhanced conductivity (3.9×10^4 S/cm) compared with AgNWs-deposited one (~ 2385 S/cm) → Inter-connected tortuous paths in pores with mean flow pore size of 0.37 μm (pore size distribution of 0.2–0.8 μm) → High stability in harsh conditions, including acid condition (pH = 2), basic condition (pH = 12), saline condition, and physical stress (i.e., ultra-sonication for 2min)	√ Increased electrostatic repulsive forces between negatively charged foulants (EPS and microalgal cells) and membrane surface (cathode) √ In-situ generation of hydrogen bubbles from membrane surface, which physically removed foulants from membrane surface √ Consistently good performance and increase in permeate flux by 480% at high electric field of 20 V/mm under continuous mode due to electric repulsion and physical pushing	Mushtaq et al. (2019)

Continued

TABLE 1 Membrane fouling control strategies during microalgae harvesting—cont'd

Methods[a]	Membrane properties[b]	Effects on fouling control[c]	
Surface-patterned membrane modification • Patterned PSf membranes prepared via adding PSf and PEG into membrane casting solution	→ Typical asymmetric structure containing finger-like macrovoids and three layers of membrane → Formation of more pronounced and uniform wave patterns on active side of membrane at higher PSf (18 wt%) and PEG concentrations (22 wt%)	√ Higher velocity and wall shear generated on the apexes than flat membrane due to the patterns in the flow channel, limiting the deposition of microalgal cells in ridge regions √ More un-fouled area in the ridge region available for water and feed passage, resulting in higher membrane flux √ Better filtration performance obtained at larger patterns	Zhao et al. (2020b)
Surface-patterned membrane modification • Negatively charged PSf membranes prepared via incorporating sPSf into PSf membrane	→ Typical asymmetric structure containing finger-like macrovoids and three layers → Increased hydrophilicity and surface charge after adding sPSf to the PSf membrane	√ The synergy between membrane vibration, surface pattern and charge √ Increased repulsive energy barrier between negatively charged microalgal cells and membrane surface, which moderated microalgal attachment √ Higher velocity and wall shear generated on the apexes, resulting in more less-fouling areas for water passage	Zhao et al. (2021a)
Combination of different fouling control methods			
• Patterned membrane filtration combined with flocculation (chitosan)	→ A typical bilayered structure containing two zones of small and narrow macrovoids	√ Formation of larger microalgal flocs and reduction of EOM due to chitosan addition √ Accumulation of microalgal flocs above valleys of the patterned membrane √ Less deposition of the flocs in valleys, resulting in more area for water passage √ Easy removal of flocs from membrane surface √ High membrane permeability of 110 L/m$^2 \cdot$ h obtained at very low CFV of 0.0025 m/s	Zhao et al. (2021b)

- Membrane vibrating system containing negatively charged and patterned PSf membranes, which was operated at high frequencies (>7 Hz)

  → A typical bilayered structure containing small and narrow macrovoids on top of cast film which possessed wider macrovoids
  → Increased pore size, porosity, hydrophilicity and surface charge after introducing sPSf

  √ Patterned membrane acting as a turbulence promotor, which provided extra vortices and turbulence on membrane surface
  √ Highest wall shear at the apexes and relatively high wall shear at the lowest area of valley region, hindering deposition of foulants (i.e., microalgal cells, EOM) in valley regions

  Zhao et al. (2021c)

- Fe–PC–CNT hollow fiber membrane as cathode and titanium mesh as anode in a filtration device when operating at constant voltage of −1.0 V

  → Effective area, 3.01 cm^2
  → Uniform outer and inner diameters, 0.8 and 0.6 mm, respectively
  → Pore size, 207 nm
  → Good hydrophilicity due to the acidification of PC and CNT
  → Contact angle at 29.38°

  √ Significantly higher pure water permeability (16 time higher) than the control one without electron-Fenton process during continuous filtration
  (1) Electric barrier around membrane, which repelled microalgal cells and EOM with negative surface charges from membrane surface (electro-chemical repulsion)
  (2) Degrading EOM on membrane surface and inside membrane by strong oxidation of •OH[d]
  (3) Regeneration of Fe^{2+} on the membrane for consistent electron-Fenton reaction process (Fe^{3+} + e$^-$ → Fe^{2+})

  Zheng et al. (2021)

---

[a] AgNWs, silver nanowires; AF, alternative feed; C–AgNWs, Ag electroplated membrane; Fe–PC–CNT, Ag electroplated membrane; Fe–PC–CNT hollow fiber membrane, electro-fenton enhanced porous carbon–carbon nanotubes hollow fiber membranes loaded with Fe^{2+}; PEG, polyethylene glycol; PES, poly(ethersulfone); PSf, polysulfone; sPSf, sulfonated polysulfone.

[b] MWCO, molecular weight cut off; PVDF, polyvinylidene difluoride; PVC, Poly(vinyl chloride); UF, ultrafiltration.

[c] BF, bottom feed; CFV, cross-flow velocity.

[d] OH, which was generated via oxygen reduction reaction (O$_2$ + H$^+$ + 2e$^-$ → H$_2$O$_2$, H$_2$O$_2$ + Fe^{2+} + H$^+$ → Fe^{3+} + •OH + H$_2$O).

33% in the TJ module, which could be considerably counterbalanced by the reduction of total specific energy consumption due to the significant increase in permeate flux induced by the great wall jet along the TJ membrane surface (Kim et al., 2019). The usage of orifices containing some holes at the inlet of a membrane module generated vigorous turbulence on the membrane surface, which inhibited the cake layer formation and enhanced membrane permeability. The specific energy demand declined (0.49 Wh/g with 8 holes and 0.44 Wh/g with 16 holes vs 1.02 Wh/g for the control one) due to the better membrane permeability despite the increased electricity consumption after installing orifices (Cho et al., 2020).

The application of the alginate-immobilized microalgae beads and/or addition of PAC is also one of the promising technologies for fouling control in MMBR or MPBR. The immobilized beads as biomass carriers mitigate membrane fouling by decreasing the concentration of suspended microalgae, hindering the growth of algae-related microorganisms, limiting the accumulation of macromolecular organics, protein-like and SMP-like substances, as well as reducing some cation-based foulants (i.e., Ca^{2+} combined with alginate) and alginate-related fouling through preparation of the immobilized beads (Cao et al., 2020; Wu et al., 2021). The immobilized beads with PAC prompted the formation of functional microbial groups to degrade foulants (i.e., genus *Opitutus* for polysaccharides degradation) and adsorption of organic matters. Moreover, the porous cake layer formed on the membrane surface owing to the presence of immobilized algal cells with a larger size served as a physical filter to retain small organic compounds for further fouling reduction. The satisfactory fouling alleviation substantially reduced total energy consumption from 132.10 J with suspended microalgae and 280.0 J with immobilized beads to 63.30 J with the immobilized beads and PAC for microalgae harvesting, leading to the low harvesting cost of $ 0.50/kg biomass (Xie et al., 2018). Table 2 compares different methods for microalgae harvesting, including some typical physical, chemical, biological, and magnetic methods.

Membrane fouling in OMPBR could be reduced by using an electric field by flocculation of microalgal cells and electrostatic repulsion between the membrane and microalgal flocs. An electric-assisted FO system was developed by Xu et al. (2021), which contained a conductive FO membrane as the cathode and a boron-doped diamond (BDD) electrode as the anode with an external voltage applied. During the electrolysis process, microbubbles (H_2 and O_2 gases) generated on electrodes inhibited the accumulation of microalgal flocs on the membrane surface as the flocs would migrate to the water surface. Furthermore, the microalgal cells in the solution formed larger flocs by electric field flocculation. As the FO membrane surface contained negative charges, microalgal flocs could not be deposited on the membrane surface due to electrostatic repulsion. Thus, the initial fouling layer on the membrane surface was not easily formed with the electric field applied (i.e., -5 V). This limited the subsequent formation of a thick and compacted cake layer on the membrane surface, reducing water flux loss by around 58% compared with that without the electric field. Another option for fouling reduction is application of a CO_2 saturated solution, which decreased pH through the release of proton (H^+) after reacting with H_2O and encouraged the formation of bubbles nucleation through pressure release. This resulted in the movement of a fouling layer and further ameliorated its interaction with the membrane surface, thereby removing colloidal and scaling fouling. Other fouling mitigation strategies include application of a feed spacer, anti-scalant-blended DS, high cross-flow velocities, osmotic backwashing, periodic air scouring, and periodical use of CO_2 saturated solution combined with hydraulic flushing (Kim et al., 2020; Ibrar et al., 2019).

TABLE 2 Comparison among different methods for microalgae harvesting.

Harvesting methods[a]	Merits	Drawbacks	Energy and cost for microalgal biomass harvesting
Physical methods			
Centrifugation	√ Rapid and reliable method √ High biomass recovery efficiency (>90%) √ Microalgal strain-independent operation √ Being applicable for almost all microalgal species √ Being efficient for large-scale application	• Energy-intensive method • Being economically incompatible owing to the high energy demand and high capital investment • Damage of cell structure by the high gravitational force and shear stresses • Being not suitable for large-scale application	→ 1 kWh/m^3 energy for harvesting of *Scenedesmus* sp.
Filtration (i.e., MF, UF, TFF, dead-end filtration)	√ No chemical additives √ High biomass harvesting √ Low energy requirement and cost √ Low impact on slurry or feed quality √ Simple operation √ Retention of intact structure and characteristics of microalgal cells by TFF √ Dead-end filtration for harvesting of large microalgal cells (diameter > 70 mm) √ High biomass recovery efficiency (70%–90%)	• Being species-dependent • Being favorable for large-size microalgae • Membrane fouling • Regular membrane cleaning and replacement required • High operating and maintenance cost • Low membrane permeability and selectivity • MF and UF for fragile cells harvesting and small-scale production of microalgae	→ 0.27 and 0.25 kWh/m^3 for *C. vulgaris* and *P. tricornutum* harvesting by submerged microfiltration, respectively → 1.43–3.87 kWh/m^3 (cost, \$0.16–0.37/kg biomass) for *Chlorella minutissima* by cross-flow microfiltration → 1.08–2.68 kWh/m^3 (cost, \$0.11–0.28 /kg biomass) for *Scenedesmus species* by cross-flow microfiltration
Gravity sedimentation	√ Simple operation √ Low cost √ Low energy required	• Time-consuming operation • Low slurry recovery • Unreliable and ineffective process owing to variable density of algal cells, low slurry recovery and requirement of further thickening • Slow sedimentation (settling rate of 0.1–2.6/cm)	

Continued

TABLE 2   Comparison among different methods for microalgae harvesting—cont'd

Harvesting methods[a]	Merits	Drawbacks	Energy and cost for microalgal biomass harvesting
Flotation (i.e., DAF, DiAF, electrolytic flotation, ODF)	√ Low cost and small footprint √ High flexibility √ High air-particle contact √ Short operational duration √ Being good for large-scale application √ Biomass recovery efficiency (50%–90%)	• Energy-intensive technology • Chemical flocculants required • Being species dependent • Being not applicable for marine microalgae harvesting • Being not suitable for large-scale application because of high cost and contamination • Breakage of flocs by oversized bubbles using DAF • Expensive equipment and high pressure drop required for generation of bubbles by DiAF • Fouling of cathodes and high power demand using electrolytic flotation • Contamination issues when employing ODF at large scale	→ High energy requirement at 7.6 kWh/m^3 for DAF
Electrical methods using electrodes (i.e., Mg electrode, Fe electrode, Al electrode, etc.)	√ Wide variety of microalgal species √ Being reliable and energy efficient √ No chemical flocculants required √ High biomass recovery efficiency >90%	• Being poorly disseminated • High cost required for large-scale application • Fouling on cathode • Damage of harvesting system	→ Higher energy consumption at higher current density, i.e., 0.2 kWh/m^3 energy at current density of 0.5 mA/cm^2, 2.28 kWh/m^3 at current density of 5.0 mA/cm^2; → 0.3–2.0 kWh/kg for electrochemical harvesting
Chemical methods			
Inorganic flocculants [metal ions (i.e., FeCl$_3$, Al$_2$(SO$_4$)$_3$, AlCl$_3$, ZnSO$_4$, etc.)] or inorganic polymers (i.e., polyelectrolyte/	√ Mature and safe technology √ High recovery efficiency (60%–100%)	• Contamination of the harvested biomass by metal salts • Contaminating downstream processes • Being not applicable in large scale	

TABLE 2 Comparison among different methods for microalgae harvesting—cont'd

Harvesting methods[a]	Merits	Drawbacks	Energy and cost for microalgal biomass harvesting
polyaluminum chloride, etc.)		• Limited recycling of culture medium • High cost per unit of microalgal harvested biomass due to the high dosage of flocculants required	
Organic polymers (i.e., chitosan/cationic starch, etc.)	√ Biodegradability √ Simple, safe and fast method √ No energy input √ High biomass recovery efficiency (75%–99%)	• Being pH dependent • High cost • Being not applicable in large scale • Secondary pollution	
Biological methods			
Bioflocculation (i.e., cocultivation with bacteria/fungi, non-flocculant microalgae with flocculant-inducing microalgae, bio-flocculant generated by microorganisms)	√ No chemical flocculants required √ Being cost-effective and energy-efficient √ Easy operation √ High recovery efficiency (>50%–98%) √ Great saving of cost for carbon source when using waste as carbon source (i.e., molasses wastewater, biogas slurry, swine wastewater)	• No clear mechanism • Slow biological flocculation process • High energy demand • Long flocculation duration • Microbiological contamination, affecting downstream process	→ Microalgae harvesting via fungal spore method, \$1.65/kg dry weight; → Microalgae harvesting via fungal pellet method, \$0.825/kg dry weight
Bioflocculation (Actinomycetes flocculation; plant-based flocculation)	√ Formation of dense network structure √ Vigorous growth and quick metabolism √ Biodegradability √ No secondary pollution for downstream process √ Being cost-effective and energy-efficient √ Easy operation	• Low flocculation efficiency • High cost as cationic quaternary amine group is introduced into some polymers	

Continued

B. Applications of membrane technology for water and wastewater treatment

TABLE 2   Comparison among different methods for microalgae harvesting—cont'd

Harvesting methods[a]	Merits	Drawbacks	Energy and cost for microalgal biomass harvesting
Autoflocculation (i.e., calcium/ phosphorus precipitate)	√ Neutralization of negative surface charge of microalgae √ No toxicity to biomass √ Being cost-effective and energy-efficient √ Easy operation √ High biomass recovery efficiency (>90%)	• Being slow and unstable • Being pH dependent	
Magnetic methods			
Magnetic nanoparticle (i.e., Fe_3O_4)	√ Being very fast and cost-effective process √ Generation of uncontaminated microalgal biomass √ High reusability √ High harvesting efficiency (>90%)	• High cost due to recycling of nanoparticles and practical applicability • Energy-intensive technology • Some effects on downstream processing	

[a] *DAF, dissolved air flotation; DiAF, dispersed air flotation; MF, microfiltration; ODF, ozonation dispersed flotation; TFF, tangential flow filtration; UF, ultrafiltration.*
References: Bilad et al. (2012), Bleeke et al. (2015), Gerardo et al. (2015), Guldhe et al. (2016), Li et al. (2020), Mathimani and Mallick (2018), Okoro et al. (2019), Singh and Patidar (2018), Vandamme et al. (2011), Yin et al. (2020).

4. Conclusions and perspectives

Microalgae harvesting faces some difficulties when employing microalgae-based processes for nutrient removal. Thus, membrane technology is employed to simultaneously harvest microalgae and remove nitrogen and phosphorus from wastewater. When operating membrane bioreactor-based systems (MPBR and OMPBR), it needs to adopt optimal operating parameters. The addition of alginate-immobilized microalgae beads and solid carriers improves MPBR performance in terms of microalgae growth, nutrient removal, and membrane fouling control. MBR combined with MPBR can treat more types of wastewaters and cultivate microalgae. Membrane modification and in situ physical cleaning methods reduce energy requirement and total operational cost due to the enhanced membrane permeability. Future studies should concentrate on the development of novel carriers and immobilized beads as well as techno-economic analyses for different membrane-based systems.

Membrane-based systems, including membrane bioreactor-based systems and membrane-based hybrid systems, have been increasingly considered for microalgae harvesting and nutrient removal. However, membrane fouling is the most challenging issue during the membrane filtration process. Thus, it is imperative to develop membrane fouling control strategies to ensure effective microalgae harvesting. Despite great efforts in the development and applications of membrane-based systems, more works for future research are required as follows:

(1) More studies need to focus on effects of light conditions on microalgae productivity and nutrient removal. The optimization of light conditions may compromise the adverse effects caused by other operating parameters (i.e., HRT, SRT, biomass concentration, temperature).

(2) More simple and cost-effective strategies should be developed to avoid the depletion or loss of essential elements (i.e., Ca^{2+}, Mg^{2+}, SO_4^{2-}) and reduce the total operation cost when employing OMPBR for microalgae harvesting.

(3) Modification of the immobilized beads is required to minimize their decomposition during the long-term operation and light shading effect even with the high microalgae cell density inside the beads.

(4) Novel media can be developed to simultaneously prompt growth of attached microalgae and alleviate membrane fouling.

(5) Distribution and properties of the attached microalgal biofilm (i.e., morphology, thickness, etc.) on carriers need to be investigated to clarify effects of the biofilm on light penetration, microalgae growth, and nutrient removal.

(6) Inhibitory thresholds of total ammonia nitrogen (including ammonium and free ammonia) need to be evaluated based on the microalgal species and cultivation conditions when feeding MPBR by the nutrient-rich AnMBR effluent.

(7) Investigations on membrane modification and physical cleaning methods should be carried out to accomplish more effective in situ fouling control.

(8) More studies should be carried out to clarify the exact mechanisms about the bioflocculation process for microalgae harvesting.

(9) When operating MPBR combined with an electrochemical process, the electric field needs to be optimized to minimize cell disruption and maintain high microbial activity.

(10) Techno-economic analyses of lab-scale or full-scale membrane-based systems need to be carried out to evaluate the economic feasibility of the proposed systems.

References

Babaei, A., Mehrnia, M.R., Shayegan, J., Sarrafzadeh, M.-H., 2016. Comparison of different trophic cultivations in microalgal membrane bioreactor containing N-riched wastewater for simultaneous nutrient removal and biomass production. Process Biochem. 51 (10), 1568–1575.

Bamba, B.S.B., Lozano, P., Ouattara, A., Elcik, H., 2020. Pilot-scale microalgae harvesting with ceramic microfiltration modules: evaluating the effect of operational parameters and membrane configuration on filtration performance and membrane fouling. J. Chem. Technol. Biotechnol. 96 (3), 603–612.

Bilad, M.R., Vandamme, D., Foubert, I., Muylaert, K., Vankelecom, I.F.J., 2012. Harvesting microalgal biomass using submerged microfiltration membranes. Bioresour. Technol. 111, 343–352.

Bilad, M.R., Arafat, H.A., Vankelecom, I.F.J., 2014. Membrane technology in microalgae cultivation and harvesting: a review. Biotechnol. Adv. 32 (7), 1283–1300.

Bleeke, F., Quante, G., Winckelmann, D., Klöck, G., 2015. Effect of voltage and electrode material on electroflocculation of Scenedesmus acuminatus. Bioresour. Biopro. 2 (1), 36–44.

Cao, S., Teng, F., Wang, T., Li, X., Lv, J., Cai, Z., Tao, Y., 2020. Characteristics of an immobilized microalgae membrane bioreactor (iMBR): nutrient removal, microalgae growth, and membrane fouling under continuous operation. Algal Res. 51, 102072.

Cho, H., Mushtaq, A., Hwang, T., Kim, H.S., Han, J.I., 2020. Orifice-based membrane fouling inhibition employing in-situ turbulence for efficient microalgae harvesting. Sep. Purif. Technol. 251, 117277.

Chong, W.C., Mohammad, A.W., Mahmoudi, E., Chung, Y.T., Kamarudin, K.F., Takriff, M.S., 2019. Nanohybrid membrane in algal-membrane photoreactor: microalgae cultivation and wastewater polishing. Chin. J. Chem. Eng. 27 (11), 2799–2806.

Eliseus, A., Bilad, M.R., Nordin, N.A.H.M., Putra, Z.A., Wirzal, M.D.H., 2017. Tilted membrane panel: a new module concept to maximize the impact of air bubbles for membrane fouling control in microalgae harvesting. Bioresour. Technol. 241, 661–668.

Fortunato, L., Lamprea, A.F., Leiknes, T., 2020. Evaluation of membrane fouling mitigation strategies in an algal membrane photobioreactor (AMPBR) treating secondary wastewater effluent. Sci. Total Environ. 708, 134548.

Gao, F., Yang, Z.H., Li, C., Zeng, G.M., Ma, D.H., Zhou, L., 2015. A novel algal biofilm membrane photobioreactor for attached microalgae growth and nutrients removal from secondary effluent. Bioresour. Technol. 179, 8–12.

Gao, F., Li, C., Yang, Z.-H., Zeng, G.-M., Feng, L.-J., Liu, J.-Z., Liu, M., Cai, H.-W., 2016. Continuous microalgae cultivation in aquaculture wastewater by a membrane photobioreactor for biomass production and nutrients removal. Ecol. Eng. 92, 55–61.

Gao, F., Peng, Y.Y., Li, C., Cui, W., Yang, Z.H., Zeng, G.M., 2018. Coupled nutrient removal from secondary effluent and algal biomass production in membrane photobioreactor (MPBR): effect of HRT and long-term operation. Chem. Eng. J. 335, 169–175.

Gao, F., Yang, Z.Y., Zhao, Q.L., Chen, D.Z., Li, C., Liu, M., Yang, J.S., Liu, J.Z., Ge, Y.M., Chen, J.M., 2021. Mixotrophic cultivation of microalgae coupled with anaerobic hydrolysis for sustainable treatment of municipal wastewater in a hybrid system of anaerobic membrane bioreactor and membrane photobioreactor. Bioresour. Technol. 337, 125457.

Gerardo, M.L., Zanain, M.A., Lovitt, R.W., 2015. Pilot-scale cross-flow microfiltration of Chlorella minutissima: a theoretical assessment of the operational parameters on energy consumption. Chem. Eng. J. 280, 505–513.

González-Camejo, J., Aparicio, S., Jiménez-Benítez, A., Pachés, M., Ruano, M.V., Borrás, L., Barat, R., Seco, A., 2020. Improving membrane photobioreactor performance by reducing light path: operating conditions and key performance indicators. Water Res. 172, 115518.

Guldhe, A., Misra, R., Singh, P., Rawat, I., Bux, F., 2016. An innovative electrochemical process to alleviate the challenges for harvesting of small size microalgae by using non-sacrificial carbon electrodes. Algal Res. 19, 292–298.

Hafiz, M.A., Hawari, A.H., Das, P., Khan, S., Altaee, A., 2020. Comparison of dual stage ultrafiltration and hybrid ultrafiltration-forward osmosis process for harvesting microalgae (Tetraselmis sp.) biomass. Chem. Eng. Process. 157, 108112.

Hu, W., Yin, J., Deng, B., Hu, Z., 2015. Application of nano TiO$_2$ modified hollow fiber membranes in algal membrane bioreactors for high-density algae cultivation and wastewater polishing. Bioresour. Technol. 193, 135–141.

Ibrar, I., Naji, O., Sharif, A., Malekizadeh, A., Alhawari, A., Alanezi, A.A., Altaee, A., 2019. A review of fouling mechanisms, control strategies and real-time fouling monitoring techniques in forward osmosis. Water 11 (4), 695.

Jafarinejad, S., 2021. Forward osmosis membrane technology for nutrient removal/recovery from wastewater: recent advances, proposed designs, and future directions. Chemosphere 263, 128116.

Jiang, W., Lin, L., Gedara, S.M.H., Schaub, T.M., Jarvis, J.M., Wang, X., Xu, X., Nirmalakhandan, N., Xu, P., 2020. Potable-quality water recovery from primary effluent through a coupled algal-osmosis membrane system. Chemosphere 240, 124883.

Kim, K., Jung, J.Y., Kwon, J.H., Yang, J.W., 2015. Dynamic microfiltration with a perforated disk for effective harvesting of microalgae. J. Membr. Sci. 475, 252–258.

Kim, D., Kwak, M., Kim, K., Chang, Y.K., 2019. Turbulent jet-assisted microfiltration for energy efficient harvesting of microalgae. J. Membr. Sci. 575, 170–178.

Kim, Y., Li, S., Ghaffour, N., 2020. Evaluation of different cleaning strategies for different types of forward osmosis membrane fouling and scaling. J. Membr. Sci. 596, 117731.

Kimura, K., Nishisako, R., Miyoshi, T., Shimada, R., Watanabe, Y., 2008. Baffled membrane bioreactor (BMBR) for efficient nutrient removal from municipal wastewater. Water Res. 42 (3), 625–632.

Lee, J.C., Baek, K., Kim, H.W., 2018. Semi-continuous operation and fouling characteristics of submerged membrane photobioreactor (SMPBR) for tertiary treatment of livestock wastewater. J. Clean. Prod. 180, 244–251.

Li, S., Hu, T., Xu, Y., Wang, J., Chu, R., Yin, Z., Mo, F., Zhu, L., 2020. A review on flocculation as an efficient method to harvest energy microalgae: mechanisms, performances, influencing factors and perspectives. Renew. Sustain. Energy Rev. 131, 110005.

Liao, Y., Bokhary, A., Maleki, E., Liao, B., 2018. A review of membrane fouling and its control in algal-related membrane processes. Bioresour. Technol. 264, 343–358.

Liu, Q., Demirel, E., Chen, Y., Gong, T., Zhang, X., Chen, Y., 2019. Improving antifouling performance for the harvesting of Scenedesmus acuminatus using Fe_2O_3 nanoparticles incorporated PVC nanocomposite membranes. J. Appl. Polym. Sci. 136, 47685.

Luo, Y., Le-Clech, P., Henderson, R.K., 2017. Simultaneous microalgae cultivation and wastewater treatment in submerged membrane photobioreactors: a review. Algal Res. 24, 425–437.

Luo, Y., Le-Clech, P., Henderson, R.K., 2018. Assessment of membrane photobioreactor (MPBR) performance parameters and operating conditions. Water Res. 138, 169–180.

Marbelia, L., Bilad, M.R., Passaris, I., Discart, V., Vandamme, D., Beuckels, A., Muylaert, K., Vankelecom, I.F.J., 2014. Membrane photobioreactors for integrated microalgae cultivation and nutrient remediation of membrane bioreactors effluent. Bioresour. Technol. 163, 228–235.

Mathimani, T., Mallick, N., 2018. A comprehensive review on harvesting of microalgae for biodiesel – key challenges and future directions. Renew. Sustain. Energy Rev. 91, 1103–1120.

Min, K.H., Kim, D.H., Ki, M.-R., Pack, S.P., 2022. Recent progress in flocculation, dewatering, and drying technologies for microalgae utilization: scalable and low-cost harvesting process development. Bioresour. Technol. 344, 126404.

Mushtaq, A., Cho, H., Ahmed, M.A., Rehman, M.S.U., Han, J.I., 2019. A novel method for the fabrication of silver nanowires-based highly electro-conductive membrane with antifouling property for efficient microalgae harvesting. J. Membr. Sci. 590, 117258.

Nawi, N.I.M., Arifin, S.N.H.M., Hizam, S.M., Rampun, E.L.A., Bilad, M.R., Elma, M., Khan, A.L., Wibisono, Y., Jaafar, J., 2020. Chlorella vulgaris broth harvesting via standalone forward osmosis using seawater draw solution. Bioresour. Technol. Rep. 9, 100394.

Nguyen, L.N., Truong, M.V., Nguyen, A.Q., Johir, M.A.H., Commault, A.S., Ralph, P.J., Semblante, G.U., Nghiem, L.D., 2020. A sequential membrane bioreactor followed by a membrane microalgal reactor for nutrient removal and algal biomass production. Environ. Sci.: Water Res. Technol. 6, 189–196.

Novoa, A.F., Fortunato, L., Rehman, Z.U., Leiknes, T., 2020. Evaluating the effect of hydraulic retention time on fouling development and biomass characteristics in an algal membrane photobioreactor treating a secondary wastewater effluent. Bioresour. Technol. 309, 123348.

Okoro, V., Azimov, U., Munoz, J., Hernandez, H.H., Phan, A.N., 2019. Microalgae cultivation and harvesting: growth performance and use of flocculants - a review. Renew. Sustain. Energy Rev. 115, 109364.

Parakh, S.K., Praveen, P., Loh, K.C., Tong, Y.W., 2020. Integrating gravity settler with an algal membrane photobioreactor for in situ biomass concentration and harvesting. Bioresour. Technol. 315, 123822.

Peng, Y.Y., Gao, F., Yang, H.L., Wu, H.W., Li, C., Lu, M.M., Yang, Z.Y., 2020. Simultaneous removal of nutrient and sulfonamides from marine aquaculture wastewater by concentrated and attached cultivation of Chlorella vulgaris in an algal biofilm membrane photobioreactor (BF-MPBR). Sci. Total Environ. 725, 138524.

Praveen, P., Heng, J.Y.P., Loh, K.C., 2016. Tertiary wastewater treatment in membrane photobioreactor using microalgae: comparison of forward osmosis & microfiltration. Bioresour. Technol. 222, 448–457.

Praveen, P., Guo, Y., Kang, H., Lefebvre, C., Loh, K.C., 2018. Enhancing microalgae cultivation in anaerobic digestate through nitrification. Chem. Eng. J. 354, 905–912.

Razak, N.N.A.N., Rahmawati, R., Bilad, M.R., Pratiwi, A.E., Elma, M., Nawi, N.I.M., Jaafar, J., Lam, M.K., 2020. Finned spacer for enhancing the impact of air bubbles for membrane fouling control in Chlorella vulgaris filtration. Bioresour. Technol. Rep. 11, 100429.

Singh, G., Patidar, S.K., 2018. Microalgae harvesting techniques: a review. J. Environ. Manage. 217, 499–508.

Solmaz, A., Işık, M., 2020. Optimization of membrane photobioreactor; the effect of hydraulic retention time on biomass production and nutrient removal by mixed microalgae culture. Biomass Bioenergy 142, 105809.

B. Applications of membrane technology for water and wastewater treatment

Uduman, N., Qi, Y., Danquah, M.K., Forde, G.M., Hoadley, A., 2010. Dewatering of microalgal cultures: a major bottleneck to algae-based fuels. J. Renew. Sustain. Energy 2, 012701.

Vandamme, D., Pontes, S.C., Goiris, K., Foubert, I., Pinoy, L.J., Muylaert, K., 2011. Evaluation of electro-coagulation-flocculation for harvesting marine and freshwater microalgae. Biotechnol. Bioeng. 108 (10), 2320–2329.

Viruela, A., Robles, Á., Durán, F., Ruano, M.V., Barat, R., Ferrer, J., Seco, A., 2018. Performance of an outdoor membrane photobioreactor for resource recovery from anaerobically treated sewage. J. Clean. Prod. 178, 665–674.

Vu, M.T., Vu, H.P., Nguyen, L.N., Semblante, G.U., Johir, M.A.H., Nghiem, L.D., 2020. A hybrid anaerobic and microalgal membrane reactor for energy and microalgal biomass production from wastewater. Environ. Technol. Innov. 19, 100834.

Vu, M.T., Nguyen, L.N., Mofijur, M., Johir, M.A.H., Ngo, H.H., Mahlia, T.M.I., Nghiem, L.D., 2022. Simultaneous nutrient recovery and algal biomass production from anaerobically digested sludge centrate using a membrane photobioreactor. Bioresour. Technol. 343, 126069.

Wang, L., Pan, B., Gao, Y., Li, C., Ye, J., Yang, L., Chen, Y., Hu, Q., Zhang, X., 2019. Efficient membrane microalgal harvesting: pilot-scale performance and techno-economic analysis. J. Clean. Prod. 218, 83–95.

Wang, Z., Lee, Y.Y., Scherr, D., Senger, R.S., Li, Y., He, Z., 2020. Mitigating nutrient accumulation with microalgal growth towards enhanced nutrient removal and biomass production in an osmotic photobioreactor. Water Res. 182, 116038.

Wu, P.H., Hsieh, T.M., Wu, H.Y., Yu, C.P., 2021. Characterization of the immobilized algae-based bioreactor with external ceramic ultrafiltration membrane to remove nutrients from the synthetic secondary wastewater effluent. Int. Biodeter. Biodegr. 164, 105309.

Xie, B., Gong, W., Yu, H., Tang, X., Yan, Z., Luo, X., Gan, Z., Wang, T., Li, G., Liang, H., 2018. Immobilized microalgae for anaerobic digestion effluent treatment in a photobioreactor-ultrafiltration system: algal harvest and membrane fouling control. Bioresour. Technol. 268, 139–148.

Xu, X., Zhang, H., Gao, T., Teng, J., 2021. Impacts of applied voltage on forward osmosis process harvesting microalgae: filtration behaviors and lipid extraction efficiency. Sci. Total Environ. 773, 145678.

Yang, L., Wang, L., Ren, S., Pan, B., Li, J., Zhang, X., Chen, Y., Hu, Q., 2019. Harvesting of Scenedesmus acuminatus using ultrafiltration membranes operated in alternative feed directions. J. Biosci. Bioeng. 128 (1), 103–109.

Yin, Z., Zhu, L., Li, S., Hu, T., Chu, R., Mo, F., Hu, D., Liu, C., Li, B., 2020. A comprehensive review on cultivation and harvesting of microalgae for biodiesel production: environmental pollution control and future directions. Bioresour. Technol. 301, 122804.

Zhao, F., Li, Z., Zhou, X., Chu, H., Jiang, S., Yu, Z., Zhou, X., Zhang, Y., 2019. The comparison between vibration and aeration on the membrane performance in algae harvesting. J. Membr. Sci. 592, 117390.

Zhao, Z., Li, Y., Muylaert, K., Vankelecom, I.F.J., 2020a. Synergy between membrane filtration and flocculation for harvesting microalgae. Sep. Purif. Technol. 240, 116603.

Zhao, Z., Li, Y., Muylaert, K., Vankelecom, I.F.J., 2020b. Optimization of patterned polysulfone membranes for microalgae harvesting. Bioresour. Technol. 309, 123367.

Zhao, Z., Muylaert, K., Szymczyk, A., Vankelecom, I.F.J., 2021a. Harvesting microalgal biomass using negatively charged polysulfone patterned membranes: influence of pattern shapes and mechanism of fouling mitigation. Water Res. 188, 116530.

Zhao, Z., Muylaert, K., Vankelecom, I.F.J., 2021b. Combining patterned membrane filtration and flocculation for economical microalgae harvesting. Water Res. 198, 117181.

Zhao, Z., Muylaert, K., Szymczyk, A., Vankelecom, I.F.J., 2021c. Enhanced microalgal biofilm formation and facilitated microalgae harvesting using a novel pH-responsive, crosslinked patterned and vibrating membrane. Chem. Eng. J. 410, 127390.

Zheng, M., Yang, Y., Qiao, S., Zhou, J., Quan, X., 2021. A porous carbon-based electro-Fenton hollow fiber membrane with good antifouling property for microalgae harvesting. J. Membr. Sci. 626, 119189.

Zou, S., Gu, Y., Xiao, D., Tang, C.Y., 2011. The role of physical and chemical parameters on forward osmosis membrane fouling during algae separation. J. Membr. Sci. 366 (1–2), 356–362.

Self-Forming Dynamic Membrane BioReactors (SFDMBRs) for wastewater treatment

Seow Wah How[a,b], Chaeyeon Kang[a], Sla Min[a], Paula Carrera[a,b], Muhammad Ahmar Siddiqui[c], Guanghao Chen[c], and Di Wu[a,b,c]

[a]Centre for Environmental and Energy Research, Ghent University Global Campus, Incheon, Republic of Korea [b]Department of Green Chemistry and Technology, Faculty of Bioscience Engineering, Ghent University, Ghent, Belgium [c]Department of Civil and Environmental Engineering, The Hong Kong University of Science & Technology, Hong Kong, China

1. Introduction

In recent years, membrane technology has gained popularity in wastewater treatment due to its low footprint, high flexibility and treatment capacity, and efficient removal of organic and inorganic matter. However, conventional membrane technologies consume a large amount of energy to maintain a consistent permeate flux. Conventional membrane technologies are also prone to clogging and may require frequent membrane maintenance, which can significantly increase the cost of operation.

The self-forming dynamic membrane bioreactor (SFDMBR) technology emerged as a potential solution to extend membrane filtration time by capitalizing on the self-forming biological matrix, known as the dynamic membrane (DM). It is composed of soluble microbial products (SMP), extracellular polymeric substances (EPS), and microorganisms. The DM deposits on the supporting material, which simultaneously filters the particulate matter and biodegrades the organic compounds in the wastewater. The DM technology was conceived in the 1960s where physical DM was studied for desalination using single-layered coated DM. This single-coated DM was formed by passing a specific colloidal material, such as powdered

Copyright © 2023 Elsevier Inc. All rights reserved.

activated carbon (PAC) and hydrolysable salts, over a porous supporting material (Marcinkowsky et al., 2002; Ye et al., 2006). Double-layer coated DMs were developed in the 1970s due to the low permeability potential and limited salt rejection of single-coated DMs (Ersahin et al., 2012). Making a double-coated DMs requires spraying two different colloidal materials through the supporting material in a stepwise approach; some of the common colloidal materials are hydrous zirconium dioxide and polyacrylate (Tanny and Johnson, 1978). DMs were later integrated with ultrafiltration (UF) and reverse osmosis (RO) to treat industrial effluents in the 1980s (Groves et al., 1983). The technology continued to evolve for protein extraction and for water purification in the 1990s (Al-Malack and Anderson, 1996). In the 2000s, more studies highlighted the potential of Self-Forming Dynamic Membrane (SFDM) to substitute the expensive UF and microfiltration (MF) systems with a relatively relaxed effluent quality (Kiso et al., 2000).

SFDMBR consists of a dynamic layer deposited on the supporting material, such as Nylon mesh, polyester mesh, and Dacron mesh, with a pore size between 10 and 200 μm. The formation of the DM is composed of four major stages (Siddiqui et al., 2021). In the first stage, an EPS gel layer known as the secondary membrane attaches freely on the supporting material. As the sludge particles accumulate, a fragile and thin DM is formed. In stage 2, the EPS gel and sludge particles begin to fill the pores and consequently make DM pore size smaller. The duration to form DM (including both stage 1 to stage 2) can vary widely from 20 h to 20 d (Ersahin et al., 2017; Zhang et al., 2010) depending on the SFDMBR configuration, supporting material mesh pore size, aeration intensity, and/or biomass concentration, to name a few (Salerno et al., 2017; Saleem et al., 2016; Rezvani et al., 2014). After complete DM formation, the membrane permeate can achieve a turbidity below 5 nominal turbidity units (NTU). Subsequently, the DM became more strongly bound in stage 3, which necessitates membrane cleaning. The last stage (stage 4) is characterized by a rapid DM reformation after cleaning (<10 min) due to the remnants of tightly bound EPS on the supporting material.

Some studies have also demonstrated the promising outcomes of using DM for desalination and wastewater treatment (Shor et al., 1968), as well as removing viral particles (Srivastava et al., 2004) and biodegradable organic and inorganic pollutants under both aerobic and anaerobic conditions. SFDMBR could be an important alternative for small and decentralized wastewater treatment systems (Ren et al., 2010). Depending on the feed characteristics and sludge properties, different membrane modules and reactor configurations were devised to attain effective wastewater treatment performance.

This book chapter introduces the common membrane modules, reactor configurations, and operating conditions of SFDMBR technologies for wastewater treatment. The microbial composition in the SFDMBR reactor, including the microorganisms that proliferate in suspended solids (SS) and/or DM biocake layer, are also detailed in this chapter. Based on the current knowledge of the operating conditions and microbial community, the strategies to optimize the SFDMBR reactor operation are summarized at the end of the chapter.

2. Dynamic membrane modules and materials

The DM module in SFDMBR can be classified into tubular (Fig. 1A) and flat sheet (Fig. 1B) configurations. The flat sheet module is easier to fabricate and operate, which is more commonly used in different lab- and pilot-scale studies; the tubular modules can achieve excellent

(A) **(B)**

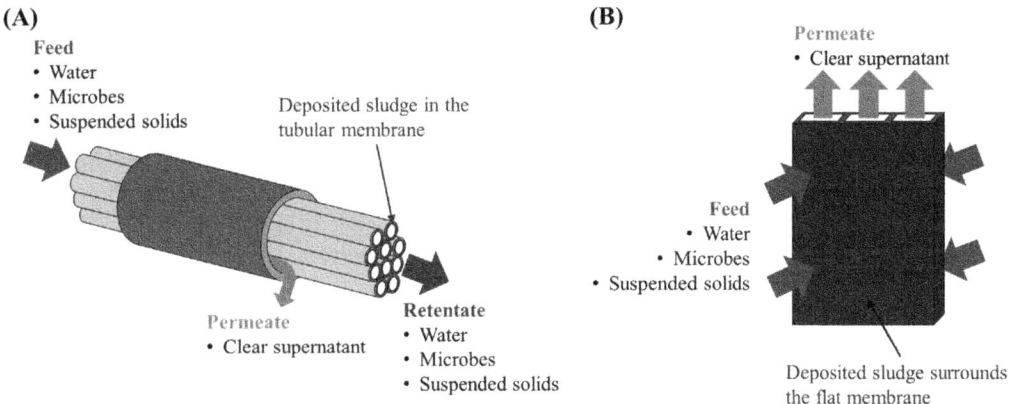

FIG. 1 (A) Tubular and (B) flat sheet membrane modules applied in SFDMBRs.

hydrodynamics with low fouling risks. However, in practice, tubular modules were seldom used due to their complex fabrication. A comparative study of the filtration characteristics of a tubular and a flat, nonwoven fabric membrane module showed that the tubular module's transmembrane pressure (TMP) remained consistent at 5 kPa (Seo et al., 2007); the flat module's TMP increased sharply to 16 kPa during the same period. The SFDM thickness in tubular thickness was also double of that in the flat sheet module due to the larger spatial area for sludge deposition (Mohan and Nagalakshmi, 2020). Nonetheless, both membrane modules were able to achieve low biological oxygen demand (BOD) and SS of 7.8 and 3.0 mg/L, respectively, for water reclamation. Studies on the effect of membrane module geometry on the DM formation are still limited to date (Pollice and Vergine, 2020), which warrants further investigations in the future.

In general, the materials for DM are meshes, woven fabrics, and nonwoven fabrics. Mesh is a permeable barrier made up of metal strands, fiber or other ductile materials that are intertwined (Ersahin et al., 2012). The woven fabric membrane is composed of knitted monofilament or multifilament yarns; however, nonwoven fabric is a thin layer of natural or manmade filaments, which have not been converted into yarns (Hutten, 2007). Some studies suggested that the flat surface of meshes was not conducive for biomass attachment on the DM (Kiso et al., 2005). Nonetheless, a review of SFDMBR technology by Alepu et al. (2016) emphasized that the suitable supporting material should be evaluated on a case-by-case basis as the current status of research is not mature enough for large-scale commercialization.

3. Reactor configuration and performance

Numerous lab- and pilot-scale studies have been conducted on SFDMBRs. The SFDMBR systems can operate under either anaerobic or aerobic conditions depending on the treatment objectives. Regarding reactor configuration, Siddiqui et al. (2021) summarized 22 studies for Anaerobic Self-Forming Dynamic Membrane BioReactor (AnSFDMBR), and Mohan and Nagalakshmi (2020) reviewed 23 aerobic SFDMBR research works. The technological descriptions of anaerobic and aerobic SFDMBRs are summarized in Tables 1 and 2, respectively.

TABLE 1 Summary of AnSFDMBR reactor configuration, operating conditions and removal performance from the literature.

Reactor type/membrane configuration	Volume (L)	Support Material	Substrate	MLSS (g/L)	Flux (LMH)	TMP (kPa)	HRT (h)	SRT (d)	COD Removal (%)	References
Upflow reactor/submerged DM	45	61-μm Dacron mesh/flat sheet	Municipal WW	NA	65	0–25	NA	NA	57.3	Zhang et al. (2010)
Upflow reactor/Submerged DM	45	61-μm Dacron mesh/flat sheet	Municipal WW	NA	65	0–25	NA	NA	63.4	Zhang et al. (2011)
Upflow reactor/submerged DM	42	61-μm Dacron mesh/flat sheet	Municipal WW	NA	60	0–35	NA	NA	81.6	Ma et al. (2013a)
Upflow reactor/submerged DM	42	61-μm Dacron mesh/flat sheet	Municipal WW	NA	60	0–35	2.2	35	80	Ma et al. (2013b)
Complete mixing/submerged DM	7	10-μm & 40-μm polypropylene woven fabric/flat sheet	Synthetic WW	17.5	9–17	50–70	NA	NA	NA	Ersahin et al. (2013)
Complete mixing/crossflow DM	1	200-μm polyamide/nylon woven mesh/flat sheet	Synthetic WW	15	1.0–7.2	0–25	1–7 d	NA	90	Alibardi et al. (2014)
Complete mixing/crossflow DM	7	10-μm polypropylene woven fabric/flat sheet	Synthetic WW	7.4±1.1	2.2	36–42	10	20	99	Ersahin et al. (2016a)
Complete mixing/crossflow DM	1	200-μm polyamide/nylon woven mesh/flat sheet	Synthetic WW	NA	1.0–7.2	0–20	NA	NA	90	Saleem et al. (2016)

7	Complete mixing/crossflow DM	10-μm polypropylene woven fabric/flat sheet	Synthetic WW	6.410	2.2	48–57	10	40	99	Ersahin et al. (2016b)
1	Complete mixing/crossflow DM	200-μm polyamide/nylon woven mesh/flat sheet	Synthetic WW	12	1.4–28	0–20	5.7–0.25	NA	90	Alibardi et al. (2016)
7	Complete mixing/Submerged and crossflow DM	10-μm polypropylene woven fabric/flat sheet	Synthetic WW	6.5	2.2	50–70 (submerged DM; 30–40 crossflow DM)	10	40	99	Ersahin et al. (2017)
7	Upflow reactor/Crossflow DM	28-μm and 46-μm polyethylene terephthalate mesh/flat sheet	Municipal WW	NA	30–100	5–40	6–3	Infinite	70	Quek et al. (2017)
14	Complete mixing/Submerged DM	150-μm polyamide/nylon mesh/tubular	Synthetic WW	3	50–150	–	0.83	NA	90	Wang et al. (2018)
15	Complete mixing/Submerged DM	30-μm polyester nonwoven fabric/flat sheet	Synthetic WW	3.5	10–30	–	NA	NA	89	Sun et al. (2018)
3.6	Upflow reactor/Submerged DM	75-μm nylon mesh	Domestic WW	7.3	22.5–180	0–35	8–1	Infinite	70–77	Yang et al. (2020)

TABLE 2 Summary of aerobic SFDMBR reactor configuration, operating conditions and removal performance from the literature.

Reactor type/ membrane configuration	Volume (L)	Support Material	Substrate	MLSS (g/L)	Flux (LMH)	TMP (kPa)	HRT (h)	SRT (d)	COD Removal (%)	References
SBR/ Submerged DM	50	50–100 μm nylon mesh/ tubular	Synthetic wastewater to represent household greywater	–	500–3000	–	24	–	≈90% for 100 μm and ≈88% for 50 μm	Khuntia et al. (2019)
Complete mixing/ Submerged DM	–	25-μm nylon mesh/tubular	Synthetic wastewater	3.0–5.0	100–650	0.6–8.2	–	40	>88	Cai et al. (2018)
Complete mixing/ Submerged DM	9	42-μm monofilament polyester mesh/flat sheet	Synthetic wastewater	3.9	–	0–70			92	Sabaghian et al. (2018)
Complete mixing/ Submerged DM	4	2-mm HDPE mesh/tubular	Synthetic wastewater	–	–	–	12	–	82.16±6.47	Sreeda et al. (2018)
Complete mixing/ Submerged DM	4	50-μm nylon mesh/flat sheet	Presettled municipal wastewater	1.7	67	0–50	8	30	≈94	Vergine et al. (2018)
Complete mixing/ Submerged DM	4.6	50-μm mesh/ flat sheet	Presettled municipal wastewater	1.5	95	0–90	6.7	16	87 to 93	Salerno et al. (2017)
Complete mixing/ Submerged DM	34	25-μm nylon mesh/flat sheet	Raw wastewater	3.0	85	–	–	30	91	Hu et al. (2016)

Configuration		Membrane material/module	Wastewater							Reference
Anoxic-oxic complete mixing/Submerged DM	19.2	Nonwoven fabric/tubular	Domestic wastewater	2.5–3.5	500–1730	5–17.3	5–8	–	–	Seo et al. (2007)
Upflow sludge microaerobic/Submerged DM	14	0.1-μm polyethylene hollow fiber/U-shaped	Raw wastewater	<6.0	7.8–9.8	0–70	7.9–5.9	46–56	81 (72–89)	Chu et al. (2006)
Complete mixing/Submerged DM	28	5-μm polypropylene nonwoven fiber/flat sheet	Synthetic domestic wastewater	–	20	–	–	–	96	Meng et al. (2005)
Complete mixing/Submerged DM	140	100-μm Dacron mesh/flat sheet	Municipal wastewater	7.5	14.8–33.3	–	3.5	56	84.2	Fan and Huang (2002)
Anoxic-oxic complete mixing/Submerged DM	750000	Polypropylene nonwoven fabric/flat sheet	Domestic wastewater	1.8	200–1000	0.5–4.9	–	–	91.6	Seo et al. (2003)
Complete mixing/Submerged DM	4	53-μm stainless steel mesh/tubular	Synthetic wastewater	3.51	20–27	0.1–1	10	Infinite	93.7	Wang et al. (2013)
Complete mixing/Submerged DM	3000	80-μm nylon mesh/flat sheet	Municipal wastewater	3.89	41.7	0.2–2.5	5	Infinite	83.3±8.4	Wang et al. (2012)
Complete mixing/Submerged DM	43.7	2-mm stainless steel mesh/tubular	Municipal wastewater	5.3–5.5	16.3	5–85	8–8.9	26.3	93–95	Zahid and El-Shafai (2011)

From the literature information, a Continuous Stirred Tank Reactor (CSTR) is the most common configuration for both AnSFDMBR and aerobic SFDMBR, while the Upflow Anaerobic Sludge Bed (UASB) reactor is adopted for AnSFDMBR operation. CSTR-type SFDMBR is subdivided into submerged, side-stream submerged, cross flow, and rotating membrane modules (Fig. 2A–D). CSTR-type submerged SFDMBR (Fig. 2A and B) has a simple configuration, consisting of a dynamic membrane submerged in a CSTR tank, but this setup is prone to severe membrane fouling, low fluxes, and high transmembrane pressure (TMP—indicating the operation energy consumption level) because of the direct contact of membrane with bulk sludge.

In an aerobic submerged membrane reactor, air bubbling is applied to shear off excess sludge attached on the membrane surface (Fig. 2B), while in the AnSFDMBR system (Fig. 2A), the biogas produced can be recirculated to the bottom to provide gas scouring. CSTR-type side-stream anaerobic/aerobic SFDMBRs (Fig. 2C and D) can form the DM rapidly within 5–10 min (Ersahin et al., 2017), and reduce the membrane fouling propensity via cross-flow filtration to shear excess sludge from the DM. Besides, Liu et al. (2016) developed a more advanced rotating membrane module (Fig. 2E) to provide better shear stress control for fouling prevention in CSTR.

Typically, SFDMBRs operating in CSTR could achieve high chemical oxygen demand (COD) removal efficiency (90%–99%) and high ammoniacal nitrogen (NH_4^+-N) nitrification efficiency (up to 99%) (Salerno et al., 2017). The total nitrogen (TN) removal efficiency is normally low when there is no anoxic zone. Some aerobic SFDMBR studies applied intermittent aeration or placed a preanoxic tank to provide the niche for denitrification, thus achieving better TN removal performance. For example, Wang et al. (2013) achieved a 80% removal of TN by applying intermittent aeration in an aerobic SFDMBR using mesh filter as the supporting material. However, incorporating a preanoxic tank before the oxic SFDMBR reactor can easily attain a high TN removal efficiency as well (Saleem et al., 2019). Besides, Integrated Fixed-Film Activated Sludge-SFDMBR (IFAS-SFDMBR) incorporated inert carriers in SFDMBR to introduce anoxic zones at the deeper layer of the biofilm adhered to the carrier (Vergine et al., 2018). The IFAS-SFDMBR achieved a lower effluent TN (28.7 ± 5.0 mg/L) when compared with that of conventional SFDMBR (35.0 ± 4.8 mg/L).

The UASB-type SFDMBRs have two common configurations, namely submerged and side-stream configurations. UASB-type submerged SFDMBR (Fig. 2F) usually requires a long start-up time (5–10 d) because of the low mixed liquor suspended solids (MLSS) at the top of the UASB where the membrane module is installed. This configuration can attain a high operating flux but a relatively lower COD removal efficiency (57%–82%) because of the poor biocake formation. However, UASB integrated with side-stream membrane tank (Fig. 2G) could achieve better COD removal (70%–90%) albeit at a slightly lower flux when compared with the UASB-type submerged SFDMBR.

4. Operating conditions

4.1 Operating flux and transmembrane pressure

SFDMBR can be operated in a constant flux mode by using a suction pump (Siddiqui et al., 2021). The flux is usually designed based on the support material pore size, filtration

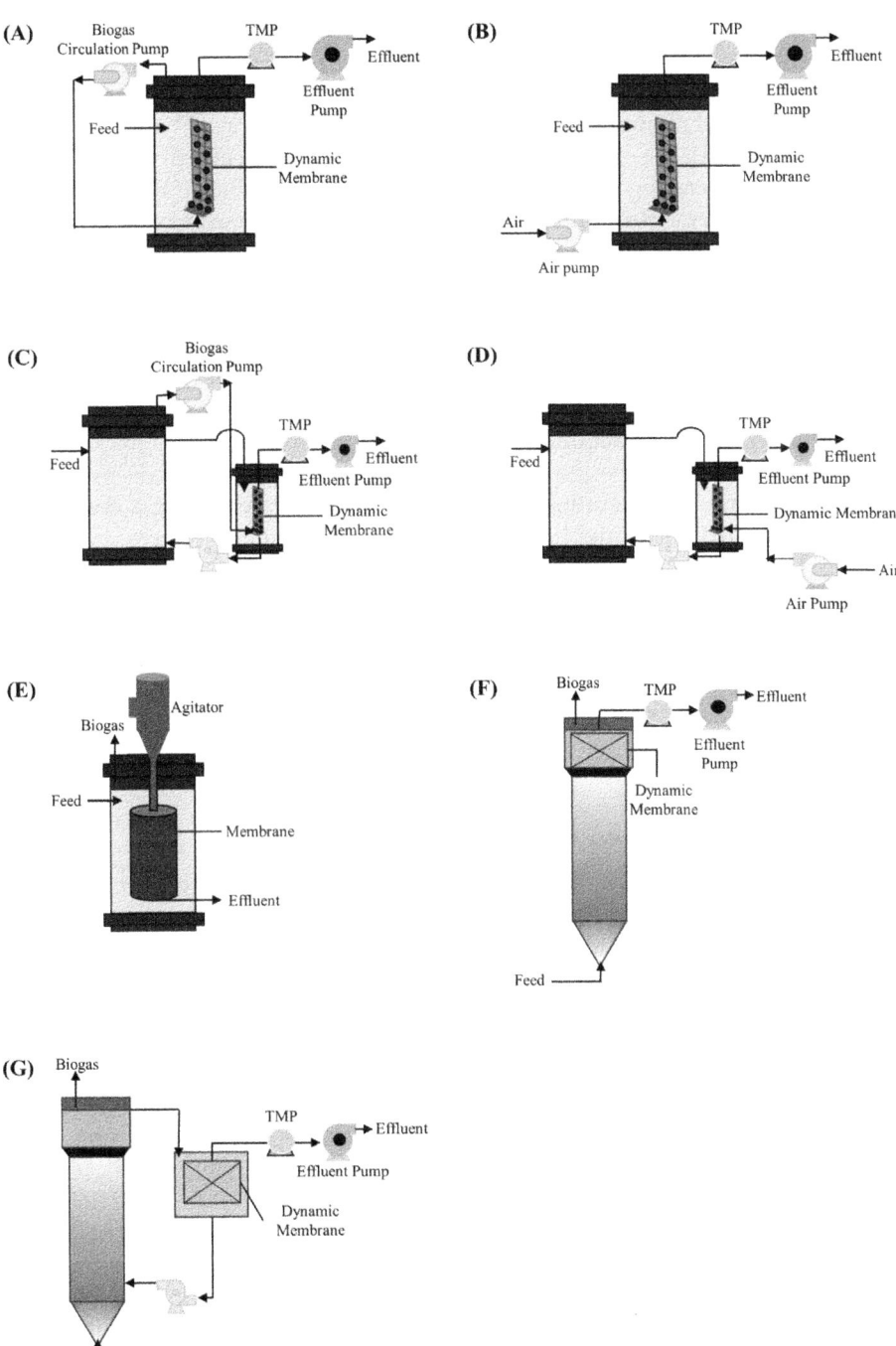

FIG. 2 Schematic diagrams of (A) anaerobic CSTR submerged, (B) aerobic CSTR submerged, (C) anaerobic CSTR side-stream submerged, (D) aerobic CSTR side-stream submerged, (E) CSTR rotating membrane, (F) UASB submerged, and (G) UASB side stream configurations.

resistance, and sludge particle size on a case-by-case basis. To be noted, the higher flux needs stronger suction energy; and after the membrane fouled, maintaining the same flux also requires stronger suction. Moreover, each SFDMBR has a critical flux below which it results in insignificant fouling and above which significant fouling can occur (Yang et al., 2019). Even for flat sheet meshes with the same pore size, the critical fluxes can vary significantly due to the difference in the operating conditions. For instance, a Membrane BioReactor (MBR) applying a relaxation strategy (9 min filtration/1 min relaxation) reported a much higher critical flux of 70 L per meter square per hour (LMH) when compared with a critical flux of 20 LMH for an MBR without relaxation (Wang et al., 2015). Operating SFDMBR at a supercritical flux above the critical flux can accelerate sludge deposition for DM formation, but it can cause frequent membrane fouling (Rezvani et al., 2014). The higher filtration resistance under supercritical flux operation may produce a denser and more compact DM. Conversely, using a subcritical flux may result in a longer DM formation time, but the DM formed is more porous and loose. This loose DM can facilitate a longer filtration time with less frequent membrane fouling (Chu et al., 2014).

Apart from constant flux operation, SFDMBR can be operated using a constant pressure mode by providing sufficient water head loss (i.e., hydraulic energy) between the water level in the SFDMBR and the effluent port. Constant pressure operation is a more economical option than constant flux operation because a suction pump is required to achieve a constant flux. The variation of flux during the constant pressure operation can indicate the operation stages correlating with DM formation and fouling. In the early stage of formation of DM, the flux reduces sharply; then, the flux plateaus during the stable operation stage; finally, a rapid decrease in the flux is observed when membrane fouling occurs.

4.2 Hydraulic and solid retention time

The hydraulic retention time (HRT) of SFDMBR systems treating domestic wastewater spans from 1 to 8 h (Tables 1 and 2). Wastewater with higher strengths and complex compositions, such as those resembling food and textile wastewater, requires a longer HRT between 2 and 10 days (Alibardi et al., 2014). The HRT should be sufficiently long to allow the degradation of biodegradable organics. For example, Yang et al. (2020) found that significant reduction in HRT from 22 to 6 h resulted in poor COD removal efficiency and a concomitant rapid increase in TMP. The rapid rise in TMP at short HRT was attributed to the accumulation of EPS and SMP, which increased the hydrophobicity of the DM and fouling propensity.

Solid retention time (SRT) can significantly influence the microbial community and predator population in the bulk sludge and biocake layer of DM. A typical SRT for SFDMBR treating domestic wastewater ranged from 20 to 40 d, while longer SRTs (125 d) were necessary for stronger wastewater. A longer SRT is beneficial for the removal of COD and NH_4^+-N, accompanied by an increase in MLSS in the SFDMBR system (Huang et al., 2019). Lengthening the SRT enables the proliferation of more slow-growing bacteria, such as nitrifiers related to *Nitrosomonas* and *Nitrospira*, thereby achieving nutrient removal from wastewater.

SRT can also alter the composition of EPS and SMP, which can affect the frequency of membrane fouling in SFDMBRs. Shorter SRTs are associated with EPS of higher protein to polysaccharide ratio (PN/PS) and being more hydrophobic in nature, which makes the DM more sticky and susceptible to membrane fouling. In contrast, longer SRTs produce thicker and more porous DM with a lower compressibility index (Mohan and Nagalakshmi, 2020). Predators, such as protozoa, metazoa, and oligochaetes, can also thrive under long SRT conditions, which help to create a more porous DM by preying on dead bacterial cells. The release of hydrolases, especially during long SRT operation, can biodegrade the excess EPS to mitigate membrane fouling.

4.3 Support materials and dynamic membrane layer formation

The support material for SFDMBR usually has a pore size between 10 and 200 μm (Saleem et al., 2017). However, the effect of the pore size on the DM formation is significant only at the beginning of the SFDMBR operation during the gel layer and DM formation. After the DM maturation, the DM layer acts as the membrane filter to produce permeate. It should be noted that a smaller pore size of the support material can improve bacterial removal. Zahid and El-Shafai (2011) found that cloth media of 200 μm (or smaller) mesh size can reduce fecal coliform count by 4 logs. However, the smaller pore size will lead to rapid rise in TMP, which necessitates more frequent cleaning. This rapid rise in TMP may be caused by the accumulation of smaller particles on the DM and the compaction of DM over time (Mohan and Nagalakshmi, 2020). The size of activated sludge floc can directly affect the formation of DM and its fouling: Large floc size is preferable to shorten the DM formation time. Floc size similar to or smaller than the pore size of the support material may cause irremovable fouling. The sludge floc size should be identified first on a case-by-case basis to ensure proper selection of support material pore size (Xie et al., 2012).

The voidage and compactness of DM can be quantified as "compressibility"—which can be measured as the ratio of weight of DM biomass-to-DM thickness (kg/m^3) (Zhang et al., 2010). The compressibility index can be used to estimate the filtration resistance of the membrane and stages of DM formation. The formation/operation of DM contains three stages—layer formation (stage 1), stable operation (stage 2), and fouling (stage 3). Compressibility increases from 20 to $40\,kg/m^3$ in stage 1 and remains approximately $40\,kg/m^3$ in stage 2, but increases to $120\,kg/m^3$ in the fouling stage (Zhang et al., 2010).

Various wastewater parameters, such as COD strength, temperature, and the presence of inorganic heavy metal ions, can affect biological activities and DM formation. For example, increasing the COD concentration from 10 to $50\,g/L$ in the feed deteriorated the DM bioreactor liquid-solid separation efficiency. High COD loading led to the accumulation of volatile fatty acids in the bioreactor, thereby reducing DM stability and leading to poor biological removal performance (Saleem et al., 2018). Most AnSFDMBRs were operated under mesophilic conditions (35°C) or ambient temperature to cut down the energy requirement for the process. In fact, higher temperatures may promote deflocculation and increase the fouling propensity (Lin et al., 2009). Lower temperatures can also cause more frequent fouling due to higher viscosity and unoxidized COD accumulation in the bioreactor (Martinez-Sosa et al., 2011; Zhang et al., 2011). In addition, inorganic heavy metal ions, such as lead, chromium, and cadmium, can inhibit bacterial activity up to 95% in a membrane bioreactor system (Saddoud et al., 2009).

4.4 Membrane fouling and cleaning

Membrane fouling is influenced by seeding sludge, substrate characteristics, membrane configuration, and reactor type (Siddiqui et al., 2021). The onset of membrane fouling can be indicated by either TMP during the constant flux operation mode or the membrane flux during constant pressure operation mode. Membrane cleaning is generally required when the TMP increases to 30–40 kPa during constant flux operation (Ma et al., 2013a; Yu et al., 2016), while the flux during the onset of membrane fouling is generally approximately 10 LMH (Sun et al., 2018). The membrane flux can vary widely between different SFDMBRs depending on the support material pore size and reactor configurations (Siddiqui et al., 2021). Liang et al. (2012) developed a fouling-related critical flux determination method by incorporating intermittent relaxation periods and parallel clean water filtration tests. Alternatively, the compressibility index can also be an indicator for membrane fouling when the index increases sharply during Stage 3 of the DM formation/operation.

Membrane foulants can be either removable (can be removed via physical cleaning) or irremovable (not able to be removed via physical cleaning, but sometimes can be removed via chemical cleaning). Typical physical cleaning methods include biogas purging, membrane relaxation, brushing, and water backwashing. Chemical cleaning can be performed using alkalis (sodium hydroxide) and oxidants (hydrogen peroxide and/or sodium hypochlorite) to remove organic foulants; while acids (hydrochloric, sulfuric, nitric, and citric acids) are used to remove inorganic and organic metal foulants. A common example of an inorganic foulant is struvite, which can be formed by chemical interactions between cations in the sludge and influent, which can significantly reduce membrane permeability (Lin et al., 2013). Some studies evaluated the efficiency of chemical cleaning with sodium hydroxide, sodium hypochlorite and surfactant (sodium dodecyl sulfate), and found that sodium hypochlorite exhibited the highest recovery of membrane permeability (99%) when compared to sodium hydroxide (87.7%) and sodium dodecyl sulfate (76.5%). Physical and cleaning methods are commonly performed together to recover membrane permeability more effectively. For example, tap water backflushing and offline membrane cleaning using 0.25 wt% sodium hypochlorite can attain a 93% flux recovery (Zhang et al., 2005).

5. Microbial composition and extracellular polymeric substances

The microbial structure in SFDMBRs differs significantly between the bulk sludge and DM because the microbial diversity is affected by the varying flux. Consequently, the microbial abundance was typically lower in the DM than in the bulk sludge. For instance, Ma et al. (2013b) found that *Proteobacteria* was the major phylum in the bulk sludge that accounted for 50% of the total read abundance. The major phyla in the DM include *Bacteroidetes*, *Firmicutes*, and *Spirochaetes*, which constituted 58% of total bacterial read abundance. Predominant classes that colonized the membrane surface include *Bacteroides*, *Clostridia*, *Gammaproteobacteria*, and *Delta-proteobacteria*.

The presence of EPS-secreting bacteria is also vital to maintain the integrity of the DM. A typical EPS-secreting bacteria is related to the family *Xanthomonadaceae* within the class *Gammaproteobacteria* (Hong et al., 2019). The EPS-secreting bacteria secrete both EPS and SMP to bind the sludge flocs together. EPS is composed of protein and polysaccharides. While the polarity of proteins and polysaccharides can differ depending on their functional groups, a sludge floc EPS polarity profiling based on reversed phase chromatography revealed that the relative hydrophobicity of the protein was higher than that of polysaccharides in EPS (Hong et al., 2018). Similarly, a Fourier transform infrared spectroscopy of sludge flocs also identified a large number of hydrophobic groups in the tightly bound EPS protein that can promote microbial aggregation (Jia et al., 2017). A higher sludge protein/polysaccharides content ratio corresponds to a higher hydrophobicity and a greater tendency of sludge flocs to deposit on the DM layer (Mohan and Nagalakshmi, 2020). Besides, SMP are soluble in wastewater, which can increase the viscosity of the wastewater. SMP can be subdivided into utilization-associated products (substrate utilization and biomass growth) and biomass-associated products (biomass decay). Compared to EPS, SMP can cause severe irremovable fouling by filling in the support material pores. The strong binding of SMP to the support material pores is attributed to the following main factors: First, SMP is a soluble EPS mainly produced from EPS release, endogenous respiration, and hydrolysis (Laspidou and Rittmann, 2002). Similar to EPS where the protein contains mainly hydrophobic amino acid groups (Jorand et al., 1998), SMP also consists of a large fraction of hydrophobic humic substances (Liang et al., 2007) and exhibits a strong tendency to deposit on the support material, causing membrane fouling (Meng et al., 2006). Second, a wide molecular distribution of SMP between 1000 and 10,000 Da enables SMP to fill in the pores of support material easily. SMP are also difficult to remove by shear forces alone because of their low transport velocity (Satyawali and Balakrishnan, 2008). Although EPS and SMP are important components of the DM to maintain its structural integrity, their production must be controlled to prevent frequent membrane fouling. A method to reduce EPS and SMP production is to add PAC, which adsorb the organics and produce less EPS than the conventional SFDMBRs (Hu et al., 2017; Skouteris et al., 2015).

Filamentous bacteria are another major concern in SFDMBRs because filamentous overgrowth can lead to more frequent membrane fouling (Ziegler et al., 2016). Some examples of common filamentous bacteria are *Chloroflexi* and *Gordonia*. Conversely, biomacromolecule-degrading bacteria are another group of bacteria that contribute to the mitigation of membrane fouling by degrading the excess EPS and SMP in the DM. The examples of biomacromolecule-degrading bacteria are *Lutibacterium* and *Cycloclasticus* (Song et al., 2020).

In anoxic/aerobic SFDMBR systems, nitrifiers and denitrifers are important functional microbes for biological nitrogen removal. Song et al. (2020) achieved biological nitrogen removal in an aerated SFDMBR system with a sponge biocarrier. The main nitrifiers detected in the system include *Nitrosomonas* and *Nitrospira*. Both *Nitrosomonas* and *Nitrospira* are common ammonia-oxidizing bacteria and nitrite-oxidizing bacteria, respectively, in municipal wastewater treatment systems. Besides, denitrifiers can thrive in the deeper anoxic zones of DM. The examples of denitrifiers present in SFDMBR systems include *Thauera*, *Desulfovibrio*, and *Sedimenticola* (Siddiqui et al., 2022; Song et al., 2020).

6. Strategies to optimize the design and operation of SFDMBR

SFDMBR optimization is mainly focused on improving the features of the dynamic membrane. Strategies have been previously proposed in the literature to shorten DM formation time, as well as to prevent the membrane's excessive compaction and/or fouling, thus reducing the membrane cleaning frequency. The most commonly applied strategies include the selection of a suitable support material during the reactor design phase, and a proper control of the SRT and flux during reactor operation.

Different support materials have been studied for SFDMBR, such as meshes (Jeison et al., 2008; Zhang et al., 2010), nonwoven fabrics (Ho et al., 2007), woven fabrics (Ersahin et al., 2013; Saleem et al., 2016), and silk (Liu et al., 2016). The woven fabric Nylon was identified as the most suitable support material because of its hydrophobicity, which helps the sludge flocs to attach on the surface. Selecting a support material with smaller pore size (10–200 μm) can prevent excessive biomass loss during the DM formation and shorten the DM formation time (Saleem et al., 2017; Siddiqui et al., 2019). Therefore, evaluating the sludge floc size distribution is critical to select a proper support material pore size and to minimize irremovable fouling in the long run.

Proper SRT control is required to prevent oversecretion of EPS, as excess EPS can result in repulsion between sludge flocs and flocculation due to the negative charges on the floc surface. For instance, parallel SFDMBR operation at 10 d, 30 d, and infinite SRT revealed that infinite SRT resulted in the highest effluent turbidity due to the accumulation of a thick and sticky biomass on the support material. The excess biomass build up reduced the food-to-microorganisms ratio, which accelerated biomass decay and leached the decay products into the effluent (Alavi Moghaddam et al., 2002). Conversely, an SRT between 10 and 30 d showed good organic carbon removal without clogging issues throughout the 4-month, lab-scale, submerged CSTR aerobic SFDMBR operation (Alavi Moghaddam et al., 2002). For sludge flocs with poor flocculation ability, multivalent cations, such as calcium and magnesium ions, or powdered activated carbon dosage can encourage flocculation (Hu et al., 2017; Higgins et al., 2004).

In addition, the SFDMBR should be operated in a subcritical operating flux range to prevent excessive DM compaction and fouling. Subcritical flux operation can encourage a more porous DM formation and subsequently lengthen the cleaning interval (Mohan and Nagalakshmi, 2020). Introducing a cross-flow stream in the SFDMBR can also reduce DM fouling by shearing excess sludge from the DM (Liao et al., 2006). For aerobic SFDMBR, cross-flow shearing can be achieved via air sparging using air pumps. The cross-flow velocity (CFV) used in aerobic SFDMBRs typically ranged from 0.011 to 0.022 m/s (Chang et al., 2007). For AnSFDMBRs, using a biogas recirculation line with a CFV of 0.015 m/s was sufficient to control the thickness of DM in AnSFDMBRs (Siddiqui et al., 2021). However, the CFV must be carefully controlled to prevent excessive shearing that can break the sludge flocs apart. These smaller sludge flocs produced from excessive shearing increase the propensity of irremovable fouling. For instance, Chang et al. (2007) found that an aeration intensity beyond 0.022 L/min caused significant filtration resistance presumably due to the smaller sludge flocs that clogged the support material pores.

7. Conclusions and perspectives

This chapter first explained the conceptualization and evolution of the SFDMBR technology from the preceding conventional membrane technologies, such as UF, RO, and MF. SFDMBR can adopt meshes, woven fabrics, and nonwoven fabrics materials in either tubular or flat sheet configurations as the support material. The SFDMBR technologies for wastewater treatment applications can be largely grouped into AnSFDMBR and aerobic SFDMBR. AnSFDMBR commonly used CSTR and UASB reactor configurations, while the aerobic SFDMBR usually applied the CSTR-type reactor. The major groups of microorganisms in SFDMBRs were EPS-secreting microorganisms, filamentous bacteria, nitrifiers, and denitrifiers. The EPS-secreting microorganisms and filamentous bacteria were important for the integrity of DM on supporting material, while the nitrifiers and denitrifiers were involved in carbon and nitrogen removal from the wastewater. The major bottlenecks of applying SFDMBR in large-scale applications were membrane fouling and cleaning. The selection of support materials with a suitable pore size, proper SRT control, and subcritical flux operation were the important considerations to mitigate the risk of frequent membrane fouling. Both CSTR- and UASB-type reactors can also employ membrane cross-flow to attain better fouling control. In addition, hybrid cleaning by combining physical and cleaning methods can be used to recover membrane permeability more effectively. With continuous development of better membrane fouling control strategies and cleaning methods, SFDMBR will emerge as a promising technology to attain a low footprint, high flexibility, and high treatment capacity wastewater treatment application in the future.

Acknowledgments

This work was supported by the Hong Kong Innovation and Technology Commission (Grant no. ITC-CNERC14EG03 and ITS/423/16FX), the Research Grants Council of the Hong Kong Special Administrative Region (Grant no. T21-604/19-R), and Ghent University, Belgium (BOF/STA/202109/022).

References

Alavi Moghaddam, M.R., Mino, T., Satoh, H., 2002. Effect of important operational parameters on performance of coarse pore filtration activated sludge process. Water Sci. Technol. 46, 229–236.

Alepu, O.E., Segun, G.A., Ikhumhen, H.O., 2016. Formation mechanism and performance of dynamic membrane technology for municipal wastewater treatment—a review. Adv. Recycl. Waste Manag. 01.

Alibardi, L., Bernava, N., Cossu, R., Spagni, A., 2016. Anaerobic dynamic membrane bioreactor for wastewater treatment at ambient temperature. Chem. Eng. J. 284, 130–138.

Alibardi, L., Cossu, R., Saleem, M., Spagni, A., 2014. Development and permeability of a dynamic membrane for anaerobic wastewater treatment. Bioresour. Technol. 161, 236–244.

Al-Malack, M.H., Anderson, G., 1996. Formation of dynamic membranes with crossflow microfiltration. J. Membr. Sci. 112, 287–296.

Cai, D., Huang, J., Liu, G., Li, M., Yu, Y., Meng, F., 2018. Effect of support material pore size on the filtration behavior of dynamic membrane bioreactor. Bioresour. Technol. 255, 359–363.

Chang, W.-K., Hu, A.Y.-J., Horng, R.-Y., Tzou, W.-Y., 2007. Membrane bioreactor with nonwoven fabrics as solid-liquid separation media for wastewater treatment. Desalination 202, 122–128.

Chu, L., Zhang, X., Yang, F., Li, X., 2006. Treatment of domestic wastewater by using a microaerobic membrane bio-reactor. Desalination 189, 181–192.

Chu, H., Zhang, Y., Zhou, X., Zhao, Y., Dong, B., Zhang, H., 2014. Dynamic membrane bioreactor for wastewater treatment: operation, critical flux, and dynamic membrane structure. J. Membr. Sci. 450, 265–271.

Ersahin, M.E., Gimenez, J.B., Ozgun, H., Tao, Y., Spanjers, H., van Lier, J.B., 2016a. Gas-lift anaerobic dynamic membrane bioreactors for high strength synthetic wastewater treatment: effect of biogas sparging velocity and HRT on treatment performance. Chem. Eng. J. 305, 46–53.

Ersahin, M.E., Ozgun, H., Dereli, R.K., Ozturk, I., Roest, K., van Lier, J.B., 2012. A review on dynamic membrane filtration: materials, applications and future perspectives. Bioresour. Technol. 122, 196–206.

Ersahin, M.E., Ozgun, H., van Lier, J.B., 2013. Effect of support material properties on dynamic membrane filtration performance. Sep. Sci. Technol. 48, 2263–2269.

Ersahin, M.E., Tao, Y., Ozgun, H., Gimenez, J.B., Spanjers, H., van Lier, J.B., 2017. Impact of anaerobic dynamic membrane bioreactor configuration on treatment and filterability performance. J. Membr. Sci. 526, 387–394.

Ersahin, M.E., Tao, Y., Ozgun, H., Spanjers, H., van Lier, J.B., 2016b. Characteristics and role of dynamic membrane layer in anaerobic membrane bioreactors. Biotechnol. Bioeng. 113, 761–771.

Fan, B., Huang, X., 2002. Characteristics of a self-forming dynamic membrane coupled with a bioreactor for municipal wastewater treatment. Environ. Sci. Technol. 36, 5245–5251.

Groves, G., Buckley, C., Cox, J., Kirk, A., Macmillan, C., Simpson, M., 1983. Dynamic membrane ultrafiltration and hyperfiltration for the treatment of industrial effluents for water reuse. Desalination 47, 305–312.

Higgins, M.J., Tom, L.A., Sobeck, D.C., 2004. Case study I: application of the divalent cation bridging theory to improve biofloc properties and industrial activated sludge system performance-direct addition of divalent cations. Water Environ. Res. 76, 344–352.

Ho, J.H., Khanal, S.K., Sung, S., 2007. Anaerobic membrane bioreactor for treatment of synthetic municipal wastewater at ambient temperature. Water Sci. Technol. 55, 79–86.

Hong, P.-N., Noguchi, M., Matsuura, N., Honda, R., 2019. Mechanism of biofouling enhancement in a membrane bioreactor under constant trans-membrane pressure operation. J. Membr. Sci. 592.

Hong, P.-N., Taing, C., Phan, P.-T., Honda, R., 2018. Polarity-molecular weight profile of extracellular polymeric substances in a membrane bioreactor: comparison between bulk sludge and cake layers. J. Water Environ. Technol. 16, 40–53.

Hu, Y., Wang, X.C., Tian, W., Ngo, H.H., Chen, R., 2016. Towards stable operation of a dynamic membrane bioreactor (DMBR): operational process, behavior and retention effect of dynamic membrane. J. Membr. Sci. 498, 20–29.

Hu, Y., Yang, Y., Wang, X.C., Hao Ngo, H., Sun, Q., Li, S., Tang, J., Yu, Z., 2017. Effects of powdered activated carbon addition on filtration performance and dynamic membrane layer properties in a hybrid DMBR process. Chem. Eng. J. 327, 39–50.

Huang, J., Wu, X., Cai, D., Chen, G., Li, D., Yu, Y., Petrik, L.F., Liu, G., 2019. Linking solids retention time to the composition, structure, and hydraulic resistance of biofilms developed on support materials in dynamic membrane bioreactors. J. Membr. Sci. 581, 158–167.

Hutten, I.M., 2007. Handbook of Nonwoven Filter Media. Elsevier, Burlington, US.

Jeison, D., Díaz, I., van Lier, J.B., 2008. Anaerobic membrane bioreactors: are membranes really necessary? Electron. J. Biotechnol. 11, 1–2.

Jia, F., Yang, Q., Liu, X., Li, X., Li, B., Zhang, L., Peng, Y., 2017. Stratification of extracellular polymeric substances (EPS) for aggregated anammox microorganisms. Environ. Sci. Technol. 51, 3260–3268.

Jorand, F., Boué-Bigne, F., Block, J.C., Urbain, V., 1998. Hydrophobic/hydrophilic properties of activated sludge exopolymeric substances. Water Sci. Technol. 37.

Khuntia, H.K., Hameed, S., Janardhana, N., Chanakya, H., 2019. Greywater treatment in aerobic bio-reactor with macropore mesh filters. J. Water Process Eng. 28, 269–276.

Kiso, Y., Jung, Y.-J., Ichinari, T., Park, M., Kitao, T., Nishimura, K., Min, K.-S., 2000. Wastewater treatment performance of a filtration bio-reactor equipped with a mesh as a filter material. Water Res. 34, 4143–4150.

Kiso, Y., Jung, Y.J., Park, M.S., Wang, W., Shimase, M., Yamada, T., Min, K.S., 2005. Coupling of sequencing batch reactor and mesh filtration: operational parameters and wastewater treatment performance. Water Res. 39, 4887–4898.

Laspidou, C.S., Rittmann, B.E., 2002. A unified theory for extracellular polymeric substances, soluble microbial products, and active and inert biomass. Water Res. 36, 2711–2720.

Liang, S., Liu, C., Song, L., 2007. Soluble microbial products in membrane bioreactor operation: behaviors, characteristics, and fouling potential. Water Res. 41, 95–101.

Liang, S., Zhao, T., Zhang, J., Sun, F., Liu, C., Song, L., 2012. Determination of fouling-related critical flux in self-forming dynamic membrane bioreactors: interference of membrane compressibility. J. Membr. Sci. 390–391, 113–120.

Liao, B.-Q., Kraemer, J.T., Bagley, D.M., 2006. Anaerobic membrane bioreactors: applications and research directions. Crit. Rev. Environ. Sci. Technol. 36, 489–530.

Lin, H., Peng, W., Zhang, M., Chen, J., Hong, H., Zhang, Y., 2013. A review on anaerobic membrane bioreactors: applications, membrane fouling and future perspectives. Desalination 314, 169–188.

Lin, H.J., Xie, K., Mahendran, B., Bagley, D.M., Leung, K.T., Liss, S.N., Liao, B.Q., 2009. Sludge properties and their effects on membrane fouling in submerged anaerobic membrane bioreactors (SAnMBRs). Water Res. 43, 3827–3837.

Liu, H., Wang, Y., Yin, B., Zhu, Y., Fu, B., Liu, H., 2016. Improving volatile fatty acid yield from sludge anaerobic fermentation through self-forming dynamic membrane separation. Bioresour. Technol. 218, 92–100.

Ma, J., Wang, Z., Xu, Y., Wang, Q., Wu, Z., Grasmick, A., 2013a. Organic matter recovery from municipal wastewater by using dynamic membrane separation process. Chem. Eng. J. 219, 190–199.

Ma, J., Wang, Z., Zou, X., Feng, J., Wu, Z., 2013b. Microbial communities in an anaerobic dynamic membrane bioreactor (AnDMBR) for municipal wastewater treatment: comparison of bulk sludge and cake layer. Process Biochem. 48, 510–516.

Marcinkowsky, A.E., Kraus, K.A., Phillips, H.O., Johnson, J.S., Shor, A.J., 2002. Hyperfiltration studies. IV. Salt rejection by dynamically formed hydrous oxide membranes. J. Am. Chem. Soc. 88, 5744–5746.

Martinez-Sosa, D., Helmreich, B., Netter, T., Paris, S., Bischof, F., Horn, H., 2011. Anaerobic submerged membrane bioreactor (AnSMBR) for municipal wastewater treatment under mesophilic and psychrophilic temperature conditions. Bioresour. Technol. 102, 10377–10385.

Meng, Z.-G., Yang, F.L., Zhang, X.W., 2005. MBR focus: do nonwovens offer a cheaper option? Filtr. Sep. 42, 28–30.

Meng, F., Zhang, H., Yang, F., Zhang, S., Li, Y., Zhang, X., 2006. Identification of activated sludge properties affecting membrane fouling in submerged membrane bioreactors. Sep. Purif. Technol. 51, 95–103.

Mohan, S.M., Nagalakshmi, S., 2020. A review on aerobic self-forming dynamic membrane bioreactor: formation, performance, fouling and cleaning. J. Water Process Eng. 37.

Pollice, A., Vergine, P., 2020. Self-forming dynamic membrane bioreactors (SFD MBR) for wastewater treatment: principles and applications. In: Current Developments in Biotechnology and Bioengineering.

Quek, P.J., Yeap, T.S., Ng, H.Y., 2017. Applicability of upflow anaerobic sludge blanket and dynamic membrane-coupled process for the treatment of municipal wastewater. Appl. Microbiol. Biotechnol. 101, 6531–6540.

Ren, X., Shon, H., Jang, N., Lee, Y.G., Bae, M., Lee, J., Cho, K., Kim, I.S., 2010. Novel membrane bioreactor (MBR) coupled with a nonwoven fabric filter for household wastewater treatment. Water Res. 44, 751–760.

Rezvani, F., Mehrnia, M.R., Poostchi, A.A., 2014. Optimal operating strategies of SFDM formation for MBR application. Sep. Purif. Technol. 124, 124–133.

Sabaghian, M., Mehrnia, M.R., Esmaieli, M., Nourmohammadi, D., 2018. Influence of static mixer on the formation and performance of dynamic membrane in a dynamic membrane bioreactor. Sep. Purif. Technol. 206, 324–334.

Saddoud, A., Abdelkafi, S., Sayadi, S., 2009. Effects of domestic wastewater toxicity on anaerobic membrane-bioreactor (MBR) performances. Environ. Technol. 30, 1361–1369.

Saleem, M., Alibardi, L., Cossu, R., Lavagnolo, M.C., Spagni, A., 2017. Analysis of fouling development under dynamic membrane filtration operation. Chem. Eng. J. 312, 136–143.

Saleem, M., Alibardi, L., Lavagnolo, M.C., Cossu, R., Spagni, A., 2016. Effect of filtration flux on the development and operation of a dynamic membrane for anaerobic wastewater treatment. J. Environ. Manag. 180, 459–465.

Saleem, M., Lavagnolo, M.C., Spagni, A., 2018. Biological hydrogen production via dark fermentation by using a side-stream dynamic membrane bioreactor: effect of substrate concentration. Chem. Eng. J. 349, 719–727.

Saleem, M., Masut, E., Spagni, A., Lavagnolo, M.C., 2019. Exploring dynamic membrane as an alternative for conventional membrane for the treatment of old landfill leachate. J. Environ. Manag. 246, 658–667.

Salerno, C., Vergine, P., Berardi, G., Pollice, A., 2017. Influence of air scouring on the performance of a Self Forming Dynamic Membrane BioReactor (SFD MBR) for municipal wastewater treatment. Bioresour. Technol. 223, 301–306.

B. Applications of membrane technology for water and wastewater treatment

Satyawali, Y., Balakrishnan, M., 2008. Treatment of distillery effluent in a membrane bioreactor (MBR) equipped with mesh filter. Sep. Purif. Technol. 63, 278–286.

Seo, G., Moon, B., Lee, T., Lim, T., Kim, I., 2003. Non-woven fabric filter separation activated sludge reactor for domestic wastewater reclamation. Water Sci. Technol. 47, 133–138.

Seo, G.T., Moon, B.H., Park, Y.M., Kim, S.H., 2007. Filtration characteristics of immersed coarse pore filters in an activated sludge system for domestic wastewater reclamation. Water Sci. Technol. 55, 51–58.

Shor, A., Kraus, K., Smith Jr., W.T., Johnson Jr., J.S., 1968. Hyperfiltration studies. XI. Salt-rejection properties of dynamically formed hydrous zirconium (IV) oxide membranes. J. Phys. Chem. 72, 2200–2206.

Siddiqui, M.A., Biswal, B.K., Saleem, M., Guan, D., Iqbal, A., Wu, D., Khanal, S.K., Chen, G., 2021. Anaerobic self-forming dynamic membrane bioreactors (AnSFDMBRs) for wastewater treatment—recent advances, process optimization and perspectives. Bioresour. Technol. 332, 125101.

Siddiqui, M.A., Dai, J., Guan, D., Chen, G., 2019. Exploration of the formation of self-forming dynamic membrane in an upflow anaerobic sludge blanket reactor. Sep. Purif. Technol. 212, 757–766.

Siddiqui, M.A., Kumar Biswal, B., Siriweera, B., Chen, G., Wu, D., 2022. Integrated self-forming dynamic membrane (SFDM) and membrane-aerated biofilm reactor (MABR) system enhanced single-stage autotrophic nitrogen removal. Bioresour. Technol. 345, 126554.

Skouteris, G., Saroj, D., Melidis, P., Hai, F.I., Ouki, S., 2015. The effect of activated carbon addition on membrane bioreactor processes for wastewater treatment and reclamation—a critical review. Bioresour. Technol. 185, 399–410.

Song, W., Lee, L.Y., You, H., Shi, X., Ng, H.Y., 2020. Microbial community succession and its correlation with reactor performance in a sponge membrane bioreactor coupled with fiber-bundle anoxic bio-filter for treating saline mariculture wastewater. Bioresour. Technol. 295, 122284.

Sreeda, P., Sathya, A., Sivasubramanian, V., 2018. Novel application of high-density polyethylene mesh as self-forming dynamic membrane integrated into a bioreactor for wastewater treatment. Environ. Technol. 39, 51–58.

Srivastava, A., Srivastava, O., Talapatra, S., Vajtai, R., Ajayan, P., 2004. Carbon nanotube filters. Nat. Mater. 3, 610–614.

Sun, F., Zhang, N., Li, F., Wang, X., Zhang, J., Song, L., Liang, S., 2018. Dynamic analysis of self-forming dynamic membrane (SFDM) filtration in submerged anaerobic bioreactor: performance, characteristic, and mechanism. Bioresour. Technol. 270, 383–390.

Tanny, G.B., Johnson, J.S., 1978. The structure of hydrous Zr(IV) oxide-polyacrylate membranes: poly(acrylic acid) deposition. J. Appl. Polym. Sci. 22, 289–297.

Vergine, P., Salerno, C., Berardi, G., Pollice, A., 2018. Sludge cake and biofilm formation as valuable tools in wastewater treatment by coupling Integrated Fixed-film Activated Sludge (IFAS) with Self Forming Dynamic Membrane BioReactors (SFD-MBR). Bioresour. Technol. 268, 121–127.

Wang, C., Chen, W.-N., Hu, Q.-Y., Ji, M., Gao, X., 2015. Dynamic fouling behavior and cake layer structure changes in nonwoven membrane bioreactor for bath wastewater treatment. Chem. Eng. J. 264, 462–469.

Wang, L., Liu, H., Zhang, W., Yu, T., Jin, Q., Fu, B., Liu, H., 2018. Recovery of organic matters in wastewater by self-forming dynamic membrane bioreactor: performance and membrane fouling. Chemosphere 203, 123–131.

Wang, Y.-K., Sheng, G.-P., Li, W.-W., Yu, H.-Q., 2012. A pilot investigation into membrane bioreactor using mesh filter for treating low-strength municipal wastewater. Bioresour. Technol. 122, 17–21.

Wang, Y.-K., Sheng, G.-P., Ni, B.-J., Li, W.-W., Zeng, R.J., Wang, Y.-Q., Shi, B.-J., Yu, H.-Q., 2013. Simultaneous carbon and nitrogen removals in membrane bioreactor with mesh filter: an experimental and modeling approach. Chem. Eng. Sci. 95, 78–84.

Xie, Y.H., Zhu, T., Xu, C.H., Nozaki, T., Furukawa, K., 2012. Treatment of domestic sewage by a metal membrane bioreactor. Water Sci. Technol. 65, 1102–1108.

Yang, T., Liu, F., Xiong, H., Yang, Q., Chen, F., Zhan, C., 2019. Fouling process and anti-fouling mechanisms of dynamic membrane assisted by photocatalytic oxidation under sub-critical fluxes. Chin. J. Chem. Eng. 27, 1798–1806.

Yang, Y., Zang, Y., Hu, Y., Wang, X.C., Ngo, H.H., 2020. Upflow anaerobic dynamic membrane bioreactor (AnDMBR) for wastewater treatment at room temperature and short HRTs: process characteristics and practical applicability. Chem. Eng. J. 383, 123186.

Ye, M., Zhang, H., Wei, Q., Lei, H., Yang, F., Zhang, X., 2006. Study on the suitable thickness of a PAC-precoated dynamic membrane coupled with a bioreactor for municipal wastewater treatment. Desalination 194, 108–120.

Yu, H., Wang, Z., Wu, Z., Zhu, C., 2016. Enhanced waste activated sludge digestion using a submerged anaerobic dynamic membrane bioreactor: performance, sludge characteristics and microbial community. Sci. Rep. 6, 20111.

Zahid, W.M., El-Shafai, S.A., 2011. Use of cloth-media filter for membrane bioreactor treating municipal wastewater. Bioresour. Technol. 102, 2193–2198.

Zhang, S., Qu, Y., Liu, Y., Yang, F., Zhang, X., Furukawa, K., Yamada, Y., 2005. Experimental study of domestic sewage treatment with a metal membrane bioreactor. Desalination 177, 83–93.

Zhang, X., Wang, Z., Wu, Z., Lu, F., Tong, J., Zang, L., 2010. Formation of dynamic membrane in an anaerobic membrane bioreactor for municipal wastewater treatment. Chem. Eng. J. 165, 175–183.

Zhang, X., Wang, Z., Wu, Z., Wei, T., Lu, F., Tong, J., Mai, S., 2011. Membrane fouling in an anaerobic dynamic membrane bioreactor (AnDMBR) for municipal wastewater treatment: characteristics of membrane foulants and bulk sludge. Process Biochem. 46, 1538–1544.

Ziegler, A.S., McIlroy, S.J., Larsen, P., Albertsen, M., Hansen, A.A., Heinen, N., Nielsen, P.H., 2016. Dynamics of the fouling layer microbial community in a membrane bioreactor. PLoS One 11, e0158811.

PART C

Membrane processes (MD-RO-FO)

Applications and challenges of membrane distillation in water reuse

My Thi Tra Ngo[a,b,c], Han Ngoc Mai Nguyen[a,b], Nguyen Cong Nguyen[d], Phuong-Thao Nguyen[a,b], and Xuan-Thanh Bui[a,b]

[a]Key Laboratory of Advanced Waste Treatment Technology & Faculty of Environment and Natural Resources, Ho Chi Minh City University of Technology (HCMUT), Ho Chi Minh City, Vietnam [b]Vietnam National University Ho Chi Minh City (VNU-HCM), Ho Chi Minh City, Vietnam [c]Graduate School of Engineering, Nagasaki University, Nagasaki, Japan [d]Faculty of Chemistry and Environment, Dalat University, Dalat, Vietnam

1. Introduction

Membrane distillation (MD) is an emerging technology that has huge potential for water reuse, particularly from high-salinity wastewater sources (Swaminathan et al., 2018; Ahmad et al., 2021). It is a thermally driven membrane separation process where water vapors are transported through a hydrophobic membrane and condensed in the distillate side. Because of this distinct mechanism, theoretically 100% rejection of nonvolatile components may be achieved (Swaminathan et al., 2018). However, conventional hydrophobic MD membranes are subjected to fast wetting and fouling caused by low-surface-tension chemicals, which are usually present in high quantities in wastewater (Yao et al., 2020). The occurrence of these chemicals reduces the membrane's hydrophobicity resulting in lower rejection than expected and potentially causing the whole process to fail (Guillen-Burrieza et al., 2016; Chamani et al., 2021). Moreover, solution for high energy consumption in circulating feed solution heating remains another challenge for MD application in wastewater reuse (Ahmed et al., 2020a). Therefore, fabrication of innovative MD membranes with desired properties such as resistance to membrane wetting and fouling, and low thermal conductivity has received massive attention (Sinha Ray et al., 2020; Gao et al., 2021; Makanjuola et al., 2021; Dai et al., 2022).

To address the wetting problem, novel membranes have been proposed in recent years for improving antiwetting and antifouling properties, especially for MD treatment of wastewater (Li et al., 2019b; Jia et al., 2021). Their surfaces have remarkable repellences toward surfactants

Copyright © 2023 Elsevier Inc. All rights reserved.

and low-surface-tension liquids. The wettability of the MD membrane can be altered by optimizing surface roughness and surface tension via modification methods supported by recent advances in materials science. Using fluorine-based solutions, omniphobic membranes have been successfully developed in the laboratory and tested with synthetic wastewater (Feng et al., 2022; Meng et al., 2022). However, the applicability of omniphobic MD membranes in water reuse remains doubtful due to health and environmental concerns regarding the use of toxic fluorine solutions. For the high-energy-consumption problem, a conventional approach is to take advantage of a low-grade heat source such as solar energy or heat waste (González et al., 2017). Limitations of this approach are the inconsistent supply of waste heat and solar energy or the unavailability of renewable energy sources. However, by incorporating photothermal materials (e.g., plasmonic metallic nanomaterials, inorganic semiconductor solar absorber materials, or carbon-based nanomaterials) into the membrane surface, membrane surface temperature can be maintained or even increased, and thus it can reduce energy consumption (Han et al., 2020; Gao et al., 2021).

Previous research on MD membranes has focused on lab-scale fabrication or modification methods to maximize water productivity and ensure water quality. Still, the produced membranes have only been tested in considerably short timeframes to evaluate the reduction of permeate flux and hydrophobicity. However, high resistance to membrane wetting-fouling and energy efficiency are crucial factors to enable long-term MD operations. In addition, the existence of chemicals of emerging concerns in wastewater can pose a viable challenge for upscaling because of membrane wetting and fouling (Yao et al., 2020). This review presents the limitations and challenges of MD applications as well as the developments of the MD membrane for water reuse with a high tolerance to harsh conditions.

2. Applications of membrane distillation in water reuse

Nowadays, membrane distillation (MD) is still in the development stage with most research conducted at the laboratory scale. With their excellent performances, MD may be applied for separating contaminants from liquid solutions in many fields, especially in water and wastewater treatment. Many studies have been focused on saltwater desalination thanks to its almost 100% rejection of nonvolatile compounds (Wang and Chung, 2015; Meng et al., 2022). More than 99.9% NaCl rejection was reported by MD membranes for seawater desalination (Donato et al., 2020; Zou et al., 2021). Furthermore, membrane distillation is capable of treating high-salinity solutions of up to $300 \, g/kg$ (Swaminathan et al., 2018). In addition to salt removal, high rejection of CECs was found using these MD membranes in wastewater treatment, such as hydrocarbon (e.g., polyphenols, >89%), household products, and industrial chemicals (e.g., benzotriazole, triclocarban, n-nonylphenol, and oxybenzone, >96%), antibiotics/prescription drugs (e.g., sulfamethoxazole, carbamazepine, diclofenac, caffeine, omeprazole, primidone, and meprobamate, >96%), pesticides/herbicide (e.g., diuron and simazine, 99%), fire retardants (e.g., tris(2-chloroethyl)phosphate, 99%), and heavy metals (e.g., boron, cobalt, cesium, strontium and so on, >99.97%) (Khayet, 2013; Liu and Wang, 2013; Wen et al., 2016; Naidu et al., 2017). Separation performance and properties of commercial MD membranes are presented in Table 1.

TABLE 1 Separation performance of existing MD membranes in water reuse.

Membrane	Membrane type/ manufacturer	Membrane properties	Wastewater sources	Operating conditions	Permeate flux	Removals	References
Flat-sheet PTFE	HP-010-30, Sumitomo Electric Industries, Ltd. Poreflon	Pore size=0.1 μm; CA=112 degree; LEP=180 kPa	Anaerobic effluent from domestic wastewater	T_f=40–70°C, T_p=10°C, $FR_f=FR_p$=0.25–1 L/min	1.41–9.22 L/m²h	97.8%–99.9% COD removal, 89.6%–96.3% ammonium removal	Jacob et al. (2015)
	Millipore, USA	Pore size=0.2 μm; CA=142 degree; LEP=280 kPa		T_f=30–60°C, T_p=20°C, FR_f=0.4 L/min, FR_p=0.3 L/min	4.5 L/m²h	>98% DOC removal, a nearly complete rejection of the residual organics	Kim et al. (2015)
Flat-sheet PVDF	Millipore, USA	Pore size=0.22 μm; CA=94 degree; LEP=204 kPa			5.3 L/m²h		
Capillary PP	Membrana GmbH, Germany	Pore size=0.2 μm; CA=134 degree; LEP=140 kPa			2.5 L/m²h		
Flat-sheet PTFE	General Electric, US	Pore size=0.2 μm; CA=139.4±1.5 degree	RO concentrate	T_f=55°C, T_p=25°C, CFV=0.6 m/s	16 kg/m²h	96%–99% removal of almost micropollutants, 50%–88% removal of salicylic acid, atrazine, triclosan bisphenol A, propylparaben, and benzophenone	Naidu et al. (2017)
	FGLP14250-EMD Millipore	Pore size=0.22 μm		T_f=50°C, T_p=20°C	16 L/m²h	≥97% EC removal	Alrehaili et al. (2020)
	GE Osmonics, Minnetonka, MN			T_f=60°C, T_p=20°C, $FR_f=FR_p$=1.5 L/min	40-5 L/m²h	>99.9% salt removal	Martinetti et al. (2009)

Continued

TABLE 1 Separation performance of existing MD membranes in water reuse—cont'd

Membrane	Membrane type/manufacturer	Membrane properties	Wastewater sources	Operating conditions	Permeate flux	Removals	References
Flat-sheet PVDF	GVHP, Millipore, USA	Pore size=0.22μm; CA=124.9 degree	Antibiotics wastewater	T_f=60°C, T_p=20°C, FR_f=FR_p=0.5 L/min	20 L/m²h	100% removal of negatively-charged and neutral antibiotics), 78% removal of Tobramycin and 86% removal of Ciprofloxacin	Guo et al. (2018)
Flat-sheet PTFE	Shanghai Minglie membrane Co., China	Pore size=0.22μm; CA=133.7 degree; LEP=271kPa	Industrial dyeing wastewater	T_f=30-60°C, T_p=20°C, FR_f=0.35 L/min, FR_p=0.25 L/min	19.8 L/m²h	90% COD removal, 94% color removal, and low EC in the permeate (1.6-399 μs/cm)	Li et al. (2018)
Capillary PP	Membrana, Germany	Pore size=0.2μm	Olive mill wastewater	T_f=50°C, T_p=15°C, FR_f=1.17 L/min, FR_p=1 L/min	6.5 kg/m²h	89% TOC removal, 99.6 % polyphenols removal, >99% EC rejection	Carnevale et al. (2016)
Flat-sheet PVDF	GVHP, Millipore, USA	Pore size=0.28μm; LEP=204kPa		T_f=40°C, T_p=20°C	5 L/m²h	89% polyphenols removal	El-Abbassi et al. (2009), El-Abbassi et al. (2013)
Flat-sheet PTFE	TF200, Gelman	Pore size=0.2μm; LEP=276kPa			7.7-9 L/m²h	99% polyphenols removal	El-Abbassi et al. (2013)
	TF200, Gelman	Pore size=0.2μm		T_f=70°C, T_p=25°C	33-23 L/m²h	>99.5% removal, fouling indexes did not exceed 2.9%	Kiai et al. (2014)
	TF450, Gelman	Pore size=0.45μm.			38-24 L/m² h	>99.5% removal, fouling index 9.8%	
	TF1000, Gelman	Pore size=1.0μm			38-18 L/m² h	>99.5% removal, fouling index 14.8%	
Flat-sheet PTFE	TF200, Gelman	Pore size=0.2μm; CA=113.6 degree; LEP=276kPa	Radioactive wastewater	T_f=55°C, T_p=21.5°C	0.155 kg/mh	Very high rejection factors, Co^{2+}, Cs^- and Sr^{2+} were not detected in the permeate	Khayet (2013)

PTFE, polytetrafluoroethylene; *PVDF*, polyvinylidene fluoride; *CA*, contact angle; *LEP*, liquid entry pressure; T_f, feed (hot) temperature; T_p, permeate (cool) temperature; *CFV*, crossflow velocity; FR_f, feed flow rate; FR_p, permeate flow rate; *EC*, electrical conductivity; *DOC*, dissolved organic carbon; *COD*, chemical oxygen demand.

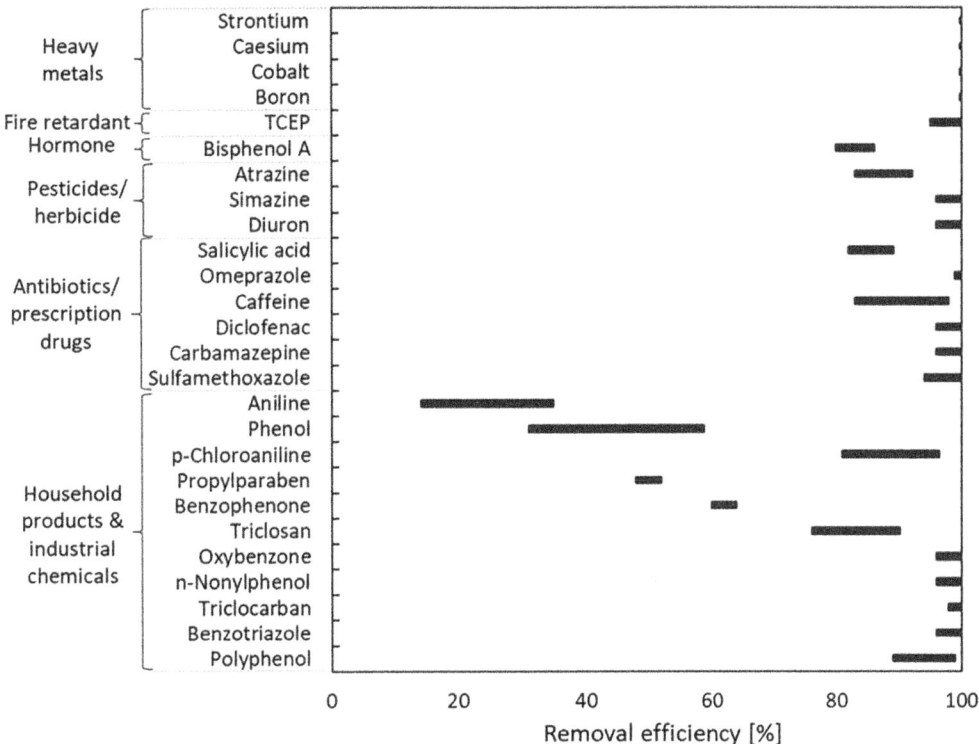

FIG. 1 Removal efficiencies of chemicals of emerging concerns by membrane distillation.

Compared to reverse osmosis, membrane distillation is far more feasible for treating solutions with high concentrations of salts (e.g., reverse osmosis concentrate) (Ngo et al., 2022). In addition, a higher rejection of toxic metals (>99.97%) in wastewater treatment was reported in MD compared to RO treatment [e.g., 96%–98% removal of cesium and strontium (Combernoux et al., 2017) and 80%–97% removal of boron (Ban et al., 2019)] (Fig. 1). Membrane distillation has also comparable removals of the charged CECs or compounds that have high volatile temperatures in comparison to RO treatment with >95% rejection of chemicals of emerging concerns (e.g., household products and industrial chemicals, antibiotics/prescription drugs, pesticides/herbicides, and fire retardants) (Fujioka et al., 2015; Fujioka et al., 2018). Further, MD requires much lower hydrostatic pressure than RO modules (Alkhudhiri et al., 2012; Tow et al., 2018). As the driving force in MD is the difference of transmembrane vapor pressure that is caused by temperature difference between the two sides of the membrane, the requirement of applied pressure is low. Therefore, MD is less susceptible to membrane fouling than other pressurized membrane filtration processes (e.g., reverse osmosis, nanofiltration, or ultrafiltration). Hence, MD has the potential to minimize a huge volume of wastewater discharged from various sources, such as municipal wastewater and landfill leachate (Arcanjo et al., 2021; Yan et al., 2021), reverse osmosis concentrate (Rajwade et al., 2020; Ngo et al., 2021), textile wastewater (Laqbaqbi et al., 2019; Fortunato et al., 2021), shale gas-produced water (Cho et al., 2018; Tavakkoli et al., 2020), olive mill, and other oily wastewaters (Carnevale et al., 2016; Kalla, 2021). In addition, MD can produce high-purity water from wastewater sources to adapt to high-quality requirements in some industries such as semiconductors and pharmaceuticals (Woldemariam et al., 2016; Noor et al., 2020) as well as to meet the standards for potable or nonpotable water by recycling municipal wastewater (Caroline Ricci et al., 2021).

3. Challenges in water-reuse applications

3.1 Membrane wetting and fouling

In MD, membranes must be hydrophobic to prevent the direct penetration of feed liquid through the microporous membrane; thus, only vapors can pass through the membrane pores (Fig. 2A). When the transmembrane hydrostatic pressure exceeds the liquid entry pressure of the membrane (Fig. 2B), pore wetting occurs. In the case of pore wetting, direct penetration of liquid reduces the rejection rate and fails the whole process (Guillen-Burrieza et al., 2016; Li et al., 2022). In long-term treatment, surface wetting happens on the membrane surface, but gaps for transporting only vapor still exist. When some membrane pores are open for liquid to pass through, the phenomenon is called partial wetting. In the case of full wetting, liquid can penetrate through almost all membrane pores (Tijing et al., 2015).

In water reuse, the wetting phenomenon happens when the feed solution has high concentrations of salts, organics, and nutrients. The deposition of these compounds on the surface and in the pores of the membrane results in the loss of hydrophobicity (Rezaei et al., 2018; Tibi et al., 2020). Moreover, the existence of surfactants or low-surface-tension liquids in wastewater can promote membrane wetting and fouling in MD treatment (Han et al., 2017). Hydrophobic compounds can also attach to the surface of MD membranes; thus, this can cause adverse effects on separation performance. While pore wetting can reduce permeate quality,

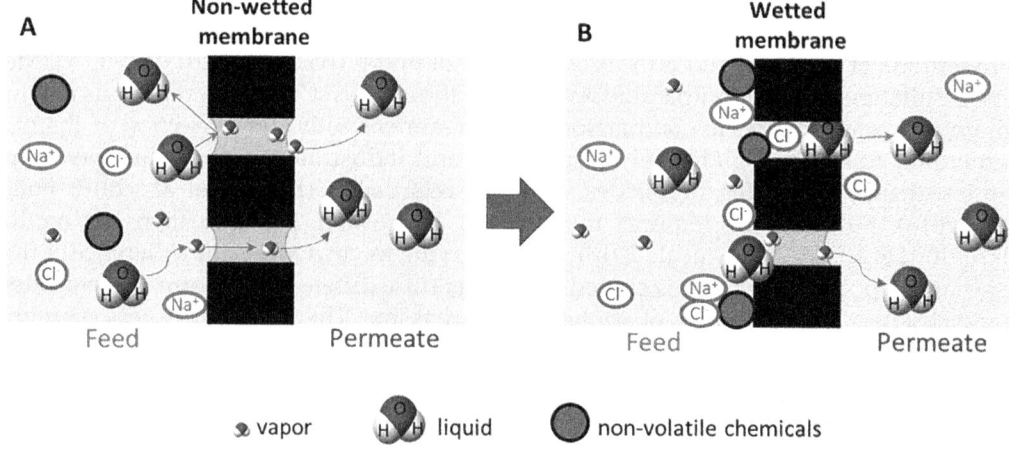

FIG. 2   Mechanisms of transportation through (A) nonwetted MD membrane and (B) wetted MD membrane.

membrane fouling happening on the feed side of the membrane can also lead to a low permeate flux (Choudhury et al., 2019). Fouling in hydrophobic membranes is affected by foulant characteristics, membrane properties, operating conditions, and feed solution characteristics (Tijing et al., 2015). In MD, membrane fouling is classified into inorganic fouling, organic fouling, and biological fouling. The presence of inorganic compounds in wastewater causes the growth of precipitates (e.g., calcium and phosphate) on the membrane surface (Khan et al., 2014), while the existence of organic compounds leads to the drop of surface tension of the solution (Rezaei et al., 2018). Hence, membrane hydrophobicity can be reduced, leading to membrane wetting and fouling. Although nonvolatile contaminants or compounds with high volatile temperatures are retained in the feed solution, the occurrence of pore wetting and membrane fouling can result in the low rejection of these contaminants. Li et al. (Li et al., 2018) reported that a direct treatment of industrial dyeing wastewater showed high removal of COD (90%) and color (94%), but serious membrane fouling was found with a decline of permeate flux from 19.8 to $\sim$0 L/m^{2}h in 48 h (Li et al., 2018). As a result of the phenomena, compounds with a higher boiling point or nonvolatile compounds, namely 2-butoxyethanol, 1-(2-butoxyethoxy)ethanol, and 2,4-dichlorobenzoic acid, have appeared in the permeate. Similarly, another research on the removal of antibiotics reported that a reduction in the rejection of Tobramycin was caused by pore wetting after 12 h operation (Guo et al., 2018). However, the growth of microorganisms in MD is limited by high salinity and high temperature of the feed solution (Gryta, 2002); thus. biological fouling could be lower than that of other membrane technologies [e.g., reverse osmosis (RO), nanofiltration (NF), or microfiltration (MF)]. However, microorganisms can still persist at the feed temperature of $<60°$C even at very low concentrations of nutrients (Tijing et al., 2015). Hence, the application of low feed temperature (e.g., $<60°$C) for low energy consumption can promote the development of biofouling in wastewater treatment.

3.2 Energy requirement

As a thermally driven separation process, thermal energy consumption in MD is the most influential factor in treatment cost. A large-scale MD plant trial of 10 m^3/h showed that heat cost accounted for about 94% of overall annual operational and maintenance expenditure. The average SEC value during direct contact MD treatment of power station wastewater was 1500 kWh/m^3 at a feed and permeate temperature of 60°C and 18–21°C, respectively (Dow et al., 2016). However, the SEC value depends on the system configuration in MD treatment. In particular, multieffect membrane distillation operating at a low feed temperature of $<80°$C showed that the SEC values were 430 kWh/m^3 for a three-stage configuration and 225 kWh/m^3 for a six-stage configuration (Morciano et al., 2020). This indicated that the design of the system configuration (e.g., multistage or semibatch configurations with heat recovery) has improved the thermal efficiency of MD operation. However, the development of thermal boundary layer reduces thermal efficiency in the MD process because of the increase in heat loss through the membrane. Therefore, energy consumption is increased to achieve stable permeate flux along the membrane length. For a countercurrent direct-contact MD system, the average additional energy was estimated to be $\sim$10 kWh/m^3 for a single pass with $\sim$2% water recovery. However, recirculation of the concentrate stream achieved a higher water recovery that resulted in a higher heating energy consumption ($350\,kWh/m^3$) (Alrehaili et al., 2020). Due to high energy consumption, membrane distillation, to date, has not been fully commercially available in water reuse.

4. Membrane development in membrane distillation

4.1 Mitigation of membrane wetting and fouling

A key aspect for the maturity of membrane technology is the existence of a wide range of commercial membranes developed for specific purposes. Unlike well-industrialized pressure-driven membrane filtration technologies, MD is still in its emerging state and thus there is a lack of suitable hydrophobic membranes specifically developed for MD. Available membranes have been fabricated from polymer materials, such as polytetrafluoroethylene (PTFE), polyvinylidene fluoride (PDVF), polypropylene (PP), polyethersulfone (PES), and polyethylene (PE) (Ravi et al., 2020; Tibi et al., 2020). Among these materials, PP, PVDF, and PTFE have been usually used in the lab-scale MD system because of their hydrophobic nature (Drioli et al., 2015). However, the presence of low-surface-tension compounds in wastewater can cause rapid wetting and serious fouling for these MD membranes. To date, many strategies have been carried out to minimize pore wetting and membrane fouling, such as pretreatment, membrane maintenance, and membrane fabrication/modification (Choudhury et al., 2019). Although pretreatment or membrane maintenance can reduce membrane fouling, the addition of further infrastructures or cleaning chemicals would lead to a large footprint or the increase of chemical demands. Therefore, modifying a hydrophobic surface or fabricating an ideal membrane becomes more attractive in recent years in order to gain high LEP and permeability, low fouling rate and thermal conductivity, and excellent mechanical strength (Ravi et al., 2020). Recently, MD membranes with omniphobic interfaces on the surface have been introduced. This omniphobic layer is capable of strongly repelling feed solutions in liquid form, even those that have low surface tensions, resulting in high wetting and fouling resistance (Lu et al., 2019). Omniphobic membranes can be classified based on membrane materials, including phase-inversed polymeric membranes, inorganic membranes, and electrospun nanofiber membranes (Lu et al., 2019). Among these membranes, the polymeric membrane is the cheapest approach to modify the omniphobic surface of conventional hydrophobic membranes (Ravi et al., 2020). Fabricating the polymeric membrane via phase inversion methods has been explored in recent years (Zheng et al., 2018; Li et al., 2019a; Lu et al., 2019). An omniphobic membrane that was prepared by spraying a coat of nanospheres on a commercial PVDF membrane showed a high contact angle of 176°, very stable flux, and high rejection of conductivity (Zheng et al., 2018). Another modification of the surface membrane by coating silica nanoparticles via a chemical bonding method (Fig. 3) in coking wastewater treatment confirmed effectiveness in wetting and fouling control (Li et al., 2019a). Despite their favorable properties, the use of fluorine-containing compounds in the modification process to form the omniphobic surface may cause human health and environmental concerns. Hence, it is necessary to develop a safe replacement for fluorine-based solutions that reduce surface tension. Graphene has been considered as a promising nanofiller material for MD membranes' enhancement, due to its high hydrophobicity, roughness, and selective sorption

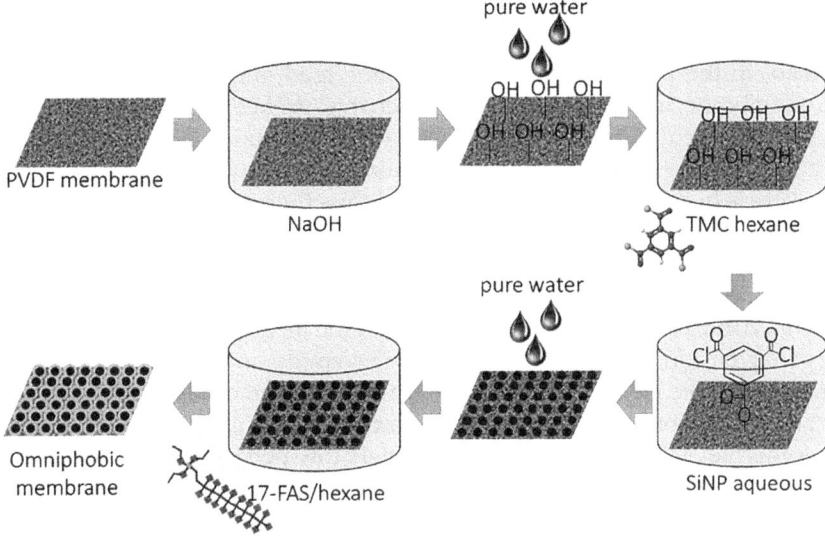

FIG. 3 Schematic illustration of the procedure for omniphobic membranes via chemical bonding method.

of water vapors (Surwade et al., 2015). Furthermore, carbon nanotubes (CNTs)-based membranes in MD have also been investigated at the laboratory scale, resulting in significant flux improvement and fouling resistance. Han et al. (2020) reported that the CNT coating layer was responsible for the excellent resistance of the MD membrane to oil fouling. As a result of the strong and repulsive force created by this layer, oil should not attach to the surface, keeping the pores open and available for transfer vapor. Although the superhydrophobicity attribute of novel MD membranes has been demonstrated to significantly improve wetting and fouling resistance on the laboratory scale, further investigations in terms of scalability for commercial applications are necessary to apply to wastewater with high concentration of both salts and organics for water reuse.

4.2 Minimization of energy consumption

In membrane distillation, one of the key issues for application potential is energy consumption for heating the feed solution. All models of MD also depend on the efficient use of heat energy to drive water evaporation; thus, there will always be vapor available at the feed-membrane interface to maintain the driving force (Bogler et al., 2017). In recent years, developments in MD membranes have been proposed to reduce the temperature polarization and improve energy efficiency by localized heating at/near the membrane surface (self-heated MD) by using photothermal, electrothermal, or induction materials (González et al., 2017; Deshmukh et al., 2018). Energy efficiency increased from <40% for traditional MD membranes to 53%–95% for novel membranes (Dongare et al., 2017; Wu et al., 2017; Anvari et al., 2019).

Among self-heated MD systems, photothermal heating is the most popular approach to date (Ahmed et al., 2020a). Unlike conventional MD systems which require preheating of the whole feed, in the photothermal approach, the feed is heated directly by solar energy at the feed-membrane interface instead of intensive preheating and thus can sustainably cut down energy consumption (Gao et al., 2021). By localizing heating to the proximity of the membrane's surface, temperature polarization phenomena can be mitigated as the interface temperature is higher than the bulk feed and thus thermal efficiency can be significantly improved. Three main types of photothermal materials, which have usually been incorporated into MD membranes, are carbon-based materials, inorganic semiconductor solar absorber materials, and plasmonic metallic nanomaterials (Wu et al., 2018; Dongare et al., 2019; Han et al., 2019; Ye et al., 2019). Depending on the incorporated photothermal materials, solar energy may be converted to heat energy via a single or a combination of the following mechanisms: plasmonic localized heating, thermal vibration of molecules, and electron-hole generation and relaxation. Despite the ability to reduce energy consumption, a major drawback of photothermal MD systems is their relatively low permeate flux, which is constrained by the low intensity of the sun radiation (flux of $0.49 \, kg/m^2 h$ under an irradiation of $0.75 \, kW/m^2$) (Jun et al., 2019). Furthermore, the productivity of photothermal MD systems is closely tied to the availability of sunlight. For unlocking the full potential of the MD system for long-term commercial operation, further considerations in terms of substitute light sources (such as infrared or UV) or enhanced photothermal conversion using photothermal materials are necessary.

In addition, electrothermal heating is a simple form of local heating where a current passes through a conductive layer, resulting in an increase in temperature (Ahmed et al., 2020a). Using electrical conductive materials [e.g., graphene or carbon nanotubes (CNTs)] on the feed side of the membrane, electromagnetic heating can maintain the temperature at the membrane surface and reduce temperature polarization, thereby minimizing energy requirement in MD. Ahmed et al. reported that combining electrothermal surface heating with MD not only reduced specific energy consumption (58%) but also increased permeate flux by 75% at feed temperatures of 40°C and 50°C (Ahmed et al., 2020b). However, due to the instabilities of the electrothermal heating method in ionizing media such as wastewater, electrothermal MD membranes have not been applied in water reuse. Therefore, a robust conductive layer is a major condition for fast electrothermal heating in water that does not interfere with membrane performance.

For another local heating method that uses induction materials, the conducting layer is produced onto a conventional MD membrane using magnetic nanoparticles (Tan et al., 2020). Within the module, electromagnetic induction is used to heat the membrane using energy created by eddy currents at the interface between the feed and membrane, thus enhancing energy efficiency. More than 84% energy efficiency ($<0.8 \, kW/m^2$) was found in previous research using Fe-CNTs as an induction material (Anvari et al., 2019). However, the commercial potential of induction heating is still limited because of its high energy input requirement. Therefore, interests in self-heated MD systems are expected to continue growing with the focus on developing new materials and fabricating techniques to produce MD membranes with desired properties including outstanding resilience to wetting and fouling, and high energy efficiency.

5. Conclusions and perspectives

Although MD has been proven to be highly effective in tackling chemicals of emerging concerns, there is still a long way for the implementation of MD systems at an industrial scale. High energy requirements for heating the feed, membrane wetting, and fouling are major challenges impeding the upscaling potential of membrane distillation in water reuse. The occurrence of surfactants and low-surface-tension liquids in wastewater can accelerate fast and serious membrane wetting and fouling in membrane distillation, resulting in worsened treatment performance or even disrupting the whole process. The realization of omniphobic membranes has brought a powerful, potential solution against membrane wetting and fouling. However, the use of fluorine-based solutions during the fabrication of an omniphobic interface has raised considerable concern regarding human and environmental health. Hence, a greener alternative to fluorine-based solutions with equal effectivity in creating omniphobic membranes remains the quest for future studies. In the same light, the emergence of membrane coating by nanoparticles, graphene, and carbon nanotubes has shown great promise in fouling mitigation.

Regarding the energy challenge, high energy consumption of membrane distillation may be compensated by the use of low-grade energy such as waste heat or solar energy. However, the dependence on abundant waste heat or renewable sources may hinder the flexibility in industrial implementations of membrane distillation due to the intermittent and unevenly distributed nature of these energy sources. Much attention has been focused on the development of a novel membrane with favorable thermal characteristics to mitigate the energy consumption of membrane distillation. In particular, several research works have reported success in incorporating photothermal, electrothermal, or induction materials into the MD membrane to achieve localized heating at the feed-membrane interface. Thanks to this new development, energy efficiency increased to 53%–95% (self-heated MD) compared to <40% of conventional MD membranes. Still, the sustainability and durability of these novel self-heated MD systems in long-term operation remains unclear as they have only been examined in a relatively short time under laboratory scale conditions while the potentially harmful impacts of the new materials used for membrane modification on human health and environments have not been well investigated. To sum up, further efforts should be made to develop highly durable MD membranes with favorable properties including high resistance to membrane wetting and fouling, and/or high energy efficiency while utilizing less toxic and biodegradable materials. Whether membrane distillation can be soon employed in real-world applications depends heavily on improving the sustainability of MD membranes.

Acknowledgments

This research was conducted under the framework of the CARE-RESCIF initiative within the International Joint Laboratory LECZ-CARE project and financially supported by Ho Chi Minh City University of Technology (HCMUT), VNU-HCM, under grant number Tc-PTN-2021-01.

References

Ahmad, N.N.R., Ang, W.L., Leo, C.P., Mohammad, A.W., Hilal, N., 2021. Current advances in membrane technologies for saline wastewater treatment: a comprehensive review. Desalination 517, 115170.

Ahmed, F.E., Lalia, B.S., Hashaikeh, R., Hilal, N., 2020a. Alternative heating techniques in membrane distillation: a review. Desalination 496, 114713.

Ahmed, F.E., Lalia, B.S., Hashaikeh, R., Hilal, N., 2020b. Enhanced performance of direct contact membrane distillation via selected electrothermal heating of membrane surface. J. Membr. Sci. 610, 118224.

Alkhudhiri, A., Darwish, N., Hilal, N., 2012. Membrane distillation: a comprehensive review. Desalination 287, 2–18.

Alrehaili, O., Perreault, F., Sinha, S., Westerhoff, P., 2020. Increasing net water recovery of reverse osmosis with membrane distillation using natural thermal differentials between brine and co-located water sources: impacts at large reclamation facilities. Water Res. 184, 116134.

Anvari, A., Kekre, K.M., Yancheshme, A.A., Yao, Y., Ronen, A., 2019. Membrane distillation of high salinity water by induction heated thermally conducting membranes. J. Membr. Sci. 589, 117253.

Arcanjo, G.S., Ricci, B.C., dos Santos, C.R., Costa, F.C.R., Silva, U.C.M., Mounteer, A.H., Koch, K., da Silva, P.R., Santos, V.L., Amaral, M.C.S., 2021. Effective removal of pharmaceutical compounds and estrogenic activity by a hybrid anaerobic osmotic membrane bioreactor – Membrane distillation system treating municipal sewage. Chem. Eng. J. 416, 129151.

Ban, S.-H., Im, S.-J., Cho, J., Jang, A., 2019. Comparative performance of FO-RO hybrid and two-pass SWRO desalination processes: boron removal. Desalination 471, 114114.

Bogler, A., Lin, S., Bar-Zeev, E., 2017. Biofouling of membrane distillation, forward osmosis and pressure retarded osmosis: principles, impacts and future directions. J. Membr. Sci. 542, 378–398.

Carnevale, M., Gnisci, E., Hilal, J., Criscuoli, A., 2016. Direct contact and vacuum membrane distillation application for the olive mill wastewater treatment. Sep. Purif. Technol. 169, 121–127.

Caroline Ricci, B., Santos Arcanjo, G., Rezende Moreira, V., Abner Rocha Lebron, Y., Koch, K., Cristina Rodrigues Costa, F., Paulinelli Ferreira, B., Luiza Costa Lisboa, F., Diniz Miranda, L., Vieira de Faria, C., Celina Lange, L., Cristina Santos Amaral, M., 2021. A novel submerged anaerobic osmotic membrane bioreactor coupled to membrane distillation for water reclamation from municipal wastewater. Chem. Eng. J. 414, 128645.

Chamani, H., Woloszyn, J., Matsuura, T., Rana, D., Lan, C.Q., 2021. Pore wetting in membrane distillation: a comprehensive review. Prog. Mater. Sci. 122, 100843.

Cho, H., Choi, Y., Lee, S., 2018. Effect of pretreatment and operating conditions on the performance of membrane distillation for the treatment of shale gas wastewater. Desalination 437, 195–209.

Choudhury, M.R., Anwar, N., Jassby, D., Rahaman, M.S., 2019. Fouling and wetting in the membrane distillation driven wastewater reclamation process – A review. Adv. Colloid Interf. Sci. 269, 370–399.

Combernoux, N., Schrive, L., Labed, V., Wyart, Y., Carretier, E., Moulin, P., 2017. Treatment of radioactive liquid effluents by reverse osmosis membranes: from lab-scale to pilot-scale. Water Res. 123, 311–320.

Dai, X., Wei, Q., Wang, Y., Li, Q., Cui, S., Nie, Z., 2022. A novel strategy to enhance the desalination stability of FAS (fluoroalkylsilane)-modified ceramic membranes via constructing a porous SiO2@ PDMS (polydimethylsiloxane) protective layer on their top. Chem. Eng. J., 134757.

Deshmukh, A., Boo, C., Karanikola, V., Lin, S., Straub, A.P., Tong, T., Warsinger, D.M., Elimelech, M., 2018. Membrane distillation at the water-energy nexus: limits, opportunities, and challenges. Energy Environ. Sci. 11, 1177–1196.

Donato, L., Garofalo, A., Drioli, E., Alharbi, O., Aljlil, S.A., Criscuoli, A., Algieri, C., 2020. Improved performance of vacuum membrane distillation in desalination with zeolite membranes. Sep. Purif. Technol. 237, 116376.

Dongare, P.D., Alabastri, A., Neumann, O., Nordlander, P., Halas, N.J., 2019. Solar thermal desalination as a nonlinear optical process. Proc. Natl. Acad. Sci. 116, 13182–13187.

Dongare, P.D., Alabastri, A., Pedersen, S., Zodrow, K.R., Hogan, N.J., Neumann, O., Wu, J., Wang, T., Deshmukh, A., Elimelech, M., 2017. Nanophotonics-enabled solar membrane distillation for off-grid water purification. Proc. Natl. Acad. Sci. 114, 6936–6941.

Dow, N., Gray, S., Li, J.-D., Zhang, J., Ostarcevic, E., Liubinas, A., Atherton, P., Roeszler, G., Gibbs, A., Duke, M., 2016. Pilot trial of membrane distillation driven by low grade waste heat: membrane fouling and energy assessment. Desalination 391, 30–42.

Drioli, E., Ali, A., Macedonio, F., 2015. Membrane distillation: recent developments and perspectives. Desalination 356, 56–84.

El-Abbassi, A., Hafidi, A., García-Payo, M.C., Khayet, M., 2009. Concentration of olive mill wastewater by membrane distillation for polyphenols recovery. Desalination 245, 670–674.

El-Abbassi, A., Hafidi, A., Khayet, M., García-Payo, M.C., 2013. Integrated direct contact membrane distillation for olive mill wastewater treatment. Desalination 323, 31–38.

Feng, H., Li, H., Li, M., Zhang, X., 2022. Construction of omniphobic PVDF membranes for membrane distillation: investigating the role of dimension, morphology, and coating technology of silica nanoparticles. Desalination 525, 115498.

Fortunato, L., Elcik, H., Blankert, B., Ghaffour, N., Vrouwenvelder, J., 2021. Textile dye wastewater treatment by direct contact membrane distillation: membrane performance and detailed fouling analysis. J. Membr. Sci. 636, 119552.

Fujioka, T., Khan, S.J., McDonald, J.A., Nghiem, L.D., 2015. Rejection of trace organic chemicals by a hollow fibre cellulose triacetate reverse osmosis membrane. Desalination 368, 69–75.

Fujioka, T., Takeuchi, H., Tanaka, H., Kodamatani, H., 2018. Online monitoring of N-nitrosodimethylamine rejection as a performance indicator of trace organic chemical removal by reverse osmosis. Chemosphere 200, 80–85.

Gao, M., Peh, C.K., Meng, F.L., Ho, G.W., 2021. Photothermal membrane distillation toward solar water production. Small Methods 5, 2001200.

González, D., Amigo, J., Suárez, F., 2017. Membrane distillation: perspectives for sustainable and improved desalination. Renew. Sust. Energ. Rev. 80, 238–259.

Gryta, M., 2002. The assessment of microorganism growth in the membrane distillation system. Desalination 142, 79–88.

Guillen-Burrieza, E., Mavukkandy, M.O., Bilad, M.R., Arafat, H.A., 2016. Understanding wetting phenomena in membrane distillation and how operational parameters can affect it. J. Membr. Sci. 515, 163–174.

Guo, J., Farid, M.U., Lee, E.-J., Yan, D.Y.-S., Jeong, S., Kyoungjin An, A., 2018. Fouling behavior of negatively charged PVDF membrane in membrane distillation for removal of antibiotics from wastewater. J. Membr. Sci. 551, 12–19.

Han, M., Dong, T., Hou, D., Yao, J., Han, L., 2020. Carbon nanotube based Janus composite membrane of oil fouling resistance for direct contact membrane distillation. J. Membr. Sci. 607, 118078.

Han, L., Tan, Y.Z., Netke, T., Fane, A.G., Chew, J.W., 2017. Understanding oily wastewater treatment via membrane distillation. J. Membr. Sci. 539, 284–294.

Han, X., Wang, W., Zuo, K., Chen, L., Yuan, L., Liang, J., Li, Q., Ajayan, P.M., Zhao, Y., Lou, J., 2019. Bio-derived ultrathin membrane for solar driven water purification. Nano Energy 60, 567–575.

Jacob, P., Phungsai, P., Fukushi, K., Visvanathan, C., 2015. Direct contact membrane distillation for anaerobic effluent treatment. J. Membr. Sci. 475, 330–339.

Jia, W., Kharraz, J.A., Sun, J., An, A.K., 2021. Hierarchical Janus membrane via a sequential electrospray coating method with wetting and fouling resistance for membrane distillation. Desalination 520, 115313.

Jun, Y.-S., Wu, X., Ghim, D., Jiang, Q., Cao, S., Singamaneni, S., 2019. Photothermal membrane water treatment for two worlds. Acc. Chem. Res. 52, 1215–1225.

Kalla, S., 2021. Use of membrane distillation for oily wastewater treatment—a review. J. Environ. Chem. Eng. 9, 104641.

Khan, M.T., Busch, M., Molina, V.G., Emwas, A.-H., Aubry, C., Croue, J.-P., 2014. How different is the composition of the fouling layer of wastewater reuse and seawater desalination RO membranes? Water Res. 59, 271–282.

Khayet, M., 2013. Treatment of radioactive wastewater solutions by direct contact membrane distillation using surface modified membranes. Desalination 321, 60–66.

Kiai, H., García-Payo, M.C., Hafidi, A., Khayet, M., 2014. Application of membrane distillation technology in the treatment of table olive wastewaters for phenolic compounds concentration and high quality water production. Chem. Eng. Process. Process Intensif. 86, 153–161.

Kim, H.-C., Shin, J., Won, S., Lee, J.-Y., Maeng, S.K., Song, K.G., 2015. Membrane distillation combined with an anaerobic moving bed biofilm reactor for treating municipal wastewater. Water Res. 71, 97–106.

Laqbaqbi, M., García-Payo, M.C., Khayet, M., El Kharraz, J., Chaouch, M., 2019. Application of direct contact membrane distillation for textile wastewater treatment and fouling study. Sep. Purif. Technol. 209, 815–825.

Li, J., Guo, S., Xu, Z., Li, J., Pan, Z., Du, Z., Cheng, F., 2019b. Preparation of omniphobic PVDF membranes with silica nanoparticles for treating coking wastewater using direct contact membrane distillation: electrostatic adsorption vs. chemical bonding. J. Membr. Sci. 574, 349–357.

Li, F., Huang, J., Xia, Q., Lou, M., Yang, B., Tian, Q., Liu, Y., 2018. Direct contact membrane distillation for the treatment of industrial dyeing wastewater and characteristic pollutants. Sep. Purif. Technol. 195, 83–91.

Li, J., Ren, L.-F., Huang, M., Yang, J., Shao, J., He, Y., 2022. Facile preparation of omniphobic PDTS-ZnO-PVDF membrane with excellent anti-wetting property in direct contact membrane distillation (DCMD). J. Membr. Sci., 120404.

Li, X., Shan, H., Cao, M., Li, B., 2019a. Facile fabrication of omniphobic PVDF composite membrane via a waterborne coating for anti-wetting and anti-fouling membrane distillation. J. Membr. Sci. 589, 117262.

Liu, H., Wang, J., 2013. Treatment of radioactive wastewater using direct contact membrane distillation. J. Hazard. Mater. 261, 307–315.

Lu, K.J., Chen, Y., Chung, T.-S., 2019. Design of omniphobic interfaces for membrane distillation—a review. Water Res. 162, 64–77.

Makanjuola, O., Lalia, B.S., Hashaikeh, R., 2021. Thermoelectric heating and cooling for efficient membrane distillation. Case Studies Therm. Eng. 28, 101540.

Martinetti, C.R., Childress, A.E., Cath, T.Y., 2009. High recovery of concentrated RO brines using forward osmosis and membrane distillation. J. Membr. Sci. 331, 31–39.

Meng, L., Mansouri, J., Li, X., Liang, J., Huang, M., Lv, Y., Wang, Z., Chen, V., 2022. Omniphobic membrane via bioinspired silicification for the treatment of RO concentrate by membrane distillation. J. Membr. Sci. 647, 120267.

Morciano, M., Fasano, M., Bergamasco, L., Albiero, A., Lo Curzio, M., Asinari, P., Chiavazzo, E., 2020. Sustainable freshwater production using passive membrane distillation and waste heat recovery from portable generator sets. Appl. Energy 258, 114086.

Naidu, G., Jeong, S., Choi, Y., Vigneswaran, S., 2017. Membrane distillation for wastewater reverse osmosis concentrate treatment with water reuse potential. J. Membr. Sci. 524, 565–575.

Ngo, M.T.T., Diep, B.Q., Sano, H., Nishimura, Y., Boivin, S., Kodamatani, H., Takeuchi, H., Sakti, S.C.W., Fujioka, T., 2021. Membrane distillation for achieving high water recovery for potable water reuse. Chemosphere, 132610.

Ngo, M.T.T., Diep, B.Q., Sano, H., Nishimura, Y., Boivin, S., Kodamatani, H., Takeuchi, H., Sakti, S.C.W., Fujioka, T., 2022. Membrane distillation for achieving high water recovery for potable water reuse. Chemosphere 288, 132610.

Noor, I.-E., Martin, A., Dahl, O., 2020. Techno-economic system analysis of membrane distillation process for treatment of chemical mechanical planarization wastewater in nano-electronics industries. Sep. Purif. Technol. 248, 117013.

Rajwade, K., Barrios, A.C., Garcia-Segura, S., Perreault, F., 2020. Pore wetting in membrane distillation treatment of municipal wastewater desalination brine and its mitigation by foam fractionation. Chemosphere 257, 127214.

Ravi, J., Othman, M.H.D., Matsuura, T., Ro'il Bilad, M., El-badawy, T.H., Aziz, F., Ismail, A.F., Rahman, M.A., Jaafar, J., 2020. Polymeric membranes for desalination using membrane distillation: a review. Desalination 490, 114530.

Rezaei, M., Warsinger, D.M., Lienhard, V.J.H., Duke, M.C., Matsuura, T., Samhaber, W.M., 2018. Wetting phenomena in membrane distillation: mechanisms, reversal, and prevention. Water Res. 139, 329–352.

Sinha Ray, S., Singh Bakshi, H., Dangayach, R., Singh, R., Deb, C.K., Ganesapillai, M., Chen, S.-S., Purkait, M.K., 2020. Recent developments in nanomaterials-modified membranes for improved membrane distillation performance. Membranes 10, 140.

Surwade, S.P., Smirnov, S.N., Vlassiouk, I.V., Unocic, R.R., Veith, G.M., Dai, S., Mahurin, S.M., 2015. Water desalination using nanoporous single-layer graphene. Nat. Nanotechnol. 10, 459–464.

Swaminathan, J., Chung, H.W., Warsinger, D.M., Lienhard, V.J.H., 2018. Energy efficiency of membrane distillation up to high salinity: evaluating critical system size and optimal membrane thickness. Appl. Energy 211, 715–734.

Tan, Y.Z., Chandrakant, S.P., Ang, J.S.T., Wang, H., Chew, J.W., 2020. Localized induction heating of metallic spacers for energy-efficient membrane distillation. J. Membr. Sci. 606, 118150.

Tavakkoli, S., Lokare, O., Vidic, R., Khanna, V., 2020. Shale gas produced water management using membrane distillation: an optimization-based approach. Resour. Conserv. Recycl. 158, 104803.

Tibi, F., Charfi, A., Cho, J., Kim, J., 2020. Fabrication of polymeric membranes for membrane distillation process and application for wastewater treatment: critical review. Process. Saf. Environ. Prot. 141, 190–201.

Tijing, L.D., Woo, Y.C., Choi, J.-S., Lee, S., Kim, S.-H., Shon, H.K., 2015. Fouling and its control in membrane distillation—a review. J. Membr. Sci. 475, 215–244.

Tow, E.W., Warsinger, D.M., Trueworthy, A.M., Swaminathan, J., Thiel, G.P., Zubair, S.M., Myerson, A.S., Lienhard, V.J.H., 2018. Comparison of fouling propensity between reverse osmosis, forward osmosis, and membrane distillation. J. Membr. Sci. 556, 352–364.

Wang, P., Chung, T.-S., 2015. Recent advances in membrane distillation processes: membrane development, configuration design and application exploring. J. Membr. Sci. 474, 39–56.

Wen, X., Li, F., Zhao, X., 2016. Removal of nuclides and boron from highly saline radioactive wastewater by direct contact membrane distillation. Desalination 394, 101–107.

Woldemariam, D., Kullab, A., Fortkamp, U., Magner, J., Royen, H., Martin, A., 2016. Membrane distillation pilot plant trials with pharmaceutical residues and energy demand analysis. Chem. Eng. J. 306, 471–483.

Wu, X., Jiang, Q., Ghim, D., Singamaneni, S., Jun, Y.-S., 2018. Localized heating with a photothermal polydopamine coating facilitates a novel membrane distillation process. J. Mater. Chem. A 6, 18799–18807.

Wu, J., Zodrow, K.R., Szemraj, P.B., Li, Q., 2017. Photothermal nanocomposite membranes for direct solar membrane distillation. J. Mater. Chem. A 5, 23712–23719.

Yan, Z., Jiang, Y., Chen, X., Lu, Z., Wei, Z., Fan, G., Liang, H., Qu, F., 2021. Evaluation of applying membrane distillation for landfill leachate treatment. Desalination 520, 115358.

Yao, M., Tijing, L.D., Naidu, G., Kim, S.-H., Matsuyama, H., Fane, A.G., Shon, H.K., 2020. A review of membrane wettability for the treatment of saline water deploying membrane distillation. Desalination 479, 114312.

Ye, H., Li, X., Deng, L., Li, P., Zhang, T., Wang, X., Hsiao, B.S., 2019. Silver nanoparticle-enabled photothermal nanofibrous membrane for light-driven membrane distillation. Ind. Eng. Chem. Res. 58, 3269–3281.

Zheng, R., Chen, Y., Wang, J., Song, J., Li, X.-M., He, T., 2018. Preparation of omniphobic PVDF membrane with hierarchical structure for treating saline oily wastewater using direct contact membrane distillation. J. Membr. Sci. 555, 197–205.

Zou, L., Zhang, X., Gusnawan, P., Zhang, G., Yu, J., 2021. Crosslinked PVDF based hydrophilic-hydrophobic dual-layer hollow fiber membranes for direct contact membrane distillation desalination: from the seawater to oilfield produced water. J. Membr. Sci. 619, 118802.

Membrane distillation technology applied in water resources

Dian Qoriati[b,c], Hismi Susane[a,b,c], Jeng-Lung Lin[b,c], Ya-Fen Wang[b,c], and Sheng-Jie You[b,c]

[a]Department of Civil Engineering, Chung Yuan Christian University, Taoyuan, Taiwan
[b]Department of Environmental Engineering, Chung Yuan Christian University, Taoyuan, Taiwan
[c]Center for Environmental Risk Management, Chung Yuan Christian University, Taoyuan, Taiwan

1. Introduction

Membrane technology for water purification has tremendously burgeoned during the last decade. Membrane distillation (MD) is a sustainable technology that has a high purification rate for numerous applications such as desalination (Mejia Mendez et al., 2018), municipal wastewater (Rajwade et al., 2020), industrial wastewater (Si et al., 2021), and recovery of other components (Amaral, 2021; Murugesan et al., 2020; Wen et al., 2021). MD is a hybrid separation technique that combines membrane separation with thermal distillation to achieve evaporation at low temperatures and pressures without heating the solution to the boiling point (Chen et al., 2021). As a nonisothermal membrane separation technique, MD involves thermally driven vapor transport through a porous hydrophobic membrane. The difference in vapor pressure between the feed and distillate streams of the membrane pores is the driving force in MD for diffusive transport, which varies exponentially with temperature differences (Murugesan et al., 2020). MD has different basic configurations such as direct contact MD (DCMD), air gap MD (AGMD), sweep gas MD (SGMD), and vacuum MD (VMD), as shown in Fig. 1, can be used for water and wastewater purification.

Copyright © 2023 Elsevier Inc. All rights reserved.

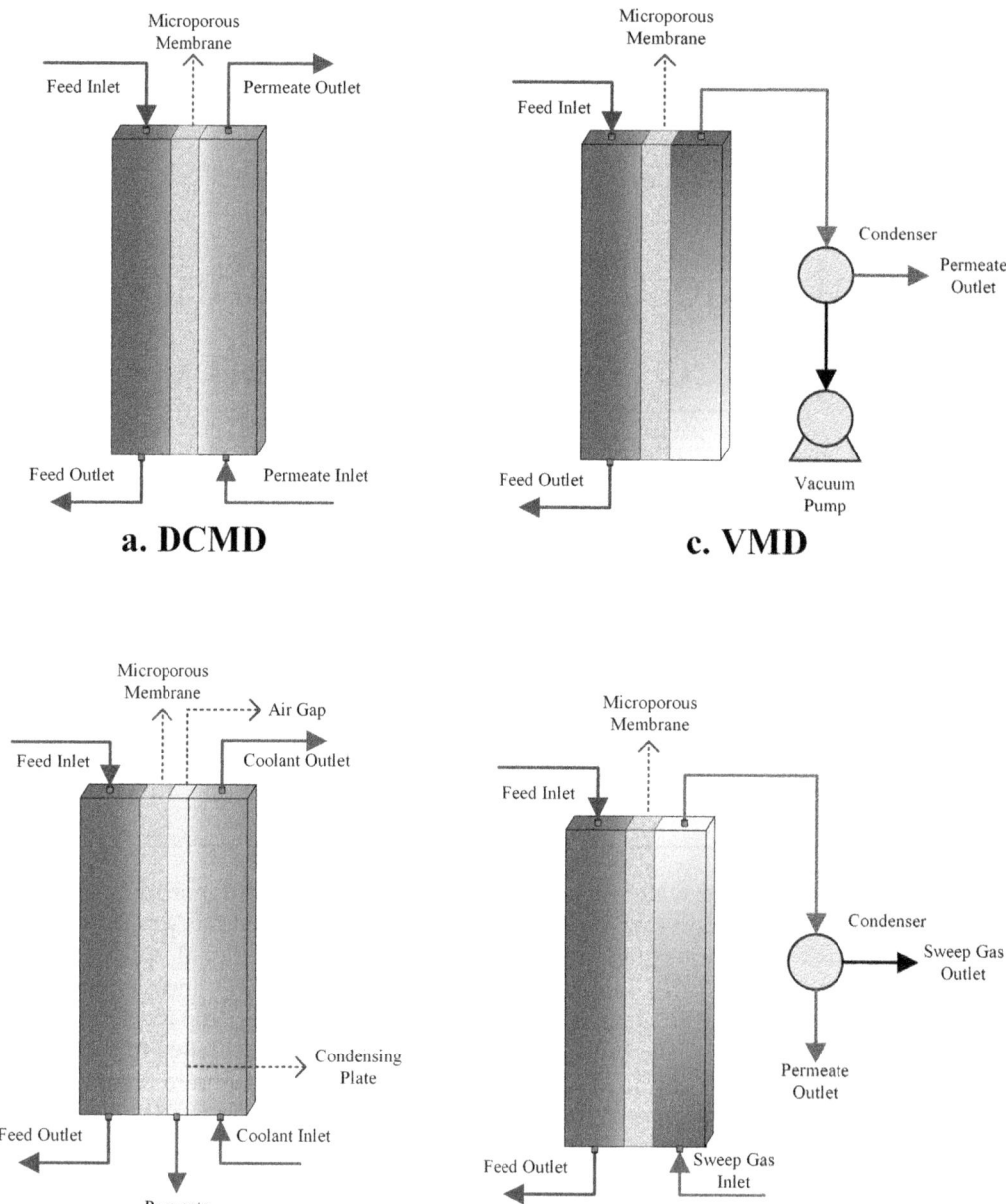

FIG. 1   Schematic representation of the four basic MD configurations: (A) DCMD, (B) AGMD, (C) VMD, and (D) SGMD.

2. Direct contact membrane distillation (DCMD)

DCMD is one of the configurations of MD in which both the heated feed solution and cold permeate solution circulated by the circulating pump is instantly connected with the membrane surface. The transmembrane temperature difference is lower than the liquid entry pressure (LEP), leading to a vapor pressure difference. Consequently, liquid–vapor interfaces are presented at the membrane pore entrances due to volatile molecules evaporating at the hot-vapor interface flowing through the membrane in the vapor phase and condensing at the cold liquid–vapor interface inside the membrane modules.

DCMD has a simple configuration and a high flux produced, which renders it worthy of consideration (Ashoor et al., 2016) to treat wastewater and recover water for solving current water scarcity issues. Furthermore, it does not require an external condenser that is relatively easier to operate than other configurations. On the one hand, the evaporator and condenser surfaces are adjacents, resulting in a low effective driving force with low thermal energy efficiency (El-Bourawi et al., 2007).

1. *Air gap membrane distillation*:

     An additional stagnant air gap interposed between the membrane and condensate part, which acts as the condensing surface at the permeate side, and a metal plate maintained at a low temperature by a coolant in the MD module is known as AGMD. The temperature difference between the heated feed and cooled permeate solutions is the driving force for the evaporation of water and volatile molecules. The volatile molecules evaporate across the membrane pores and the air gap to eventually condense over a cold permeate surface. The presence of an air gap to address heat loss due to conduction through the membrane results in a relatively low efficiency of the MD process. AGMD has high thermal energy use and heat recovery efficiency due to the vaporization and recovery of the latent heat condensation. However, it exhibits a lower flux due to the interposed air gap leading to a higher mass transfer resistance (Salmanli et al., 2022).

2. *Vacuum membrane distillation*:

     As the driving force in VMD, vacuum pressure on the permeate side below the equilibrium vapor pressure eliminates all air from the membrane pores. Vacuum pressure increases membrane permeability and separates the volatile solvent from the heated feed solution. The application of vacuum pressure is owing to the vapor pressure differences between the feed and permeate sections that occur through the diminished pressure generated (Yang et al., 2021). VMD has a higher flux than the other configuration due to the resistance of mass transfer in the boundary layer, higher partial pressure gradient (Abu-Zeid et al., 2015), and lower heat loss. In contrast, VMD has a high probability of membrane wetting owing to vacuum conditions and membrane fouling. On the other hand, VMD has a higher salt rejection rate (Abu-Zeid et al., 2015) that can be reached for nonvolatile solutes, but it has minimum selectivity of volatile components (Ding et al., 2006). VMD requires to provide a vacuum pump and an external condenser that can increase energy costs from the operation.

3. *Sweep gas membrane distillation*:

     The stream of cold, inert gas is applied in the permeate section in the MD module to swipe out the vaporized molecules and condensation away from the permeate chamber of

TABLE 1 The advantages and disadvantages among MD configurations.

Configuration	Advantages	Disadvantages
DCMD	Simple in its design High gained output ratio (GOR) (Drioli et al., 2015; Meindersma et al., 2006)	High-temperature polarization Relatively large conductive heat losses (Meindersma et al., 2006)
AGMD	Higher thermal resistance Smaller heat losses The most energy-efficient (Meindersma et al., 2006; Zhani et al., 2015)	Higher mass transfer resistance and lower transmembrane fluxes are lower than DCMD (Meindersma et al., 2006; Zhani et al., 2015)
VMD	Small conductive heat losses High transmembrane flux Lower temperature polarization (Drioli et al., 2015; Izquierdo-gil and Jonsson, 2003)	Higher probability of pore wetting More prone to fouling (Drioli et al., 2015)
SGMD	Lower temperature polarization	Additional equipment which has complexity, where heat recovery is difficult (Drioli et al., 2015)

the membrane pore. Then the vaporized permeate heads into an external condenser and condensed into liquid outside the membrane module. To solve issues in different MD configurations, inert gases such as humid air, dry N_2, dry air, and air are used. It minimizes conductive heat loss with minimal mass transfer resistance due to the gas flow on the permeate side, low pore wetting risk, and high evaporation efficiency (Said et al., 2020). Thus, it is to be noted that SGMD is a promising and potential configuration with an excellent perspective for future research and development (Table 1).

MD and other membrane separation technologies are frequently misinterpreted to have a similar mechanism, as VMD with PV and DCMD with OD. Nevertheless, there are fundamental distinctions, practically. The driving force for PV and VMD is provided by applying a vacuum and low pressure on the permeate side. The basic difference between VMD and PV is the involvement of the membrane in separation. VMD employs a porous and hydrophobic membrane, with the transport mechanism based on flow through the pores. In contrast, PV requires a compact and selective membrane, with the separation being based on each component's relative solubility and diffusivity in the membrane material.

The difference in vapor pressure between the feed and permeate solutions of the porous hydrophobic membrane is the driving force in both DCMD and OD. The distinction is the way to generate propulsion. OD has the draw solution on the downstream side as a known osmotic agent (Curcio et al., 2010). The hypertonic solution has been widely used such as NaCl, $CaCl_2$, $MgCl_2$, $MgSO_4$, K_2HPO_4, and KH_2PO_4, and several organic liquids such as glycerol or polyglycols (Laqbaqbi et al., 2017) flows on the stripping side removing the volatile compound. OD was performed at ambient temperature and required less energy which is frequently applied to concentrate an aqueous solution of volatile and nonvolatile compounds and concentrate fruits and vegetable juices and dealcoholization of binary mixtures, i.e., water-ethanol (Esteras-Saz et al., 2021). In recent studies, MD combined with OD, also known as osmotic membrane distillation (OMD), was expanded in order to increase the difference in transmembrane water vapor pressure. Both transmembrane temperature and concentration gradient are the driving forces that are produced by osmotic solution on the downstream side of the membrane. Generally, a brine solution utilized on the permeate side was proven to raise the vapor pressure driving force so that mass transport diffuses through the membrane will be higher than a single configuration (Khayet and Matsuura, 2011a). However, the regeneration of the osmotic agent to recover its performance while also achieving freshwater should be extended (Table 2).

TABLE 2 The differences among membrane distillation (MD), osmotic distillation (OD), and pervaporation (PV).

	MD	OD	PV
Principle	A thermal separation process that allows only vapor molecules to transport through a microporous hydrophobic membrane	The presence of an osmotic agent in the stripping stream through a hydrophobic microporous membrane contactor to concentrate production. The aqueous streams on both membrane sides do not come into touch through the pores due to the membrane's hydrophobicity; therefore, a liquid–vapor interface forms at each pore edge. Between the feed phase-air, there is a vapor–liquid equilibrium distribution.	The feed stream's liquid mixture evaporates through a polymer membrane, and the permeate is removed by the stream of a sweep gas or vacuum pump (Polena and Marie, 2021)
Driving force	The difference in vapor pressures generated by the temperature difference across the hydrophobic membrane, the vacuum (VMD), and the sweep gas (SGMD)	The gradient of partial vapor pressures from a concentration gradient produced using an extracting solution in the permeate side	Mass transfer: solution-diffusion mechanism
Advantages	– High separation efficiency for macromolecules, inorganic ions, and nonvolatile compounds (Ji et al., 2021; Kiai et al., 2014) – Low energy consumption (Ji et al., 2021) – Lower fouling and concentration polarization (Kiai et al., 2014) – Relatively low operating temperature compared to other thermal desalination techniques – The operating pressure of the MD process is less compared to the pressure-driven membrane separation processes	Performs at normal ambient temperature Highly suitable for extracting liquid foods and dilute aqueous solution of nonvolatile solutes Less energy required compared to MD (Laqbaqbi et al., 2017)	Low power consumption No chemicals required

3. Mechanism of membrane distillation (MD) transport

MD is a concurrent process that incorporates both heat and mass transfer processes. Heat and mass are simultaneously transported through the hydrophobic membrane. The heat and mass transport mechanisms in the MD module involve liquid evaporation at the heated feed side, diffusion of vapor through the membrane pores, and condensation of vapor at the permeate side (Olatunji and Camacho, 2018). The transport of gases and vapors across porous membranes varies depending on the MD configurations.

3.1 Mass transfer

Mass transfer proceeds through the pores of the membrane as water flows from the hot feed to the surface of the hydrophobic membrane as shown in Fig. 2. Mass transfer occurs first through the boundary layer at the feed side and second across the membrane. The models most often applied in MD are the dusty gas (DG) model and Fick's law model (Olatunji and Camacho, 2018). The DG model is significantly extended since it employs the average pore size to predict mass transport of volatile water molecules and assumes an average temperature across the membrane (Hitsov et al., 2017). Fick's law model is extensively emphasized in order to consider that vaporized molecules diffuse through pores in the air-filled membrane (Hitsov et al., 2015). Mass transport in MD has a specific difference due to the variation of the driving force in each configuration. The water evaporates and flows via the membrane pore to the cold side of the membrane. The water vapor subsequently mixes with the cold water and condenses (in DCMD), or it condenses on the cold plate after crossing

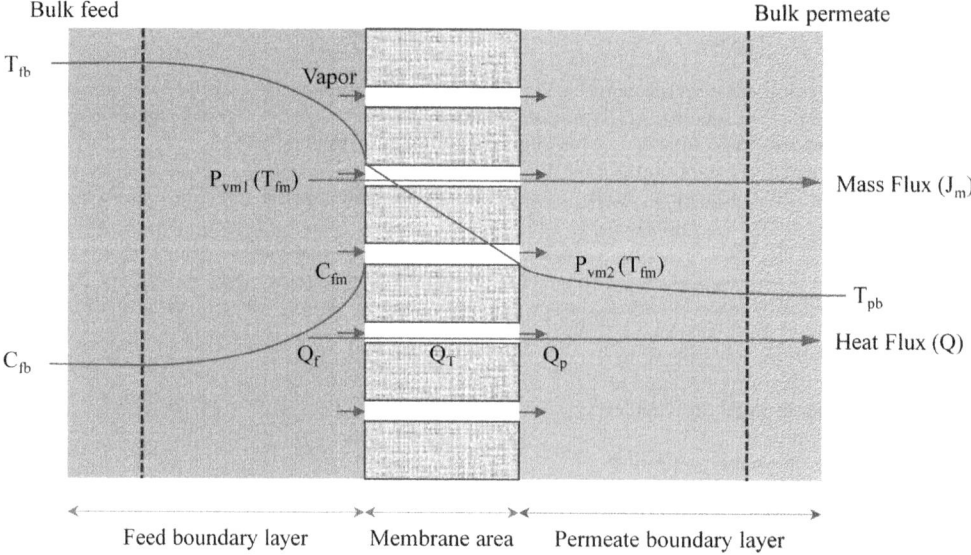

FIG. 2   Heat and mass transfer profiles in MD.

the air gap (in AGMD), or a vacuum pump is employed to transport the steam to the condenser (in VMD), or it is swamped with gas in the condenser (in SGMD) (Olatunji and Camacho, 2018).

3.2 Heat transfer

When the vapor molecules diffuse through the membrane matrix and its pores, heat is transported across the membrane by conduction in the form of sensible heat and latent heat. It has occurred in all MD configurations, but the heat transport by conduction is negligible in VMD (Hitsov et al., 2015; Olatunji and Camacho, 2018). Heat is transferred across the membrane by latent heat accompanying the vapor or gas flux and thermal conduction across the membrane material and the gases within its pores. Along the MD process, the convective heat transmission from the bulk feed is diffused via the feed boundary layer in the hot feed membrane surface. The heat conduction occurs from the vapor–liquid which is evaporated across the microporous membrane. Lastly, the convective heat transmission is passed through the membrane surface of the permeate side (Al-Anezi et al., 2012). The major difference among MD configurations is the condition of heat transport at the permeate side. In DCMD, the convective heat diffuses from the cold side of the membrane surface to the permeate bulk via the membrane boundary layer; via the vacuum side (in VMD), the convective heat transfer passes via the air gap and the heat conduction passes via the air (in AGMD) or conductive and convective heat transfer diffuses via the sweep gas (in SGMD) (Olatunji and Camacho, 2018).

4. Membrane used and characterization in membrane distillation

A fundamental requirement for the MD process is that the membranes used should be hydrophobic and have low surface energy and be constructed of polymers or inorganic materials. Thus, only vapors and noncondensable gases remain in the pores. In general, MD membranes must fulfill multiple requirements: high thermal stability, excellent chemical resistance to various feed solutions, high hydrophobicity and porosity, high LEP and permeability, narrow pore size distribution, and low thermal conductivity (Curcio et al., 2010).

MD membranes should have a large mean pore size of the porous structure. The distillate flux is related to the pore size and porosity and inversely related to the membrane thickness and tortuosity of the membrane pore. In order to prevent membrane wetting, membranes should have a high LEP value. Furthermore, supported membranes with a specific pore size demonstrated greater flux than unsupported membranes with the same pore size. Hamzah and Leo (2016, 2017) carried out the application of DCMD in synthetic wastewater to remove phenol compounds (10 wt%) using unsupported membrane $PVDF/TiO_2$ and supported membrane $PVDF/TiO_2/LiCl$ 2%. The supported and unsupported membranes had the same membrane properties such as a pore size of 0.47 µm, a thickness of 168.5 µm, and an effective area of 1.26×10^{-3} m^2. Both the supported and unsupported membranes were conducted by the same parameter condition in the MD process, which were running for 6.7 h with a feed flow rate of 18 L/h at feed temperature and the permeate temperature 40°C and 20°C, respectively. Supported and unsupported membranes showed high performance with the rejection factor of phenol reached by 99.9%. However, there is a difference in the permeate flux produced. The permeate flux of PVDF/TiO$_2$/LiCl 2% and PVDF/TiO$_2$ membranes gained 5–6 and 2.5 kg/m^{2}h, respectively. LiCl was doped into a superhydrophobic membrane to improve the pore size of the membrane without altering the hydrophobicity of the membrane surface. A decent MD membrane should have a low membrane resistance to mass transfer and thermal conductivity, a high LEP of distilled water or feed solution, and high thermal stability and durability.

Commercial polymeric membranes such as polypropylene (PP), polyvinylidene-fluoride (PVDF), and polytetrafluoroethylene (PTFE) are commonly used in both industrial and laboratory scales. Each membrane has a variety of advantageous characteristics. PP is chemically resistant and hydrophobic; PVDF has high-temperature resistance and is inherently hydrophobic, and PTFE has high-temperature and chemical (acid) resistance, cannot be irradiated, and is inherently hydrophobic (Curcio et al., 2010).

The prediction of membrane performance in MD could be presented with the assistance of notable information about the thermal conductivity of the membrane, pore size, distribution, pore tortuosity, porosity or empty volume fraction, and membrane surface properties. Membrane development can be identified by providing appropriate characterization to determine the performance of MD membranes for various MD applications.

To characterize MD membranes, two physical methods are commonly used. Firstly, the liquid and gas flow tests allow the determination of various permeation-related parameters. Additionally, scanning electron microscopy (SEM), scanning emission field electron microscopy (FESEM), and atomic force microscopy (AFM) could be used to directly obtain the morphological and structural properties of membranes.

5. Determination of pore size

Gas permeation test: This technique is appropriate to measure not only the mean pore size but also the effective porosity, which is defined as the ratio between the porosity and the effective pore length that considers the pore tortuosity (Khayet et al., 2002). It consists of measuring only the gas flow rate through a dry porous membrane at different transmembrane pressures.

Wet/dry flow method: Wet/dry flow or liquid displacement methods can assess the maximum pore size, the mean pore size, and the pore size distribution of MD membranes using a combined bubble point and gas permeation test (Khayet et al., 2002). It involves the displacement of a liquid from a wetted membrane using a purified gas.

Mercury porosimetry: The same principles are applied to this method of determining membrane pore size as the bubble pressure method. Nevertheless, mercury is employed here as the nonwetting liquid for porous materials. Mercury is forced into dry membrane pores, and the volume of mercury that enters the pores is measured at each pressure applied.

Electron microscopy: This method incorporates several electronic microscopy techniques, such as scanning electron microscopy (SEM), transmission electron microscopy (TEM), field emission scanning electron microscopy (FESEM), and others, that allow for direct visualization of the morphological structure of porous membranes (top and bottom surfaces for flat sheet membranes, internal and external surfaces of capillaries and hollow fiber membranes, as well as membrane cross section) (Khayet et al., 2002). To acquire pore size distributions and porosities, computerized image analysis of the corresponding micrographs is typically performed.

Atomic force microscopy (AFM): This method creates high-resolution methodologies for investigating the surface morphology of membrane surfaces at the nanometer scale. It is employed in the air to obtain explicit three-dimensional topographic pictures of the membrane surface as well as cross sections to atomic-level resolution or in liquids by scanning the sharp tip of the microscopic cantilever over the surface (Hitsov et al., 2017).

The membrane modules used for the MD process vary according to the function required. The plate and frame module are widely used in desalination and water treatment. Furthermore, the flat sheet membrane module is frequently used on the laboratory scale due to its being the simplest and easiest to operate and maintain, while the packing density, defined as the ratio of membrane area to packing volume, is low, requiring membrane support (Hitsov et al., 2015). Several hollow fibers are bundled and sealed inside a shell tube in the hollow fiber module, which has been used in MD. The main advantages of the hollow fiber module are its extremely high packing density (Khayet and Matsuura, 2011b) leading to exhibiting a large surface resulting in a hollow fiber having a high volume of water productivity.

On the other hand, the main advantage of hollow fiber is a high fouling tendency, and it is a challenging step for maintenance. As a consequence of the penetration of feed solution through membrane pores and its presence on the permeate side, the membrane module should be cleaned. Physical and chemical cleaning was performed, which showed high water flux (Peng et al., 2015). Besides, numerous pretreatment processes such as adsorption, coagulation/flocculation, ultrafiltration, etc. were carried out in several areas such as desalination and wastewater treatment.

The tubular module, which is tube-shaped and fits between two cylindrical chambers (hot and cold fluid chambers), is more appealing in the commercial field because it has a low fouling tendency and a large effective area and is easy to clean. The tubular module has a low packing density and a high operating cost. Flat sheet membranes are assembled in spiral-wound modules. The feed flows axially across the membrane surface, whereas the permeate flows radially to the center and exits through the collection tube. The spiral-wound membrane has a good packing density, an average fouling tendency, and an acceptable energy consumption (Hitsov et al., 2015).

6. Effect of operation factor for membrane distillation performance

6.1 Feed temperature

The driving force in the MD process is the temperature difference between the feed and permeate sides of the membrane that influences the difference in the vapor partial pressure. The exponential increase in water vapor pressure is attributed to the rise in feed temperature.

In other words, feed temperature is a crucial operational parameter that determines the permeate flux (Mittal et al., 2021; Xu et al., 2016; Alsaadi et al., 2013; Guillén-Burrieza et al., 2011). The enhanced temperature difference between the feed and permeate sides of the membrane will favorably affect the diffusion factor (Reza et al., 2018). Diffusion is the mechanism for transporting water through the boundary layer. The diffusion factor of volatile molecules is the governing mass transfer mechanism and the membrane permeability. The previous study discovered the relevant relationship between the experimental and theoretical data using the mass transfer coefficient. The rise in feed bulk temperature leads to an increase in the water vapor flux, causing an increase in the water vapor diffusion coefficient via the stagnant air (Qtaishat et al., 2008). When the temperature of the hot fluid rises, the vapor pressure gradient rises faster than when the temperature of the cold fluid drops. The distillate flux will increase at a steady temperature difference as the feed temperature rises, as shown in Table 3, indicating that the vapor pressure is more susceptible to high temperatures.

TABLE 3   Effect of temperature on the permeate flux.

Solution	T_f (°C)	Flux (J) (kg/m^{2}h)	References
Artificial seawater	40–70	≈1–7	Banat and Simandl (1998)
EtAc (5 wt%)	57	≈1.8–3.0	Al-Gharabli et al. (2018)
NaCl (3%)	5–45	0.5–6	Hsu et al. (2002)
NaCl (2 mol/L)	17.5–31	≈2.88–25.2	Martínez-Díez et al. (1999)
Pure water	40–70	≈5.8–18.7	Phattaranawik et al. (2003)
Pure water	36–66	≈5.4–36	Yun et al. (2006)
Pure water	40–70	≈7–33 L/m^{2}h	Srisurichan et al. (2006)
NaCl (24.6 wt%)	43–68	≈6.1–28.8	Yun et al. (2006)
NaCl (5%)	5–45	1–42	Hsu et al. (2002)
Sugar	61–81	≈18–38	Schofield et al. (1990)
Orange juices	24–45	30×10^3–108×10^3	Calabro et al. (1994)
Phenol (0.87 wt%)	40	10.62	El-Abbassi et al. (2012)
Phenol (10 wt%)	40	2.5	Hamzah and Leo (2016)
Phenol (10 wt%)	47	≈5–6	Hamzah and Leo (2017)
Phenol (0.213 wt%)	70	23 L/m^{2}h	Kiai et al. (2014)
Phenol (0.003 wt%)	45	9.57	Veleva et al. (2021)
Phenol (0.51–2.51 µg/L)	60	1.45	Ramos et al. (2009)
TCA	50–60	≈0.050–0.2	Wu et al. (2006)
NaCl (35%)	7–60	3.2	Murugesan et al. (2020)
Pure water	40–70	≈7–33 L/m^{2}h	Khayet et al. (2000)

The vapor pressure of the water could be predicted by Antoine's equation as given in the following equations:

$$p_{v,f} = e^{A - \frac{B}{C+T_m}} \tag{1}$$

where $p_{v,f}$ is the vapor pressure of water (Pa) for feed solutions, T_m is the feed temperature in the membrane surface (K), A, B, and C are Antoine's constants for water ($A = 231,963$, $B = 3816.44$, and $C = -46.13$). Based on Antoine's equation, the exponential relationship between water vapor pressure and temperature, the increasing temperature, the vapor pressure increases exponentially (Santos et al., 2021).

6.2 Feed concentration

When the feed concentration rises, the vapor pressure is reduced linearly along with the permeate flux, as shown in Table 4 (Reza et al., 2018; Xu et al., 2016). Likewise, both the reduction in the water activity and reduction in the coefficient of mass transfer attributable to concentration polarization and drop in the coefficient of heat transport leading to decline in the membrane surface temperature all contribute to flow reduction. Moreover, membrane fouling could occur at high solute concentrations in the feed (Hitsov et al., 2017).

TABLE 4   Effect of concentration on permeate flux.

Solution	Concentration (g/L)	Flux (J) (kg/m²h)	References
HNO_3	2–6M	0.9–2.1 L/m²h	Matheswaran et al. (2007)
NaCl	1 w/v%	4.11 L/m²h	Eryildiz et al. (2021)
Boron	2.6–6.2	32–26 L/m²h	Eryildiz et al. (2021)
NaCl	0–116.8	32.4–25.2	Martínez-Díez et al. (1999)
NaCl	0–5290	44–63	Schofield et al. (1990)
NaCl	0–24.6 wt%	36–28.8	Yun et al. (2006), Eryildiz et al. (2021)
H_2SO_4	2M	15	Eryildiz et al. (2021)
NaOH	0.1M	16	Martínez-Díez et al. (1999)
NaCl	0.7	15.3 L/m²h	Schofield et al. (1990)
NaCl	0.7	32	Ji et al. (2021)
NaCl	3.5 wt%	11	Ji et al. (2021)
NaCl	35	10	Woo et al. (2017)
NaCl	100–300	10.7–7	Lin et al. (2015)
2-BE/water	2	≈3.25 L/m²h	Chew et al. (2017)
EG/water	2	≈2.8 L/m²h	Tong et al. (2016)
IPA/water	2	≈3.1 L/m²h	Matheswaran et al. (2007)
Acetic acid/water	2	≈3.25 L/m²h	Yao et al. (2018)

TABLE 5   Effect of flowrate on permeate flux.

Solution	Flowrate (l/min)	J (kg/m^{2}h)	References
NaCl (35 g/L)	1	$\approx$7.5 L/m^{2}h	Matheswaran et al. (2007)
SDS (10 mg/L)	150 mL/min	10	Tong et al. (2016)
NaCl	120 L/h	$\approx$30.4–38.4 L/m^{2}h	Lin et al. (2015)
TDS (15 mg/L)	2	15.3 L/m^{2}h	Woo et al. (2017)
NaCl (3.5 wt%)	12 L/h	11	Chew et al. (2017)
Simulated seawater	0.7	3	Banat and Simandl (1998)
Juice (100 g/L)	5	2.9	Calabro et al. (1994)
NaCl (1 M)	5 kg/min	$\approx$3.24–3.96	Khayet (2011), Khayet et al. (2000)
	0.07–0.21	$\approx$5.4–5.76	Yun et al. (2006)
NaCl (17.7 wt%)	0.056–0.33	$\approx$25.9–29.5	Bandini and Sarti (1999)
Acetone (5 wt%)	0.1–2.6 L/min	12.6–21.6	Safavi and Mohammadi (2009)
NaCl (300 g/L)	0.015–0.03 L/s	7–9.1	

6.3 Feed flow rate

The increase in flow rate is proportional to the increase in permeate flux, as shown in Table 5. In the feed side, the thickness of the boundary layer is lower and the mixing temperature is increased. The asymptotic proportion of the distillate flux was attained as caused by a change in the laminar flow regime to turbulence as the flow rate increased (Reza et al., 2018). Similarly, the Reynolds number and heat transfer coefficient contribute to the decrease in temperature polarization.

6.4 Running time operation

The most challenging aspect of commercializing MD is ensuring membrane stability (Reza et al., 2018). The findings reveal that the permeate flux rose after membrane solidification during the first hours of the test. As the feed flow reduces, it settles into a steady state. Flux reduction was linked to partial pore wetting and fouling (Guillen-Burrieza et al., 2014). Over long-term running, the optimal duration was expanded and membrane deterioration was prevented. Overall, the permeate flux decreased by 16% of continuous operation over the 100 days with synthetic seawater (McGaughey et al., 2017) since certain fouling occurred over long periods. Moreover, the small feed volume passed through the membrane led to a slight flux increase ($<$1 L/m^{2}h) and additionally guaranteed that the distillate conductivity rose after 50 days.

7. Membrane distillation application for water resources

Water scarcity has been and continues to be a significant issue worldwide. Water-stressed countries have 2.3 billion people, with 733 million living in high and critically water-stressed countries (UN Water, 2021). Furthermore, less than 3% is freshwater, and agriculture uses 72% of all water withdrawals, municipalities 16% for families and services, and industries 12% (UN Water, 2021). Most of the world's water is seawater, which is unfit for human consumption due to its high salinity. Given this problem, solutions must be searched for, and desalination may significantly contribute, given the high population density close to the coasts (Zapata-Sierra et al., 2022).

One of the emerging desalination technologies in water reclamation is MD. MD has high separation efficiency, lower heating requirements and its modular design makes it simple and compact (Ahmed et al., 2021). Although MD is mainly used for desalination, it has other applications like removing any solutes from water, concentrating nonvolatile compounds, volatile compounds extraction, and purification of wastewater. The MD process is mainly applied at the laboratory scale (Khayet and Matsuura, 2011c). There have been various pilot plant uses proposed for seawater desalination. One of the cases is seawater desalination for potable water provision on remote islands in Vietnam (Duong et al., 2021).

Desalination to produce high purity water from brackish water, saline water, brine water, and seawater to be potable water has been the dominant MD use. One of the reasons is that the rejection of nonvolatile solutes in aqueous solutions is over 100%. As the permeate product is unspoiled, it is likewise essential for application in the medical and pharmaceutical sectors. In fact, only water molecules flow via membrane pores as in a solution with nonvolatile components (Khayet and Matsuura, 2011d).

MD has been extensively applied to wastewater treatment at a laboratory scale to generate a hazardousless permeate to the environment or recover valuable substances. MD has been rigorously tested for treating petrochemicals (Santos et al., 2020; Veleva et al., 2021), textiles (Fortunato et al., 2021; Leaper et al., 2019), and wastewater contaminated with heavy metals, oil–water emulsions (Kalla, 2021), and radioactive wastewater solutions (Chen et al., 2019; Jia et al., 2017, 2021; Nie et al., 2021).

MD provides a unique opportunity to improve the performance of these brackish water reverse osmosis (BWRO) plants by recovering a cleaner permeate and minimizing the amount of concentrate that must be discharged. However, adequate processing of the BWRO concentrate is required to ensure that this technology performs as expected (Zhang et al., 2021). MD also can improve the quality of the secondary wastewater treatment effluent (Liu et al., 2020). The hybrid technology of MD and postcoagulation has been proposed to treat the secondary effluent further to gain the high quality of reclaimed water. The coagulation-MD hybrid treatment process proved could reclaim water production from the purifying secondary effluent. Two coagulants, poly-titanium tetrachloride (PTC) and poly-aluminum chloride (PAC), were utilized for secondary effluent pretreatment, followed by DCMD to purify further the coagulated effluent for enhancing the effluent quality and mitigating the membrane fouling.

MD has the ability to treat contaminated biological wastewater. This was shown by Ruiz-Aguirre et al.'s (2017) research on treating highly resistant spores, *Bacillus subtilis* (*B. subtilis*) and Clostridium sp. spores, from wastewater. Results showed that MD could eliminate spores from the wastewater altogether and produce permeate, which can be directly discharged into the river or used in agriculture. In this experiment, they also investigate the effect of MD as pretreatment for the photo-Fenton process. The treatment of MD's concentrate with photo-Fenton reduces spore concentration in a shorter amount of time than spores treated without the MD process. These results are essential because *B. subtilis* and Clostridium sp. spores are highly resistant to tertiary treatment. The MD—photo-Fenton treatment can be the solution for very resistant endospores. Moreover, it could be qualified to remove slightly resistant pathogens with the same conditions.

MD can recover freshwater from protein-rich discharge from the meat processing and rendering industries. Two distinct PTFE (native hydrophobic and hydrophilic coated) membranes were used in batch and continuous experiments to process water samples of blood stick from an industrial plant. MD has been shown to produce a concentrated high strength stick to water with a concentration factor of 20 for potential reuse and recover clean water with a salinity and TOC rejection of >99% and TN rejection of >95%. Volatile fatty acids transferred through the membrane into the permeate were explored in depth in MD investigations, with concentrations of 30–160 mg/L in the permeate being proven to be more linked to hydrophobic surface diffusion than volatile diffusion. MD for the valorization of the meat industry effluent is feasible under the right conditions and with suitable membranes. Yet the reclaimed water quality owing to fatty acids, as well as other practical problems such as membrane cleaning, must be considered (Muster-Slawitsch et al., 2021).

Despite MD being an innovative, eco-friendly technology, the study for use of MD process for treating oily wastewater is limited, presumably due to the poor performance experienced. Han et al. (2017) aimed to comprehend the major challenges by conducting a comprehensive investigation of the impact of the critical components (especially, oil, surfactant, and salt) in such feeds. This study shows that NaCl and Sodium dodecyl sulfate (SDS) in the feed are primarily responsible for the deterioration of the MD process rather than the oil itself. Further research is required since the results might only be specific to the oil, salt, and surfactant studied. Understanding the interactions between NaCl and SDS may indeed be useful in providing information on the contact strength as well as how to avoid such issues.

As mentioned above and shown in Table 6, MD has a high removal efficiency in eliminating solute and pollutants to gain freshwater. To be utilized commercially, MD's applicability has so far been confined to a pilot plant and still confronts significant obstacles. The notable drawbacks of MD are the low permeate flux and high energy consumption compared to other technologies (Yadav et al., 2021).

The primary source of energy consumption in MD systems is feed heating and permeate condensation. Depending on the configuration, energy is also used for vacuum and peristaltic pumps. In terms of MD's configuration, DCMD and AGMD have better energy efficiency compared to SGMD and VMD. Both AGMD and DCMD designs considered retrieving the energy to improve efficiency. As in AGMD, the gap's spacing highly affects energy efficiency (Ahmed et al., 2020). One of the solutions to improve energy efficiency is to apply low-grade heat energy sources in MD. Some of the recommended energy sources are solar energy, geothermal energy, and waste heat recovered from ships, industries, and power stations (Yadav et al., 2021).

The last 5 years have shown a shift to exploring improvement strategies in MD's process to improve energy efficiency using integrated heat recovery, renewable energy MD, hybrid MD

TABLE 6 Rejection factor and energy consumption of membrane distillation for water reclaim.

Configuration	Application	Membrane material	Parameter	Permeate flux (kg/m²h)	Rejection factor (%)	Energy consumption (kWh/m³)	GOR	Reference
SGMD	Desalination	CNIM	$A_m = 12.50\,cm^2$; $T_f = 90°C$	19.2	99			Bhadra et al. (2013)
MS-AGMD	Desalination	PP	$A_m = 1.38\,m^2$; $T_f = 95°C$	3.9			± 3.4	Geng et al. (2015)
AGMD	Desalination	PTFE, PVDF, PP	$A_m = 0.005\,m^2$; $T_f = 80°C$	4–5	95–100			Xu et al. (2016a)
AGMD	Desalination	Alumina, LP3 (T-PFS)	$T_f = 80°C$	33	99.99			García-Fernández et al. (2017)
MS-AGMD	Desalination	PTFE	$A_m = 0.0074\,m^2$ $T_f = 50$–$90°C$	±15	99.97	10–30	0.45–0.6	Khalifa et al. (2017)
AGMD	Heavy metal	PVDF	$A_m = 36.88\,m^2$	20	99.36			Attia et al. (2017)
DCMD	Desalination	SBS	$T_f = 60°C$	10	99.70			Duong et al. (2018)
MS-AGMD	Desalination	PTFE	$A_m = 0.0074\,m^2$ $T_f = 50$–$90°C$	30	97.7–99.97	5.77–9.3		Khalifa and Alawad (2018)
MS-WGMD	Desalination	PTFE	$A_m = 0.0074\,m^2$ $T_f = 50$–$90°C$	60	97.7–99.97	5–6.96		Khalifa and Alawad (2018)
DCMD	Dye wastewater	MfSNPS/PVDF/PTFE	$A_m = 0.0125\,m^2$ $T_f = 50°C$	0.0041	99			Khumalo et al. (2019)
AGMD	Dye wastewater	PTFE	$A_m = 7.16\,cm^2$ $T_f = 70°C$	~12.4	100			Leaper et al. (2019)
DCMD	Dye wastewater	PTFE	$A_m = 7.60\,cm^2$ $T_f = 70°C$	12.8	99			Leaper et al. (2019)
Solar driven VMD	Desalination		$A_m = \sim 0.2\,m^2$ $T_f = \sim 60°C$	5		4–6	~0.4–0.5	Li et al. (2019)

Continued

TABLE 6 Rejection factor and energy consumption of membrane distillation for water reclaim—cont'd

Configuration	Application	Membrane material	Parameter	Permeate flux (kg/m²h)	Rejection factor (%)	Energy consumption (kWh/m³)	GOR	Reference
VMD	Radioactive	ceramic nanofiltration membrane	$T_f = 60\text{-}80°C$	>20	>99.9			Chen et al. (2019)
VA-AGMD	Heavy metal	PP	$T_f = 70°C$	135.23	99			Eryildiz et al. (2020)
VA-AGMD	Heavy metal	PTFE	$T_f = 70°C$	82.36	99			Eryildiz et al. (2020)
VMD/V-MEMD	Desalination	PP	$A_m = 0.016 \, m^2$ $T_f = \sim 90°C$	~5.1	>99.99	263	~2.6	Li et al. (2020)
DCMD	Organic matter	PTFE	$A_m = 12 \, cm^2$ $T_f = 40\text{-}60°C$	5.06, 9.07, 11.82	~93			Liu et al. (2020)
VMD	Radioactive	PTFE	$T_f = 75°C$	7.3	>99.99			Nie et al. (2021)
VMD	Radioactive	PTFE	$A_m = 8.2 \, m^2$ $T_f = 90°C$	6.82			0.8-1	Jia et al. (2021)
AGMD	Desalination	LDPE	$A_m = 25.9 \, m^2$ $T_f = 70°C$	0.6		87		Duong et al. (2021)

A_m = effective membrane area; T_f = feed temperature; T_c = coolant temperature; low-density polyethylene (LDPE); vacuum assisted air gap membrane distillation (VA-AGMD); vacuum multi effect membrane distillation (VMD/V-MEMD); multistage air gap membrane distillation (MS-AGMD); Multi Effect Air Gap Membrane Distillation (ME-AGMD); polypropylene (PP); polytetrafluroethylene (PTFE); polyvinylidene fluoride (PVDF); styrene-butadiene-styrene (SBS); methyl functionalized mesoporous silica nanoparticles (MfSNPs); carbon nanotube immobilized membrane (CNIM); low-density polyethylene (LDPE).

systems, alternative heating techniques (Ahmed et al., 2021). The current study examines the experimental and numerical feasibility of an integrated solar membrane distillation prototype for potable water and thermal energy production (with membrane distillation modules placed directly into evacuated solar tubes). This attractive combination of an evacuated tube solar collector and a membrane distillation unit, according to Li et al. (2019), is a novel method that marries two well-developed technologies into an efficient yet relatively low-cost hybrid energy-water production system.

A novel hybrid system is presented that combines DCMD with an alkaline fuel cell to reuse the alkaline fuel cell's exhaust heat for water purification. Calculation results illustrated that the power output density and energy efficiency of the hybrid system were 144.58% and 144.55%, respectively, larger than the standalone alkaline fuel cell. Moreover, extensive parametric studies indicated that the performance of the hybrid system benefits the feedwater temperature, convective heat transfer coefficients, water flow velocities, and hydrodynamic membrane pore radius. In contrast, the permeate temperature and hydrodynamic membrane thickness negatively influence the hybrid system performance (Zhao et al., 2021).

MD has a promising potential as a water reclamation process from wastewater. MD has a high removal efficiency for a large range of wastewater, from organic matter to radioactive substances. However, as compared to other membrane separation technologies, for instance membrane bioreactor (MBR) technology, it is applied less on an industrial scale, although in terms of production quality both have a high effluent quality. Moreover, for the membrane used, MD's membrane is more resistant to fouling compared to MBR. Kiss and Kattan Readi (2018) have reviewed that Low permeate flow, high energy efficiency, unclear economic cost, and a restricted membrane render MD unsuitable for industrial-scale applications.

MD has a low permeate flux that is caused by concentration polarization and membrane fouling. Concentration polarization occurs when the solute concentration adjacent to the membrane surface is higher than on the bulk feed. This affects the measurement of solute concentration due to mass and heat transfer transpiring at the same time. The effect of Membrane fouling in MD is lower compared to the pressure-driven membrane technology; however, it reduces the membrane surface effective area which led to the poor MD performance. To increase the permeate flux, these problems should be minimized (Kebria and Rahimpour, 2012).

High energy consumption is one of the major problems in MD. MD's operating temperature is low, but the thermal energy takes a big part of the energy consumption. Thermal energy is necessary to operate MD. The energy consumption is related to thermal efficiency. One of the reasons for the thermal efficiency lows is because of temperature polarization. Thus, the energy consumption is high due to temperature and concentration polarization. The high energy consumption led to high economic cost (Ahmed et al., 2020). MD's cost could be competitive if the system integrated with waste heat or renewable energy (Kebria and Rahimpour, 2012). Using a hydrophobic membrane to resist solute from the permeate is the process of MD. However, to produce a membrane with reliable antifouling properties, tolerance of chlorine charge and great mechanical strength are challenging. Moreover, this membrane is not as commercially available as hydrophilic membrane, especially for a large-scale system (Kebria and Rahimpour, 2012).

Despite the many advantages that MD has, it is still challenging to apply MD on an industrial scale. The reason is the low water production flux, high energy consumption, and limitation of hydrophobic membrane on a large scale. More research is still needed to confront this challenge. In recent years, one of the ways is by integrated MD with other technology as last stage treatment and to apply waste heat or renewable energy to the MD operation.

8. Conclusions and perspectives

MD is a sustainable and promising separation technology dealing with high salinity water and wastewater to produce reclaimed water. Numerous investigations have been performed in the recent decade to determine ways to enhance the performance of MD. The subject matter ranged from the modified membrane, extended module, and modified configuration to models and system design performance and evaluation. DCMD accounts for the vast majority of MD's fundamental configuration studies, followed by VMD and AGMD, and SGMD requires less MD configuration investigation.

Despite both fouling and scaling issues being less of a concern in MD than in pressure-driven separation processes. Nevertheless, a factor to consider for enhanced MD desalination due to the decrement in distillate water production and increment in treatment costs. Feedwater and foulant parameters, membrane qualities, and operational circumstances play an important role in fouling and scaling. Fouling is also caused by organic, inorganic, and biological causes, resulting in decreased MD performance.

Researchers experimented with membrane modification to prevent membrane issues by using a technique such as blending and electrospinning. These techniques are one of the effective processes to enhance membrane performance. They also innovated a novel material for membranes, such as carbon-based materials, followed by introducing styrene-butadiene-styrene (SBS), carbon nanotubes (CNTs), graphene, and graphene oxide used due to their particular characteristics. Developing new nanomaterials and polymers will increase membrane hydrophobicity. It also can improve penetration rates, salt rejection, and stability, as well as encourage exploration and growth in high-performance and multifunctional membranes in membrane distillation.

MD has emerged as a viable method for generating high-quality freshwater in recent years. However, both theoretical and experimental computations are fraught with uncertainty. Trials on large-scale pilot plants with continuous monitoring of the various processes are still needed to solve the current obstacles in MD. Several developments in technology have occurred recently concerning MD modules and configurations, including integration with electromagnetic (e.g., electrically powered self-contained direct contact membrane distillation (EPSCD), electricity-assisted membrane distillation, integrated with a solar system/self-heating (e.g., the self-heating-membrane distillation (SHMD), Solar thermal membrane distillation, Photothermal membrane distillation, a solar-powered vacuum membrane distillation coupled with a liquid ring vacuum pump, solar-powered tubular direct contact membrane distillation (TDCMD) systems, a solar thermal-photovoltaic vacuum membrane distillation (STPVMD) system, integrated with another approach which is more known as hybrid systems (e.g., Membrane distillation crystallization, VMD - Osmotic Distillation (OD), integrated forward osmosis-membrane distillation (FO-MD)). The essential configuration modification is also developed, such as vacuum-assisted air gap membrane distillation (V-AGMD), passive multistage membrane distillation, submerged vacuum membrane distillation (S-VMD), air sparged-DCMD (or AS-DCMD), a hollow fiber-based multieffect VMD (HF V-MEMD), air gap diffusion distillation (AGDD), vacuum multieffect membrane distillation (VMEMD).

MD's advantage, coupled with other separation techniques, is achieving zero liquid discharge (ZLD). Using hybrid multistep processes, the rejected (concentrated) brine can be minimized to solid waste. It means that it is easier to dispose of than focused brine, and most of the water is desalinated. It has been determined that this method of water recovery is feasible both technically and economically. In terms of integration with solar energy, the location of the system's installation is critical to the system's overall operation and performance. Even though numerous studies have been conducted to evaluate the performance of solar-powered MD, more large-scale pilot plants are required to demonstrate the feasibility of using MD for sustainable water production. Although these configuration methods appear to have the potential to improve MD efficiency, most of them were developed in the lab. More theoretical and experimental investigations, focusing on their module materials and cost-effective materials, are still needed.

References

Abu-Zeid, M.A.E.R., Zhang, Y., Dong, H., Zhang, L., Chen, H.L., Hou, L., 2015. A comprehensive review of vacuum membrane distillation technique. Desalination 356, 1–14. https://doi.org/10.1016/j.desal.2014.10.033.

Ahmed, F.E., Lalia, B.S., Hashaikeh, R., Hilal, N., 2020. Alternative heating techniques in membrane distillation: a review. In: Desalination. Vol. 496. Elsevier B.V., https://doi.org/10.1016/j.desal.2020.114713.

Ahmed, F.E., Khalil, A., Hilal, N., 2021. Emerging desalination technologies: current status, challenges and future trends. In: Desalination. vol. 517. Elsevier B.V., https://doi.org/10.1016/j.desal.2021.115183.

Al-Anezi, A.A.-H., Sharif, A.O., Sanduk, M.I., Khan, A.R., 2012. Experimental investigation of heat and mass transfer in tubular membrane distillation module for desalination. ISRN Chem. Eng. 2012, 1–8. https://doi.org/10.5402/2012/738731.

Al-Gharabli, S., Kujawski, W., El-Rub, Z.A., Hamad, E.M., Kujawa, J., 2018. Enhancing membrane performance in removal of hazardous VOCs from water by modified fluorinated PVDF porous material. J. Membr. Sci. 556, 214–226. https://doi.org/10.1016/j.memsci.2018.04.012.

Alsaadi, A.S., Ghaffour, N., Li, J., Gray, S., Francis, L., Maab, H., Amy, G.L., 2013. Modeling of air-gap membrane distillation process : a theoretical and experimental study. J. Membr. Sci. 445, 53–65. https://doi.org/10.1016/j.memsci.2013.05.049.

Amaral, M.C.S., 2021. Sustainable ammonia resource recovery from landfill leachate by solar-driven modified direct contact membrane distillation. Sep. Purif. Technol. 264 (September 2020). https://doi.org/10.1016/j.seppur.2021.118356.

Ashoor, B.B., Mansour, S., Giwa, A., Dufour, V., Hasan, S.W., 2016. Principles and applications of direct contact membrane distillation (DCMD): a comprehensive review. Desalination 398, 222–246. https://doi.org/10.1016/j.desal.2016.07.043.

Attia, H., Alexander, S., Wright, C.J., Hilal, N., 2017. Superhydrophobic electrospun membrane for heavy metals removal by air gap membrane distillation (AGMD). Desalination 420 (July), 318–329. https://doi.org/10.1016/j.desal.2017.07.022.

Banat, F.A., Simandl, J., 1998. Desalination by membrane distillation: a parametric study. Sep. Sci. Technol. 33 (2), 201–226. https://doi.org/10.1080/01496399808544764.

Bandini, S., Sarti, G.C., 1999. Heat and mass transport resistances in vacuum membrane distillation per drop. AICHE J. 45 (7), 1422–1433. https://doi.org/10.1002/aic.690450707.

Bhadra, M., Roy, S., Mitra, S., 2013. Enhanced desalination using carboxylated carbon nanotube immobilized membranes. Sep. Purif. Technol. 120, 373–377. https://doi.org/10.1016/j.seppur.2013.10.020.

Calabro, V., Jiao, B.L., Drioli, E., 1994. Theoretical and experimental study on membrane distillation in the concentration of Orange juice. Ind. Eng. Chem. Res. 33 (7), 1803–1808.

Chen, X., Chen, T., Li, J., Qiu, M., Fu, K., Cui, Z., Fan, Y., Drioli, E., 2019. Ceramic nanofiltration and membrane distillation hybrid membrane processes for the purification and recycling of boric acid from simulative radioactive waste water. J. Membr. Sci. 579, 294–301. https://doi.org/10.1016/j.memsci.2019.02.044.

Chen, L., Chen, Z., Wang, Y., Mao, Y., Cai, Z., 2021. Effective treatment of leachate concentrate using membrane distillation coupled with electrochemical oxidation. Sep. Purif. Technol. 267 (March), 118679. https://doi.org/10.1016/j.seppur.2021.118679.

Chew, N.G.P., Zhao, S., Loh, C.H., Permogorov, N., Wang, R., 2017. Surfactant effects on water recovery from produced water via direct-contact membrane distillation. J. Membr. Sci. 528 (October 2016), 126–134. https://doi.org/10.1016/j.memsci.2017.01.024.

Curcio, E., di Profio, G., Drioli, E., 2010. Membrane distillation and osmotic distillation. Compr. Membr. Sci.

Ding, Z., Liu, L., Li, Z., Ma, R., Yang, Z., 2006. Experimental study of ammonia removal from water by membrane distillation (MD): the comparison of three configurations. J. Membr. Sci. 286 (1–2), 93–103. https://doi.org/10.1016/j.memsci.2006.09.015.

Drioli, E., Ali, A., Macedonio, F., 2015. Membrane distillation: recent developments and perspectives. Desalination 356, 56–84. https://doi.org/10.1016/j.desal.2014.10.028.

Duong, H.C., Chuai, D., Woo, Y.C., Shon, H.K., Nghiem, L.D., Sencadas, V., 2018. A novel electrospun, hydrophobic, and elastomeric styrene-butadiene-styrene membrane for membrane distillation applications. J. Membr. Sci. 549 (September 2017), 420–427. https://doi.org/10.1016/j.memsci.2017.12.024.

Duong, H.C., Tran, L.T.T., Truong, H.T., Nelemans, B., 2021. Seawater membrane distillation desalination for potable water provision on remote islands—a case study in Vietnam. Case Stud. Chem. Environ. Eng. 4, 100110. https://doi.org/10.1016/j.cscee.2021.100110.

El-Abbassi, A., Kiai, H., Hafidi, A., García-Payo, M.C., Khayet, M., 2012. Treatment of olive mill wastewater by membrane distillation using polytetrafluoroethylene membranes. Sep. Purif. Technol. 98, 55–61. https://doi.org/10.1016/j.seppur.2012.06.026.

El-Bourawi, M.S., Khayet, M., Ma, R., Ding, Z., Li, Z., Zhang, X., 2007. Application of vacuum membrane distillation for ammonia removal. J. Membr. Sci. 301 (1–2), 200–209. https://doi.org/10.1016/j.memsci.2007.06.021.

Eryildiz, B., Yuksekdag, A., Korkut, S., Zeytuncu, B., Pasaoglu, M.E., Koyuncu, I., 2020. Effect of operating parameters on removal of boron from wastewater containing high boron concentration by vacuum assisted air gap membrane distillation. J. Water Process Eng. 38. https://doi.org/10.1016/j.jwpe.2020.101579.

Eryildiz, B., Yuksekdag, A., Korkut, S., Koyuncu, İ., 2021. Performance evaluation of boron removal from wastewater containing high boron content according to operating parameters by air gap membrane distillation. Environ. Technol. Innov. 22. https://doi.org/10.1016/j.eti.2021.101493.

Esteras-Saz, J., de la Iglesia, Ó., Peña, C., Escudero, A., Téllez, C., Coronas, J., 2021. Theoretical and practical approach to the dealcoholization of water-ethanol mixtures and red wine by osmotic distillation. Sep. Purif. Technol. 270. https://doi.org/10.1016/j.seppur.2021.118793.

Fortunato, L., Elcik, H., Blankert, B., Ghaffour, N., Vrouwenvelder, J., 2021. Textile dye wastewater treatment by direct contact membrane distillation: membrane performance and detailed fouling analysis. J. Membr. Sci. 636. https://doi.org/10.1016/j.memsci.2021.119552.

García-Fernández, L., Wang, B., García-Payo, M.C., Li, K., Khayet, M., 2017. Morphological design of alumina hollow fiber membranes for desalination by air gap membrane distillation. Desalination 420 (July), 226–240. https://doi.org/10.1016/j.desal.2017.07.021.

Geng, H., Wang, J., Zhang, C., Li, P., Chang, H., 2015. High water recovery of RO brine using multi-stage air gap membrane distillation. Desalination 355, 178–185. https://doi.org/10.1016/j.desal.2014.10.038.

Guillén-Burrieza, E., Blanco, J., Zaragoza, G., Alarcón, D., Palenzuela, P., Ibarra, M., Gernjak, W., 2011. Experimental analysis of an air gap membrane distillation solar desalination pilot system. J. Membr. Sci. 379, 386–396. https://doi.org/10.1016/j.memsci.2011.06.009.

Guillen-Burrieza, E., Ruiz-Aguirre, A., Zaragoza, G., Arafat, H.A., 2014. Membrane fouling and cleaning in long term plant-scale membrane distillation operations. J. Membr. Sci. 468, 360–372. https://doi.org/10.1016/j.memsci.2014.05.064.

Hamzah, N., Leo, C.P., 2016. Fouling prevention in the membrane distillation of phenolic-rich solution using superhydrophobic PVDF membrane incorporated with TiO2 nanoparticles. Sep. Purif. Technol. 167, 79–87. https://doi.org/10.1016/j.seppur.2016.05.005.

Hamzah, N., Leo, C.P., 2017. Membrane distillation of saline with phenolic compound using superhydrophobic PVDF membrane incorporated with TiO2 nanoparticles: separation, fouling and self-cleaning evaluation. Desalination 418 (March), 79–88. https://doi.org/10.1016/j.desal.2017.05.029.

Han, L., Tan, Y.Z., Netke, T., Fane, A.G., Chew, J.W., 2017. Understanding oily wastewater treatment via membrane distillation. J. Membr. Sci. 539, 284–294. https://doi.org/10.1016/j.memsci.2017.06.012.

Hitsov, I., Maere, T., de Sitter, K., Dotremont, C., Nopens, I., 2015. Modelling approaches in membrane distillation: a critical review. Sep. Purif. Technol. 142, 48–64. https://doi.org/10.1016/j.seppur.2014.12.026.

Hitsov, I., Eykens, L., de Schepper, W., de Sitter, K., Dotremont, C., Nopens, I., 2017. Full-scale direct contact membrane distillation (DCMD) model including membrane compaction effects. J. Membr. Sci. 524 (September 2016), 245–256. https://doi.org/10.1016/j.memsci.2016.11.044.

Hsu, S.T., Cheng, K.T., Chiou, J.S., 2002. Seawater desalination by direct contact membrane distillation. Desalination 143 (3), 279–287. https://doi.org/10.1016/S0011-9164(02)00266-7.

Izquierdo-Gil, M.A., Jonsson, G., 2003. Factors affecting flux and ethanol separation performance in vacuum membrane distillation (VMD). J. Membr. Sci. 214, 113–130.

Ji, H., Zhang, G., Teng, L., Xing, J., Jia, X., Luo, H., 2021. Fabrication of aramid-coated asymmetric PVDF membranes towards acidic and alkaline solutions concentration via direct contact membrane distillation. Appl. Surf. Sci. 562 (March), 150185. https://doi.org/10.1016/j.apsusc.2021.150185.

Jia, F., Li, J., Wang, J., 2017. Recovery of boric acid from the simulated radioactive wastewater by vacuum membrane distillation crystallization. Ann. Nucl. Energy 110, 1148–1155. https://doi.org/10.1016/j.anucene.2017.07.024.

Jia, X., Lan, L., Zhang, X., Wang, T., Wang, Y., Ye, C., Lin, J., 2021. Pilot-scale vacuum membrane distillation for decontamination of simulated radioactive wastewater: system design and performance evaluation. Sep. Purif. Technol. 275. https://doi.org/10.1016/j.seppur.2021.119129.

Kalla, S. (2021). Use of membrane distillation for oily wastewater treatment—a review. In J. Environ. Chem. Eng. (Vol. 9, issue 1). Elsevier Ltd. doi:https://doi.org/10.1016/j.jece.2020.104641.

Kebria, M.R.S., Rahimpour, A., 2012. Membrane distillation: basics, advances, and applications. In: Abdelrasoul, A. (Ed.), Advances in Membrane Technologies, p. 13.

Khalifa, A.E., Alawad, S.M., 2018. Air gap and water gap multistage membrane distillation for water desalination. Desalination 437 (November 2017), 175–183. https://doi.org/10.1016/j.desal.2018.03.012.

Khalifa, A.E., Alawad, S.M., Antar, M.A., 2017. Parallel and series multistage air gap membrane distillation. Desalination 417 (April), 69–76. https://doi.org/10.1016/j.desal.2017.05.003.

Khayet, M., 2011. Membranes and theoretical modeling of membrane distillation: a review. Adv. Colloid Interface Sci. 164 (1–2), 56–88. https://doi.org/10.1016/j.cis.2010.09.005.

Khayet, M., Matsuura, T., 2011a. Direct contact membrane distillation. In: Membrane Distillation., https://doi.org/10.1016/b978-0-444-53126-1.10010-7.

Khayet, M., & Matsuura, T. (2011b). Formation of hollow fibre MD membranes. Membrane Distillation, Md, 59–87. doi:https://doi.org/10.1016/b978-0-444-53126-1.10004-1.

Khayet, M., Matsuura, T., 2011c. Introduction to membrane distillation. In: Membrane Distillation. Elsevier, pp. 1–16, https://doi.org/10.1016/b978-0-444-53126-1.10001-6.

Khayet, M., Matsuura, T., 2011d. Thermally induced phase separation for MD membrane formation. In: Membrane Distillation., https://doi.org/10.1016/b978-0-444-53126-1.10005-3.

Khayet, M., Godino, P., Mengual, J.I., 2000. Nature of flow on sweeping gas membrane distillation. J. Membr. Sci. 170 (2), 243–255. https://doi.org/10.1016/S0376-7388(99)00369-5.

Khayet, M., Feng, C.Y., Khulbe, K.C., Matsuura, T., 2002. Preparation and characterization of polyvinylidene fluoride hollow fiber membranes for ultrafiltration. Polymer 43 (14), 3879–3890. https://doi.org/10.1016/S0032-3861(02)00237-9.

Khumalo, N.P., Nthunya, L.N., De Canck, E., Derese, S., Verliefde, A.R., Kuvarega, A.T., Mamba, B.B., Mhlanga, S.D., Dlamini, D.S., 2019. Congo red dye removal by direct membrane distillation using PVDF/PTFE membrane. Sep. Purif. Technol. 211 (September 2018), 578–586. https://doi.org/10.1016/j.seppur.2018.10.039.

Kiai, H., García-Payo, M.C., Hafidi, A., Khayet, M., 2014. Application of membrane distillation technology in the treatment of table olive wastewaters for phenolic compounds concentration and high quality water production. Chem. Eng. Process. Process Intensif. 86, 153–161. https://doi.org/10.1016/j.cep.2014.09.007.

Kiss, A.A., Kattan Readi, O.M., 2018. An industrial perspective on membrane distillation processes. J. Chem. Technol. Biotechnol. 93 (8), 2047–2055. https://doi.org/10.1002/jctb.5674.

Laqbaqbi, M., Sanmartino, J.A., Khayet, M., García-Payo, C., Chaouch, M., 2017. Fouling in membrane distillation, osmotic distillation and osmotic membrane distillation. Appl. Sci. (Switzerland) 7 (4). https://doi.org/10.3390/app7040334.

Leaper, S., Abdel-Karim, A., Gad-Allah, T.A., Gorgojo, P., 2019. Air-gap membrane distillation as a one-step process for textile wastewater treatment. Chem. Eng. J. 360, 1330–1340. https://doi.org/10.1016/j.cej.2018.10.209.

C. Membrane processes (MD-RO-FO)

Li, Q., Beier, L.J., Tan, J., Brown, C., Lian, B., Zhong, W., Wang, Y., Ji, C., Dai, P., Li, T., le Clech, P., Tyagi, H., Liu, X., Leslie, G., Taylor, R.A., 2019. An integrated, solar-driven membrane distillation system for water purification and energy generation. Appl. Energy 237, 534–548. https://doi.org/10.1016/j.apenergy.2018.12.069.

Li, Q., Omar, A., Cha-Umpong, W., Liu, Q., Li, X., Wen, J., Wang, Y., Razmjou, A., Guan, J., Taylor, R.A., 2020. The potential of hollow fiber vacuum multi-effect membrane distillation for brine treatment. Appl. Energy 276 (June), 115437. https://doi.org/10.1016/j.apenergy.2020.115437.

Lin, P.J., Yang, M.C., Li, Y.L., Chen, J.H., 2015. Prevention of surfactant wetting with agarose hydrogel layer for direct contact membrane distillation used in dyeing wastewater treatment. J. Membr. Sci. 475, 511–520. https://doi.org/10.1016/j.memsci.2014.11.001.

Liu, X., Tian, C., Sun, W., Zhao, Y., Shih, K., 2020. Secondary effluent purification towards reclaimed water production through the hybrid post-coagulation and membrane distillation technology: a preliminary test. J. Clean. Prod. 271. https://doi.org/10.1016/j.jclepro.2020.121797.

Martínez-Díez, L., Florido-Díaz, F.J., Vázquez-González, M.I., 1999. Study of evaporation efficiency in membrane distillation. Desalination 126 (1–3), 193–198. https://doi.org/10.1016/S0011-9164(99)00174-5.

Matheswaran, M., Kwon, T.O., Kim, J.W., Moon, I.S., 2007. Factors affecting flux and water separation performance in air gap membrane distillation. J. Ind. Eng. Chem. 13 (6), 965–970.

McGaughey, A.L., Gustafson, R.D., Childress, A.E., 2017. Effect of long-term operation on membrane surface characteristics and performance in membrane distillation. J. Membr. Sci. 543 (August), 143–150. https://doi.org/10.1016/j.memsci.2017.08.040.

Meindersma, G.W., Guijt, C.M., de Haan, A.B., 2006. Desalination and water recycling by air gap membrane distillation. Desalination 187 (February 2005), 291–301. https://doi.org/10.1016/j.desal.2005.04.088.

Mejia Mendez, D.L., Castel, C., Lemaitre, C., Favre, E., 2018. Membrane distillation (MD) processes for water desalination applications. Can dense selfstanding membranes compete with microporous hydrophobic materials? Chem. Eng. Sci. 188, 84–96. https://doi.org/10.1016/j.ces.2018.05.025.

Mittal, S., Gupta, A., Srivastava, S., Jain, M., 2021. Chemical engineering and processing - process intensification artificial neural network based modeling of the vacuum membrane distillation process : effects of operating parameters on membrane fouling. Chem. Eng. Process. Process Intensif. 164 (April), 108403. https://doi.org/10.1016/j.cep.2021.108403.

Murugesan, V., Rana, D., Matsuura, T., Lan, C.Q., 2020. Optimization of nanocomposite membrane for vacuum membrane distillation (VMD) using static and continuous flow cells: effect of nanoparticles and film thickness. Sep. Purif. Technol. 241 (February). https://doi.org/10.1016/j.seppur.2020.116685.

Muster-Slawitsch, B., Dow, N., Desai, D., Pinches, D., Brunner, C., Duke, M., 2021. Membrane distillation for concentration of protein-rich waste water from meat processing. J. Water Process Eng. 44. https://doi.org/10.1016/j.jwpe.2021.102285.

Nie, X., Hu, X., Liu, C., Xia, X., Dong, F., 2021. Decontamination of uranium contained low-level radioactive wastewater from UO2 fuel element industry with vacuum membrane distillation. Desalination 516. https://doi.org/10.1016/j.desal.2021.115226.

Olatunji, S.O., Camacho, L.M., 2018. Heat and mass transport in modeling membrane distillation configurations: a review. Front. Energy Res. 6 (130), 1–18. https://doi.org/10.3389/fenrg.2018.00130.

Peng, Y., Ge, J., Li, Z., Wang, S., 2015. Effects of anti-scaling and cleaning chemicals on membrane scale in direct contact membrane distillation process for RO brine concentrate. Sep. Purif. Technol. 154, 22–26. https://doi.org/10.1016/j.seppur.2015.09.007.

Phattaranawik, J., Jiraratananon, R., Fane, A.G., 2003. Effect of pore size distribution and air flux on mass transport in direct contact membrane distillation. J. Membr. Sci. 215 (1–2), 75–85. https://doi.org/10.1016/S0376-7388(02)00603-8.

Polena, J., Marie, Š., 2021. Pervaporation of dichloromethane-cyclopentane and methylal-cyclopentane mixtures through membranes from chloroprene rubber. J. Appl. Polymer. https://doi.org/10.1002/app.51320.

Qtaishat, M., Matsuura, T., Kruczek, B., Khayet, M., 2008. Heat and mass transfer analysis in direct contact membrane distillation. Desalination 219 (1–3), 272–292. https://doi.org/10.1016/j.desal.2007.05.019.

Rajwade, K., Barrios, A.C., Garcia-segura, S., Perreault, F., 2020. Pore wetting in membrane distillation treatment of municipal wastewater desalination brine and its mitigation by foam fractionation. Chemosphere 257, 127214. https://doi.org/10.1016/j.chemosphere.2020.127214.

Ramos, R., Vinatea, L., Seiffert, W., Beltrame, E., Silva, J.S., da Costa, R.H.R., 2009. Treatment of shrimp effluent by sedimentation and oyster filtration using *Crassostrea gigas* and *C. rhizophorae*. Braz. Arch. Biol. Technol. 52 (3), 775–783. https://doi.org/10.1590/S1516-89132009000300030.

Reza, M., Kebria, S., Rahimpour, A., 2018. Membrane Distillation: Basics, Advances, and Applications. pp. 1–21, https://doi.org/10.5772/intechopen.86952.

Ruiz-Aguirre, A., Polo-López, M.I., Fernández-Ibáñez, P., Zaragoza, G., 2017. Integration of membrane distillation with solar photo-Fenton for purification of water contaminated with *Bacillus* sp. and *Clostridium* sp. spores. Sci. Total Environ. 595, 110–118. https://doi.org/10.1016/j.scitotenv.2017.03.238.

Safavi, M., Mohammadi, T., 2009. High-salinity water desalination using VMD. Chem. Eng. J. 149 (1–3), 191–195. https://doi.org/10.1016/j.cej.2008.10.021.

Said, I.A., Chomiak, T., Floyd, J., Li, Q., 2020. Sweeping gas membrane distillation (SGMD) for wastewater treatment, concentration, and desalination: a comprehensive review. Chem. Eng. Proc. Process Intensif. 153 (January), 107960. https://doi.org/10.1016/j.cep.2020.107960.

Salmanli, O.M., Yuksekdag, A., Koyuncu, I., 2022. Boron removal by using vacuum assisted air gap membrane distillation (VAGMD). Environ. Technol. Innov. 26, 102395. https://doi.org/10.1016/j.eti.2022.102395.

Santos, P.G., Scherer, C.M., Fisch, A.G., Rodrigues, M.A.S., 2020. Petrochemical wastewater treatment: Water recovery using membrane distillation. J. Clean. Prod. 267. https://doi.org/10.1016/j.jclepro.2020.121985.

Santos, P.G., Scherer, C.M., Fisch, A.G., Rodrigues, M.A.S., 2021. Membrane distillation: pre-treatment effects on fouling dynamics. Membranes 11 (12), 1–8. https://doi.org/10.3390/membranes11120958.

Schofield, R.W., Fane, A.G., Fell, C.J.D., Macoun, R., 1990. Factors affecting flux in membrane distillation. Desalination 77 (C), 279–294. https://doi.org/10.1016/0011-9164(90)85030-E.

Si, Z., Han, D., Xiang, J., 2021. Experimental investigation on the mechanical vapor recompression evaporation system coupled with multiple vacuum membrane distillation modules to treat industrial wastewater. Sep. Purif. Technol. 275 (June), 119178. https://doi.org/10.1016/j.seppur.2021.119178.

Srisurichan, S., Jiraratananon, R., Fane, A.G., 2006. Mass transfer mechanisms and transport resistances in direct contact membrane distillation process. J. Membr. Sci. 277 (1–2), 186–194. https://doi.org/10.1016/j.memsci.2005.10.028.

Tong, D., Wang, X., Ali, M., Lan, C.Q., Wang, Y., Drioli, E., Wang, Z., Cui, Z., 2016. Preparation of Hyflon AD60/PVDF composite hollow fiber membranes for vacuum membrane distillation. Sep. Purif. Technol. 157, 1–8. https://doi.org/10.1016/j.seppur.2015.11.026.

UN Water, 2021. Summary Progress Update 2021: SDG 6-Water and Sanitation for all.

Veleva, I., Vanoppen, M., Hitsov, I., Phukan, R., Wyseure, L., Dejaeger, K., Cornelissen, E.R., Verliefde, A.R.D., 2021. Selection of membranes and operational parameters aiming for the highest rejection of petrochemical pollutants via membrane distillation. Sep. Purif. Technol. 259. https://doi.org/10.1016/j.seppur.2020.118143.

Wen, L., Elfa, L., Li, X., 2021. Contactless membrane distillation for effective ammonia recovery from waste sludge: a new configuration and mass transfer mechanism. J. Membr. Sci. 638 (June), 119733. https://doi.org/10.1016/j.memsci.2021.119733.

Woo, Y.C., Chen, Y., Tijing, L.D., Phuntsho, S., He, T., Choi, J.S., Kim, S.H., Shon, H.K., 2017. CF4 plasma-modified omniphobic electrospun nanofiber membrane for produced water brine treatment by membrane distillation. J. Membr. Sci. 529 (January), 234–242. https://doi.org/10.1016/j.memsci.2017.01.063.

Wu, B., Tan, X., Li, K., Teo, W.K., 2006. Removal of 1,1,1-trichloroethane from water using a polyvinylidene fluoride hollow fiber membrane module: vacuum membrane distillation operation. Sep. Purif. Technol. 52 (2), 301–309. https://doi.org/10.1016/j.seppur.2006.05.013.

Xu, J., Singh, Y.B., Amy, G.L., Ghaffour, N., 2016. Effect of operating parameters and membrane characteristics on air gap membrane distillation performance for the treatment of highly saline water. J. Membr. Sci. 512, 73–82. https://doi.org/10.1016/j.memsci.2016.04.010.

Yadav, A., Labhasetwar, P.K., Shahi, V.K., 2021. Membrane distillation using low-grade energy for desalination: a review. J. Environ. Chem. Eng. 9 (5), 105818. https://doi.org/10.1016/j.jece.2021.105818.

Yang, S., Abdalkareem Jasim, S., Bokov, D., Chupradit, S., Nakhjiri, A.T., El-Shafay, A.S., 2021. Membrane distillation technology for molecular separation: a review on the fouling, wetting and transport phenomena. J. Mol. Liq., 118115. https://doi.org/10.1016/j.molliq.2021.118115.

Yao, M., Chul, Y., Tijing, L.D., Choi, J., Kyong, H., 2018. Effects of volatile organic compounds on water recovery from produced water via vacuum membrane distillation. Desalination 440 (May 2017), 146–155. https://doi.org/10.1016/j.desal.2017.11.012.

Yun, Y., Ma, R., Zhang, W., Fane, A.G., Li, J., 2006. Direct contact membrane distillation mechanism for high concentration NaCl solutions. Desalination 188 (1–3), 251–262. https://doi.org/10.1016/j.desal.2005.04.123.

Zapata-Sierra, A., Cascajares, M., Alcayde, A., Manzano-Agugliaro, F., 2022. Worldwide research trends on desalination. In: Desalination. vol. 519. Elsevier B.V., https://doi.org/10.1016/j.desal.2021.115305.

Zhang, Z., Lokoare, O.R., Gusa, A.V., Vidic, R.D., 2021. Pretreatment of brackish water reverse osmosis (BWRO) concentrate to enhance water recovery in inland desalination plants by direct contact membrane distillation (DCMD). Desalination 508. https://doi.org/10.1016/j.desal.2021.115050.

Zhani, K., Zarzoum, K., Ben Bacha, H., Koschikowski, J., Pfeifle, D., (2015). Autonomous solar powered membrane distillation systems: state of the art. 3994(December). doi:https://doi.org/10.1080/19443994.2015.1117821.

Zhao, Q., Zhang, H., Hu, Z., Li, Y., 2021. An alkaline fuel cell/direct contact membrane distillation hybrid system for cogenerating electricity and freshwater. Energy 225. https://doi.org/10.1016/j.energy.2021.120303.

Membrane distillation for wastewater treatment: Recent advances in process optimization and membrane modification

Helen Julian[a,b], Pri Januar Gusnawan[a], Vita Wonoputri[a], and Tjandra Setiadi[a]

[a]Department of Chemical Engineering, Faculty of Industrial Technology, Institut Teknologi Bandung, Bandung, Indonesia [b]Department of Food Engineering, Faculty of Industrial Technology, Institut Teknologi Bandung, Sumedang, Indonesia

1. Introduction

Clean water scarcity due to industrialization and population growth has inspired the search for alternative water sources. Even though seawater desalination has been acknowledged as an applicative and successful strategy for clean water provision, its application in inland areas is limited. Wastewater treatment is one of the promising strategies to produce clean water that can meet the high requirement of potable water. In addition, wastewater treatment and reuse would also decrease the volume of wastewater to be further treated and discharged to the environment. There are diverse wastewater sources, such as textile, oil, metallurgical, food industry, sewage wastewater, etc., with various contaminants. Common conventional strategies for wastewater treatment involve physical, chemical, and electrical methods (e.g., adsorption, coagulation/flocculation, chemical, and biological oxidation) (Kalla, 2021). Limitations of the conventional processes occur due to the high cost, inefficient output, triggering of secondary pollutants, and product contaminations (Padaki et al., 2015). Therefore, it is essential to explore advanced water purification methods for wastewater treatment, such as membrane technology.

Copyright © 2023 Elsevier Inc. All rights reserved.

Pressure-based membranes have been widely studied for wastewater treatment in recent years due to their low-energy consumption, compactness, and straightforward operation and maintenance. Ultrafiltration (UF) and reverse osmosis (RO) are at the frontier of membrane technology for wastewater treatment. However, limitations occur and need to be considered. As the separation principle in UF is based on molecule sieving, high-purity product water is challenging to achieve without further purification steps. However, in RO application, high feed pressure is required, and the water recovery is limited due to the osmotic pressure of the feed. Among other membrane technologies, membrane distillation (MD) has shown great potential for wastewater treatment, and the study of MD for wastewater treatment has increased rapidly over the past decade.

Membrane distillation is a separation process driven by a vapor pressure gradient between the hot feed and permeate streams. Employing a hydrophobic membrane as the barrier, water in the feed solution travels as vapor through the membrane's pores. To generate water vapor, the feed solution does not necessarily need to be heated to the boiling point as the process can be carried out at feed temperature as low as 40°C (Alkhudhiri et al., 2012). Compared to the existing conventional pressure-driven membrane separation process, the energy requirement for MD operation is lower, particularly when low-grade heat of renewable energy is readily available for the process (Julian et al., 2021a). As the driving force for vapor transport is the vapor pressure gradient, the need for high feed pressure to overcome the osmotic pressure, as is required in RO, can be eliminated, and MD can be operated at extremely high feed concentrations. The ability of MD to concentrate the feed solution until crystallization occurs has been investigated in the application of zero liquid discharge desalination and sucrose production (Choi et al., 2019; Julian et al., 2021b). In wastewater application, this feature highlights the ability of MD to recover a significant amount of clean water as the permeate. Due to membrane hydrophobicity, 100% theoretical solute rejection can be achieved in MD operation. In the case of membrane bioreactors for wastewater treatment, this indicates MD's ability to produce high-quality water while maintaining the concentration of materials in the bioreactor.

Various types of wastewaters, such as textile, radioactive, produced water, municipal, pharmaceutical, oil and gas, and petroleum, have been treated by MD (Asif et al., 2018; Caroline Ricci et al., 2021; Chen et al., 2019; Zhang et al., 2021; Zou et al., 2020). To date, most studies on MD for wastewater treatment were conducted on a laboratory scale, with the exception of the pilot trial to treat pharmaceutical (Woldemariam et al., 2016), radioactive (Jia et al., 2021), and textile wastewater (Dow et al., 2017). Excellent solute rejections were achieved in the treatment of various wastewaters. In textile wastewater treatment, 100% color removal (Fortunato et al., 2021), 98.15% chemical oxygen demand (COD) rejection (Shirazi et al., 2020b), and 97.93% biological oxygen demand (BOD) rejection (Shirazi et al., 2020a) were reported. Other studies reported the complete removal of CFX and CTX antibiotics (Guo et al., 2018), more than 99% boron removal (Eryildiz et al., 2020), and 99% trace organic compound removal (Asif et al., 2017) in the respective wastewater treatments.

Despite MD's superior separation performance, fouling and wetting are the major drawbacks, which may lead to poor process efficiency and impede the industrial application of MD for wastewater treatment. This chapter reviews the fundamental principles of MD and factors affecting MD performance. Recent applications of MD to treat various types of wastewater are comprehensively reviewed and the major challenges for each application are highlighted. Strategies to overcome the challenges, such as membrane modification and process optimization, are discussed. Lastly, advancement of MD's integration with other processes to achieve better performance is elaborated.

2. Membrane distillation: Basic principles and module configurations

In MD operation, mass and heat transfer cooccur. In terms of mass transfer, the water vapor partial pressure difference is the driving force, while the membrane wall and the associated fluid that contact with the membrane is the resistance. The mass transfer can be written as follows (Bird et al., 2006):

$$J_w = K(p_w^h - p_w^C) \tag{1}$$

where J_w is the permeate (water vapor) flux across the membrane (kg/m^2/s), K is the overall mass transfer coefficient (s/m), p_w^h is water vapor partial pressure on the hot side (Pa), and p_w^C is water vapor partial pressure on the cold side (Pa). Raoult's law shows that in an ideal system, the partial water pressure is the multiplication of the water mole fraction in the vapor phase, y_W, with the total pressure, P, or the multiplication of the water mole fraction in the liquid phase, x_W, with the water-saturated vapor pressure, p_W^{sat} (Smith et al., 2018).

$$p_W = y_W P = x_W p_W^{sat} \tag{2}$$

The vapor pressure itself is proportional to the temperature in exponential relation as shown by Antoine equation:

$$p_W^{sat}(\text{kPa}) = \exp\left(16.262 - \frac{3799.89}{T - 46.8}\right) \tag{3}$$

The temperature corresponds to the energy carried by the wastewater, Q (kJ), the sensible energy. It is proportional to the amount of water, m (kg), heat capacity, C_p (kJ/kg/K), and the temperature differences, ΔT (K) (Smith et al., 2018).

$$Q = m \cdot C_p \cdot \Delta T \tag{4}$$

The sensible heat is transformed into the latent heat of evaporation so that the water evaporates. The vapor partial pressure difference across the membrane is generated, and the vapor transfer across the membrane.

There are four common configurations of the MD process, namely direct contact membrane distillation (DCMD), air gap membrane distillation (AGMD), sweep gas membrane distillation (SGMD), and vacuum membrane distillation (VMD). The schematic diagrams of various MD configurations are shown in Fig. 1. In all configurations, the hot feed solution is continuously circulated and directly contacts the membrane surface. The difference in each configuration is determined by the condition of the water vapor in the permeate stream. In DCMD, a cold permeate stream is circulated and is in direct contact with the opposite membrane side that contains the hot feed solution. The temperature difference between the feed and permeate stream creates vapor pressure difference, thus induces the transport of water vapor from the feed side to the permeate side. In AGMD, the membrane and a cool condensing plate are separated by a stagnant air gap. Therefore, the water vapor from the feed have to pass across the air gap before being condensed at the surface of the condensing plate. In SGMD, the permeate side is filled with cold, inert, sweep gases that collect and carry water vapor to an external condenser. In VMD, the vapor pressure difference is created by applying vacuum pressure on the permeate side. The water vapor then travels across the membrane and condenses outside the membrane module.

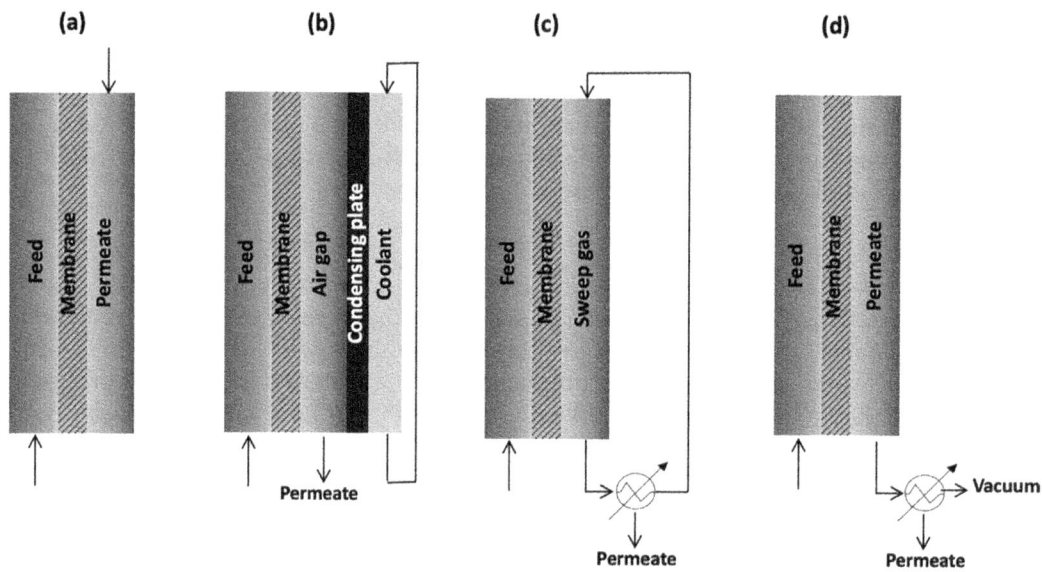

FIG. 1   Schematic set-up of (A) direct contact membrane distillation (DCMD), (B) air gap membrane distillation (AGMD), (C) sweep gas membrane distillation (SGMD), and (D) vacuum membrane distillation (VMD).

3. Factors affecting MD performance and challenges in MD for wastewater treatment

In wastewater treatment, water is the separation objective that needs to be purified from unwanted components. The MD performance is determined by the amount and quality of water vapor that can be transferred across an area of the membrane wall from the hot side (evaporation side) to the cold side (condenser side). In this section, the factors affecting MD performance based on the mass and energy transfer theory discussed in the previous section are analyzed, and the challenges of MD for wastewater treatment application are identified. The challenges for specific applications will be discussed in the following sections.

3.1 Feed-side factor and challenges

A high water-vapor partial pressure difference is needed to treat the wastewater by MD. Eq. (1) shows that the water vapor partial pressure in the hot streamside needs to be as high as possible. However, Eqs. (1), (2) show that MD's permeate flux will naturally be low relative to pressure-based membrane processes. It is attributed to the limitation of the water-vapor pressure driving force. The typical hot stream of MD temperature is 85°C. At this temperature, the water vapor pressure is only 0.58 bar. Higher partial pressure is possible, but it is also limited by the boiling point, as shown in Eq. (3).

Eq. (2) indicates that the maximum partial pressure in the hot side can be achieved if the hot stream contains only water. However, due to the wastewater containing different components, the water concentration in the wastewater is less than 1, which proportionally reduces the partial pressure. Furthermore, since water evaporation occurs in the water and membrane interface, the water concentration decreases faster than in the bulk. Consequently, the concentration of the components in the interface is higher than in the bulk, which is called concentration polarization (Luis, 2018). The concentration polarization reduces the water vapor partial pressure that leads to permeate flux reduction. The polarization may also lead to materials accumulation that causes fouling. Maintaining the hot feed stream turbulence is essential to make the stream well mixed and minimize the polarization.

The wastewater may contain oil, organics, inorganics, solid, and microorganisms (Henze, 2008). The characteristics of these components may also become problematic when operating the MD system. The oil is immiscible but has a lower density than water. The oils may stick to the membrane surface, block the membrane pore, and cause fouling. Similarly, the heavy organics may also interact with the hydrophobic membrane surface by nonpolar-nonpolar adsorption, leading to fouling initiation. Fouling is also risky in wastewater with high solid organics due to membrane pore blockage. Membrane surface modification with oleophobic nanocomposite coatings was reported to have antioil fouling properties (Wang et al., 2016a; Zuo and Wang, 2013).

The light organic contents in wastewater typically have vapor pressure higher than water or are more volatile (Smith et al., 2018). In the DCMD process, the organic vapors may pass through the membrane wall and contaminate the cold stream. Besides, the volatile organics may also condense early inside the membrane pore. It is worth noting that volatile organics is easier to be vaporized; but, at the same time, it is also easy to be condensed. The condensate may be accumulated until it reaches the other side of the membrane wall, totally plugging the membrane pore; consequently, no separation occurs anymore since the hot and cold fluids mixed each other. Despite the additional mass transfer resistance, other MD modules' configuration of AGMD, SGMD, or VMD modules is an excellent solution to overcome these problems. In these modules, vapor condensation takes place separated from the membrane surface, and therefore there is no contact between the membrane surface and the condense permeate (Alkhudhiri et al., 2012). Miscible organics with lower surface tension than water reduce the wastewater surface tension. It increases the tendency of membrane wetting and leads to poor separation performance of MD. Tailoring membrane materials and structures are needed to address the problem (Liao et al., 2013; Razmjou et al., 2012).

Some organic solvents such as 2,5-dimethylfuran (DMF), N-methyl-2-pyrrolidone (NMP), and dimethylacetamide (DMAc) potentially dissolve in or swell the polymeric membrane, leading to deformation of the membrane structure (Hansen and Charles, 2007). Oxidative organics, along with a high-temperature environment, may cause membrane failure due to polymer aging that changes the mechanical properties (Celina, 2013). To overcome the problem, identifying wastewater composition is critical so that an adequate membrane material can be wisely selected. A pretreatment process such as filtration or flotation that separates "membrane-dangerous" organics may be installed to avoid MD operational problems and prolong MD usage time.

The inorganics content in wastewaters typically has very low vapor pressure and dissolves very well due to ionization (Yaws, 1995). The inorganic ion content changes the water properties to follow colligative properties that lower the vapor pressure (McQuarrie et al., 2011). Operation at higher temperatures is needed to offset the reduction of vapor pressure. The inorganic may also be precipitated when the concentration reaches its supersaturation point due to water evaporation (Mersmann, 2001). The precipitation leads to inorganic fouling that blocks the membrane pore. The precipitate crystal may also break the membrane when the crystal structure is sharp and is dragged by the flow (He et al., 2009). To solve the problem, controlling the hot stream flow rate and recycle ratio is significant in maintaining the water concentration and in mitigating the supersaturation condition. Additional precipitator units may also be installed before the MD module to separate the precipitated crystal in front.

3.2 Temperature, thermal energy, and heat loss

According to Antoine's equation shown in Eqs. (2), (3), the component vapor pressure is proportional to temperature in an exponential relation. Therefore, the mass transfer flux is maximum when the hot stream can be maintained at the highest temperature possible. Fig. 2 shows that there is convective heat resistance (R_H) in the interface between the hot stream and the membrane. The resistance causes the temperature in the interface to be lower than in bulk. Regarding that, the vapor partial pressure is naturally reduced.

Furthermore, the concentration polarization in the interface increases the convective heat resistance, leading to a further temperature reduction in the interface. This phenomenon refers to temperature polarization (Termpiyakul et al., 2005). Temperature polarization can reduces the overall driving force for water vapor to transport across the membrane. In fact, temperature polarization may cause up to 80% driving force reduction on the MD process (El-Bourawi et al., 2006). Maintaining the turbulent flow rate in the hot side is the key to reducing temperature and concentration polarizations and ensuring that the flow is well mixed (Bird et al., 2006).

Fig. 2 also shows that the sensible energy carried by the hot streamflow is ideally only converted to the latent heat to evaporate the water. However, in reality, sensible heat loss results in thermal deficiency resistance (Zou et al., 2020). The heat may release into the

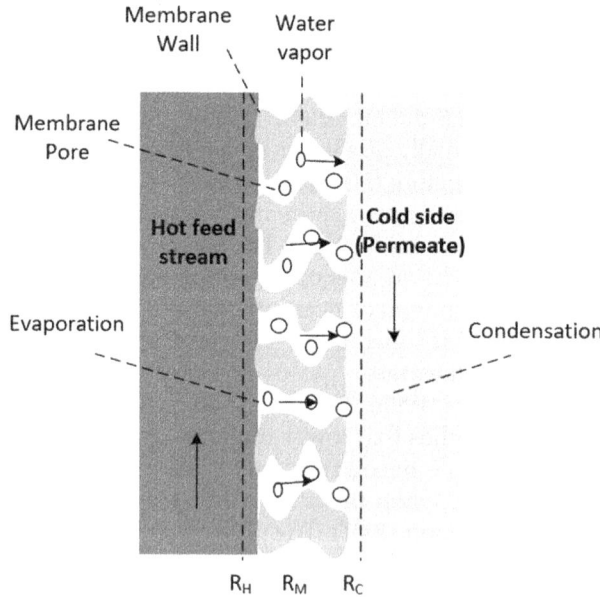

FIG. 2   Membrane distillation mass transfer schematic.

environment around the module. Proper insulation that covers the membrane module and piping system is required. Also, since the membrane wall is thin, the membrane thermal resistance is poor. The heat from the hot stream may conductively move via the membrane wall to heat the cold stream.

Consequently, the hot stream temperature decreases while the cold side temperature increases. It reduces the mass transfer driving flux along with the membrane module. Modifying the membrane structure to have a better heat insulation property is one effort to overcome the challenge. Applying Janus membrane, which has two different layers of hydrophobic-hydrophilic, maybe the alternative. The hydrophilic layer in the cold stream side will get wetted by the water from the cold stream. As water has a good heat resistance property, the structure significantly improves the membrane heat resistance (Zou et al., 2020). Despite that, the additional layer makes the membrane thicker and forms an unstable structure on the layer interface, increasing membrane resistance. Besides, the layer may easily get skinned by the flow as two different materials with opposite characteristics are cast. Changing the membrane configuration to AGMD, SGMD, or VMD may also reduce the heat loss as the condensation occurs separately from the membrane surface, and the associated air gap also adds more heat resistance.

Typically, the hot feed stream temperature for MD operation is 60–85°C. However, the wastewater effluent's typical temperature is around the ambient condition of 25–35°C, which needs to be heated if the treatment using MD is desired. The specific heat capacity of pure water is approximately 4182 kJ/kg/K (Smith et al., 2018). As the wastewater contains different components, the specific heat capacity may be lower depending on the component concentration. For instance, if the wastewater has 6% salinity, the heat capacity is reduced by 8% to become 3850 kJ/kg/K, which requires less heating energy (Lienhard et al., 2012). Despite that, the energy requirement is still categorized as energy extensive because the heat capacity is high compared to other fluids (Smith et al., 2018). Fortunately, since the temperature requirement is relatively low, low-grade heat may be utilized as an energy source such as industrial waste heat, solar heat, shallow geothermal, or municipal waste combustion. The hot feed stream recycling is also a method to conserve energy during the MD operation while improving permeate recovery.

3.3 Cold stream side factor and challenges

According to Eq. (1), a high partial pressure difference can be achieved if the partial pressure in the cold side is low. If the fluid contains only water, $x_i = 1$, like presumably in the cold stream of DCMD, Eq. (2) may be simplified as

$$p_w^c = p_W^{sat} \tag{5}$$

It shows that the partial pressure equals the water vapor pressure in the cold stream.

Eqs. (3) and (5) show that to have high permeate flux in DCMD, the cold stream flow temperature needs to be maintained as low as possible. Controlling the cold stream at a high flow rate may be conducted to maintain the temperature. However, it is limited in that a high flow rate may lead to membrane wetting which elevates the mass transfer resistance.

In AGMD, SGMD, or VMD modules, the approach to evaluating the water vapor partial pressure is different because there is no fluid on the cold side. For these cases, the partial pressure can be written as

$$p_w^c = y_W P \tag{6}$$

The low partial pressure on the cold side is achieved by lowering the total pressure, P, or the vapor concentration on the cold side, y_W. The total pressure is reduced to vacuum in the VMD module by installing a vacuum pump to suck up the permeate. The mechanical strength of the membrane is a limitation when performing the VMD module. The pressure is also limited by the liquid entry pressure of the membrane; otherwise, the hot stream may enter the pore and increase the mass transfer resistance. In AGMD and SGMD, the low water vapor partial pressure is maintained by decreasing the vapor concentration. In AGMD, it is performed by condensing the vapor in a surface separated from the membrane, while in SGMD, it is maintained by sweeping the vapor using noncondensable gases such as air, nitrogen, or helium.

3.4 Membrane factor and challenges

Ideally, the membrane condition during MD operation is completely gas-filled, where the pore is dry, and the surface is clean and unaffected by the fluid condition, as shown in Fig. 2. At this condition, only three mass transfer resistances involve the hot stream resistance, R_H, the permeate resistance, R_C, and the membrane resistance, R_M. It is worth noting that the R_C is MD-module dependent.

Despite there being different proposed models (Hitsov et al., 2015), the mass transfer in the membrane R_m is dependent on the porosity, thickness, and tortuosity, which can be defined as

$$R_m = \frac{\tau \cdot \delta}{D_m \cdot \varepsilon} \tag{7}$$

where D_m is the effective water vapor diffusion coefficient in the gas-filled membrane pores (m^2/s), δ is the membrane thickness (m), ε is the membrane porosity, and τ is the membrane tortuosity. The D_m can be determined by the summation of bulk diffusion and Knudsen diffusion coefficients.

$$\frac{1}{D_m} = \frac{1}{D_g} + \frac{3}{d_p} \sqrt{\frac{\pi M_w}{8RT}} \tag{8}$$

where D_g is the diffusion coefficient of water in the air (m^2/s), M_w is water molecular weight (kg/kmol), d_p is membrane pore diameter (m), R is the universal gas constant (J/kmol/K), and T is the temperature (K) (Feron and Jansen, 2002). Eqs. (7) and (8) show that the minimum membrane resistance is reached when the membrane has low tortuosity, low thickness, high porosity, large pore size, and is operated at high temperature. Low tortuosity and low thickness represent a short distance and low curvature of the pore length, while high porosity and large pore size relate to the evaporation area (Alkhudhiri et al., 2012). Those properties are determined during membrane fabrication from materials' selection, casting method, additive selection, layer configuration, and fabrication condition (Gusnawan, 2020). However, the properties are in contrast with the membrane mechanical strength, which determines durability and usage time of operation. The membrane with low tortuosity, low thickness, high porosity, and high pore size tends to have low mechanical strength. A membrane design and fabrication optimization are critical to producing a high-performance membrane with sufficient mechanical strength. Currently, a porosity of 60%–80%, a mean pore size of 0.1–1 µm, and thickness of 60–180 µm are often used in MD (Alkhudhiri et al., 2012).

Another limitation to the membrane design is the liquid entry pressure (LEP). The LEP is the minimum pressure required that allows the liquid to enter the membrane pore. It describes the membrane resistance to wetting where high LEP means more resistance to wetting. Wetting is undesirable since it blocks the pore, increases the mass transfer resistance, and leads to poor separation performance. LEP is typically defined by the Laplace-Young equation as follows:

$$LEP = -\frac{2B\gamma \cos\theta}{r_p} \tag{9}$$

where γ is surface tension (N/m), θ is the contact angle between the liquid and membrane surface (degrees), r_p is the membrane pore radius (m), and B is the irregularities factor (If the pore is homogenous and cylindrical, $B=1$) (Franken et al., 1987). From Eq. (9), the membrane with a large pore size tends to have a low LEP and get wetted, which contradicts the high flux membrane property. Also, the membrane with high surface tension that produces a high liquid contact angle will be less likely to get wetted. Typically, a surface with a water contact angle of more than 90 degrees is considered hydrophobic.

As was explained in the previous section, the organic content in the wastewater tends to produce a lower contact angle on the membrane surface due to their low surface tension properties. The effect is amplified in MD applications due to relatively high-temperature operation. Surface tensions decrease at higher temperatures (Palmer, 1976). For wastewater treatment using MD, a membrane that provides high surface tension is required to minimize the contact and mitigate the interaction between liquid and the surface. In conjunction with tailoring the structure, selecting membrane materials with low surface energy (high surface tension) is essential to designing the membrane. Commonly used membrane materials are polypropylene (PP), polyvinylidene fluoride (PVDF), and polytetrafluoroethylene (PTFE) which has a low surface energy of 30, 30.3, and 19.1 mJ/m², respectively, and water contact angle of 100–118 degrees, 92–100 degrees, 113–133.5 degrees, respectively (Gusnawan, 2020). The contact angle properties may also be improved by coating and developing a mixed matrix structure. Membrane coating or mixing with lower surface energy materials such as silicon-based and hydrophobic nano-silica was also conducted to produce a superhydrophobic membrane with a water contact angle of more than 140 degrees (Liao et al., 2013; Razmjou et al., 2012). However, the limitation is mechanical properties and stability. Mixing the membrane matrix with inorganics, such as silica, produced polymer solidification in the particle interface, resulting in a more brittle membrane (Shimekit et al., 2011). The stable coating is still a challenge regarding the shear force from the liquid flow rate. Omniphobic membranes, made of low-surface energy material with specific re-entrant structure, have shown good performance in repelling high- and low-surface-tension liquids, as are found in wastewater. Another membrane modification involves integrating materials of opposing wettability, referred to as a Janus membrane. The studies on the application of Janus membranes for wastewater treatment have also shown promising results (Chew et al., 2019; Zhu et al., 2018).

4. Application of MD for wastewater treatment

MD has been applied to treat various wastewaters (Table 1). It is worth noting that the quality of the product water and the MD performance differs from one process to the other, depending on the source and composition of the treated wastewater.

4.1 Textile wastewater

The textile industry requires a significant amount of water for the production process, particularly the dyeing process (Marcuccl et al., 2001). Only a tiny amount of dye is fixed in the cloth; then, the remaining dye is disposed of as wastewater needs to be treated before being discharged to the environment. The textile wastewater is a complex mixture of dyes, surfactants, leveling agents, softeners, and salts, making the treatments complicated (Yang et al., 2021). Conventional treatments for textile wastewater consist of adsorption (Herrera-González et al., 2019), chemical oxidation (Gebregiorgis et al., 2018), the combination of coagulation/flocculation (Dotto et al., 2019), and activated sludge biological treatment (Amar et al., 2009). Conventional treatments can remove most biological oxygen demand (BOD), chemical oxygen demand (COD), suspended solids, and organic molecules in the wastewater. However, the processes efficiency is low, and some processes produce additional waste, which poses another challenge. In addition, biological treatment may also exacerbate the toxicity of the wastewater due to the various pollutants, pH, and organic concentrations in the textile wastewater (Paździor et al., 2019).

The treatment of textile wastewater by MD has gained much attention, particularly as the textile wastewater is discharged in the temperature range of 50–80°C (Mokhtar et al., 2014). The high wastewater temperature is beneficial for MD operation as no additional feed heating is required. However, it is worth noting that most studies in MD for textile wastewater treatment were conducted using a synthetic dye solution (Table 1), which cannot represent the actual challenges in the real textile wastewater treatment. Fouling and wetting have been reported as the major challenges in textile wastewater treatment using MD. Fouling may occur due to the deposition of various compounds on the membrane surface and lead to pore blocking and reduced flux. Surfactants in textile wastewater can induce wetting by forming hydrophobic-hydrophobic interactions with the membrane and hydrophilic-hydrophilic interactions with water. In addition, the presence of organic compounds also leads to wetting due to the reduction of wastewater's surface tension.

The severity of fouling and wetting depends on the composition of the textile wastewater being treated. More than 98.7% color rejection was achieved by treating polyester dyeing wastewater using the DCMD setup. Even though fouling occurred, as indicated by the flux reduction, wetting was not detected during the operation. However, the high concentration of cationic and nonionic surfactants in the cotton and viscose dyeing wastewater led to severe wetting in operation using a similar DCMD setup (Ramlow et al., 2020). The dye charge has also been reported to impact MD performance, with the cationic dyes increasing the fouling

TABLE 1 Selected wastewater treatment by MD.

MD configuration	MD material	Wastewater source	Major/ challenging constituents	MD performance	Ref.
DCMD	PVDF	Dye solution	Dye	Excellent dye rejection of 99.87%	Mokhtar et al. (2014)
DCMD	PDMS-PVDF	Dye solution	Dye	Complete color rejection, 50% increase of permeate flux compared to the pristine PVDF, excellent antifouling properties	An et al. (2017)
DCMD	PTFE	Dye solution	Dye	High flux of 30–34 LMH and excellent dye rejection in 5 days tests	An et al. (2016)
VMD	PP	Dye solution	Dye	High flux of 40 LMH can be achieved with 100% rejection. Membrane swelling occurred due to the interaction of dye and membrane material	Criscuoli et al. (2008)
DCMD	PTFE, flat sheet, pore size: 0.2 mm	Olive mill wastewater	Residual oil, TOC, COD	84% flux reduction in 30 h. Negligible polyphenol detected in permeate	El-Abbassi et al. (2013)
DCMD	PVDF/PEG/TiO$_2$	Synthetic seawater + mineral oil	Oil and salts	No fouling and wetting were detected for 24 h of operation. Permeate conductivity was less than 10 mS/cm	Zuo and Wang (2013)
DCMD	PDA/SiNPs-PVDF	Saline crude oil emulsion 1000 ppm wt%	Crude oil and salts	Negligible flux reduction for 12 h of operation. the salt rejection rate of over 99.9%	Wang et al. (2016b)
DCMD	Fluorinated silica (F-SiO$_2$)@PS poly(vinylidene fluoride-co-hexafluoropropylene) (PVDF-HFP)/ PS and SiO$_2$@ polyacrylonitrile (PAN) (Janus membrane)	1000 ppm lubricating oil in 3.5wt% NaCl solution	Lubricating oil and salts	Stable water flux of 25.42 LMH and a high salt rejection ratio of 100%. No oil droplets visible in the permeate after 30 h of operation	Zhu et al. (2018)

Continued

TABLE 1 Selected wastewater treatment by MD—cont'd

MD configuration	MD material	Wastewater source	Major/challenging constituents	MD performance	Ref.
DCMD	Fluorocarbon surfactant (FS)/fluorinated alkyl silane (FAS)/Si-amino functionalized PVDF membrane	SDS in 3.5 wt% NaCl solution, 1.0% v/v mineral oil and SDS in 3.5 wt% NaCl solution	Surfactant, oil, and salts	A stable flux of 27 LMH for 48 h of operation and 99.99% SDS rejection. No traceable pollutants in the permeate stream	Li et al. (2019)
DCMD	SiNPs/perfluorodecyltrichlorosilane-PVDF	Shale gas produced water	Oil, grease, surfactants, organic compounds, salts	Modified PVDF membrane exhibit lower permeate flux, delayed flux decline, and reduced salt passage	Boo et al. (2016)
DCMD	PVDF	Simulated radioactive wastewater	Cs^+, Sr^{2+}, Co^{2+}	Excellent rejection of Cs^+, Sr^{2+}, Co^{2+}, high-quality permeate to be reused or discharged directly	Liu and Wang (2013)
DCMD	SMMs/PES	Simulated radioactive wastewater	^{60}Co, ^{137}Cs, ^{85}Sr	Modified membrane showed a low radioactivity decontamination factor of more than 888	Khayet (2013)
VMD	PP	Simulated radioactive wastewater	Cs^+, Sr^{2+}, Co^{2+}	Decontamination factor of 7600, 8900, and 7800 for Cs^+, Co^{2+}, and Sr^{2+}, respectively	Wen et al. (2016)
DCMD, pilot	PTFE	Simulated radioactive wastewater	^{60}Co, ^{137}Cs, ^{65}Zn, ^{110}Ag, ^{133}Ba, ^{134}Cs, ^{170}Tm, ^{192}Ir	Decontamination factor of 4336.5 for ^{60}Co, 43.8 for ^{137}Cs, and infinity for the rest of the compounds	Zakrzewska-Trznadel et al. (1999)
VMD, pilot	PTFE	Simulated radioactive wastewater	^{133}Cs	Concentration factor of 110.3, high-quality water production rate of 8 m^3/day, decontamination factor of $10^{4.85}$	Jia et al. (2021)
AGMD, pilot	PTFE with PP support	Municipal wastewater	Various pharmaceutical residues	High removal of the pharmaceutical residues	Woldemariam et al. (2016)

and wetting potential (Laqbaqbi et al., 2019). A pilot trial has been conducted to achieve zero liquid discharge (ZLD), at which the textile wastewater is treated to produce fresh water and concentrated dye solution simultaneously. The concentrated dye can be reused during the dyeing process to reduce the operational production cost. Rapid wetting was observed due to surfactants and biomolecules in the wastewater. Pretreatment by foam fractionation was required to mitigate the wetting in the long-term trial (Dow et al., 2017). Wetting was also detected in another study using a PTFE membrane, and fractionation or ozonation should be conducted to pretreat the textile wastewater (Zhang et al., 2021).

Using a noninvasive fouling characterization, namely optical coherence tomography (OCT), the evolution and spatial distribution of fouling in textile wastewater treatment was investigated. The results showed that the fouling deposition is highly dependent on the feed temperature and velocity. Of all the conditions tested, fouling formation was more severe at high operating temperature (80°C) and low feed flow rate (15 L/h). Moreover, fouling thickness was found to increase along with the membrane cell, where thicker fouling was observed at the outlet of the cell (Fig. 3A) (Fortunato et al., 2021). Fouling and wetting are disadvantageous for wastewater treatment as they lead to poor separation performance, increased cleaning frequency, operational downtime, and energy demand (Naidu et al., 2017;

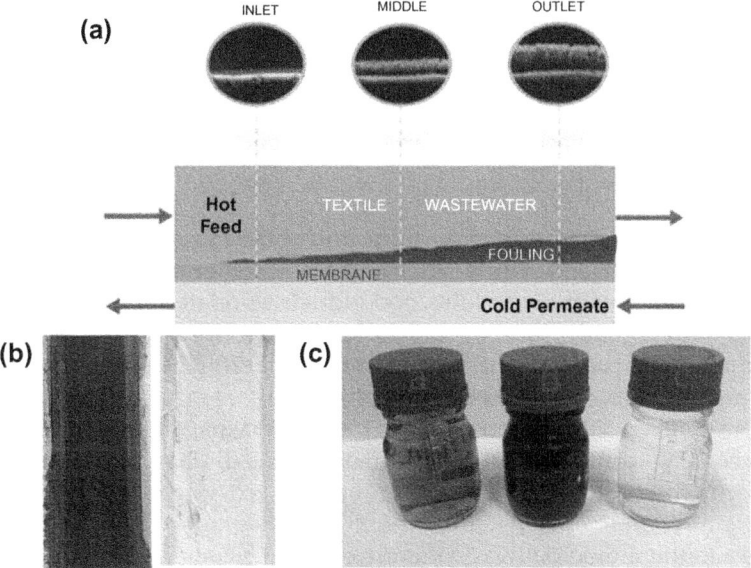

FIG. 3   (A) Spatial distribution of fouling layer in synthetic dye solution treatment. (B)The surface of PDMS/PVDF (purple) and pristine PVDF (white) after 24 h treatment of single dye solution with water flushing. (C) Distillate of PVDF 0.2 mm, PVDF 0.45 mm, and PTFE membranes after 5 days of dye solution treatment. *Reproduced from (A) Fortunato, L., Elcik, H., Blankert, B., Ghaffour, N., Vrouwenvelder, J., 2021. Textile dye wastewater treatment by direct contact membrane distillation: membrane performance and detailed fouling analysis. J. Membr. Sci. 636, 119552. https://doi. org/10.1016/j.memsci.2021.119552, with permission from Elsevier; (B) An, A.K., Guo, J., Lee, E.J., Jeong, S., Zhao, Y., Wang, Z., Leiknes, T.O., 2017. PDMS/PVDF hybrid electrospun membrane with superhydrophobic property and drop impact dynamics for dyeing wastewater treatment using membrane distillation. J. Membr. Sci. 525, 57–67. https://doi.org/10.1016/j. memsci.2016.10.028, with permission from Elsevier; (C) An, A.K., Guo, J., Jeong, S., Lee, E.J., Tabatabai, S.A.A., Leiknes, T.O., 2016. High flux and antifouling properties of negatively charged membrane for dyeing wastewater treatment by membrane distillation. Water Res. 103, 362–371. https://doi.org/10.1016/j.watres.2016.07.060, with permission from Elsevier.*

Warsinger et al., 2015). An effort to mitigate the fouling and wetting in textile wastewater treatment has been focused on modifying membrane material and structure. A thin, hydrophilic layer of agarose hydrogel was attached to the hydrophobic Teflon membrane to mitigate wetting in the treatment of synthetic textile wastewater. The synthetic wastewater consisted of NaCl and surfactants (sodium dodecyl sulfonate (SDS), Tween 20, and Tween 85). In 24h of the test, no wetting occurred when the concentration of SDS or Tween 20 was 10 mg/L. It was due to the entrapment of the surfactants by the hydrogel. However, wetting was detected in a higher surfactant concentration test (Lin et al., 2015).

Superhydrophobic polydimethylsiloxane (PDMS)/PVDF membrane with antifouling properties treated synthetic dye solutions. The membrane's unique surface structure and negative charge resulted in a strong repulsive force between dye and membrane, thus increasing the chance of dye-dye binding. It led to a loose fouling layer that can be easily cleaned by water flushing (Fig. 3B). In addition, the PDMS/PVDF membrane exhibit 50% higher permeates flux than the pristine PVDF membrane, which enhances the process productivity (An et al., 2017). Similar results were reported in the study using PVDF and PTFE membrane to treat synthetic dye solution. Negatively charged PVDF and PTFE membrane repulsed negatively charged dye, which then formed aggregated flake away from the membrane surface and led to superior dye rejection (Fig. 3C). In long-term tests, higher flux and better fouling control were observed when the PTFE membrane was employed due to its higher porosity and hydrophobicity (An et al., 2016).

4.2 Oily wastewater

Oily wastewater can be discharged from shale oil and gas, petroleum, metallurgy field, restaurants, domestic effluents, meat processing, vegetable oil refinery, etc. (Kalla, 2021). Produced water consists mainly of oil, which constitutes of petroleum hydrocarbon (PHCs). Dissolved solids (calcium, magnesium, sodium, and phosphorus), toxic heavy metals (nickel, chromium, cadmium, lead, and boron), and radioactive materials are also present in the produced water. The main constituents of the food industries-related wastewater are BOD, COD, surfactants, and oil and grease. Moreover, the pH of the food-industries wastewater is also an essential parameter being used for further treatment. The effluent of vegetable oil industries is mostly acidic and consists of oil and colloidal particles.

However, the main constituents of meat-processing wastewater are suspended solids, oil, and grease. According to many countries' regulations, the discharge of oily wastewater to the environment is strictly prohibited, affecting the aquatic ecosystem. In addition, as oily wastewater also contains components that are difficult to be degraded biologically, the disposal of oily wastewater to the municipal wastewater treatment plants is also not permitted.

The treatment of olive mill wastewater was conducted using a commercial PTFE membrane in DCMD configuration to obtain clean water and phenolic-rich concentrate. Pretreatment of the wastewater by microfiltration (MF) before entering the MD resulted in delayed fouling and prolonged operation time. Hydroxytyrosol, the main phenolic compound in the olive mill wastewater, was concentrated more than 2 times after 40h of MD treatment. However, reduction of permeate flux was observed as the phenolic concentration in the wastewater increased. Rapid flux reduction occurred within the first 10h of operation. It reached the low value of 1.44 LMH after 30h of operation due to the higher feed concentration and fouling (El-Abbassi et al., 2013). The oily wastewater treatment by MD has been hindered due to the severe fouling and wetting that occurred in MD operation using a commercially available hydrophobic membrane. Thus, membrane materials with specifically tailored properties are required to treat oily wastewater.

Development of antioil fouling MD was conducted by modifying the surface wettability of a PVDF membrane via plasma-induced grafting of polyethylene glycol (PEG) and titanium dioxide (TiO_2) particles on the membrane surface to form a hydrophilic surface. In the synthetic seawater test containing 0.01 wt% of mineral oil, the modified PVDF membrane exhibited antioil fouling and wetting due to membrane hydrophilicity and smaller pore sizes (Zuo and Wang, 2013). Oil-fouling mitigation can also be conducted by modifying the membrane's surface charge, creating a composite membrane with in-air hydrophilic and underwater oleophobic surfaces (Fig. 4A). Composite PVDF membranes were modified by coating a layer of negatively charged polydopamine/silica nanoparticles (PDA/SiNPs) or positively charged poly(diallyl dimethyl-ammonium chloride)/SiNPs (PDDA/SiNPs). Both modified membranes showed better antioil fouling than the pristine PVDF membrane in a

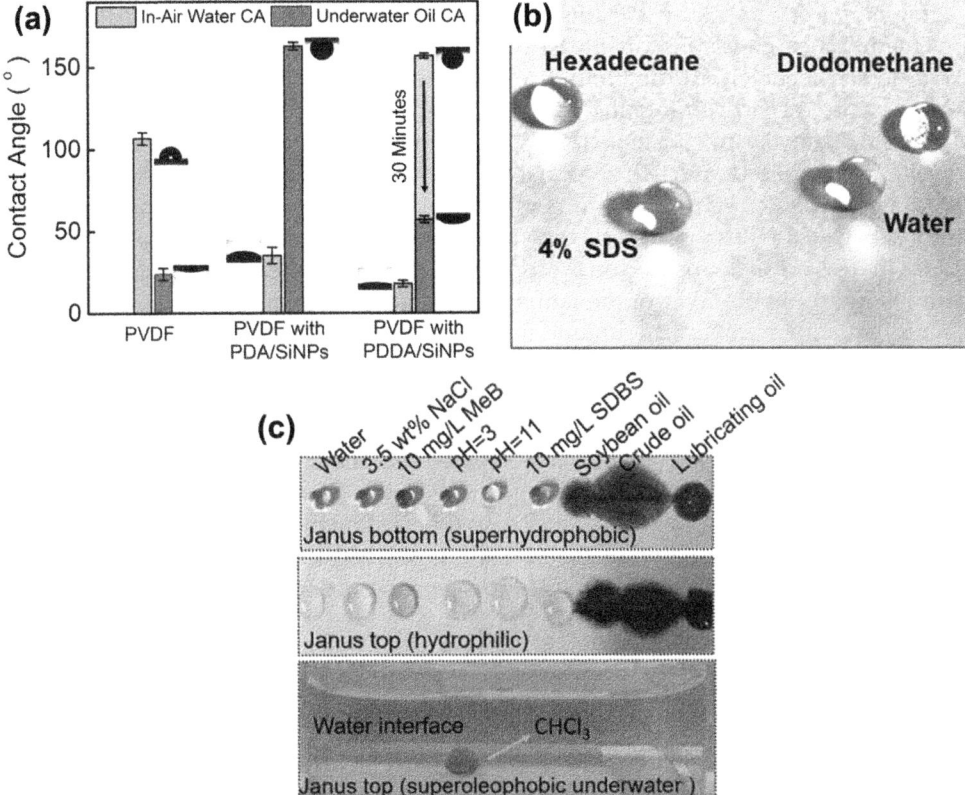

FIG. 4 (A) In-air water contact angle and underwater oil contact angle of pristine and nanocomposite PEG-TiO2-PVDF membranes. (B) Visualization of various droplets on the fluorinated SiNPs-PS-PVDF membrane. (C) Droplets on the top and bottom surface of the Janus membrane. *Reproduced from (A) Wang, Z., Jin, J., Hou, D., Lin, S., 2016b. Tailoring surface charge and wetting property for robust oil-fouling mitigation in membrane distillation. J. Membr. Sci. 516, 113–122. https://doi.org/10.1016/j.memsci.2016.06.011, with permission from Elsevier; (B,C) Zhu, Z., Liu, Z., Zhong, L., Song, C., Shi, W., Cui, F., Wang, W., 2018. Breathable and asymmetrically superwettable Janus membrane with robust oil-fouling resistance for durable membrane distillation. J. Membr. Sci. 563, 602–609. https://doi.org/10.1016/j.memsci.2018.06.028.*

test using saline crude oil emulsion; however, the PDA/SiNP-PVDF membrane was superior with no fouling occurrence after 12 h of the test (Wang et al., 2016b).

Another promising strategy to mitigate fouling and wetting in wastewater treatment is utilizing a hierarchically structured omniphobic membrane. An omniphobic PVDF membrane was prepared by spray coating SiNPs-polystyrene (PS) onto the membrane surface, followed by the adhesion of the particle on the membrane support and fluorination to tailor the membrane omniphobicity. The membrane exhibits a high contact angle for high and low surface energy components, such as water, 4% SDS, hexadecane, and diiodomethane (Fig. 4B). In contrast to the pristine PVDF membrane, the omniphobic PVDF can maintain a stable flux and low permeate conductivity for 1000 min of the DCMD test using an emulsion of SDS, hexadecane, and NaCl in water at a concentration of 240, 2400, and 10,000 mg/L, respectively (Zheng et al., 2018). Omniphobic membranes have also been fabricated by coating amino-functionalized PVDF with a solution consisting of fluorocarbon surfactant (FD), fluorinated alkyl silane (FAS), and SiNPs. Compared to pristine PVDF, the modified membrane showed superior antifouling and antiwetting properties against surfactants, oil, and hydrophilic organic pollutants (Li et al., 2019).

Janus membranes, consisting of at least two layers with asymmetric wettability properties, have also been extensively studied to address fouling and wetting issues. An ultrathin Janus skin layer was recently integrated with a hydrophobic nanofibrous membrane via electrospinning and electrospraying (Fig. 4C). Excellent underwater superoleophobicity of 164 degrees was reported and led to robust oil resistance during the test using 1000 ppm oil in saline water emulsion. While the commercial PVDF membrane flux instantaneously reduced once the test began, the flux on the test using a Janus membrane was stable for 30 h of operation. It indicates there is no oil deposition on the membrane surface that can accumulate and block the pores on the hydrophobic layer of the Janus membrane (Zhu et al., 2018). In another research fabricating Janus membranes, the unique feature of carbon nanotubes was explored. Layers of hydrophilic multiwalled carbon nanotubes (M-CNT) functionalized with carboxylic groups and polyvinyl alcohol (PVA) were sprayed on top of a hydrophobic PVDF membrane. Cross-linking was then performed to obtain a hydrophilic top layer with enhanced flux and antifouling properties. The modified membrane showed excellent underwater oleophobicity, a smoother surface, and smaller membrane pores than the pristine PVDF. However, the flux of the PVDF-M-CNT membrane was higher than the pristine membrane due to the high water permeability and heat conduction of the M-CNT. In the test using a hexadecane emulsion in NaCl solution, superior antifouling and antiwetting properties were observed (Han et al., 2020).

4.3 Radioactive wastewater

The treatment of radioactive wastewaters is crucial in avoiding life-threatening radioactive contamination. Treatment for processing radioactive wastewater before disposal to the environment includes chemical precipitation, sedimentation, adsorption, ion exchange, and thermal evaporation (Zhang et al., 2019). Those treatments, however, need to be conducted in sequence to remove the contaminants. In addition, high operational costs and the generation of secondary waste requiring further treatment limit the application of the treatment mentioned above (Khayet, 2013).

MD has been proposed to treat low- and intermediate-level radioactive wastewater; however, the study of MD for this particular application is still lacking. Preliminary economic analysis indicated that MD could be economically attractive for treating radioactive wastewater in small-capacity plants utilizing waste heat from the nuclear reactor's cooling system (Zakrzewska-

Trznadel et al., 1999). Low-level radioactive wastewater has significant volume and low activity; thus, concentration is required for efficient further treatment and disposal. An MD pilot plant has been constructed for this application using a spiral-wound PTFE membrane in DCMD configuration. A model solution containing cobalt-60 (^{60}Co) and cesium-137 (^{137}Cs) with a total specific activity of 5000 Bq/dm^3) was used. All radioisotopes can be removed from the permeate stream, enabling direct disposal of the permeate stream to the environment or utilization as process water (Zakrzewska-Trznadel et al., 1999). Another study with promising results was also conducted using a PP membrane in DCMD configuration to treat actual wastewater consisting of chromium-51 (^{51}Cr), ^{60}Co, ^{137}Cs, and other β γ emitters (Chmielewski et al., 1997). Effective nuclides' separation was reported with high decontamination factors of 6000, 3700, and 8300 for Cs$^+$, Sr^{2+}, and Co^{2+}, respectively (Wen et al., 2016). The removal of Sr^{2+} and Co^2 individually has also been performed using the PP membrane in separate VMD tests, with removals of over 99.6% were achieved for each experiment (Jia et al., 2017, 2018). The pilot plant for low radioactive wastewater treatment has been tested and is shown in Fig. 5.

Even though MD has been proven to treat radioactive wastewater with high separation and decontamination factors, the selection of membrane material and module is essential. For industrial application of MD for radioactive wastewater treatment, the membrane has to show resistance to ionizing radiation; it is applicable in treating various radioactive chemicals in wastewater. Hydrophobic/hydrophilic composite membrane was fabricated by integrating fluorinated surface-modifying macromolecules (SMMs) into polysulfone (PS) or polyethersulfone (PES) as the polymer matrix. The DCMD test was conducted with a radioactive model solution containing ^{60}Co, ^{137}Cs, and ^{85}Sr. The modified membrane exhibited lower permeate flux, lower radionuclides' adsorption, and higher rejection than the TF200 commercial membrane. Complete radioactive chemical rejection was performed

FIG. 5   Pilot-plant for low-level radioactive wastewater treatment. *Reproduced from Jia, F., Li, J., Wang, J., Sun, Y., 2017. Removal of strontium ions from simulated radioactive wastewater by vacuum membrane distillation. Ann. Nucl. Energy 103, 363–368. https://doi.org/10.1016/j.anucene.2017.02.003, with permission from Elsevier.*

in a one-stage operation, and the permeate activity is similar to the natural background, indicating no wetting during the test (Khayet, 2013). Recently, Korolkov et al. increased the contact angle of track-etched poly(ethylene terephthalate (PET)) by UV-induced grafting of styrene. The modified membrane was tested in separating various salts (Cs, Mo, Sr, Sb, Al, Ca, Fe, Mg, K, and Na) and decontaminating radioisotopes (^{60}Co, ^{137}Cs, and ^{241}Am). Rejection in the range of 90%–100% was achieved for most salts, and a decontamination factor of 1727 for ^{137}Cs was reported (Korolkov et al., 2019).

4.4 Other wastewaters

Pharmaceutical industries discharge specific wastewaters that can be toxic and harmful to the environment. The wastewater may contain antibiotics, analgesics, steroids, antidepressants, antipyretics, stimulants, antimicrobials, hormones, antiinflammatory drugs, β-blockers, lipid modulators, contrast agents, and impotence drugs (Bu et al., 2013). These pharmaceutical residues can also be found in domestic wastewater. Removal of pharmaceutical residues has been conducted by various methods, but the results were not satisfactory. The application of MD for pharmaceutical residue removal is of interest due to the complete rejection of nonvolatile solutes in MD. The water produced by MD-treated sewage treatment plants containing pharmaceutical and dissolved organic matters has to meet the drinking water standards in Korea (Jeong et al., 2021). A carbon nanotube-immobilized membrane (CNIM) with PVDF as the polymer matrix treated synthetic pharmaceutical wastewater containing ibuprofen, diphenhydramine, acetaminophen, and dibucaine. The CNIM showed a higher flux and superior enrichment factor than the plain PVDF. Enrichment factor, which represents the ratio of compound concentration in the membrane module's outlet to that in the inlet, for all compounds of interest, was more than 10 (Gethard et al., 2012).

An MD pilot plant treated municipal wastewater containing pharmaceutical residues (Fig. 6). High removal efficiency was obtained for 19 detected pharmaceuticals in the wastewater. However, the energy consumption for feed heating in MD was high, making MD not competitive with pressure-driven membranes. Integration of MD with district heating may reduce the energy requirement, though this should be supported by module improvement to minimize the parasitic heat loss during the operation (Woldemariam et al., 2016). Fouling also occurred during the treatment of pharmaceutical wastewater and impacted the effectiveness of compound removal. The severity of fouling in pharmaceutical wastewater treatment depends on the interaction between the membrane surface and the compounds. Stable flux and complete rejection of antibiotics were reported when the membrane material and compounds to be separated share similar charges (Guo et al., 2018). Engineering of membrane surface charge and energy by coating PVDF membrane with perfluorooctyl-triethoxysilane (FTES)—TiO$_2$ nanoparticles resulted in stable performance even though the membrane and antibiotics have opposite charges. As expected, no visible foulant layer was formed when the membrane treated cefotaxime. On the contrary, a thick foulant layer (around 326–408 μm) and blockages of pore happened during the treatment of positively charged tobramycin. Consequently, only 78% rejection was obtained during the process (Guo et al., 2020).

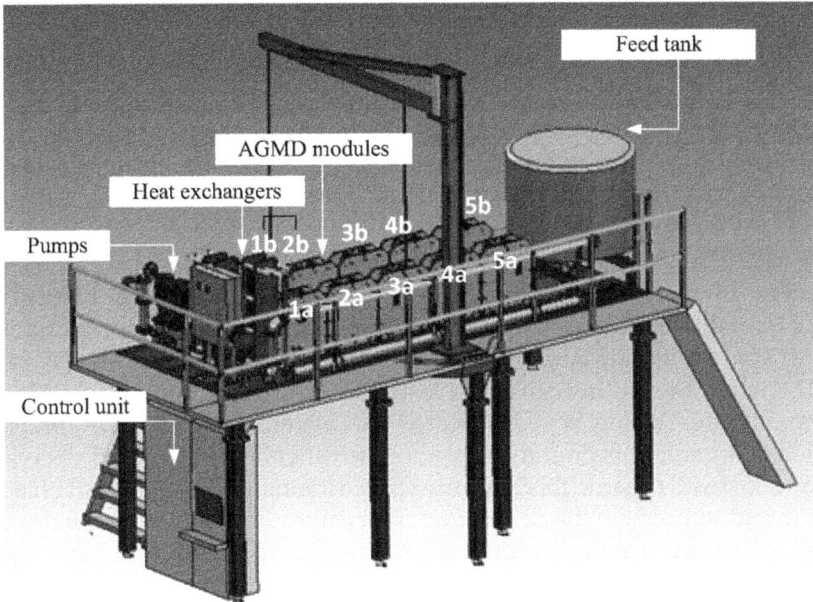

FIG. 6 Schematic setup of AGMD pilot-plant in Hammarby Sjöstadsverk for municipal wastewater treatment. *Reproduced from Woldemariam, D., Kullab, A., Fortkamp, U., Magner, J., Royen, H., Martin, A., 2016. Membrane distillation pilot plant trials with pharmaceutical residues and energy demand analysis. Chem. Eng. J. 306, 471–483. https://doi.org/10.1016/j.cej.2016.07.082, with permission from Elsevier.*

5. MD configuration for wastewater treatment

5.1 Stand-alone MD

Stand-alone MD is widely applied for wastewater treatment. The specific application of stand-alone MD has been discussed in Section 4. Direct contact membrane distillation (DCMD) is the most used setup among all MD configurations due to ease of design and operation. Thus, most literature focused on the application of DCMD on different wastewater. DCMD, based on PTFE and PVDF, has been used to treat five different samples of dyeing wastewater. The difference in sample compositions and concentration resulted in slightly different COD and color removal efficiency, but higher than 89% removal are achieved. PTFE performs slightly better than PVDF, attributed to enhanced hydrophobicity and reduced membrane wettability (Li et al., 2018). Petrochemical wastewater that originated from an electrodialysis-reversal pilot plant can be treated using PTFE (functional)-PP (support) DCMD, where more than 95%, 80%, and 81% removal was reported for TDS, TOC, and turbidity, respectively (Santos et al., 2020). Wastewater from fermentation processes that contain high organic concentrations (10,000–90,000 mg/L COD) also can be treated using DCMD. After 12 days of the process, an overall organic rejection of over 95% was achieved. However, decreasing permeate flux (from 8.7 to 4.3 LMH) and increased concentration of COD and TOC in permeate water are observed due to membrane fouling and wetting (Wu et al., 2018).

Other types of MD configurations also have been used. A direct comparison of DCMD and VMD processes in olive mill wastewater treatment showed that both configurations reported high rejections (>99.6%) at all the feed temperatures tested (30°C, 40°C, and 50°C). However, the VMD configuration has a significantly higher permeate flux (19 kg/m^2/h) compared to DCMD (6.5 kg/m^2/h) (Carnevale et al., 2016). The performance of AGMD in treating simulated textile wastewater was compared with DCMD, and the result showed that higher flux and better performance (in terms of permeate conductivity, color removal, and carbon content rejection) was observed in AGMD. Overall, color removal in all AGMD tests can reach 100% even after 70 h of operation. Conversely, only 99% color removal after 8 h of operation was reported for DCMD (Leaper et al., 2019).

Although MD is more fouling resistant than other membrane-based processes such as reverse osmosis, it has remained one of the main challenges in the industrial application of MD. It is especially applicable in configurations that resulted in more conduction losses such as DCMD or configurations with higher permeate flux such as VMD (Warsinger et al., 2015). In MD, membrane fouling can occur as organic fouling or inorganic scaling (Yan et al., 2018). MD membrane scaling was dominated by calcium-, magnesium-, phosphate-, and silicon-related inorganic compounds; however, acidification of the feed decreased scaling formation, and instead, organic fouling formation dominated. At higher pH feed, inorganic scaling was found to precede the deposition of organics on the membrane, forming severe organic fouling. The occurrence of fouling increases the possibility of membrane wetting, which can reduce the permeate quality (Yan et al., 2019). Membrane wetting is mainly influenced by scaling than by other membrane characteristics such as pore size (Yan et al., 2018).

5.2 High-retention MD: Membrane distillation bioreactor and osmotic membrane bioreactor-MD

The combinations of MD with other processes to increase the overall performance also have been studied (Table 2). For instance, pressure-driven microfiltration (MF) or ultrafiltration (UF) commonly used in membrane bioreactors can be replaced by MD. In this system, high-grade product water can be obtained in a single step due to its ability to provide complete retention to nonvolatile components such as salts, microorganisms, and nonvolatile organics. Moreover, the organic retention times in this system can be much higher than the hydraulic retention times (Phattaranawik et al., 2008), allowing even higher degradation of organic components in the wastewater (Goh et al., 2013). Due to the elevated feed temperature used in MD, heat-tolerant microbes have to be used in the bioreactor. Moreover, this proves advantageous as thermophilic bacteria have higher COD (chemical oxygen demand) removal efficiency and low sludge net yield than mesophilic ones (Phattaranawik et al., 2008).

Incorporating MD with membrane bioreactor has been done to both aerobic and anaerobic bioreactors. Aerobic membrane distillation bioreactor (MDBR) is especially appealing in eliminating trace organic contaminants (TrOCs). Wijekoon et al. (2014a) reported that aerobic thermophilic bioreactors alone can remove most of the TrOCs in the water sample but cannot eliminate compounds containing electron-withdrawing functional groups such as diclofenac, triclosan, or carbamazepine. However, an aerobic thermophilic MDBR system can eliminate all 25 TrOCs, with total organic carbon and total nitrogen removal reaching >99% and 96%, respectively (Wijekoon et al., 2014b). The removal of TrOCs also has been tested in an

TABLE 2 Hybrid MD configuration for wastewater treatment.

MD configuration	Membrane material and structure	Feed solution	Flux	Remarks	Ref.
MDBR-aerobic	Porous PVDF (Millipore Durapore GVHP), pore size of 0.22 μm	Synthetic wastewater with 0.67 g COD/L and 0.04 g total nitrogen (TN)/L	6.8 LMH (8% lower than MD). A 20% flux reduction occurred in 13 days	Fouling reduce membrane hydrophobicity and accelerate wetting. Wetting can be delayed by 1.7–3.6 times by reducing organic content and nutrient concentration. TOC removal of more than 94%	Goh et al. (2013)
MDBR-aerobic thermophilic	Porous PTFE (GE, Minnetonka, MN), pore size of 0.22 μm	Synthetic wastewater containing 100 mg/L glucose, 100 mg/L peptone, 17.5 mg/L KH_2PO_4, 17.5 mg/L $MgSO_4$, 10 mg/L $FeSO_4$, 225 mg/L CH_3COONa, and 35 mg/L urea with the addition of TrOCs at concentration of 5 μg/L each	Initial flux = 4 LMH Flux declined to 1.2 LMH after 10 days and stable until 35 days	Most trace organic compounds (TrOC) were removed (>95%), with exception to those contain electron withdrawing functional group (0%–52%)	Wijekoon et al. (2014b)
MDBR-anaerobic	PTFE top layer and PP support layer, pore size of 0.2 μm	Synthetic wastewater with addition of TrOC	Initial MD flux = 13.2 LMH. Feed conductivity increased from 3 to 55 mS/cm in 30 days of operation	Produce biogas 0.3–0.5 L/g COD_{added} Overall removal of TrOC = 76%. Potential phosphorus recovery from AnMBR effluent	Song et al. (2018)
OMBR-MD (NaCl as the draw solution)	FO membrane: cellulose triacetate (CTA) MD membrane: PTFE	Synthetic wastewater mimicking domestic wastewater	OMBR flux was in the range of 1–4 LMH for 200 days of operation. Initial conductivity was 5 mS/cm, then increased to 15 mS/cm	The OMBR operation were altered between aerobic/anoxic condition. NH_4^+-N removal of 90.2%. COD removal of 98.4%	Morrow et al. (2018)
OMBR-MD (NaCl as the draw solution)	FO membrane: polyamide (replaced in every 7 days). MD membrane: PE hollow fiber	Synthetic wastewater with added supernatant from anaerobic digester in local sewage treatment plant	OMBR flux was around 10 LMH. MF flux = 1.2 LMH	Anaerobic fluidized bed bioreactor was used. Phosphorus removal of 99%. Ammonia nitrogen removal of 80%. Reverse salt diffusion occurred but did not affect the performance	Kwon et al. (2021)

Continued

TABLE 2 Hybrid MD configuration for wastewater treatment—cont'd

MD configuration	Membrane material and structure	Feed solution	Flux	Remarks	Ref.
OMBR-MD (FO and MD membrane were integrated in one module)	FO membrane: cellulose triacetate (Hydration Technologies, Albany, United States). MD membrane: PTFE (Sterlitech, United States)	Real domestic wastewater with addition of 7 pharmaceutical active compounds (PhAC)	Initial FO flux was 5.2 LMH, then decrease to around 1 LMH. Initial MD membrane was 2 LMH, then decreased to 1 LMH	Single module integrating FO and MD membrane was used for 22 days of operation. PhAC removal = >96%. Dissolved organic carbon removal = 97.1%	Caroline Ricci et al. (2021)
Coagulation/flocculation-MD or MF-MD	MD membrane: PTFE (TF200, Gelman). MF membrane: CM-CELFA Membrantrenntechnik AG, model P-28	Industrial olive mill water	Using MF as the pretreatment, initial MD flux was at 9 LMH, then decreased to 1.44 LMH	Phenolic compounds in olive mill effluent were concentrated more than two times	El-Abbassi et al. (2013)
NF-MD	Self-fabricated ceramic NF and MD membrane	Synthetic radioactive wastewater	NF flux = 178 LMH. MD flux = 20 LMH	NF was used to remove the boric acid from the nuclide. Boric acid was concentrated by MD from 1 to 107 g/L. Two-stage NF retention of Co^{2+} and Ag^+ were 99.9% and 95%, respectively	Chen et al. (2019)

anaerobic MDBR system. Anaerobic membrane bioreactors are more energy efficient than aerobic ones; however, they often have a lower removal capacity of nutrients (nitrogen and phosphorus) and TrOCs. Using an anaerobic MDBR hybrid system, high removal of COD and phosphate throughout the testing period was achieved. The system also achieved 76% removal of all 26 selected TrOCs, although low ammonium removal (decreases from 90% to 60% after 30 days of operation) was observed (Song et al., 2018).

Degradation of TrOCs in the MDBR system that uses an enzyme (laccase) as the biocatalyst also has been reported. Laccase is an oxidoreductase enzyme that can degrade pollutants such as aromatic hydrocarbons, aliphatic amines, and trace organic contaminants. This enzyme is especially useful in degrading TrOCs recalcitrant to the activated sludge process. The application of laccase in wastewater treatment is challenging due to the requirement of a bioreactor that can prevent enzyme washout along with the treated effluents. By using MDBR, laccase can be retained together with the TrOCs. In a continuous MDBR process, around 94%–99% removal of all 30 TrOCs tested was achieved (Asif et al., 2018). On the other hand, only partial removal (54%–70%) was performed in a stand-alone MD system (Wijekoon et al., 2014a).

Although MDBR has superior performance in wastewater reclamation, the system is even more susceptible to fouling than MD, mainly due to the inclusion of biomass and biomolecules (Goh et al., 2013). The exact steps involved in fouling formation in MDBR have not been studied yet; however, fouling in an MBR process usually starts with the deposition of the organic fouling layer. This triggers the attachment of microbes, which is then followed by microbial growth, forming biofouling. During their growth, microbes excrete extracellular polymeric substances (EPS), which in turn interact with inorganic ions. This interaction causes inorganic salts' crystallization, forming inorganic fouling (Herzberg 2009 EST). Therefore, organic, inorganic, and biofouling can cooccur in MDBR.

Comparison between fouled MD and MDBR showed that MD has 19% higher flux than MDBR at similar fouling thickness (Goh et al., 2013). It indicates that fouling other characteristics, such as structure, composition, and porosity, strongly affects the system's performance. For porous fouling such as calcium carbonate, higher thermal resistance is observed. However, higher thermal and hydraulic resistance occurs when nonporous organic fouling is formed (Gryta, 2008). As thermal and hydraulic resistance increase, a reduction in flux follows. The occurrence of fouling can be minimized by selecting appropriate feed pretreatment and process condition. For instance, 50% flux reduction was observed after 7 days of operation when a feed temperature of 65°C was used; however, only minimal flux reduction is observed at a lower feed temperature (45°C) (Gryta, 2012).

Contrary to fouling formation, membrane wetting is slightly delayed in the MDBR system compared to the MD system. It was reported that membrane wetting in the MDBR system happens after 22 days of operation, compared to MD wetting that occurs after only 9 days. This is attributed to the higher TOC and TN in MD operation, which causes faster formation and higher concentration of foulants formed on the membrane surface (Goh et al., 2013).

Another high-retention MD that has been used is an osmotic membrane bioreactor with MD, commonly known as an osmotic membrane bioreactor (OMBR). In OMBR, the forward osmosis (FO) membrane is used in place of conventional MF or UF. The FO membrane is advantageous as it consumes low energy, has high rejection, and low fouling prospensity (Nguyen et al., 2016). Treated wastewater from the bioreactor is used as feed, and a highly concentrated solution of NaCl, $CaCl_2$, sucrose, $MgCl_2$, CH_3COONa, or $Mg(CH_3COO)_2$ is used as draw solution (DS) on the permeate side. This resulted in different osmotic pressure gradients, which act as the driving force for water transport (Achilli et al., 2010). DS concentration slowly diluted during the process, causing lower water flux across the membrane. MD can then be used to regenerate DS, recover the FO driving force, and produce high-quality water in the permeate side of MD (Fig. 7) (Morrow et al., 2018; Nguyen et al., 2016).

OMBR-MD has been used with numerous types of bioreactors. Morrow et al. (2018) used aerobic/anoxic OMBR-MD and reported 90.2% removal of NH_4^+-N and 98.4% COD removal. The use of attached growth biofilm-OMBR with MD to remove nutrients from wastewater while minimizing biofouling formation also has been proposed. During the 60 days of operation, stable water flux of $3.62 L/m^2h$ and high nutrient removal is obtained (Nguyen et al., 2016). These studies reported higher than 99.7% salt rejection in the MD process.

An anaerobic OMBR coupled with a hybrid FO-MD submerged module that has been used for municipal wastewater reclamation showed high removal of organic matter (91%), phosphorus (95%), and ammonium nitrogen (71%) (Caroline Ricci et al., 2021). Similarly, around 80% ammonium nitrogen and 99% phosphorus removal were observed for hybrid FO-MD combined with an anaerobic fluidized bed bioreactor (AFBR). However, ammonium nitrogen accumulated in the DS, resulting in a reverse salt flux through the FO membrane (Fig. 8) (Kwon et al., 2021). Moreover, these FO-MD systems are still susceptible to fouling formation. During the process, salinity build-up inside the bioreactor increased EPS production and caused biofouling formation on the FO membrane. Consequently, organic matter accumulated in the DS in a closed-loop system, resulting in fouling formation in the MD system (Caroline Ricci et al., 2021). Observations using scanning electron microscopy revealed that the inorganic scale dominated the membrane's fouling structure on the membrane (Kwon et al., 2021).

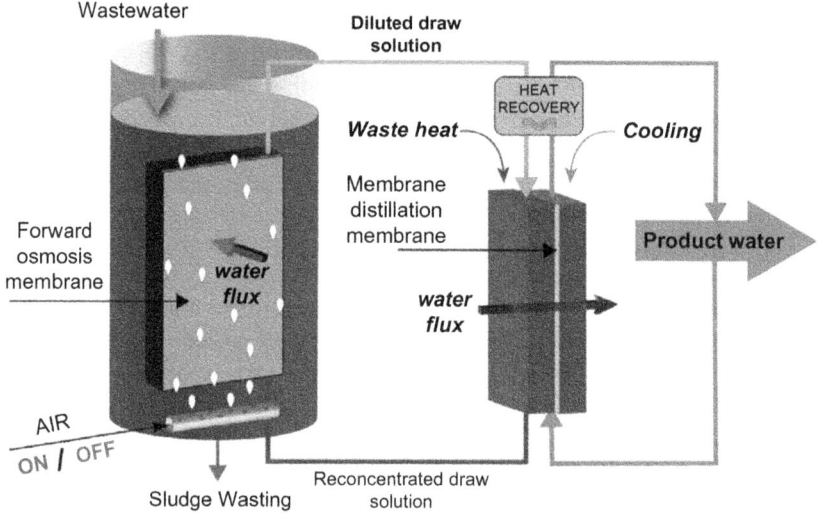

FIG. 7 Schematic of OMBR-MD laboratory setup. *Reproduced from Morrow, C.P., Furtaw, N.M., Murphy, J.R., Achilli, A., Marchand, E.A., Hiibel, S.R., Childress, A.E., 2018. Integrating an aerobic/anoxic osmotic membrane bioreactor with membrane distillation for potable reuse. Desalination 432, 46–54 (September 2017) https://doi.org/10.1016/j.desal.2017.12.047, with permission from Elsevier.*

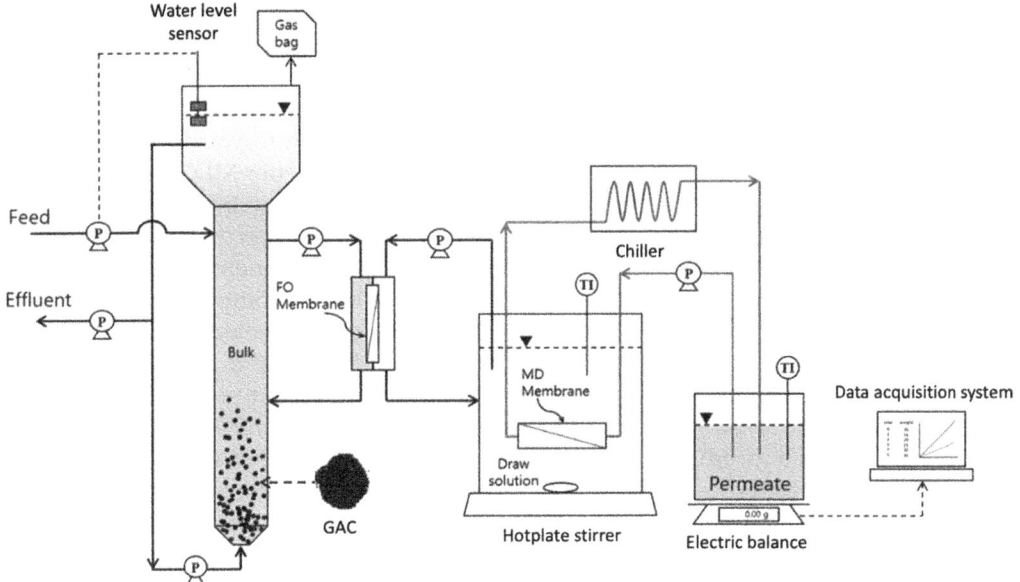

FIG. 8 Laboratory setup of AFBR-FO-MD hybrid system. *Reproduced from Kwon, D., Bae, W., Kim, J., 2021. Hybrid forward osmosis/membrane distillation integrated with anaerobic fluidized bed bioreactor for advanced wastewater treatment. J. Hazard. Mater. 404 (Part A), 124160. https://doi.org/10.1016/j.jhazmat.2020.124160, with permission from Elsevier.*

5.3 Other hybrid MD configurations

Some applications incorporate other processes, especially membrane-based processes, before MD is done to increase MD performance and mitigate fouling formation. MF or UF can potentially be used to remove particulate and large molecules commonly present in wastewater (Warsinger et al., 2015). In olive mill wastewater treatment, two different pretreatments before the DCMD process are tested, coagulation/flocculation and MF. Pretreatment using MF resulted in higher permeate flux of 7.7 LMH compared to crude olive mill wastewater (5.6 LMH) or coagulation/flocculation-treated olive mill wastewater (6.9 LMH) (El-Abbassi et al., 2013). A hybrid system containing ceramic nanofiltration (NF) and VMD was used to concentrate boric acid from simulated radioactive wastewater. The ceramic NF membrane was mainly used to remove radionuclides from the feed, followed by the VMD process to obtain purified boric acid solution. The NF membrane effectively removed simulated ionic radionuclides using Co^{2+} and Ag^+ ions. In the subsequent VMD process, permeate flux was maintained above 20 LMH, while boric acid concentration increased from 1 to 107 g/L (Chen et al., 2019). Although the incorporation of MF/UF showed potential, research on this topic is still limited.

6. Conclusions and perspectives

Wastewater treatment by MD has been conducted for freshwater production, nutrient recovery, and wastewater volume reduction prior to disposal. Feed temperature is crucial in the

MD operation as it determines the driving force for water vapor transport. The concentration and temperature polarization adversely affect the permeate flux's driving force. In addition, sensible heat loss to the environment and parasitic heat conduction across the membrane material result in energy inefficiency and, hence, need to be minimized. These challenges deter the commercialization of MD for wastewater treatment. To push forward the industrial application of MD for wastewater treatment, strategies that target better MD's performance and energy utilization are important. Optimization of the flow rate may limit concentration polarization and temperature polarization. In contrast, proper insulation of the membrane module, modification of membrane structure, and optimization of membrane thickness can improve the energy efficiency. As wastewater consists of various components with different characteristics, fouling is complex and challenging in wastewater treatment. Specifically tailored membrane materials and structure and the choice of MD configuration are required to address the fouling. Wetting is also critical in wastewater treatment using MD as it leads to poor MD performance. While the hydrophobic membrane is used in MD, organic content in the wastewater tends to produce a lower contact angle on the membrane surface due to their low surface tension properties. Omniphobic and Janus membranes have been acknowledged to delay the wetting due to their reentrant structure and opposing wettability, respectively.

Various wastewater types, such as textile, oil and gas, radioactive, pharmaceutical, and municipal, have been treated by MD, highlighting promising results. Excellent nonvolatile solute rejections were reported. In many studies, membranes with modified structure and material were utilized to address fouling and wetting. However, it is worth noting that almost all experiments were conducted on a laboratory scale using a synthetic feed solution. Further study on the MD applicability to treat real wastewater at higher capacity is essential. Aside from the stand-alone configuration, hybrid MDs are widely studied. High-retention MDs, consisting of MDBR and OMBR-MD, have been studied to treat various wastewaters. Despite the excellent solute rejection, low flux, fouling, and wetting were still reported. However, the application of modified membranes in high-retention MDs is still lacking. Studies using specifically tailored membranes in long-term operation using real wastewater are needed to improve and commercialize the high-retention MD performance.

References

Achilli, A., Cath, T.Y., Childress, A.E., 2010. Selection of inorganic-based draw solutions for forward osmosis applications. J. Membr. Sci. 364 (1–2), 233–241. https://doi.org/10.1016/j.memsci.2010.08.010.

Alkhudhiri, A., Darwish, N., Hilal, N., 2012. Membrane distillation: a comprehensive review. Desalination 287, 2–18. https://doi.org/10.1016/j.desal.2011.08.027.

Amar, N.B., Kechaou, N., Palmeri, J., Deratani, A., Sghaier, A., 2009. Comparison of tertiary treatment by nanofiltration and reverse osmosis for water reuse in denim textile industry. J. Hazard. Mater. 170 (1), 111–117. https://doi.org/10.1016/j.jhazmat.2009.04.130.

An, A.K., Guo, J., Jeong, S., Lee, E.J., Tabatabai, S.A.A., Leiknes, T.O., 2016. High flux and antifouling properties of negatively charged membrane for dyeing wastewater treatment by membrane distillation. Water Res. 103, 362–371. https://doi.org/10.1016/j.watres.2016.07.060.

An, A.K., Guo, J., Lee, E.J., Jeong, S., Zhao, Y., Wang, Z., Leiknes, T.O., 2017. PDMS/PVDF hybrid electrospun membrane with superhydrophobic property and drop impact dynamics for dyeing wastewater treatment using membrane distillation. J. Membr. Sci. 525, 57–67. https://doi.org/10.1016/j.memsci.2016.10.028.

Asif, M.B., Nguyen, L.N., Hai, F.I., Price, W.E., Nghiem, L.D., 2017. Integration of an enzymatic bioreactor with membrane distillation for enhanced biodegradation of trace organic contaminants. Int. Biodeterior. Biodegrad. 124, 73–81. https://doi.org/10.1016/j.ibiod.2017.06.012.

Asif, M.B., Hai, F.I., Kang, J., van de Merwe, J.P., Leusch, F.D.L., Price, W.E., Nghiem, L.D., 2018. Biocatalytic degradation of pharmaceuticals, personal care products, industrial chemicals, steroid hormones and pesticides in a membrane distillation-enzymatic bioreactor. Bioresour. Technol. 247, 528–536. https://doi.org/10.1016/j.biortech.2017.09.129 (August 2017).

Bird, B.R., Stewart, W.E., Lightfoot, E.N., 2006. Transport Phenomena. John Wiley & Sons.

Boo, C., Lee, J., Elimelech, M., 2016. Omniphobic polyvinylidene fluoride (PVDF) membrane for desalination of shale gas produced water by membrane distillation. Environ. Sci. Technol. 50 (22), 12275–12282. https://doi.org/10.1021/acs.est.6b03882.

Bu, Q., Wang, B., Huang, J., Deng, S., Yu, G., 2013. Pharmaceuticals and personal care products in the aquatic environment in China: a review. J. Hazard. Mater. 262, 189–211. https://doi.org/10.1016/j.jhazmat.2013.08.040.

Carnevale, M.C., Gnisci, E., Hilal, J., Criscuoli, A., 2016. Direct contact and vacuum membrane distillation application for the olive mill wastewater treatment. Sep. Purif. Technol. 169, 121–127. https://doi.org/10.1016/j.seppur.2016.06.002.

Caroline Ricci, B., Santos Arcanjo, G., Rezende Moreira, V., Abner Rocha Lebron, Y., Koch, K., Cristina Rodrigues Costa, F., Paulinelli Ferreira, B., Luiza Costa Lisboa, F., Diniz Miranda, L., Vieira de Faria, C., Celina Lange, L., Cristina Santos Amaral, M., 2021. A novel submerged anaerobic osmotic membrane bioreactor coupled to membrane distillation for water reclamation from municipal wastewater. Chem. Eng. J. 414. https://doi.org/10.1016/j.cej.2021.128645 (January).

Celina, M.C., 2013. Review of polymer oxidation and its relationship with materials performance and lifetime prediction. Polym. Degrad. Stab. 98 (12), 2419–2429. https://doi.org/10.1016/j.polymdegradstab.2013.06.024.

Chen, X., Chen, T., Li, J., Qiu, M., Fu, K., Cui, Z., Fan, Y., Drioli, E., 2019. Ceramic nanofiltration and membrane distillation hybrid membrane processes for the purification and recycling of boric acid from simulative radioactive waste water. J. Membr. Sci. 579, 294–301. https://doi.org/10.1016/j.memsci.2019.02.044 (February).

Chew, N.G.P., Zhang, Y., Goh, K., Ho, J.S., Xu, R., Wang, R., 2019. Hierarchically structured janus membrane surfaces for enhanced membrane distillation performance. ACS Appl. Mater. Interfaces 11 (28), 25524–25534. https://doi.org/10.1021/acsami.9b05967.

Chmielewski, A.G., Harasimowicz, M., Zakrzewska-Trznadel, G., 1997. Purification of radioactive wastes by low temperature evaporation (membrane distillation). Sep. Sci. Technol. 32 (1–4), 709–720. https://doi.org/10.1080/01496399708003225.

Choi, Y., Naidu, G., Nghiem, L.D., Lee, S., Vigneswaran, S., 2019. Membrane distillation crystallization for brine mining and zero liquid discharge: opportunities, challenges, and recent progress. Environ. Sci. Water Res. Technol. 5 (7), 1202–1221. Royal Society of Chemistry https://doi.org/10.1039/c9ew00157c.

Criscuoli, A., Zhong, J., Figoli, A., Carnevale, M.C., Huang, R., Drioli, E., 2008. Treatment of dye solutions by vacuum membrane distillation. Water Res. 42 (20), 5031–5037. https://doi.org/10.1016/j.watres.2008.09.014.

Dotto, J., Fagundes-Klen, M.R., Veit, M.T., Palácio, S.M., Bergamasco, R., 2019. Performance of different coagulants in the coagulation/flocculation process of textile wastewater. J. Clean. Prod. 208, 656–665. https://doi.org/10.1016/j.jclepro.2018.10.112.

Dow, N., Villalobos García, J., Niadoo, L., Milne, N., Zhang, J., Gray, S., Duke, M., 2017. Demonstration of membrane distillation on textile waste water assessment of long term performance, membrane cleaning and waste heat integration. Environ. Sci. Water Res. Technol. 3 (3), 433–449. https://doi.org/10.1039/c6ew00290k.

El-Abbassi, A., Hafidi, A., Khayet, M., García-Payo, M.C., 2013. Integrated direct contact membrane distillation for olive mill wastewater treatment. Desalination 323, 31–38. https://doi.org/10.1016/j.desal.2012.06.014.

El-Bourawi, M.S., Ding, Z., Ma, R., Khayet, M., 2006. A framework for better understanding membrane distillation separation process. J. Membr. Sci. 285 (1–2), 4–29. https://doi.org/10.1016/j.memsci.2006.08.002.

Eryildiz, B., Yuksekdag, A., Korkut, S., Zeytuncu, B., Pasaoglu, M.E., Koyuncu, I., 2020. Effect of operating parameters on removal of boron from wastewater containing high boron concentration by vacuum assisted air gap membrane distillation. J. Water Process Eng. 38. https://doi.org/10.1016/j.jwpe.2020.101579.

Feron, P.H.M., Jansen, A.E., 2002. CO_2 separation with polyolefin membrane contactors and dedicated absorption liquids: performances and prospects. In. Sep. Purif. Technol. 27. www.elsevier.com/locate/seppur.

Fortunato, L., Elcik, H., Blankert, B., Ghaffour, N., Vrouwenvelder, J., 2021. Textile dye wastewater treatment by direct contact membrane distillation: membrane performance and detailed fouling analysis. J. Membr. Sci. 636, 119552. https://doi.org/10.1016/j.memsci.2021.119552.

Franken, A., Nolten, J.A.M., Mulder, M.H.V., Bargeman, D., Smolders, C.A., 1987. Wetting criteria for the applicability of membrane distillation. J. Membr. Sci. 33, 315–328. Elsevier Science Publishers B.V.

Gebregiorgis, T., van Hullebusch, E.D., Hagos, K., 2018. Decolourization of real textile wastewater by the combination of photocatalytic and biological oxidation processes. In: Advances in Science, Technology and Innovation. Springer Nature, pp. 115–117, https://doi.org/10.1007/978-3-319-70548-4_40.

Gethard, K., Sae-Khow, O., Mitra, S., 2012. Carbon nanotube enhanced membrane distillation for simultaneous generation of pure water and concentrating pharmaceutical waste. Sep. Purif. Technol. 90, 239–245. https://doi.org/10.1016/j.seppur.2012.02.042.

Goh, S., Zhang, J., Liu, Y., Fane, A.G., 2013. Fouling and wetting in membrane distillation (MD) and MD-bioreactor (MDBR) for wastewater reclamation. Desalination 323, 39–47. https://doi.org/10.1016/j.desal.2012.12.001.

Gryta, M., 2008. Fouling in direct contact membrane distillation process. J. Membr. Sci. 325 (1), 383–394. https://doi.org/10.1016/j.memsci.2008.08.001.

Gryta, M., 2012. Polyphosphates used for membrane scaling inhibition during water desalination by membrane distillation. Desalination 285, 170–176. https://doi.org/10.1016/j.desal.2011.09.051.

Guo, J., Farid, M.U., Lee, E.J., Yan, D.Y.S., Jeong, S., Kyoungjin An, A., 2018. Fouling behavior of negatively charged PVDF membrane in membrane distillation for removal of antibiotics from wastewater. J. Membr. Sci. 551, 12–19. https://doi.org/10.1016/j.memsci.2018.01.016.

Guo, J., Fortunato, L., Deka, B.J., Jeong, S., An, A.K., 2020. Elucidating the fouling mechanism in pharmaceutical wastewater treatment by membrane distillation. Desalination 475. https://doi.org/10.1016/j.desal.2019.114148.

Gusnawan, P.J., 2020. Soybean-Based Biosolvent for Flue-Gas CO_2 Capture via Hollow Fiber Membrane Contactor. New Mexico Institute of Mining and Technology ProQuest Dissertations Publishing, United States.

Han, M., Dong, T., Hou, D., Yao, J., Han, L., 2020. Carbon nanotube based Janus composite membrane of oil fouling resistance for direct contact membrane distillation. J. Membr. Sci. 607. https://doi.org/10.1016/j.memsci.2020.118078.

Hansen, Charles, M., 2007. Hansen Solubility Parameters A User's Handbook Second Edition. CRC Press, Taylor & Francis Group, Boca Raton, United States.

He, F., Sirkar, K.K., Gilron, J., 2009. Studies on scaling of membranes in desalination by direct contact membrane distillation: $CaCO_3$ and mixed $CaCO_3$/$CaSO_4$ systems. Chem. Eng. Sci. 64 (8), 1844–1859. https://doi.org/10.1016/j.ces.2008.12.036.

Henze, M., 2008. Biological Wastewater Treatment: Principles, Modelling and Design. IWA Pub.

Herrera-González, A.M., Caldera-Villalobos, M., Peláez-Cid, A.A., 2019. Adsorption of textile dyes using an activated carbon and crosslinked polyvinyl phosphonic acid composite. J. Environ. Manag. 234, 237–244. https://doi.org/10.1016/j.jenvman.2019.01.012.

Hitsov, I., Maere, T., de Sitter, K., Dotremont, C., Nopens, I., 2015. Modelling approaches in membrane distillation: a critical review. Sep. Purif. Technol. 142, 48–64. Elsevier https://doi.org/10.1016/j.seppur.2014.12.026.

Jeong, S., Song, K.G., Kim, J., Shin, J., Maeng, S.K., Park, J., 2021. Feasibility of membrane distillation process for potable water reuse: a barrier for dissolved organic matters and pharmaceuticals. J. Hazard. Mater. 409. https://doi.org/10.1016/j.jhazmat.2020.124499.

Jia, F., Li, J., Wang, J., Sun, Y., 2017. Removal of strontium ions from simulated radioactive wastewater by vacuum membrane distillation. Ann. Nucl. Energy 103, 363–368. https://doi.org/10.1016/j.anucene.2017.02.003.

Jia, F., Yin, Y., Wang, J., 2018. Removal of cobalt ions from simulated radioactive wastewater by vacuum membrane distillation. Prog. Nucl. Energy 103, 20–27. https://doi.org/10.1016/j.pnucene.2017.11.008.

Jia, X., Lan, L., Zhang, X., Wang, T., Wang, Y., Ye, C., Lin, J., 2021. Pilot-scale vacuum membrane distillation for decontamination of simulated radioactive wastewater: system design and performance evaluation. Sep. Purif. Technol. 275. https://doi.org/10.1016/j.seppur.2021.119129.

Julian, H., Nurgirisia, N., Sutrisna, P.D., Wenten, I.G., 2021a. Advances in seawater membrane distillation (SWMD) towards stand-alone zero liquid discharge (ZLD) desalination. Rev. Chem. Eng., 196–207. De Gruyter Open Ltd https://doi.org/10.1515/revce-2020-0073.

Julian, H., Rizqullah, H., Siahaan, M.A., Wenten, I.G., 2021b. Cane sugar crystallization using submerged vacuum membrane distillation crystallization (SVMDC). J. Food Sci. Technol. 58 (6), 2368–2376. https://doi.org/10.1007/s13197-020-04749-z.

Kalla, S., 2021. Use of membrane distillation for oily wastewater treatment—a review. J. Environ. Chem. Eng. 9 (1). https://doi.org/10.1016/j.jece.2020.104641. Elsevier Ltd.

Khayet, M., 2013. Treatment of radioactive wastewater solutions by direct contact membrane distillation using surface modified membranes. Desalination 321, 60–66. https://doi.org/10.1016/j.desal.2013.02.023.

Korolkov, I.V., Yeszhanov, A.B., Zdorovets, M.V., Gorin, Y.G., Güven, O., Dosmagambetova, S.S., Khlebnikov, N.A., Serkov, K.V., Krasnopyorova, M.V., Milts, O.S., Zheltov, D.A., 2019. Modification of PET ion track membranes for membrane distillation of low-level liquid radioactive wastes and salt solutions. Sep. Purif. Technol. 227. https://doi.org/10.1016/j.seppur.2019.115694.

Kwon, D., Bae, W., Kim, J., 2021. Hybrid forward osmosis/membrane distillation integrated with anaerobic fluidized bed bioreactor for advanced wastewater treatment. J. Hazard. Mater. 404 (Part A), 124160. https://doi.org/10.1016/j.jhazmat.2020.124160.

Laqbaqbi, M., García-Payo, M.C., Khayet, M., el Kharraz, J., Chaouch, M., 2019. Application of direct contact membrane distillation for textile wastewater treatment and fouling study. Sep. Purif. Technol. 209, 815–825. https://doi.org/10.1016/j.seppur.2018.09.031.

Leaper, S., Abdel-Karim, A., Gad-Allah, T.A., Gorgojo, P., 2019. Air-gap membrane distillation as a one-step process for textile wastewater treatment. Chem. Eng. J. 360, 1330–1340. https://doi.org/10.1016/j.cej.2018.10.209 (August 2018).

Li, F., Huang, J., Xia, Q., Lou, M., Yang, B., Tian, Q., Liu, Y., 2018. Direct contact membrane distillation for the treatment of industrial dyeing wastewater and characteristic pollutants. Sep. Purif. Technol. 195, 83–91. https://doi.org/10.1016/j.seppur.2017.11.058 (November 2017).

Li, X., Shan, H., Cao, M., Li, B., 2019. Facile fabrication of omniphobic PVDF composite membrane via a waterborne coating for anti-wetting and anti-fouling membrane distillation. J. Membr. Sci. 589. https://doi.org/10.1016/j.memsci.2019.117262.

Liao, Y., Wang, R., Fane, A.G., 2013. Engineering superhydrophobic surface on poly(vinylidene fluoride) nanofiber membranes for direct contact membrane distillation. J. Membr. Sci. 440, 77–87. https://doi.org/10.1016/j.memsci.2013.04.006.

Lienhard, J.H., Antar, M.A., Bilton, A., Blanco, J., Zaragoza, G., 2012. Solar desalination. In: Annual Review of Heat Transfer. vol. 15., https://doi.org/10.1504/IJND.2007.015802.

Lin, P.J., Yang, M.C., Li, Y.L., Chen, J.H., 2015. Prevention of surfactant wetting with agarose hydrogel layer for direct contact membrane distillation used in dyeing wastewater treatment. J. Membr. Sci. 475, 511–520. https://doi.org/10.1016/j.memsci.2014.11.001.

Liu, H., Wang, J., 2013. Treatment of radioactive wastewater using direct contact membrane distillation. J. Hazard. Mater. 261, 307–315. https://doi.org/10.1016/j.jhazmat.2013.07.045.

Luis, P., 2018. Introduction. In: Fundamental Modeling of Membrane Systems: Membrane and Process Performance. Elsevier, pp. 1–23, https://doi.org/10.1016/B978-0-12-813483-2.00001-0.

Marcuccl, M., Nosenzo, G., Capannelli, G., Ciabattp, I., Corrieri, D., Ciardelli, G., 2001. Treatment and reuse of textile effluents based on new ultrafiltration and other membrane technologies. Desalination 138. www.elsevier.com/locate/desal.

McQuarrie, D.A., Rock, P.A., Gallogly, E.B., 2011. General Chemistry. University Science Books, Mill Valley, CA, United States.

Mersmann, A., 2001. Crystallization Technology Handbook. Marcel Dekker.

Mokhtar, N.M., Lau, W.J., Ismail, A.F., 2014. The potential of membrane distillation in recovering water from hot dyeing solution. J. Water Process Eng. 2, 71–78. https://doi.org/10.1016/j.jwpe.2014.05.006.

Morrow, C.P., Furtaw, N.M., Murphy, J.R., Achilli, A., Marchand, E.A., Hiibel, S.R., Childress, A.E., 2018. Integrating an aerobic/anoxic osmotic membrane bioreactor with membrane distillation for potable reuse. Desalination 432, 46–54. https://doi.org/10.1016/j.desal.2017.12.047 (September 2017).

Naidu, G., Shim, W.G., Jeong, S., Choi, Y.K., Ghaffour, N., Vigneswaran, S., 2017. Transport phenomena and fouling in vacuum enhanced direct contact membrane distillation: experimental and modelling. Sep. Purif. Technol. 172, 285–295. https://doi.org/10.1016/j.seppur.2016.08.024.

Nguyen, N.C., Nguyen, H.T., Chen, S.S., Ngo, H.H., Guo, W., Chan, W.H., Ray, S.S., Li, C.W., Hsu-Te, H., 2016. A novel osmosis membrane bioreactor-membrane distillation hybrid system for wastewater treatment and reuse. Bioresour. Technol. 209, 8–15. https://doi.org/10.1016/j.biortech.2016.02.102.

Padaki, M., Surya Murali, R., Abdullah, M.S., Misdan, N., Moslehyani, A., Kassim, M.A., Hilal, N., Ismail, A.F., 2015. Membrane technology enhancement in oil-water separation. A review. Desalination 357, 197–207. Elsevier https://doi.org/10.1016/j.desal.2014.11.023.

Palmer, S.J., 1976. The effect of temperature on surface tension. Phys. Educ. 11, 119.

Paździor, K., Bilińska, L., Ledakowicz, S., 2019. A review of the existing and emerging technologies in the combination of AOPs and biological processes in industrial textile wastewater treatment. Chem. Eng. J. 376. https://doi.org/10.1016/j.cej.2018.12.057.

Phattaranawik, J., Fane, A.G., Pasquier, A.C.S., Bing, W., 2008. A novel membrane bioreactor based on membrane distillation. Desalination 223 (1–3), 386–395. https://doi.org/10.1016/j.desal.2007.02.075.

Ramlow, H., Machado, R.A.F., Bierhalz, A.C.K., Marangoni, C., 2020. Direct contact membrane distillation applied to wastewaters from different stages of the textile process. Chem. Eng. Commun. 207 (8), 1062–1073. https://doi.org/10.1080/00986445.2019.1640683.

Razmjou, A., Arifin, E., Dong, G., Mansouri, J., Chen, V., 2012. Superhydrophobic modification of TiO_2 nanocomposite PVDF membranes for applications in membrane distillation. J. Membr. Sci. 415–416, 850–863. https://doi.org/10.1016/j.memsci.2012.06.004.

Santos, P.G., Scherer, C.M., Fisch, A.G., Rodrigues, M.A.S., 2020. Petrochemical wastewater treatment: water recovery using membrane distillation. J. Clean. Prod. 267. https://doi.org/10.1016/j.jclepro.2020.121985.

Shimekit, B., Mukhtar, H., Murugesan, T., 2011. Prediction of the relative permeability of gases in mixed matrix membranes. J. Membr. Sci. 373 (1–2), 152–159. https://doi.org/10.1016/j.memsci.2011.02.038.

Shirazi, M.M.A., Bazgir, S., Meshkani, F., 2020a. A novel dual-layer, gas-assisted electrospun, nanofibrous SAN4-HIPS membrane for industrial textile wastewater treatment by direct contact membrane distillation (DCMD). J. Water Process Eng. 36. https://doi.org/10.1016/j.jwpe.2020.101315.

Shirazi, M.M.A., Bazgir, S., Meshkani, F., 2020b. A dual-layer, nanofibrous styrene-acrylonitrile membrane with hydrophobic/hydrophilic composite structure for treating the hot dyeing effluent by direct contact membrane distillation. Chem. Eng. Res. Des. 164, 125–146. https://doi.org/10.1016/j.cherd.2020.09.030.

Smith, J.M., van Ness, H.C., Abbott, M.M., Swihart, M.T., 2018. Introduction to Chemical Engineering Thermodynamics. *McGraw-Hill Education*, https://doi.org/10.1021/ed027p584.3 (Issue 8).

Song, X., Luo, W., McDonald, J., Khan, S.J., Hai, F.I., Price, W.E., Nghiem, L.D., 2018. An anaerobic membrane bioreactor—membrane distillation hybrid system for energy recovery and water reuse: removal performance of organic carbon, nutrients, and trace organic contaminants. Sci. Total Environ. 628–629, 358–365. https://doi.org/10.1016/j.scitotenv.2018.02.057.

Termpiyakul, P., Jiraratananon, R., Srisurichan, S., 2005. Heat and mass transfer characteristics of a direct contact membrane distillation process for desalination. Desalination 177 (1–3), 133–141. https://doi.org/10.1016/j.desal.2004.11.019.

Wang, Z., Hou, D., Lin, S., 2016a. Composite membrane with underwater-oleophobic surface for anti-oil-fouling membrane distillation. Environ. Sci. Technol. 50 (7), 3866–3874. https://doi.org/10.1021/acs.est.5b05976.

Wang, Z., Jin, J., Hou, D., Lin, S., 2016b. Tailoring surface charge and wetting property for robust oil-fouling mitigation in membrane distillation. J. Membr. Sci. 516, 113–122. https://doi.org/10.1016/j.memsci.2016.06.011.

Warsinger, D.M., Swaminathan, J., Guillen-Burrieza, E., Arafat, H.A., Lienhard, V., J. H., 2015. Scaling and fouling in membrane distillation for desalination applications: a review. Desalination 356, 294–313. https://doi.org/10.1016/j.desal.2014.06.031.

Wen, X., Li, F., Zhao, X., 2016. Filtering of low-level radioactive wastewater by means of vacuum membrane distillation. Nucl. Technol. 194 (3), 379–386. https://doi.org/10.13182/NT15-74.

Wijekoon, K.C., Hai, F.I., Kang, J., Price, W.E., Cath, T.Y., Nghiem, L.D., 2014a. Rejection and fate of trace organic compounds (TrOCs) during membrane distillation. J. Membr. Sci. 453, 636–642. https://doi.org/10.1016/j.memsci.2013.12.002.

Wijekoon, K.C., Hai, F.I., Kang, J., Price, W.E., Guo, W., Ngo, H.H., Cath, T.Y., Nghiem, L.D., 2014b. A novel membrane distillation-thermophilic bioreactor system: biological stability and trace organic compound removal. Bioresour. Technol. 159, 334–341. https://doi.org/10.1016/j.biortech.2014.02.088.

Woldemariam, D., Kullab, A., Fortkamp, U., Magner, J., Royen, H., Martin, A., 2016. Membrane distillation pilot plant trials with pharmaceutical residues and energy demand analysis. Chem. Eng. J. 306, 471–483. https://doi.org/10.1016/j.cej.2016.07.082.

Wu, Y., Kang, Y., Zhang, L., Qu, D., Cheng, X., Feng, L., 2018. Performance and fouling mechanism of direct contact membrane distillation (DCMD) treating fermentation wastewater with high organic concentrations. J. Environ. Sci. (China) 65, 253–261. https://doi.org/10.1016/j.jes.2017.01.015.

Yan, Z., Yang, H., Yu, H., Qu, F., Liang, H., van der Bruggen, B., Li, G., 2018. Reverse osmosis brine treatment using direct contact membrane distillation (DCMD): effect of membrane characteristics on desalination performance and the wetting phenomenon. Environ. Sci. Water Res. Technol. 4 (3), 428–437. https://doi.org/10.1039/c7ew00468k.

Yan, Z., Yang, H., Qu, F., Zhang, H., Rong, H., Yu, H., Liang, H., Ding, A., Li, G., van der Bruggen, B., 2019. Application of membrane distillation to anaerobic digestion effluent treatment: identifying culprits of membrane fouling and scaling. Sci. Total Environ. 688, 880–889. https://doi.org/10.1016/j.scitotenv.2019.06.307.

Yang, G., Zhang, J., Peng, M., Du, E., Wang, Y., Shan, G., Ling, L., Ding, H., Gray, S., Xie, Z., 2021. A mini review on antiwetting studies in membrane distillation for textile wastewater treatment. Processes 9 (2), 1–16. MDPI AG https://doi.org/10.3390/pr9020243.

Yaws, C.L., 1995. Handbook of Vapor Pressure. Gulf Publishing Company.

Zakrzewska-Trznadel, G., Harasimowicz, M., Chmielewski, A.G., 1999. Concentration of radioactive components in liquid low-level radioactive waste by membrane distillation. J. Membr. Sci. 163.

Zhang, X., Gu, P., Liu, Y., 2019. Decontamination of radioactive wastewater: state of the art and challenges forward. Chemosphere 215, 543–553. Elsevier Ltd https://doi.org/10.1016/j.chemosphere.2018.10.029.

Zhang, J., Mirza, N.R., Huang, Z., Du, E., Peng, M., Shan, G., Wang, Y., Pan, Z., Ling, L., Xie, Z., 2021. Evaluation of direct contact membrane distillation coupled with fractionation and ozonation for the treatment of textile effluent. J. Water Process Eng. 40. https://doi.org/10.1016/j.jwpe.2020.101789.

Zheng, R., Chen, Y., Wang, J., Song, J., Li, X.M., He, T., 2018. Preparation of omniphobic PVDF membrane with hierarchical structure for treating saline oily wastewater using direct contact membrane distillation. J. Membr. Sci. 555, 197–205. https://doi.org/10.1016/j.memsci.2018.03.041.

Zhu, Z., Liu, Z., Zhong, L., Song, C., Shi, W., Cui, F., Wang, W., 2018. Breathable and asymmetrically superwettable Janus membrane with robust oil-fouling resistance for durable membrane distillation. J. Membr. Sci. 563, 602–609. https://doi.org/10.1016/j.memsci.2018.06.028.

Zou, L., Gusnawan, P., Zhang, G., Yu, J., 2020. Novel Janus composite hollow fiber membrane-based direct contact membrane distillation (DCMD) process for produced water desalination. J. Membr. Sci. 597. https://doi.org/10.1016/j.memsci.2019.117756.

Zuo, G., Wang, R., 2013. Novel membrane surface modification to enhance anti-oil fouling property for membrane distillation application. J. Membr. Sci. 447, 26–35. https://doi.org/10.1016/j.memsci.2013.06.053.

Membrane distillation-liquid desiccant air-conditioning for thermal comfort in buildings

Hung Cong Duong[a,b], Hai Thuong Cao[a], Long Duc Nghiem[b], Ashley Joy Ansari[c], Ngoc Lieu Le[d,e], Thi Oanh Doan[f], and Nguyen Cong Nguyen[g]

[a]Le Quy Don Technical University, Hanoi, Vietnam [b]Centre for Technology in Water and Wastewater, School of Civil and Environmental Engineering, University of Technology Sydney, Ultimo, NSW, Australia [c]Strategic Water Infrastructure Laboratory, School of Civil Mining and Environmental Engineering, University of Wollongong, Wollongong, NSW, Australia [d]International University, Ho Chi Minh City, Vietnam [e]Vietnam National University, Ho Chi Minh City, Vietnam [f]Hanoi University of Natural Resources and Environment, Hanoi, Vietnam [g]Faculty of Chemistry and Environment, Dalat University, Dalat, Vietnam

1. Introduction

The energy efficiency of air-conditioning systems in buildings remains an important aspect in the efforts toward a zero-emission buildings industry. Carbon dioxide (CO_2) emissions from the building sector accounted for 37% of total global energy-related CO_2 emissions in 2020 (UNEP, 2021). These emissions originate from the energy required for activities such as air-conditioning, space heating and/or cooling, ventilation, water heating, and lighting, in addition to other site-specific greenhouse gas emissions (Röck et al., 2020). In fact, air-conditioning and space cooling contributed to almost 5% of total energy consumption worldwide in 2020 and there is a projected 75% increase in buildings floor space worldwide over the next 30 years (UNEP, 2021). Furthermore, healthy and comfortable indoor building

Copyright © 2023 Elsevier Inc. All rights reserved.

environments are essential for the productivity of societies, and this signifies the important role that innovative and energy-efficient air-conditioning technologies must play.

Conventional air-conditioning systems operate using vapor compression refrigeration, which has the benefit of a small footprint; however, they exhibit limited energy efficiency in hot temperatures and humid conditions (Pasqualin et al., 2022; Su et al., 2022). To overcome this challenge, the liquid desiccant air-conditioning (LDAC) process has been developed and has attracted significant research attention over the past 10 years (Qi et al., 2020; Gurubalan and Simonson, 2021; Pasqualin et al., 2022; Su et al., 2022). This air-conditioning method employs a liquid desiccant solution for the absorption of moisture from the supply air. Because of its hygroscopic nature, the liquid desiccant solution effectively dehumidifies the air without the need for overcooling below the dew point temperature as required in conventional vapor compression-based air conditioners. Therefore, the LDAC process offers an energy-efficient alternative to conventional air-conditioning systems (Duong et al., 2017, 2019).

To sustain the LDAC process, the diluted liquid desiccant solution (after dehumidification) requires reconcentration, which can be achieved using a regeneration process. The regeneration step is regarded as the most critical component of the LDAC process. Indeed, a majority of the LDAC energy consumption is attributed to the regeneration step. As a result, it is essential to explore innovative technologies that are low-cost and energy-efficient for the regeneration of liquid desiccant solutions in the LDAC process.

Membrane distillation (MD) has been explored as a robust and potentially low-cost treatment method for various impaired water sources including seawater and saline solutions. The MD process embodies several advantages that make it a promising option for the strategic treatment of hypersaline waters such as liquid desiccant solutions used in the LDAC process (Duong et al., 2020a, 2020b). These advantages distinguish MD from other desalination processes based on membrane separation or conventional thermal distillation. Therefore, in recent years, extensive studies have been conducted to elucidate the technical feasibility and examine the performance of the MD process for regeneration of liquid desiccant solutions (Pasqualin et al., 2022).

This chapter aims to provide a comprehensive analysis of the combined MD-LDAC process for providing thermal comfort in buildings. In the first section of the chapter, the working principles of the LDAC process are outlined, and the key bottlenecks are discussed, including the selection and regeneration of appropriate liquid desiccant solutions. Then, the fundamentals of the MD process are presented in the context related to the regeneration of liquid desiccant solutions in LDAC systems. Finally, recent works on the combination of MD with the LDAC process for energy-efficient and low CO_2 emission thermal comfort provision in buildings are critically discussed and analyzed to shed light on the future direction of this innovative air-conditioning technology.

2. Liquid desiccant air-conditioning process

2.1 Working principles of the liquid desiccant air-conditioning process

The function of the LDAC process is to supply cool air with low humidity to buildings. To perform this function, LDAC removes excess moisture and heat from the supply air using a liquid desiccant solution as a dehumidifying medium (Gurubalan et al., 2019; Gurubalan and Simonson, 2021). In the LDAC operation, the liquid desiccant solution absorbs moisture and heat when in contact with the supply air and subsequently desorbs them into the exhaust air. Therefore, the LDAC process consists of at least an air dehumidifier and a liquid desiccant regenerator. In some cases, the LDAC process also entails a complementary cooler to further cool the dried air before distributing it to the buildings (Fig. 1) (Duong et al., 2019).

In the LDAC process, the water vapor pressure difference between the liquid desiccant solution and the air plays as the driving force to transfer the moisture (Gurubalan et al., 2019; Gurubalan and Simonson, 2021). In particular, the liquid desiccant solution has a low water vapor pressure because of its high desiccant concentration and low temperature. When the hot and humid supply air passes through the dehumidifier, the liquid desiccant solution absorbs moisture from, and hence dries, the supply air when they are in contact in the dehumidifier. In tandem with the moisture absorption, heat is transferred from the supply air to the liquid desiccant solution. As the result of moisture and heat transfer, the liquid desiccant solution dilutes and warms up, elevating its water vapor pressure and hence reducing the driving force for the moisture transfer from the supply air. This inevitably reduces the air dehumidification efficiency of the LDAC process. Therefore, the diluted and warm liquid desiccant solution eventually needs to be regenerated in the regenerator before being returned to the dehumidifier (Fig. 1).

The regeneration of liquid desiccant solution in the regenerator is fundamentally a reversion of air dehumidification whereby water is desorbed or removed from the diluted liquid desiccant solution to reconcentrate it. During this process, the diluted liquid desiccant solution is heated to raise its water vapor pressure and hence facilitate the water transfer from the liquid into the exhaust air. Subsequently, the reconcentrated liquid desiccant solution is cooled down before returning to the air dehumidifier. The changes in moisture content,

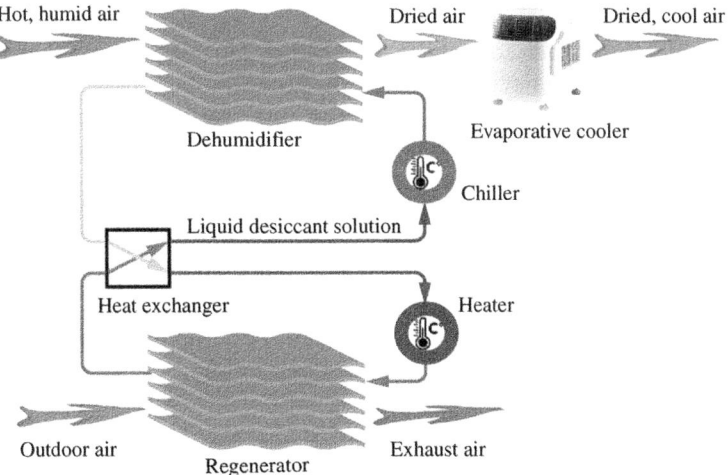

FIG. 1 The schematic diagram of an LDAC system, consisting of an air dehumidifier, a liquid desiccant solution regenerator, and an evaporative cooler. *From Duong, H.C., Ansari, A.J., Nghiem, L.D., Cao, H.T., Vu, T.D., Nguyen, T.P., 2019. Membrane processes for the regeneration of liquid desiccant solution for air conditioning. Curr. Pollut. Rep. 5, 308–318.*

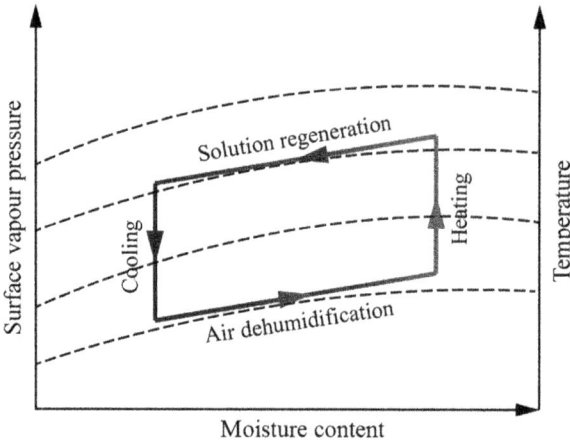

FIG. 2   Changes in moisture content, surface water vapor pressure, and temperature of liquid desiccant solution during the LDAC process. *Modified from Gurubalan, A., Maiya, M.P., Geoghegan, P.J., 2019. A comprehensive review of liquid desiccant air conditioning system. Appl. Energy 254, 113673.*

temperature, and surface water vapor pressure of the liquid desiccant solution during the LDAC process are illustrated in Fig. 2.

In theory, the air dehumidifier and liquid desiccant solution regenerator have similar working principles, except for the direction of moisture transfer from the air to the liquid desiccant solution and vice versa. As a result, they also have similar structures and designs, which can be classified into two types, namely direct contact and indirect contact. In the direct contact design, air and liquid desiccant solution come into direct contact, hence achieving a high mass and heat transfer between them. The direct contact type might be equipped with packing fills to enlarge the contact area and thus boost the heat and mass transfers. Nevertheless, the carryover of desiccant droplets into the outlet air is an intrinsic drawback of these systems. In order to overcome this problem, the indirect contact model is designed, whereby a membrane is implemented between the air and liquid desiccant solution (Zhao et al., 2015, 2017; Yong et al., 2017). This book chapter focuses on indirect contact liquid desiccant solution regenerators based on MD, which will be discussed further in later sections.

To achieve complete air-conditioning, the LDAC process can be coupled with a complementary cooler to control the temperature of the dried air after air dehumidification (Fig. 1). This cooler is responsible for only the sensible cooling of the dried air. The most used complementary coolers include vapor compression, vapor absorption, and evaporative cooling systems. This coupled system enables the independent control of the moisture and temperature (i.e., the latent and sensible heat) of the air. Therefore, it can achieve dried and cool air without the need for overcooling the air to below its dew point temperature as required in conventional vapor compression-based air conditioners. This attribute renders the LDAC process an energy-efficient alternative to most commercial conventional air conditioners.

2.2 Liquid desiccant solutions used in the liquid desiccant air-conditioning process

Liquid desiccants with a hygroscopic nature and low surface water vapor pressure are used in the LDAC process. The most commonly used liquid desiccant solutions include glycol-based solutions, halide salt solutions, and ionic liquids. Among these liquid desiccants, glycol-based solutions are the first ones to be explored for LDAC systems; however, these solutions are volatile and highly viscous and hence render economic and environmental consequences of LDAC systems (Lowenstein, 2008). On the other hand, solutions of halide salts (e.g., chloride or bromide) with lithium, potassium, calcium, and magnesium are much less volatile and less viscous while offering a noticeably low water vapor pressure. As a result, halide salt solutions have been extensively demonstrated as liquid desiccant solutions in many studies on the LDAC process. Nevertheless, as compared to glycol-based solutions, halide salt solutions are more corrosive and costlier, inevitably raising the operational costs of LDAC systems (Gurubalan and Simonson, 2021). Recently, ionic liquids have been proposed for the LDAC process because of their low water vapor pressure, low viscosity, and less corrosive impact to metals (Su et al., 2022). Some preliminary studies have demonstrated the feasibility of ionic liquids for LDAC systems, paving the way for their potential future applications (Luo et al., 2012; Watanabe et al., 2019).

The properties of liquid desiccant solutions exert profound impacts on the performance of the LDAC process. The typically required characteristics of liquid desiccant solutions are tabulated in Table 1. Some of the most vital properties of liquid desiccant solutions include equivalent specific humidity (i.e., surface water vapor pressure), viscosity, density, thermal conductivity, cost, and safety. These properties determine the performance efficiency and operational cost of not only the air dehumidifier but also the regenerator that compose the LDAC process.

TABLE 1 Typical characteristics of liquid desiccant solutions used in LDAC systems (Giampieri et al., 2018).

Classification	Property
Absorption characteristics	Equivalent specific humidity
	Heat of absorption
	Diffusion coefficient
	Energy storage capacity
Transport characteristics	Specific heat capacity
	Viscosity
	Density
	Thermal conductivity
	Surface tension
Economic and environmental characteristics	Cost
	Safety
	Material compatibility

Pure or single liquid desiccant solutions hardly possess all the desired properties listed in Table 1. In practice, mixed liquid desiccant solutions consisting of two or more desiccants are commonly used. For example, the mixture of calcium chloride and lithium chloride might offer similar absorption properties as compared to a single lithium chloride solution but at a lower cost. Moreover, additives can be added to liquid desiccant solutions to enhance their transport properties such as specific heat capacity, density, and thermal conductivity (Table 2). The additives include nanoparticles of copper (Ali et al., 2004), iron (Kang et al., 2008), silver (Wu et al., 2010), ferric and aluminum oxide (Yang et al., 2011), and carbon nanotubes (Kang et al., 2008). Besides nanoparticles, surfactants have been employed to reduce the surface tension of liquid desiccant solutions and hence improve their wettability on the contact surface (Kang et al., 2007; Cihan et al., 2017; Wen et al., 2018). It is noteworthy that studies on the LDAC process using additive-containing liquid desiccant solutions mostly focus on demonstrating the enhanced air dehumidification efficiency but fail to assess their subsequent regeneration. The altered properties of the mixed liquid desiccant solutions that are caused by the additives might pose considerable challenges to their regeneration stage. Indeed, the regeneration of liquid desiccant solutions plays a pivotal role in the LDAC process as it mainly accounts for the total energy consumption. The next section will analyze several methods that have been applied for regeneration of liquid desiccant solutions in the LDAC process.

TABLE 2   The additives and surfactants added to improve the properties of liquid desiccant solutions.

Additives	Solution(s)	Improved properties	Reference
Copper nanoparticles	NA	Thermal conductivity	Ali et al. (2004)
Iron nanoparticles	LiBr	Diffusion coefficient Thermal conductivity	Kang et al. (2008)
Silver nanoparticles	NH_3/H_2O	Specific heat capacity	Wu et al. (2010)
Ferric and aluminum oxides	NH_3/H_2O	Viscosity Diffusion coefficient	Yang et al. (2011)
Carbon nanotubes	LiBr	Thermal conductivity Diffusion coefficient	Kang et al. (2008)
Surfactants	LiCl	Surface tension	Kang et al. (2007), Cihan et al. (2017), and Wen et al. (2018)

2.3 Regeneration of liquid desiccant in the liquid desiccant air-conditioning process

Regeneration of liquid desiccant solution is an integral stage of the LDAC process. The air dehumidification efficiency of the LDAC process is directly dependent on the moisture absorption capacity of the liquid desiccant solution, which is a function of several factors, most notably including the solution concentration and temperature. During the air dehumidification stage, the moisture absorption and heat transfer from the supply air to the liquid desiccant solution result in a decrease in the concentration but an increase in the temperature of the solution. These changes raise the surface water vapor pressure (i.e., equivalent specific humidity) of the liquid desiccant solution, leading to a reduction in the water vapor pressure difference between it and the supply air, which is the driving force of the moisture transfer. Consequently, regeneration of the weak liquid desiccant solution to reconcentrate and cool it down is required to restore its moisture absorption capacity and the air dehumidification efficiency of the LDAC process.

In most current LDAC systems, the weak liquid desiccant solution is regenerated using the conventional thermal evaporation method. In principle, the weak liquid desiccant solution is heated up to about 70–90°C to raise its surface water vapor pressure. The hot solution is then circulated through the regenerator with a warm and dry scavenging airstream. Inside the regenerator, these two fluids can be arranged in the counter- or cross-flow currents to increase their contact. Due to its higher water vapor pressure, the desiccant solution releases moisture into the scavenging air and hence is reconcentrated. After flowing through the regenerator, the reconcentrated desiccant solution is cooled down to further reduce its surface water vapor pressure. The concentrated and cool desiccant solution with regenerated moisture absorption capacity is subsequently returned to the strong solution storage for the next cycle of air dehumidification.

Thermal evaporation regenerators are classified into three different types according to the structure of a solution/air contactor used inside the regenerators. The function of this contactor is to promote the contact and to regulate the residence of the liquid desiccant solution and scavenging airstream inside the regenerators. These three types are spray tower, packed bed tower, and falling film tower. As visualized by their names, these regenerators have different internal structures and principles of contact, and therefore they exhibit different operational costs and regeneration efficiency.

In all the three aforementioned types of regenerators, direct contact between the liquid desiccant solution and scavenging air is maintained to facilitate the heat and mass transfer and hence achieve high moisture desorption rates. However, this type of contact poses a considerable risk of desiccant carryover. When the desiccant solution and the scavenging airstreams come into direct contact under their counter- or cross-flow mode, tiny desiccant droplets at the surface of the solution can be swept into and carried over with the air. Desiccant carryover is considered a bottleneck of the thermal evaporation regeneration method. It not only leads to the desiccant solution replenishment (i.e., thus increasing the operational cost) for the LDAC process, but also causes detrimental problems to building equipment and occupants due to the corrosive nature of desiccant solutions.

To mitigate the problems associated with desiccant carryover, indirect contact regenerators based on membrane processes have been trialed. In these systems, a semipermeable membrane is used to facilitate the separation of water from the liquid desiccant solution, hence reconcentrating it. Several membrane processes, including reverse osmosis (RO), electrodialysis (ED), forward osmosis (FO), and MD, have been integrated into the desiccant solution regenerators. The advantages and drawbacks of these membrane processes for regeneration of liquid desiccant solution application have been thoroughly discussed in a recent comprehensive review (Duong et al., 2019). A great number of experimental and simulation investigations have been conducted to elucidate the feasibility of these integrated systems. The proposed membrane processes have been demonstrated to effectively eradicate the issue of desiccant carryover owing to their excellent membrane selectivity. However, it appears that the high concentration of liquid desiccant solutions used in LDAC systems presents a serious challenge to the pressure-driven, electrically driven, and osmotically driven membrane processes. Their water flux is heavily dependent on the salt concentration and the osmotic pressure of the liquid desiccant solution. On the other hand, MD can be a potential alternative since it is a thermally driven membrane process and its performance is less affected by the high concentration of liquid desiccant solutions. The MD process also enables the usage of waste heat and solar thermal energy since heating is its main energy input. Thus, it is considered an ideal process to be coupled with LDAC for cost-effective and environmentally friendly thermal comfort provision in buildings. The next section provides fundamentals of the MD process with detailed discussions on its attributes that render it destined for regeneration of liquid desiccant solutions in LDAC systems.

3. Membrane distillation process

3.1 Fundamentals and configurations of the membrane distillation process

MD has been emerging as a viable process for treatment of challenging saline waters including liquid desiccant solutions used in the LDAC process (Duong et al., 2020a, 2020b; Qasim et al., 2021). MD is fundamentally a hybrid process that combines membrane separation with thermal distillation. In the MD process, a hydrophobic microporous membrane is deployed to separate hot saline water from cold and fresh distillate. The hydrophobic nature and microporous structure of the membrane allows for the transfer of water vapor and volatile compounds while preventing the penetration of liquid water through the membrane. As a result, when there is a transmembrane temperature difference between the two sides of the membrane, water evaporates at the membrane surface on the feed side (i.e., hot saline water), diffuses through the membrane pores in vapor form, and then condenses into liquid water (distillate) on the permeate side of the membrane (i.e., cold distillate). As only water vapor and volatile compounds can transfer through its membrane, the MD process can theoretically achieve a complete salt rejection, which is superior to the salt rejections offered by most pressure-driven membrane desalination processes including RO and nanofiltration (NF) (Alkhudhiri et al., 2012). Therefore, the MD process can provide an ideal solution to the desiccant carryover issue during the regeneration of liquid desiccant solutions used in LDAC systems (Duong et al., 2019).

One critical attribute of the MD process is its workability with hypersaline waters such as liquid desiccant solutions. Unlike in RO and NF, in the MD process the driving force for mass transfer (i.e., transfer of water) is the water vapor pressure difference between the feed and permeate sides of the membrane, which can be induced by maintaining a transmembrane temperature gradient via heating and cooling and/or applying vacuum or sweeping gas. In other words, the driving force of the MD process is not strictly subject to the salinity and osmotic pressure of the feed water as in RO and NF. Given this attribute, the MD process has been explored for strategic treatment of hypersaline waters including brines from the RO and NF desalination, draw solutions in FO, and diluted liquid desiccant solutions from LDAC systems.

Besides its complete salt rejection and workability with the hypersalinity, the MD process for regeneration of liquid desiccant solutions facilitates the utilization of low-grade waste heat and solar energy in LDAC systems. The MD process is operated without the need for heating the feed water to its boiling point temperature as required in conventional thermal distillation methods. The MD desalination process has practically been efficiently operated at a feed temperature as low as 40°C. This mild feed temperature can be effectively sourced from low-grade waste heat and/or solar thermal energy, thus reducing the energy cost of the LDAC process when it is coupled with MD for liquid desiccant solution regeneration. It is noteworthy that the energy consumption of liquid desiccant regeneration contributes around 75% to the overall energy consumption of the LDAC process.

The MD process is operated in four basic configurations depending on the method applied to generate the transmembrane water vapor pressure difference. They include direct contact membrane distillation (DCMD), air gap membrane distillation (AGMD), vacuum membrane distillation (VMD), and sweeping gas membrane distillation (SGMD) (Fig. 3). These configurations share the same structure of the feed channel but have a distinct permeate channel arrangement. As a result, they have different attributes that render their suitability for specific desalination applications.

Among the four basic configurations, DCMD is the simplest one and most used for various MD applications. In the DCMD configuration, a cold distillate is circulated on the permeate side and is in direct contact with the membrane. The transmembrane water vapor pressure gradient is solely induced by the temperature difference between the hot saline feed water and the cold distillate, and the condensation of water vapor to distillate occurs right at the permeate membrane surface inside the membrane module (Fig. 3). This offers simplicity to the systems and processes based on DCMD as they require only two water circulating pumps and heating and cooling sources for their operation. Therefore, this configuration is utilized in most lab-scale studies on MD for diverse desalination applications, including regeneration of liquid desiccant solutions used in the LDAC process. However, the direct contact arrangement in the DCMD configuration leads to significant heat conduction across the membrane from the hot feed to the cold distillate. The conductive heat transferred across the membrane is considered the heat loss in the MD process. As a result, the energy efficiency of the DCMD configuration is much lower than that of other configurations. Thus, DCMD is not a favorable configuration with respect to energy efficiency, which is a critical factor for large-scale desalination applications. Indeed, only a few studies on pilot scale seawater desalination using the DCMD configuration have been reported in the literature (Song et al., 2008; Dow et al., 2016).

In the AGMD configuration, an air gap is maintained between the membrane surface and the cold water by deploying a condensate film in the permeate channel. Water vapor after crossing the membrane pores diffuses through the air gap and condenses to distillate on

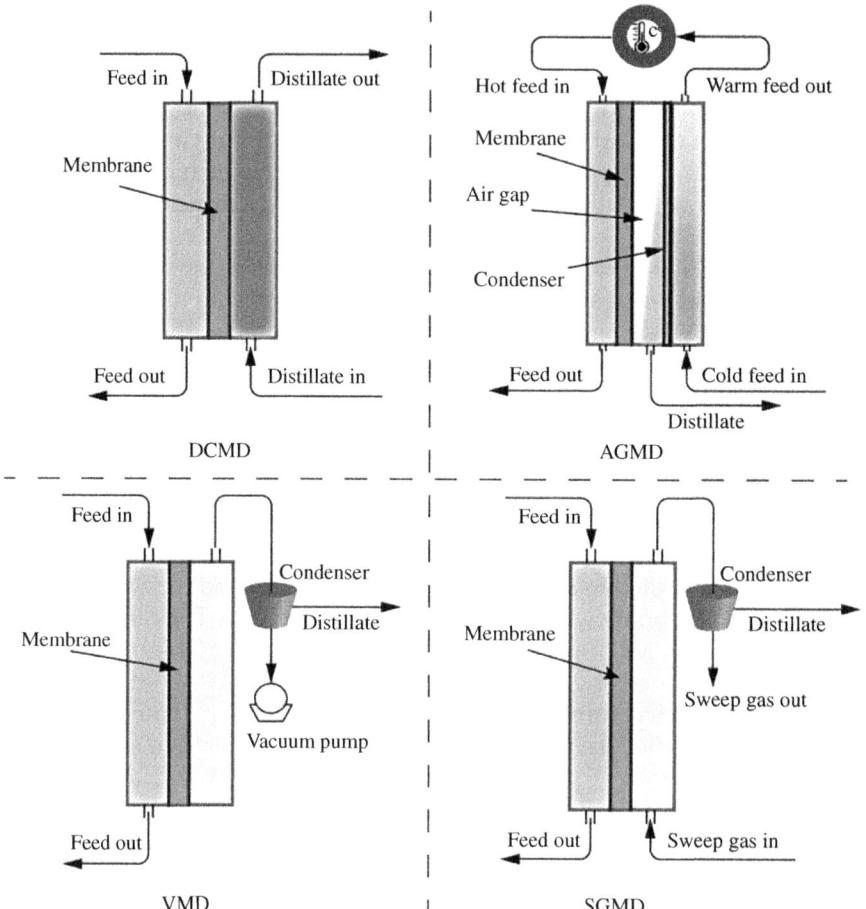

FIG. 3 The four basic configurations of the MD process: (A) direct contact membrane distillation (DCMD), (B) air gap membrane distillation (AGMD), (C) vacuum membrane distillation (VMD), and (D) sweeping gas membrane distillation (SGMD).

the surface of the condensate film. Because of the air gap, the heat loss due to conduction through the membrane is reduced; however, the air gap also adds resistance to the mass transfer in the AGMD configuration. As a result, the AGMD configuration has an improved thermal efficiency but lower water flux as compared to the DCMD one. The air gap created in the AGMD configuration also enables the recovery of the latent heat of condensation to preheat the feed water, hence reducing the process energy consumption (Duong et al., 2016a, 2016b). With these attributes, the AGMD configuration is widely used for pilot or large-scale seawater desalination applications whereby low energy consumption is a decisive factor (Duong et al., 2021b). Nevertheless, the limited driving force and low water flux restrict the application of the AGMD configuration for treatment of hypersaline waters such as liquid desiccant solutions. Literally none of the reported studies on MD regeneration of liquid desiccant solutions used in LDAC systems are based on the AGMD configuration.

The drawbacks of both DCMD (i.e., low thermal efficiency) and AGMD (i.e., limited water flux) are mitigated in the VMD and SGMD configurations. In VMD and SGMD, vacuum or sweeping gas is applied on the permeate side of the membrane to generate the transmembrane water vapor pressure gradient. In other words, the driving force for water vapor transfer in VMD and SGMD can be induced by both the temperature and pressure difference. Therefore, these two configurations allow to achieve a higher driving force and hence improve the water flux as compared to DCMD and AGMD. The vacuum pressure and sweeping gas also help reduce the heat conduction through the membrane, thus improving the thermal efficiency of the configurations. Nevertheless, these two types of design require additional equipment and external condensers for water vapor condensation to distillate, hence making them more complex than DCMD and AGMD. Given these attributes, VMD and SGMD are best suited for applications in which hypersaline solutions need to be treated to recover the concentrated solutions and the attainment of distillate (freshwater) is not the ultimate purpose. Regeneration of liquid desiccant solutions in the LDAC process might be one kind of these applications.

Besides the four basic configurations, the MD process has been applied in several improved or modified configurations such as permeate gap membrane distillation (PGMD), material gap membrane distillation (MGMD), multieffect membrane distillation (MEMD), and multieffect vacuum membrane distillation (MEVMD) (Drioli et al., 2015). Compared to the basic ones, the modified configurations exhibit enhanced performance in terms of water flux and thermal efficiency, but with the cost of process complexity. It is noteworthy that these modified configurations have been explored and investigated for seawater desalination application, but not yet for regeneration of liquid desiccant solutions in LDAC systems. Until now, most reported studies on MD regeneration of LDAC liquid desiccant solutions rely on DCMD and VMD configurations. However, it is envisaged that modified configurations will be dominant in pilot- or large-scale MD regeneration of liquid desiccant solutions.

3.2 Performance of the membrane distillation process

The MD process has been explored for various applications such as desalination of seawater or waste brine for freshwater (distillate) production and reconcentration/regeneration of saline solutions used in the membrane crystallization (MC), forward osmosis (FO), and LDAC processes. For these applications, the performance of the MD process is commonly assessed with regard to water flux, salt rejections, and energy consumption. These performance indexes are strictly regulated by the heat and mass transfer through the MD membrane.

3.2.1 Water flux and polarization effects of the membrane distillation process

In the MD process, water transfers in vapor form through the membrane from the feed to the permeate. This transfer of water vapor is proportional to the water vapor pressure difference across the membrane and is expressed as:

$$J = C_m \left(P_{m \cdot f} - P_{m \cdot p} \right) \tag{1}$$

where J is the rate of water vapor transfer, which is termed as water flux (L/m^2 h); C_m is the membrane permeability (L/m^2 h Pa); $P_{m \cdot f}$ and $P_{m \cdot p}$ are the water vapor pressure at the membrane feed and permeate surfaces, respectively. While the water vapor pressure at the feed membrane surface $P_{m \cdot f}$ is a function of the feed water salt concentration and temperature, the water vapor at the permeate membrane surface $P_{m \cdot p}$ is induced by different methods depending on the configuration of the MD process. For example, $P_{m \cdot p}$ is the partial water vapor pressure of the cold distillate in DCMD and AGMD, the sweeping gas in SGMD, and the vacuum in VMD.

The transfer of water vapor through the membrane pores is also controlled by the properties of the membrane, represented by the membrane permeability C_m. Indeed, C_m is a complex function of membrane properties (e.g., pore size, porosity, pore tortuosity, and membrane thickness) and the temperature and pressure of water vapor inside the membrane pores. In the MD literature, different equations are proposed for the calculation of C_m, depending on the mechanism of water vapor transfer through the membrane pores. Three mechanisms are largely used for the modeling of water vapor transfer through the MD membrane pores, namely Knudsen diffusion, ordinary molecular diffusion, and Poiseuille flow (Alkhudhiri et al., 2012). The transfer of water vapor inside the MD membrane pores might occur following more than one mechanism; however, for simplicity's sake, the calculation of C_m might be based on the dominant mechanism(s). For example, in the DCMD process using the membrane with a small pore size (i.e., pore radius d_p < the mean free path of the water vapor molecule λ), Knudsen diffusion is the dominant mechanism, and C_m is calculated as follows:

$$C_m = \frac{2\pi}{3} \frac{1}{RT} \left(\frac{8RT}{\pi M_w}\right)^{1/2} \frac{r^3}{\tau \delta} \tag{2}$$

where r, τ, and δ are the membrane pore size, pore tortuosity, and membrane thickness.

For the DCMD process with the membrane of large pore sizes (i.e., $d_p > 100\lambda$), the ordinary molecular diffusion represents the transfer of water vapor inside the pores. In this case, C_m is calculated using Eq. (3) as below:

$$C_m = \frac{\pi}{RT} \frac{PD}{P_{air}} \frac{r^2}{\tau \delta} \tag{3}$$

In the AGMD process, stagnant air exists in the membrane pores; therefore, ordinary molecular diffusion is the best suited mechanism for the calculation of C_m. In VMD and SGMD, due to the transmembrane pressure gradient induced by the vacuum or sweeping gas, Knudsen diffusion and/or the Poiseuille flow are the dominating mechanisms for the transfer of water vapor molecules. It is noteworthy that C_m exhibits higher values under the order of Poiseuille flow, Knudsen diffusion, and ordinary molecular diffusion. As a result, under the same operating conditions, higher water flux is achieved for the VMD and SGMD configurations compared to that for the DCMD configuration. Among the DCMD processes, higher water flux is associated with the membrane of small pore sizes in which the Knudsen diffusion is dominant (Khayet et al., 2006). Of particular note, the AGMD configuration achieves the lowest water flux due to its ordinary molecular mechanism.

In the MD process, the transfer of water vapor through the membrane always occurs simultaneously with the transfer of heat from the feed water to the permeate. This simultaneous

Hot feed

Cold permeate

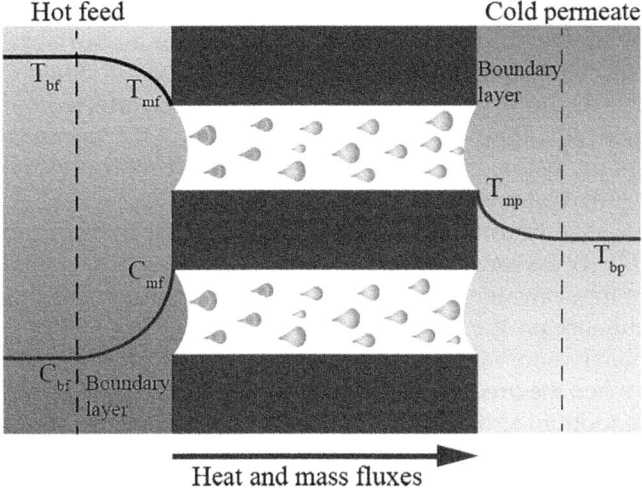

Heat and mass fluxes

FIG. 4 The formation of the boundary layers and temperature and concentration polarization effects in the DCMD process.

mass and heat transfer across the membrane inevitably leads to the formation of concentration and temperature boundary layers adjacent to the membrane surfaces in the feed and permeate streams. Due to these layers, the temperature and salt concentrations in the bulk streams are different from those at the membrane surfaces (Fig. 4). These effects are termed temperature and concentration polarization. Both the temperature and concentration polarization effects render the water vapor pressure difference between the two membrane surfaces smaller than that between the bulk feed and permeate streams, thus negatively affecting the water flux of the MD process. In the MD literature, the temperature and concentration polarization effects are quantified using the temperature polarization coefficient (ψ) and the concentration polarization coefficient (Φ), respectively. For the DCMD process, ψ and Φ are calculated as follows:

$$\psi = \frac{T_{m \cdot f} - T_{m \cdot p}}{T_{b \cdot f} - T_{b \cdot p}} \tag{4}$$

$$\Phi = \frac{C_{m \cdot f}}{C_{b \cdot f}} \tag{5}$$

where $T_{m \cdot f}$, $T_{m \cdot p}$, $T_{b \cdot f}$, and $T_{b \cdot p}$ are the temperatures at the feed and permeate membrane surfaces and in the bulk feed and permeate, respectively; $C_{m \cdot f}$ and $C_{b \cdot f}$ are the salt concentrations at the feed membrane surface and in the bulk feed solution.

The magnitudes of temperature and concentration polarization depend on the configurations and the operating conditions of the MD process. The ideal MD process is expected to possess temperature and polarization coefficients of unity (i.e., equal to 1), which presents the condition of no temperature and concentration polarization effects. In practice, polarization effects always occur in the MD process; therefore, their coefficients deviate from unity. For example, the temperature polarization coefficient is in the range of 0.4–0.7 for the DCMD configuration depending on the feed temperature and concentration and the feed and distillate velocity. Improving the fluid turbulence near the membrane surfaces by raising the fluid velocity and using spacers helps minimize the boundary layers and hence mitigate the polarization effects in the MD process. Polarization effects are also affected by water flux through the membrane. Specifically, they are exacerbated when the MD process is operated at a high water flux (Duong et al., 2015a, 2015b, 2015c). As a result, flux control is recommended as a method to mitigate polarization effects in the MD process (Duong et al., 2015a, 2015b, 2015c).

It is noteworthy that concentration polarization is deemed to have a negligible impact on water flux in the MD process for seawater desalination applications. In these applications, the salt concentration of the seawater feed (i.e., around 32,000 ppm) exerts little effect on its water vapor pressure and hence on the driving force for the mass transfer of the MD process. Experimental studies have reported that the water flux of seawater MD desalination processes reduced negligibly when the process water recovery and the feed salt concentration increased at a low level (Hickenbottom and Cath, 2014; Duong et al., 2015a, 2015b, 2015c). Therefore, the impact of concentration polarization on the MD water flux can be ignored. However, in the MD process of hypersaline solutions such as liquid desiccants, the impact of salt concentration on the water flux is significant. In this context, the impact of concentration polarization on water flux must be seriously considered.

3.2.2 Salt rejection and wettability of the membrane in the membrane distillation process

The MD process achieves its separation efficiency via evaporation of liquid water on the feed side and condensation of water vapor on the permeate side of the membrane (Chamani et al., 2021). The transfer of water through the MD membrane occurs in vapor form. Under this ideal condition, only water vapor and volatile compounds are allowed to transfer from the feed water to the permeate. Thus, the MD process can completely retain dissolved salts and other involatile contaminants in the feed water and achieve a complete salt rejection. The ideal condition and the resulting complete salt rejection can be achieved only when all the membrane pores are dry during the MD process. However, in practice, some membrane pores might be wetted, and then liquid water and dissolved salts can penetrate the wetted pores from the feed water to the permeate, leading to compromised salt rejection and contamination of the permeate.

The salt rejection of the MD process is calculated as follows:

$$\text{Rejection} = \left(1 - \frac{C_f}{C_p}\right) \times 100\% \tag{6}$$

where C_f and C_p are the salt concentrations of the feed and permeate, respectively. For dilute saline solutions like seawater, salt concentration is linearly proportional to electrical conductivity. As a result, the salt rejection of the MD process is assessed by measuring the electrical conductivity of the feed and permeate. Most MD processes reported in the literature achieve excellent salt rejection (i.e., >99.9%) under stable conditions (i.e., low water recovery and low feed water organic contents). However, when operating under stressed conditions like high water recovery and added organic matters in the feed, membrane pore wetting occurs, and consequently the MD salt rejection deteriorates.

The nonwet condition of the MD membrane pores is induced by the hydrophobic nature and microstructure of the membrane. Due to its small pore size and surface hydrophobicity, the membrane helps maintain the liquid-vapor interfaces at the entrances of the pores (Fig. 4), thus preventing liquid water from penetrating through the pores when the hydrostatic pressure does not exceed the liquid entry pressure (*LEP*). *LEP* is a critical parameter regarding membrane pore wetting in the MD process, where the membrane with a high value of *LEP* is more resistant to membrane pore wetting. Indeed, *LEP* is affected not only by the membrane properties, but also by the characteristics of the feed water as expressed in the equation below:

$$LEP = -\frac{2B\gamma \cos\theta}{r_{max}} \tag{7}$$

where γ is the surface tension of the feed water; B is the pore geometry coefficient; θ is the contact angle of the membrane surface; and r_{max} is the maximum pore radius. Thus, any factors that reduce the value of *LEP* or increase the hydrostatic pressure of the feed and permeate streams might cause the risk of membrane pore wetting in the MD process. These factors include the presence of surfactants and organic matters in the feed water and the formation of foul/scale layers on the membrane surface. As a result, the pretreatment of feed water and process operating condition control are often required to prevent membrane pore wetting in the MD process.

The risk of membrane pore wetting is low for MD regeneration of liquid desiccant solutions. There are two reasons for this. Firstly, compared to wastewater or seawater, liquid desiccant solutions often contain negligible contents of organic matters and surfactants as they are prepared by dissolving desiccant salts (e.g., LiCl, LiBr, and $CaCl_2$) in clean water. Secondly, their high concentration of salts increases the surface tension of the solutions (i.e., γ), hence raising the value of *LEP* expressed in Eq. (7). Owing to these characteristics, the risk of membrane pore wetting during MD regeneration of liquid desiccant solutions might be lower than that experienced in the MD process for treatment of wastewater or seawater. However, when operating the MD process of liquid desiccant solutions under extreme conditions (i.e., high feed concentration, high water recovery, and low feed velocity), membrane fouling and/or scaling might occur and alter the interaction between the feed water and membrane surface. As a result, the risk of membrane pore wetting might be heightened for these circumstances.

3.2.3 *Thermal efficiency and energy consumption of the membrane distillation process*

Thermal efficiency and energy consumption are critical performance indices of the MD process. As a thermally driven separation process, MD requires heating and cooling to facilitate the evaporation of liquid water and condensation of water vapor to achieve its separation efficiency, while electrical energy is consumed only for water circulation. Therefore, thermal energy for heating and cooling is the major energy input of the MD process. Indeed, previous pilot studies have demonstrated that the thermal energy consumption of the MD process is several orders of magnitude higher than its electrical energy input (Duong et al., 2015a, 2015b, 2015c, 2016a, 2016b).

It is noteworthy that most MD studies regarding thermal efficiency and energy consumption have focused on thermal energy for heating of the feed water while ignoring the cooling of the permeate. Actually, cooling is inevitable for the MD process to generate the driving force for the mass transfer and to obtain distillate (i.e., freshwater). However, most MD studies assume that cooling can be supplied by mild temperature cooling sources available on-site (i.e., cool seawater or air). As a result, thermal energy consumption of the MD process is largely calculated based on heating consumed by the feed water loops.

MD is considered a low and thermally efficient desalination process. In this process, the transfer of water always occurs simultaneously with the transfer of heat through the membrane. Of the total heat transfer through the MD membrane, the heat transfer associated with the mass transfer (i.e., latent heat) is useful, while the heat transfer caused by conduction via the membrane is deemed to be the heat loss. Thermal efficiency is an important parameter presenting the ratio between the useful heat and the heat loss of the MD process, and it is calculated as follows:

$$\Pi = \frac{J\Delta H_v}{J\Delta H_v + \frac{k_m}{\delta}\left(T_{m\cdot f} - T_{m\cdot p}\right)} \tag{8}$$

where $J\Delta H_v$ is the latent heat associated with water transfer, and $\frac{k_m}{\delta}\left(T_{m\cdot f} - T_{m\cdot p}\right)$ is the heat conducted through the membrane. The thermal efficiency of the MD process depends on various factors, including membrane properties and operating conditions, but it can be as low as 0.6 for the DCMD configuration whereby the cold distillate is in direct contact with the membrane (Fane et al., 1987). Therefore, methods to improve thermal efficiency in the MD process have been centered on reducing the conductive heat loss through the membrane and promoting the latent heat transfer. These methods include increasing the membrane porosity, optimizing the membrane thickness, operating the process at high water flux, and utilizing modified MD configurations.

The specific thermal energy consumption (STEC) of the MD process for desalination applications is calculated as follows:

$$STEC = \frac{E_{input}}{Q_{distillate}} \tag{9}$$

where STEC is measured in kWh/m³; E_{input} is the heating required on the feed water loops (kJ); and $Q_{distillate}$ is the flow rate of the obtained distillate (L/h). It is interesting to note that the thermal energy consumption of most MD processes reported in the literature is based on the volume of the obtained distillate. This is because the MD process is largely applied for desalination applications in which freshwater (i.e., distillate) is the primary product. For MD regeneration of liquid desiccant solutions, the attainment of freshwater is not the ultimate purpose, but the increased concentration of liquid desiccants is. Thus, the specific thermal energy consumption of MD regeneration of liquid desiccant solutions might not be comparable to that of seawater MD desalination or wastewater MD treatment. This will be discussed further in the next section.

4. Membrane distillation regeneration of liquid desiccant solutions for energy-efficient liquid desiccant air-conditioning

The MD process was first proposed and investigated for desalination of seawater for freshwater provision. However, recent years have witnessed a gradual change in MD applications from seawater desalination to strategic treatment of challenging saline waters such as liquid desiccant solutions used in LDAC systems. The change in MD applications can be attributed to the distinguishing characteristics of this process, namely excellent salt rejection, compatibility with hypersalinity, and the utilization of low-grade waste heat and solar energy. Moreover, the low thermal efficiency and high energy consumption of the MD process as compared to other mature desalination technologies (e.g., RO and conventional thermal distillation) are also factors that orientate MD to its strategic applications.

The MD process has been combined with LDAC systems for liquid desiccant solution regeneration with two distinct purposes: (1) mitigating the desiccant carryover issue of the conventional regeneration method and (2) reducing the energy cost of LDAC systems. To achieve these targets, most studies on MD regeneration of liquid desiccant solutions used in LDAC systems have focused on demonstrating the technical feasibility of MD for reconcentration of liquid desiccant solutions using renewable energy. In this context, the technical feasibility of MD is assessed with respect to water flux, membrane salt rejection, membrane fouling/scaling formation, thermal efficiency, and thermal energy consumption. This section provides an in-depth discussion on notable demonstrations of MD for regeneration of liquid desiccant solutions.

Duong et al. (2017) systematically investigated the regeneration of a lithium chloride (LiCl) liquid desiccant solution by a lab-scale DCMD process. The experimental results proved that the DCMD operation at the feed and distillate temperatures of 65°C and 25°C, respectively, could raise the concentration of the LiCl solution from 20% to 29% without any problems with membrane fouling/scaling and membrane wetting. The performance indices of this process, including water flux, regeneration capacity (i.e., the LiCl concentration increase from the inlet to the outlet), and energy consumption, were profoundly affected by the operating conditions such as feed temperature and concentration and water circulation rates. Operating the DCMD process at higher feed temperatures and water circulation rates was beneficial because it enhanced the water flux and regeneration capacity but reduced the process energy consumption. On the other hand, increasing LiCl concentration negatively affected the performance of the DCMD process as this led to linear reduction in water flux and regeneration capacity but a hyperbolic increase in the process energy consumption. The results reported by Duong et al. (2017) imply that despite its workability with high salinity, the high concentration of the LiCl desiccant solution remains a considerable technical challenge to the DCMD regeneration process with respect to energy consumption.

In another study, Duong et al. (2018) demonstrated the regeneration capacity of the DCMD process using three different liquid desiccant solutions prepared from single LiCl, $CaCl_2$, and mixed LiCl/$CaCl_2$. The reported results manifest that the type of salts used in preparing the liquid desiccant solutions exerted huge impacts on the regeneration capacity of the DCMD process. Among the three solutions investigated in the work (Duong et al., 2018), the LiCl solution is the most hygroscopic, and hence the LDAC system using it as a desiccant can achieve the highest air dehumidification efficiency. However, the LiCl solution is the most challenging one to be regenerated in the DCMD process. Indeed, the DCMD process exhibited the lowest water flux and regeneration capacity when using the LiCl solution feed as compared to the single $CaCl_2$ and mixed LiCl/$CaCl_2$ solutions (Fig. 5). The solutions of LiCl, $CaCl_2$, and mix LiCl/$CaCl_2$ could be reconcentrated to 29%, 40%, and >50%, respectively. Moreover, the presence of desiccant salts (e.g., LiCl and $CaCl_2$) at high concentration in liquid desiccant solutions caused a discernible concentration polarization effect to the DCMD process. In the

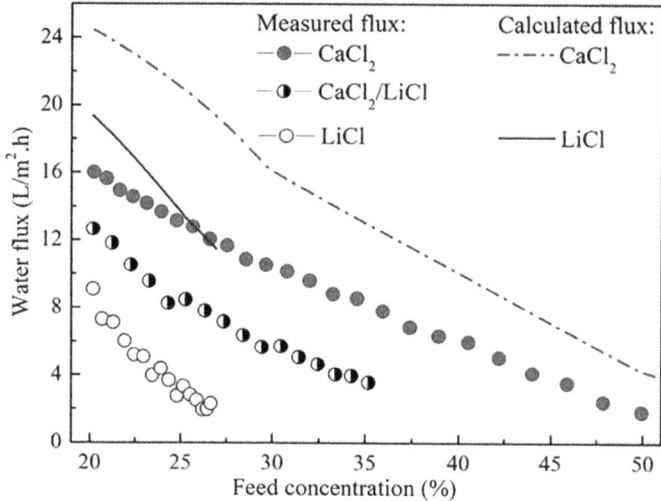

FIG. 5 The impacts of desiccant salts on the water flux of the DCMD process during the regeneration of liquid desiccant solutions prepared from single LiCl, CaCl₂ and mixed LiCl/CaCl₂. *From Duong, H.C., Álvarez, I.R.C., Nguyen, T.V., Nghiem, L.D., 2018. Membrane distillation to regenerate different liquid desiccant solutions for air conditioning. Desalination 443, 137–142.*

DCMD process of the single LiCl and CaCl₂ solutions, the experimentally measured water flux was reduced by nearly a half as compared with the calculated water flux which excluded the concentration polarization effect.

The impacts of concentration and temperature polarization effects on water flux and regeneration capacity during the DCMD process of LiCl solution were insightfully elucidated by analyzing the heat and mass transfer through the membrane using computer simulation (Duong et al., 2020a). In a recent work, Duong et al. (2020a) used a descriptive heat and mass transfer model to quantitatively describe the heat and mass transfer and to examine the negative impacts of temperature and concentration polarization on the water flux and LiCl concentration increase during its DCMD regeneration. The simulation results reveal significant impacts of both temperature and concentration polarization on the transmembrane water vapor pressure gradient (ΔP_m), rendering it only two-thirds of the water vapor pressure difference between the bulk feed and distillate (ΔP_b) (Fig. 6). Together with the reduced membrane mass transfer coefficient, the reduction in ΔP_m caused by the temperature and concentration polarization effects inevitably led to the decline in the DCMD process water flux. The simulation results also signify that during the DCMD regeneration of LiCl desiccant solution, temperature polarization was dominating as compared to concentration polarization; nevertheless, the negative impact of concentration polarization on the water flux and regeneration capacity of the DCMD process was discernible, given the high concentration of the LiCl solution.

Liu et al. (2022) also investigated the performance of DCDM for the regeneration of liquid desiccant solutions, but with a particular interest in the tendency of membrane scaling and membrane wetting. The experimental results reported by the authors demonstrate that

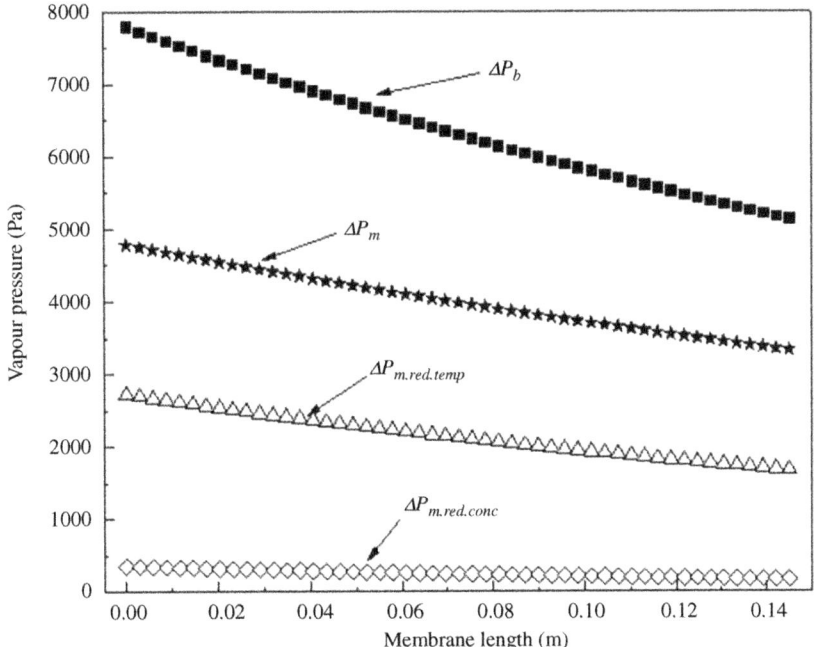

FIG. 6   Reduction in the water vapor pressure difference between the feed and permeate membrane surfaces as compared to that between the bulk feed and permeate along the membrane leaf inside a DCMD membrane module. *From Duong, H.C., Ansari, A.J., Cao, H.T., Nguyen, N.C., Do, K.-U., Nghiem, L.D., 2020. Membrane distillation regeneration of liquid desiccant solution for air-conditioning: insights into polarisation effects and mass transfer. Environ. Technol. Innov. 19, 100941.*

elevating the feed temperature improved the water flux and the regeneration capacity of the DCMD process but compromised the membrane surface hydrophobicity, hence incurring the risk of membrane wetting. High water flux resulting from the elevating feed temperature also aggravated the concentration polarization effect and thus promoted the formation of loose and dense scale layers of LiCl crystals on the membrane surface. This, together with the reduced membrane surface hydrophobicity, heightened the risk of membrane wetting. As a result, the DCMD process of the LiCl solution at the feed temperature of 65°C and feed concentration of >30% exhibited a negative water flux whereby the transfer of water was reversed from the distillate to the feed. The results reported in the work by Liu et al. (2022) indicate that membrane scaling and the resultant membrane wetting might be a challenge for the DCMD regeneration of liquid desiccant solution at high feed temperatures and concentrations. Thus, further studies should be prioritized to shed light on membrane scaling and wetting in long-term and durable MD regeneration of liquid desiccant solutions used in LDAC systems.

It is noteworthy that the DCMD configuration is commonly used for studies on MD regeneration of liquid desiccant solutions due to its simple arrangement and ease of operation. However, as the driving force for the mass transfer in this configuration is induced only by the temperature difference between the hot feed and the cold distillate, the DCMD process achieves a limited water flux and thus a low regeneration capacity, particularly when treating liquid desiccant solutions of high concentrations. For example, the DCMD process trialed in Duong et al. (2017) could raise the LiCl concentration to only 29% at the feed and distillate temperature of 65°C and 25°C, respectively. For higher LiCl concentrations, the DCMD process is required to operate at higher feed temperatures. However, these conditions are not beneficial to the DCMD process in terms of energy consumption, and, as discussed previously, it might aggravate the polarization effects and membrane scaling/wetting. As a result, researchers are more interested in using VMD for regeneration of liquid desiccant solutions with high salt concentrations.

Zhou et al. (2018, 2019, 2020a, 2020b) have extensively investigated VMD regeneration of liquid desiccant solutions using both experimental and simulation approaches. For example, Zhou et al. (2018) developed a mathematical model based on the heat and mass balance to simulate the VMD and traditional thermal (TH) regeneration processes. The VMD and TH processes were compared with respect to their regeneration capacity and energy consumption. The simulation results prove that the VMD regeneration method was superior to the TH one when the feed solution temperature was below 60°C, which is the temperature that can be supplied by low-grade energy. Moreover, the VMD regeneration consumed energy up to 37% less than the TH process. In experimental works (Zhou et al., 2020a, 2020b), Zhou et al. demonstrated that the VMD process using hollow fiber membranes with the effective length of 0.52 m could increase the LiCl concentration from 30% to a maximum of 31.2% after one cycle through the membrane module. The feed temperature and the initial LiCl concentration were the most critical operating parameters for the VMD process to achieve its target regeneration capacity (i.e., amount of concentration increase) with a given membrane length. To further examine the effects of the hollow fiber membrane length on the regeneration capacity, in a recent study Zhou et al. (2020a) developed a simulation model for pilot scale membranes (with the membrane length ranging from 0 to 0.6 m). However, in this pilot scale simulation, the authors had to rely on the DCMD configuration for simplicity in modeling.

VMD has been used for regeneration of liquid desiccant solutions used in LDAC systems for both cooling and freshwater supplies, particularly for desert greenhouses (Lefers et al., 2016, 2018, 2019). Lefers et al. (2019) simulated a combined air dehumidification and VMD process using triple-pore hollow fiber membranes for humidity and temperature control and water irrigation of a greenhouse. The VMD regeneration process powered by solar energy could achieve a water flux from 2.8 to 7.0 L/m² h when operated at the feed temperature of 30°C and vacuum pressure of 3 mbar. The freshwater obtained from the VMD process had an electrical conductivity of <18 μS/cm; therefore, it was suitable for makeup irrigation water. In a recent study, Lefers et al. (2018) examined a lab-scale system for regeneration of calcium and magnesium chloride desiccant solutions to confirm the technical feasibility of the VMD process. The VMD process at the feed temperature of 50°C and vacuum pressure of 25 mbar achieved a water flux of 8 L/m²·h when the initial feed concentration was 30%. The lab-scale experimental results also demonstrated that the VMD process exhibited a comparable water flux when the $CaCl_2$ or $MgCl_2$ solution of the same concentration was used as the feed.

It is important to note that to date, most studies on MD regeneration of liquid desiccant solutions used for LDAC systems have been limited to lab-scale demonstrations and investigation. There have been only a few MD studies on this topic at the pilot scale level (Zhou et al., 2020a, 2020b; Duong et al., 2021a) as compared to a great number of pilot MD demonstrations for seawater desalination. This reveals the fact that the MD application for liquid desiccant solution regeneration is just at its initial stage and more future research should be devoted to this application. Moreover, as discussed above, even though MD has attracted increasing attention for regeneration of liquid desiccant solution, most MD studies have focused on the feasibility demonstration and elucidation of heat and mass transfer and polarization effects. To facilitate the realization of MD for liquid desiccant solution regenerations, pilot studies and large-scale demonstrations with focuses on membrane scaling/ membrane wetting and energy consumption are needed.

5. Conclusions and perspectives

Liquid desiccant air-conditioning (LDAC) has been deemed an environmentally friendly and energy-efficient process to meet the increasing demand for thermal comfort in buildings in the context of reduced CO_2 emission. The LDAC process essentially entails the air dehumidification and liquid desiccant solution regeneration, and the efficiency of the LDAC process is directly affected by the nature of liquid desiccant solution and the method to regenerate it. Membrane distillation (MD), which is a hybrid membrane separation process, has been explored for the regeneration of liquid desiccant solution due to its distinguishing attributes such as excellent salt rejection, workability with extreme salt concentrations, and the capability of using low-grade waste heat and solar thermal energy. However, the application of MD for regeneration of liquid desiccant solutions has been only at its initial phase, and most MD studies on this topic have been focused on technical feasibility demonstrations using lab-scale DCMD and VMD processes. These demonstrations have paved the way for progress on the MD regeneration of liquid desiccant solutions. Nevertheless, extensive MD studies on this topic at the pilot or large-scale operations are urgently needed to fully realize MD for the regeneration of liquid desiccant solution applications.

References

Ali, A., Vafai, K., Khaled, A.R.A., 2004. Analysis of heat and mass transfer between air and falling film in a cross flow configuration. Int. J. Heat Mass Transf. 47, 743–755.

Alkhudhiri, A., Darwish, N., Hilal, N., 2012. Membrane distillation: a comprehensive review. Desalination 287, 2–18.

Chamani, H., Woloszyn, J., Matsuura, T., Rana, D., Lan, C.Q., 2021. Pore wetting in membrane distillation: a comprehensive review. Prog. Mater. Sci. 122, 100843.

Cihan, E., Kavasoğulları, B., Demir, H., 2017. Enhancement of performance of open liquid desiccant system with surface additive. Renew. Energy 114, 1101–1112.

Dow, N., Gray, S., Li, J.-D., Zhang, J., Ostarcevic, E., Liubinas, A., Atherton, P., Roeszler, G., Gibbs, A., Duke, M., 2016. Pilot trial of membrane distillation driven by low grade waste heat: membrane fouling and energy assessment. Desalination 391, 30–42.

Drioli, E., Ali, A., Macedonio, F., 2015. Membrane distillation: recent developments and perspectives. Desalination 356, 56–84.

Duong, H.C., Chivas, A.R., Nelemans, B., Duke, M., Gray, S., Cath, T.Y., Nghiem, L.D., 2015a. Treatment of RO brine from CSG produced water by spiral-wound air gap membrane distillation—a pilot study. Desalination 366, 121–129.

Duong, H.C., Cooper, P., Nelemans, B., Cath, T.Y., Nghiem, L.D., 2015b. Optimising thermal efficiency of direct contact membrane distillation by brine recycling for small-scale seawater desalination. Desalination 374, 1–9.

Duong, H.C., Gray, S., Duke, M., Cath, T.Y., Nghiem, L.D., 2015c. Scaling control during membrane distillation of coal seam gas reverse osmosis brine. J. Membr. Sci. 493, 673–682.

Duong, H.C., Cooper, P., Nelemans, B., Cath, T.Y., Nghiem, L.D., 2016a. Evaluating energy consumption of air gap membrane distillation for seawater desalination at pilot scale level. Sep. Purif. Technol. 166, 55–62.

Duong, H.C., Duke, M., Gray, S., Cooper, P., Nghiem, L.D., 2016b. Membrane scaling and prevention techniques during seawater desalination by air gap membrane distillation. Desalination 397, 92–100.

Duong, H.C., Hai, F.I., Al-Jubainawi, A., Ma, Z., He, T., Nghiem, L.D., 2017. Liquid desiccant lithium chloride regeneration by membrane distillation for air conditioning. Sep. Purif. Technol. 177, 121–128.

Duong, H.C., Álvarez, I.R.C., Nguyen, T.V., Nghiem, L.D., 2018. Membrane distillation to regenerate different liquid desiccant solutions for air conditioning. Desalination 443, 137–142.

Duong, H.C., Ansari, A.J., Nghiem, L.D., Cao, H.T., Vu, T.D., Nguyen, T.P., 2019. Membrane processes for the regeneration of liquid desiccant solution for air conditioning. Curr. Pollut. Rep. 5, 308–318.

Duong, H.C., Ansari, A.J., Cao, H.T., Nguyen, N.C., Do, K.-U., Nghiem, L.D., 2020a. Membrane distillation regeneration of liquid desiccant solution for air-conditioning: insights into polarisation effects and mass transfer. Environ. Technol. Innov. 19, 100941.

Duong, H.C., Ansari, A.J., Hailemariam, R.H., Woo, Y.C., Pham, T.M., Ngo, L.T., Dao, D.T., Nghiem, L.D., 2020b. Membrane distillation for strategic water treatment applications: opportunities, challenges, and current status. Curr. Pollut. Rep. 6, 173–187.

Duong, H.C., Nghiem, L.D., Ansari, A.J., Vu, T.D., Nguyen, K.M., 2021a. Assessment of pilot direct contact membrane distillation regeneration of lithium chloride solution in liquid desiccant air-conditioning systems using computer simulation. Environ. Sci. Pollut. Res. 29, 41941–41952.

Duong, H.C., Tran, L.T.T., Truong, H.T., Nelemans, B., 2021b. Seawater membrane distillation desalination for potable water provision on remote islands—a case study in Vietnam. Case Stud. Chem. Environ. Eng. 4, 100110.

Fane, A.G., Schofield, R.W., Fell, C.J.D., 1987. The efficient use of energy in membrane distillation. Desalination 64, 231–243.

Giampieri, A., Ma, Z., Smallbone, A., Roskilly, A.P., 2018. Thermodynamics and economics of liquid desiccants for heating, ventilation and air-conditioning—an overview. Appl. Energy 220, 455–479.

Gurubalan, A., Simonson, C.J., 2021. A comprehensive review of dehumidifiers and regenerators for liquid desiccant air conditioning system. Energy Convers. Manag. 240, 114234.

Gurubalan, A., Maiya, M.P., Geoghegan, P.J., 2019. A comprehensive review of liquid desiccant air conditioning system. Appl. Energy 254, 113673.

Hickenbottom, K.L., Cath, T.Y., 2014. Sustainable operation of membrane distillation for enhancement of mineral recovery from hypersaline solutions. J. Membr. Sci. 454, 426–435.

Kang, B.H., Kim, K.H., Lee, D.Y., 2007. Fluid flow and heat transfer on a falling liquid film with surfactant from a heated vertical surface. J. Mech. Sci. Technol. 21, 1807.

Kang, Y.T., Kim, H.J., Lee, K.I., 2008. Heat and mass transfer enhancement of binary nanofluids for H_2O/LiBr falling film absorption process. Int. J. Refrig. 31, 850–856.

Khayet, M., Matsuura, T., Mengual, J.I., Qtaishat, M., 2006. Design of novel direct contact membrane distillation membranes. Desalination 192, 105–111.

Lefers, R., Bettahalli, N.M.S., Nunes, S.P., Fedoroff, N., Davies, P.A., Leiknes, T., 2016. Liquid desiccant dehumidification and regeneration process to meet cooling and freshwater needs of desert greenhouses. Desalin. Water Treat. 57, 23430–23442.

Lefers, R., Bettahalli, N.M.S., Fedoroff, N., Nunes, S.P., Leiknes, T., 2018. Vacuum membrane distillation of liquid desiccants utilizing hollow fiber membranes. Sep. Purif. Technol. 199, 57–63.

Lefers, R.M., Srivatsa Bettahalli, N.M., Fedoroff, N.V., Ghaffour, N., Davies, P.A., Nunes, S.P., Leiknes, T., 2019. Hollow fibre membrane-based liquid desiccant humidity control for controlled environment agriculture. Biosyst. Eng. 183, 47–57.

Liu, J., Albdoor, A.K., Lin, W., Hai, F.I., Ma, Z., 2022. Membrane fouling in direct contact membrane distillation for liquid desiccant regeneration: effects of feed temperature and flow velocity. J. Membr. Sci. 642, 119936.

Lowenstein, A., 2008. Review of liquid desiccant technology for HVAC applications. HVAC&R Res. 14, 819–839.

Luo, Y., Shao, S., Qin, F., Tian, C., Yang, H., 2012. Investigation on feasibility of ionic liquids used in solar liquid desiccant air conditioning system. Sol. Energy 86, 2718–2724.

Pasqualin, P., Lefers, R., Mahmoud, S., Davies, P.A., 2022. Comparative review of membrane-based desalination technologies for energy-efficient regeneration in liquid desiccant air conditioning of greenhouses. Renew. Sust. Energ. Rev. 154, 111815.

Qasim, M., Samad, I.U., Darwish, N.A., Hilal, N., 2021. Comprehensive review of membrane design and synthesis for membrane distillation. Desalination 518, 115168.

Qi, R., Dong, C., Zhang, L.-Z., 2020. A review of liquid desiccant air dehumidification: from system to material manipulations. Energy Build. 215, 109897.

Röck, M., Saade, M.R.M., Balouktsi, M., Rasmussen, F.N., Birgisdottir, H., Frischknecht, R., Habert, G., Lützkendorf, T., Passer, A., 2020. Embodied GHG emissions of buildings—the hidden challenge for effective climate change mitigation. Appl. Energy 258, 114107.

Song, L., Ma, Z., Liao, X., Kosaraju, P.B., Irish, J.R., Sirkar, K.K., 2008. Pilot plant studies of novel membranes and devices for direct contact membrane distillation-based desalination. J. Membr. Sci. 323, 257–270.

Su, W., Lu, Z., She, X., Zhou, J., Wang, F., Sun, B., Zhang, X., 2022. Liquid desiccant regeneration for advanced air conditioning: a comprehensive review on desiccant materials, regenerators, systems and improvement technologies. Appl. Energy 308, 118394.

UNEP, 2021. 2021 Global Status Report for Bulidings and Constructions: Towards a Zero-emission, Efficient and Resilient Buildings and Construction Sector.

Watanabe, H., Komura, T., Matsumoto, R., Ito, K., Nakayama, H., Nokami, T., Itoh, T., 2019. Design of ionic liquids as liquid desiccant for an air conditioning system. Green Energy Environ. 4, 139–145.

Wen, T., Lu, L., Zhong, H., 2018. Investigation on the dehumidification performance of $LiCl/H_2O$-MWNTs nanofluid in a falling film dehumidifier. Build. Environ. 139, 8–16.

Wu, W., Pang, C., Sheng, W., Chen, S., Wu, R., 2010. Enhancement on NH_3/H_2O bubble absorption in binary nanofluids by mono nano Ag. CIESC J. 61.

Yang, L., Du, K., Niu, X.F., Cheng, B., Jiang, Y.F., 2011. Experimental study on enhancement of ammonia–water falling film absorption by adding nano-particles. Int. J. Refrig. 34, 640–647.

Yong, W.F., Ho, Y.X., Chung, T.-S., 2017. Nanoparticles embedded in amphiphilic membranes for carbon dioxide separation and dehumidification. ChemSusChem 10, 4046–4055.

Zhao, B., Peng, N., Liang, C., Yong, W.F., Chung, T.-S., 2015. Hollow fiber membrane dehumidification device for air conditioning system. Membranes 5.

Zhao, B., Yong, W.F., Chung, T.-S., 2017. Haze particles removal and thermally induced membrane dehumidification system. Sep. Purif. Technol. 185, 24–32.

Zhou, J., Zhang, X., Sun, B., Su, W., 2018. Performance analysis of solar vacuum membrane distillation regeneration. Appl. Therm. Eng. 144, 571–582.

Zhou, J., Zhang, X., Su, W., Sun, B., 2019. Performance analysis of vacuum membrane distillation regenerator in liquid desiccant air conditioning system. Int. J. Refrig. 102, 112–121.

Zhou, J., Noor, N., Wang, F., Zhang, X., 2020a. Simulation and experiment on direct contact membrane distillation regenerator in the liquid dehumidification air-conditioning system. Build. Environ. 168, 106496.

Zhou, J., Wang, F., Noor, N., Zhang, X., 2020b. An experimental study on liquid regeneration process of a liquid desiccant air conditioning system (LDACs) based on vacuum membrane distillation. Energy 194, 116891.

Reverse osmosis (RO) membrane development and industrial applications

Nirenkumar Pathak, Umakant Badeti, Weonjung Sohn, Sherub Phuntsho, and Ho Kyong Shon

School of Civil and Environmental Engineering, University of Technology, Sydney, NSW, Australia

1. Introduction

In the future decades, rapid population expansion, urbanization, and industrialization will increase global demand for fresh water, necessitating the creation of new water sources. While numerous possibilities exist for supplementing freshwater sources, these options alone will not be sufficient to meet this demand. Although seawater desalination has the potential to provide a plentiful and consistent supply of fresh water cleansed from the vast oceans, it should be explored only after all other options have been exhausted (Elimelech and Phillip, 2011). Desalination of water has developed into a scientifically and economically viable alternative for addressing the issues connected with growing water scarcity in many parts of the world (Ghaffour et al., 2015).

Currently, more than 20 different technologies are employed to desalinate salt water. Three basic categories of commercial desalination techniques exist: thermal processes, membrane separations, and emerging technologies (Feria-Díaz et al., 2021). Thermal or phase change separation processes such as multistage flash (MSF), multiple effect distillation (MED), freeze desalination, and vapor compression (VC) are equilibrium-governed separation processes, whereas membrane separation processes such as reverse osmosis (RO) and electrodialysis reversal (EDR) are rate-governed separation processes. Distillation and freezing are processes that remove pure water from salty brine in the form of water vapor or ice (Gibbons, 1988; Liyanaarachchi et al., 2014). The International Desalination Association (IDA) reports that

Copyright © 2023 Elsevier Inc. All rights reserved.

the world's desalination plant capacity reached 99.8 million m^3/day in 2017, with over 18,500 plants installed in 150 countries (Feria-Díaz et al., 2021).

Although thermal techniques have been employed in the past, the reduced cost of membrane materials and the flexibility to tailor them have resulted in a fast increase in the market share of membrane-based desalination over the last several decades (Ahmed et al., 2019). RO is the world's fastest growing technology, with an expected market value of 9227 million USD in 2022. RO has garnered considerable interest as a result of its distinct benefits over other processes in terms of simplicity, scalability, compact footprint, and energy efficiency (Dai et al., 2020; Feria-Díaz et al., 2021).

Today's seawater reverse osmosis (SWRO) facilities require about 3–6 kWh of electricity to create 1 m^3 of product water, compared to the previous generation. A lot of energy is needed for phase change processes, which makes them more expensive. MSF and MED processes require 10–16 and 6–12 kWh/m^3, respectively, in terms of energy consumption. The unit production cost (UPC) of water produced by the MSF and MED processes ranges from $0.60 to $1.17 per unit. It is interesting to note that the UPC for RO is 0.45–0.95 US dollars with a combined energy demand of 3–6 kWh/m^3. Approximately 40%–60% of water can be recovered using a single-stage RO process (Liyanaarachchi et al., 2014).

2. Membrane separation characteristics

Membrane separation is defined by the simultaneous retention of species and the flow of product across the semipermeable membrane. Membrane performance is determined by its high selectivity and flux; the membrane materials' mechanical, chemical, and thermal durability; defect-free manufacturing; minimum fouling; and decent compatibility in working condition (Nath, 2017; Singh, 2015).

In mild operating conditions, membranes that exhibit high selectivity can include porous/dense materials; anisotropic or symmetric ceramic and metal flicks made of homogeneous, heterogeneous, or composite materials; and ionic membranes (Singh, 2015). Membrane/membrane processes can be classified according to a number of different classification schemes as shown in Fig. 1. Both lengthy channels and a porous sponge-like structure of 100–200 μm thickness are possible in symmetric membranes (Dai et al., 2016). In asymmetric membrane, the nonidentical membranes form two or more structural planes. Distinct chemical or structural layers are formed in composite membrane (Nath, 2017). While convective flow is used in porous membranes to transport tiny molecules through openings, sorption and diffusion are used in nonporous membranes (Singh, 2015). In RO membranes, a dense or nonporous matrix (about 0.0001–0.001 μm) selectively remove low molecular weight species such as inorganic solids (including salt ions, minerals, and metal ions) and organic molecules (Dai et al., 2016; Figoli and Criscuoli, 2017).

2.1 RO transport mechanisms

Solute and solvent transport through membranes can be described using a variety of phenomenological and mechanistic models. Solution-diffusion (SD) and preferentially

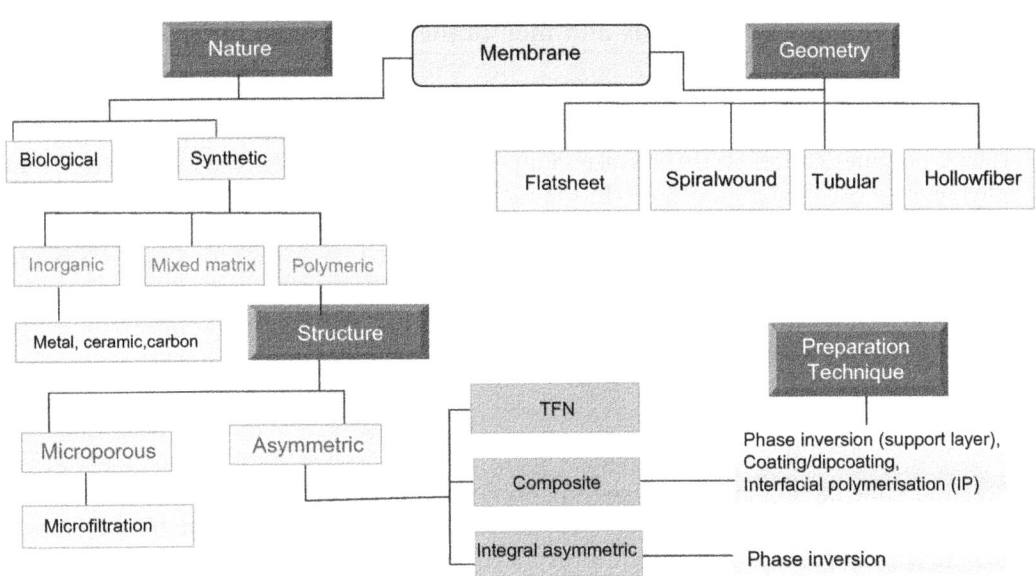

FIG. 1   Membrane classification.

sorption-capillary flow (PSCF) models can be used to explain water transport over RO membranes (Singh, 2015; Wang et al., 2014). In contrast to the SD model, which assumes a pore-free membrane surface, the preferentially sorption-capillary flow model was created by Sourirajan and others, which relied on hypothesis that the reverse osmosis membrane surface has pores (Singh, 2015).

The solution-diffusion (SD) model is the most widely recognized model at the moment since it does not need pores. Thus, the SD model holds true for nonporous membranes such as the RO membrane, where separation occurs as due to differences in permeate solubility and diffusivity (Wang et al., 2014). To summarize, the model entails three processes for the transfer of both the solvent (water) and the solute (salt) across the membrane:

1) adsorption onto the membrane,
2) diffusion across the membrane,
3) desorption from the membrane.

According to this model, solvent and solute are adsorbed at one side of the membrane, diffuse through it, and desorbed at the other side (Ismail and Matsuura, 2018). Thus, RO can be seen as a diffusion-controlled process in which the mass transfer of permeate via RO membranes is regulated by diffusion, a process known as solution–diffusion process. Permeate dissolves in the membrane material and then diffuses through it (Wenten, 2016). The chemical potential between the feed and permeate sides of the membrane is what drives mass transfer. Due to its simplicity, the solution-diffusion model has been effectively applied (Ismail and Matsuura, 2018). When it comes to membrane performance, selectivity vs productivity are the two factors to consider. The lower the permeate flux or productivity, the higher the selectivity (Singh, 2015).

3. Membrane materials and membrane preparation techniques

Reverse osmosis membranes have been made using a variety of membrane materials and manufacturing methods. Cellulose acetate, polyamide, and "thin film composites" made via interfacial polymerization on the surface of a porous support are the most often used materials for reverse osmosis membranes. In the majority of commercial RO membranes, the interfacial polymerization process is used (Abetz et al., 2006). Sintering, stretching, track etching, sol–gel process, pHase inversion, and solution coating are some of the more prevalent methods for preparing membranes (Figoli and Criscuoli, 2017; Nath, 2017). Because of the low flux that results from the thicker materials, dense symmetric membranes are seldom employed in purification operations (Figoli and Criscuoli, 2017; Nath, 2017).

3.1 Solution casting

A casting knife having metal edge sitting on two runners, configured to provide an accurate slit between edge and plate, is used to distribute an even film of a suitable polymer solution across a flat plate. It is then left to dry, and the solvent evaporates to produce a thin, homogenous polymer layer. The dried film can be removed from the glass once the solvent has evaporated completely. Soaking the cast film in a swelling nonsolvent, such as water or alcohol, is generally enough to remove it from the plate (Baker, 2012; Nath, 2017). The melt extrusion procedure is used when casting solvents fail to dissolve polymers. At little above the polymer's melting point, two heated plates are used to compress the polymer between them for 1–5 min at a pressure of 2000–5000 pounds per square inch. Phase inversion and coating techniques are used to create asymmetric porous membranes (Baker, 2012; Figoli and Criscuoli, 2017; Nath, 2017).

3.2 Phase inversion

The phase inversion approach is widely used to manufacture polymeric membranes in both laboratory and industrial settings. The term "phase inversion" denotes the transition from a liquid phase to a solid phase. The evaporation of the solvent in the polymer solution causes a phase separation from a homogenous starting solution. This phase separation might be caused due to the changes in temperature, the use of nonsolvents, or other factors. Different morphologies, such as open or dense structures, can be obtained based on the approach used to prepare polymeric membranes (Baker, 2012; Figoli and Criscuoli, 2017). In order to prepare porous or nonporous membranes, one must manage the initial stage of the phase transition. As a result, phase inversion is a highly adaptable approach that may produce a wide range of morphologies. Phase inversion is the most common method for obtaining commercial membranes (Nath, 2017).

The phase inversion procedure can be broken down into four different types based on the driving force: nonsolvent-induced phase separation (NIPS), evaporation-induced phase separation (EIPS), vapor-induced phase separation (VIPS), and thermally induced phase separation (TIPS) (Ji et al., 2021). For example, in NIPS process, the cast polymer solution film is immersed in a nonsolvent bath. Nonsolvent enters the film, displacing solvent; the film is enriched in nonsolvent, causing phase separation (Baker, 2012). By manipulating the solvent-nonsolvent system's chemical composition and temperature difference, it is possible to maintain tuning of water flux and rejection by changing the asymmetric structure and thickness of active layer. In EIPS process, demixing occurs as solvent departs from the polymer, which raises the concentration of the polymer above its limit. On exposure of dope solution to the environment where solvent vapor does not exist, the polymer solution adsorbs the nonsolvent and induces the phase separation called VIPS. However, alternative phase inversion procedures, for example, evaporation-induced phase inversion and vapor-induced phase separation are often used in conjunction with NIPS to generate integrally skinned asymmetric (ISA) membranes (Ji et al., 2021).

3.3 Coating

Fluxes in dense polymeric membranes such as RO, where diffusion is the primary mode of transport, tend to be low. The effective membrane thickness must be decreased as much as feasible in order to enhance the flow. Preparation of composite membranes, consisting of two distinct materials, can accomplish this. Thin, dense structures with great (intrinsic) selectivity and relatively high flux are typically created using coating processes. A thin layer of a highly selective membrane material is put on top of a sublayer that is more or less permeable. The thin top layer determines the true selectivity, whereas the porous sublayer serves only as a foundation. A phase-separated asymmetric membrane is frequently used as the primary base material in all composite membranes. Dip coating, plasma polymerization, interfacial polymerization, and in situ polymerization are just a few of the coating methods that can be utilized to create composite membranes (Nath, 2017).

There is a sort of composite membrane with an asymmetrical structure that has a thin, dense top layer supported by a porous bottom layer. Unlike dense homogeneous polymer films, composite membranes have a regular thickness (20–200 µm) that eliminates their poor permeate rate properties. The advantage of a composite membrane is that each layer may be tuned to get the best performance in terms of selectivity, penetration rate, and thermal stability. A thin coating of a very dilute polymer solution atop water or mercury was used to create the first composite membranes. As the solvent evaporated, a very thin polymeric coating was formed on the surface. The polymeric film was then placed on top of a porous support. The composite membranes produced by this method are referred to as thin film composite membranes had mechanical stability issues and were not appropriate for large-scale manufacture (Nath, 2017). TFC NF and RO membranes are manufactured using IP, which is the most common method used in industry (Ji et al., 2021). Today, most RO membranes are made of a composite material that was invented in 1970 by a company called FilmTec, which is now part of DuPont (Nath, 2017).

3.4 Interfacial polymerization (IP)

In 1959, Wittbecker and Morgan came up with the idea of interfacial polymerization. At first, it refers to the polycondensation of diamine and diacid monomers, which makes polyamide and hydrogen chloride. Interfacial polymerization has become a strong and effective

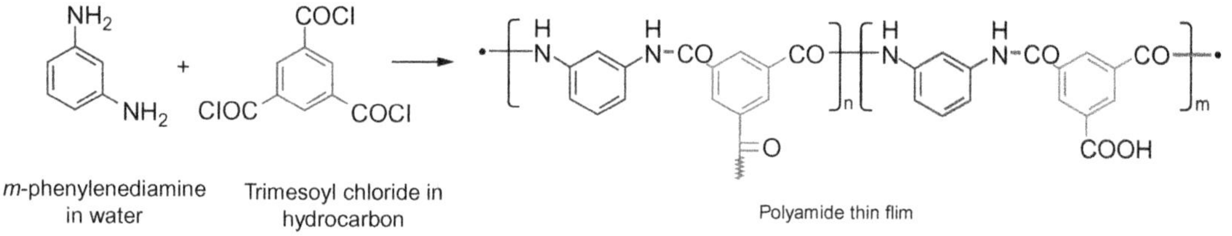

FIG. 2 Standard commercial polyamide membrane derived from m-phenylenediamine (MPD) and trimesoyl chloride (TMC) via interfacial polymerization. *Reproduced with permission from Lau, W., Ismail, A., Misdan, N., Kassim, M., 2012. A recent progress in thin film composite membrane: a review. Desalination 287, 190–199.*

way to make a wide range of useful polymer materials (Nath, 2017). In reverse osmosis (RO) and nanofiltration (NF) membranes, a thin film composite (TFC) structure is frequently used, in which a polyamide thin film with a thickness of 10–300 nm is formed in situ on top of a porous substrate via an interfacial polymerization (IP) reaction. A typical IP reaction begins with the impregnation of the substrate with a water solution of an amine monomer [e.g., m-phenylenediamine (MPD) or piperazine (PIP), followed by the application of an organic solution (water-insoluble) of an acid chloride monomer (e.g., trimesoyl chloride (TMC)] to form the polyamide thin film (Peng et al., 2022) as shown in Fig. 2 (Lau et al., 2012).

Polymeric supports are impregnated with diamine aqueous solution and then exposed to triacid chloride organic solution in the standard IP procedure. Polymerization reactions occur primarily at the solvent-nonsolvent surface, where diamine monomers permeate through water phase to the solvent phase and quickly interacts with triacid chlorides due to their low solubility in water and moderate solubility in organic solvents. PA asymmetric nanofilms containing amine and carboxylic groups on either side of the Janus reaction zone (between 20 and 200 nm in total thickness) are formed as the result of this reaction. Researchers concluded that nonhomogeneous and very fine thickness of polyamide film is accomplished due to negative charge of top layer followed by thick positively charged support layer (Freger, 2003). Diffusion limits are commonly used to describe the IP reaction (Ji et al., 2021; Ukrainsky and Ramon, 2018). Improved chemical compositions, improved membrane architectures, and reduced thickness of selective layers have all been employed by researchers over the years to further enhance this magnificent technology (Ji et al., 2021).

4. Reverse osmosis membrane development

4.1 First-generation RO membranes

Around 1920, the first evidence of reverse osmosis (RO) began to emerge. However, it went undiscovered for 30 years until Reid and his colleagues rediscovered it (Singh, 2015). Because of the enormous pore size of microporous membranes making them unsuitable for desalination, the Office of Saline Water (in the United States) began investigating RO (also known as "hyperfiltration") for desalinating salt water in the early 1950s. Despite Reid and Berton's discovery of polymeric membranes with great salt rejection in the mid-1950s, they were unable to cast ideal thin membranes (thickness about 6.0 μm). As a result, the low water fluxes through the membranes made them unusable. Much of the early research focused on salt water purification (about 35,000 mg/L), comprising membranes with solute retention above 99.3% to create produced water having salt levels below 500 mg/L. Early membranes could achieve this desired performance only at very high pressure (up to 100 bar) (Baker, 2012). It has been more than half a century since membrane technology was first developed in a laboratory setting. When World War II came to an end, the first important use of membranes occurred in the testing of drinking water (Singh, 2015). Fig. 3 depicts the evolution of the reverse osmosis membrane from its infancy to its most recent development.

Loeb and Sourirajan (L-S) invented the first asymmetric cellulose acetate (CA) membrane for RO systems in the 1960s. The phase-inversion-produced Loeb-Sourirajan RO membranes possessed actual skin thickness of 100–200 nm, allowing for standard fluxes (2–20 lmh) at tolerable feed pressures (30–60 bar g) for brine. These asymmetric membranes showed that high flux is largely due to their exceptional thinness and porosity (Feria-Díaz et al., 2021; Loeb and Sourirajan, 1962). The Loeb-Sourirajan RO membrane was developed for seawater desalination using the Preferential Sorption-Capillary Flow (PS-CF) model, and apertures are essential in order to pass solvent through reverse osmosis membrane (Ismail and Matsuura, 2018). In 1964, Loeb and Sourirajan introduced the first RO desalination technique. Because of this, it has made significant advancements, establishing it as the industry's top technology for desalination operations (Feria-Díaz et al., 2021).

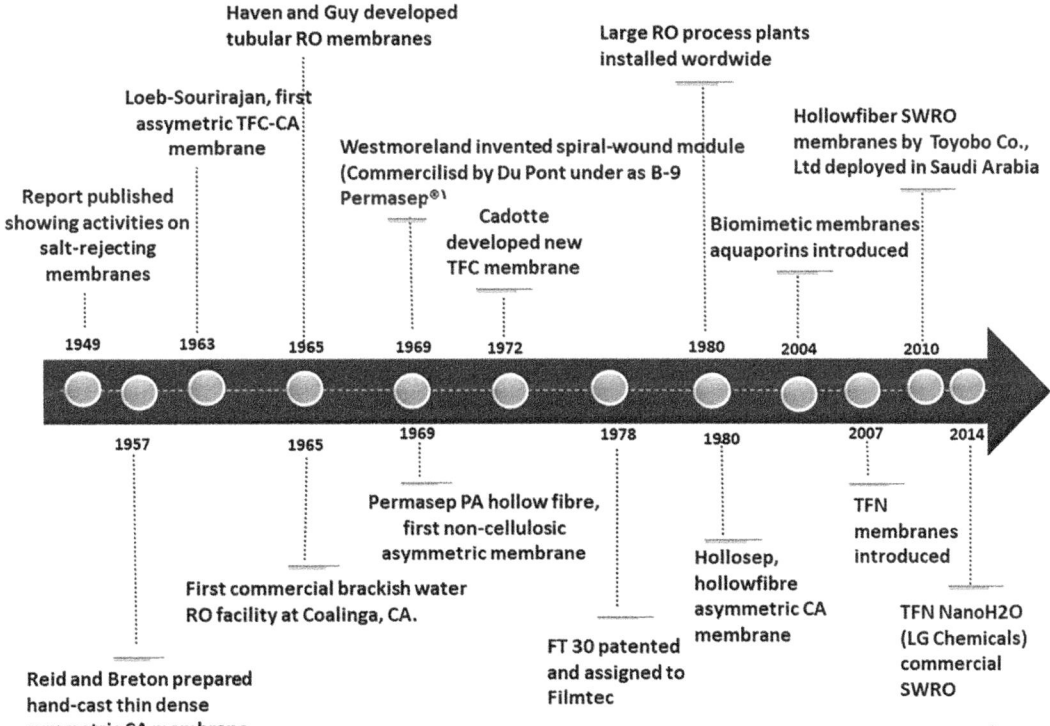

FIG. 3   Reverse osmosis membrane development.

Then, in the 1970s, Cadotte and coworkers created a novel membrane comprised of a thin film composite (TFC), which has dominated market for several years (Cadotte et al., 1980; Feria-Díaz et al., 2021). The introduction of a casting machine (1 m width) in 1989 enabled reproducible fabrication of flat sheet membranes (Abetz et al., 2006) as shown in Fig. 4.

Shortly after L–S membrane technology was invented, RO membrane modules were developed. In the mid-1960s, Haven and Guy invented the tubular RO membrane. Spiral-wound modules first developed by Westmoreland and then by Bray in the late 1960s were found to be more efficient than tube-in-shell modules. Rolled-up spiral-wound module resembles a plate-and-frame structure. Original modules contained a single membrane, whereas spiral-wound membranes are more common today which incorporate many membrane leaves (Singh, 2015).

Even though the oldest RO membranes were cellulose acetate, they had a restricted chemical, thermal, and mechanical strength. To this day, the most commonly used cellulose esters are a mixture of cellulose acetate and cellulose triacetate each in the composition of 40%. The CA membranes are biodegradable and hydrolyze when operated outside of a pH range of 4–6. Cellulose acetate hydrolyzes gradually in water, reducing membrane's acetyl content and impairing its function. Plasticizing CA membranes occurs when polar organic solvents are absorbed into the membranes, swelling them. However, CA membranes are more resistant to fouling than PA membranes, and they are more resistant to chlorine. Many big seawater RO plants use Toyobo hollow fiber CTA membranes (Singh, 2015).

FIG. 4   Membrane casting machine with 1 m width, production of membranes using nonwoven material as support. *Reproduced with permission from Abetz, V., Brinkmann, T., Dijkstra, M., Ebert, K., Fritsch, D., Ohlrogge, K., et al., 2006. Developments in membrane research: from material via process design to industrial application. Adv. Eng. Mater. 8(5), 328–358.*

4.2 Polymeric membranes—Integrally skinned asymmetric (ISA) membranes

Membranes composed of synthetic materials can be created from a wide variety of materials. However, not all materials are appropriate for pressure-driven membrane processes. Membranes should exhibit a high degree of permeability and selectivity while maintaining a sufficient mechanical stability (Singh, 2015). To ensure mechanical stability, the majority of polymeric membranes are produced on a nonwoven backing material (Marchetti et al., 2014). Surface morphologies and features of membranes such as smoothness, charge, and affinity to water have a significant impact on permeability, selectivity, and scaling (Singh, 2015). Membranes can be engineered to remove specific inorganic and organic pollutants, including trihalomethanes (Gibbons, 1988). Polymers exhibit a variety of useful features for separations, and their modification can improve membrane selectivity. Because chemical and thermal stability are desired, polymers having high glass transition temperatures, Tg, melting points, and crystallinity are preferred (Singh, 2015). Polymeric membranes are classified into two types: integrally skinned asymmetric (ISA) membranes and thin film composite (TFC) membranes (Marchetti et al., 2014).

Asymmetric membranes are typically composed of a dense layer of thickness 100–1000 nm which is reinforced by a very porous support layer of thickness 100–200 μm. When it comes to membrane selectivity, the dense layer is superior. The chemical nature, the size of pores (0.4–1 nm), and the thickness of the skin layer determine the separation qualities. The porous substructure is supposed to offer mechanical support for the thin and fragile selective layer and is thought to have little impact on the membrane's separation ability (Wang et al., 2014). Membranes can last up to 7 years, depending on the membrane composition and the quality of the feed water, and are typically constructed of cellulose acetate, aromatic polyamide, polyimide, polysulfones, or thin film composites. Seawater membranes are routinely replaced every 3–5 years (Gibbons, 1988).

4.3 Thin film composite (TFC) membranes

Only a few soluble polymers are capable of forming asymmetric structures in a single step, and even fewer are commercially viable in terms of the correct balance of permeability and salt rejection. The CA asymmetric membrane's middle transition layer becomes denser under pressure. A two-step casting procedure was developed, allowing the microporous support film and barrier layer to be individually optimized to provide desired selectivity and permeability while providing outstanding mechanical strength and compression resistance (Lau et al., 2012; Lee et al., 2011). Nonsolvent induced phase separation (NIPS) is used to generate an integrally skinned membrane for the porous support layer in TFC membranes. Both interfacial polymerization and dip coating are commonly used to create the skin layer (Qasim et al., 2019; Wang et al., 2014). As a result, the surface properties of a PA film are different from those of the PA dense layer because the polymer density is not evenly distributed. Although the PA dense layer is strongly negatively charged because acyl chloride groups are not entirely converted to amide during the manufacturing process, direct titration experiments have shown that the dense layer of composite PA NF membranes contains both positive and negative fixed charges (Wang et al., 2014).

Fig. 5 shows the cross-sectional structural differences in mechanical robustness between the integrally skinned asymmetric and the thin film composite membrane, each are made from

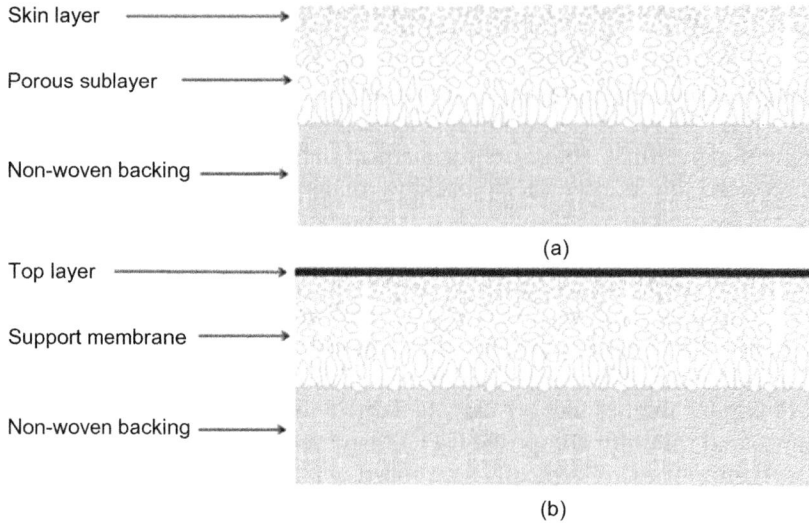

Skin layer

Porous sublayer

Non-woven backing

(a)

Top layer

Support membrane

Non-woven backing

(b)

FIG. 5   Illustration of membrane structures: (A) an integrally skinned asymmetric membrane and (B) a TFC membrane. *Reproduced with permission from Marchetti, P., Jimenez Solomon, M. F., Szekely, G., Livingston, A.G., 2014. Molecular separation with organic solvent nanofiltration: a critical review. Chem. Rev. 114(21), 10735–10806.*

nonwoven cloth (Marchetti et al., 2014). When it comes to membrane manufacture and chemical makeup, these two types of membranes are very different. In situ formation of an ISA membrane's selective layer occurs during phase separation, where it gradually transforms into nondense support layer having same chemical structure. In comparison to an ISA membrane, a TFC membrane's postdesigned selective layer has a diverse chemical composition as compared to integrally skinned asymmetric support, allowing for more precise selection (Ji et al., 2021).

Currently, the most often utilized polymeric membrane material for RO is polyamide (PA). Table 1 summarizes the most frequently used and newly described monomers for the manufacture of thin film composite membranes (Lau et al., 2012).

Compaction effect under pressure, surface fouling, and chemical degradation and oxidation are the main limitations of TFC membranes. Fouling on the surface is classified as colloidal particulate fouling and bacterial biofouling. As the water pressure increases, the polymers slightly reorganize themselves into a more compact structure, resulting in a reduced porosity and ultimately restricting the efficiency of the system meant to utilize them. Largely, the higher the pressure, the more compact the material (Nath, 2017).

4.4 Novel membranes

4.4.1 Nanocomposite membranes and mixed matrix membranes

Nanomaterials including as graphene, graphene oxide, zeolites, silica nanoparticles, polyhedral oligomeric silsesquioxane nanofiller, silver nanoparticles, and carbon nanotubes have recently been explored for improving TFC RO polymer membrane water flux, salt rejection, and anti(*bio*)fouling capabilities (Choi et al., 2022; Figoli and Criscuoli, 2017; Ng et al., 2022;

TABLE 1 The commonly used monomers and newly reported monomers for thin film composite membrane preparation.

Amine monomer (abbreviation)	Chemical structure	Molecular weight	Acyl chloride monomer (abbreviation)	Chemical structure	Molecular weight
Piperazine (PIP)		86.14	Trimesoyl chloride (TMC)		265.48
M-Phenylenediamine (MPD)		108.10	Isophthaloyl chloride (IPC)		203.02
P-Phenylenediamine (PPD)		108.10	5-isocyanato-isophthaloyl chloride (ICIC)		244.04
Sulfonated cardo poly(arylene ether sulfone) (SPES-NH$_2$)		774.71	mm-Biphenyl tetraacyl chloride (mm-BTEC)		404.03

Continued

TABLE 1 The commonly used monomers and newly reported monomers for thin film composite membrane preparation—cont'd

Amine monomer (abbreviation)	Chemical structure	Molecular weight	Acyl chloride monomer (abbreviation)	Chemical structure	Molecular weight
3,5-diamino-N-(4-aminophenyl)benzamide (DABA)		242.27	om-Biphenyl tetraacyl chloride (om-BTEC)		404.03
Triethanolamine (TEOA)		149.19	op-Biphenyl tetraacyl chloride (op-BTEC)		404.03
Methyl-diethanolamine (MDEOA)		119.16	Cyclohexane-1,3,5-tricarbonyl chloride (HTC)		271.53
1,3-cyclohexanebis(methylamine) (CHMA)		142.24	5-chloroformyloxy-isophthaloyl chloride (CFIC)		281.48

m-
phenylenediamine-
4-methyl (MMPD)

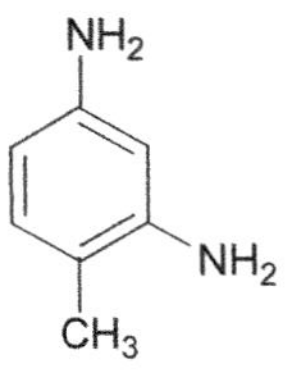

122.17

Hexafluoroalcohol-m-
Phenylenediamine
(HFA-MPD)

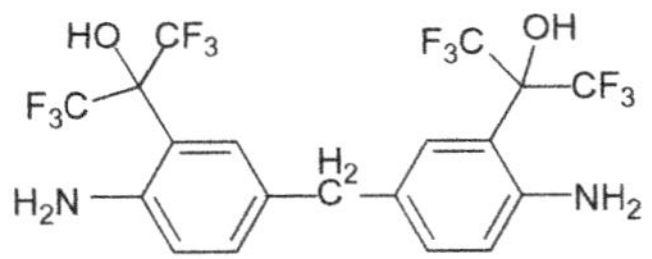

530.31

Reproduced with permission from Lau, W., Ismail, A., Misdan, N., Kassim, M., 2012. A recent progress in thin film composite membrane: a review. Desalination, 287, 190–199.

Wenten, 2016; Yang et al., 2020). Additionally, by utilizing nanostructured materials, energy costs and membrane area requirements are reduced, pretreatment operations are simplified, module repair expenditures are reduced, single pass operation is feasible, and plant size is increased (Fajardo-Diaz et al., 2022; Wenten, 2016; Zhao et al., 2021).

Monomers (MPD or TMC) can be used to integrate nanomaterials into polyamide rejection layers to create thin film nanocomposite membranes (TFN). For the first time, researchers introduced thin film nanocomposite in 2007 by inserting very fine nanoparticles into the polyamide rejection layer of an RO membrane. Water permeance was increased by 81% but salt rejection remained practically unchanged as compared to the original membrane (Yang et al., 2020). Zeolite nanoparticles were added to the thick film of a PA to boost water permeation by about double at the same time maintained similar salt retention. It was deduced that the zeolites' pores worked as a preferred channels for solvent, however have pore radius smaller to retain solutes (Jeong et al., 2007; Wang et al., 2014). Other researchers, working in parallel, used a polysulfone (PSF) substrate to include silica and zeolite nanoparticles, resulting in a TFCn structure that has increased water flux and robustness. To back this up, researchers found that adding hydrophilic and porous zeolite nanoparticles to an FO membrane enhanced water flux and lessened concentration polarization in the membrane's substrate. Thin film nanocomposite membranes are also commercially viable. Up to a 200% increase in water permeance has been observed in recent research for TFN and TFCn membranes. This improvement in A/B selectivity is less visible because of the clump of nanoparticles or weakened stability of active layer, but it still can be improved (Ma et al., 2013; Pendergast et al., 2010; Yang et al., 2020).

There are many advantages to incorporating a porous support into a TFC membrane fabrication process that go beyond simply providing structural support (Ji et al., 2021). Recently, a new form of TFN membrane (TFNi) with an interlayer of nanomaterials has appeared (Gonzales et al., 2021). In addition to a significant increase in water flow (up to an order of magnitude), these new TFNi membranes also demonstrate improved selectivity. Nanoparticles, one- and two-dimensional materials, and interfacial coatings are all examples of nanomaterials that can be used to create these kinds of advancements. While improving polyamide rejection layer formation, nanostructured interlayers also provide a more efficient water transport channel, giving TFNi membranes the ability to elucidate the trade-off between membrane permeability and selectivity. TFNi membranes may be able to help with crucial industrial usages such as treatment of elements and trace organics from water matrices (Yang et al., 2020). As previously stated, interlayer formation on prefabricated support membranes has been shown to be a highly effective method for producing both extremely selective and highly permeable interlayered-TFC membranes, referred to as i-TFC membranes. Various membrane configurations are shown in Fig. 6 (Ji et al., 2021). Contrary to popular belief, interlayer-based TFC PA membranes are still in their infancy and they were designed for flat sheets. More study is needed to extend the interlayer to commercial hollow fibers and spiral wound membranes (Elimelech and Phillip, 2011).

PVA membranes were modified with a conductive type of networked cellulose (NC), made by incorporating carbon nanostructures in NC, and are known as NC-CNS. As compared to NC alone, employing NC-CNS increased membrane mechanical strength and, when fed with 25,000 mg/L NaCl, increased RO flux by 93% (Ahmed et al., 2019). The high-performance TFC PA membranes are made possible by organic interlayers (such as PDA,

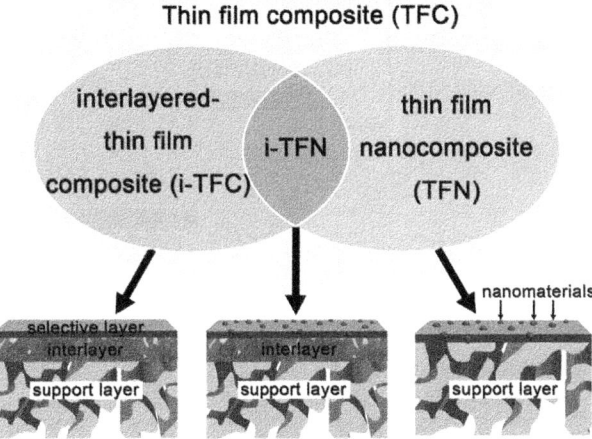

FIG. 6 Definitions and cross-sectional structures of TFN, i-TFC, and i-TFN membranes. *Reproduced with permission from Ji, C., Zhai, Z., Jiang, C., Hu, P., Zhao, S., Xue, S., et al., 2021. Recent advances in high-performance TFC membranes: a review of the functional interlayers. Desalination 500, 114869.*

TA, and other organic materials) and 1D and 2D nanomaterial interlayers. TFC PA membranes made with organic interlayers are the most promising for commercial scale-up among the various types of interlayers (R. Dai et al., 2020). PVA-interlayered PA membrane preparation was tested on a pilot NF production line to confirm its viability even further. A 30 lmh/bar flux and 99% Na_2SO_4 rejection were observed in laboratory-scale and pilot-scale PVA-interlayered PA membranes (Dai et al., 2020).

As shown in Fig. 7, a suitable nanofibrous interlayer for the IP process was created by using polydopamine impregnated single-walled carbon nanotubes (PD/SWCNTs) on a polyether sulfone MF membrane (Zhu et al., 2016). This newly developed interlayer had uniform and fine pores, and smooth hydrophilic surface. By carefully controlling the quantities of monomers, very thin and perfectly formed polyamide active layer up to 10 nm could be obtained. Water permeance up to 32 lmh/bar and divalent ion rejection of 95.9% are both demonstrated by the TFC membrane, which are the outcome of this process (Ji et al., 2021; Zhu et al., 2016).

Even though nanostructured materials have the potential to revolutionize membrane technologies, there are still many obstacles to be solved, including high costs, scale-up issues, and concerns about safety health and well-being when employing nanocomposites. Because of this, scientists are still looking for the best ways to develop materials with adequate selectivity and permeability while keeping their costs down (Wenten, 2016).

For water treatment, organic–inorganic hybrid membranes have garnered a lot of attention as the next-generation membrane materials (Ma et al., 2022; Wang et al., 2014). The organic and inorganic components of mixed-matrix membranes are intertwined. Inorganic particles are distributed within polymer matrix membranes used in aqueous separations. It has been traditional to add micron-scale inorganic fillers to PA layers to create superior fluid channels to enhance transfer of specific compounds (e.g., zeolites and silicalite) (Mahajan et al., 2002). As single filler

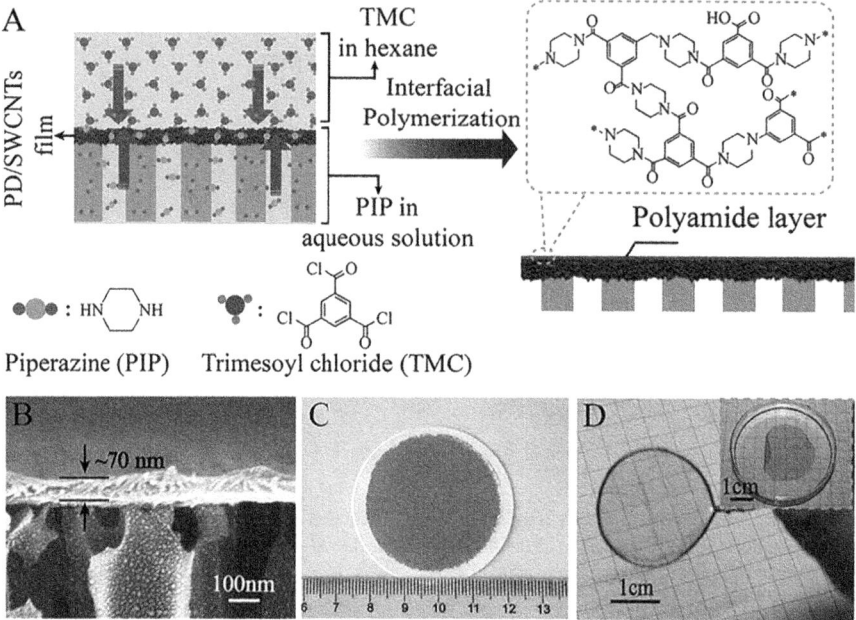

FIG. 7 Fabrication of TFC membrane with ultrathin (12 nm) PA selective layer with the assistance of a single-walled carbon nanotube built interlayer. (A) An illustration of the PD/SWCNTs interlayer assisted IP process, chemical structures of the monomers and the resulting PA selective layer, (B) the cross-sectional SEM image showing a PES membrane with a ~70-nm thick PD/SWCNTs interlayer, (C) a photo of the interlayered support membrane, and (D) a photo of PA/PD/SWCNTs composite. *Reproduced with permission from Zhu, Y., Xie, W., Gao, S., Zhang, F., Zhang, W., Liu, Z., Jin, J., 2016. Single-walled carbon nanotube film supported nanofiltration membrane with a nearly 10 nm thick polyamide selective layer for high-flux and high-rejection desalination. Small 12(36), 5034–5041.*

material benefits by imparting improvement to a whole membrane, mixed matrices theoretically give membrane production an additional degree of flexibility (Zimmerman et al., 1997). Mixing matrices has been used to enhance organic membranes' mechanical and chemical resilience, as well as their interface functioning, such as reducing fouling and enhancing selectivity. Unfortunately, very few scientists have investigated the transport of solvents and solutes in mixed matrix materials (Wang et al., 2014). A TFN membrane study used UiO-66, a MOF nanoparticle filler, to remove boron from a polyamide (PA)-based TFN RO membrane. The UiO-66 TFN membrane has a higher boron rejection (91.2%) than the benchmark TFC membrane at 0.05% (w/v) blending concentration (82.3%) (Liu et al., 2019).

4.4.2 Biomimetic membranes

Due to the exceptional water transport capabilities of biological membranes, interest in developing biomimetic membranes for desalination has grown. Aquaporin (AQP) or synthetic water nanochannels can be included into lipid bilayers or amphiphilic block copolymers to create biomimetic membranes (Lim et al., 2021). In the cell membrane, aquaporins are proteins that serve as water channels. As many as 10^9 water molecules per second can be transported through each aquaporin channel whereas all other solutes are completely rejected (Luo et al., 2018). Specially engineered water channels in the PA selective layer of

TABLE 2   Table Key properties of the aquaporin and two conventional FO membranes (average ± standard deviation of triplicate measurements).

Parameters	Aquaporin	CTA	TFC
Pure water permeability (L/m² h-bar)	2.09 ± 0.02	0.84 ± 0.03	2.50 ± 0.25
Salt (NaCl) permeability (L/m² h)	0.07 ± 0.01	0.32 ± 0.06	0.19 ± 0.03
Membrane structural parameter (μm)	301 ± 36	575 ± 28	245 ± 35
Observed NaCl rejection (%)	99.9 ± 0.1	92.0 ± 1.4	98.0 ± 0.2
Contact angle (°)	74.5 ± 8.9	60.4 ± 5.2	42.3 ± 3.2
Zeta potential at pH8 (mV)	−16.4 ± 2.3	−4.5 ± 0.4	−14.2 ± 0.5

Reproduced with permission from Luo, W., Xie, M., Song, X., Guo, W., Ngo, H.H., Zhou, J.L., Nghiem, L.D., 2018. Biomimetic aquaporin membranes for osmotic membrane bioreactors: membrane performance and contaminant removal. Bioresour. Technol., 249, 62–68.

RO membranes could lead to a change in the way water is moved through the membrane via fast single-file transport. Accordingly, AQP-based biomimetic membranes should theoretically be capable of high permeability and selectivity (Lim et al., 2021; Luo et al., 2018). As shown in Table 2, the aquaporin FO membrane was evaluated and compared to other commercially available membranes, such as the CTA and TFC FO membranes (Luo et al., 2018). The use of aquaporins in membrane technology has been the subject of nearly 30 patent applications since 2004 (Tang et al., 2015).

The first AQP-based biomimetic membrane for SWRO desalination was developed by incorporating aquaporin proteins into the polyamide selective layer of a hollow fiber TFC FO membrane. This membrane significantly increased the water flux through the membrane (100% flux enhancement) while decreasing the reverse salt flux (Li et al., 2017; Luo et al., 2018). Using an interfacial polymerization approach, other researchers successfully built an aquaporin-based biomimetic membrane. Workers discovered that the resultant membrane AMB-wild exhibited excellent stability at 10 bar pressure. When operated under 5 bar pressure, a 4 lmh/bar permeability and very high salt retention (approximately 97%) were found. The membrane outperformed commercial RO membranes (BW30 and SW30HR) in terms of separation performance, revealing the significant potential of interfacially polymerized ABM membranes (Ismail et al., 2019). Among the next-generation RO membranes now under development, it was suggested that biomimetic membranes show greatest promise for fabricating membranes with ultrahigh selectivity, and are anticipated to yield unparalleled improvements in SWRO (Lim et al., 2021). Although aquaporin membranes are now used in the laboratory, industrial-scale manufacture of these membranes (for full-scale applications) could be easily produced in the near future (Ismail et al., 2019).

5. Reverse osmosis membrane applications

RO is now a dominant technology because of its wide range of purification and concentration applications. It has become more affordable to use RO technology for commercial use as a result of advancements in polymers scheme, module and plant design, pretreatment, and energy reclamation (Wenten, 2016). RO has been a prominent tool for desalination; RO units are currently employed in a wide range of other fields as well. These include semiconductors, food processing and manufacture, power generation, pharmaceuticals, desalination, and biotechnology (Malaeb and Ayoub, 2011).

5.1 Water treatment—Seawater and brackish water desalination

Currently, RO is most commonly used to turn salt water and brackish water into potable water. Many coastal communities are increasingly relying on RO-based salt water desalination to meet their water needs (Yang et al., 2020). Desalination plants using RO membranes now account for 60% of all plants in operation. According to IDE technologies, the world's largest SWRO desalination plant was inaugurated at Sorek, Israel, in 2013 with a production capacity of 624,000 m^3/day potable water. The plant harvest water at a specific energy usage of 4 kWh/m^3 besides boron concentrations of 0.3 mg/L at maximum capacity (Wenten, 2016).

Fig. 8 illustrates how a typical desalination plant's process is handled to obtain the required high-quality permeate water (Kaya et al., 2015). There is an open intake with a feed water tank (not shown in Fig. 8) and pumping prior to physical pretreatment with first-stage sand filters, second-stage 5-μm cartridge filters, and two pressure vessels of about 1 m in length, each containing a single 2.5″ diameter membrane element. The previously chlorinated salt water in the feed tank is sent to a low-pressure pump and then to a sand filter and a cartridge filter.

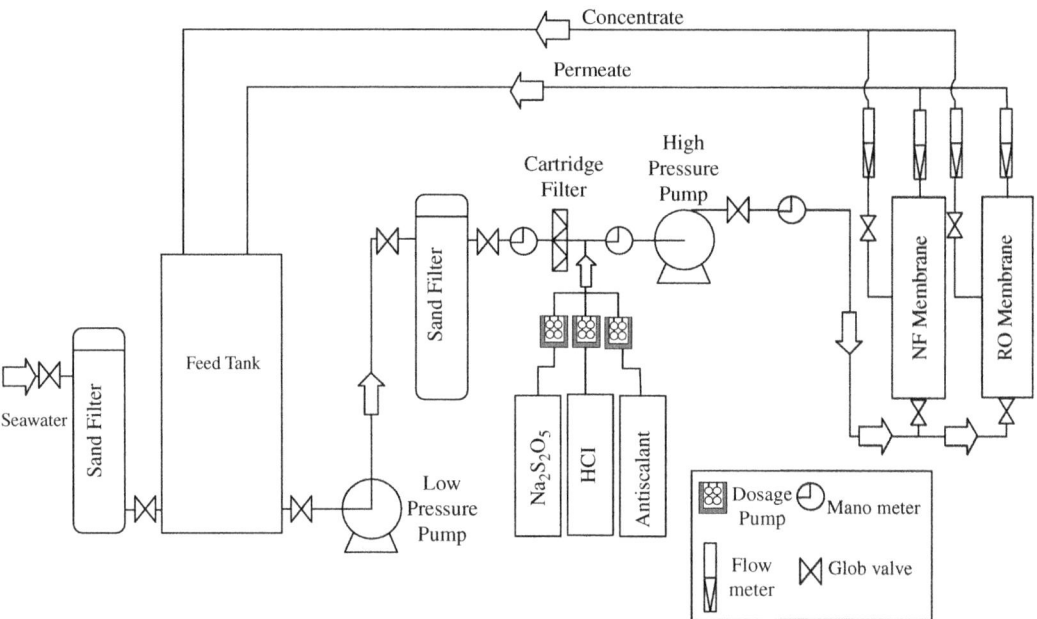

FIG. 8 Desalination process diagram (Kaya et al., 2015). *Reproduced with permission from Kaya, C., Sert, G., Kabay, N., Arda, M., Yüksel, M., Egemen, Ö., 2015. Pre-treatment with nanofiltration (NF) in seawater desalination—preliminary integrated membrane tests in Urla, Turkey. Desalination 369, 10–17.*

Seawater that has been physically and chemically treated is sent to the SWRO membrane via a high-pressure pump for desalination. Following pretreatment, a high-pressure pump precedes a single-stage reverse osmosis system. In some desalination plant arrangements the brine is routed straight to the Pelton turbine, and the resulting product is stored in a tank of potable water (Kaya et al., 2015; Leon and Ramos, 2021).

There is an average concentration of 4.5 mg/L of boron in Earth's oceans (World Health Organization, 2009). As the global demand for fresh water increases, so do the criteria for acceptable and allowed boron levels in drinking water. For example, the WHO recommends a maximum boron level of 2.4 mg/L while the EU recommends a boron level of 1.0 mg/L (Hu et al., 2016). Most commercial SWRO membranes only reject 40%–80% of boron because of the tiny elemental radius as well as neutral charge of boron under ambient pH levels (pKa = 9.25) (Yang et al., 2020). In order to achieve adequate boron removal, several desalination facilities employ a two-pass reverse osmosis process that includes second pass and a pH control. Adding a second-pass RO to seawater desalination dramatically increases the overall cost, SEC, and footprint (Yang et al., 2020).

LG Chem installed a NanoH₂O innovative TFN seawater RO membranes at Pembroke desalination facility in Malta in early 2016. The project's most difficult requirement was to keep boron concentrations in the permeate stream below 0.9 ppm for 5 years. They retained outstanding boron rejection performance after 2 years of operation without any further pH adjustment of the input water, which is around 6.7. The presence of nanoparticles in the polyamide layer increased water permeability without increasing salt permeability. The resultant membranes have a salt rejection rate of up to 99.85%. This study implies that the nanomaterial in TFN membranes improves boron rejection. In theory, the nanomaterial binds to boron species covalently and forms internal coordination complexes that are stable over a wide pH range (Lagartos et al., 2019). As shown in another study, on the PA layer, sulfonate functional groups can create charge-aggregate cavities that chelate boric acid, preventing it from passing through the SWRO membrane. In one stage of the RO process, boron removal was 90.6% under neutral circumstances. Therefore, the penetrated water met the WHO drinking water criteria. Its exceptional performance is due in part to the highly aggregated charges' strong interactions with ions and chelating impact with boron acid (Hu et al., 2016; Zhao et al., 2022).

5.2 Municipal and trade effluent treatment and water reclamation

Dissolved solids and hazardous pollutants can be removed by RO in municipal wastewater treatment (Liyanaarachchi et al., 2014; Wenten, 2016). The membrane-based sewage treatment and reclamation emerged as a more promising option to desalination thanks to its low energy requirement, as a result the Water Factory in Southern California, US installed earliest treatment unit. Membrane-based water reuse is often less concerned with TDS removal because the TDS level of normal sewage is substantially less as compared to seawater. However, because of ill effect on flora and fauna organic micropollutants, disinfection by-products, herbicides, and antibiotics are of higher concern than other pollutants. The typical hydrophobic trace organics and other small molecular weight cut-off compounds, despite being well optimized for TDS rejection, typically demonstrate insufficient removal by commercial TFC membranes. Virus eradication and membrane integrity are also important considerations when reusing potable water (Yang et al., 2020). Water that is reclaimed from secondary or tertiary wastewater effluents can be used for drinking or irrigation (Liyanaarachchi et al., 2014; Wenten, 2016). Table 3 lists RO uses for wastewater treatment (Wenten, 2016).

Total suspended solids, total dissolved compounds, heavy metals, sulfur, nitrogen, and phosphorous compounds and peculiar trade effluents can all be removed by RO and NF membrane systems in effluent treatment and reclamation. These applications require a high degree of membrane selectivity in order to guarantee excellent treatment efficiency and dependability (Yang et al., 2020). The characteristics that affect RO performance for removing phenol from water have been studied by researchers. Even in areas with shallow

TABLE 3　Applications of RO in wastewater treatment (Wenten, 2016).

Wastewater	Target of treatment	Remarks	References
Municipal sewage effluent	Make-up water for high-pressure steam production	RO combined with ion-exchange process	Wethern and Katzaras (1995)
Municipal wastewater	Water reclamation	Integrated MF–RO system	Ghayeni et al. (1996)
Municipal wastewater	Potable water production	Pretreatment required: fast mixing, coagulation, flocculation and sedimentation using Fe III, cationic polyelectrolyte and a sanitizing agent	Abdel-Jawad et al. (1997)
Municipal wastewater	Drinking water production	Integrated MF–RO–UV system. Chlorine is dosed before and after the MF to control biofouling.	Wintgens et al. (2005)
Dumpsite leachate	Zero discharge	Integrated RO–NF–crystallization–high-pressure RO system. A 97% water recovery and $\sim 8.3\,kWh/m^3$ permeate energy consumption.	Rautenbach et al. (1997)
Landfill leachate	Purification, discharge	High-pressure RO (120 bar) and NF in combination with a controlled crystallization process. Permeate recovery rates of more than 95%.	Peters (1998)
Electroplating wastewater	Purification, discharge	Pretreatment: cartridge depth filtration, pH adjustment and UF. Initial copper concentration of 340 mg/L was reduced to 4 mg/L.	Chai et al. (1997)
Sugary wastewater	Sugar recovery	Combined RO–EDI system. A more than 99% rejection of sugar on RO membrane.	Widiasa and Wenten (2007)
Wastewater of dairy industry	Water reuse	A 95% water recovery.	Vourch et al. (2008)
Tannery wastewater	Water reuse	Integrated microfilter–UF–RO system. Flux of RO is 40 lmh at 30 bar. More than 98% rejection of salts and COD rejection.	Fababuj-Roger et al. (2007)
Olive mill wastewater	Water reuse	Centrifuge–UF–RO system. COD removal: $\sim 96\%$; conductivity removal: $\sim 95\%$; permeate flux: ~ 13 lmh	Coskun et al. (2010)

lmh, $1m^{-2}h^{-1}$; MF, microfiltration; NF, nanofiltartion; UF, ultrafiltration.
Reproduced with permission from Wenten, I.G., 2016. Reverse osmosis applications: prospect and challenges. Desalination 391, 112–125.

groundwater, the permeate from secondary wastewater treatment was shown to be suitable for land application (Malaeb and Ayoub, 2011; Wenten, 2016). Ultrafiltration (UF) and reverse osmosis (RO) were found to be effective in treating olive mill wastewater. Aquatic receptors and irrigation can be used for concentrate disposal (Rahimi et al., 2021). Total hardness reduction from textile effluent is substantially higher with RO (90%) than with NF (75%) (Malaeb and Ayoub, 2011).

Pesticides and other pollutants from the irrigation system may enter surface water. Surface water must be cleaned of impurities before it can be utilized for drinking water. Using RO, surface water may be made safe to drink by removing both dissolved solids and contaminants (Cornelissen et al., 2021). Some study has focused on low pressure RO (LPRO) membranes. Researchers found that LPRO effectively removed pollutants from water. Commercial RO membranes are capable of removing 99.8% of refractory compounds (Wenten, 2016).

5.3 Food, dairies, and other industrial applications

In the dairy and sugar industries, RO has been used to concentrate heat-sensitive products, and in the alcoholic beverage business, it has been used to dealcoholize drinks. The performance and parameters of RO which affect the process have been evaluated in early research for a variety of fruit extractions (Braddock et al., 1988; Merson and Morgan, 1968; Wenten, 2016). According to these early research, RO membranes may preserve fruit juice's aromas, polyamide membranes offers greater permeate and selectivity in terms of taste and other ingredients, albeit membrane performance is influenced by working circumstances as well as module configurations (Wenten, 2016). RO preconcentration of sugar juice has been analyzed in terms of energy consumption, and the results are encouraging. The RO-evaporator system was compared to a conventional evaporator in terms of electricity usage. The RO-evaporator system outperformed to traditional system having saved 33% on energy. Two-stage reverse osmosis membrane efficiently thickened syrup up to 20°Brix by minimal sugar loss (Madaeni and Zereshki, 2010). RO has been used to remove ethanol from fermented beverages such as beer, wine, and cider (Ivić et al., 2022; Ramsey et al., 2021). RO is able to produce dealcoholized beverages without affecting their taste or quality. Low operating temperature and high aroma content made the RO process an excellent choice for reducing the ethanol content of beer to less than 0.5% volumetric alcohol content (v/v) (Catarino et al., 2006, 2007).

The RO has the potential to be used in the dairy industry for milk and whey concentration. The valuable components of milk can be preserved while reducing nearly 70% of the moisture. As a result, nutritive constituent destruction in process might be evaded due to the low-temperature operating conditions. The cost of transportation can be reduced if milk is preconcentrated before being transported (Hiddink et al., 1980; Kumar et al., 2013; Wenten, 2016). Because of its high organic load, whey has been a significant problem in dairy because it is a by-product of the cheese making process and contains 100,000 mg/L COD. Ultrafiltration, nanofiltration, and reverse osmosis were used to treat two different whey samples to produce a cleaner discharge and recover the whey proteins for repurposing. In terms of cleaner discharge and the recovery of proteins in whey for reuse, the NF + RO combination produced the best results out of the applied cascade operations. Both protein and lactose could be recovered using this NF + RO combination, with protein being recovered in the first stage and lactose being recovered in the second (Yorgun et al., 2008).

6. Conclusions and perspectives

It is already common practice to use reverse osmosis to purify seawater all around the world. Desalination of seawater, despite significant developments in desalination technology, is still more energy-intensive than conventional methods for treating fresh water. Desalination plants that use seawater on a large scale could have an impact on the environment. More polluted water necessitates the use of pretreatment to combat corrosion and biofilm growth, scale, and particle deposition, among other things. In terms of fouling resistance and high throughput, RO technology has experienced breakthroughs in membrane material, as well as module and process design, pretreatment, and energy recovery, all of which have reduced prices and piqued the interest of businesses looking for commercial applications.

The interlayer-based TFC PA membranes, contrary to popular opinion, are still in the early stages of development and were originally intended for flat sheets. The interlayer can be extended to commercial hollow fibers and spiral wrapped membranes, but more research is needed. Membrane modules and spacers can also be made using 3D printing, which has experienced an uptick in interest in recent years. Since 3D printing methods are expensive and have limitations on scalability as well as resolution, they can only be used in the production of membranes. When designing a new membrane, molecular simulations can help determine how to increase permeability and selectivity. Despite a rise in publications on the synthesis of novel membranes, these membranes rarely work well in operating plants. In the industrial sector, more study on membrane long-term and real-water performance is required. As little is known about the potential effects of nanoparticles on human and environmental health, they should be used with caution.

Additional research should be conducted to determine the potential synergies between enhanced antifouling capabilities, higher water permeability, and maintaining salt rejections. Industrial desalination cannot occur without conducting cross-disciplinary research.

References

Abdel-Jawad, M., Ebrahim, S., Al-Atram, F., Al-Shammari, S., 1997. Pretreatment of the municipal wastewater feed for reverse osmosis plants. Desalination 109 (2), 211–223.

Abetz, V., Brinkmann, T., Dijkstra, M., Ebert, K., Fritsch, D., Ohlrogge, K., et al., 2006. Developments in membrane research: from material via process design to industrial application. Adv. Eng. Mater. 8 (5), 328–358.

Ismail, A.F., Khulbe, K.C., Matsuura, T., 2019. Reverse Osmosis. Elsevier, Amsterdam, Netherlands.

Ahmed, F.E., Hashaikeh, R., Hilal, N., 2019. Fouling control in reverse osmosis membranes through modification with conductive carbon nanostructures. Desalination 470, 114118.

Baker, R.W., 2012. Membrane Technology and Applications, third ed. John Wiley & Sons, West Sussex, UK. ISBN: 9780470743720.

Braddock, R., Nikdel, S., Nagy, S., 1988. Composition of some organic and inorganic compounds in reverse osmosis-concentrated citrus juices. J. Food Sci. 53 (2), 508–512.

Cadotte, J.E., Petersen, R., Larson, R., Erickson, E., 1980. A new thin-film composite seawater reverse osmosis membrane. Desalination 32, 25–31.

Catarino, M., Mendes, A., Madeira, L., Ferreira, A., 2006. Beer dealcoholization by reverse osmosis. Desalination 200 (1–3), 397–399.

Catarino, M., Mendes, A., Madeira, L.M., Ferreira, A., 2007. Alcohol removal from beer by reverse osmosis. Sep. Sci. Technol. 42 (13), 3011–3027.

Chai, X., Chen, G., Po-Lock, Y., Mi, Y., 1997. Pilot scale membrane separation of electroplating waste water by reverse osmosis. J. Membr. Sci. 123 (2), 235–242.

Choi, P.J., Lim, S., Shon, H., An, A.K., 2022. Incorporation of negatively charged silver nanoparticles in outer-selective hollow fiber forward osmosis (OSHF-FO) membrane for wastewater dewatering. Desalination 522, 115402.

Cornelissen, E.R., Harmsen, D.J.H., Blankert, B., Wessels, L.P., van der Meer, W.G.J., 2021. Effect of minimal pretreatment on reverse osmosis using surface water as a source. Desalination 509, 115056.

Coskun, T., Debik, E., Demir, N.M., 2010. Treatment of olive mill wastewaters by nanofiltration and reverse osmosis membranes. Desalination 259 (1–3), 65–70.

Dai, Z., Ansaloni, L., Deng, L., 2016. Recent advances in multi-layer composite polymeric membranes for CO2 separation: a review. Green Energy Environ. 1 (2), 102–128.

Dai, R., Li, J., Wang, Z., 2020. Constructing interlayer to tailor structure and performance of thin-film composite polyamide membranes: a review. Adv. Colloid Interface Sci., 102204.

Elimelech, M., Phillip, W.A., 2011. The future of seawater desalination: energy, technology, and the environment. Science 333 (6043), 712–717.

Fababuj-Roger, M., Mendoza-Roca, J., Galiana-Aleixandre, M., Bes-Pia, A., Cuartas-Uribe, B., Iborra-Clar, A., 2007. Reuse of tannery wastewaters by combination of ultrafiltration and reverse osmosis after a conventional physical-chemical treatment. Desalination 204 (1–3), 219–226.

Fajardo-Diaz, J.L., Morelos-Gomez, A., Cruz-Silva, R., Matsumoto, A., Ueno, Y., Takeuchi, N., et al., 2022. Antifouling performance of spiral wound type module made of carbon nanotubes/polyamide composite RO membrane for seawater desalination. Desalination 523, 115445.

Feria-Díaz, J.J., Correa-Mahecha, F., López-Méndez, M.C., Rodríguez-Miranda, J.P., Barrera-Rojas, J., 2021. Recent desalination technologies by hybridization and integration with reverse osmosis: a review. Water 13 (10), 1369.

Figoli, A., Criscuoli, A., 2017. Sustainable Membrane Technology for Water and Wastewater Treatment. Springer, Singapore. Ebook ISBN 9789811056239.

Freger, V., 2003. Nanoscale heterogeneity of polyamide membranes formed by interfacial polymerization. Langmuir 19 (11), 4791–4797.

Ghaffour, N., Bundschuh, J., Mahmoudi, H., Goosen, M.F., 2015. Renewable energy-driven desalination technologies: a comprehensive review on challenges and potential applications of integrated systems. Desalination 356, 94–114.

Ghayeni, S.S., Madaeni, S., Fane, A., Schneider, R., 1996. Aspects of microfiltration and reverse osmosis in municipal wastewater reuse. Desalination 106 (1–3), 25–29.

Gibbons, U., 1988. Using desalination technologies for water treatment. In: Paper presented at the Recommended by US Congress, Office of Technology Assessment, OTA-BP-O-46 (Washington, DC: US Government Printing Office).

Gonzales, R.R., Zhang, L., Guan, K., Park, M.J., Phuntsho, S., Abdel-Wahab, A., et al., 2021. Aliphatic polyketone-based thin film composite membrane with mussel-inspired polydopamine intermediate layer for high performance osmotic power generation. Desalination 516, 115222.

Hiddink, J., De Boer, R., Nooy, P., 1980. Reverse osmosis of dairy liquids. J. Dairy Sci. 63 (2), 204–214.

Hu, J., Pu, Y., Ueda, M., Zhang, X., Wang, L., 2016. Charge-aggregate induced (CAI) reverse osmosis membrane for seawater desalination and boron removal. J. Membr. Sci. 520, 1–7.

Ismail, A.F., Matsuura, T., 2018. Progress in transport theory and characterization method of reverse osmosis (RO) membrane in past fifty years. Desalination 434, 2–11.

Ivić, I., Kopjar, M., Buljeta, I., Pichler, D., Mesić, J., Pichler, A., 2022. Influence of reverse osmosis process in different operating conditions on phenolic profile and antioxidant activity of conventional and ecological cabernet sauvignon red wine. Membranes 12 (1), 1–76.

Jeong, B.-H., Hoek, E.M., Yan, Y., Subramani, A., Huang, X., Hurwitz, G., et al., 2007. Interfacial polymerization of thin film nanocomposites: a new concept for reverse osmosis membranes. J. Membr. Sci. 294 (1–2), 1–7.

Ji, C., Zhai, Z., Jiang, C., Hu, P., Zhao, S., Xue, S., et al., 2021. Recent advances in high-performance TFC membranes: a review of the functional interlayers. Desalination 500, 114869.

Kaya, C., Sert, G., Kabay, N., Arda, M., Yüksel, M., Egemen, Ö., 2015. Pre-treatment with nanofiltration (NF) in seawater desalination—preliminary integrated membrane tests in Urla, Turkey. Desalination 369, 10–17.

Kumar, P., Sharma, N., Ranjan, R., Kumar, S., Bhat, Z., Jeong, D.K., 2013. Perspective of membrane technology in dairy industry: a review. Asian Australas. J. Anim. Sci. 26 (9), 1347.

Lagartos, A., Rozenbaoum, E., Oruc, M., Hyung, H., Armas, J.C., S. D., 2019. Long-term boron rejection of thin-film nanocomposite membrane at Pembroke desalination Plant in Malta: a case study. Desalin. Water Treat. 157, 274–280.

Lau, W., Ismail, A., Misdan, N., Kassim, M., 2012. A recent progress in thin film composite membrane: a review. Desalination 287, 190–199.

Lee, K.P., Arnot, T.C., Mattia, D., 2011. A review of reverse osmosis membrane materials for desalination—development to date and future potential. J. Membr. Sci. 370 (1–2), 1–22.

Leon, F., Ramos, A., 2021. Performance analysis of a full-scale desalination plant with reverse osmosis membranes for irrigation. Membranes 11 (10), 774.

Li, X., Loh, C.H., Wang, R., Widjajanti, W., Torres, J., 2017. Fabrication of a robust high-performance FO membrane by optimizing substrate structure and incorporating aquaporin into selective layer. J. Membr. Sci. 525, 257–268.

Lim, Y.J., Goh, K., Kurihara, M., Wang, R., 2021. Seawater desalination by reverse osmosis: current development and future challenges in membrane fabrication–a review. J. Membr. Sci., 119292.

Liu, L., Xie, X., Qi, S., Li, R., Zhang, X., Song, X., Gao, C., 2019. Thin film nanocomposite reverse osmosis membrane incorporated with UiO-66 nanoparticles for enhanced boron removal. J. Membr. Sci. 580, 101–109.

Liyanaarachchi, S., Shu, L., Muthukumaran, S., Jegatheesan, V., Baskaran, K., 2014. Problems in seawater industrial desalination processes and potential sustainable solutions: a review. Rev. Environ. Sci. Biotechnol. 13 (2), 203–214.

Loeb, S., Sourirajan, S., 1962. Sea water demineralization by means of an osmotic membrane. In: Saline Water Conversion—II. vol. 38. ACS Publications, Washington, DC, pp. 117–132. Chapter 9, 20036, US, eISBN: 9780841222045.

Luo, W., Xie, M., Song, X., Guo, W., Ngo, H.H., Zhou, J.L., Nghiem, L.D., 2018. Biomimetic aquaporin membranes for osmotic membrane bioreactors: membrane performance and contaminant removal. Bioresour. Technol. 249, 62–68.

Ma, N., Wei, J., Qi, S., Zhao, Y., Gao, Y., Tang, C.Y., 2013. Nanocomposite substrates for controlling internal concentration polarization in forward osmosis membranes. J. Membr. Sci. 441, 54–62.

Ma, X., Wan, X., Fang, Z., Li, Z., Wang, X., Hu, Y., et al., 2022. Orientational seawater transportation through Cu (TCNQ) nanorod arrays for efficient solar desalination and salt production. Desalination 522, 115399.

Madaeni, S., Zereshki, S., 2010. Energy consumption for sugar manufacturing. Part I: evaporation versus reverse osmosis. Energ. Conver. Manage. 51 (6), 1270–1276.

Mahajan, R., Vu, D.Q., Koros, W.J., 2002. Mixed matrix membrane materials: an answer to the challenges faced by membrane based gas separations today? J. Chin. Inst. Chem. Eng. 33 (1), 77–86.

Malaeb, L., Ayoub, G.M., 2011. Reverse osmosis technology for water treatment: state of the art review. Desalination 267 (1), 1–8.

Marchetti, P., Jimenez Solomon, M.F., Szekely, G., Livingston, A.G., 2014. Molecular separation with organic solvent nanofiltration: a critical review. Chem. Rev. 114 (21), 10735–10806.

Merson, R., Morgan, J.A., 1968. Juice concentration by reverse osmosis. Food Technol. 22 (5), 97.

Nath, K., 2017. Membrane separation processes, second ed. PHI Learning Pvt. Ltd., New Delhi, India. ISBN: 8120352912.

Ng, Z.C., Lau, W.J., Lai, G.S., Meng, J., Gao, H., Ismail, A.F., 2022. Facile fabrication of polyethyleneimine interlayer-assisted graphene oxide incorporated reverse osmosis membranes for water desalination. Desalination 526, 115502.

Pendergast, M.T.M., Nygaard, J.M., Ghosh, A.K., Hoek, E.M., 2010. Using nanocomposite materials technology to understand and control reverse osmosis membrane compaction. Desalination 261 (3), 255–263.

Peng, L.E., Yang, Z., Long, L., Zhou, S., Guo, H., Tang, C.Y., 2022. A critical review on porous substrates of TFC polyamide membranes: mechanisms, membrane performances, and future perspectives. J. Membr. Sci. 641, 119871.

Peters, T.A., 1998. Purification of landfill leachate with reverse osmosis and nanofiltration. Desalination 119 (1–3), 289–293.

Qasim, M., Badrelzaman, M., Darwish, N.N., Darwish, N.A., Hilal, N., 2019. Reverse osmosis desalination: a state-of-the-art review. Desalination 459, 59–104.

Rahimi, B., Afzali, M., Farhadi, F., Alamolhoda, A.A., 2021. Reverse osmosis desalination for irrigation in a pistachio orchard. Desalination 516, 115236.

Ramsey, I., Yang, Q., Fisk, I., Ayed, C., Ford, R., 2021. Assessing the sensory and physicochemical impact of reverse osmosis membrane technology to dealcoholize two different beer styles. Food Chem. X 10, 100121.

Rautenbach, R., Vossenkaul, K., Linn, T., Katz, T., 1997. Waste water treatment by membrane processes—new development in ultrafiltration, nanofiltration and reverse osmosis. Desalination 108 (1–3), 247–253.

Singh, R. (2015). Membrane Technology and Engineering for Water Purification—Application, Systems Design and Operation. (second ed.), Elsevier, Oxford, United Kingdom, ISBN 978–0–444-63362-0.

Tang, C., Wang, Z., Petrinić, I., Fane, A.G., Hélix-Nielsen, C., 2015. Biomimetic aquaporin membranes coming of age. Desalination 368, 89–105.

Ukrainsky, B., Ramon, G.Z., 2018. Temperature measurement of the reaction zone during polyamide film formation by interfacial polymerization. J. Membr. Sci. 566, 329–335.

Vourch, M., Balannec, B., Chaufer, B., Dorange, G., 2008. Treatment of dairy industry wastewater by reverse osmosis for water reuse. Desalination 219 (1–3), 190–202.

Wang, J., Dlamini, D.S., Mishra, A.K., Pendergast, M.T.M., Wong, M.C., Mamba, B.B., et al., 2014. A critical review of transport through osmotic membranes. J. Membr. Sci. 454, 516–537.

Wenten, I.G., 2016. Reverse osmosis applications: prospect and challenges. Desalination 391, 112–125.

Wethern, M., Katzaras, W., 1995. Reverse osmosis treatment of municipal sewage effluent for industrial reuse. Desalination 102 (1–3), 293–299.

Widiasa, I., Wenten, I.G., 2007. Combination of reverse osmosis and electrodeionization for simultaneous sugar recovery and salts removal from sugary wastewater. E-rea 11 (2), 91–97.

Wintgens, T., Melin, T., Schäfer, A., Khan, S., Muston, M., Bixio, D., Thoeye, C., 2005. The role of membrane processes in municipal wastewater reclamation and reuse. Desalination 178 (1–3), 1–11.

World Health Organization (2009). Boron in drinking-water: Background document for development of WHO Guidelines for Drinking-water Quality. Retrieved from https://apps.who.int/iris/handle/10665/70170.

Yang, Z., Sun, P.-F., Li, X., Gan, B., Wang, L., Song, X., et al., 2020. A critical review on thin-film nanocomposite membranes with interlayered structure: mechanisms, recent developments, and environmental applications. Environ. Sci. Technol. 54 (24), 15563–15583.

Yorgun, M., Balcioglu, I.A., Saygin, O., 2008. Performance comparison of ultrafiltration, nanofiltration and reverse osmosis on whey treatment. Desalination 229 (1–3), 204–216.

Zhao, D.L., Zhao, Q., Chung, T.-S., 2021. Fabrication of defect-free thin-film nanocomposite (TFN) membranes for reverse osmosis desalination. Desalination 516, 115230.

Zhao, Q., Zhao, D.L., Feng, F., Chung, T.-S., Chen, S.B., 2022. Thin-film nanocomposite reverse osmosis membranes incorporated with citrate-modified layered double hydroxides (LDHs) for brackish water desalination and boron removal. Desalination 527, 115583.

Zhu, Y., Xie, W., Gao, S., Zhang, F., Zhang, W., Liu, Z., Jin, J., 2016. Single-walled carbon nanotube film supported nanofiltration membrane with a nearly 10 nm thick polyamide selective layer for high-flux and high-rejection desalination. Small 12 (36), 5034–5041.

Zimmerman, C.M., Singh, A., Koros, W.J., 1997. Tailoring mixed matrix composite membranes for gas separations. J. Membr. Sci. 137 (1–2), 145–154.

Potentialities of membrane distillation and membrane crystallization

E. Drioli, F. Alessandro, and F. Macedonio

Institute on Membrane Technology, National Research Council of Italy (CNR-ITM), Rende, Italy

1. Introduction

Membrane contactors are systems in which hydrophobic porous membranes do not perform the function of a separating septum but, at the level of their pores, allow direct contact between two phases (e.g., liquid–liquid or liquid–gas) without dispersion of one into the other, facilitating the transport of mass and energy. Contactor-based systems can be used for separating miscible and nonmiscible liquids, for transporting components of gas mixtures in liquid phase, etc. As a result, they are widely used in air and water purification (Grossi et al., 2020; Li et al., 2022), semiconductors (Drioli et al., 2015), and pharmaceuticals (Yadav et al., 2021).

Membrane technology is increasingly emerging in the industrial field as an alternative (or complement) approach to traditional separation and concentration techniques. This is because membrane processes are more cost-effective in terms of energy consumption, separation efficiency, environmental sustainability, and process management (Bernardo et al., 2009; Siagian et al., 2019). In fact, today, membranes are being used to make production processes increasingly innovative, in the context of a circular economy that responds to environmental and human needs. Membrane processes differ according to the size and physical state of the substances present in mixtures to be processed. Many membrane operations, including microfiltration (MF), ultrafiltration (UF), nanofiltration (NF), reverse osmosis (RO), dialysis, electrodialysis (ED), etc., are commonly used. Emerging membrane processes involve membrane distillation (MD) and membrane crystallization (MCr).

Membrane distillation separates volatile solutes from nonvolatile solutes such as macromolecules, salts, and colloids. In the process, a membrane is in contact with a solution having a temperature in the range of 30–90°C. Since the membrane is hydrophobic, it does not allow

Copyright © 2023 Elsevier Inc. All rights reserved.

the passage of water in the liquid phase but of the water vapor in which it is converted at the solution-membrane interface due to the effect of a thermal gradient. The vapor thus diffuses into the membrane until it reaches the permeate side, where, thanks to a cooling system, it is condensed to return again to the liquid state (Capizzano et al., 2022; Macedonio and Drioli, 2019; Ursino et al., 2021; Zhao et al., 2021). Nonvolatile solutes are concentrated in the residual volume upstream of the membrane. Academic interest in MD is driven by the versatility of processes. MD is a promising technology, for instance, for the desalination of high saline waters (Macedonio et al., 2021; Soukane et al., 2021). In addition, MD technology has the potential to produce "green" results and provide solutions to environmental problems.

Membrane crystallization is a process very similar to the membrane distillation. It allows us to obtain solutes in a crystalline form by gradual concentration (until supersaturation) of the solution containing them. The membrane supports nucleation for crystalline growth and controls the saturation conditions (Drioli et al., 2012). Examples of application of this technique are the crystallization of inorganic salts, organic acids, recovery of crystals from the sea, and separations in the biomedical field, such as the crystallization of proteins (Di Profio et al., 2014; Polino et al., 2019).

Before discussing MD and MCr application and research trends, MD and MCr transport principles are discussed in the following sections.

2. Membrane distillation

2.1 Process description

Membrane distillation is a process that was described for the first time in the late 1960s (Bodell, 1963; Findley, 1967) and then investigated worldwide as a potential alternative to conventional separation methods. As mentioned before, this is a thermally driven membrane process in which a heated aqueous solution is placed in contact with one side (feed/retentate side) of a hydrophobic microporous membrane that inhibits the mass transfer of the liquid. Water and volatile solutes evaporate at the liquid–vapor interface that forms at the entrance of the nonwetting pores, to then condense on the distillate/permeate side (Fig. 1). Ideally, MD allows complete rejection of the salts and nonvolatile components of the feed.

Other benefits of MD with respect to conventional processes are:

- lower operating temperatures;
- lower operating pressure and reduced influence of concentration;
- minor issues related to membrane fouling.

As regards transport mechanisms, water vapor crosses the microporous membrane by means of a combination of three transport mechanisms: Knudsen diffusion, Poiseuille flow, and molecular diffusion. As a consequence of these effects, water vapor reaches a cold surface on which it condenses.

The Knudsen number, defined as the ratio between the mean free path of gas and the pore diameter of the membrane, is a dimensionless quantity used with the aim of characterizing the physical nature of flow through membrane pores. In particular, if $K_n > 1$, the mass

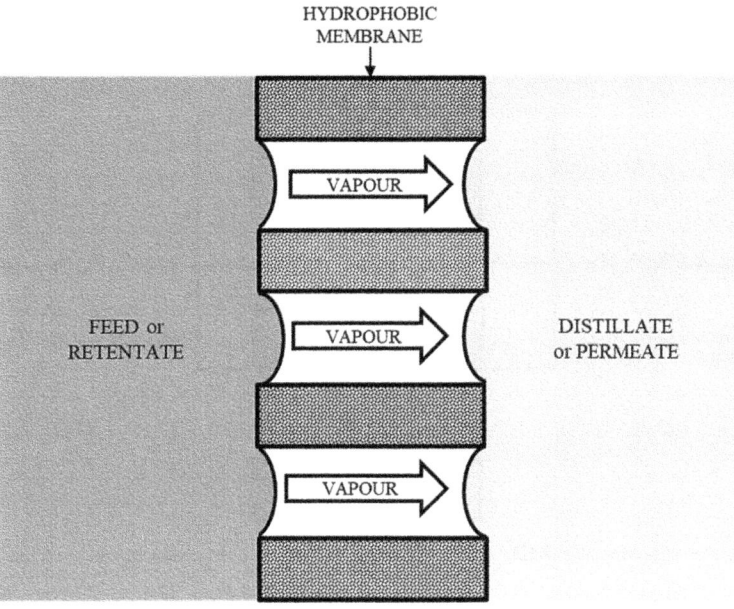

FIG. 1   General scheme of the MD process.

transport follows Knudsen diffusion, whereas when $0.01 \ll K_n < 1$, the transport mechanism is described by a combination of Knudsen and molecular diffusion, and finally, if $K_n \ll 0.01$, the mass transport follows molecular diffusion.

In line with the mass transport model, in the MD system the water vapor flux J across the membrane can be expressed by the following equation:

$$J = C \cdot \Delta P \tag{1}$$

where ΔP is the vapor pressure difference between the warm and cold sides of membrane and C is a membrane distillation coefficient associated with the membrane structure and flow parameters of the aqueous solution.

Fig. 2 illustrates the possible mass transfer resistances in the MD process with an electrical analogy. This approach considers the mass transfer in MD in terms of serial resistances upon transfer between the bulks of two phases in contact with the membrane.

Boundary layers can contribute to the overall mass transfer resistance and the rate is generally limited by the molecular diffusion. The mass transfer resistance inside the pores of the membrane depends on the transfer of momentum (viscous resistance), collision between diffusing molecules (molecular resistance), or on the collision between the molecules and the membrane walls (Knudsen resistance).

Furthermore, in MD resistance related to surface diffusion is present but is usually negligible because the surface area is smaller than the pore area.

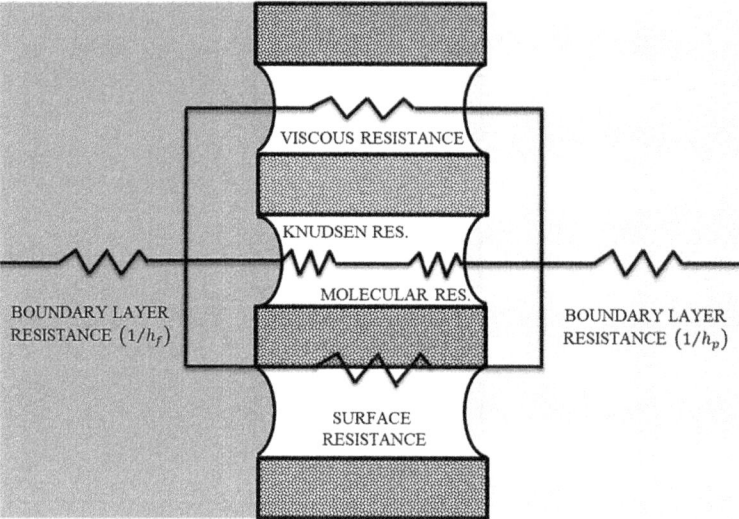

FIG. 2   Mass-transfer resistances in MD.

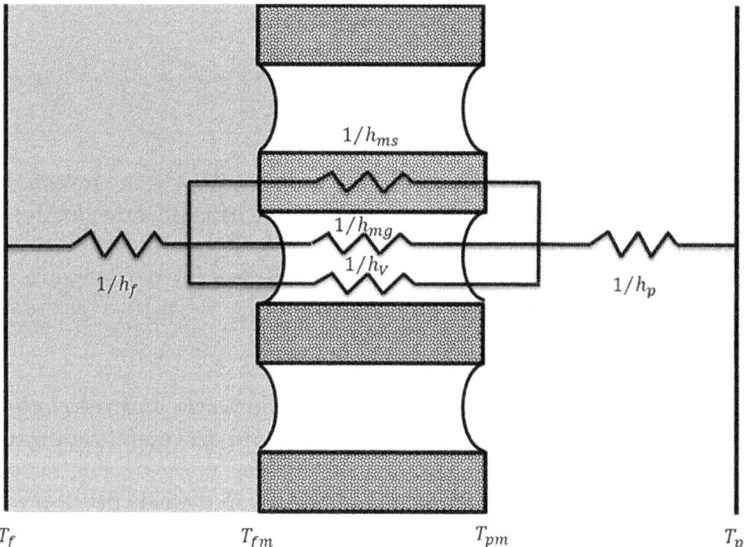

FIG. 3   Heat-transfer resistances in MD.

Fig. 3 shows a general schema of heat transfer that can be modified based on the MD configuration considered. Unlike other processes such as direct osmosis, in MD the vapor mass transfer is coupled with heat transfer across the membrane and across boundary layers next to the
two sides of the membrane. In fact, in the MD process, heat transfer and mass transfer are closely
related, so much so that the rate of heat transfer depends on the mass flow and vice versa.

Heat is first conducted from the heated feed solution of temperature T_f, across the thermal boundary layer, to the membrane surface of temperature T_{fm}, at a rate of

$$Q_f = h_f \cdot (T_f - T_{fm}) \tag{2}$$

where h_f is the heat transfer coefficient.

The total heat transfer across the membrane can be defined as $Q = Q_V + Q_m$ where

$$Q_V = h_V \cdot \Delta T_m = N \cdot \Delta H_V \tag{3}$$

(N is the rate of mass transfer and ΔH_V is the heat of vaporization) is the heat transferred across the membrane due to liquid evaporation at the surface of the membrane.

$$Q_m = h_m \cdot \Delta T_m \tag{4}$$

where $h_m = \varepsilon \cdot h_{mg} + (1 - \varepsilon)h_{ms}$ (ε is the membrane porosity and h_{mg} and h_{ms} represent the heat transfer coefficients of the vapor within the membrane pores and the solid membrane material, respectively) is the heat transferred through the membrane material and via the vapor that fills the pores.

Because no mass transfer is associated with the conduction, it can be considered a heat loss mechanism. Finally, the vapor condenses at the liquid–vapor interface and heat is removed from the surface of the membrane onto the cold side through the thermal boundary layer at a rate of

$$Q = h_p \cdot \Delta T_p. \tag{5}$$

In the MD process, polarization phenomena take place that lead to a decrease in the real flux with respect to that produced in the event that these phenomena do not exist.

The commonly measures used to quantify the heat transfer and mass resistance within the boundary layer are given by the temperature polarization coefficient (TPC) and concentration polarization coefficient (CPC), respectively. These are defined as follows:

$$\text{TPC} = \frac{T_{fm} - T_{pm}}{T_f - T_p} \tag{6}$$

$$\text{CPC} = \frac{C_{B_m}}{C_{Bb}} \tag{7}$$

where T_{fm} is the temperature of the feed at the membrane surface, T_{pm} is the temperature of the permeate at the membrane surface, T_f is the temperature of the feed in the bulk, T_p is the temperature of the permeate in the bulk, C_{B_m} is the salt concentration at the hot membrane surface, and C_{Bb} is the salt concentration in the bulk (feed side).

Generally, the MD process is limited by heat transfer across the boundary layers on either sides of the membrane and not by mass transfer across the membrane due to the fact heat transfer across the boundary layers often represents rate-limiting steps. In addition to the transport resistance produced by boundary layers and the membrane, the most important phenomena that can cause a reduction in transmembrane flux or damage the membrane are the membrane wetting and fouling (both will be described later in the chapter).

2.2 Membrane distillation configuration

The most common MD systems can be classified into four basic configurations (Fig. 4A–D) depending on the methods used to induce a vapor pressure difference across the membrane and the type of permeate collection on the cold side (Davey et al., 2021; Khayet and Matsuura, 2011; Okati et al., 2022; Ruiz-Aguirre et al., 2019; Shaheen et al., 2022).

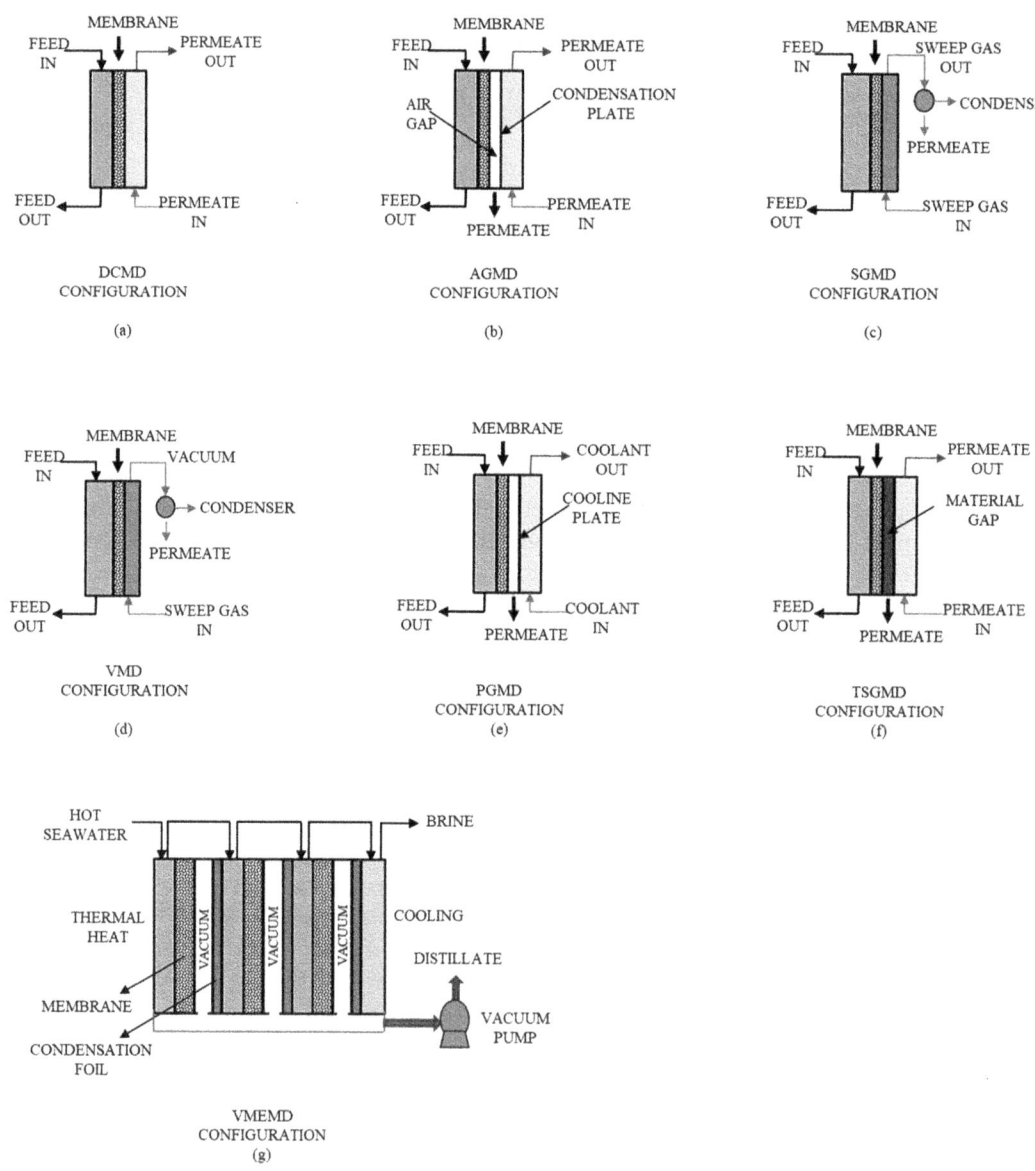

FIG. 4   Different types of MD configurations.

(1) *Direct contact membrane distillation* (DCMD)

DCMD is the simplest configuration of a membrane distillation system in which both the solutions, feed and permeate, are in direct contact with the membrane. The temperature at the feed side is higher than that at the permeate side, and this difference creates a vapor pressure difference which induces volatile molecules of the feed solution to evaporate at the liquid–vapor interface, to diffuse across the membrane, until they condense directly into an aqueous solution flowing on the cold permeate side (Baghbanzadeh et al., 2016). DCMD is capable of producing a relatively high amount of permeate flux, without the use of an external condenser. DCMD is widely used in desalination processes (Bonyadi et al., 2022; Essalhi et al., 2021; Marques Lisboa et al., 2021), concentration of aqueous solutions in the food industry (Quist-Jensen et al., 2016; Tundis et al., 2021), boron removal (Wen et al., 2016), and acid production (Wang et al., 2022). The main drawback of this configuration is the heat lost by conduction, which increases the energy required to maintain the feed solutions at temperatures hot enough to drive evaporation (Parani and Oluwafemi, 2021).

(2) *Air gap membrane distillation* (AGMD)

AGMD represent the most versatile technique in which only the feed solution is directly in contact with the hot side of the membrane. A thin air gap separates the hydrophobic membrane from the cold condensing surface providing the possibility of condensing the permeate vapors on the cold surface rather than directly in a cold liquid. This corresponds to a considerable reduction of the amount of heat lost by conduction through the membrane (Noamani et al., 2020). However, in AGMD systems, with respect to the DCMD or VMD configurations (Ansari et al., 2022; Noamani et al., 2020; Yang et al., 2019), the fluxes are reduced because the air gap creates an additional resistance to mass transfer. AGMD can be used for both DCMD applications and for removing volatile components as traces of aqueous alcohol solutions.

(3) *Sweeping gap membrane distillation* (SGMD)

The SGMD configuration combines the low conductive heat loss of AGMD with the reduced mass transfer resistance of DCMD. In SGMD, an inert gas is used to sweep the vapor at the permeate side of the membrane, maintaining the gradient necessary for transport (Jiang et al., 2022; Ko et al., 2018; Lee et al., 2022). The main advantage of this process is that the mass transport resistance of the air gap is substantially reduced. However, because a small volume of permeate diffuses into a large sweep gas volume, large external condensers are necessary to condense the vapor water and collect the permeate. SGMD is mostly used for the removal of volatile organic compounds or dissolved gases from aqueous solutions (Jiang et al., 2022; Xie et al., 2009).

(4) *Vacuum membrane distillation* (VMD)

In the VMD configuration, a pump is used to create a vacuum at the permeate side of the membrane (Ko et al., 2018). To facilitate vapor transport across the membrane, the applied vacuum pressure must be lower than the saturation pressure of volatile molecules in the feed solution. The vapor is transferred continuously from the vacuum chamber to the external condenser (Jia et al., 2021). One of the major advantages of VMD with respect to other configurations is the negligible conductive heat loss through the membrane. However, the main issues are the complicated set-up and the possible wetting of membranes due to the fact that the variation of pressures at the membrane interface is higher in VMD than in other MD processes. VMD is typically used for the desalination of seawater or saline solutions (Chang et al., 2021), and for the removal of volatile components from aqueous solutions.

Recently, new configurations (Fig. 4E–G) with improved energy efficiency and better permeation flow have been proposed.

- *Permeate-gap membrane distillation* (PGMD)

   An improvement of the direct contact-MD configuration is PGMD, a process in which a third channel is introduced by an additional nonpermeable foil. One of the main advantages of this system is the separation of the distillate from the coolant. This implies that the cold feed water can be used as a coolant. Consequently, the sensible heat losses are minimized by the presence of the distillate channel. The main disadvantage is due to the reduction of the effective temperature difference across the membrane, which tends to lower the permeation rate (Alquraish et al., 2021; Macedonio and Drioli, 2019).

- *Material-gap membrane distillation* (MGMD)

   MGMD is an evolution of AGMD in which the gap between the membrane and the condensation surface is filled with different materials such as polyurethane (sponge), polypropylene mesh, and deionized water (Cai et al., 2020; Francis et al., 2013). This new MD module design involves an increase of 200%–800% of water vapor flux compared to the AGMD, where the presence of stagnant air between the membrane and the condensation plate produces a decrease of permeation flux in relation to other MD processes.

- *Thermostatic sweeping gas membrane distillation* (TSGMD)

   TSGMD is a hybrid configuration, which combines both SGMD and AGMD systems. In TSGMD, the driving force is increased by including a cold wall to the system, which decreases the temperature of the sweeping gas. TSGMD exhibits some benefits over DCMD in terms of better selectivity performance and smaller temperature polarization, although at a lower permeate flux. TSGMD can be used for distilled water production, concentration of the nonvolatile solutes in the feed aqueous solution, and concentration of contaminants in aqueous solutions (Rivier et al., 2002).

- *Multieffect membrane distillation* (MEMD)

   MEMD is a modified configuration of AGMD realized to increase the internal heat recovery. The condensation surface is placed above the cold feed solution that acts as a coolant to condense the permeated vapor and heat the solution.

- *Vacuum multieffect membrane distillation* (VMEMD)

   VMEMD represents a modified form of VMD that combines the benefits of thermal multieffect distillation with the membrane filtration step. As a first step, water is separated from its pollutants by evaporation and later the produced steam is filtered by a microporous and hydrophobic membrane. The typical V-MEND is composed of a heater, multiple evaporation–condensation stages, and an external condenser (Zhao et al., 2013). The VMEMD configuration can be used in a wide range of applications. It can be applied to ultrapure water applications like in the pharmaceutical industry or for groundwater purification, moisture removal, brine concentration, and solar-energy driven desalination.

2.3 Membrane characteristics

The main obstacle in the widespread application of MD is the unavailability of appropriate membranes for MD systems. The performances of the MD processes depend on the structural characteristics and physical–chemical properties of the membranes used. Thickness, porosity, tortuosity, mean pore size, and pore distribution are included among the physical properties of MD membranes while liquid entry pressure, thermal stability, thermal conductivity, and fouling rate represent some of the physical–chemical parameters.

2.3.1 Thickness

Membrane thickness is an important parameter of membrane efficiency. As in any other process, membrane thickness is inversely proportional to the permeate flux. In fact, as the thickness of the membrane increases, the permeate flux is reduced because the mass transfer resistance increases.

Usually, thicknesses of 10–300 µm are used for MD membranes.

2.3.2 Porosity and tortuosity

Membrane porosity is the volume fraction of the pores of the membrane. Membranes with greater porosity have a larger surface area for evaporation. So, a membrane with high porosity has higher permeate flux and lower conductive heat loss.

Membrane porosity can be determined by a gravimetric method (Feng et al., 2004), measuring the weight of liquid contained in the membrane pores. The porosity of the membranes (ε) can be calculated using the following equation:

$$\varepsilon = \frac{(W_1 - W_2)/\rho_k}{(W_1 - W_2)/\rho_k + W_2/\rho_P} \cdot 100\% \tag{8}$$

where W_1 is the weight of the wet membrane, W_2 is the weight of the dry membrane, ρ_k is the density of the liquid, and ρ_P is the polymer density.

In general, membranes used in MD systems have porosities as high as 70%–80%.

Tortuosity (τ) is a property of a porous material, usually defined as the deviation of the pore structure from the cylindrical shape. Consequently, the higher the tortuosity value, the lower the permeate flux. The best relationship has been suggested by Macki-Meares (Srisurichan et al., 2006), where:

$$\tau = \frac{(2 - \varepsilon)^2}{\varepsilon} \tag{9}$$

2.3.3 Mean pore size and pore distribution

Membranes with pore sizes between 100 nm and 1 µm are typically used in MD systems (Woods et al., 2011). The permeate flow rate increases depending on the size of the membrane pores (Woods et al., 2011). Based on the pore size and mean free path of the transferred molecules, it is possible to determine the mechanism of mass transfer and calculate the permeate flow. The mean pore size is used to estimate the vapor flux: the pore size should be small to avoid liquid penetration whereas a large pore size is required for a high permeate flow. In this case, even if the membrane is very hydrophobic, wetting occurs in pores and the membrane selectivity decreases. Therefore, an optimal pore size should be determined for each operating condition.

2.3.4 Liquid entry pressure

The liquid entry pressure (LEP) is defined as the minimum transmembrane pressure that a feed solution requires to penetrate the hydrophobic membrane. LEP is a very important parameter for MD membranes, where the desired property of the membrane is to avoid the wetting of the pores. The higher the LEP, the better it is in preventing membrane wetting.

LEP can be estimated as follows (Franken et al., 1987):

$$LEP_W = -\frac{2B\gamma_l \cos\theta}{r_{max}} \tag{10}$$

where B is a geometric factor determined by pore structure (for instance, $B=1$ for cylindrical pores), γ_l is the liquid surface tension, θ is the liquid/solid contact angle between water and the membrane surface, and r_{max} is the maximum pore size.

When the hydrostatic pressure on the feed side of an MD membrane is greater than LEP, liquid penetrates the membrane pores and this pore membrane penetration is known as wetting of a membrane. In its simplest form, the wettability of a liquid droplet on a flat smooth surface is commonly determined by Young's equation. In reality, however, the surfaces contain some roughness and for this reason Wenzel's theory (Wenzel, 1936) was proposed, according to which the contact angle is correlated with surface roughness through the following equation:

$$\cos\theta_w = r^* \frac{(\gamma_{sv} - \gamma_{sl})}{\gamma_{lv}} \tag{11}$$

where r^* represents the surface roughness factor ($r^* > 1$), θ is the contact angle, γ_{sv}, γ_{sl}, and γ_{lv} are solid/vapor, solid/liquid, and liquid/vapor interfacial tensions, respectively.

Once wetting takes place, the membrane begins to lose its hydrophobicity locally.

If the wetting is partial and the majority of the pores are dry, the MD process can be continued. In the case of full wetting, the MD membrane no longer acts as a barrier, resulting in a viscous flow of liquid water through the membrane pores, incapacitating the MD process (Rezaei and Samhaber, 2016; Rezaei et al., 2017).

2.3.5 Thermal stability and thermal conductivity

An MD membrane is thermally stable if high temperatures do not degrade or decompose its material.

The thermal behavior is influenced by the material, thickness, and porosity of the membrane. In order to reduce the conductive heat losses and the thermal polarization phenomena, membranes with low thermally conductive are preferable. Thinner membranes are more conductive than thicker ones. Moreover, high porosity reduces the conductive heat losses, as the vapor has lower thermal conductivity than the solid membrane material, and improves the permeability.

2.3.6 *Fouling rate*

Membrane fouling is a process in which organic and inorganic materials are deposited on the surface or in the pores of the membrane. It is an important factor that influences membrane wettability and reduces its performance by lowering the flux (An et al., 2016).

Fouling is a more relevant problem in pressure-driven processes than in MD because in the latter process the pores are larger than, for example, the pores present in UF and RO and also because of its low operating pressure that induces the deposition of foulants on the membrane surface to be less compact and only slightly affect transport resistance.

This unwanted phenomenon can be tackled through various strategies including prefiltration, membrane cleaning, and modification of the membrane surface.

2.4 Membrane materials

The internal structure of the membranes is determined by the method and the type of material used for their preparation. There are two main categories of membrane structure: symmetric and asymmetric. Symmetrical membranes can exhibit a porous or a dense homogeneous structure. Porous membranes are structures having straight or sponge-like pores, widely employed in membrane contactors. Asymmetric membranes are comprised of a highly porous structure (having a typical thickness of 100–200 μm) that supports a very thin, dense, or porous skin layer (typically less than 1 μm). This latter provides the membrane the desired selectivity while the porous substructure supplies the necessary mechanical properties and offers the advantage of low resistance to the mass transport through the membrane.

As previously mentioned, MD technology requires the use of hydrophobic membranes. Over the past few decades, among hydrophobic materials, fluoropolymers have attracted significant attention due to their high thermal stability, low surface tension, and good chemical resistance.

An important fluoropolymer used for fabricating MD membranes is poly(vinylidene fluoride) (PVDF). PVDF is a semicrystalline polymer with typical crystallinity between 35% and 70%. The crystalline phase of PVDF has five distinct crystal polymorphs: α (phase II), β (phase I), γ (phase III), δ, and ε. The α form is the most common nonpolar phase formed of $(CH_2–CF_2)_n$ chains in a monoclinic crystallographic form while the polar β phase is the most thermodynamically stable form. This latter is of special interest due to its pyro- and piezoelectric properties.

PVDF shows good thermal stability, mechanical and chemical resistance, and it can be dissolved in most common solvents such as N-methyl-2-pyrrolidone (NMP), N,N-dimethyl acetamide (DMAc), and N,N-dimethyl formamide (DMF). It has a glass transition temperature (T_g) in the range from −40°C to −30°C and a melting point between 155°C and 192°C.

As regards the synthesis processes, nonsolvent-induced phase separation (NIPS) is the mainly used technique for the production of PVDF membranes for MD applications. In NIPS, a polymer solution is immersed in a nonsolvent bath, inducing phase separation of the solution into a polymer-rich (membrane matrix) phase and a polymer-poor phase (membrane pores).

Other techniques used to synthesize PVDF membranes are thermally induced phase separation (TIPS) and electrospinning. In the TIPS process, membrane formation is induced by cooling the polymer solution. Some of the advantages of the TIPS process are the simplicity of the process, high porosity, and the capability to form narrow pore distribution. Recently, the electrospinning process has also attracted considerable attention mainly due to the fact that it is a simple process with which to produce submicron-scale and nanoscale fibers. Compared to the NIPS techniques, electrospun membranes show superior porosity, high specific surface area, and high strength-to-weight ratio (Liu et al., 2011).

To reduce fouling or increase wetting resistance, various modifications have been suggested such as surface coating, blending, and pore filling (Zhong et al., 2021). A disadvantage can be represented by the fact that the introduction of such modifications can involve a decrease of the membrane permeability.

Another polymer widely used for the MD membrane production is polytetrafluoroethylene (PTFE), better known by the trade name Teflon. It is a thermoplastic polymer and it is formed by carbon chains having two fluorine atoms for each carbon atom. The most important properties of PTFE such as high chemically stability, high heat resistance, and strong hydrophobicity derive from the strong C—C and C—F bonds and the carbon backbone, protected from chemical attacks by a uniform and continuous covering formed by fluorine atoms. PTFE is electrically inert; so, its volume and surface resistivity are high. It is insoluble in common solvents at room temperature. Additionally, PTFE is highly crystalline (92%–98% crystallinity) and the crystallites have a high melting point (342°C), which makes the processability of PTFE difficult. Therefore, the commonly used phase inversion or melt spinning techniques for membrane synthesis cannot be exploited for PTFE. Currently, PTFE membranes are prepared using complicated extrusion, rolling, stretching, and sintering methods, which make difficult the control of the porous structure. Consequently, further efforts are required to create new methods to produce PTFE membranes with increasingly performing characteristics.

Properties such as low cost, good thermal stability, chemical resistance, and mechanical strength make polypropylene (PP) suitable for membrane applications. PP consists of a repeating unit, the $-CH_2CH(CH_3)$ monomer. It exists in both semicrystalline and amorphous forms but the isostatic PP, a semicrystalline form, is the most used for membrane preparation on a commercial scale. Moreover, compared with PVDF, PP has high surface energy and hence low hydrophobicity.

The PP membrane is fabricated using the following three main techniques: TIPS, stretching, and track etching. The stretching method is a solvent-free technique and is applicable for semicrystalline polymers. It consists of four fundamental processes: the molten polymer film is extruded into a film (1), which is subsequently annealed (2); annealing is followed by cold and hot stretching (3) that generates and enlarges the pore size; to avoid pore closure, heat setting is performed as the last step (4). The track-etching technique is mainly based on the use of heavy ion accelerators.

The advantage offered by this technique is the possibility of varying the size and density of the pores in a controllable manner so that a membrane with the required characteristics of transport can be produced.

Future research is expected to be directed toward investigating novel preparation methods for PP membranes using, in particular, less explored techniques such as electrospinning. Furthermore, the use of fillers such as carbon nanotubes (CNTs) can also be investigated to create membranes having specific properties.

PVDF, PTFE, and PP membranes demonstrate some drawbacks in terms of thermal and chemical stabilities that can affect their lifetime (He et al., 2019). As a result, hydrophobic ceramic membranes have attracted considerable interest due to their outstanding characteristics, which allow long-term operation without significant performance deterioration. However, ceramic membranes are generally hydrophilic in nature (Cerneaux et al., 2009; He et al., 2019; Tai et al., 2020; Wu et al., 2013). Indeed, TiO_2, alumina, silica, and zirconia, the materials mainly used for their synthesis, present hydroxyl groups at the surface, which confer hydrophilic character to the ceramic membrane. Hence, several membrane surface modifications are required to acquire hydrophobic properties for MD applications. The most widely used hydrophobization method is direct grafting using silane agents (Hubadillah et al., 2019; Kujawa et al., 2013a,b). Nevertheless, the long-lasting character of the various applied modifications is still a difficulty that must be overcome. In addition, ceramic membranes are expensive; hence, studies should be directed toward exploring cost-effective ceramic materials.

Recently, significant efforts have occurred in developing nanomaterial-based MD membranes.

Due to their exceptional mechanical strength, chemical resistance, and thermal properties, carbon nanotubes (CNTs) have gained significant attention. CNTs consist of rolled-up sheets of single-layer carbon atoms (graphene). The main features that make CNTs an emerging nanomaterial in water desalination are their large specific surface area, high aspect ratio, ease of functionalization, high transport of water molecules, and the possibility to change the water-membrane interaction favoring the preferential transport of vapors through the pores (Aroon et al., 2010; Jamed et al., 2019; Roy et al., 2014). Furthermore, CNT-based membranes exhibit outstanding porosity and hydrophobicity, basic requisites for MD applications. CNTs used as fillers have proven to significantly improve membrane performance in terms of strength, rejection, and permeability. In 2010, the first evidence was provided regarding the potential of CNT membranes for desalination by MD (Dumée et al., 2010). A significant increase in flow enhancement is observed when CNTs-based membranes are used in MD applications. The main drawbacks are related to the long-term operation, synthesis, processing of CNTs, and scale-up approaches.

One of the most important two-dimensional (2D) materials used in MD membranes is graphene. Since its isolation by exfoliation from graphite samples (Novoselov et al., 2004), graphene is considered among the most promising materials in almost every field of applied science and technology. Graphene has become a valuable and useful nanomaterial due to its exceptionally high tensile strength, electrical conductivity, transparency, and the fact that it is the thinnest two-dimensional material in the world. These exceptional properties derive from the unique behavior of its π bands, which give rise to so-called Dirac cones at the K points of the Brillouin zone (BZ). Graphene and its derivatives [such as graphene oxide (GO) and reduced graphene oxide (rGO)] can be used to improve the performance of polymeric membranes or to synthesize polymer nanocomposite membranes for many separation applications. The use of graphene in MD membranes allows to reach higher flux, higher rejection, improved antifouling, enhancement of mechanical properties, and superior heat recovery. In addition, graphene-based membranes exhibit exceptional selective permeability toward various components. For example, these membranes can selectively permeate some metal ions present in a solution containing different types of ions (Sun et al., 2013) or have an excellent selectivity toward various gases (Nidamanuri et al., 2020). Thanks to these features, it is possible to apply graphene-based membranes in desalination applications including MD (Frappa et al., 2020; Gontarek-Castro et al., 2022; Mao et al., 2020; Xu et al., 2020). However, the high thermal conductivity of graphene represents the main drawback for its practical application in the MD process.

Following the extensive investigation of graphene, enormous effort has been directed toward other 2D materials "beyond graphene." MXenes (Meng et al., 2021), zeolites (Donato et al., 2020), and metal–organic framework (Wu et al., 2021) nanosheets have emerged as promising 2D materials for high-performance membranes. Two-dimensional material membranes, due to their well-defined transport channels and ultralow thickness, offer an ultralow resistance to mass transport. Outstanding performances have been observed for liquid and gas separation applications. Recently, the use of a dichalchogenide material (Bi_2Te_3) for desalination of water by MD was proposed, for the first time (Frappa et al., 2021). The authors demonstrated that the use of Bi_2Te_3 (exfoliated and included in PVDF membranes) increases freshwater production, contrasting with the loss of conductive heat.

The main current challenges in implementing on a commercial scale of these 2D materials-based membranes include limited available exfoliation techniques for exfoliating the high aspect ratio and intact nanoporous monolayers from bulk crystals, drilling of the pores with required characteristics in the membrane matrix, and scaling up of these atomic-scale membranes into real-scale separation devices (Liu et al., 2016). In recent times, an important patent (Politano et al., 2018) was deposited describing the production of a nanocomposite membrane with 2D crystals obtained through exfoliation of layered materials by a wet-jet milling technique.

2.5 Applications

Membrane distillation is a promising technology for desalination and wastewater treatment (Gryta, 2021; Kalla, 2021; Niknejad et al., 2021; Yadav et al., 2022). It is a process not limited by concentration polarization phenomena as in the case of RO, and which offers the enticement of operating at atmosphere pressure and relatively low temperatures, with the theoretical ability to achieve 100% salt rejection. In addition, lower operating temperatures with respect to those usually used in conventional distillation methods make the MD method also interesting for processing pharmaceutical compounds, juices, dairy products, etc. (Rehman et al., 2022). Membrane distillation, either as a stand-alone process or in integration with other processes, can be considered as a potential candidate for the removal of heavy metals from drinking water (Attia et al., 2017) and recovery of phosphorus from agricultural, household, and industrial runoffs (Simoni et al., 2021). It has also been used in membrane bioreactors (MDBR) for the treatment of industrial and municipal wastewater to effectively capture small contaminants (Ricci et al., 2021).

In addition, MD can also be used in integrated systems with pressure-retarded osmosis (PRO) or reverse electrodialysis (RED) for utilizing a salinity gradient for energy production.

However, a particular drawback of MD operated with conventional membranes is represented by the temperature polarization phenomenon, which reduces the feed temperature at the membrane surface, involving a decrease of the driving force to water evaporation.

Overcoming the polarization temperature issue is crucial, for example, for a future integration of membrane distillation into solar energy systems. Recently, materials that exhibit photothermal effect have attracted notable attention due to their ability to absorb and convert natural or artificial radiation into heat. Experimental studies have demonstrated that the use of these materials as fillers can increase the thermal efficiency of the MD process (Dongare et al., 2017; Huang et al., 2019; Tan et al., 2018; Wu et al., 2018). Innovative nanostructured membranes that include Ag nanoparticles in a PVDF matrix were proposed (Politano et al., 2017) for seawater desalination. In this work, the authors have proved that photothermal plasmonic heating generated by Ag nanoparticles considerably improved the MD performance, increasing the transmembrane flux by 9–11 times for 0.5 M NaCl and pure water. Nonetheless, the high cost of nanoparticles is a limit for the employment of noble metals in photothermal membranes. In perspective, further materials that exhibit interesting photothermal properties, such as carbon-based materials and nanostructures based on layered materials (black phosphorus, Bi_2Se_3, and MXenes), can be suitable candidates for thermoplasmonic MD.

3. Membrane crystallization

3.1 Process description

Membrane crystallization (MCr) has been proposed as one of the most promising extensions of the MD technology. This innovative process combines the concepts of mass transfer and heat transport across porous membranes and the theory of heterogeneous nucleation. Essentially, the working principle is the same as that of membrane distillation. In MCr, a solution containing a nonvolatile solute is in contact, through a microporous hydrophobic membrane, with the permeate side. The vapor pressure gradient created across the porous membrane induces the evaporation of the volatile components from the feed side toward the distillate side where, finally, they recondense. In a membrane crystallizer, the continuous removal of solvent from the feed solution increases the solute concentration, thus attaining a supersaturated solution. Therefore, in MCr, the membrane is a physical support that allows the controlled removal of solvent from/to the crystallizing solution generating a supersaturated environment where crystals may nucleate and grow. The degree and rate of the supersaturation might be regulated with high precision operating on the process parameters (temperature, concentration, flow rate, etc.) and using membranes with suitable physicochemical properties. As described earlier, the membranes might be made of polymeric or inorganic materials or a combination of both in a hybrid or composite system.

MCr can be operated in all four basic MD configurations: DCMD, SGMD, AGMD, and VMD.

From a process design point of view, crystallization can take place directly in the membrane module (Fig. 5A) (Edwie and Chung, 2013; Lakerveld et al., 2010), where supersaturation occurs, or in the crystallizer (Fig. 5B) (Curcio et al., 2005; Kieffer et al., 2009). In the latter case, the membrane module is used to generate the supersaturation, or simply to concentrate the solid phase. In 2010, a new design of MCr was proposed (Di Profio et al., 2009) in which crystallization is induced by using an antisolvent. This approach operates in two configurations:

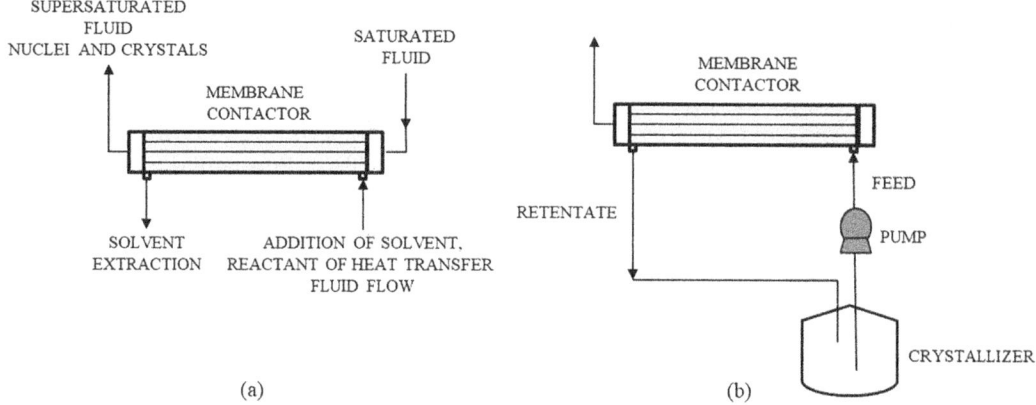

FIG. 5   Schematic diagrams of the set-up used for the MCr experiments in which crystallization can take place directly in the membrane module (A) or in the crystallizer (B).

(1) solvent/antisolvent demixing, where the solvent is removed at a higher flow rate than the antisolvent so that phase inversion promotes supersaturation; and

(2) antisolvent addition, in which the antisolvent is dosed into the crystallizing solution.

Congruent with the general concept of membrane crystallization, in these two configurations, solvent and antisolvent migration occur in the phase vapor and it is controlled by the membrane structure, acting on the operative process parameters.

This mechanism is different than that observed when forcing the liquid phases through the pores and the possibility to accurately dose the antisolvent in the vapor phase through the porous membrane entails an increasingly efficient control of the solution composition during the process, involving a consequent improvement of the final crystal properties.

A well-documented advantage of membrane crystallization processes over conventional methods is represented by the accelerated rate of the crystallization process, as proved by induction time (the time elapsed from the attainment of a given supersaturation up to the formation of critical nuclei) and nucleation/growth rate measurements. This involves the production of narrow size and high-quality crystals in terms of purity and size distribution (Fig. 6).

Conversely, the main concernment regarding MCr is related to the membrane stability. Indeed, in the case of membrane wetting, the process collapses because of the direct dispersion of one phase into the other without any selectivity. This issue can be reduced by operating with moderate transmembrane fluxes and using membrane materials that are sufficiently hydrophobic.

3.2 Heterogeneous nucleation

One of the main features of the MCr process is that the membrane operates not only as a support for the solvent evaporation rate, but also promotes heterogeneous nucleation starting at low supersaturation ratios, depending on the surface characteristics of the membrane.

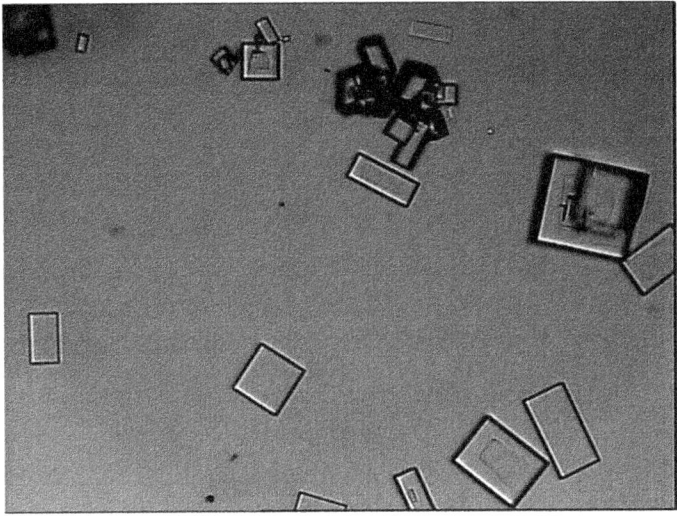

FIG. 6 Sodium chloride crystals as obtained via MCr.

The influence of morphological parameters (contact angle, porosity, and roughness) on the nucleation is shown in the following paragraph.

3.2.1 Effect of contact angle

A crystallizing solution can be imagined as a certain number of solute particles moving among the molecules of solvent and colliding with each other, so that a number of them converge into clusters. Nevertheless, an energetic barrier must be crossed in order to induce the formation of stable nuclei.

The energy barrier for a nucleus is located in correspondence to its critical size.

The presence of foreign surfaces, in particular polymeric membranes, decreases the work required to produce critical nuclei and increases locally the probability of nucleation with respect to other locations in the system. This phenomenon is defined as heterogeneous nucleation.

According to the classical nucleation theory (CNT), the reduction of the Gibbs free energy $\Delta G^{*}_{heterogeneous}$ due to heterogeneous nucleation is associated to the contact angle θ between the solution and the solid substrate:

$$\frac{\Delta G^{*}_{heterogeneous}}{\Delta G^{*}_{homgeneous}} = \left(\frac{1}{2} - \frac{3}{4} \cos\theta + \frac{1}{4} \cos^{3}\theta \right) \tag{12}$$

where $\Delta G^{*}_{homgeneous}$ is the value of the Gibbs energy threshold for homogeneous nucleation.

Because of the dependence of the energy barrier to heterogeneous nucleation on the contact angle and other relevant morphological properties, the heterogeneous nucleation rate can be controlled by tuning the interactions between the crystallizing solution and the membrane surface.

A mathematical relationship exists between ΔG^* and the nucleation rate:

$$N = \Gamma e^{-\frac{\Delta G^*}{kT}} \tag{13}$$

where Γ is a preexponential kinetic factor, k is the Boltzmann constant, and T is the absolute temperature.

3.2.2 Effect of membrane porosity

Young's equation for surface tensions is not applicable since it is strictly valid for ideal and nonporous surfaces. For a porous membrane, a modified form of the Young equation correlating the surface porosity (ε) to the measured and equilibrium contact angle (θ) is used:

$$(\gamma_s - \gamma_i) = \gamma_L \left[\cos\theta + \frac{4\varepsilon(1 + \cos\theta)}{(1 - \varepsilon)(1 - \cos\theta)} \right] \tag{14}$$

in which γ_L, γ_i, and γ_s are the surface tensions of the nucleus-liquid, nucleus-substrate, and liquid-substrate interface, respectively.

The resulting energetic barrier to heterogeneous nucleation occurring on a porous membrane $\Delta G^*_{heterogeneous}$ is

$$\frac{\Delta G^*_{heterogeneous}}{\Delta G^*_{homgeneous}} = \frac{1}{4}(2 + \cos\theta)(1 - \cos\theta)^2 \left[1 - \varepsilon \frac{(1 + \cos\theta)^2}{(1 - \cos\theta)^2} \right]^3 \tag{15}$$

If $\varepsilon = 0$, Eq. (15) reduces to a mathematical form describing the heterogeneous nucleation on solid surfaces expressed by Eq. (12).

In general, the nucleation rate increases on highly porous substrates because the energetic barrier decreases at higher porosity.

3.2.3 Effect of membrane roughness

As stated in Section 3.2.2, the Young equation is only valid for solid, ideal, and smooth surfaces. Therefore, the wetting behavior of rough surfaces is described by the Wenzel equation. The Gibbs free energy barrier of a critical nucleus developing on a roughness membrane is

$$\frac{\Delta G^*_{heterogeneous}}{\Delta G^*_{homgeneous}} = \frac{1}{4}Y^3 \left(\cos^2\theta - 3\cos\theta + 2 \right) \tag{16}$$

where

$$Y = \frac{r^* \cos\theta(1 + \cos\theta) - 2}{\cos\theta(1 + \cos\theta) - 2} \tag{17}$$

and r^* is the Wenzel roughness factor and θ is the equilibrium contact angle.

It was proved that the nucleation rate increases with the roughness of the surface. For example, the nucleation rate of calcium sulfate dehydrate was increased by the roughness of stainless-steel surfaces where it was deposited (Gunn, 1980).

3.3 Applications

The large spectrum of applications offered by the membranes and the future perspectives shown in the literature present the MCr technology as a promising and competitive alternative to conventional crystallizers in many fields of science and engineering. Nowadays, many products used as additives in the fields of cosmetics, personal care, and pharmaceutical products are formulated in the form of crystalline powders. MCr can represent a valid alternative to the production of solid crystalline substances. Further applications can be found in the fields of microelectronics, nonlinear optics, and sensing, where the solid state represents a fundamental concept for the realization of single crystal semiconductor devices. The use of crystalline materials is also increasing in the field of heterogeneous catalysis and the controlled release of active substances. Membrane crystallization is also recognized as an innovative and efficient method for enhancing the crystallization of biomolecules, allowing the production of crystals with controlled shape, size distribution, and polymorphism. This last is an important phenomenon in the pharmaceutical industry, as the various structures of polymorphs have different physicochemical properties which can have huge influence on bioavailability, filtration, tableting, etc. Experimental investigation carried out on glycine and paracetamol have highlighted the interesting potential of MCr (Di Profio et al., 2007a, b, 2013). In both cases, membrane-assisted crystallization allowed the selective production of a specific polymorph depending on the transmembrane flux of solvent. In addition, different nucleation mechanisms for different ranges of supersaturation have been identified. Specifically, at high supersaturation, nucleation appears prevalently homogeneous, whereas at low supersaturation heterogeneous nucleation mechanism prevails. Carbamazepine is an example where the formation of metastable forms is promoted at higher supersaturation rates (Caridi et al., 2012).

Further applications are wastewater treatment for the recovery of high-purity silver (Tang et al., 2010) or sodium sulfate (Li et al., 2014), CO_2 capture (Luis et al., 2013), synthesis of $BaSO_4$ and $CaCO_3$ particles (Jia et al., 2003; Zhou et al., 2014), and recovery of antibiotics and polystyrene microparticles (Drioli et al., 2014).

In line with the process intensification (PI) strategy (Tian et al., 2018), an important field where MCr technology is expected to give a fundamental contribution is seawater desalination. Although RO water desalination is considered as the most cost-effective solution to potable water demand, a fundamental problem, relating this process, is represented by the environmental aspects of brine discharge from RO plants. A possible solution might be represented by the complete redesign of a desalination system by introducing MCr operations also. Compared to the RO process, MCr is not subject to notable restrictions due to the osmotic pressure of highly concentrated brines. Thus, it can be employed when high permeate recovery factors or brine recovery is requested. In principle, membrane crystallization is a process through which the goal of creating desalination plants with zero-discharge liquids could be achieved. Despite this, the application of MCr on an industrial scale requires the development of proper membranes and an accurate evaluation of the benefits/advantages/disadvantages with respect to conventional processes.

4. Conclusions and perspectives

Membrane contactors, especially MD and MCr, can provide new opportunities to design and optimize innovative productions. Some of the most interesting developments are related to the possibility of integrating new membrane devices together with traditional membrane systems, with important advantages related to a synergic integration between them. Notwithstanding their high potential, MD and MCr still fail to meet all expectations. In order to overcome the existing limits, it is necessary to carry out an increasingly systematic analysis of the possible advantages or disadvantages concerning the introduction of an innovative membrane system, an increasingly appropriate choice of materials and operating conditions, and an increasingly simpler modeling for a facilitated scale-up.

Therefore, the preparation of new and specific microporous membranes with high hydrophobicity and stability, with a narrow pore distribution and better morphological properties, represents a crucial aspect in improving MD and MCr performances. As already expressed in this chapter, the main future applications for MD and MCr concern various fields including desalination of seawater, wastewater treatment, and agro-food and pharmaceutical and biotechnological industries, where membranes with high physico-chemical stability are required.

References

Alquraish, M.M., Mejbri, S., Abuhasel, K.A., Zhani, K., 2021. Experimental investigation of a pilot solar-assisted permeate gap membrane distillation. Membranes 11 (5), 336.

An, A.K., Guo, J., Jeong, S., Lee, E.J., Tabatabai, S.A.A., Leiknes, T., 2016. High flux and antifouling properties of negatively charged membrane for dyeing wastewater treatment by membrane distillation. Water Res. 103, 362–371.

Ansari, A., Galogahi, F.M., Thiel, D.V., Helfer, F., Millar, G., Soukane, S., Ghaffour, N., 2022. Downstream variations of air-gap membrane distillation and comparative study with direct contact membrane distillation: a modelling approach. Desalination 526, 115539.

Aroon, M.A., Ismail, A.F., Montazer-Rahmati, M.M., Matsuura, T., 2010. Effect of raw multi-wall carbon nanotubes on morphology and separation properties of polyimide membranes. Sep. Sci. Technol. 45 (16), 2287–2297.

Attia, H., Shirin, A., Wright, C.J., Hilal, N., 2017. Superhydrophobic electrospun membrane for heavy metals removal by air gap membrane distillation (AGMD). Desalination 420, 318–329.

Baghbanzadeh, M., Lan, C.Q., Rana, D., Matsuura, T., 2016. Membrane distillation. In: Nanostructured Polymer Membranes. vol. 1. John Wiley & Sons, Inc., Hoboken, NJ, pp. 419–455.

Bernardo, P., Drioli, E., Golemme, G., 2009. Membrane gas separation: a review/state of the art. Ind. Eng. Chem. Res. 48 (10), 4638–4663.

Bodell, B.R., 1963. Silicone rubber vapor diffusion in saline water distillation. US Patent 285,032.

Bonyadi, E., Niknejad, A.S., Ashtiani, F.Z., Bazgir, S., Kargari, A., 2022. A well-designed polystyrene/polycarbonate membrane for highly saline water desalination using DCMD process. Desalination 528, 115604.

Cai, J., Yin, H., Guo, F., 2020. Transport analysis of material gap membrane distillation desalination processes. Desalination 481, 114361.

Capizzano, S., Frappa, M., Macedonio, F., Drioli, E., 2022. A review on membrane distillation in process engineering: design and exergy equations, materials and wetting problems. Front. Chem. Sci. Eng. 16, 592–613.

Caridi, A., Di Profio, G., Caliandro, R., Guagliardi, A., Curcio, E., Drioli, E., 2012. Selecting the desired solid form by membrane crystallizers: crystals or Cocrystals. Cryst. Growth Des. 12 (7), 4349–4356.

Cerneaux, S., Strużyńska, I., Kujawski, W.M., Persin, M., Larbot, A., 2009. Comparison of various membrane distillation methods for desalination using hydrophobic ceramic membranes. J. Membr. Sci. 337 (1–2), 55–60.

Chang, Y.S., Ooi, B.S., Ahmad, A.L., Leo, C.P., Low, S.C., 2021. Vacuum membrane distillation for desalination: scaling phenomena of brackish water at elevated temperature. Sep. Purif. Technol. 254, 117572.

Curcio, E., Simone, S., Di Profio, G., Drioli, E., Cassetta, A., Lamba, D., 2005. Membrane crystallization of lysozyme under forced solution flow. J. Membr. Sci. 257 (1–2), 134–143.

Davey, C.J., Liu, P., Kamranvand, F., Williams, L., Jiang, Y., Parker, A., Tyrrel, S., McAdam, E.J., 2021. Membrane distillation for concentrated Blackwater: influence of configuration (air gap, direct contact, vacuum) on selectivity and water productivity. Sep. Purif. Technol. 263, 118390.

Di Profio, G., Tucci, S., Curcio, E., Drioli, E., 2007a. Controlling polymorphism with membrane-based crystallizers: application to form I and II of paracetamol. Chem. Mater. 19 (10), 2386–2388.

Di Profio, G., Tucci, S., Curcio, E., Drioli, E., 2007b. Selective glycine polymorph crystallization by using microporous membranes. Cryst. Growth Des. 7 (3), 526–530.

Di Profio, G., Stabile, C., Caridi, A., Curcio, E., Drioli, E., 2009. Antisolvent membrane crystallization of pharmaceutical compounds. J. Pharm. Sci. 98 (12), 4902–4913.

Di Profio, G., Reijonen, M.T., Caliandro, R., Guagliardi, A., Curcio, E., Drioli, E., 2013. Insights into the polymorphism of glycine: membrane crystallization in an electric field. Phys. Chem. Chem. Phys. 15 (23), 9271–9280.

Di Profio, G., Polino, M., Nicoletta, F.P., Belviso, B.D., Caliandro, R., Fontananova, E., Filpo, G.D., Curcio, E., Drioli, E., 2014. Tailored hydrogel membranes for efficient protein crystallization. Adv. Funct. Mater. 24, 1582–1590.

Donato, L., Garofalo, A., Drioli, E., Alharbi, O., Aljlil, S.A., Criscuoli, A., Algieri, C., 2020. Improved performance of vacuum membrane distillation in desalination with zeolite membranes. Sep. Purif. Technol. 237, 116376.

Dongare, P.D., Alabastri, A., Pedersen, S., Zodrow, K.R., Hogan, N.J., Neumann, O., Wu, J., Wang, T., Deshmukh, A., Elimelech, M., Li, Q., Nordlander, P., Halas, N.J., 2017. Nanophotonics-enabled solar membrane distillation for off-grid water purification. Proc. Natl. Acad. Sci. 114, 6936–6941.

Drioli, E., Di Profio, G., Curcio, E., 2012. Progress in membrane crystallization. Curr. Opin. Chem. Eng. 1 (2), 178–182.

Drioli, E., Carnevale, M.C., Figoli, A., Criscuoli, A., 2014. Vacuum membrane dryer (VMDr) for the recovery of solid microparticles from aqueous solutions. J. Membr. Sci. 472, 67–76.

Drioli, E., Ali, A., Macedonio, F., 2015. Membrane distillation: recent developments and perspectives. Desalination 356, 56–84.

Dumée, L.F., Sears, K., Schütz, J., Finn, N., Huynh, C., Hawkins, S., Duke, M., Gray, S., 2010. Characterization and evaluation of carbon nanotube Bucky-paper membranes for direct contact membrane distillation. J. Membr. Sci. 351 (1–2), 36–43.

Edwie, F., Chung, T.S., 2013. Development of simultaneous membrane distillation–crystallization (SMDC) technology for treatment of saturated brine. Chem. Eng. Sci. 98, 160–172.

Essalhi, M., Khayet, M., Tesfalidet, S., Alsultan, M., Tavajoh, N., 2021. Desalination by direct contact membrane distillation using mixed matrix electrospun nanofibrous membranes with carbon-based nanofillers: a strategic improvement. Chem. Eng. J. 426, 131316.

Feng, C., Shi, B., Li, G., Wu, Y., 2004. Preparation and properties of microporous membrane from poly (vinylidene fluoride-co-tetrafluoroethylene)(F2.4) for membrane distillation. J. Membr. Sci. 237 (1–2), 15–24.

Findley, M.E., 1967. Vaporization through porous membranes. Ind. Eng. Chem. Process Des. Develop. 6 (2), 226–230.

Francis, L., Ghaffour, N., Alsaadi, A.A., Amy, G.L., 2013. Material gap membrane distillation: a new design for water vapor flux enhancement. J. Membr. Sci. 448, 240–247.

Franken, A.C.M., Nolten, J.A.M., Mulder, M.H.V., Bargeman, D., Smolderset, C.A., 1987. Wetting criteria for the applicability of membrane distillation. J. Membr. Sci. 33 (3), 315–328.

Frappa, M., Del Rio Castillo, A.E., Macedonio, F., Politano, A., Drioli, E., Bonaccorso, F., Pellegrini, V., Gugliuzza, A., 2020. Few-layer graphene for advanced composite PVDF membranes dedicated to water distillation: a comparative study. Nanoscale Adv. 2, 4728–4739.

Frappa, M., Del Rio Castillo, A.E., Macedonio, F., Di Luca, G., Drioli, E., Gugliuzza, A., 2021. Exfoliated Bi2Te$_3$-enabled membranes for new concept water desalination: freshwater production meets new routes. Water Res. 203, 117503.

Gontarek-Castro, E., Di Luca, G., Lieder, M., Gugliuzza, A., 2022. Graphene-coated PVDF membranes: effects of multi-scale rough structure on membrane distillation performance. Membranes 12 (5), 511.

Grossi, L.B., Alvim, C.B., Alvares, C.M.S., Martins, M.F., Amaral, M.C.S., 2020. Purifying surface water contaminated with industrial failure using direct contact membrane distillation. Sep. Purif. Technol. 233, 116052.

Gryta, M., 2021. Application of polypropylene membranes hydrophilized by plasma for water desalination by membrane distillation. Desalination 515, 115187.

Gunn, D.J., 1980. Effect of surface roughness on the nucleation and growth of calcium sulphate on metal surfaces. J. Cryst. Growth 50 (2), 533–537.

He, Z., Lyu, Z., Gu, Q., Zhang, L., Wang, J., 2019. Ceramic-based membranes for water and wastewater treatment. Colloids Surf. A Physicochem. Eng. Asp. 578, 123513.

Huang, Q., Gao, S., Huang, Y., Zhang, M., Xiao, C., 2019. Study on photothermal PVDF/ATO nanofiber membrane and its membrane distillation performance. J. Membr. Sci. 582, 203–210.

Hubadillah, S.K., Tai, Z.S., Othman, M.H.D., Harun, Z., Jamalludin, M.R., Rahman, M.A., Jaafar, J., Ismail, A.F., 2019. Hydrophobic ceramic membrane for membrane distillation: a mini review on preparation, characterization, and applications. Sep. Purif. Technol. 217, 71–84.

Jamed, M.J., Alhathal Alanezi, A., Alsalhy, Q.F., 2019. Effects of embedding functionalized multi-walled carbon nanotubes and alumina on the direct contact poly (vinylidene fluoride-co-hexafluoropropylene) membrane distillation performance. Chem. Eng. Commun. 206 (8), 1035–1057.

Jia, Z., Liu, Z., He., F., 2003. Synthesis of Nanosized $BaSO_4$ and $CaCO_3$ particles with a membrane reactor: effects of additives on particles. J. Colloid Interface Sci. 266 (2), 322–327.

Jia, X., Lan, L., Zhang, X., Wang, T., Wang, Y., Ye, C., Lin, J., 2021. Pilot-scale vacuum membrane distillation for decontamination of simulated radioactive wastewater: system design and performance evaluation. Sep. Purif. Technol. 275, 119129.

Jiang, H., Straub, A.P., Karanikola, V., 2022. Ammonia recovery with sweeping gas membrane distillation: energy and removal efficiency analysis. ACS ES &T Eng. 2 (4), 617–628.

Kalla, S., 2021. Use of membrane distillation for oily wastewater treatment–a review. J. Environ. Chem. Eng. 9 (1), 104641.

Khayet, M., Matsuura, T., 2011. Membrane Distillation: Principles and Applications. Elsevier, Great Britain.

Kieffer, R., Mangin, D., Puel, F., Charcosset, C., 2009. Precipitation of barium sulphate in a hollow fiber membrane contactor, Part I: Investigation of particulate fouling. Chem. Eng. Sci. 64 (8), 1759–1767.

Ko, C.C., Ali, A., Drioli, E., Tung, K.L., Chen, C.H., Chen, Y.R., Macedonio, F., 2018. Performance of ceramic membrane in vacuum membrane distillation and in vacuum membrane crystallization. Desalination 440, 48–58.

Kujawa, J., Kujawski, W., Koter, S., Jarzynka, K., Rozicka, A., Bajda, K., Cerneaux, S., Persin, M., Larbot, A., 2013a. Membrane distillation properties of TiO_2 ceramic membranes modified by perfluoroalkylsilanes. Desalin. Water Treat. 51, 1352–1361.

Kujawa, J., Kujawski, W., Koter, S., Rozicka, A., Cerneaux, S., Persin, M., Larbot, A., 2013b. Efficiency of grafting of Al_2O_3, TiO_2 and ZrO_2 powders by perfluoroalkylsilanes. Colloids Surf. A Physicochem. Eng. Asp. 420, 64–73.

Lakerveld, R., Kuhn, J., Kramer, H.J., Jansens, P.J., Grievink, J., 2010. Membrane assisted crystallization using reverse osmosis: influence of solubility characteristics on experimental application and energy saving potential. Chem. Eng. Sci. 65 (9), 2689–2699.

Lee, T., Min, C., Naidu, G., Huang, Y., Shon, H.K., Kim, S.H., 2022. Optimizing the performance of sweeping gas membrane distillation for treating naturally heated saline groundwater. Desalination 532, 115736.

Li, W., Van der Bruggen, B., Luis, P., 2014. Integration of reverse osmosis and membrane crystallization for sodium Sulphate recovery. Chem. Eng. Process. Process Intensif. 85, 57–68.

Li, Q., Charlton, A.J., Omar, A., Dang, B., Le-Clech, P., Scott, J., Taylor, R.A., 2022. A novel concentrated solar membrane-distillation for water purification in a building integrated design. Desalination 535, 115828.

Liu, F., Hashim, N.A., Liu, Y., Abed, M.R.M., Li, K., 2011. Progress in the production and modification of PVDF membranes. J. Membr. Sci. 375, 1–27.

Liu, G., Jin, W., Xu, N., 2016. Two-dimensional material membranes: a new family of high-performance separation membranes. Angew. Chem. Int. Ed. 55 (43), 13384–13397.

Luis, P., Van Aubel, D., Van der Bruggen, B., 2013. Technical viability and exergy analysis of membrane crystallization: closing the loop of CO_2 sequestration. Int. J. Greenhouse Gas Contr. 12, 450–459.

Macedonio, F., Drioli, E., 2019. Membrane distillation development. In: Sustainable Water and Wastewater Processing. Elsevier, pp. 133–159.

Macedonio, F., Criscuoli, A., Drioli, E., 2021. Membrane distillation pilot units for seawater desalination. In: Advances in Water Desalination Technologies, pp. 227–261.

Mao, Y., Huang, Q., Meng, B., Zhou, K., Liu, G., Gugliuzza, A., Drioli, E., Jin, W., 2020. Roughness-enhanced hydrophobic graphene oxide membrane for water desalination via membrane distillation. J. Membr. Sci. 611, 118364.

Marques Lisboa, K., Busson de Moraes, D., Naveira-Cotta, C.P., Machado, C.R., 2021. Analysis of the membrane effects on the energy efficiency of water desalination in a direct contact membrane distillation (DCMD) system with heat recovery. Appl. Therm. Eng. 182, 116063.

Meng, B., Liu, G., Mao, Y., Liang, F., Liu, G., Jin, W., 2021. Fabrication of surface-charged MXene membrane and its application for water desalination. J. Membr. Sci. 623, 119076.

Nidamanuri, N., Li, Y., Li, Q., Dong, M., 2020. Graphene and graphene oxide-based membranes for gas separation. Eng. Sci. 9 (9), 3–16.

Niknejad, A.S., Bazgir, S., Kargari, A., 2021. Desalination by direct contact membrane distillation using a superhydrophobic nanofibrous poly (methyl methacrylate) membrane. Desalination 511, 115108.

Noamani, S., Niroomand, S., Rastgar, M., Azhdarzadeh, M., Sadrzadeh, M., 2020. Modeling of air-gap membrane distillation and comparative study with direct contact membrane distillation. Ind. Eng. Chem. Res. 59 (50), 21930–21947.

Novoselov, K.S., Geim, A.K., Morozov, S.V., Jiang, D.E., Zhang, Y., Dubonos, S.V., Grigorieva, I.V., Firsov, A.A., 2004. Electric field effect in atomically thin carbon films. Science 306 (5696), 666–669.

Okati, V., Moghadam, A.J., Farzaneh-Gord, M., Moein-Jahrom, M., 2022. Thermo-economical and environmental analyses of a direct contact membrane distillation (DCMD) performance. J. Clean. Prod. 340, 130613.

Parani, S., Oluwafemi, O.S., 2021. Membrane distillation: recent configurations, membrane surface engineering, and applications. Membranes 11 (12), 934.

Polino, M., Portugal, C.A.M., Di Profio, G., Coelhoso, I.M., Crespo, J.G., 2019. Protein crystallization by membrane-assisted technology. Cryst. Growth Des. 19 (8), 4871–4883.

Politano, A., Argurio, P., Di Profio, G., Sanna, V., Cupolillo, A., Chakraborty, S., Arafat, H.A., Curcio, E., 2017. Photothermal membrane distillation for seawater desalination. Adv. Mater. 29, 1603504.

Politano, A., Bonaccorso, F., Del Rio Castillo, A.E., Drioli, E., Gugliuzza, A., Macedonio, F., Pellegrini, V., 2018. Italian Patent: A procedure to fabricate a nanocomposite membrane with bidimensional crystals obtained through exfoliation of layered materials by wet-jet milling technique. acceptance: 2020/11/22, IT102018000020641.

Quist-Jensen, C.A., Macedonio, F., Conidi, C., Cassano, A., Aljlil, S., Alharbi, O., Drioli, E., 2016. Direct contact membrane distillation for the concentration of clarified orange juice. J. Food Eng. 187, 37–43.

Rehman, W.U., Sarwar, B., Saqib, S., Mukhtar, A., Younas, M., Rezakazemi, M., 2022. Applications of membrane contactors in food industry. In: Membrane Contactor Technology: Water Treatment, Food Processing, Gas Separation, and Carbon Capture. Wiley-VCH GmbH, pp. 219–245, https://doi.org/10.1002/9783527831036.

Rezaei, M., Samhaber, W.M., 2016. Wetting behaviour of Superhydrophobic membranes coated with nanoparticles in membrane distillation. Chem. Eng. Trans. 47, 373–378.

Rezaei, M., Warsinger, D.M., Lienhard V, J.H., Samhaber, W.M., 2017. Wetting prevention in membrane distillation through superhydrophobicity and recharging an air layer on the membrane surface. J. Membr. Sci. 530, 42–52.

Ricci, B.C., Arcanjo, G.S., Moreira, V.R., Lebron, Y.A.R., Koch, K., Costa, F.C.R., Ferreira, B.P., Lisboa, F.C., Miranda, L.D., de Faria, C.V., Lange, L.C., Amaral, M.C.S., 2021. A novel submerged anaerobic osmotic membrane bioreactor coupled to membrane distillation for water reclamation from municipal wastewater. Chem. Eng. J. 414, 128645.

Rivier, C.A., García-Payo, M.C., Marison, I.W., Von Stockar, U., 2002. Separation of binary mixtures by thermostatic sweeping gas membrane distillation: I. theory and simulations. J. Membr. Sci. 201 (1–2), 1–16.

Roy, S., Bhadra, M., Mitra, S., 2014. Enhanced desalination via functionalized carbon nanotube immobilized membrane in direct contact membrane distillation. Sep. Purif. Technol. 136, 58–65.

Ruiz-Aguirre, A., Andrés-Mañas, J.A., Zaragoza, G., 2019. Evaluation of permeate quality in pilot scale membrane distillation systems. Membranes 9, 69.

Shaheen, A., AlBadi, S., Zhuman, B., Taher, H., Banat, F., AlMarzooqi, F., 2022. Photothermal air gap membrane distillation for the removal of heavy metal ions from wastewater. Chem. Eng. J. 431, 133909.

Siagian, U.W., Raksajati, A., Himma, N.F., Khoiruddin, K., Wenten, I.G., 2019. Membrane-based carbon capture technologies: membrane gas separation vs. membrane contactor. J. Nat. Gas Sci. Eng. 67, 172–195.

Simoni, G., Kirkebæk, B.S., Quist-Jensen, C.A., Christensen, M.L., Ali, A., 2021. A comparison of vacuum and direct contact membrane distillation for phosphorus and ammonia recovery from wastewater. J. Water Process Eng. 44, 102350.

Soukane, S., Elcik, H., Alpatova, A., Orfi, J., Ali, E., AlAnsary, H., Ghaffour, N., 2021. Scaling sets the limits of large scale membrane distillation modules for the treatment of high salinity feeds. J. Clean. Prod. 287, 125555.

Srisurichan, S., Jiraratananon, R., Fane, A.G., 2006. Mass transfer mechanisms and transport resistances in direct contact membrane distillation process. J. Membr. Sci. 277 (1–2), 186–194.

Sun, P., Zhu, M., Wang, K., Zhong, M., Wei, J., Wu, D., Xu, Z., Zhu, H., 2013. Selective ion penetration of graphene oxide membranes. ACS Nano 7 (1), 428–437.

Tai, Z.S., Abd Aziz, M.H., Othman, M.H.D., Mohamed Dzahir, M.I.H., Hashim, N.A., Koo, K.N., Hubadillah, S.K., Ismail, A.F., Rahman, M.A., Jaafar, J., 2020. Ceramic membrane distillation for desalination. Separ. Purif. Rev. 49 (4), 317–356.

Tan, Y.Z., Wang, H., Han, L., Tanis-Kanbur, M.B., Pranav, M.V., Chew, J.W., 2018. Photothermal-enhanced and fouling-resistant membrane for solar-assisted membrane distillation. J. Membr. Sci. 565, 254–265.

Tang, B., Yu, G., Fang, J., Shi, T., 2010. Recovery of high-purity silver directly from dilute effluents by an emulsion liquid membrane-crystallization process. J. Hazard. Mater. 177 (1–3), 377–383.

Tian, Y., Demirel, S.E., Hasan, M.F., Pistikopoulos, E.N., 2018. An overview of process systems engineering approaches for process intensification: state of the art. Chem. Eng. Proc. Process Intensif. 133, 160–210.

Tundis, R., Conidi, C., Loizzo, M.R., Sicari, V., Romeo, R., Cassano, A., 2021. Concentration of bioactive phenolic compounds in olive mill wastewater by direct contact membrane distillation. Molecules 26, 1808.

Ursino, C., Ounifi, I., Di Nicolò, E., Cheng, X.Q., Shao, L., Zhang, Y., Drioli, E., Criscuoli, A., Figoli, A., 2021. Development of non-woven fabric-based ECTFE membranes for direct contact membrane distillation application. Desalination 500, 114879.

Wang, Y., Yu, H., Yang, X., Liu, L., Xu, S., He, H., Zhang, Y., He, T., 2022. Concentrating phosphoric acid by direct contact membrane distillation using a low-cost polyethylene separator. Desalination 530, 115664.

Wen, X., Li, F., Zhao, X., 2016. Removal of nuclides and boron from highly saline radioactive wastewater by direct contact membrane distillation. Desalination 394, 101–107.

Wenzel, R.N., 1936. Resistance of solid surfaces to wetting by water. Ind. Eng. Chem. 28 (8), 988–994.

Woods, J., Pellegrino, J., Burch, J., 2011. Generalized guidance for considering pore-size distribution in membrane distillation. J. Membr. Sci. 368 (1–2), 124–133.

Wu, Z., Faiz, R., Li, T., Kingsbury, B.F., Li, K., 2013. A controlled sintering process for more permeable ceramic hollow fibre membranes. J. Membr. Sci. 446, 286–293.

Wu, X., Jiang, Q., Ghim, D., Singamaneni, S., Jun, Y.S., 2018. Localized heating with a photothermal polydopamine coating facilitates a novel membrane distillation process. J. Mater. Chem. A 6, 18799–18807.

Wu, X.Q., Mirza, N.R., Huang, Z., Zhang, J., Zheng, Y.M., Xiang, J., Xie, Z., 2021. Enhanced desalination performance of aluminium fumarate MOF-incorporated electrospun nanofiber membrane with bead-on-string structure for membrane distillation. Desalination 520, 115338.

Xie, Z., Duong, T., Hoang, M., Nguyen, C., Bolto, B., 2009. Ammonia removal by sweep gas membrane distillation. Water Res. 43 (6), 1693–1699.

Xu, Z., Yan, X., Du, Z., Li, J., Cheng, F., 2020. Effect of oxygenic groups on desalination performance improvement of graphene oxide-based membrane in membrane distillation. Sep. Purif. Technol. 251, 117304.

Yadav, A., Yadav, P., Labhasetwar, P.K., Shahi, V.K., 2021. CNT functionalized ZIF-8 impregnated poly(vinylidene fluoride-co-hexafluoropropylene) mixed matrix membranes for antibiotics removal from pharmaceutical industry wastewater by vacuum membrane distillation. J. Environ. Chem. Eng. 9 (6), 106560.

Yadav, P., Farnood, R., Kumar, V., 2022. Superhydrophobic modification of electrospun nanofibrous Si@PVDF membranes for desalination application in vacuum membrane distillation. Chemosphere 287, 132092.

Yang, C., Peng, X., Zhao, Y., Wang, X., Cheng, L., Wang, F., Li, Y., Li, P., 2019. Experimental study on VMD and its performance comparison with AGMD for treating copper-containing solution. Chem. Eng. Sci. 207, 876–891.

Zhao, K., Heinzl, W., Wenzel, M., Büttner, S., Bollen, F., Lange, G., Heinzl, S., Sarda, N., 2013. Experimental study of the memsys vacuum-multi-effect-membrane-distillation (V-MEMD) module. Desalination 323, 150–160.

Zhao, S., Jiang, C., Fan, J., Hong, S., Mei, P., Yao, R., Liu, Y., Zhang, S., Li, H., Zhang, H., Sun, C., Guo, Z., Shao, P., Zhu, Y., Zhang, J., Guo, L., Ma, Y., Zhang, J., Feng, X., Wang, F., Wu, H., Wang, B., 2021. Hydrophilicity gradient in covalent organic frameworks for membrane distillation. Nat. Mater. 20 (11), 1551–1558.

Zhong, L., Wang, Y., Liu, D., Zhu, Z., Wang, W., 2021. Recent advances in membrane distillation using electrospun membranes: advantages, challenges, and outlook. Environ. Sci.: Water Res. Technol. 7 (6), 1002–1019.

Zhou, J., Cao, X., Yong, X., Wang, S.Y., Liu, X., Chen, Y.-L., Zheng, T., Ouyang, P.K., 2014. Effects of various factors on biogas purification and Nano-CaCO$_3$ synthesis in a membrane reactor. Ind. Eng. Chem. Res. 53 (4), 1702–1706.

Forward osmosis: Principle and applications in sustainable water and energy development

*Duc-Viet Nguyen[a], Thanh-Tin Nguyen[b],
Rusnang Syamsul Adha[b], Lei Zheng[c], Xuan-Thanh Bui[d,e],
Xiaoli Ma[f], and Hoang Nhat Phong Vo[g]*

[a]Graduate School of Water Resources, Sungkyunkwan University, Suwon, Republic of Korea
[b]School of Earth Sciences and Environmental Engineering, Gwangju Institute of Science and Technology, Gwangju, Republic of Korea [c]Chongqing Institute of Green and Intelligent Technology, Chinese Academy of Sciences, Chongqing, China [d]Vietnam National University Ho Chi Minh City (VNU-HCM), Ho Chi Minh City, Vietnam [e]Key Laboratory of Advanced Waste Treatment Technology & Faculty of Environment and Natural Resources, Ho Chi Minh City University of Technology (HCMUT), Ho Chi Minh City, Vietnam [f]Department of Materials Science and Engineering, University of Wisconsin-Milwaukee, Milwaukee, WI, United States [g]Queensland Alliance for Environmental Health Sciences (QAEHS), The University of Queensland, Woolloongabba, QLD, Australia

1. Introduction

The world population is expected to reach the figure of 8.4–9.8 billion around the middle of the 21st century, while the global water cycle is changing due to global warming. On the other hand, water consumption has risen sharply in recent decades at a faster rate than the population growth which leads to water scarcity. The current world water consumption is expected to be around $4600 \, km^3$ per year, with a predicted growth of 20%–30% between 5500 and $6000 \, km^3$ per year by 2050 (Burek et al., 2016). The clean water demand is growing for various purposes including drinking, irrigation, industrial application. However, the

Copyright © 2023 Elsevier Inc. All rights reserved.

water sources need to be treated before being used, and seawater desalination technology has filled the gap in the market for its unlimited amount of resources (Bhojwani et al., 2019). On the other side, the sustainable development to produce a limited amount of waste is also a highly important strategy for the environmental protection and efficiency while solving the issue of freshwater resources which has become increasingly limited (Blandin et al., 2016a,b).

Seawater desalination and water reuse are two major methods of supplying clean water to address the issue of water shortage. Nowadays, the cost of producing water for contemporary reverse osmosis (RO) plants in seawater reverse osmosis (SWRO) plants ranges from 0.5 to 2 USD/m^3, mostly depending on local energy prices (Elimelech and Phillip, 2011). Nonetheless, due to thermodynamic constraints, additional major reductions in SWRO energy usage will be small. To minimize the cost of generated water, the utilization of contaminated water sources (e.g., primary or secondary wastewater effluent) might be a viable option for recovery and reuse (Valladares Linares et al., 2016). In this case, the water production cost ranges from 0.4 to 1.26 USD/m^3 (Valladares Linares et al., 2016), depending on the degree of treatment for reuse (direct/indirect potable or nonpotable reuse, industrial water or irrigation). In most cases, pressure membrane filtering methods are used for water reclamation. However, these technologies, particularly RO, have a large energy requirement. On the other hand, if wastewater is fed directly into an RO process, it is predicted that irreversible fouling of the RO membrane would develop (Li et al., 2007). Currently, the forward osmosis membrane process is developing with the promise of overcoming the constraints of pressure-driven membrane processes in order to provide energy-efficient and sustainable water treatment (Wang et al., 2018).

2. Forward osmosis membrane process

2.1 Forward osmosis concept

Forward osmosis (FO) is a process in which water is moved across a semipermeable membrane by the action of a gradient from low osmotic pressure to high osmotic pressure (e.g., the driving force is naturally created by the difference in osmotic pressure between the DS and the FS). Fig. 1 illustrates this approach. FO has the following benefits over traditional

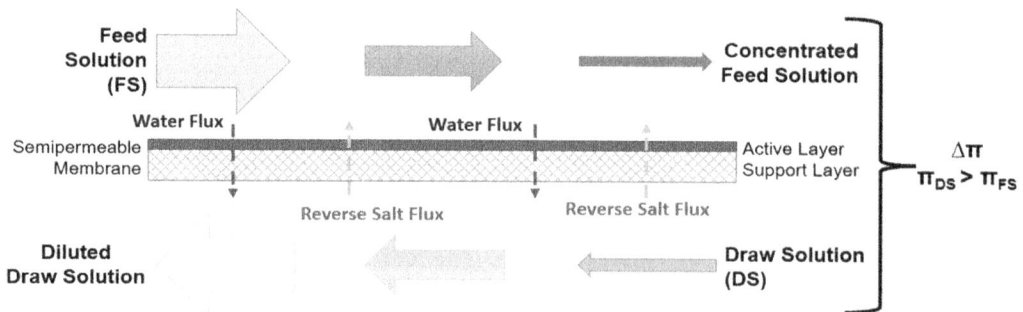

FIG. 1   FO process with the membrane active layer facing the feed solution.

pressure-driven membrane processes (Haupt and Lerch, 2018): (i) lower energy requirement, (ii) reduced membrane fouling potential due to lack of compression (e.g., induced by applied hydraulic pressure), (iii) simultaneous treatment of two solutions in one treatment step, and (iv) treatment of liquids that are incompatible with other membrane processes.

In addition to the "conventional" forward osmosis procedure, there are similar FO-related processes that involve physical pressure. The two forms are pressure-enhanced osmosis (PEO) and pressure-retarded osmosis (PRO).

FO water flow and solute flux are given as Eqs. (1) and (2), respectively:

$$J_v = A \left(\pi_{draw} - \pi_{support} \right) \tag{1}$$

$$J_s = B \left(C_{draw} - C_{support} \right) \tag{2}$$

wherein A and B are the water and solute transfer coefficients, respectively; J_v is the volumetric weight of water; J_s denotes the solute mass flow. C_{draw} and π_{draw} represent the concentration and osmotic pressure of the DS, respectively.

The real driving force during an FO process, on the other hand, is much lower than the osmotic pressure gradient between the bulk FS and the bulk DS. This is due to the internal concentration polarization (ICP) phenomenon within the membrane support layer (Fig. 2). In particular, this impact might be either DS dilution on the DS side (Fig. 2A) or solute buildup on the FS side. This is known as dilutive ICP in the FO mode (i.e., the active layer-facing-feed solution (AL-FS). In particular, greater CFV over the surface can reduce external concentration polarization (ECP), but ICP is directly related to the properties of the substrate layer, such as the support layer, which can be defined using a structural parameter (S):

$$S = \frac{\tau * l}{\varepsilon} \tag{3}$$

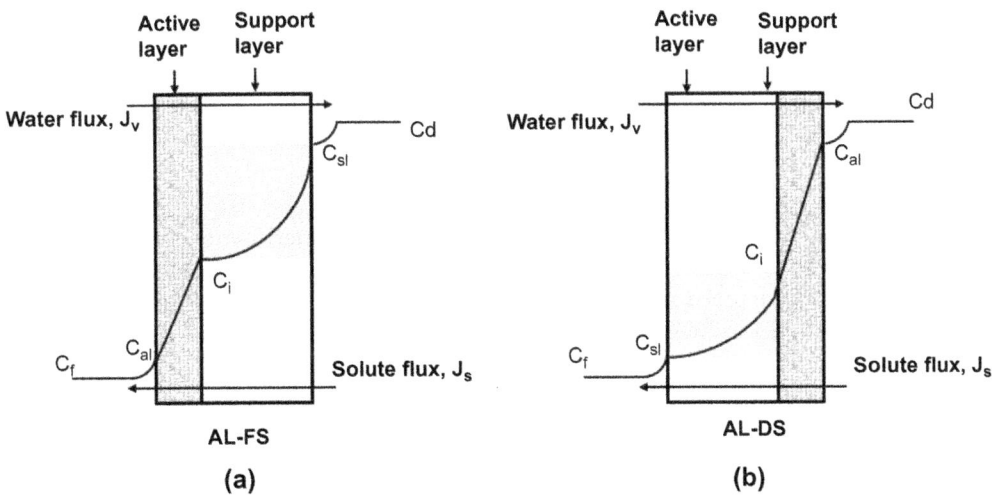

FIG. 2   External and Internal concentration polarization (ICP) in the FO membrane cross section. (A) AL-FS orientation; and (B) AL-DS orientation.

whereas τ is tortuosity, l is the substrate layer thickness, and ε is the substrate porosity.

A previous work constructed a model to validate the link between FO water flow, J_v (in AL-DS and AL-FS mode), solute flux, S, and A, B values to get a better understanding of the interplay between these parameters (Tang et al., 2010).

The flux equation for the AL-facing FW configuration can be described as

$$J_v = K_m \, ln \left(\frac{A\pi_{draw} + B}{A\pi_{feed} + J_v + B} \right) \tag{4}$$

whereas the flux equation for the AL-facing DS configuration can be described as

$$J_v = K_m \, ln \left(\frac{A\pi_{draw} - J_v + B}{A\pi_{feed} + B} \right) \tag{5}$$

where K_m is the mass transfer coefficient.

For the solute flux, it can be described as

$$\frac{J_s}{J_v} = \frac{B \left(C_{draw} - C_{support} \right)}{A \left(\pi_{draw} - \pi_{support} \right)} \tag{6}$$

By applying the van't Hoff equation, the solute flux can be derived:

$$J_s = \frac{B}{A.\beta R_g T} J_v \tag{7}$$

2.2 Commercial forward osmosis membranes

Several materials, as is well known, are used in FO membranes. As a first-generation material, the Hydration Technology Innovation (HTI) successfully synthesized cellulose triacetate (CTA). This membrane is resistant to chlorine and is less prone to thermal, chemical, and biological degradation. The success of the CTA membrane has resulted in fast growth in the FO membrane design due to an improved understanding of the needed properties of FO membranes. Thin-film composite (TFC) membranes for FO were said to outperform CTA membranes. As a result, research on the TFC membrane is necessary to provide membrane substrates with ideal features to decrease internal concentration polarization (ICP) or strong selective layers for high FO performance and fouling control.

With reference to the commercial FO membrane used for experiments of lab scale, a recent review paper (Haupt and Lerch, 2018) indicated that most of the studies on FO used the flat-sheet type. Self-manufactured membranes accounted for 20% of the total, while commercial membranes accounted for 76%. HTI, a company that is no longer in business, supplied the majority of the commercial membranes (57%). Aquaporin (Denmark), Toray (Korea), and FTS are the other commercial providers.

On the other hand, some pilot scales were implemented with different FO membrane modules: HTI CTA FO ($20.2\,m^2$) (Phuntsho et al., 2016), HTI CTA FO ($9\,m^2$) (Kim et al., 2015, 2017a,b), Toray TFC FO module ($15\,m^2$) (Kim et al., 2017a,b), Toray PA-TFC FO module ($15\,m^2$) (Kook et al., 2018). The spiral wrapped FO element and a plate-and-frame FO element are shown in Fig. 3.

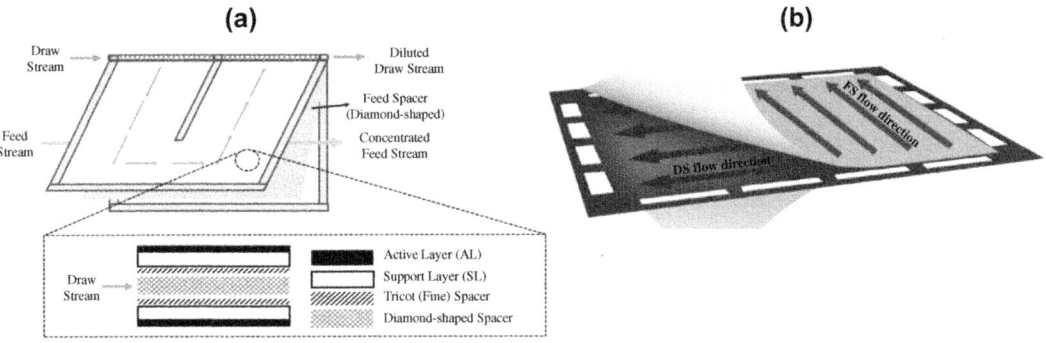

FIG. 3 Flow channel configuration in (A) the spiral-wound FO element (Kook et al., 2018) and (B) the plate frame FO element (Song et al., 2018).

TABLE 1 The water permeability of commercial membranes.

Manufacturer	Membrane types	Water permeability (LMH/bar)	Reference
Hydration Technology Inc.	CTA—Plate frame	0.82 0.45	Nguyen et al. (2019) Li et al. (2018a,b)
	CTA—Spiral wound	1.60	Kim et al. (2017a,b)
	TFC—Plate frame	1.47 1.30	Ortega-Bravo et al. (2016) Blandin et al. (2016b)
Porifera	TFC—Plate frame	2.00	Blandin et al. (2016b)
Toyobo	CTA—Hollow fiber	0.27–0.55	Shibuya et al. (2015)
Toray chemical Korea	PA-TFC—Spiral wound	8.82 8.90	Nguyen et al. (2019) Kim et al. (2017a,b)
	PA-TFC—Plate frame	6.47	Volpin et al. (2018)

There were two commercial membrane generations: CTA (low permeability), e.g., HTI, Toyobo, and PA-TFC membrane (high permeability) (e.g., Toray Chemical Korea) (Table 1).

3. Critical factors controlling the performance of the forward osmosis membrane

3.1 pH

pH is a significant factor that affects not only membrane properties but also the feed and draw solution, thus ultimately influencing the overall performance (i.e., water flux, removal efficiency, membrane fouling). (Ruengruehan et al., 2016) found that membrane fouling was influenced by pH fluctuations. Due to the protonation of carboxylic functional groups in octanoic acid molecules, the flow drop was more pronounced at low pH (4.0) than at high pH (9.0). In terms of solute selectivity, pH has no impact on the FO membrane selectivity

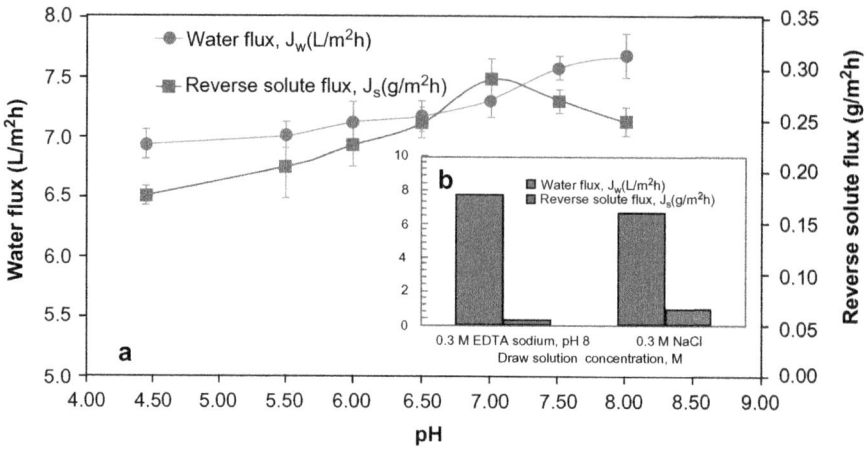

FIG. 4   Influence of pH on (A) the water flux (Ruengruehan et al., 2016) (B) the RSF of the FO system (Hau et al., 2014).

(Ruengruehan et al., 2016). In another work, (Zhu et al., 2018) investigated the effects of pH on the removal efficiency of three different naphthenic (NA) compounds by an FO membrane. The elimination behavior of Merichem NAs was unaffected by changing the pH from 6 to 9. The overall rejection of three NA model compounds was greater than 95% at pH 9. These results, however, contradicted recent findings that the water flux drop was pH-independent (Zhu et al., 2018).

pH variations in the DS, in addition to the FS, had a significant impact on the performance of the FO. The change in pH of the draw solute (EDTA sodium salt) caused the effects on the < affects? > membrane water flux and RSF (Hau et al., 2014). Their study indicated that pH range 4–9 affected the water flux and RSF. In comparison to standard inorganic salts, good solubility and high charged species of EDTA at pH 8 not only provided effective osmotic pressure to pull water from the discarded sludge, but also dramatically reduced salt leakage (Fig. 4) (Hau et al., 2014).

3.2 Temperature

The diffusion coefficient, density, viscosity, and osmotic pressure of a solution alter as its temperature changes. As a result, temperature has an effect on the performance of the FO membrane. Several studies have been conducted to assess the influence of temperature on FO. A previous study looked at how temperature affected water flow, recovery, concentration factor, and membrane scaling. Water flow, recovery, and concentration factor all rose as temperature climbed; however, higher temperatures resulted in more severe membrane scaling.

Feng et al. (2018) reported that increasing the FS temperature from 20°C to 40°C increased the water flux by 15.74% (i.e., from 7.5 to 9.0 LMH). Meanwhile, at a high temperature of 40°C, the flux increased by 47.46% (i.e., from 8 to 12 LMH) (Fig. 5). The DS (NH_4HCO_3) diffusion coefficient increases as the DS temperature rises, but the DS viscosity decreases. As a result,

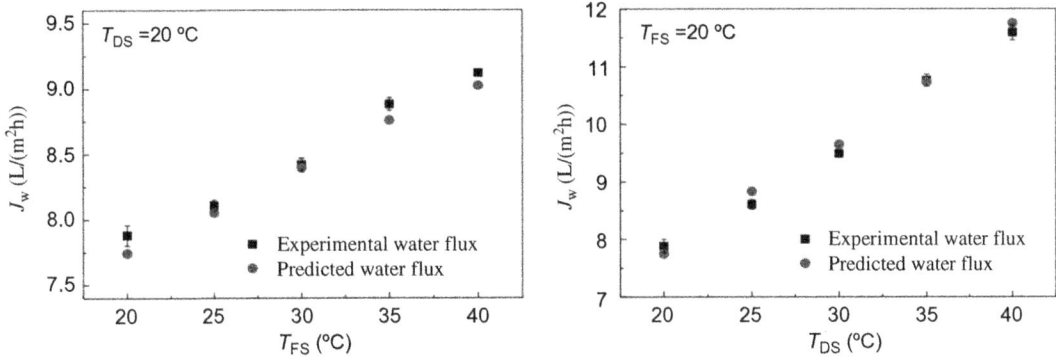

FIG. 5 The influence of temperature on the FO water flux (Feng et al., 2018).

the interface between the active and support layers has a greater DS concentration and osmotic pressure. Their study also reported that the water flux increased by 15.74%, 47.46% at a temperature of 40°C of the FS and DS solutions, respectively (Fig. 5).

3.3 Membrane properties

The kind of FO membrane utilized can have a considerable influence on the FO system performance, including water and solute flow, membrane fouling, and cleaning. To date, many FO materials have been proposed. They are cellulose triacetate (CTA) FO, thin film composite (TFC) FO, and aquaporin FO membrane. Because of the commercial dominance of the cellulose triacetate (CTA) FO membranes, the CTA membranes were employed in the bulk of the previous FO fouling research (Chung et al., 2011; Lutchmiah et al., 2014; Yu et al., 2017). TFC membranes outperform CTA membranes in terms of pH stability, hydrolysis resistance, and biological degradation (Cath et al., 2006; Gu et al., 2013). In terms of membrane fouling, previous research indicated that TFC FO membranes had a greater fouling propensity than CTA FO membranes under moderate fouling circumstances due to their rougher surface. As a result, the water flow through the TFC membrane reduced dramatically (Gu et al., 2013). While FO is supposed to be more fouling resistant in the AL-FS orientation, significant fouling can develop even at low flux, especially for TFC membranes or unstable feed solutions (Gu et al., 2013).

As shown in Fig. 6, the water flux of TFC membrane exhibited a much higher decline rate than that of CTA membrane. For the osmotic membrane bioreactor (OMBR) system, the decline in water flux is attributed to the combination of membrane fouling and reduced driving force caused by salinity (Wang et al., 2016). Furthermore, as compared to the CTA FO, the TFC FO had a lower salt accumulation due to the lower water flux and less reverse salt transport.

3.4 Draw solution

Various types of draw solution (DS) were utilized to create an osmotic pressure difference across the membrane (Chekli et al., 2012). As a draw solution, sugar solutions were used in a few cases, while salt solutions were used in the majority of cases. The draw solution used is

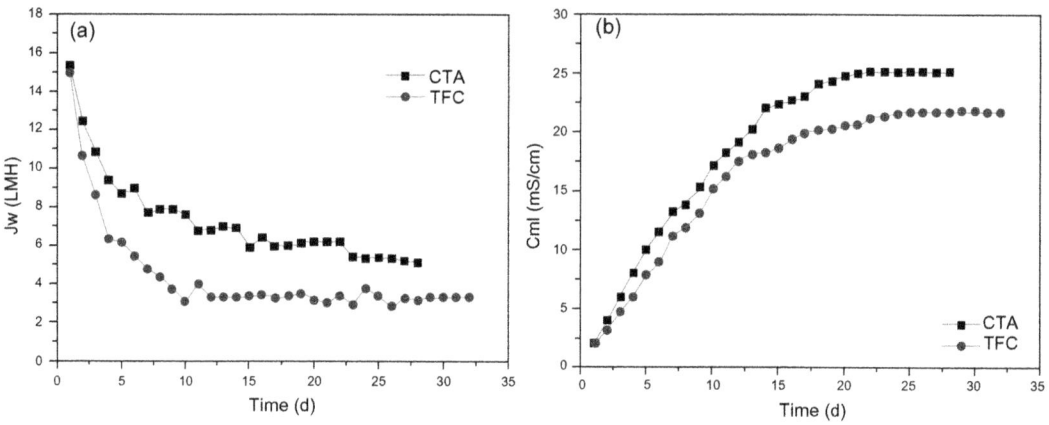

FIG. 6 Variations of water flux (A) and the conductivity of FS (B) during the operation of the FO system (Wang et al., 2016).

important to the success of the FO process. The literature describe the conditions for an excellent FO draw solution (Cornelissen et al., 2008). A good draw solution is water soluble and has a low molecular weight, which results in strong osmotic pressures. In addition, the draw solution must be nontoxic, chemically compatible with the membrane material, and simply and cheaply removed from the permeate (Viet et al., 2019).

Table 2 depicts the use of several types of DS in FO systems. The bulk of recent research has employed NaCl as the DS; however, seawater, $MgCl_2$, and KCl have also been commonly

TABLE 2   Draw solutions and system configurations in FO systems.

No.	FO membrane	Config.	Feed solution	Draw solution	Reference
1	CTA (HTI)	AL-FS	Raw sewage	0.5M NaCl	Luo et al. (2016b)
2	CTA (HTI)	AL-FS	Wastewater	0.75M NaCl, KCl, CH_3COONa	Pathak et al. (2018)
3	CTA	AL-FS	Wastewater	48.4 g/L $MgCl_2$; 49 g/L NaCl	Qiu and Ting (2014a,b)
4	CTA (HTI)	AL-FS	Synthetic municipal ww	Fertilizers	Wang et al. (2017b)
5	CTA	AL-FS	Wastewater	1.5M NaCl; $NaNO_3$; $MgSO_4$; $ZnSO_4$	Cornelissen et al. (2008)
6	TFC (Toray)	AL-FS	Wastewater	35 g/L synthetic seawater	Blandin et al. (2018)
7	TFC and CTA (HTI)	AL-FS	Wastewater	150 mg/L PAA	Yang et al. (2019)
8	CTA (HTI)	AL-FS	Dyeing ww	1M $MgCl_2$	Li et al. (2018a,b)
9	CTA (HTI)	AL-FS	Wastewater	0.6M NaOAc; 0.3M EDTA-2Na; 0.5M NaCl	Luo et al. (2016a)

TABLE 2   Draw solutions and system configurations in FO systems—cont'd

No.	FO membrane	Config.	Feed solution	Draw solution	Reference
10	CTA (HTI)	AL-FS	Municipal ww	Seawater brine	Qiu et al. (2015)
11	TFC and CTA (HTI)	AL-FS	Wastewater	Mixed 1,5 M $MgCl_2$ and 1.5 mM Triton X-114	Nguyen et al. (2016a)
12	CTA (HTI)	AL-FS	Wastewater (phenol)	1.5 M $MgCl_2$	Praveen et al. (2015)
13	Hollow fiber	AL-FS	Wastewater	1 M NaCl	Ding et al. (2016)
14	TFC (HTI)	AL-FS	Wastewater	Seawater	Luo et al. (2017a,b)
15	CTA (HTI)	AL-FS	Wastewater	EDTA-2Na	Nguyen et al. (2016b)

used (Viet et al., 2019). Previous work indicated that using a monovalent salt (NaCl) could induce a decrease in organic matter removal and an increase in salt leakage compared to using divalent draw solutes (Abdelrasoul et al., 2018; Yu et al., 2017). Cornelissen et al. (2008) investigated the effect of DS types on water flux and salt flux. Their results implied that differences in electrolyte diffusivity in the porous membrane structure were the primary factor controlling the performance of the FO membrane. Bivalent electrolytes ($MgSO_4$ and $ZnSO_4$) have much lower diffusion coefficients than monovalent electrolytes (NaCl and $NaNO_3$). As a result, bivalent DSs had reduced water fluxes (Cornelissen et al., 2008). Furthermore, because divalent cations are bigger and less likely to pass across the membrane, the indicative salt flow values for $MgSO_4$ and $ZnSO_4$ solutions were lower ($5 g/m^2 h$) than for monovalent DSs. Lower bivalent electrolyte diffusion coefficients obstruct back diffusion inside the porous substructure, resulting in more severe internal concentration polarization (ICP). Because nitrate retention is lower than chloride retention, the larger indicated salt flow for the $NaNO_3$ solution was explained.

3.5 System configuration

Asymmetric FO membranes generally have two separate layers: an active layer (AL) and a support layer (SL). The former is in charge of substance selection and rejection, while the latter is in charge of mechanical support (Hoover et al., 2011; Shaffer et al., 2014). The vast majority of the experiments were conducted in the FO (AL-FS) modality. This is because fouling caused by the AL-FS mode is more reversible, allowing for a reduction in operating expenses and cleaning cycles (Matin et al., 2021). However, as the support layer became more exposed to the feed, it developed a larger fouling proclivity. This is owing to the decreased hydrodynamic shear force within the membrane's porous structure, which results in increased foulant deposition on the membrane surface. Furthermore, the PRO mode had a greater specific RSF value than the FO mode (Alturki et al., 2012).

3.6 Cross-flow velocity

Water flux is generally known to increase as cross-flow velocity increases. This fact is thanks to the decrease in the ECP effect (Suh and Lee, 2013). Suh and Lee (2013) discovered that as cross-flow velocity (CFV) increased, the proportion of external concentration polarization (CPs) decreased. At higher cross-flow velocities, permeated water mixes faster in the bulk draw solution. The proportion of dilutive ECP decreased from 32.5% to 15.5%. However, the water flux increased only from 13.7 to 15.2 LMH, and the rate of increase in water flux was only 11.0% when the cross-flow velocity significantly increased from 0.1 to 1.7 L/min. The reason for this fact is that the ICP increase cancels out the dilutive ECP decrease. ICP's contribution to the support layer increased from 46.0% to 60.6% (Suh and Lee, 2013). Similarly, Wang et al. (2017a,b,c) investigated the effect of CFV (from 8 to 26 cm/s) on the FO operation. Their work noted that the water flux was greater at higher velocities (Wang et al., 2017a). The water flux difference in the FO mode, on the other hand, was minor at three CFVs. This finding differed from a previous study. In detail, the water flux was approximately 6.74 LMH at a CFV of 8.5 cm/s, the flux value was only 2.22% and 6.68% higher at a cross-flow velocity of 17 and 25.5 cm/s, respectively (Wang et al., 2017a).

4. Applications in sustainable water and energy development

4.1 Desalination

In contrast to pressure-driven systems, forward osmosis (FO) processes rely on the osmotic flow of water over a semipermeable membrane from a dilute feed solution into a concentrated draw solution (Cath et al., 2006). Because FO does not require high pressure for separation, it has the potential to use less energy than RO systems to create water. As a result, FO has recently gained popularity as a novel method for wastewater treatment, food processing, and salty and brackish water desalination (Choi et al., 2009). (Kwon et al., 2016) investigated the use of an FO-MD system for saline solution desalination. The FO process was used to extract pure water from salty water, whereas the MD process was used to manufacture freshwater and recover draw solutes from DS (Kwon et al., 2016). As the DS concentration increased, so did the flow of the FO process. The flow of the MD process, on the other hand, decreased as DS concentrations increased. The flux of the FO/MD system was greatest when 4 M KCl was used at 60°C. Even while the DS temperatures were constant, the water flow between the salts varied. In addition, the ability of forward osmosis for osmotic dilution before RO contributes to a significant reduction of energy consumption (Lee et al., 2020a,b). Below are some examples of the application of FO in desalination.

4.1.1 Glucose forward osmosis

Glucose was used by Kravath and Davis as a draw solution to desalinate seawater (Kravath and Davis, 1975). Water was transferred from saltwater to a glucose draw solution, resulting in a drinkable glucose solution. There was no recoverable product water. The idea may be used for emergency water supply. The other techniques employed the similar idea, but they used nutrients and fructose as draw solutions (Qasim et al., 2015).

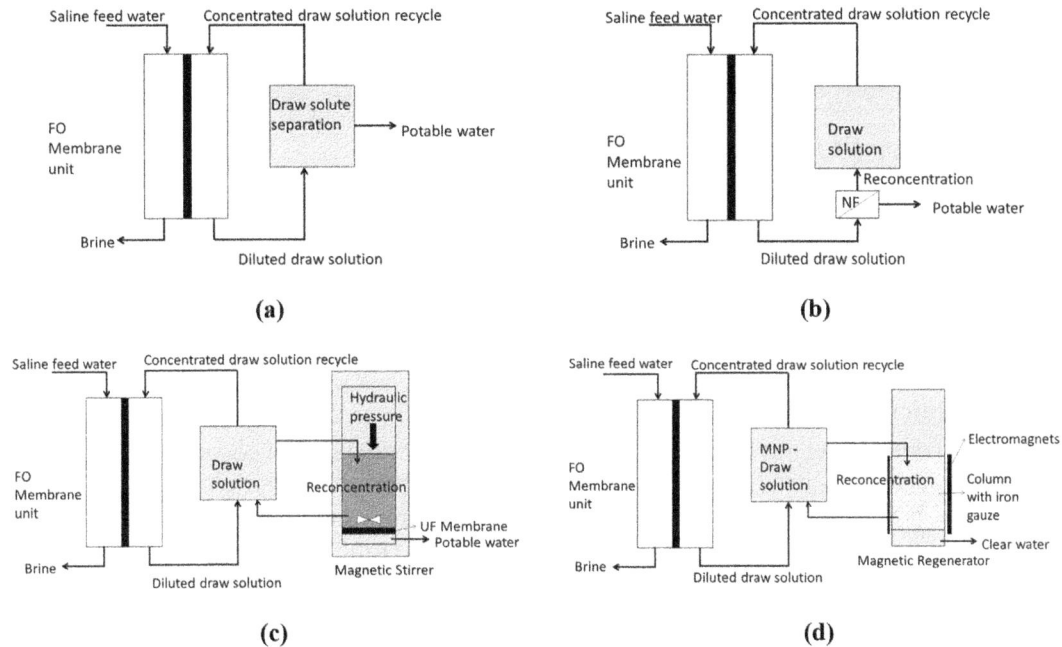

FIG. 7   (A) The schematic of ammonia-carbon dioxide FO; (B) The schematic of the FO-NF desalination; (C) The schematic FO-UF desalination; (D) The schematic diagram of Magnetic regeneration of draw solute.

4.1.2 Ammonia-carbon dioxide forward osmosis

The process employed ammonium bicarbonate as a draw solution to extract water from saline feed water across the FO membrane (McCutcheon et al., 2005). Ammonium bicarbonate is a very soluble draw solution that generates a very high osmotic pressure. This resulted in a high permeate recovery. The process was helped by mild heating, which decomposed ammonium bicarbonate into ammonia and carbon dioxide gases that can be separated from the water molecule (Fig. 7A).

4.1.3 FO-Nanofiltration with a divalent draw solution

Typically, the FO desalination process applies both osmotic dilution and freshwater regeneration from a diluted draw solution. Flat-sheet triacetate (CTA) membranes from Hydration Technology were used. The process employed divalent Na_2SO_4 and $MgSO_4$ as draw solution and brackish water as feed solution. This is the reason why the selection of draw solution and the freshwater generation method is very important. Under high osmotic pressure difference between the feed solution and draw solution, water is transported across the semipermeable FO membrane. Potable water was then generated from the NF separation process of the diluted draw solution. The pressure applied for NF was 4–10 bar (Fig. 7B).

4.1.4 FO-ultrafiltration regeneration

In this process, DI water and seawater were used as FS (Ling and Chung, 2011). Following water extraction from the feed solution in the FO process, the diluted DS containing hyperhydrophilic nanoparticles was filtered. At a transmembrane pressure of 5 bar, the UF membrane was used (TMP). The concentration polarization effect was assumed to be mitigated by magnetic stirring (Fig. 7C).

4.1.5 FO-magnetic field regeneration

The concept used for FO desalination in this process is similar to the previous method. However, the draw solution containing magnetic nanoparticles (MNPs) is regenerated differently by using a magnetic field (Ge et al., 2011). Iron gauze is packed into a plastic column. The magnetized iron attracts MNPs and allows water molecules to pass. To remove the magnetic field, iron gauze is demagnetized causing the trapped MNPs to be easily washed away by DI water (Fig. 7D).

4.1.6 FO-microgel desalination

This FO is produced in a batch procedure. A copolymer microgel of Nisopropylacrylamide and acrylic acid was used by (Hartanto et al., 2015). The authors employed a copolymer microgel of Nisopropylacrylamide and acrylic acid as the draw solution and saltwater as the feed solution. In the FO process, the microgel was employed to absorb water. Water was then extracted by centrifugation at 40°C and 12,000 rpm for 10 min.

4.2 Wastewater treatment and water reuse

Several studies have been conducted on the use of FO in water and wastewater treatment. Table 3 contains the detailed information. Wang et al. (2016), for example, tested a pilot scale method for municipal wastewater treatment using a spiral-wound FO module. The concentration factor (CF) of this FO system was 8 in the AL-FS mode with 0.5 M NaCl as the draw solution, according to the authors. Long-term operation at a concentration factor of 5 at an average flow of 6 LMH may achieve (99.8 ± 0.6%) chemical oxygen demand and (99.7 ± 0.5%) total phosphorus rejection rates. In comparison, just 48.1%–10.5% and 67%–8.3% of ammonium and total nitrogen were rejected, respectively. The fouling cake layer induced a promoted concentration polarization, thus reducing in water flux. Fouling also resulted in a cake-enhanced concentration polarization effect, which promoted the ammonium rejection rate with an increasing cycle operating duration (Wang et al., 2016). Similarly, Lampi and Shetji (2014) investigated an FO system for the treatment of landfill leachate wastewater. Their work demonstrated that the FO-RO system can achieve water recoveries of up to 90%. The hybrid system for treating landfill leachate involves a three FO as primary pretreatment driven by a 9% sodium chloride osmotic draw solution, then polished and reconcentrated in a closed loop by a multistage RO regeneration step. Over the course of a week, the average water recovery rate is greater than 93%. Furthermore, the system rejects most contaminants by more than 99%. The required total dissolved solids concentration of the reverse osmosis process permeate is less than 100 mg/L. The final permeate from HTI's FO/RO system has a total dissolved solids concentration of 5 mg/L (Lampi and Shetji, 2014).

Aside from wastewater treatment, FO has been demonstrated to be a capable process for the concentration of activated sludge. Hau et al. (2014) investigated an FO system with EDTA as a draw solution for high nutrient sludge concentration. After 16 h of operation with the

TABLE 3 FO application in desalination and wastewater treatment and reuse.

No	FO Application	Mem. type	Config.	Target FS	Draw solution	Temp	pH	CFV (cm/s)	Js (LMH)	Removal efficiency (%)				Reference
										DOC	TP	TN	NH4	
1	Desalination	CTA	AL-FS	DI	NaCl 4M	–	–	–	3	–	–	–	–	Choi et al. (2009)
2	Desalination	CTA	AL-FS	Seawater	NaCl 4M	25	–	–	2.5	–	–	–	–	Kwon et al. (2016)
3	Desalination	TFC	AL-FS	DI	NaCl 0.5M	–	–	–	25	–	–	–	–	Hawari et al. (2016)
4	WW treatment and reuse	CTA	AL-FS	Municipal ww	NaCl 0.5M	20	–	20	6	100	100	70	50	Wang et al. (2016)
5	WW treatment and reuse	TFC	AL-FS	Synthetic ww	NaCl 4M	20	–	–	5	–	–	–	–	Mi and Elimelech (2010)
6	WW treatment and reuse	CTA	AL-FS	Synthetic ww	NaCl 5M	20	–	–	4	–	–	–	–	Lee et al. (2010)
7	WW treatment and reuse	CTA	AL-FS	Synthetic ww	Seawater	–	7	8.5	3	–	–	–	–	Yu et al. (2017)
8	WW treatment and reuse	CTA	AL-FS	Municipal ww	NaCl 0.5M	–	–	9	–	–	–	–	–	Xie et al. (2014)
9	WW treatment and reuse	CTA	AL-FS	Municipal ww	Seawater	22	–	0.14	5	–	–	–	–	Zhang et al. (2014)
10	WW treatment and reuse	CTA	AL-FS	Landfill leachate	NaCl 9%	–	–	–	–	95	95	95	95	Lampi and Shetji (2014)
11	WW treatment and reuse	CTA	AL-FS	Oily ww	NaCl 9%	–	–	–	–	–	–	–	–	Lampi and Shetji (2014)

Continued

TABLE 3 FO application in desalination and wastewater treatment and reuse—cont'd

No	FO Application	Mem. type	Config.	Target FS	Draw solution	Temp	pH	CFV (cm/s)	Js (LMH)	Removal efficiency (%)				Reference
										DOC	TP	TN	NH4	
12	WW treatment and reuse	CTA	AL-FS	Digested sludge	NaCl 1M	20	–	0.4	2.5	–	–	–	–	Husnain et al. (2015)
13	WW treatment and reuse	CTA	AL-FS	Municipal ww	Seawater	20	7.1	9	8	–	–	–	–	Xie et al. (2013)
14	WW treatment and reuse	CTA	AL-FS	Municipal ww	Seawater	–	–	–	22.5	–	–	–	90	Volpin et al. (2018)
15	WW treatment and reuse	TFC	AL-FS	WWTP effluent	NaCl 0.5M	–	–	–	18	–	–	–	–	Bell et al. (2016)
16	WW treatment and reuse	CTA	AL-FS	Naphthenhic acids	NaCl 1M	21	–	14	6.5	–	–	–	–	Zhu et al. (2018)
17	WW treatment and reuse	CTA	AL-FS	Sludge	EDTA 0.3M	25	–	2.5	6.5	–	99	–	98	Hau et al. (2014)
18	WW treatment and reuse	CTA	AL-FS	Rainwater	Cooling water 5000 µS/cm	–	–	4.5	1.5	–	–	–	–	Wang et al. (2014b)
19	WW treatment and reuse	TFC	AL-FS	WWTP effluent	NaCl 0.6M	25	–	–	12.5	–	–	–	–	Im et al. (2021)

CTA FO membrane and EDTA DS 0.3 M, the final concentration of sludge was 32,000 mg/L. This was linked to the formation of multibarrier layers of sludge on the membrane surface as well as the steric impact of the FO membrane (Hau et al., 2014).

4.3 Concentration of materials

The forward osmosis (FO) process is capable of generating high-quality reused water from a wide range of wastewater sources. Aside from the purifying function in wastewater treatment, a unique osmosis-driven phenomena with various characteristics makes FO a prospective use in the commercial concentration process. The characteristics are as follows: (1) high water recovery rate based on a dense membrane with ideal rejection of most compounds; (2) simple operation without the need for pressure vessels and high pressure pumps; (3) ambient concentration process minimized additional energy input; and (4) less compact fouling layer requiring simple cleaning procedure. Thus, FO can be a preferred option for the emerging concentration of streams and an alternative to the established concentration process. During the recent decade, not only scientific research publications but also commercial products employing FO as a concentration technique have risen dramatically (Blandin et al., 2020). Several types of concentration application (beverage and dairy industry, algae harvesting) will be introduced as follows.

4.3.1 *Beverage concentration*

Concentrated beverage greatly reduces the cost in terms of storage, package, handling, and transportation. The thermal process is the most widely utilized technique for shelf -life extension and food preservation and concentration in the commercial beverage company (Sant'Anna et al., 2012). However, flavors and aromas are quickly damaged in the typical high-heat processing. Besides, the high energy demand also inevitably increases the investment of the process. Membrane technology provides the promising supplement for the current concentration process. Various hydraulic pressure-driven membrane technologies have been therefore attempted to separate different molecular weight compounds due to their corresponding pore size. During the concentration process, the flavors and aroma can be better preserved without thermal degradation. Nevertheless, the concentration ratio appears to be the major obstacle for these membrane processes. For example, reverse osmosis only achieved 25–30 brix due to the high osmotic pressure of concentrated solution, which still lacks the competence compared to the evaporation of 45–65 brix (Jiao et al., 2004).

Contrary to these conventional membrane processes, a broad range of beverage concentrations is successfully realized by FO (Kim et al., 2019; Wang et al., 2022; Sant'Anna et al., 2016). With the development of FO membranes, intense studies have demonstrated that the fruit juice is able to be concentrated within high acid and color rejection. Nayak et al. (2011) observed that the content of betalains increased 57.1-fold in beetroot juice and anthocyanin increased 6.8-fold in grape juice (Nayak et al., 2011). The previous works have conducted several studies by using membrane technology from reverse osmosis to FO to concentrate tomato juice (Petrotos et al., 1998; Dova et al., 2007). Despite the low flux ($4.5 \, \mathrm{L \, m^{-2} \, h^{-1}}$), FO achieved positive results. Advanced FO membrane companies (Aquaporin and Porifera) subsequently aim to approach varieties of beverage

TABLE 4   Fruit juice concentrated by forward osmosis.

Application	Membrane type	Draw solution	Initial conc. (brix)	Final conc. (brix)	Reference
Grapefruit	TFC	2M NaCl	Pure juice	22	Kim et al. (2019)
Beetroot	CTA	6M NaCl	2.3	52	Nayak et al. (2011)
Sweetlime	CTA	6M NaCl	11	50	Chanukya and Rastogi (2017)
Orange	CTA	4M NaCl	8	10.5	Garcia-Castello and McCutcheon (2011)
Tomato	TFC	4M NaCl	5.5	16	Petrotos et al. (2010)

concentrations by their own designed systems. For example, the juice concentration ratio can reach as high as 65–70 brix without any loss of flavor, aroma, or nutrients. Examples of fruit juice concentration by FO are summarized in Table 4. Besides fruit juice concentration, some researchers have tried to separate alcohol from water to reduce shipped and stored volumes without losing flavor, which is beneficial for the beer or wine industry. Zhang et al. (2013) found that ethanol concentration can increase up to 90 wt% in both membrane orientations. This technique, on the other hand, suffers from some ethanol loss due to reverse salt diffusion. As part of the FO concentration procedures in alcoholic drinks, FO membrane vendors support both alcohol dehydration and dealcoholization, as well as tunability to application demands.

Several factors should be considered during the concentration process. Reverse salt diffusion will inevitably interfere with the product quality. To date, however, scarce adverse impact has been reported when the most selected draw solution, namely NaCl, was used. In addition, membrane fouling and associated cake-enhanced concentration polarization tend to exacerbate the performance. With the progress of membrane material and cleaning methods, less severe fouling greatly alleviates the blockage of water permeation and improves the concentration ratio (Chanukya and Rastogi, 2017).

4.3.2 Dairy concentration

In the dairy industry, large quantities of milk and whey proteins are processed to powder after the concentration process. Evaporation is the most used concentration process which can get rid of moisture in the milk (85.5%–89.5%) to obtain the final product containing little moisture (3%–5%) (Schuck et al., 2015). However, conventional evaporation used to consume nearly 15% of the total energy consumption of the whole process, which is considered as the energy-intensive unit. Furthermore, thermal concentration inevitably degrades the ingredients and digestibility of dairy heat-sensitive products. Hydraulic membrane technology, such as reverse osmosis, has been used as the preconcentration step due to less energy consumption than evaporation (Ramírez et al., 2006).

Since FO can be operated at ambient condition, which appears a possible alternative to replace the conventional evaporation. Chen et al. (2019) investigated the concentration of skim milk and whey by a pilot scale FO system. Results showed that the concentration ratio achieved 2.5 for both skim milk and fresh whey. Moderate fouling was observed when the flux was below $10 \, \text{L} \, \text{m}^{-2} \, \text{h}^{-1}$, in which, citric acid followed by enzyme-based solution cleaning can completely remove the fouling layer. In addition, significant energy saving was achieved by using FO instead of evaporation. This pilot system consumed <8.5 and < 10 kWh of electrical energy to remove one tone of water for skim milk and whey, respectively, which was much smaller than that for the vacuum evaporation process (17–70 kWh). On the other hand, some previous studies have conducted successive trials for whey recovery by the hybrid FO system (Aydiner et al., 2013, 2014). The high concentration of whey (25%–35%) and removal of water presented these hybrid systems as competitive and promising in the dairy industry.

4.3.3 *Microalgae harvesting*

As a sustainable biomass, microalgae gradually show great potential in the aspects of biofuel, food supplement, and carbon neutralization (Barros et al., 2015). The biorefinery for processing microalgae consists of several steps including cultivation, harvesting, and purification. Dewatering and thickening could accumulate the microalgae concentration from 6 to $150 \, \text{g/L}$, which enables the biorefinery to be economically competitive. However, a major dilemma regarding economic feasibility still remains in a large-scale processing of microalgae. Multiple techniques, including centrifugation, membrane filtration, flotation, and sedimentation, have been applied to harvest microalgae.

Due to its intrinsic property, FO is currently served as a recent microalgae harvesting method, which hopefully meets the global carbon neutralization. Ye et al. (2018) have used the TFC membrane to concentrate *Scenedesmus acuminatus* 20 times despite the appearance of fouling. Of particular note, most of the membrane fouling was reversible realized by simple physical flushing and the water flux was recovered back to 97% of the pristine water flux. Draw solution recovery was avoided by coupling with microalgae cultivation to obtain the highly concentrated biomass ($120 \, \text{g/L}$) (Ryu et al., 2020). However, biomass cultivating conditions and unexpected polymeric substances would intensify the fouling boosted by the reverse salt diffusion (Larronde-Larretche and Jin, 2016, 2017).

4.4 Resource recovery

4.4.1 *FO membrane-based hybrid systems for water recovery*

One application of FO for resource recovery is to recover fresh water and reconcentrate the draw solution. To make that application possible, FO needs to be in combination with other separation processes to form a hybrid system. There are a variety of options such as pressure-driven (e.g., NF and RO) (Nguyen et al., 2015), thermal-driven (e.g., membrane distillation) (Zhang et al., 2014), and electric-driven processes (e.g., electrodialysis) (Zhang et al., 2013). The operational principle of the hybrid systems is that wastewater is pretreated by the FO part and the resources in the foulant-free solution are recovered by the subsequent systems. Hence, the hybrid system is able to produce a more quality effluent compared to other single membrane processes.

One issue of the FO hybrid system for water recovery is the accumulation of contaminants in the draw solution (Luo et al., 2016a,b,c). The FO membrane is unable to retain all the dissolved contaminants which would increasingly accumulate in the draw solution as part of a closed-loop system. Some contaminants such as ammonium and phosphate accumulated in the FO-RO system (Luo et al., 2016a,b,c). In addition, accumulation of trace organic pollutants are also unavoidable in the FO system driven by the discrepancy of rejection capability of the FO and draw solute (D'Haese et al., 2013).

The accumulation of contaminants in the FO hybrid system is a barrier for its wide application in water recovery. For example, the accumulation of organic contaminants and ammonium in the Ae-OMBR-RO system has had a negative impact on water quality (Luo et al., 2016a,b,c). The accumulation of contaminants in an FO-RO system was also modeled and the trace organic contaminants in the draw solute were higher than the feed due to cumulative accumulation (D'Haese et al., 2013). The quality of water recovery is compromised, and membrane fouling becomes worse.

4.4.2 Nutrient recovery using FO

Another application of FO in terms of resource recovery is nutrient recovery. The rejection solute of the FO system contains high-quality nutrient products from wastewater. Phosphorus is one of the nutrients of interest which frequently appears at high concentration in the rejection. There is a wide range of phosphorus levels reported in the literature which resulted from various types of feed solutions and operating conditions. The wastewater sources for phosphorus recovery are also diverse such as waste activated sludge (Luo et al., 2016a,b,c) and digested sludge centrate (Holloway et al., 2007). In addition, the hybrid FO system for phosphorus recovery appears in the form of several configurations such as Ae-OMBR and direct FO filtration; however, the principle is to concentrate phosphorus at the first stage by FO and subsequently use conventional techniques to precipitate phosphorus (e.g., struvite, calcium phosphate). The FO system is particularly suitable for nutrient recovery application in three ways: (i) reduce chemical for precipitation of phosphorus in the subsequent stage, (ii) reuse the reverse flux for nutrient recovery, and (iii) automatically increase the pH of the feed solution. The recovery of struvite needs a significant amount of chemicals (i.e., magnesium salt and ammonium) to surpass the stoichiometric ratio of struvite precipitation. The phosphorus solution of the FO system is able to facilitate the precipitation kinetics and reduce the burden of chemicals for the precipitation step. The reverse solute flux is one problematic issue of the FO which can be in turn used for nutrient recovery. This approach can be strategically developed by using $MgCl_2$ as the draw solution and, through the reverse magnesium flux, the feed solution is enriched. Regarding the bidirectional, it has been reported in numerous studies worldwide, for example, Ansari et al. (2016) have shown the benefit of the bidirectional transport of Mg^{2+}/Ca^{2+} and proton (H^+) for phosphorus precipitation.

Like the water recovery application, there is a range of FO-based system configurations used for nutrient recovery. Ae-OMBR is one of the configurations that have shown an effective recovery of nutrients (Holloway et al., 2015). On a broader scale, the direct preconcentration processes of the anaerobic digestion effluent also showed a satisfying result and is applicable for integrating with the available wastewater treatment infrastructure. This approach maximized the nutrient recovery efficiency by minimizing the loss of nutrient which happened by biomass uptake. For the aerobic process, there is likely less phosphorus available for recovery as it has been consumed extensively by the aerobic activated sludge.

The application of the FO system for nutrient recovery remains a huge potential and keeps attracting the interest of researchers (Ding et al., 2014). However, there are some technical challenges that need to be addressed such as membrane fouling, scaling, and precipitation.

Membrane scaling is a critical constraint for the wider application of the FO process due to the influence on the membrane performance and longevity. The problem stems from the precipitation of phosphate minerals on the membrane surface. At the moment, there is no evidence that membrane scaling in the nutrient recovery application can cause a significant problem. One argument can come from the short-term studies which are not enough to come up with a conclusion. The cake layer formation in the feed solution is another concern, but it can be eradicated by membrane flushing (Ansari et al., 2016; Ning and Troyer, 2007).

4.5 Agricultural fertigation

Apart from wastewater treatment and reclamation, FO is also capable of water reuse application. By utilizing the concentrated fertilizers as DS, the diluted DS after FO could be used directly for agricultural purposes such as irrigation [this so-called fertilizer drawn FO (FDFO)]. This facilitates the use of membrane in simultaneous wastewater treatment and reuse. Recent studies have reported the performance of FO membranes in water recovery for agricultural fertigation. Multiple types of wastewater as well as the fertilizer draw solution have been investigated in previous studies. For instance, municipal wastewater has been treated by FO in the studies of several groups around the world (Adnan et al., 2019; Phuntsho et al., 2014; Wang et al., 2017b); the concentrated fertilizers (i.e., ammonium-based fertilizers) were diluted for fertigation purpose. The authors demonstrated that during the FO operation, $(NH_4)_2SO_4$ (SOA) showed the longest filtration running time, at around 17 days compared to other types of fertilizers. This facilitated the use of SOA as the draw solution in the process (Adnan et al., 2019). In addition, MAP ($NH_4H_2PO_4$) was also another potential candidate for the use of fertilizers as DS in FO processes. Wang et al. (2017a,b,c) reported that MAP was one of the most suitable fertilizers for the FDFO processes because it resulted in a high water flux as well as a low reverse salt flux compared with other fertilizers. In addition to domestic wastewater, brackish water and seawater were also utilized for water extraction by FO for fertigation. Phuntsho et al. (2012) published a study on the use of blended fertilizer (urea and SOA) as the draw solution in FOFO with brackish water as the feed solution. The authors concluded that the blended solution of KCl and $NH_4H_2PO_4$ as the concentrated DS can lead to a lower final concentration and ratio of N/P/K compared to single solution, supporting a higher potential of diluted fertilizer for direct irrigation (Phuntsho et al., 2012). However, achieving the final and acceptable concentration of nutrients is still a challenge for this process and this may require further dilution.

Other types of wastewater such as landfill leachate or coal seam gas were also utilized as feed solution in FDFO for water extraction for fertigation (Kim et al., 2017a,b; Li et al., 2017). Over 90% water recovery was achieved during the FDFO process using these types of feed. Kim et al. (2017a,b) reported that KNO_3 was the worst fertilizer for treatment of the coal seam gas effluent with a high flux decline because the feed solution conductivity increased critically due to RSF. Besides, $Ca(NO_3)_2$ should not be used in this process owing to its high scaling potential that is caused by calcium ions. By contrast, a very high potential candidate (NH_4HCO_3) for FDFO treating landfill leachate wastewater was reported by Li et al. (2017). The authors demonstrated that 92% water recovery was reached by FDFO with NH_4HCO_3. The permeate water quality was also very high with an extremely low concentration of contaminants such as polycyclic aromatic hydrocarbons (PAHs), facilitating direct use of the diluted fertilizer draw solution for agricultural irrigation. The most recent studies by Viet and Jang (2022) and Manzoor et al. (2022) reported the use of a fertilizer draw solution index in supporting the selection of the appropriate fertilizer for use in the FDFO system. The very first recovery of water from textile wastewater for direct irrigation with a very high water flux of around 9.7 LMH was demonstrated.

Though FDFO has been intensively investigated in recent years, there are still several limitations of this process that need to be paid attention to in future research, such as the final concentration of diluted fertilizers was still higher than the demand of plants, or the selection of the best fertilizer candidate for each specific purpose was also another challenge of this FO application (Table 5).

TABLE 5 FO applications in fertigation.

FO	Config.	Target FS	Draw solution	Temp	CFV	Ave water flux (LMH)	Water recovery (%)	Reference
CTA	AL-FS	Groundwater	$(NH_4)_2SO_4$ 1 M	–	–	6	–	Phuntsho et al. (2013)
CTA	AL-FS	Seawater	NH_4NO_3 2 M	25	8.5 cm/s	5	–	Phuntsho et al. (2011)
CTA	AL-FS	Brackish water	Blended fertilizers (urea + SOA)	–	–	2.16	–	Phuntsho et al. (2012)
TFC	AL-FS	Landfill leachate	NH_4HCO_3 3 M	25	8.4 cm/s	3	92	Li et al. (2017)
CTA	AL-FS	Synthetic ww	$(NH_4)_2SO_4$ 0.25 M	25	0.5 L/min	2.58	–	Adnan et al. (2019)
TFC	AL-FS	Coal seam gas	$Ca(NO_3)_2$ 1 M	25	8.5 cm/s	–	90	Kim et al. (2017a,b)
CTA	AL-FS	Synthetic ww	KCl 1 M	25	8.5 cm/s	7.5	–	Phuntsho et al. (2014)
CTA	AL-FS	Domestic ww	$NH_4H_2PO_4$ 2 M	22	0.2 L/min	7	97	Wang et al. (2017b)
TFC	AL-FS	Domestic ww	MAP	25	8 cm/s	7.6	–	Viet and Jang (2022)
TFC	AL-FS	Textile ww	SOA	36	–	9.7	–	Manzoor et al. (2022)

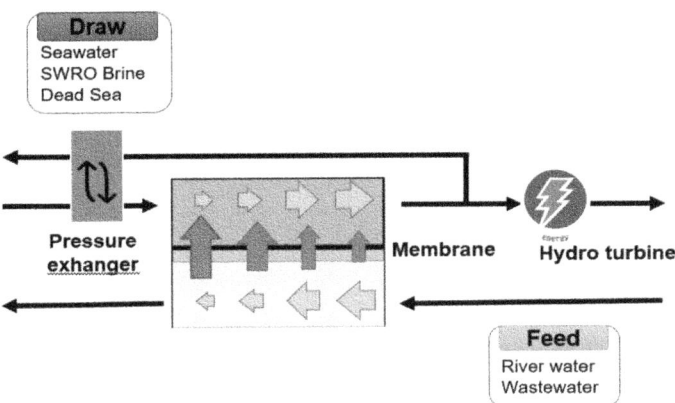

FIG. 8   The schematic of a PRO process including a hydroturbine and a pressure exchanger.

4.6 Energy harvesting

PRO has emerged as a promising technology for harvesting energy from a salinity gradient. This concept is illustrated in the schematic diagram (Fig. 8). PRO extracts the Gibbs free energy of process by providing water to flow spontaneously through a membrane against hydraulic pressure from a low osmotic FS to a high osmotic DS. The Gibbs free energy is converted to diluted brine hydraulic pressure, which can then be converted to mechanical energy by a pressure exchanger or electrical energy by a hydroturbine. A recent review paper provided comprehensive information about the current status of PRO technologies in membrane fabrication, module design, process optimization, and commercialization of PRO (Lee et al., 2020). This process favors generating a high-power density compared to another salinity energy [i.e., Reverse electrodialysis (RED)]. A past work highlighted that the PRO process exhibited the energy harvesting of 2.4–38 W/m^2 while RED only attained a low value of 0.77–1.2 W/m^2 (Yip and Elimelech, 2014).

Table 6 summarizes the outcomes of energy harvesting using the commercial FO membrane in the operation of the PRO process. The membranes developed by HTI achieved a power density of 14.1 W/m^2 at 20.7 bar when 1.0 M NaCl was used as the draw solution. In another work, Han et al. (2013) developed a new generation of the TFC membrane using polyimide support. This membrane possesses a sponge-like structure that can withstand a hydraulic pressure of 15 bar. At a pressure of 15 bar and a draw solution of 1.0 M NaCl, the fabricated membrane achieved a power density of 12.0 W/m^2. The most recent study published by Fang et al. (2018) demonstrated a power density of around 3.29 W/m^2 was extracted by a CTA FO membrane system with a small pressure gradient (i.e., <4 bar). However, most of the published studies on PRO have been utilizing DI water as a feed solution. This may cause multiple differences in the real situations where various types of feed could be the potential source for the PRO process. Further studies using different types of feed solution such as wastewater indicated that groundwater would be necessary to improve the practical possibility of the PRO process.

TABLE 6   Application of PRO to energy harvesting.

Membrane	Feed solution	Draw solution	A (L/m² h.bar)	B (L/m² h)	Pressure difference ΔP (bar)	Power density (W/m²)	Reference
TFC-200	DI water	1.0 M NaCl	5.3	2.0	15.0	12.0	Han et al. (2013)
TFC-B-5	DI water	1.0 M NaCl	1.0	0.18	9.0	6.0	Li and Chung (2013)
TFC-2	DI water	1.0 M NaCl	0.9	–	22.0	12.9	Wei et al. (2016)
HTI-CTA	DI water	1.0 M NaCl	0.7	0.4	9.7	5.1	Achilli et al. (2009a,b)
HTI-TFC	DI water	1.0 M NaCl	2.5	0.39	20.7	14.1	Straub et al. (2013)
HTI OsMem 2521 FO-CTA- MS-P-3H	0–1000	4000–60,000 mg/L	–	–	0–4	0.97–3.29	Fang et al. (2018)
HTI 2521 spiral wound FO membrane module	500	10,000–60,000 mg/L	–	–	0–11.5	0.16–0.8	She et al. (2012)

5. Conclusions and perspectives

Over the past decade, the FO (PRO) membrane has emerged as a potential technology for sustainable water and energy development. The FO process is increasingly used for domestic and industrial wastewater treatment, desalination, resource recovery, concentration of materials, and agricultural fertigation on laboratory and pilot scales thanks to its outstanding characteristics featuring high-quality treated water, low energy consumption, and less fouling for the FO mode and energy harvesting for the PRO mode. However, the key drawback of FO is its inherent low flux and high reverse salt flux (RSF). These critical obstacles render the FO membrane costly, increasing the capital cost of applications. Either operating in the PRO mode or using a higher RSF caused serious fouling of the FO membrane, which resulted in increased operating costs. Considering the application of PRO to energy harvesting, the inefficiency of the module design is surprising. These critical factors make it impossible to apply FO membranes in full-scale conditions. Although a recent new generation of the TFC membrane has the potential to enhance water flux, the FO membrane is still costly. It is necessary to study and fabricate more generation membranes using surface modification, organic and inorganic additives, and various fabrication techniques in the future. The ideal FO membrane should possess high flux, low RSF, and sufficient ion selectivity. The fabricated membranes must be translated to module manufacturing in order to make the FO process feasible for desalination, water treatment, resource recovery, and material concentration.

Acknowledgments

This research is funded by Vietnam National University Ho Chi Minh City (VNU-HCM) under grant number NCM2021-20-01." We acknowledge the support of time and facilities from Ho Chi Minh City University of Technology (HCMUT), VNU-HCM for this study.

References

Abdelrasoul, A., Doan, H., Lohi, A., Cheng, C., 2018. Fouling in forward osmosis membranes: mechanisms, control and challenges. Intech Open. https://doi.org/10.5772/intechopen.72644.

Achilli, A., Cath, T.Y., Childress, A.E., 2009a. Power generation with pressure retarded osmosis: an experimental and theoretical investigation. J. Membr. Sci. 343, 42–52. https://doi.org/10.1016/j.memsci.2009.07.006.

Achilli, A., Cath, T.Y., Marchand, E.A., Childress, A.E., 2009b. The forward osmosis membrane bioreactor: a low fouling alternative to MBR processes. Desalination 239, 10–21. https://doi.org/10.1016/j.desal.2008.02.022.

Adnan, M., Khan, S.J., Manzoor, K., Hankins, N.P., 2019. Performance evaluation of fertilizer draw solutions for forward osmosis membrane bioreactor treating domestic wastewater. Process. Saf. Environ. Prot. 127, 133–140. https://doi.org/10.1016/j.psep.2019.05.006.

Alturki, A., Mcdonald, J., Khan, S.J., Hai, F.I., Price, W.E., Nghiem, L.D., 2012. Performance of a novel osmotic membrane bioreactor (OMBR) system: flux stability and removal of trace organics. Bioresour. Technol. 113, 201–206. https://doi.org/10.1016/j.biortech.2012.01.082.

Ansari, A.J., Hai, F.I., Price, W.E., Nghiem, L.D., 2016. Phosphorus recovery from digested sludge centrate using seawater-driven forward osmosis. Sep. Purif. Technol. 163, 1–7.

Aydiner, C., Topcu, S., Tortop, C., Kuvvet, F., Ekinci, D., Dizge, N., Keskinler, B., 2013. A novel implementation of water recovery from whey: "forward- reverse osmosis" integrated membrane system. Desalin. Water Treat. 51, 786–799. https://doi.org/10.1080/19443994.2012.693713.

Aydiner, C., Sen, U., Topcu, S., Ekinci, D., Altinay, A.D., Koseoglu-Imer, D.Y., Keskinler, B., 2014. Techno-economic viability of innovative membrane systems in water and mass recovery from dairy wastewater. J. Membr. Sci. 458, 66–75. https://doi.org/10.1016/j.memsci.2014.01.058.

Barros, A.I., Gonçalves, A.L., Simões, M., Pires, J.C.M., 2015. Harvesting techniques applied to microalgae: a review. Renew. Sustain. Energy Rev. 41, 1489–1500. https://doi.org/10.1016/j.rser.2014.09.037.

Bell, E.A., Holloway, R.W., Cath, T.Y., 2016. Evaluation of forward osmosis membrane performance and fouling during long-term osmotic membrane bioreactor study. J. Membr. Sci. 517, 1–13. https://doi.org/10.1016/j.memsci.2016.06.014.

Bhojwani, S., Topolski, K., Mukherjee, R., Sengupta, D., El-Halwagi, M.M., 2019. Technology review and data analysis for cost assessment of water treatment systems. Sci. Total Environ. 651, 2749–2761. https://doi.org/10.1016/j.scitotenv.2018.09.363.

Blandin, G., Verliefde, A.R.D., Comas, J., Rodriguez-Roda, I., Le-Clech, P., 2016a. Efficiently combining water reuse and desalination through forward osmosis-reverse osmosis (FO-RO) hybrids: a critical review. Membranes (Basel). https://doi.org/10.3390/membranes6030037.

Blandin, G., Vervoort, H., D'Haese, A., Schoutteten, K., Van den Bussche, J., Vanhaecke, L., Myat, D.T., Le-Clech, P., Verliefde, A.R.D., 2016b. Impact of hydraulic pressure on membrane deformation and trace organic contaminants rejection in pressure assisted osmosis (PAO). Process. Saf. Environ. Prot. 102, 316–327. https://doi.org/10.1016/j.psep.2016.04.004.

Blandin, G., Gautier, C., Sauchelli Toran, M., Monclús, H., Rodriguez-Roda, I., Comas, J., 2018. Retrofitting membrane bioreactor (MBR) into osmotic membrane bioreactor (OMBR): a pilot scale study. Chem. Eng. J. 339, 268–277. https://doi.org/10.1016/j.cej.2018.01.103.

Blandin, G., Ferrari, F., Lesage, G., Le-Clech, P., Héran, M., Martinez-Lladó, X., 2020. Forward osmosis as concentration process: review of opportunities and challenges. Membranes (Basel). 10, 1–40. https://doi.org/10.3390/membranes10100284.

Burek, P., Langan, S., Cosgrove, W., Fischer, G., Kahil, T., Magnuszewski, P., Satoh, Y., Tramberend, S., Wada, Y., Wiberg, D., 2016. The Water Futures and Solutions Initiative of IIASA. pp. 23–26.

Cath, T.Y., Childress, A.E., Elimelech, M., 2006. Forward osmosis: principles, applications, and recent developments. J. Membr. Sci. 281, 70–87. https://doi.org/10.1016/j.memsci.2006.05.048.

Chanukya, B.S., Rastogi, N.K., 2017. Ultrasound assisted forward osmosis concentration of fruit juice and natural colorant. Ultrason. Sonochem. 34, 426–435. https://doi.org/10.1016/j.ultsonch.2016.06.020.

Chekli, L., Phuntsho, S., Shon, H., Vigneswaran, S., Kandasamy, J.K., Chanan, A.P., 2012. A review of draw solutes in forward osmosis process and their use in modern applications. Desalin. Water Treat. 43 (1–3), 167–184. https://doi.org/10.1080/19443994.2012.672168.

Chen, G.Q., Artemi, A., Lee, J., Gras, S.L., Kentish, S.E., 2019. A pilot scale study on the concentration of milk and whey by forward osmosis. Sep. Purif. Technol. 215, 652–659. https://doi.org/10.1016/j.seppur.2019.01.050.

Choi, Y.J., Choi, J.S., Oh, H.J., Lee, S., Yang, D.R., Kim, J.H., 2009. Toward a combined system of forward osmosis and reverse osmosis for seawater desalination. Desalination 197, 239–246. https://doi.org/10.1016/j.desal.2008.12.028.

Chung, T.-S., Zhang, S., Wang, K.Y., Ling, M.M., 2011. Forward osmosis processes: yesterday, today and tomorrow. Desalination 287, 78–81. https://doi.org/10.1016/j.desal.2010.12.019.

Cornelissen, E.R., Harmsen, D., de Korte, K.F., Ruiken, C.J., Qin, J.J., Oo, H., Wessels, L.P., 2008. Membrane fouling and process performance of forward osmosis membranes on activated sludge. J. Membr. Sci. 319, 158–168. https://doi.org/10.1016/j.memsci.2008.03.048.

D'Haese, A., Le-Clech, P., Van Nevel, S., Verbeken, K., Cornelissen, E.R., Khan, S.J., Verliefde, A.R.D., 2013. Trace organic solutes in closed-loop forward osmosis applications: influence of membrane fouling and modeling of solute build-up. Water Res. 47 (14), 5232–5244.

Ding, Y., Tian, Y., Li, Z., Liu, F., You, H., 2014. Characterization of organic membrane foulants in a forward osmosis membrane bioreactor treating anaerobic membrane bioreactor effluent. Bioresour. Technol. 167, 137–143.

Ding, Y., Tian, Y., Liu, J., Li, N., Zhang, J., Zuo, W., Li, Z., 2016. Investigation of microbial structure and composition involved in membrane fouling in the forward osmosis membrane bioreactor treating anaerobic bioreactor effluent. Chem. Eng. J. 286, 198–207. https://doi.org/10.1016/j.cej.2015.10.084.

Dova, M.I., Petrotos, K.B., Lazarides, H.N., 2007. On the direct osmotic concentration of liquid foods: part II. Development of a generalized model. J. Food Eng. 78, 431–437. https://doi.org/10.1016/j.jfoodeng.2005.10.011.

Elimelech, M., Phillip, W.A., 2011. The future of seawater desalination: energy, technology, and the environment. Science. https://doi.org/10.1126/science.1200488.

Fang, L.F., Cheng, L., Jeon, S., Wang, S.Y., Takahashi, T., Matsuyama, H., 2018. Effect of the supporting layer structures on antifouling properties of forward osmosis membranes in AL-DS mode. J. Membr. Sci. 552, 265–273. https://doi.org/10.1016/j.memsci.2018.02.028.

Feng, L., Xie, L., Suo, G., Shao, X., Dong, T., 2018. Influence of temperature on the performance of forward osmosis using ammonium bicarbonate as draw solute. Trans. Tianjin Univ. 24, 571–579. https://doi.org/10.1007/s12209-018-0159-1.

Garcia-Castello, E.M., McCutcheon, J.R., 2011. Dewatering press liquor derived from orange production by forward osmosis. J. Membr. Sci. 372, 97–101. https://doi.org/10.1016/j.memsci.2011.01.048.

Ge, Q., Su, J., Chung, T.S., Amy, G., 2011. Hydrophilic superparamagnetic nanoparticles: synthesis, characterization, and performance in forward osmosis processes. Ind. Eng. Chem. Res. 50, 382–388. https://doi.org/10.1021/ie101013w.

Gu, Y., Wang, Y.N., Wei, J., Tang, C.Y., 2013. Organic fouling of thin-film composite polyamide and cellulose triacetate forward osmosis membranes by oppositely charged macromolecules. Water Res. 47, 1867–1874. https://doi.org/10.1016/j.watres.2013.01.008.

Han, G., Zhang, S., Li, X., Chung, T.S., 2013. High performance thin film composite pressure retarded osmosis (PRO) membranes for renewable salinity-gradient energy generation. J. Membr. Sci. 440, 108–121. https://doi.org/10.1016/j.memsci.2013.04.001.

Hartanto, Y., Yun, S., Jin, B., Dai, S., 2015. Functionalized thermo-responsive microgels for high performance forward osmosis desalination. Water Res. 70, 385–393. https://doi.org/10.1016/j.watres.2014.12.023.

Hau, N.T., Nguyen, N.C., Guo, W., Ngo, H.H., Huang, K.Z., Chen, S.-S., 2014. Exploration of EDTA sodium salt as novel draw solution in forward osmosis process for dewatering of high nutrient sludge. J. Membr. Sci. 455, 305–311. https://doi.org/10.1016/j.memsci.2013.12.068.

Haupt, A., Lerch, A., 2018. Forward osmosis application in manufacturing industries: a short review. Membranes (Basel). 8. https://doi.org/10.3390/membranes8030047.

Hawari, A.H., Kamal, N., Altaee, A., 2016. Combined influence of temperature and flow rate of feeds on the performance of forward osmosis. Desalination 398, 98–105. https://doi.org/10.1016/j.desal.2016.07.023.

Holloway, R.W., Childress, A.E., Dennett, K.E., Cath, T.Y., 2007. Forward osmosis for concentration of anaerobic digester centrate. Water Res. 41 (17), 4005–4014.

Holloway, R.W., Wait, A.S., Fernandes da Silva, A., Herron, J., Schutter, M.D., Lampi, K., Cath, T.Y., 2015. Long-term pilot scale investigation of novel hybrid ultrafiltration-osmotic membrane bioreactors. Desalination 363, 64–74. https://doi.org/10.1016/j.desal.2014.05.040.

Hoover, L.A., Phillip, W.A., Tiraferri, A., Yip, N.Y., Elimelech, M., 2011. Forward with osmosis: emerging applications for greater sustainability. Environ. Sci. Technol. 45, 9824–9830. https://doi.org/10.1021/es202576h.

Husnain, T., Mi, B., Riffat, R., 2015. A combined forward osmosis and membrane distillation system for Sidestream treatment. J. Water Resour. Prot. 7, 1111–1120. https://doi.org/10.4236/jwarp.2015.714091.

Im, S.J., Viet, N.D., Jang, A., 2021. Real-time monitoring of forward osmosis membrane fouling in wastewater reuse process performed with a deep learning model. Chemosphere 130047. https://doi.org/10.1016/j.chemosphere.2021.130047.

Jiao, B., Cassano, A., Drioli, E., 2004. Recent advances on membrane processes for the concentration of fruit juices: a review. J. Food Eng. 63, 303–324. https://doi.org/10.1016/j.jfoodeng.2003.08.003.

Kim, J.E., Phuntsho, S., Lotfi, F., Shon, H.K., 2015. Investigation of pilot-scale 8040 FO membrane module under different operating conditions for brackish water desalination. Desalin. Water Treatment. https://doi.org/10.1080/19443994.2014.931528.

Kim, J., Blandin, G., Phuntsho, S., Verliefde, A., Le-Clech, P., Shon, H., 2017a. Practical considerations for operability of an 8″ spiral wound forward osmosis module: hydrodynamics, fouling behaviour and cleaning strategy. Desalination 404, 249–258. https://doi.org/10.1016/j.desal.2016.11.004.

Kim, Y., Woo, Y.C., Phuntsho, S., Nghiem, L.D., Shon, H.K., Hong, S., 2017b. Evaluation of fertilizer-drawn forward osmosis for coal seam gas reverse osmosis brine treatment and sustainable agricultural reuse. J. Membr. Sci. 537, 22–31. https://doi.org/10.1016/j.memsci.2017.05.032.

Kim, D.I., Gwak, G., Zhan, M., Hong, S., 2019. Sustainable dewatering of grapefruit juice through forward osmosis: improving membrane performance, fouling control, and product quality. J. Membr. Sci. 578, 53–60. https://doi.org/10.1016/j.memsci.2019.02.031.

Kook, S., Lee, C., Nguyen, T.T., Lee, J., Shon, H.K., Kim, I.S., 2018. Serially connected forward osmosis membrane elements of pressure-assisted forward osmosis-reverse osmosis hybrid system: process performance and economic analysis. Desalination. https://doi.org/10.1016/j.desal.2018.09.019.

Kravath, R.E., Davis, J.A., 1975. Desalination of sea water by direct osmosis. Desalination 16, 151–155. https://doi.org/10.1016/S0011-9164(00)82089-5.

Kwon, Y.N., Kim, M.J., Lee, Y.T., 2016. Application of a FO/MD-combined system for the desalination of saline solution. Desalin. Water Treat. 57, 14347–14354. https://doi.org/10.1080/19443994.2015.1066714.

Lampi, K., Shetji, J., 2014. Forward osmosis industrial wastewater treatment: landfill leachate and oil and gas produced waters. In: Proceedings of International Forward Osmosis Summit 2014. Australia, pp. 1–13, https://doi.org/10.13140/RG.2.1.1652.1444.

Larronde-Larretche, M., Jin, X., 2016. Microalgae (*Scenedesmus obliquus*) dewatering using forward osmosis membrane: influence of draw solution chemistry. Algal Res. 15, 1–8. https://doi.org/10.1016/j.algal.2016.01.014.

Larronde-Larretche, M., Jin, X., 2017. Microalgal biomass dewatering using forward osmosis membrane: influence of microalgae species and carbohydrates composition. Algal Res. 23, 12–19. https://doi.org/10.1016/j.algal.2016.12.020.

Lee, S., Boo, C., Elimelech, M., Hong, S., 2010. Comparison of fouling behavior in forward osmosis (FO) and reverse osmosis (RO). J. Membr. Sci. 365, 34–39. https://doi.org/10.1016/j.memsci.2010.08.036.

Lee, C., Chae, S.H., Yang, E., Kim, S., Kim, J.H., Kim, I.S., 2020. A comprehensive review of the feasibility of pressure retarded osmosis: recent technological advances and industrial efforts towards commercialization. Desalination 491, 114501. https://doi.org/10.1016/j.desal.2020.114501.

Lee, C., Nguyen, T.-T., Adha, R.S., Shon, H.K., Kim, I.S., 2020a. Influence of hydrodynamic operating conditions on organic fouling of spiral-wound forward osmosis membranes: fouling-induced performance deterioration in FO-RO hybrid system. Water Res. 185, 116154. https://doi.org/10.1016/j.watres.2020.116154.

Lee, C., Thanh, N., Syamsul, R., Kim, I.S., 2020b. Performance analysis of serially-connected membrane element for pressure- assisted forward osmosis: wastewater reuse and seawater desalination. Desalin. Water Treat. 183, 104–113. https://doi.org/10.5004/dwt.2020.25260.

Li, Q., Xu, Z., Pinnau, I., 2007. Fouling of reverse osmosis membranes by biopolymers in wastewater secondary effluent: role of membrane surface properties and initial permeate flux. J. Membr. Sci. https://doi.org/10.1016/j.memsci.2006.12.027.

Li, J., Niu, A., Lu, C., Zhang, J., Junaid, M., Strauss, P.R., Xiao, P., Wang, X., Ren, Y., Pei, D., 2017. A novel forward osmosis system in land fi ll leachate treatment for removing polycyclic aromatic hydrocarbons and for direct fertigation. Chemosphere 168, 112–121. https://doi.org/10.1016/j.chemosphere.2016.10.048.

Li, X., Chung, T.S., 2013. Effects of free volume in thin-film composite membranes on osmotic power generation. Sep. Mater. Devices Processes 59 (12), 4749–4761. https://doi.org/10.1002/aic.14217.

Li, J.Y., Ni, Z.Y., Zhou, Z.Y., Hu, Y.X., Xu, X.H., Cheng, L.H., 2018a. Membrane fouling of forward osmosis in dewatering of soluble algal products: comparison of TFC and CTA membranes. J. Membr. Sci. 552, 213–221. https://doi.org/10.1016/j.memsci.2018.02.006.

Li, F., Xia, Q., Gao, Y., Cheng, Q., Ding, L., Yang, B., Tian, Q., Ma, C., Sand, W., Liu, Y., 2018b. Anaerobic biodegradation and decolorization of a refractory acid dye by a forward osmosis membrane bioreactor. Environ. Sci. Water Res. Technol. 4, 272–280. https://doi.org/10.1039/c7ew00400a.

Ling, M.M., Chung, T.S., 2011. Desalination process using super hydrophilic nanoparticles via forward osmosis integrated with ultrafiltration regeneration. Desalination 278, 194–202. https://doi.org/10.1016/j.desal.2011.05.019.

Luo, W., Hai, F.I., Price, W.E., Elimelech, M., Nghiem, L.D., 2016a. Evaluating ionic organic draw solutes in osmotic membrane bioreactors for water reuse. J. Membr. Sci. 514, 636–645. https://doi.org/10.1016/j.memsci.2016.05.023.

Luo, W., Hai, F.I., Price, W.E., Guo, W., Ngo, H.H., Yamamoto, K., Nghiem, L.D., 2016b. Phosphorus and water recovery by a novel osmotic membrane bioreactor-reverse osmosis system. Bioresour. Technol. 200, 297–304. https://doi.org/10.1016/j.biortech.2015.10.029.

Luo, W., Xie, M., Hai, F.I., Price, W.E., Nghiem, L.D., 2016c. Biodegradation of cellulose triacetate and polyamide forward osmosis membranes in an activated sludge bioreactor: observations and implications. J. Membr. Sci. 510, 284–292. https://doi.org/10.1016/j.memsci.2016.02.066.

Luo, W., Phan, H.V., Li, G., Hai, F.I., Price, W.E., Elimelech, M., Nghiem, L.D., 2017a. An osmotic membrane bioreactor-membrane distillation system for simultaneous wastewater reuse and seawater desalination: performance and implications. Environ. Sci. Technol. 51, 14311–14320. https://doi.org/10.1021/acs.est.7b02567.

Luo, W., Phan, H.V., Xie, M., Hai, F.I., Price, W.E., Elimelech, M., Nghiem, L.D., 2017b. Osmotic versus conventional membrane bioreactors integrated with reverse osmosis for water reuse: biological stability, membrane fouling, and contaminant removal. Water Res. 109, 122–134. https://doi.org/10.1016/j.watres.2016.11.036.

Lutchmiah, K., Verliefde, A.R.D., Roest, K., Rietveld, L.C., Cornelissen, E.R., 2014. Forward osmosis for application in wastewater treatment: a review. Water Res. 58, 179–197. https://doi.org/10.1016/j.watres.2014.03.045.

Manzoor, K., Khan, S.J., Khan, A., Abbasi, H., Zaman, W.Q., 2022. Woven-fiber microfiltration couple with anaerobic forward osmosis membrane bioreactor treating textile wastewater: use of fertilizer draw solutes for direct fertigation. Biochem. Eng. J. 181, 108385. https://doi.org/10.1016/j.bej.2022.108385.

Matin, A., Laoui, T., Falath, W., Farooque, M., 2021. Fouling control in reverse osmosis for water desalination & reuse: current practices & emerging environment-friendly technologies. Sci. Total Environ. 765, 142721. https://doi.org/10.1016/j.scitotenv.2020.142721.

McCutcheon, J.R., McGinnis, R.L., Elimelech, M., 2005. A novel ammonia-carbon dioxide forward (direct) osmosis desalination process. Desalination 174, 1–11. https://doi.org/10.1016/j.desal.2004.11.002.

Mi, B., Elimelech, M., 2010. Organic fouling of forward osmosis membranes: fouling reversibility and cleaning without chemical reagents. J. Membr. Sci. 348, 337–345. https://doi.org/10.1016/j.memsci.2009.11.021.

Nayak, C.A., Valluri, S.S., Rastogi, N.K., 2011. Effect of high or low molecular weight of components of feed on transmembrane flux during forward osmosis. J. Food Eng. 106, 48–52. https://doi.org/10.1016/j.jfoodeng.2011.04.006.

Nguyen, H.T., Nguyen, N.C., Chen, S.-S., Ngo, H.H., Guo, W., Li, C.-W., 2015. A new class of draw solutions for minimizing reverse salt flux to improve forward osmosis desalination. Sci. Total Environ. 538, 129–136.

Nguyen, N.C., Chen, S.S., Nguyen, H.T., Ray, S.S., Ngo, H.H., Guo, W., Lin, P.H., 2016a. Innovative sponge-based moving bed-osmotic membrane bioreactor hybrid system using a new class of draw solution for municipal wastewater treatment. Water Res. 91, 305–313. https://doi.org/10.1016/j.watres.2016.01.024.

Nguyen, N.C., Nguyen, H.T., Chen, S.-S., Ngo, H.H., Guo, W., Chan, W.H., Ray, S.S., Li, C.-W., Hsu, H.-T., 2016b. A novel osmosis membrane bioreactor-membrane distillation hybrid system for wastewater treatment and reuse. Bioresour. Technol. 209, 8–15. https://doi.org/10.1016/j.biortech.2016.02.102.

Nguyen, T., Kook, S., Lee, C., Field, R.W., Kim, I.S., 2019. Critical flux-based membrane fouling control of forward osmosis : behavior, sustainability, and reversibility. J. Membr. Sci. 570–571, 380–393. https://doi.org/10.1016/j.memsci.2018.10.062.

Ning, R.Y., Troyer, T.L., 2007. Colloidal fouling of RO membranes following MF/UF in the reclamation of municipal wastewater. Desalination 208 (1), 232–237.

Ortega-Bravo, J.C., Ruiz-Filippi, G., Donoso-Bravo, A., Reyes-Caniupán, I.E., Jeison, D., 2016. Forward osmosis: evaluation thin-film-composite membrane for municipal sewage concentration. Chem. Eng. J. 306, 531–537. https://doi.org/10.1016/j.cej.2016.07.085.

Pathak, N., Li, S., Kim, Y., Chekli, L., Phuntsho, S., Jang, A., 2018. Assessing the removal of organic micropollutants by a novel baffled osmotic membrane bioreactor-microfiltration hybrid system. Bioresour. Technol. 262, 98–106. https://doi.org/10.1016/j.biortech.2018.04.044.

Petrotos, K.B., Quantick, P., Petropakis, H., 1998. A study of the direct osmotic concentration of tomato juice in tubular membrane—module configuration. I. The effect of certain basic process parameters on the process performance. J. Membr. Sci. 150, 99–110. https://doi.org/10.1016/S0376-7388(98)00216-6.

Petrotos, K.B., Tsiadi, A.V., Poirazis, E., Papadopoulos, D., Petropakis, H., Gkoutsidis, P., 2010. A description of a flat geometry direct osmotic concentrator to concentrate tomato juice at ambient temperature and low pressure. J. Food Eng. 97, 235–242. https://doi.org/10.1016/j.jfoodeng.2009.10.015.

Phuntsho, S., Shon, H.K., Hong, S., Lee, S., Vigneswaran, S., 2011. A novel low energy fertilizer driven forward osmosis desalination for direct fertigation: evaluating the performance of fertilizer draw solutions. J. Membr. Sci. 375, 172–181. https://doi.org/10.1016/j.memsci.2011.03.038.

Phuntsho, S., Shon, H.K., Majeed, T., El Saliby, I., Vigneswaran, S., Kandasamy, J., Hong, S., Lee, S., 2012. Blended fertilizers as draw solutions for fertilizer-drawn forward osmosis desalination. Environ. Sci. Technol. 46, 4567–4575. https://doi.org/10.1021/es300002w.

Phuntsho, S., Hong, S., Elimelech, M., Shon, H.K., 2013. Forward osmosis desalination of brackish groundwater: meeting water quality requirements for fertigation by integrating nanofiltration. J. Membr. Sci. 436, 1–15. https://doi.org/10.1016/j.memsci.2013.02.022.

Phuntsho, S., Lotfi, F., Hong, S., Shaffer, D.L., Elimelech, M., Shon, H.K., 2014. Membrane scaling and flux decline during fertiliser-drawn forward osmosis desalination of brackish groundwater. Water Res. 57, 172–182. https://doi.org/10.1016/j.watres.2014.03.034.

Phuntsho, S., Kim, J.E., Johir, M.A.H., Hong, S., Li, Z., Ghaffour, N., Leiknes, T.O., Shon, H.K., 2016. Fertiliser drawn forward osmosis process: pilot-scale desalination of mine impaired water for fertigation. J. Membr. Sci. https://doi.org/10.1016/j.memsci.2016.02.024.

Praveen, P., Nguyen, D.T.T., Loh, K.C., 2015. Biodegradation of phenol from saline wastewater using forward osmotic hollow fiber membrane bioreactor coupled chemostat. Biochem. Eng. J. 94, 125–133. https://doi.org/10.1016/j.bej.2014.11.014.

Qasim, M., Darwish, N.A., Sarp, S., Hilal, N., 2015. Water desalination by forward (direct) osmosis phenomenon: a comprehensive review. Desalination 374, 47–69. https://doi.org/10.1016/j.desal.2015.07.016.

Qiu, G., Ting, Y.P., 2014a. Direct phosphorus recovery from municipal wastewater via osmotic membrane bioreactor (OMBR) for wastewater treatment. Bioresour. Technol. 170, 221–229. https://doi.org/10.1016/j.biortech.2014.07.103.

Qiu, G., Ting, Y.P., 2014b. Short-term fouling propensity and flux behavior in an osmotic membrane bioreactor for wastewater treatment. Desalination 332, 91–99. https://doi.org/10.1016/j.desal.2013.11.010.

Qiu, G., Law, Y.M., Das, S., Ting, Y.P., 2015. Direct and complete phosphorus recovery from municipal wastewater using a hybrid microfiltration-forward osmosis membrane bioreactor process with seawater brine as draw solution. Environ. Sci. Technol. 49, 6156–6163. https://doi.org/10.1021/es504554f.

Ramírez, C.A., Patel, M., Blok, K., 2006. From fluid milk to milk powder: energy use and energy efficiency in the European dairy industry. Energy 31, 1984–2004. https://doi.org/10.1016/j.energy.2005.10.014.

Ruengruehan, K., Kim, H., Hai Yen, L.T., Jang, A., Lee, W., Kang, S., 2016. Fatty acids fouling on forward osmosis membrane: impact of pH. Desalin. Water Treat. 57, 7531–7537. https://doi.org/10.1080/19443994.2015.1030118.

Ryu, H., Kim, K., Cho, H., Park, E., Chang, Y.K., Han, J.I., 2020. Nutrient-driven forward osmosis coupled with microalgae cultivation for energy efficient dewatering of microalgae. Algal Res. 48, 101880. https://doi.org/10.1016/j.algal.2020.101880.

Sant'Anna, V., Marczak, L.D.F., Tessaro, I.C., 2012. Membrane concentration of liquid foods by forward osmosis: process and quality view. J. Food Eng. 111, 483–489. https://doi.org/10.1016/j.jfoodeng.2012.01.032.

Sant'Anna, V., Gurak, P.D., de Vargas, N.S., da Silva, M.K., Marczak, L.D.F., Tessaro, I.C., 2016. Jaboticaba (*Myrciaria jaboticaba*) juice concentration by forward osmosis. Sep. Sci. Technol. 51, 1708–1715. https://doi.org/10.1080/01496395.2016.1168845.

Schuck, P., Jeantet, R., Tanguy, G., Méjean, S., Gac, A., Lefebvre, T., Labussière, E., Martineau, C., 2015. Energy consumption in the processing of dairy and feed powders by evaporation and drying. Drying Technol. 33, 176–184. https://doi.org/10.1080/07373937.2014.942913.

Shaffer, D.L., Werber, J.R., Jaramillo, H., Lin, S., Elimelech, M., 2014. Forward osmosis: where are we now? Desalination 356, 271–284. https://doi.org/10.1016/j.desal.2014.10.031.

She, Q., Jin, X., Tang, C.Y., 2012. Osmotic power production from salinity gradient resource by pressure retarded osmosis: effects of operating conditions and reverse solute diffusion. J. Membr. Sci. 401–402, 262–273. https://doi.org/10.1016/j.memsci.2012.02.014.

Shibuya, M., Yasukawa, M., Takahashi, T., Miyoshi, T., Higa, M., Matsuyama, H., 2015. Effects of operating conditions and membrane structures on the performance of hollow fiber forward osmosis membranes in pressure assisted osmosis. Desalination 365, 381–388. https://doi.org/10.1016/j.desal.2015.03.005.

Song, M., Im, S.J., Jeong, S., Jang, A., 2018. Evaluation of an element-scale plate-type forward osmosis: effect of structural parameters and operational conditions. Desalination. https://doi.org/10.1016/j.desal.2017.12.010.

Straub, A.P., Yip, N.Y., Elimelech, M., 2013. Raising the Bar: increased hydraulic pressure allows unprecedented high power densities in pressure-retarded osmosis. Environ. Sci. Technol. Lett. 1, 55–59. https://doi.org/10.1021/ez400117d.

Suh, C., Lee, S., 2013. Modeling reverse draw solute flux in forward osmosis with external concentration polarization in both sides of the draw and feed solution. J. Membr. Sci. 427, 365–374. https://doi.org/10.1016/j.memsci.2012.08.033.

Tang, C.Y., She, Q., Lay, W.C.L., Wang, R., Fane, A.G., 2010. Coupled effects of internal concentration polarization and fouling on flux behavior of forward osmosis membranes during humic acid filtration. J. Membr. Sci. 354, 123–133. https://doi.org/10.1016/j.memsci.2010.02.059.

Valladares Linares, R., Li, Z., Yangali-Quintanilla, V., Ghaffour, N., Amy, G., Leiknes, T., Vrouwenvelder, J.S., 2016. Life cycle cost of a hybrid forward osmosis—low pressure reverse osmosis system for seawater desalination and wastewater recovery. Water Res. 88, 225–234. https://doi.org/10.1016/j.watres.2015.10.017.

Viet, N.D., Jang, A., 2022. Fertilizer draw solution index in osmotic membrane bioreactor for simultaneous wastewater treatment and sustainable agriculture. Chemosphere 296, 134002. https://doi.org/10.1016/j.chemosphere.2022.134002.

Viet, N.D., Cho, J., Yoon, Y., Jang, A., 2019. Enhancing the removal efficiency of osmotic membrane bioreactors: a comprehensive review of influencing parameters and hybrid configurations. Chemosphere 236, 124363. https://doi.org/10.1016/j.chemosphere.2019.124363.

Volpin, F., Fons, E., Chekli, L., Kim, J.E., Jang, A., Shon, H.K., 2018. Hybrid forward osmosis-reverse osmosis for wastewater reuse and seawater desalination: understanding the optimal feed solution to minimise fouling. Process. Saf. Environ. Prot. 117, 523–532. https://doi.org/10.1016/j.psep.2018.05.006.

Wang, W., Zhang, Y., Mariem, E.-A., Wang, X., Yang, H.Y., Xie, Y., 2014b. Effects of pH and temperature on forward osmosis membrane flux using rainwater as the makeup for cooling water dilution. Desalination 351, 70–76. https://doi.org/10.1016/j.desal.2014.07.025.

Wang, X., Zhao, Y., Yuan, B., Wang, Z., Li, X., Ren, Y., 2016. Comparison of biofouling mechanisms between cellulose triacetate (CTA) and thin-film composite (TFC) polyamide forward osmosis membranes in osmotic membrane bioreactors. Bioresour. Technol. 202, 50–58. https://doi.org/10.1016/j.biortech.2015.11.087.

Wang, C., Gao, B., Zhao, P., Li, R., Yue, Q., Shon, H.K., 2017a. Exploration of polyepoxysuccinic acid as a novel draw solution in the forward osmosis process. RSC Adv. 7, 30687–30698. https://doi.org/10.1039/c7ra04036a.

Wang, J., Pathak, N., Chekli, L., Phuntsho, S., Kim, Y., Li, D., Shon, H.K., 2017b. Performance of a novel fertilizer-drawn forward osmosis aerobic membrane bioreactor (FDFO-MBR): mitigating salinity build-up by integrating microfiltration. Water (Switzerland) 9, 1–13. https://doi.org/10.3390/w9010021.

Wang, X., Wang, C., Tang, C.Y., Hu, T., Li, X., Ren, Y., 2017c. Development of a novel anaerobic membrane bioreactor simultaneously integrating microfiltration and forward osmosis membranes for low-strength wastewater treatment. J. Membr. Sci. 527, 1–7. https://doi.org/10.1016/j.memsci.2016.12.062.

Wang, H., Zhang, Y., Ren, S., Pei, J., Li, Z., 2022. Athermal concentration of apple juice by forward osmosis: process performance and membrane fouling propensity. Chem. Eng. Res. Des. 177, 569–577. https://doi.org/10.1016/j.cherd.2021.11.023.

Wang, Y.N., Goh, K., Li, X., Setiawan, L., Wang, R., 2018. Membranes and processes for forward osmosis-based desalination: recent advances and future prospects. Desalination 434, 81–99. https://doi.org/10.1016/j.desal.2017.10.028.

Wei, J., Li, Y., Setiawan, L., Wang, R., 2016. Influence of macromolecular additive on reinforced flat-sheet thin film composite pressure-retarded osmosis membranes. J. Membr. Sci. 511, 54–64. https://doi.org/10.1016/j.memsci.2016.03.046.

Xie, M., Nghiem, L.D., Price, W.E., Elimelech, M., 2013. A forward osmosis-membrane distillation hybrid process for direct sewer mining: system performance and limitations. Environ. Sci. Technol. 47 (23), 13486–13493. https://doi.org/10.1021/es404056e.

Xie, M., Nghiem, L.D., Price, W.E., Elimelech, M., 2014. Relating rejection of trace organic contaminants to membrane properties in forward osmosis: measurements, modelling and implications. Water Res. 49, 265–274. https://doi.org/10.1016/j.watres.2013.11.031.

Yang, Y., Song, H., He, Z., 2019. Mitigation of solute buildup by using a biodegradable and reusable polyelectrolyte as a draw solute in an osmotic membrane bioreactor. Environ. Sci. Water Res. Technol. 5 (1), 19–27. https://doi.org/10.1039/c8ew00556g.

Ye, J., Zhou, Q., Zhang, X., Hu, Q., 2018. Microalgal dewatering using a polyamide thin film composite forward osmosis membrane and fouling mitigation. Algal Res. 31, 421–429. https://doi.org/10.1016/j.algal.2018.02.003.

Yip, N.Y., Elimelech, M., 2014. Comparison of energy efficiency and power density in pressure retarded osmosis and reverse electrodialysis. Environ. Sci Technol. 48 (18), 11002–11012. https://doi.org/10.1021/es5029316.

Yu, Y., Lee, S., Maeng, S.K., 2017. Forward osmosis membrane fouling and cleaning for wastewater reuse. J. Water Reuse Des. 7 (2), 111–120. https://doi.org/10.2166/wrd.2016.023.

Zhang, X., Ning, Z., Wang, D.K., Costa, J.C.D.D., 2013. A novel ethanol dehydration process by forward osmosis. Chem. Eng. J. 232, 397–404. https://doi.org/10.1016/j.cej.2013.07.106.

Zhang, X., Ning, Z., Wang, D.K., Costa, J.C.D.D., 2014. Processing municipal wastewaters by forward osmosis using CTA membrane. J. Membr. Sci. 468, 269–275. https://doi.org/10.1016/j.memsci.2014.06.016.

Zhu, S., Li, M., Gamal El-Din, M., 2018. The roles of pH and draw solute on forward osmosis process treating aqueous naphthenic acids. J. Membr. Sci. 549, 456–465. https://doi.org/10.1016/j.memsci.2017.12.029.

Application of forward osmosis membrane technology in nutrient recovery and water reuse

*Hau Thi Nguyen[a], Nguyen Cong Nguyen[a], Shiao-Shing Chen[b],
Huu Hao Ngo[c], Xuan-Thanh Bui[d,e],
and Phuong-Thao Nguyen[d,e]*

[a]Faculty of Chemistry and Environment, Dalat University, Dalat, Vietnam [b]Institute of
Environmental Engineering and Management, National Taipei University of Technology, Taipei,
Taiwan, Republic of China [c]Centre for Technology in Water and Wastewater, School of Civil and
Environmental Engineering, University of Technology Sydney, Sydney, NSW, Australia
[d]Key Laboratory of Advanced Waste Treatment Technology & Faculty of Environment and
Natural Resources, Ho Chi Minh City University of Technology (HCMUT), Ho Chi Minh City,
Vietnam [e]Vietnam National University Ho Chi Minh City (VNU-HCM), Ho Chi Minh City,
Vietnam

1. Introduction

Forward osmosis (FO) is a physical phenomenon that allows water to permeate through semipermeable membranes due to the gradient of osmotic pressure between the draw solution and feed solution. Many types of wastewaters (domestic or industrial) are used as the feed solution and the high concentration of organic and inorganic salts is used as the draw solution which can generate high osmotic pressure (Hau et al., 2014; Nguyen et al., 2016). In general, polymeric materials are used to produce the semipermeable FO membrane whereby only water can pass through; meanwhile, trace organic, virus, and heavy metal ions are retained on the membrane (Alihemati et al., 2020), as shown in Fig. 1.

As compared to the hydraulic pressure-driven reverse osmosis (RO), the osmotic pressure-driven process has lower membrane fouling (Jin et al., 2012; Nguyen et al., 2013). As

Copyright © 2023 Elsevier Inc. All rights reserved.

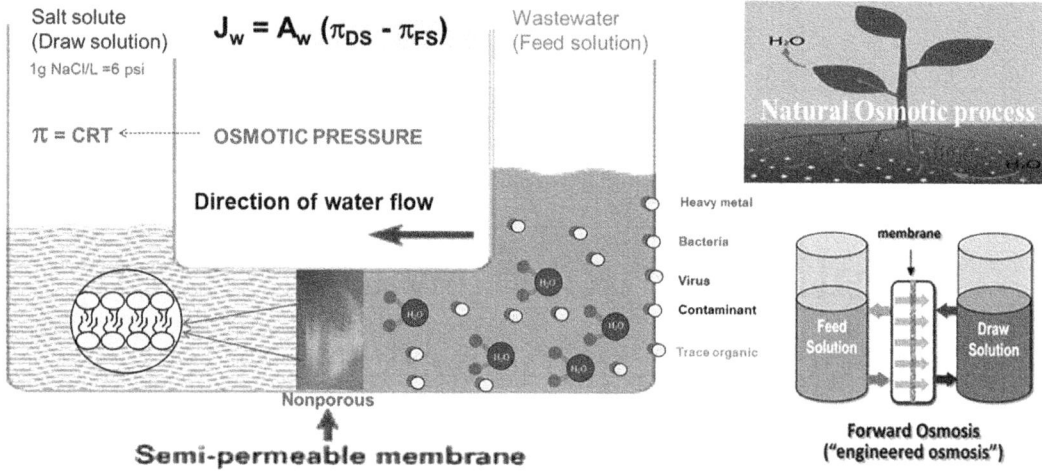

FIG. 1   Forward osmosis process concept.

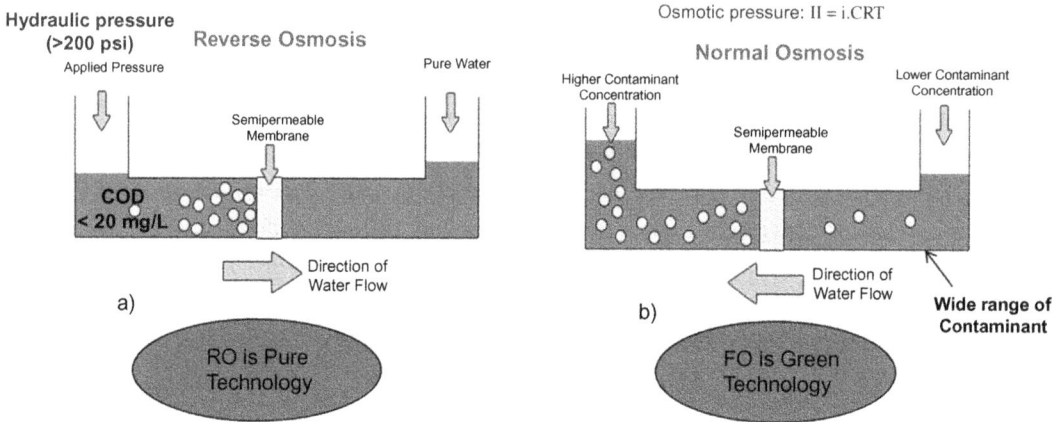

FIG. 2   Different operational mechanisms between (A) RO and (B) FO.

illustrated in Fig. 2, hydraulic pressure is not required during the FO operation; hence, the manufacture of FO modules is simple and cheap. Moreover, the investment cost of the FO process is lower than that of the RO. Clearly, the less membrane fouling of FO makes it a perfect technology to be integrated with biological processes, which are destined for wastewater treatment with high organic concentration. However, a posttreatment technology should be coupled with the FO process to recover the diluted draw solution for reuse and produce purified water (Ricci et al., 2021; Nguyen et al., 2015).

The relationships between the permeate water flux (J_w), different hydraulic pressure (Δp), and different osmotic pressure ($\Delta \pi$) and in the RO and FO processes are calculated as follows (Lonsdale et al., 1965; Baker, 2004):

$$J_w = A \cdot (\Delta p - \Delta \pi) \tag{1}$$

TABLE 1 The differences between RO and FO technology.

Factor	Forward osmosis (FO)	Reverse osmosis (RO)
Technology	Green technology	Pure technology
Driving force	Osmotic pressure	External hydraulic pressure
Field	Water treatment Food processing Fertilization Desalination	Water treatment Desalination
Operating conditions	$P \sim$ atmospheric, pH 6–11 Draw solution: Seawater, inorganic and organic salt Feed solution: Impaired water, seawater	$P \sim$ 10–70 bar, pH 6–7 Feedwater: Brackish and seawater
Membrane property		
Physical morphology	Membrane with low torturous sublayer and dense active layer	Membrane with low torturous sublayer and dense active layer Good mechanical stability
Membrane requirement	High water flux High solute removal	High water flux Robust for high-pressure operation
Target performance	High contaminant removal (>95%) High recovery	High contaminant removal (>95%) Low water recovery
Challenges	ICP Recovery of draw solution	Energy consumption Operating cost

where A is the membrane water permeability (L/m^2 h bar). In the RO process, a Δp higher than $\Delta \pi$ is applied to the feed solution to induce the transfer of water from the feed solute to the permeate. In the FO process, Δp is zero and water diffuses from the feed solution to the higher osmotic pressure draw solution. In one variant process of FO, a positive Δp smaller than $\Delta \pi$ is applied on the draw solute side to induce work when water transfers from the feed solution to the draw solution. The differences between FO and RO are summarized in Table 1.

2. Key factors affecting the performance of FO

The permeation of water across the FO membrane could be affected by the characteristics of the feed and draw solutions and by the properties and orientations of the FO membrane, as shown in Figs. 3 and 4. In FO-based processes, the concentration polarization (CP) phenomenon from the feed and draw solutions has a significant impact on water recovery efficiency because the CP appearing on the FO membrane raises the osmotic pressure at the active layer, leading to prevention of the permeate flow. Hence, there have been many studies on the structure of the FO membrane to decrease the impact of ICP (Loeb et al., 1997; McCutcheon and Elimelech, 2006, 2007). Moreover, recent studies have developed new draw solutions with high osmotic pressures and relatively easy recovery (McCutcheon et al., 2005).

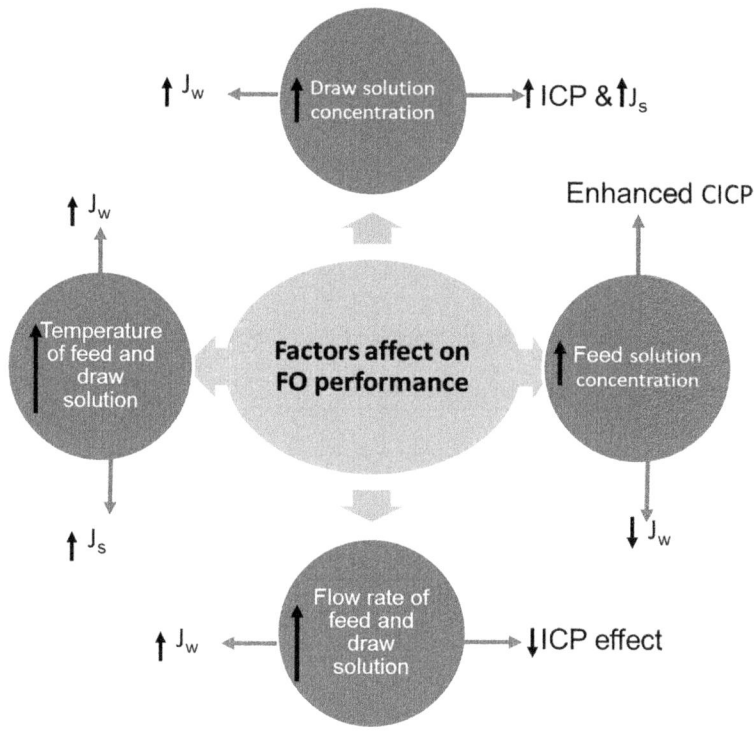

FIG. 3 The effect of operational condition on FO performance (CICP: concentrative internal concentration polarization).

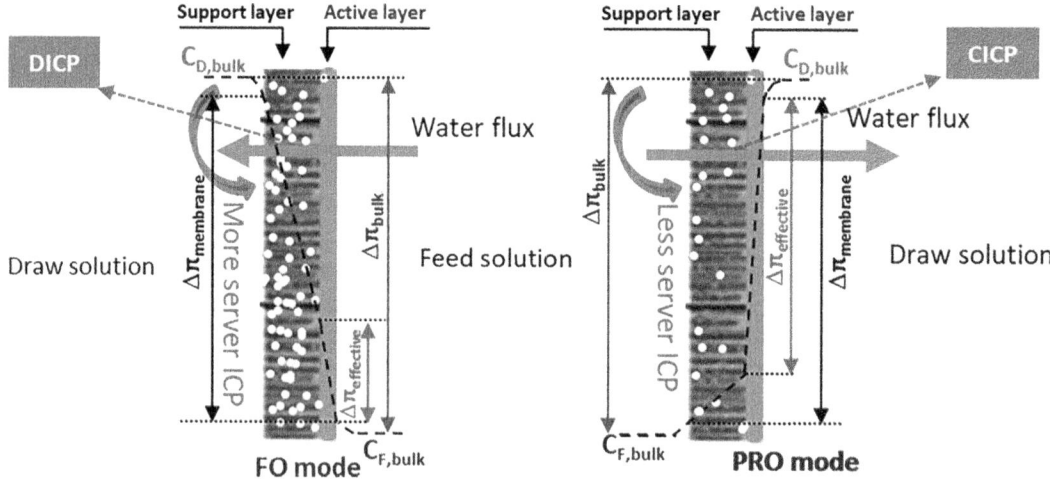

FIG. 4 The effect of FO and the PRO model on FO performance (DICP: diluted internal concentration polarization).

2.1 Membrane type and membrane orientation

The pollutant removal and water recovery efficiency of FO-based systems depend on the properties of the FO membrane. A suitable FO membrane should simultaneously have characteristics including low salt transport, high resistance to a wide range of pH, high water permeability, and good chemical stability (Chekli et al., 2016; Shon et al., 2015). Basically, there are two kinds of FO membranes that are often used in the FO-based system, namely the thin-film composite (TFC) polyamide FO and cellulose triacetate (CTA) FO membranes. The properties of the TFC FO and CTA FO membranes are shown in Table 2.

As compared to the TFC FO membranes, the CTA FO membranes were more popular in the hybrid FO-based systems. The main reason is that the chemical stability of the CTA FO membrane is better than that of the TFC membrane. Nguyen et al. (2020) employed the CTA-ES membrane in the FO-based hybrid process and achieved an average water flux of 4.01 LMH during the 45-day anaerobic osmotic membrane bioreactor (AnOsMBR) operation. TFC FO membranes have been developed more recently and TFC FO membranes presented higher efficiency in removing micropollutant compounds as compared to CTA membrane because the active layer of TFC FO membrane was more negatively charged than that of the CTA FO membrane.

Li et al. (2017) recently combined the aquaporin protein with the active layer of the TFC FO membrane to produce new aquaporin-based TFC FO membranes. The previous research showed that aquaporin-based TFC FO membranes could achieve higher permeate flux than conventional FO CTA or TFC membranes (2015). The reason is that the active layer of the CTA membrane is less hydrophilic than that of the aquaporin FO membrane. Luo et al. (2018) studied the performance of OsMBR when using the conventional CTA FO membrane and

TABLE 2 Properties of forward osmosis membranes.

Factor	FO CTA-ES	FO CTA-NW	FO TFC-ES
Pore size	0.37 nm (Xie et al., 2014; Kazner et al., 2014)	0.37 nm (Xie et al., 2014; Kazner et al., 2014)	0.37 nm (Tiraferri et al., 2011)
Water flux in FO mode	$9 L/m^2 h$	$4 L/m^2 h$	$20 L/m^2 h$
Contact angle	78 degrees	57 degrees	70 degrees
Maximum operating temperature	71 °C	71 °C	71 °C
Maximum chlorine	2 ppm	2 ppm	<0.1 ppm
Thickness (μm)	40 μm (Cath et al., 2006)	<50 μm (Cath et al., 2006)	<50 μm (Cartinella et al., 2006)
S value (mm)	480 μm (Kong et al., 2014)	700 μm (Kong et al., 2014)	–
pH range	3–8	3–8	2–11

aquaporin FO membrane and a higher water flux ($J_w = 15.6$ LMH) was obtained with the aquaporin FO membrane as compared to the CTA FO membrane ($J_w = 5.5$ LMH). Regarding salt leakage, the aquaporin FO membrane has lower reverse salt flux than both the CTA and TFC FO membranes. In terms of nutrient and organic removal, conventional FO and aquaporin-based membranes in the OsMBR system obtained a high rejection of NH_4^+-N and TOC at 90% and 80%, respectively (Wang et al., 2016).

Besides, the membrane orientations also have a significant effect on the FO performance. Jin et al. (2011) demonstrated that the feed solution facing the active layer achieved a better rejection than the draw solution facing the active layer when using CTA membranes. The active layer was faced with draw solute orientation, however, recorded a higher rejection of NH_4^+ than the feed solution facing the active layer when using TFC membrane. The reason is that the TFC membrane exhibits low retention of NH_4^+, meanwhile, the active layer facing with draw solution orientation shows a higher water flux which creates a more diluted draw solution. Moreover, the highly negative charge at the active layer on the TFC membrane attracted more NH_4^+ cations in the feed solution, when membrane orientation was the active layer facing with feed solution, consequently, a lower retention of NH_4^+. In contrast with NH_4^+ cations, NO_2^- and NO_3^- anions were rejected with higher efficiency by the TFC FO membrane as compared with the CTA FO membrane. The reason is the stronger electrostatic repulsion between the negative charge TFC membrane and negative charge of NO_2^- and NO_3^- anions.

2.2 The influence of the draw solution on FO-based performance

The main point of a successful hybrid FO-based application is a suitable draw solution. In FO-based systems, draw solutions affect significantly on bacterial activity, water recovery, and operational cost. Basically, a good draw solution must have high solubility, high osmotic pressure, high charge, and large particle size to reduce salt accumulation, nontoxic to bacterial activity and easy recovery (Ricci et al., 2021; Nguyen et al., 2015, 2016, 2020; Tan and Ng, 2010), as shown in Fig. 5.

Among single-inorganic salts, NaCl was used as a draw solution in many FO-based studies since it is cheap. However, Nguyen et al. (2013) showed that the high salt leakage of NaCl raised salt accumulation in the reactor; consequently, it affected bacteria performance since most bacteria were more sensitive under conditions of high salinity (Moussa et al., 2006; Osaka et al., 2008). Qiu and Ting (2014) recorded that the conductivity concentration in the bioreactor rose to 33 mS/cm within the 20-day OsMBR operation when using NaCl as a draw solute. This phenomenon could explain that the hydrated radii of Cl^- (0.33 nm) and Na^+ (0.36 nm) were small and easy to transfer through the membranes (Nguyen et al., 2013). Uygur and Kargi (Dinçer and Kargi, 2001; Uygur and Kargi, 2004) demonstrated that the removal efficiency of PO_4^{3-} and NH_4^+ decreased to 62% and 20%, respectively, as the concentration of salinity in the bioreactor was 5%.

To reduce salt accumulation in hybrid FO-based systems, organic draw solutions were explored in many studies. Nguyen et al. (2015) and Bowden et al. (2012) investigated the effect of organic draw solutions on the hybrid FO-based performance. Yang et al. (2014) demonstrated that polyacrylic acid (PAA) used as a draw solution in the FO process could reduce salt

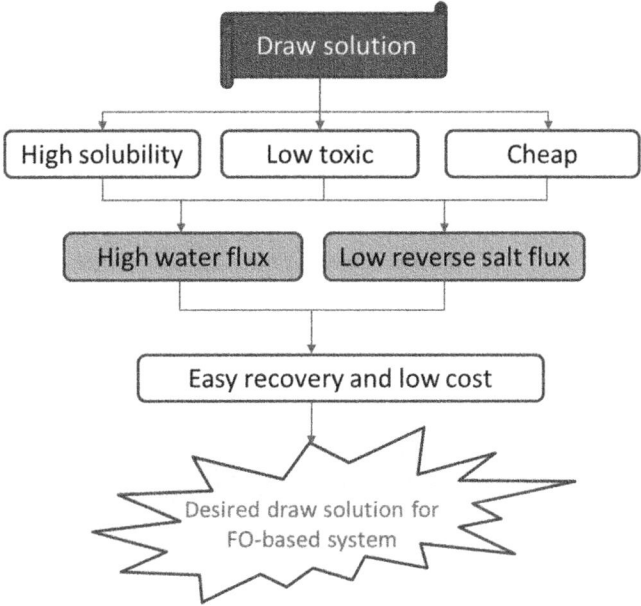

FIG. 5 Selection of the desired draw solution for the FO-based system.

leakage (98.7%), increase nutrient removal (58.4%), and achieve a higher water recovery as compared with the NaCl draw solution. The reason is that the low reverse salt flux of the organic draw solution could enhance FO performance significantly (Nawaz et al., 2013; Bowden et al., 2012).

Although organic salts are considered as potential draw solutes, their high biodegradation and high replenishment cost prevent their application in FO-based systems (Shon et al., 2015). To solve this problem, Nguyen et al. (2020) explored a mixture of Na_3PO_4 and EDTA-2Na as the draw solution in hybrid FO-based systems for treating municipal wastewater. The result recorded the lowest specific reverse salt flux ($J_s/J_w = 0.08\,g/L$) of the mixed $0.2\,M$ Na_3PO_4/$0.25\,M$ EDTA-2Na draw solution. The reason is that the size of high valent charge such as $HEDTA^{3-}$, $NaEDTA^{3-}$, HPO_4^{2-} in the mixed draw solution is large, leading to a decline in reverse salt flux. In addition, the mixed high charge draw solution could produce a high water flux because the predominant compositions in the mixed draw solution at pH 8 are $HEDTA^{3-}$, $NaEDTA^{3-}$, HPO_4^{2-} ions; consequently, high index of ideal Van't Hoff and high water flux. Clearly, a mixed Na_3PO_4 and EDTA-2Na was an appropriate draw solution for hybrid FO-based systems to reduce conductivity in the reactor.

2.3 Membrane fouling and scaling

Generally, the FO-based technology is considered to lower scaling and fouling more than other technologies [MBR, reverse osmotic (RO), membrane distillation (MD)] since it uses osmotic pressure as a driving force instead of hydraulic pressure (Ricci et al., 2021). However, under unfavorable conditions, scaling and fouling present an important role in the long-term hybrid FO-based operation, such as a reduced membrane life span, decreased water quality

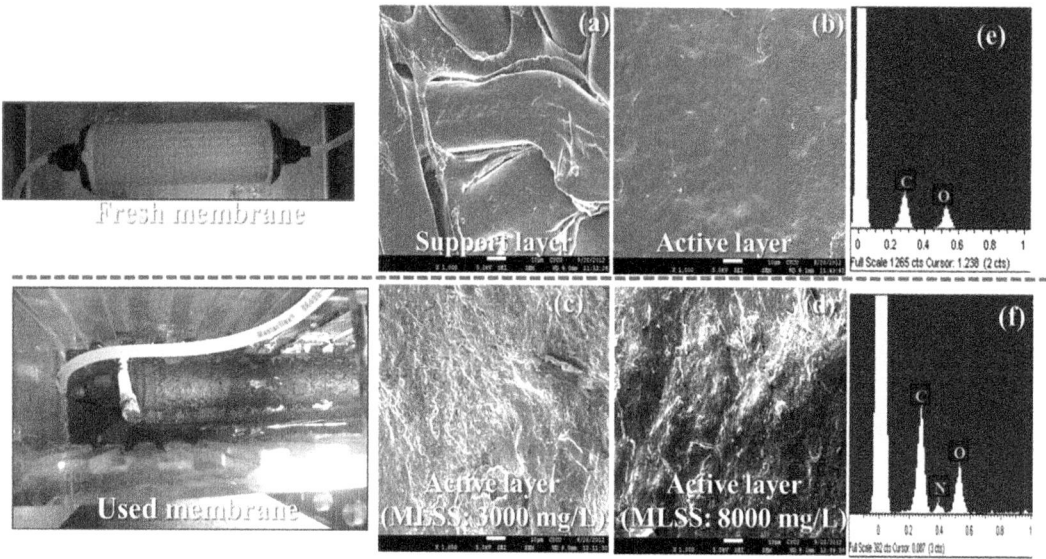

FIG. 6   Membrane fouling in the FO-based system: SEM graphs of (A) original support layer membrane; (B) original active layer of the membrane; (C) fouled layer of the used membrane with low sludge concentration; (D) fouled layer of the used membrane with high sludge concentration.

(Chang et al., 2019; Qiu and Ting, 2014), as shown in Fig. 6. Crystallization and particulate fouling are the two main mechanisms playing important roles during scale formation on FO membrane surfaces. Previous studies found that Mg^{2+}, Ca^{2+}, CO_3^{2-}, and SO_4^{2-} were the main components leading to scaling of the FO membrane. In the OsMBR system, there have been some scalants commonly presented in the draw and feed solution of the OsMBR system including $CaSO_4$, $CaCO_3$, SiO_2, and $Ca_3(PO_4)_2$.

In addition, scalant cake layers may also enhance to attract biological and organic materials; consequently, more severe membrane fouling as shown in Fig. 6. Zhu et al. (2018) demonstrated that the reason for the reduced water flux is membrane fouling in three steps. Firstly, foulants are deposited on the FO membrane surface; then fouling layers are formed on the surface of the FO membrane. Finally, biofilm is condensed on the surface of the FO membrane. Achilli et al. (2009) showed that approximately 38% of the water flux declined in the OMBR system due to membrane fouling when using the CTA FO membrane. In addition, Li et al. (2016) demonstrated that the fouling phenomenon in the OsMBR system also reduced the water flux by 45% when using the TFC membrane. Li et al. (2016) also recorded that the raising of specific reverse salt flux in OsMBR is due to fouling layers on the FO membrane surfaces. Basically, the higher negative charged fouling layer is attracted by the higher reverse cation flux in the FO-based technology.

3. Hybrid FO-based system for nutrient recovery

Basically, the FO technology could highly concentrate nutrients in reactors; then it integrated with other processes (e.g., chemical precipitation) to harvest nutrients as a fertilizer

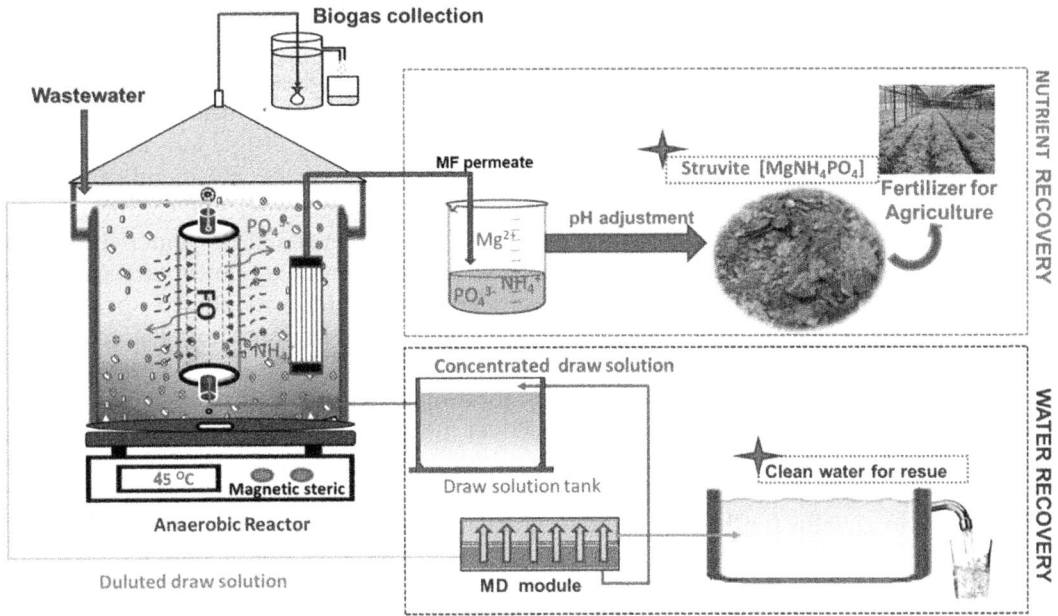

FIG. 7 A hybrid FO-based system for nutrient and water recovery.

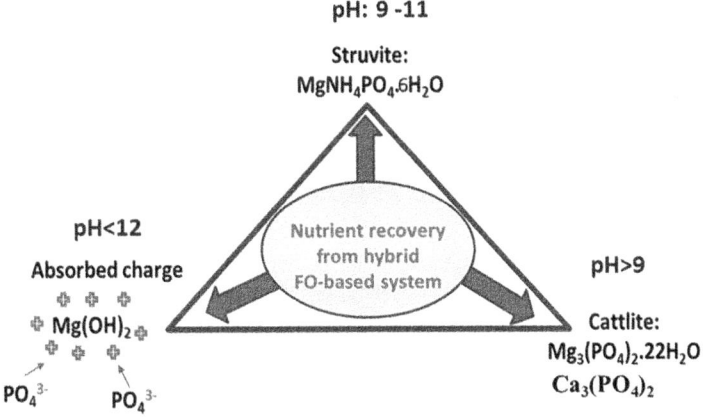

FIG. 8 Nutrient recovery form from a hybrid FO-based system.

for agricultural reuse (Fig. 7). Both phosphate and ammonium in the feed solution are concentrated in the reactor of FO-based systems due to the high FO membrane rejection; hence, it offers suitable conditions for NH_4^+ and PO_4^{3-} recovery in the form calcium phosphate, magnesium phosphate, or struvite (Nguyen et al., 2020), as shown in Fig. 8. Typically, the struvite is formed based on the equation as follows:

$$Mg^{2+} + PO_4^{3-} + NH_4^+ + 6H_2O \rightarrow MgNH_4PO_4{\bullet}6H_2O \quad K_{sp} = 7.1 \times 10^{-14} \tag{2}$$

Moreover, the high retention of Ca^{2+} and Mg^{2+} ions by the FO process is also useful for the recovery of nutrients since it provides favorable components (Mg^{2+}, NH_4^+, PO_4^{3-}) for precipitation of $MgNH_4PO_4 \cdot 6H_2O$. Recently, there have been many studies on the hybrid FO-based system to recover phosphorus from wastewater to prevent the eutrophication phenomenon in water sources (Qiu and Ting, 2014; Chang et al., 2019). NH_4^+, PO_4^{3-}, Mg^{2+} ions in wastewater are concentrated in an anaerobic reactor due to the barrier of the FO membrane in the AnOsMBR system. Membrane filtration (MF) is used to extract the enriched solution, to control pH to recover phosphorus in the form of struvite. Nguyen et al. (2020) demonstrated that the phosphorus recovery from the MF/AnOsMBR hybrid system was most effective as the solution pH was adjusted to 9.5 and concentrated wastewater was stirred speed at 180 rpm in 90 min, which 1.3% (w/w) of nitrogen and 17.4% (w/w) of phosphorus were confined in constituents of struvite precipitation (Fig. 9).

To save cost for the precipitation reactor in formed struvite, Wu et al. (2018) used 0.5 M $MgCl_2$ as a draw solute in the FO process to recover nutrients from digested swine wastewater. The reverse-fluxed Mg^{2+} ions from the draw solution to the feed solution enhanced the struvite precipitation. As a result, 93% NH_4^+ and 99% PO_4^{3-} were recovered. The preliminary estimation of total value for the recovered struvite and water in this study was $1.35/m^3$. Qiu et al. (2015) used a hybrid MF-FO to enrich the PO_4^{3-} concentration in the wastewater to 70 mg/L and the experiment results recorded that the recovery rate of PO_4^{3-} was 98%. Besides, a microbial recovery cell (MRC) was coupled with AnOsMBR to recover phosphorus. The MRC system was used to drive mineral salts and nutrients from the bioreactor then concentrate them in a separated space for forming a high nutrient stream (Hou et al., 2017). Hou et al. (2017) recorded that the concentration of phosphorus was 280 mg/L in MRC-AnOMBR, then the phosphorus reacted with ammonium, magnesium, and calcium to form $MgNH_4PO_4.6H_2O$. Consequently, the recovery of NH_4^+ and PO_4^{3-} by MRC-AnOsMBR were 45% and 65%, respectively. Zou et al. (2017) demonstrated that a hybrid microbial fuel cells–FO (MFCs-FO) system could efficiently recover both PO_4^{3-} and NH_4^+ from the synthetic sidestream centrate. The first reason is that the activity of bacteria from MFCs could produce electricity; hence, it consumed less energy than MECs. Secondly, the value of feed solution pH

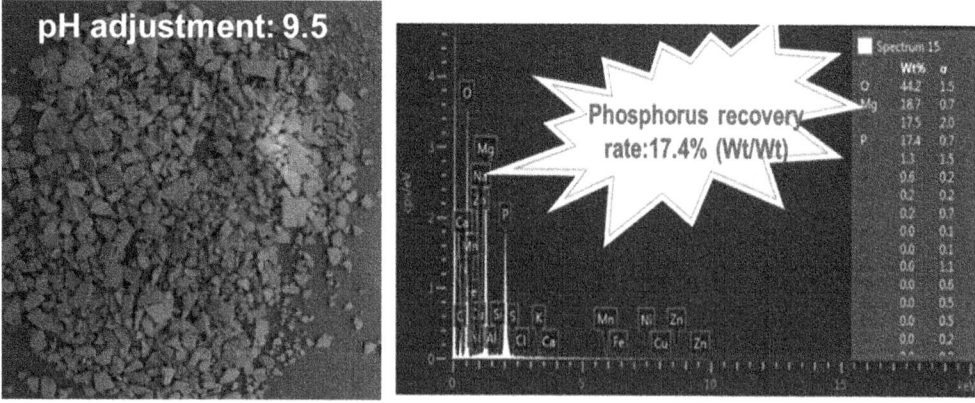

FIG. 9   Phosphorus recovery from the Sponge-AnOsMBR/MD hybrid system.

was raised due to the reactor of cathode in MFCs. Consequently, NH_4^+ was concentrated by the MFCs and PO_4^{3-} was enriched by the FO membrane. A previous study recorded that 79.5% of PO_4^{3-}-P and 99.7% of NH_4^+-N were recovered in hybrid MFCs-FO systems when adding Mg^{2+} into the reactor to form the struvite precipitation. A specific energy consumption rate was approximately $1.17\,kWh\,kg^{-1}$ recovered NH_4^+-N. Nevertheless, the biggest challenge of hybrid MFCs-FO systems is optimum energy consumption and decreasing nutrient loss.

4. Advanced hybrid FO-based system for wastewater treatment and water reuse

Because of the low energy consumption and membrane fouling, FO has been considered an effective technology for wastewater treatment and water reclamation. The efficiency of nutrient recovery and water reuse from hybrid FO-based systems is shown in Table 3. Hancock et al. (2013) investigated the FO/RO pilot scale for wastewater treatment and the results showed that high PO_4^{3-} rejection (>99%) was achieved and the water quality after treatment of the whole system met the standard of drinking water of the Environmental Protection Agency. Utilizing the advantages of the FO process, many researchers have combined the FO process with other biological processes to treat wastewater for reuse, known as OsMBR or AnOsMBR. AnOsMBR technology combines the FO membrane with an anaerobic bioreactor which has more benefits than the MBR technology, such as lower energy consumption, lower sludge production, and higher organic and nutrient removal (Ricci et al., 2021; Chang et al., 2019; Nguyen et al., 2020; Wenhui Lee et al., 2021; Gao et al., 2021). In a hybrid AnOsMBR system, pollutant compounds could be treated both anaerobic microorganism community and high retention of FO membrane, consequently, the water quality after AnOsMBR is better than after AnMBR processing (Chang et al., 2019; Nguyen et al., 2020). Nguyen et al. (2020) showed that the hybrid AnOsMBR system could achieve a good COD removal rate of more than 99% most of the time. During a 45-day AnOsMBR operation, the effluent concentration of COD was less than $5\,mg/L$, while AnMBR can remove 90%–98% of COD (Radjenovic et al., 2007; Luo et al., 2017). The reason is that the FO membrane with a small pore size of 0.37 nm in the AnOsMBR system can retain all contaminants of the influent wastewater, but the conventional AnMBR with MF and Ultrafiltration membranes (UF) cannot remove all organic compounds from wastewater.

As compared with the traditional AnMBR, the AnOsMBR hybrid system could enhance PO_4^{3-} and NH_4^{4+} removal efficiency significantly, since the nutrient removal efficiency depends on the pore size of the FO membrane in AnOsMBR. More specially, in almost all of the experiments, the NH_4^+-N removal efficiency was lower than that of PO_4^{3-}-P because of the hydrated radius effect. A previous study recorded that the hydrated radius of NH_4^+ was smaller than that of PO_4^{3-} (Kiriukhin and Collins, 2002); consequently, the smaller hydrated radius has lower rejection efficiency. Moreover, the charge repulsion phenomenon also affects pollutant removal. Nguyen et al. (2013) showed that the CTA FO membrane is negatively charged (Fig. 10) when an active layer of the membrane faces feed solution (at pH 7.0); hence, NH_4^+ cations were easily attracted to the surface membrane and passed through the membrane due to the electrostatic attraction force. Meanwhile, PO_4^{3-} anions were pushed due to electrostatic repulsions (Fig. 10).

TABLE 3 Comparison of efficiency of nutrient recovery and water reuse between hybrid FO-based systems.

FO membrane	Hybrid system	Draw solution	Feed solution	Operating time	J_w (LMH)	Nutrient removal/ recovery	Water quality
CTA-ES (Nguyen et al., 2020)	Moving sponge-AnOMBR/MD	0.2 M Na$_3$PO$_4$ coupled with 0.25 M EDTA-2Na	Real municipal wastewater	45 day	3.1–4.5	Phosphorus recovery was achieved 82% at pH 9.5	NH$_4^+$-N = 0.95 mg/L PO$_4^{3-}$-P = 0.83 mg/L
CTA-ES (Cong Nguyen et al., 2021)	Dynamic-OsMBR/NF	Mixed 0.1 M EDTA-2Na/0.1 M Na$_2$CO$_3$/0.9 mM Triton114	Dalat municipal wastewater	40 days	2.6–2.4	PO$_4^{3-}$-P removal = 93%; NH$_4^+$-N removal = 99%	NH$_4^+$-N = 3.57 mg/L; COD = 16.2 mg/L; SS = 0.9 mg/L; TDS = 425 mg/L; PO$_4^{3-}$-P = 0.25 mg/L
CTA-NW (Chang et al., 2019)	AnOsMBR-MD	1.5 M MgSO$_4$	Synthetic wastewater	46 days	1.45–1.05	Nutrient removal >99.9%	COD <25 mg/L; PO$_4^{3-}$ <1.7 mg/L; NH$_4^+$ <9 mg/L
CTA (Qiu et al., 2015)	MF-FOMBR	Municipal wastewater	Seawater brine	98 days	9.1–6.5	>90% phosphorus recovery was achieved at pH 9.0	90% removal of TOC; 99% removal of NH$_4^+$-N; 97.9% of PO$_4^{3-}$-P
CTA (Wu et al., 2018)	FO	Digested swine wastewater	0.5 M MgCl$_2$	24 h	3.2–1.9	>99% phosphate recovery	>93% NH$_4^+$-N removal
TFC (Schneider et al., 2019)	FO	Real and synthetic anaerobically digested effluents	0.66 M MgCl$_2$	45 days	3.3–1.9	–	96.95% rejections of NH$_4^+$-N 95.87% rejection of TP
Cellulose acetate (CA) (Luo et al., 2016)	OsMBR-RO	Raw sewage from the Wollongong Wastewater Treatment Plant	0.5 M NaCl	60 days	5.5–2.5	The recovered precipitate contained 15%–20% (wt/wt) of phosphorus at pH 10	TOC of 10 mg/L; NH$_4^+$-N <1 mg/L; PO$_4^{3-}$-P <0.5 mg/L
TFC (Chang et al., 2019)	UAS-OsMBR	Synthetic wastewater	116.6 g/L of MgSO$_4$	40 days	0.8–0.3	Phosphorus recovered at approximately 57–105 mg/L in the form of struvite as the pH varies from 9 to 12	COD, PO$_4^{3-}$, and NH$_4^+$ removal were all more than 95%

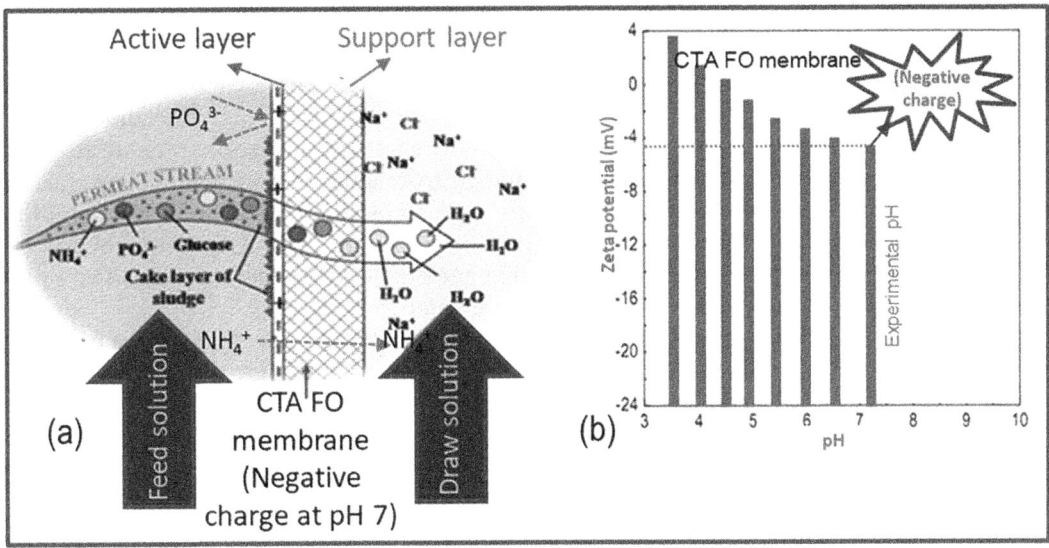

FIG. 10 (A) Schematic diagram representing the effect of charge repulsion effect on pollutants removal in the FO process; (B) variations of pH versus charge of the CTA FO membrane.

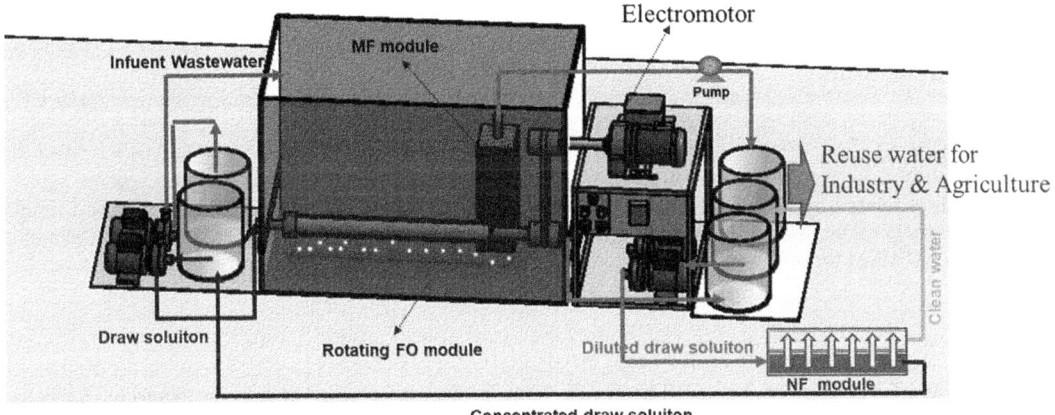

FIG. 11 The lab scale Dynamic-OsMBR/NF hybrid system for wastewater treatment and reuse.

Recently, to reduce membrane fouling and enhance water quality, Nguyen et al. (2021) have shown that the Dynamic-OsMBR/NF hybrid system could achieve a high contaminant removal (almost >98%), as shown in Fig. 11. It is important to note that the concentrations of TDS, PO_4^{3-}-P, NH_4^+-N, and COD in the permeate stream during the Dynamic-OsMBR/NF operation were as low as 429 ± 6 mg/L, 0.25 ± 0.03 mg/L, 3.57 ± 0.28 mg/L, and 16.21 ± 0.58 mg/L, respectively, which was appropriate for water reclamation.

5. Conclusions and perspectives

The hybrid FO-based technology has been considered a good solution for water reuse and nutrient recovery from a wide range of wastewater sources due to its high rejection capacity. There were more studies regarding hybrid FO-based systems, and some outstanding progress achieved that could transfer the FO technology from laboratory-scale to large-scale applications. Despite these advantages, there are still some challenges such as membrane fouling in long-term operation, high energy consumption for draw solution recovery, low FO water flux as compared to MBR water flux that need to be solved before it could be a practical system in a commercial application for wastewater treatment.

To improve the progress of the hybrid FO-based process, the development of a suitable draw solution with low reverse salt flux and easy regeneration is significantly necessary in nutrient recovery and water reuse. More investigation into the full scale of hybrid FO-based systems for nutrient recovery should be conducted to compare with other technologies in terms of environmental footprint, economic analyses as well as energy consumption. Scaling and fouling present an important role in the long-term FO-based operation with real wastewater, such as reduced membrane life span and decreased water quality; hence, further research on scaling and fouling in FO-based systems should be carried out to understand more fouling mechanisms and cleaning strategies for fouled membranes. Furthermore, novel FO membrane designs such as a rotating FO and multiangle tubular FO modules should be employed to enhance water flux and reduce membrane fouling during nutrient and water recovery. To improve nutrient and water recovery, researchers should focus on fabricating innovative FO membranes and understanding in detailed conditions for chemical precipitation of struvite form.

Acknowledgments

This research is funded by Vietnam National University Ho Chi Minh City (VNU-HCM) under grant number NCM2021-20-01. We acknowledge the support of time and facilities from Ho Chi Minh City University of Technology (HCMUT), VNU-HCM for this study.

References

Achilli, A., Cath, T.Y., Marchand, E.A., Childress, A.E., 2009. The forward osmosis membrane bioreactor: a low fouling alternative to MBR processes. Desalination 239, 10–21.

Alihemati, Z., Hashemifard, S.A., Matsuura, T., Ismail, A.F., Hilal, N., 2020. Current status and challenges of fabricating thin film composite forward osmosis membrane: a comprehensive roadmap. Desalination 491, 114557.

Baker, R.W., 2004. Membrane transport theory. In: Membrane Technology and Applications. John Wiley & Sons, Ltd.

Bowden, K.S., Achilli, A., Childress, A.E., 2012. Organic ionic salt draw solutions for osmotic membrane bioreactors. Bioresour. Technol. 122, 207–216.

Cartinella, J.L., Cath, T.Y., Flynn, M.T., Miller, G.C., Hunter, K.W., Childress, A.E., 2006. Removal of natural steroid hormones from wastewater using membrane contactor processes. Environ. Sci. Technol. 40, 7381–7386.

Cath, T.Y., Childress, A.E., Elimelech, M., 2006. Forward osmosis: principles, applications, and recent developments. J. Membr. Sci. 281, 70–87.

Chang, H.-M., Chen, S.-S., Lu, M.-Y., Duong, C.C., Nguyen, N.C., Chang, W.-S., Ray, S.S., 2019. Mesophilic microfiltration–anaerobic osmotic membrane bioreactor–membrane distillation hybrid system for phosphorus recovery. J. Chem. Technol. Biotechnol. 94, 1230–1239.

Chekli, L., Phuntsho, S., Kim, J.E., Kim, J., Choi, J.Y., Choi, J.-S., Kim, S., Kim, J.H., Hong, S., Sohn, J., Shon, H.K., 2016. A comprehensive review of hybrid forward osmosis systems: performance, applications and future prospects. J. Membr. Sci. 497, 430–449.

Dinçer, A.R., Kargi, F., 2001. Salt inhibition kinetics in nitrification of synthetic saline wastewater. Enzym. Microb. Technol. 28, 661–665.

Gao, T., Zhang, H., Xu, X., Teng, J., Lu, M., 2021. Enhanced water and energy recovery from anaerobic osmotic membrane bioreactors treating waste activated sludge based on the draw solution concentration and temperature regulation. Chem. Eng. J. 417, 129325.

Hancock, N., Xu, P., Roby, M., Gomez, J., Cath, T., 2013. Towards direct potable reuse with forward osmosis: technical assessment of long-term process performance at the pilot scale. J. Membr. Sci. 445, 34–46.

Hau, N.T., Chen, S.-S., Nguyen, N.C., Huang, K.Z., Ngo, H.H., Guo, W., 2014. Exploration of EDTA sodium salt as novel draw solution in forward osmosis process for dewatering of high nutrient sludge. J. Membr. Sci. 455, 305–311.

Hou, D., Lu, L., Sun, D., Ge, Z., Huang, X., Cath, T.Y., Ren, Z.J., 2017. Microbial electrochemical nutrient recovery in anaerobic osmotic membrane bioreactors. Water Res. 114, 181–188.

Jin, X., Tang, C.Y., Gu, Y., She, Q., Qi, S., 2011. Boric acid permeation in forward osmosis membrane processes: modeling, experiments, and implications. Environ. Sci. Technol. 45, 2323–2330.

Jin, X., Shan, J., Wang, C., Wei, J., Tang, C.Y., 2012. Rejection of pharmaceuticals by forward osmosis membranes. J. Hazard. Mater. 227-228, 55–61.

Kazner, C., Jamil, S., Phuntsho, S., Shon, H.K., Wintgens, T., Vigneswaran, S., 2014. Forward osmosis for the treatment of reverse osmosis concentrate from water reclamation: process performance and fouling control. Water Sci. Technol. 69, 2431–2437.

Kiriukhin, M.Y., Collins, K.D., 2002. Dynamic hydration numbers for biologically important ions. Biophys. Chem. 99, 155–168.

Kong, F.-x., Yang, H.-w., Wu, Y.-q., Wang, X.-m., Xie, Y.F., 2014. Rejection of pharmaceuticals during forward osmosis and prediction by using the solution–diffusion model. J. Membr. Sci. 476, 410–420.

Li, F., Cheng, Q., Tian, Q., Yang, B., Chen, Q., 2016. Biofouling behavior and performance of forward osmosis membranes with bioinspired surface modification in osmotic membrane bioreactor. Bioresour. Technol. 211, 751–758.

Li, Z., Linares, R.V., Bucs, S., Fortunato, L., Hélix-Nielsen, C., Vrouwenvelder, J.S., Ghaffour, N., Leikness, T., Amy, G., 2017. Aquaporin based biomimetic membrane in forward osmosis: chemical cleaning resistance and practical operation. Desalination 420, 208–215.

Loeb, S., Titelman, L., Korngold, E., Freiman, J., 1997. Effect of porous support fabric on osmosis through a Loeb-Sourirajan type asymmetric membrane. J. Membr. Sci. 129, 243–249.

Lonsdale, H.K., Merten, U., Riley, R.L., 1965. Transport properties of cellulose acetate osmotic membranes. J. Appl. Polym. Sci. 9, 1341–1362.

Luo, W., Hai, F.I., Price, W.E., Guo, W., Ngo, H.H., Yamamoto, K., Nghiem, L.D., 2016. Phosphorus and water recovery by a novel osmotic membrane bioreactor–reverse osmosis system. Bioresour. Technol. 200, 297–304.

Luo, W., Phan, H.V., Xie, M., Hai, F.I., Price, W.E., Elimelech, M., Nghiem, L.D., 2017. Osmotic versus conventional membrane bioreactors integrated with reverse osmosis for water reuse: biological stability, membrane fouling, and contaminant removal. Water Res. 109, 122–134.

Luo, W., Xie, M., Song, X., Guo, W., Ngo, H.H., Zhou, J.L., Nghiem, L.D., 2018. Biomimetic aquaporin membranes for osmotic membrane bioreactors: membrane performance and contaminant removal. Bioresour. Technol. 249, 62–68.

McCutcheon, J.R., Elimelech, M., 2006. Influence of concentrative and dilutive internal concentration polarization on flux behavior in forward osmosis. J. Membr. Sci. 284, 237–247.

McCutcheon, J.R., Elimelech, M., 2007. Modeling water flux in forward osmosis: implications for improved membrane design. AICHE J. 53, 1736–1744.

McCutcheon, J.R., McGinnis, R.L., Elimelech, M., 2005. A novel ammonia-carbon dioxide forward (direct) osmosis desalination process. Desalination 174, 1–11.

Moussa, M.S., Sumanasekera, D.U., Ibrahim, S.H., Lubberding, H.J., Hooijmans, C.M., Gijzen, H.J., Van Loosdrecht, M.C.M., 2006. Long term effects of salt on activity, population structure and floc characteristics in enriched bacterial cultures of nitrifiers. Water Res. 40, 1377–1388.

Nawaz, M.S., Gadelha, G., Khan, S.J., Hankins, N., 2013. Microbial toxicity effects of reverse transported draw solute in the forward osmosis membrane bioreactor (FO-MBR). J. Membr. Sci. 429, 323–329.

Nguyen, N.C., Chen, S.-S., Yang, H.-Y., Hau, N.T., 2013. Application of forward osmosis on dewatering of high nutrient sludge. Bioresour. Technol. 132, 224–229.

Nguyen, H.T., Nguyen, N.C., Chen, S.-S., Li, C.-W., Hsu, H.-T., Wu, S.-Y., 2015b. Innovation in draw solute for practical zero salt reverse in forward osmosis desalination. Ind. Eng. Chem. Res. 54, 6067–6074.

Nguyen, N.C., Nguyen, H.T., Chen, S.-S., Ngo, H.H., Guo, W., Chan, W.H., Ray, S.S., Li, C.-W., Hsu, H.-T., 2016. A novel osmosis membrane bioreactor-membrane distillation hybrid system for wastewater treatment and reuse. Bioresour. Technol. 209, 8–15.

Nguyen, C.N., Duong, H.C., Chen, S.-S., Nguyen, H.T., Ngo, H.H., Guo, W., Le, H.Q., Duong, C.C., Trang, L.T., Le, A.H., Bui, X.T., Nguyen, P.D., 2020. Water and nutrient recovery by a novel moving sponge—anaerobic osmotic membrane bioreactor—membrane distillation (AnOMBR-MD) closed-loop system. Bioresour. Technol. 312, 123573.

Nguyen, C.N., Nguyen, H.T., Duong, H.C., Chen, S.-S., Le, H.Q., Duong, C.C., Trang, L.T., Chen, C.-K., Nguyen, P.D., Bui, X.T., Guo, W., Ngo, H.H., 2021. A breakthrough dynamic-osmotic membrane bioreactor/nanofiltration hybrid system for real municipal wastewater treatment and reuse. Bioresour. Technol. 342, 125930.

Osaka, T., Shirotani, K., Yoshie, S., Tsuneda, S., 2008. Effects of carbon source on denitrification efficiency and microbial community structure in a saline wastewater treatment process. Water Res. 42, 3709–3718.

Qiu, G., Ting, Y.-P., 2014. Direct phosphorus recovery from municipal wastewater via osmotic membrane bioreactor (OMBR) for wastewater treatment. Bioresour. Technol. 170, 221–229.

Qiu, G., Law, Y.-M., Das, S., Ting, Y.-P., 2015. Direct and complete phosphorus recovery from municipal wastewater using a hybrid microfiltration-forward osmosis membrane bioreactor process with seawater brine as draw solution. Environ. Sci. Technol. 49, 6156–6163.

Radjenovic, J., Petrovic, M., Barceló, D., 2007. Analysis of pharmaceuticals in wastewater and removal using a membrane bioreactor. Anal. Bioanal. Chem. 387, 1365–1377.

Ricci, C., Bárbara, G.S., Arcanjo, V.R., Moreira, Y.A., Lebron, R., Koch, K., Costa, F.C.R., Ferreira, B.P., Lisboa, F.L.C., Miranda, L.D., Vieira, C., de Faria, L., Lange, C., Amaral, M.C.S., 2021. A novel submerged anaerobic osmotic membrane bioreactor coupled to membrane distillation for water reclamation from municipal wastewater. Chem. Eng. J. 414, 128645.

Schneider, C., Rajmohan, R.S., Zarebska, A., Tsapekos, P., Hélix-Nielsen, C., 2019. Treating anaerobic effluents using forward osmosis for combined water purification and biogas production. Sci. Total Environ. 647, 1021–1030.

Shon, H.K., Chekli, L., Phuntsho, S., Kim, J., Cho, J., 2015. Draw Solutes in Forward Osmosis Processes. Forward Osmosis.

Tan, C.H., Ng, H.Y., 2010. A novel hybrid forward osmosis—nanofiltration (FO-NF) process for seawater desalination: draw solution selection and system configuration. Desalin. Water Treat. 13, 356–361.

Tiraferri, A., Yip, N.Y., Phillip, W.A., Schiffman, J.D., Elimelech, M., 2011. Relating performance of thin-film composite forward osmosis membranes to support layer formation and structure. J. Membr. Sci. 367, 340–352.

Uygur, A., Kargi, F., 2004. Salt inhibition on biological nutrient removal from saline wastewater in a sequencing batch reactor. Enzym. Microb. Technol. 34, 313–318.

Wang, X., Zhao, Y., Yuan, B., Wang, Z., Li, X., Ren, Y., 2016. Comparison of biofouling mechanisms between cellulose triacetate (CTA) and thin-film composite (TFC) polyamide forward osmosis membranes in osmotic membrane bioreactors. Bioresour. Technol. 202, 50–58.

Wenhui Lee, L., Zhu, X., Liu, Z., Gao, Y., Chen, C., Huang, X., 2021. Probing the key foulants and membrane fouling under increasing salinity in anaerobic osmotic membrane bioreactors for low-strength wastewater treatment. Chem. Eng. J. 413, 127450.

Wu, Z., Zou, S., Zhang, B., Wang, L., He, Z., 2018. Forward osmosis promoted in-situ formation of struvite with simultaneous water recovery from digested swine wastewater. Chem. Eng. J. 342, 274–280.

Xie, M., Nghiem, L.D., Price, W.E., Elimelech, M., 2014. Relating rejection of trace organic contaminants to membrane properties in forward osmosis: measurements, modelling and implications. Water Res. 49, 265–274.

Yang, J., Li, Y., Li, H.B., Cao, H.B., Zhang, Y., 2014. Forward osmosis performance of sodium polyacrylate as draw solution. Guocheng Gongcheng Xuebao/Chin. J. Process Eng. 14, 395–401.

Zhu, S., Li, M., El-Din, M.G., 2018. The roles of pH and draw solute on forward osmosis process treating aqueous naphthenic acids. J. Membr. Sci. 549, 456–465.

Zou, S., Qin, M., Moreau, Y., He, Z., 2017. Nutrient-energy-water recovery from synthetic sidestream centrate using a microbial electrolysis cell—forward osmosis hybrid system. J. Clean. Prod. 154, 16–25.

Index

Note: Page numbers followed by *f* indicate figures and *t* indicate tables.

CPI Antony Rowe
Eastbourne, UK
February 17, 2023